# The Biochemistry of the Nucleic Acids

## TENTH EDITION

Roger L. P. Adams
John T. Knowler
David P. Leader
*Department of Biochemistry*
*University of Glasgow*

LONDON    NEW YORK
Chapman and Hall

*First published in 1950 by Methuen and Co. Ltd*
*Second edition 1953, Third edition 1957, Fourth edition 1960,*
*Fifth edition 1965, Sixth edition 1969, Seventh edition 1972*

*Published by Chapman and Hall Ltd*
*11 New Fetter Lane, London EC4P 4EE*
*First published as a Science Paperback 1972*
*Eighth edition 1976, Ninth edition 1981, Tenth edition 1986*

*Published in the USA by Chapman and Hall*
*29 West 35th Street, New York, NY 10001*

© 1986 Chapman and Hall

*Printed in Great Britain at the*
*University Press, Cambridge*

ISBN 0 412 27270 9 (cased)
0 412 27280 6 (Science Paperback)

British Library Cataloguing in Publication Data

Adams, R.L.P.
  The biochemistry of the nucleic acids. —
10th ed.
  1. Nucleic acids
  I. Title   II. Knowler, J. T.   III. Leader,
David P.
  574.87'328      QP620

  ISBN 0-412-27270-9
  ISBN 0-412-27280-6 Pbk

Library of Congress Cataloging in Publication Data

Adams, R. L. P. (Roger Lionel Poulter)
  The biochemistry of the nucleic acids.

  Rev. ed. of: The Biochemistry of the nucleic
acids / R. L. P. Adams. 9th ed.
  Includes bibliographical references and index.
  1. Nucleic acids. I. Knowler, John T., 1942—
II. Leader, David P., 1943–   . III. Title.
QP620.B56 1986      574.87'328      86-9749
ISBN 0-412-27270-9
ISBN 0-412-27280-6 (pbk.)

# Contents

# Preface

When the first edition of this book was published in 1950, it set out to present an elementary outline of the state of knowledge of nucleic acid biochemistry at that time and it was the first monograph on the subject to appear since Levene's book on Nucleic Acids in 1931. The fact that a tenth edition is required after thirty five years and that virtually nothing of the original book has been retained is some measure of the speed with which knowledge has advanced in this field.

As a result of this vast increase in information it becomes increasingly difficult to fulfil the aims of providing an introduction to nucleic acid biochemistry and satisfying the requirements of advanced undergraduates and postgraduates in biochemistry, genetics and molecular biology. We have attempted to achieve these aims by concentrating on those basic aspects not normally covered in the general biochemistry textbooks and by providing copious references so that details of methodology can readily be retrieved by those requiring further information.

The first seven editions emerged from the pen of J. N. Davidson who died in September 1972 shortly after completing the seventh edition. The subsequent editions have been produced by various colleagues who have tried to retain something of the character and structure of the earlier editions while at the same time introducing new ideas and concepts and eliminating some of the more out-dated material.

With each new edition very extensive revisions, not only in the content of individual chapters but also in general organization and layout, have been required. With a large amount of additional material to present, the book has grown in size, but every effort has been made to keep the increase within bounds by excluding non-essential detail. In a field in which new developments are occurring so rapidly it is inevitable that new knowledge will accumulate more quickly than it can be embodied in a new edition but we have endeavoured to incorporate into this edition material published up to the date of completion of the manuscript in December 1985.

It is a pleasure to express our thanks to those who have allowed us to reproduce figures and diagrams, especially those who have provided original photographs.

We are particularly grateful to the secretarial staff in the Biochemistry Department at the University of Glasgow for the cheerful and tireless efforts they have put in, typing and retyping the manuscript; and to the artists of the Medical Illustrations Unit.

*R.L.P.A.*
*J.T.K.*
*D.P.L.*
*December 1985*

# Abbreviations and nomenclature

The abbreviations employed in this book are those approved by the Commission on Biochemical Nomenclature (CBN) of the International Union of Pure and Applied Chemistry (IUPAC) and the International Union of Biochemistry (IUB).

## Nucleosides

| | |
|---|---|
| A | adenosine |
| G | guanosine |
| C | cytidine |
| U | uridine |
| $\psi$ | 5-ribosyluracil (pseudouridine) |
| I | inosine |
| X | xanthine |
| rT | ribosylthymine (ribothymidine) |
| N | unspecified nucleoside |
| R | unspecified purine nucleoside |
| Y | unspecified pyrimidine nucleoside |
| dA | 2'-deoxyribosyladenine |
| dG | 2'-deoxyribosylguanine |
| dC | 2'-deoxyribosylcytosine |
| dT or T | 2'-deoxyribosylthymine (thymidine) |

## Minor nucleosides (when in sequence)

| | |
|---|---|
| $m^1A$ | 1-methyladenosine |
| $m_2^6A$ | $N^6$-dimethyladenosine |
| iA | $N^6$-isopentenyladenosine |
| $m^5C$ | 5-methylcytidine |
| $ac^4C$ | $N^4$-acetylcytidine |
| $m^1G$ | 1-methylguanosine |
| $m^2G$ | $N^2$-methylguanosine |
| $m_2^2G$ | $N^2$-dimethylguanosine |
| $m^1I$ | 1-methylinosine |
| Cm | 2'-$O$-methylcytidine |
| Gm | 2'-$O$-methylguanosine |
| Um | 2'-$O$-methyluridine |
| D | 5,6-dihydrouridine |

| | |
|---|---|
| mcm$^5$U | 5-(methoxycarboxylmethyl)uridine |
| mcm$^5$s$^2$U | 5-(methoxycarboxylmethyl)-2-thiouridine |
| mnm$^5$s$^2$U | 5-(methylaminomethyl)-2-thiouridine |
| mo$^5$U | 5-methoxyuridine |
| cmo$^5$U | 5-(carboxymethoxyuridine) |
| Q | Queosine |
| yW | Wybutosine |

## Nucleotides

| | |
|---|---|
| AMP | adenosine 5′-monophosphate |
| GMP | guanosine 5′-monophosphate |
| CMP | cytidine 5′-monophosphate |
| UMP | uridine 5′-monophosphate |
| dAMP | 2′-deoxyribosyladenine 5′-monophosphate |
| dGMP | 2′-deoxyribosylguanine 5′-monophosphate |
| dCMP | 2′-deoxyribosylcytosine 5′-monophosphate |
| dTMP | 2′-deoxyribosylthymine 5′-monosphosphate |
| 2′-AMP, 3′-AMP, 5′-AMP etc | 2′-, 3′- and 5′-phosphates of adenosine etc. |
| ADP etc. | 5′-(pyro) diphosphates of adenosine etc. |
| ATP etc. | 5′-(pyro) triphosphates of adenosine etc. |
| ddTTP etc. | 2′, 3′-dideoxyribosylthymine 5′-triphosphate |
| araCTP | 1-$\beta$-D-arabinofuranosylcytosine 5′-triphosphate |

## Polynucleotides

| | |
|---|---|
| DNA | deoxyribonucleic acid |
| cDNA | complementary DNA |
| mtDNA | mitochondrial DNA |
| RNA | ribonucleic acid |
| mRNA | messenger RNA |
| rRNA | ribosomal RNA |
| tRNA | transfer RNA |
| nRNA | nuclear RNA |
| hnRNA | heterogeneous nuclear RNA |
| snRNA | small nuclear RNA |
| Alanine tRNA or tRNA$^{Ala}$ etc. | transfer RNA that normally accepts alanine |
| Alanyl-tRNA$^{Ala}$ or Ala-tRNA$^{Ala}$ or Ala-tRNA | transfer RNA that normally accepts alanine with alanine residue covalently linked |
| poly(N), or (N)$_n$ or (rN)$_n$ | polymer of ribonucleotide N |
| poly(dN) or (dN)$_n$ | polymer of deoxyribonucleotide N |
| poly(N-N′), or r(N-N′)$_n$ or(rN-rN′)$_n$ | copolymer of–N–N′–N–N′–in regular, alternating, *known* sequence |

| poly(A).poly(B) or | |
|---|---|
| $(A)_n.(B)_n$ | two chains, generally or completely associated |
| poly(A), poly(B) or | |
| $(A)_n, (B)_n$ | two chains, association unspecified or unknown |
| poly(A) + poly(B) or | |
| $(A)_n + (B)_n$ | two chains, generally or completely unassociated |

## Miscellaneous

| | |
|---|---|
| RNase, DNase | ribonuclease, deoxyribonuclease |
| $P_i$, $PP_i$ | inorganic orthophosphate and pyrophosphate |
| nt | nucleotide |
| bp | base pair |
| mt | mitochondrial |

## Amino acids

| | |
|---|---|
| Ala or A | alanine |
| Arg or R | arginine |
| Asn or N | asparagine |
| Asp or D | aspartic acid |
| Cys or C | cysteine |
| Gln or Q | glutamine |
| Glu or E | glutamic acid |
| Gly or G | glycine |
| His or H | histidine |
| Ile or I | isoleucine |
| Leu or L | leucine |
| Lys or K | lysine |
| Met or M | methionine |
| fMet | formylmethionine |
| Phe or F | phenylalanine |
| Pro or P | proline |
| Ser or S | serine |
| Thr or T | threonine |
| Trp or W | tryptophan |
| Tyr or Y | tyrosine |
| Val or V | valine |

In naming enzymes, the recommendations of the Nomenclature Committee of the International Union of Biochemistry (1984) are followed as far as possible. The numbers recommended by the Commission are inserted in the text after the name of each enzyme.

# 1

# Introduction

The fundamental investigations which led to the discovery of the nucleic acids were made by Friedrich Miescher [1] (1844–95), who may be regarded as the founder of our knowledge of the chemistry of the cell nucleus. In early work carried out in 1868, in the laboratory of Hoppe-Seyler in Tübingen, he isolated the nuclei from pus cells obtained from discarded surgical bandages and showed the presence in them of an unusual phosphorus-containing compound that he called 'nuclein' and which we now know to have been nucleoprotein. It was subsequently shown that nucleic acids were normal constituents of all cells and tissues and Miescher's investigations of the nucleic acids were continued by Altman, who in 1889 described a method for the preparation of protein-free nucleic acids from animal tissues and from yeast [2].

On hydrolysis, the nucleic acid from thymus glands was found to yield the purine bases adenine and guanine, the pyrimidine bases cytosine and thymine, a deoxypentose and phosphoric acid. The nucleic acid from yeast on the other hand yielded on hydrolysis adenine, guanine, cytosine, uracil, a pentose sugar and phosphoric acid. Yeast nucleic acid therefore differed from thymus nucleic acid in containing uracil in place of thymine and a pentose in place of a deoxypentose. This led to the impression that deoxypentose nucleic acid was characteristic of animal tissues, and pentose nucleic acid was characteristic of plant tissues.

It was not long before the validity of this concept was questioned but final proof that ribonucleic acid is a general constituent of animal, plant and bacterial cells was not forthcoming until the early 1940s as a consequence of the ultraviolet spectrophotometric studies of Caspersson [3], the histochemical observations of Brachet [4] and the chemical analysis of Davidson [5, 6].

It took a surprisingly long time also to establish the nature of the sugars in deoxypentose and pentose nucleic acids but they now form the basis of the names deoxyribonucleic acid (DNA) and ribonucleic acid (RNA). The elucidation of the detailed structure of nucleosides and nucleotides can largely be attributed to Todd and his collaborators (for review see [7]), who established the nature of the glycosidic linkage between the sugar residues and the purine or pyrimidine bases and the nature of the phosphate ester bonds. Their work taken together with the studies of Cohn and his colleagues [8] provided final confirmation of the nature of the 3′,5′-internucleotide linkage in both DNA and RNA and made it possible for clear concepts as to the primary structure of the two types of nucleic acids to be put forward.

These advances established the biology of the nucleic acids on a new foundation. The use of new techniques in cytochemistry and cell fractionation showed that DNA and RNA are normal constituents of all cells, whether plant or animal, DNA being confined mainly to the nucleus while RNA is found also in the cytoplasm [5, 6, 9, 10].

The development of techniques of subcellular fractionation and for the isolation of nuclei made possible chemical measurements of the distributions of DNA and RNA amongst the subcellular fractions of various cell types, and led ultimately to the recognition of RNA in the nuclear, ribosomal and soluble fractions of cells and to the demonstration of the constancy in the average amount of DNA per nucleus in the somatic cells of any given species [11].

The presence of the bases in approximately equimolar proportions led to the development of the tetranucleotide hypothesis for both DNA and RNA, in which both nucleic acids were considered to be polymeric structures containing equivalent amounts of mononucleotides derived from each of the four purine and pyrimidine bases linked together in repeating units. It was only when methods for the quantitative analysis of the nucleic acids had been developed that this hypothesis was finally abandoned as a consequence of the demonstration that the various nucleotides did not necessarily occur in equimolar proportions [12].

In the early 1950s Chargaff [13] drew attention to certain regularities in the composition of DNA, namely that the sum of the purines was equal to the sum of the pyrimidines, that the sum of the amino bases (adenine and cytosine) was equal to the sum of the keto bases (guanine and thymine) and that adenine and thymine, and guanine and cytosine, were present in equivalent amounts. These observations were to be of crucial importance in the subsequent interpretation of X-ray crystallographic analyses which were performed by Astbury [14], Pauling and Corey [15], Wilkins and colleagues [16] and Franklin and Gosling [17]. The two sets of data were brilliantly combined by Watson and Crick [18] in their now famous double-helical structure made up of specifically hydrogen-bonded base pairs which suggested 'a possible copying mecha-

nism for the genetic material'.

For a while after the elucidation of the structure of the double helix, it was thought that there was one sort of RNA, ribosomal RNA (rRNA), which carried the genetic message from the nucleus to the site of protein synthesis in the cell cytoplasm. By the early 1960s, however, it was realized that rRNA was too stable and too constant in base composition to fulfil this function and in 1961 Jacob and Monod [19] proposed a short-lived messenger molecule (mRNA). Support for the concept was not long in coming. Phage infection of bacteria had been shown to be followed by the synthesis of an RNA complementary to the DNA and it was further demonstrated that this RNA associated with bacterial ribosomes and became the template for phage protein synthesis [20]. Thus was established the concept of the flow of information symbolized as:

$$\text{DNA} \xrightarrow{\textit{transcription}} \text{RNA} \xrightarrow{\textit{translation}} \text{protein}$$

The third major species of RNA, tRNA, was first proposed by Crick [21] as an adapter molecule required for the insertion of amino acids against an RNA template. It was also this species which formed the subject of many of the breakthroughs in the analysis of RNA structure beginning with the elucidation of the primary sequence of a yeast tRNA for alanine [22] and followed nine years later by the elucidation of the three-dimensional structure of the yeast tRNA for phenylalanine [23, 24].

An enzyme catalysing the synthesis of DNA-like polymers from deoxyribonucleoside 5′-triphosphates was first identified in *Escherichia coli* by Kornberg and his collaborators [25]. This enzyme was named DNA polymerase and is now recognized as one of a family of enzymes concerned in the replication and repair of DNA molecules [26].

RNA polymerases, catalysing the synthesis of polyribonucleotides from ribonucleoside 5′-triphosphates, were identified almost simultaneously by several groups around 1960 (for review see [27]) so that within the space of a very few years understanding of the whole field of nucleic acid biosynthesis and function underwent a complete transformation.

The first phase of nucleic acid chemistry and biochemistry provided a broad structural and mechanistic understanding of genetic replication and gene expression. In order to obtain a detailed understanding of genetic organization and the regulation of gene expression – especially in eukaryotes – structural analysis of individual genes was required. The revolution that realized what at one time had seemed an impossible dream arose from the forging together of the enzymology of the restriction endonucleases [28] and the genetics of bacteria and bacteriophages, The cloning of recombinant DNA enabled the isolation of large quantities of individual genes, the nucleotide sequences of which became soluble through the chemical and enzymic techniques developed by Maxam and Gilbert [29] and Sanger [30].

The application of these techniques to prokaryotic genomes has been exciting and rewarding enough. However in the study of the eukaryotic genome it has enabled a quantum leap to be made. It has revealed the unexpected as in the discovery of introns and the unravelling of the nucleic acid rearrangements in the field of immunology. But, most excitingly, it is holding the promise of the key to understanding the basic biological problems of differentiation and development. Today there is such an integration of different disciplines in attacking problems related to nucleic acids that it is unlikely that we would have chosen the present title if this were the first rather than the tenth edition of this book. Nevertheless, although we hope this edition reflects the spirit of the times, its central pole is still biochemistry and the contribution this discipline has made, and is continuing to make, to the fascinating study of gene expression.

## REFERENCES

1 Miescher, F. (1879), *Die histochemischen und physiologischen Arbeiten*, Leipzig.
2 Altmann, R. (1889), *Arch. Anat. Physiol.*, 524.
3 Caspersson, T. (1950), *Cell Growth and Cell Function*, Norton, New York.
4 Brachet, J. (1950), *Chemical Embryology*, Interscience, New York.
5 Davidson, J. N. and Waymouth, C. (1944), *Biochem. J.*, **38**, 39.
6 Davidson, J. N. and Waymouth, C. (1944–5), *Nut. Abs. Rev.*, **14**, 1.
7 Brown, D. M. and Todd, A. R. (1955), in *The Nucleic Acids* (ed. E. Chargaff and J. N. Davidson), Academic Press, New York, Vol.1, p. 409.
8 Cohn, W. E. (1956), in *Currents in Biochemical Research* (ed. D. E. Green), Interscience, New York, p.460.
9 Feulgen, R. and Rossenbeck, H. (1924), *Hoppe-Seyler's Zeitschr.*, **135**, 203.
10 Behrens, M. (1938), *Hoppe-Seyler's Zeitschr.*, **253**, 185.
11 Vendrely, R. (1955), *The Nucleic Acids* (ed. E. Chargaff and J. N. Davidson), Academic Press, New York, Vol. 2, p. 155.
12 Chargaff, E. (1950), *Experientia*, **6**, 201.
13 Chargaff, E. (1955), in *The Nucleic Acids* (ed. E. Chargaff and J. N. Davidson), Academic Press, New York, Vol. 1, p.307.
14 Astbury, W. T. (1947), *Symp. Soc. Exp. Biol.*, **1**, 66.
15 Pauling, L. and Corey, R. B. (1953), *Proc. Natl. Acad. Sci. USA*, **39**, 84.
16 Wilkins, M. F. H., Stokes, A. R. and Wilson, H. R. (1953), *Nature (London)*, **171**, 738.
17 Franklin, R. E. and Gosling, R. G. (1953), *Nature (London)*, **171**, 740; **172**, 156.
18 Watson, J. D. and Crick, F. H. C. (1953), *Nature (London)*, **171**, 737.
19 Jacob, F. and Monod, J. (1961), *J. Mol. Biol.*, **3**, 318.
20 Brenner, S., Jacob, F. and Meselson, M. (1961), *Nature (London)*, **190**, 576.
21 Crick, F. (1957), *Biochem. Soc. Symp.*, **14**, 25.
22 Holley, R. W., Apgar, J., Everett, G. A., Madison, J. T., Marguisee, M., Merrill, S. H., Penswick, J. R. and Zamir, A. (1965), *Science*, **147**, 1462.

23  Robertus, J. D., Ladner, J. E., Finch, J. T., Rhodes, D., Brown, R. S., Clark, B. F. E. and Klug, A. (1974), *Nature (London)*, **250**, 546.

24  Quigley, G. J., Wang, A. H. J., Seeman, N. C., Suddath, F. L., Rich, A., Sussman, J. L. and Kim, S. H. (1975), *Proc. Natl. Acad. Sci. USA*, **72**, 4866.

25  Lehmann, I. R., Bessman, M. J., Simms, E. S. and Kornberg, A. (1958), *J. Biol. Chem.*, **233**, 163.

26  Kornberg, A. (1980), *DNA Replication*, Freeman, San Francisco.

27  Smellie, R. M. S. (1963), in *Progress in Nucleic Acid Research* (ed. J. N. Davidson and W. E. Cohn), Academic Press, New York, Vol. 1, p. 27.

28  Kelly, T. J. and Smith, H. O. (1970), *J. Mol. Biol.*, **51**, 393.

29  Maxam, A. M. and Gilbert, W. (1977), *Proc. Natl. Acad. Sci. USA*, **74**, 560.

30  Sanger, F., Air, G. M., Barrell, B. G., Brown, N. L., Coulson, A. R., Fiddes, J. C., Hutchison, C. A., Slocombe, P. M. and Smith, M. (1977), *Nature (London)*, **265**, 687.

<h1>2</h1>

<h1>The structure of the
nucleic acids</h1>

## 2.1 MONOMERIC COMPONENTS

Before any account is given of the structure of
the nucleic acids proper it is necessary to
describe the structures of their component parts.
Nucleic acids are high-molecular-weight polymeric compounds which on complete hydrolysis
yield pyrimidine and purine bases, a sugar
component and phosphoric acid. Partial
hydrolysis yields compounds known as nucleosides and nucleotides. Each of these component
parts will be discussed in turn.

### 2.1.1 Pyrimidine bases

The pyrimidine bases (Fig. 2.1) are derivatives of
the parent compound pyrimidine, and the bases
found in the nucleic acids are cytosine found in
both RNA and DNA, uracil found in RNA and
thymine and 5-methylcytosine found in DNA. In
certain of the bacterial viruses cytosine is
replaced by 5-hydroxymethylcytosine or glucosylated derivatives of 5-hydroxymethylcytosine.

The pyrimidine bases can undergo keto–enol
tautomerism as shown for uracil in Fig. 2.2.

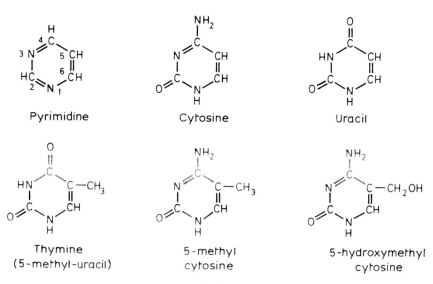

Pyrimidine

Cytosine

Uracil

Thymine
(5-methyl-uracil)

5-methyl
cytosine

5-hydroxymethyl
cytosine

**Fig. 2.1**

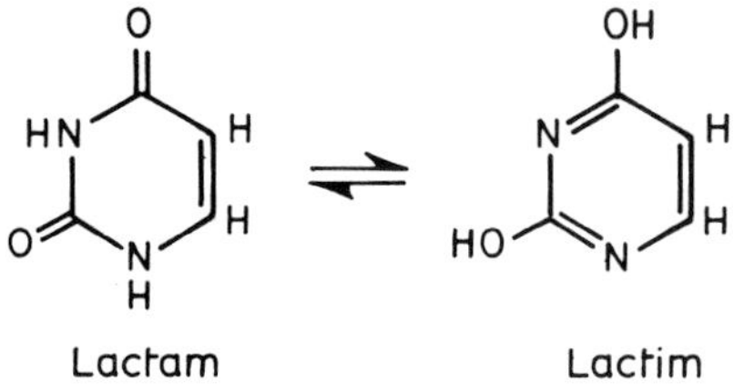

**Fig. 2.2**

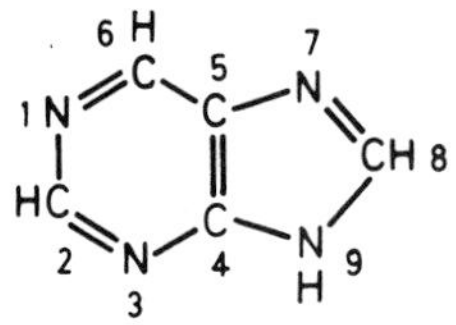

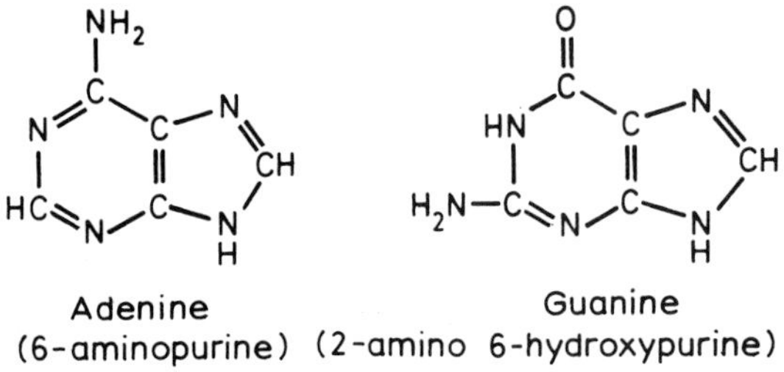

Lactam ⇌ Lactim

**Fig. 2.3**

Adenine
(6-aminopurine)

Guanine
(2-amino 6-hydroxypurine)

### 2.1.2 Purine bases

Both types of nucleic acids contain the purine bases, adenine and guanine. They are derivatives of the parent compound purine which is formed by the fusion of a pyrimidine ring and an iminazole ring. It should be noted that the style of numbering of the pyrimidine ring in the purines differs from that used for pyrimidines themselves. Adenine and guanine have the structures shown in Fig. 2.3.

Other naturally occurring purine derivatives include hypoxanthine, xanthine and uric acid (Fig. 2.4).

As in the pyrimidines so with the purines the bases can exist in two tautomeric forms.

Certain 'minor bases' are also found in small amounts in some nucleic acids [1–4]. For example, 'transfer' RNA (tRNA) which is discussed in Sections 2.9, 9.5.2 and Chapter 11 contains a wide variety of methylated bases, including thymine [3]. These unusual bases comprise less than 5% of the total base content of the tRNA and vary in relative amounts from species to species. Some of the minor bases in RNA are listed in Table 2.1.

The phenylalanine transfer RNA from yeast contains a most unusual base known as Wybutosine, the structure of which is shown in Fig. 2.5 [5, 6, 119].

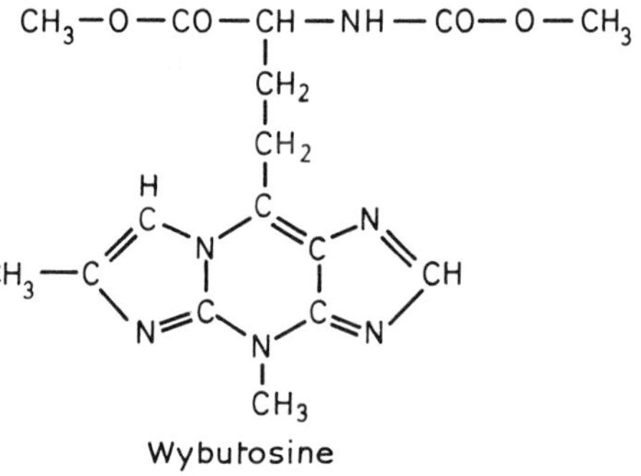

Wybutosine

**Fig. 2.5**

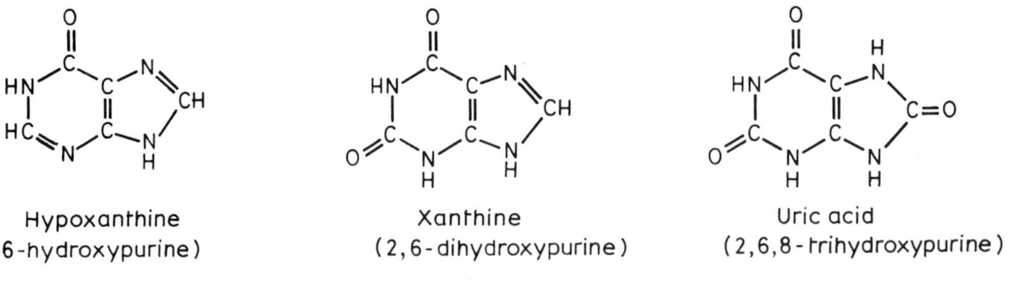

Hypoxanthine
(6-hydroxypurine)

Xanthine
(2,6-dihydroxypurine)

Uric acid
(2,6,8-trihydroxypurine)

**Fig. 2.4**

**Table 2.1** Some of the more important minor bases in RNA.

| | |
|---|---|
| 1-methyladenine | dihydrouracil |
| 2-methyladenine | 5-hydroxyuracil |
| 6-methyladenine | 5-carboxymethyluracil |
| 6,6-dimethyladenine | 5-methyluracil (thymine) |
| 6-isopentenyladenine | 5-hydroxymethyluracil |
| 2-methylthio-6-isopentenyladenine | 2-thiouracil |
| 6-hydroxymethylbutenyladenine | 3-methyluracil |
| 6-hydroxymethylbutenyl-2-methylthioadenine | 5-methylamino-2-thiouracil |
| 1-methylguanine | 5-methyl-2-thiouracil |
| 2-methylguanine | 5-uracil-5-hydroxyacetic acid |
| 2,2-dimethylguanine | 3-methylcytosine |
| 7-methylguanine | 4-methylcytosine |
| 2,2,7-trimethylguanine | 5-methylcytosine |
| hypoxanthine | 5-hydroxymethylcytosine |
| 1-methylhypoxanthine | 2-thiocytosine |
| xanthine | 4-acetylcytosine |
| 6-aminoacyladenine | |
| 7-(4,5-*cis*-dihydroxyl-1-cyclopenten-3-ylaminomethyl)-7-deazaguanosine(Q) | |

## 2.1.3 Pentose and deoxypentose sugars

The main sugar component of RNA is D-ribose which in polynucleotides occurs in the D-ribofuranose form. In DNA this sugar is replaced by 2-deoxyribose also in the D-furanose form (Fig. 2.6). This apparently small difference between the two types of nucleic acid has wide-ranging effects on both their chemistry and structure since the presence of the bulky hydroxyl group on the 2 position of the sugar not only limits the range of possible secondary structures available to the RNA molecule but also makes it more susceptible to chemical and enzymic degradation.

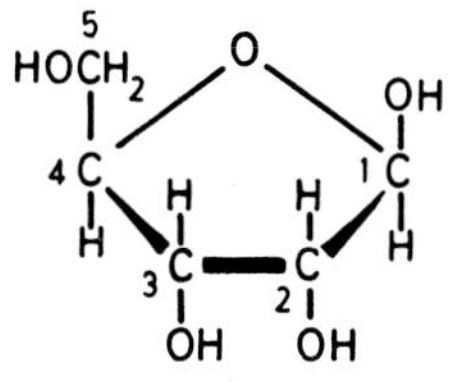

Some RNAs, notably ribosomal RNAs, contain very small amounts of 2-$O$-methyl ribose.

When the pentose sugars occur in nucleic acids or nucleotides the carbon atoms are numbered as 1', 2', 3' etc. to avoid confusion with the numbering of the ring atoms of the bases.

β-D-ribofuranose

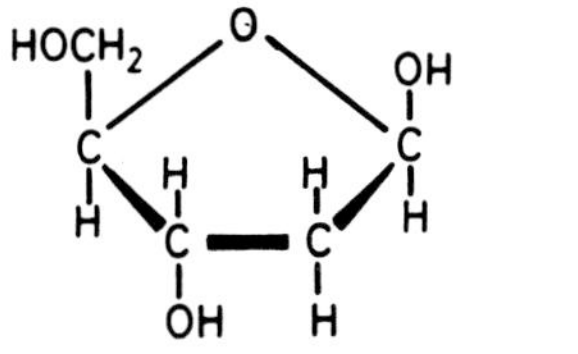

β-D-2-deoxyribofuranose

**Fig. 2.6**

## 2.1.4 Nucleosides

When a purine or a pyrimidine base is linked to ribose or deoxyribose the resulting compound is known as a nucleoside. Thus adenine condenses with ribose to form the nucleoside adenosine, guanine forms guanosine, cytosine forms cytidine and uracil forms uridine (Fig. 2.7). These ribonucleosides can be formed on partial hydrolysis of RNA. The ribonucleoside from hypoxahtnine is named inosine. The nucleosides derived from 2-deoxyribose are known as deoxyribonucleosides – deoxyadenosine,

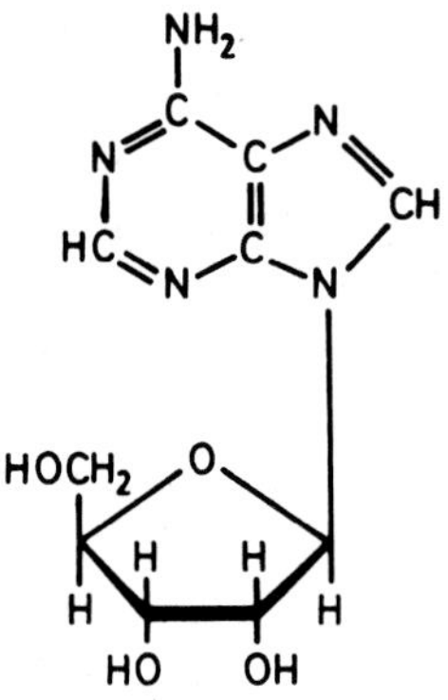

Adenosine
(9-β-D-ribofuranosyl adenine)

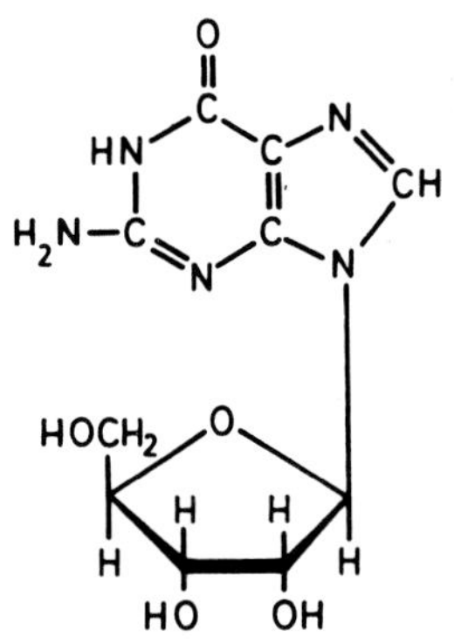

Guanosine
(9-β-D-ribofuranosyl guanine)

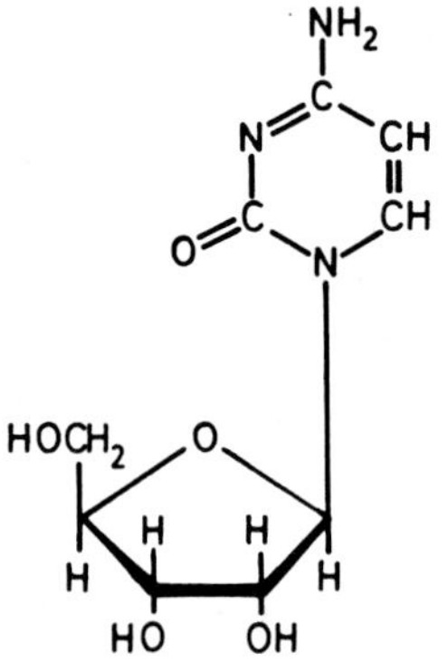

Cytidine
(1-β-D-ribofuranosyl cytosine)

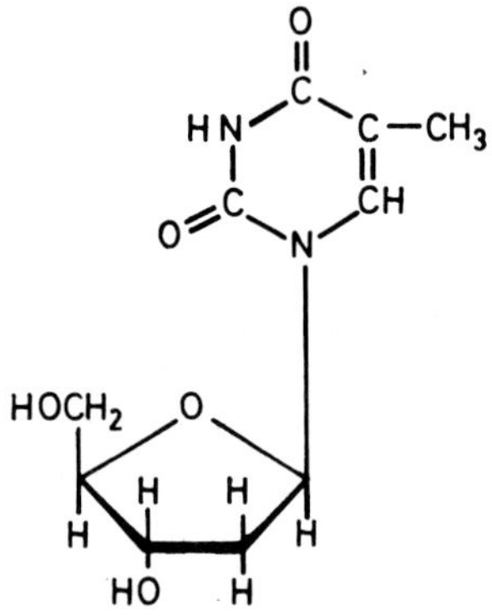

Thymidine
(1-β-D-2-deoxyribo-furanosylthymine)

**Fig. 2.7**

deoxyguanosine, deoxycytidine, deoxythymidine and so on.

In addition to the nucleosides listed above several others are found in very small amounts in certain classes of nucleic acids. These are listed in Table 2.2

The nucleoside 5-ribosyluracil has been obtained in small amounts from the digestion products of RNA, particularly tRNA, and has been named pseudouridine (Fig. 2.8).

**Table 2.2** Minor nucleosides in RNA [7].

| | |
|---|---|
| 1-ribosylthymine | 2′-*O*-methyluridine |
| 5-ribosyluracil (pseudouridine) | 2′-*O*-methylcytidine |
| 2-ribosylguanine | 2′-*O*-methylpseudouridine |
| 2′-*O*-methyladenosine | 2′-*O*-methyl-4-methylcytidine |
| 2′-*O*-methylguanosine | |

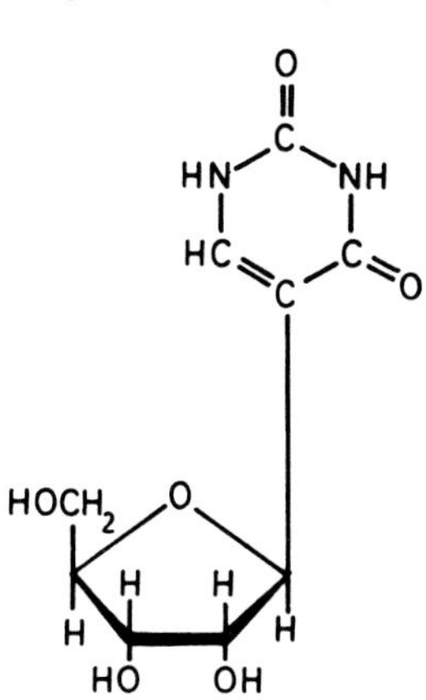

Pseudouridine ($\psi$)
(5-$\beta$-D-ribofuranosyluracil)

**Fig. 2.8**

### 2.1.5 Nucleotides

The structures of nucleotides are dealt with in more specialized terms by Hutchinson [8] and Michelson [9]. They are all phosphoric acid esters of the nucleosides. Those derived from ribonucleosides are usually referred to as ribonucleotides and those from deoxyribonucleosides as deoxyribonucleotides. These terms are sometimes abbreviated to riboside, ribotide, deoxyriboside and deoxyribotide.

Since the ribonucleosides have three free hydroxyl groups on the sugar ring, three possible ribonucleoside monophosphates can be formed. Adenosine, for example, can give rise to three monophosphates (adenylic acids), adenosine 5'-phosphate, adenosine 3'-phosphate and adenosine 2'-phosphate.

In the same way guanosine, cytidine and uridine can give rise to three guanosine monophosphates (guanylic acids), three cytidine monophosphates (cytidylic acids) and three uridine monophosphates (uridylic acids) respectively. These are frequently referred to by the abbreviations shown in the table of abbreviations at the beginning of the book.

The ribonucleoside 5'-phosphates may be further phosphorylated at position 5' to yield 5'-di- and -tri-phosphates. Thus adenosine 5'-phosphate (AMP) yields adenosine 5'-diphosphate (ADP) and adenosine 5'-triphosphate (ATP) (Fig. 2.10). Adenosine 5'- and guanosine 5'-tetraphosphate have also been described.

Similarly the other ribonucleoside 5'-phosphates yield such di- and tri-phosphates as GDP, CDP, UDP, GTP, CTP and UTP. The 5'-monophosphates of adenosine, guanosine, cytidine and uridine together with the corresponding di- and tri-phosphates all occur in the free state in the cell as do the deoxyribonucleoside 5'-phosphates which are referred to as dAMP, dADP, dATP, dTMP, dTDP, dTTP etc.

Ribonucleoside 3',5'-diphosphates and 2',3'-cyclic monophosphates can be formed on hydrolysis of RNA molecules, and ribonucleoside 3',5'-cyclic monophosphates of adenine and guanine occur in many tissues where they play multiple roles in the regulation of metabolic pathways (Fig. 2.11) (see [107] for review).

## 2.2 THE PRIMARY STRUCTURE OF THE NUCLEIC ACIDS

Nucleic acids are polymers made up of hundreds, thousands or millions of nucleotides coupled together by phosphodiester linkages. In the case of DNA where C-4' in the sugar is occupied in ring formation and C-2' carries no hydroxyl group, only the hydroxyl groups at positions 3' and 5' are available for internucleotide linkages. The primary structure of the polynucleotide chain in DNA is shown in Fig. 2.12.

In the case of RNA where there is a hydroxyl group at the 2' position, it is possible to postulate a 2',5' linkage rather than the 3',5' linkage which occurs in DNA. However, hydrolysis with phosphodiesterase from snake venom yields nucleoside 5'-monophosphates and hydrolysis with phosphodiesterase from spleen yields the nucleoside 3'-monophosphates, suggesting that the internucleotide linkage in most RNA is identical to that in DNA.

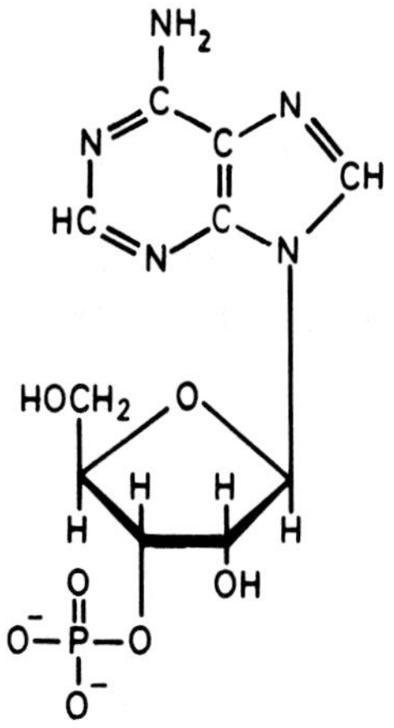

Adenosine
3'-phosphate

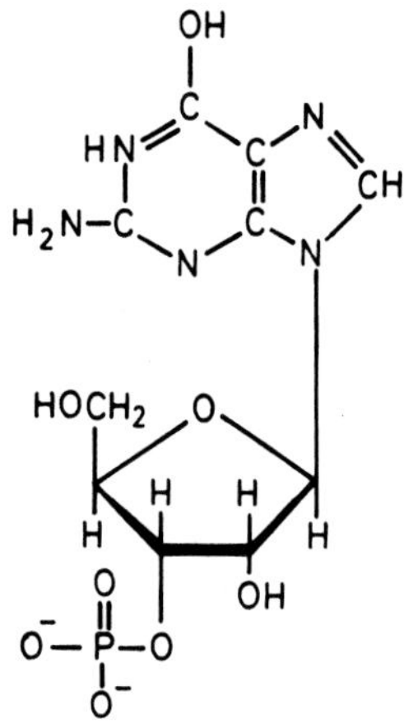

Guanosine
3'-phosphate

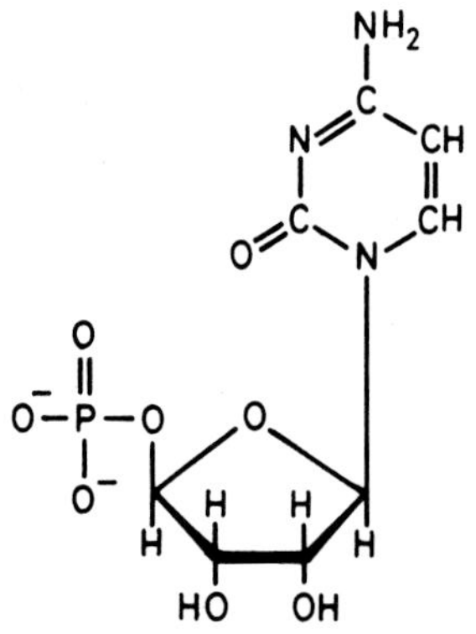

Cytidine
5'-phosphate

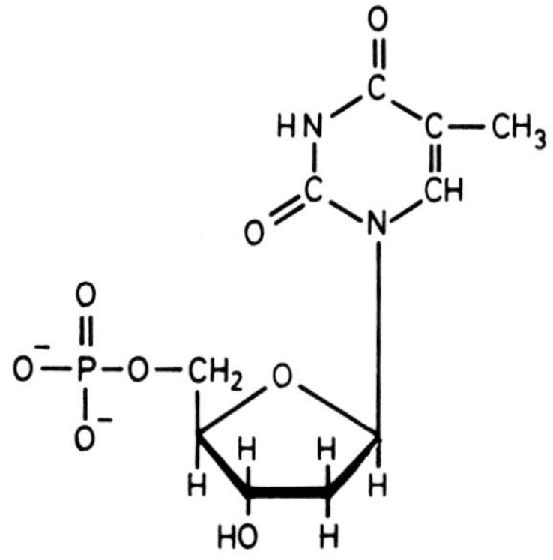

Thymidine
5'-phosphate

**Fig. 2.9**

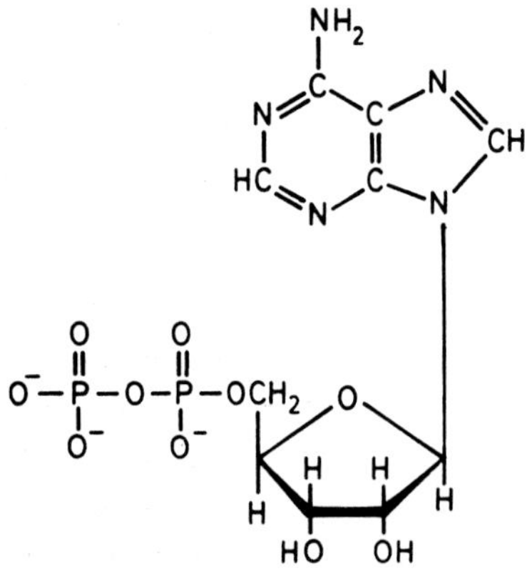

Adenosine diphosphate (ADP)

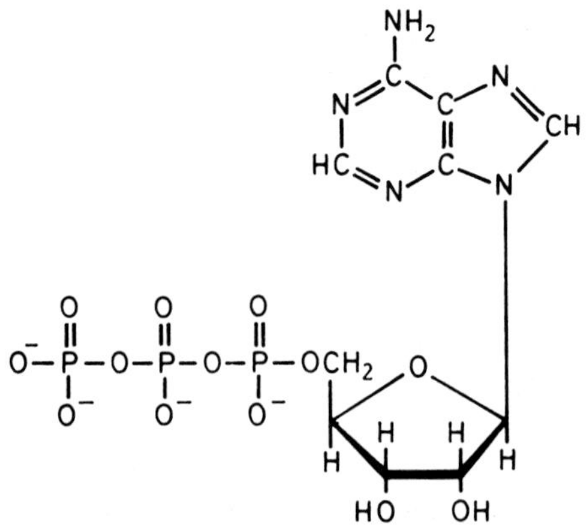

Adenosine triphosphate (ATP)

**Fig. 2.10**

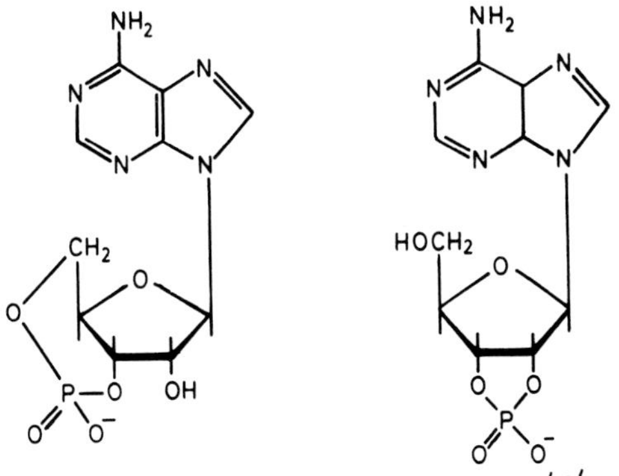

Adenosine 3′ 5′-cyclic
phosphate (cAMP)

Adenosine 2′ 3′-cyclic
phosphate

**Fig. 2.11**

## 2.3 SHORTHAND NOTATION

The representation of polynucleotide chains by complete formulae is clumsy and it has become customary to use the schematic system illustrated in Fig. 2.12, where the chain, shown in full on the left, is abbreviated as on the right. The vertical line denotes the carbon chain of the sugar with the base attached at C-1′. The diagonal line from the middle of the vertical line indicates the phosphate link at C-3′ while that at the end of the vertical line remote from the base denotes the phosphate link at C-5′. This system may be used for either RNA or DNA.

To further simplify the representation of specific polynucleotides the following shorthand notation is commonly used. It was originally suggested by Heppel, Ortiz and Ochoa [117] and is now embodied in the Rules of the CBN (see Preface). A phosphate group is denoted by p; when placed to the right of the nucleoside symbol (e.g. Cp), the phosphate is attached to the C-3′ of the ribose moiety; when placed to the left of the nucleoside symbol (e.g. pC), the phosphate is attached to the C-5′ of the ribose moiety. Thus, UpUp (or U-Up) is a dinucleotide with a phosphodiester bond between the C-3′ of the first uridine and the C-5′ of a second uridine which also has a second phosphate group attached to its C-3′ group. UpU or U-U would be

the dinucleoside monophosphate, uridylyl (3′5′) uridine. The letter p *between* nucleoside residues may be replaced by a hyphen, and is often omitted altogether to save space when long nucleotide sequences are being reported.

The examples in Fig. 2.13 illustrate the method.

The letters A, G, C and U represent adenosine, guanosine, cytidine and uridine respectively. The prefix d (e.g. dA) may be used to indicate a deoxyribonucleoside, but is usually omitted in the case of (deoxy) thymidine. Thymine ribonucleoside is represented as rT to make the distinction.

Cyclic-terminal nucleotides may be represented by using the cyclic-p to indicate a 2′:3′-phosphoryl group or by means of the symbol >p. Thus, U-cyclic-p or U>p is uridine 2′:3′-phosphate and UpU-cyclic-p or UpU>p is the cyclic-terminal dinucleotide.

## 2.4 BASE COMPOSITION ANALYSIS OF DNA

The common monomeric units of DNA are the four deoxyribonucleotides containing the bases adenine, cytosine, guanine and thymine. Many DNAs, however, contain small amounts of other bases, e.g. 5-methylcytosine, which is particularly abundant in wheatgerm DNA (Table 2.3). In a few bacteriophages, one of the common pyrimidine bases is completely replaced by a different pyrimidine base, e.g. in T2, T4 and T6 5-hydroxymethylcytosine completely replaces cytosine, and in PBS 1 (a phage which attacks *Bacillus subtilis*) uracil replaces thymine.

Methods for determining the molar proportions of bases by hydrolysis and chromatography are discussed in detail by Bendich [10]. More recent methods involving high-performance liquid chromatography are now available [11].

The results of the analysis of a number of DNAs as shown in Table 2.3 reveal wide variations in the molar proportions of bases in DNAs from different species although the

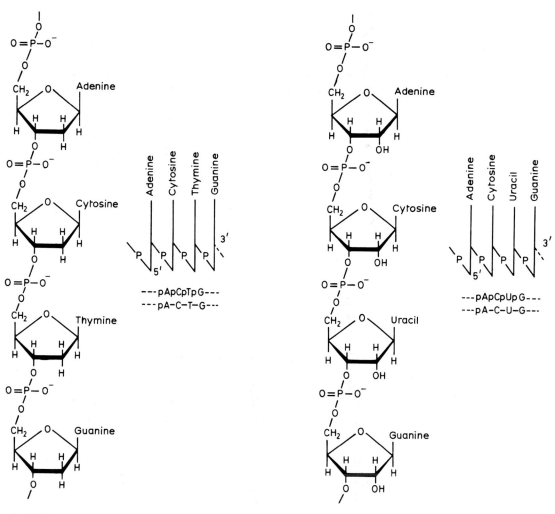

**Fig. 2.12** A section of the polynucleotide chain in DNA (on the left) and RNA (on the right). The shorthand notations are shown alongside.

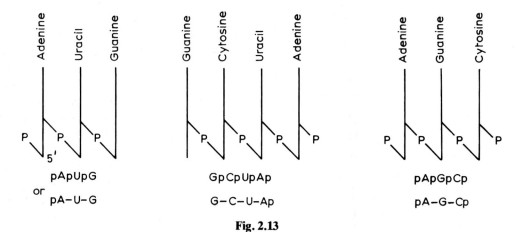

**Fig. 2.13**

**Table 2.3**  Molar proportions of bases (as moles of nitrogenous constituents per 100 g-atoms of P) in DNAs from various sources (data from various authors).

| Source of DNA | Adenine | Guanine | Cytosine | Thymine | 5-Methylcytosine |
|---|---|---|---|---|---|
| Bovine thymus | 28.2 | 21.5 | 21.2 | 27.8 | 1.3 |
| Bovine spleen | 27.9 | 22.7 | 20.8 | 27.3 | 1.3 |
| Bovine sperm | 28.7 | 22.2 | 20.7 | 27.2 | 1.3 |
| Rat bone marrow | 28.6 | 21.4 | 20.4 | 28.4 | 1.1 |
| Herring testes | 27.9 | 19.5 | 21.5 | 28.2 | 2.8 |
| *Paracentrotus lividus* | 32.8 | 17.7 | 17.3 | 32.1 | 1.1 |
| Wheatgerm | 27.3 | 22.7 | 16.8 | 27.1 | 6.0 |
| Yeast | 31.3 | 18.7 | 17.1 | 32.9 | – |
| *E. coli* | 26.0 | 24.9 | 25.2 | 23.9 | – |
| *M. tuberculosis* | 15.1 | 34.9 | 35.4 | 14.6 | – |
| φX174 | 24.3 | 24.5 | 18.2 | 32.3 | – |

DNAs from the different organs and tissues of any one species are essentially the same. Extensive tables showing the molar proportions of bases have been published [12].

It was Chargaff [13] who first drew attention to certain regularities in the composition of DNA. The sum of the purines is equal to the sum of the pyrimidines; the sum of the amino bases (adenine and cytosine) is equal to the sum of the keto (oxo) bases (guanine and thymine); adenine and thymine are present in equimolar amounts, and guanine and cytosine are also found in equimolar amounts. This equivalence of A and T and of G and C is of the utmost importance in relation to the formation of the DNA helix and may be referred to as Chargaff's rule. There are two major deviations from the rule in Table 2.3. (a) In wheatgerm DNA guanine and cytosine are not present in equimolar amounts but this is explained by the scarcity of cytosine being compensated for by the presence of 5-methylcytosine. (b) In φX174 DNA and in the DNA of several similar small coliphages the adenine is not equimolar with thymine nor is guanine with cytosine. This is because φX174 DNA is single-stranded.

## 2.5 MOLECULAR WEIGHT OF DNA (see also Appendix Section A.1.3)

The molecular weights of DNA molecules are very difficult to determine accurately by the methods of classical chemistry since they range from $10^6$ to more than $10^{10}$ (Table 2.4). Conventional analytical equilibrium ultracentrifugation is unsatisfactory since the available instruments are not stable at speeds low enough to balance centrifugal forces against diffusion forces. The best absolute methods of DNA molecular-weight determination are light scattering on low-angle instruments [14–16], equilibrium analytical ultracentrifugation in caesium chloride gradients [17] and viscoelastic relaxation [18]. Most other methods have an empirical basis and rely on the few light-scattering experiments which have been performed for their calibration. These more empirical methods include the measurement of intrinsic viscosity [19], sedimentation rate [19], electron microscopy [20], electrophoresis [21] and radioautography [22]. Generally a combination of two or more of these methods can be used [23]. The advent of the laser has led to the development of a new technique for determining

**Table 2.4**  DNA molecular weights.

| Source | Mol. wt. | Length | Number of kilobase pairs (kbp) |
|---|---|---|---|
| Bacteriophage $\phi$X174 | $1.6 \times 10^6$ | 0.6 $\mu$m | – |
| Polyoma virus | $3 \times 10^6$ | 1.5 $\mu$m | 4.5 |
| Mouse mitochondria | $9.5 \times 10^6$ | 4.9 $\mu$m | 14 |
| Bacteriophage $\lambda$ | $33 \times 10^6$ | 17 $\mu$m | 50 |
| Bacteriophage T2 or T4 | $1.3 \times 10^8$ | 67 $\mu$m | 200 |
| Mycoplasma PPLO strain H-39 | $4 \times 10^8$ | 200 $\mu$m | 600 |
| *H. influenzae* chromosome | $8 \times 10^8$ | 400 $\mu$m | 1 200 |
| *E. coli* chromosome | $3 \times 10^9$ | 1.5 mm | 4 500 |
| *Drosophila melanogaster* chromosome | $43 \times 10^9$ | 20 mm | 70 000 |

diffusion coefficients of large molecules from the Doppler shift in the wavelength of the scattered laser light. This technique together with conventional analytical velocity centrifugation to determine sedimentation coefficients should lead to reliable absolute molecular-weight determinations [24]. The determination of the molecular weight of DNA has also been complicated by the difficulties experienced in the preparation of whole DNA molecules since high-molecular-weight DNA is very susceptible to hydrodynamic shearing forces and to the action of contaminating nucleases. Thus all molecular-weight determinations on an unknown DNA must be carried out in nuclease-free (heat-sterilized) conditions, without pipetting or any other manipulation which puts shear stress on the DNA; any methods which might themselves shear the DNA, such as viscosity determinations, must be used with caution. The size of large DNA molecules from individual chromosomes of eukaryotic cells has recently been determined by viscoelastic relaxation, and it appears that in *Drosophila* at least the DNA is contained in one complete molecule per chromosome [18]. The molecular weights of DNAs from a variety of sources are shown in Table 2.4.

In more recent years it has become customary to describe the size of DNA molecules in terms of kilobase pairs (kbp), rather than daltons or molecular weight. A molecular weight of one million corresponds approximately to 1.5 kilobase pairs.

## 2.6 THE SECONDARY STRUCTURE OF DNA

### 2.6.1 The basic structures

X-ray diffraction has been extensively employed in the study of the molecular architecture of DNA by Astbury [25] and later by Franklin and Gosling [26] and on a very extensive scale by Wilkins and his colleagues [27–29]. Using early information obtained by this technique and chemical observations, Watson and Crick [30–32] in 1953 put forward the view that the DNA molecule is *double stranded* and in the form of a right-handed helix with the two polynucleotide chains wound round the same axis and held together by hydrogen bonds between the bases.

By making scale models they were able to show that the bases could fit in if they were arranged in pairs of one pyrimidine and one purine. The equivalence of adenine and thymine, and guanine and cytosine in most naturally occurring DNA molecules, first observed by Chargaff [13], suggested that the most likely hydrogen-bonding configuration of base pairs be

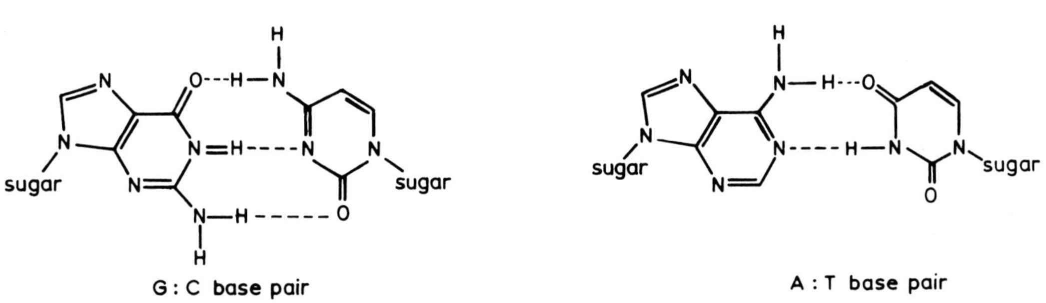

G : C base pair                    A : T base pair

**Fig. 2.14**   The normal base-pairing arrangement found in DNA.
(The dashed lines indicate hydrogen bonds).

of these two types (Fig. 2.14). These base-pairing arrangements have been confirmed as the only ones possible in the B-form of DNA (see Section 2.6.2). Although other base-pairing arrangements have since been suggested [33] they have only been shown to occur in RNA where they are involved in secondary and tertiary structure stabilization (Section 2.9).

The most important consequence of the base-pairing configuration found in DNA is that the order in which the bases occur in one chain automatically determines the order in which they occur in the other, complementary chain. Apart from this essential condition, there are no restrictions on the sequence of the bases along the chains.

The shorthand method of describing two DNA strands has been to have the strand of $5' \rightarrow 3'$ polarity on the top line of the sequence with the complementary strand of opposite polarity lying below. For example:

$$5' - AGGTC - 3'$$
$$3' - TCCAG - 5'$$

The pairs of bases are flat and may be stacked one above the other like a pile of plates so that the molecule is readily represented as a spiral staircase with the base pairs forming the treads (Fig. 2.15). The two polynucleotide chains are of opposite polarity in the sense that the internucleotide linkage in one strand is $3' \rightarrow 5'$ while in the other it is $5' \rightarrow 3'$. The two helices are

right handed (i.e. (a) turn in a clockwise direction as viewed from the near end or (b) turn in the direction of the fingers of the right hand when the thumb indicates the line of sight) and cannot be separated without unwinding. The pitch of the duplex is 3.4 nm and since there are 10 base pairs in each turn of the helix, there is a distance of 0.34 nm between each base pair. The diameter of the helix is 2 nm.

The base–sugar linkage is normally *anti* with the C8 of the purine or the C6 of the pyrimidine ring over the sugar (Fig. 2.16). The sugar pucker is variable but a base pair usually has the sugar attached to the purine in the C2' *endo* configuration and that attached to the pyrimidine in the C3' *endo* configuration. There is a lot of water associated with DNA. Up to 72 molecules of water per 12 base pairs are found largely in the minor groove where they form a spine.

The stacked bases form a hydrophobic core but the amino and keto groups point into the grooves and allow interaction with the solvent. The sugar phosphate backbone points out to give hydrophilic ridges with one negative charge per phosphate.

While the basic model put forward by Watson and Crick remains close to the accepted structure of the DNA molecule in solution, the more refined X-ray diffraction studies of Wilkins and his colleagues demonstrated that DNA fibres can have three possible structures (Table 2.5) with the B structure corresponding closely to the

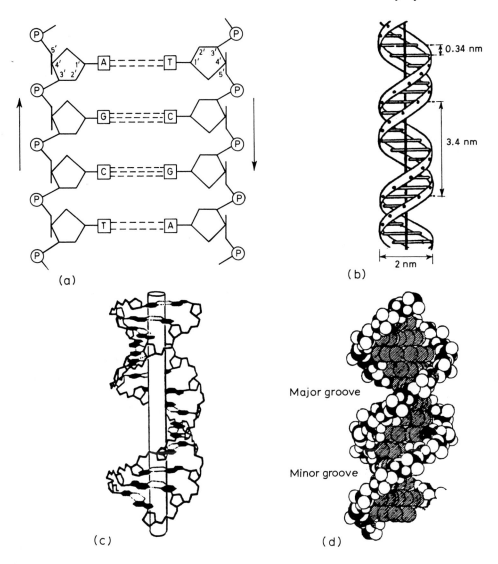

**Fig. 2.15** Various diagrammatic ways of representing DNA: (a) showing polarity and base pairing but no helical twist; (b) showing helical twist and helix parameters but not base pairs; (c) showing helix and base pairs; (d) space-filling representation showing major and minor grooves.

original Watson–Crick model. The A and C structures are also right-handed double helices but differ in the pitch and in the number of bases per turn. In both A and C conformations the bases are not flat but are tilted. Their biological significance is not clear but the A conformation is believed to be very close to the structure adopted by double-stranded RNA and by DNA–RNA hybrids [34]. Because of the presence of the extra 2′ hydroxyl group, RNA appears to be unable to adopt the B conformation. Thus when it is engaged as a template for making RNA (Chapter 10), the DNA molecule must adopt the A conformation. Various organic solvents [35]

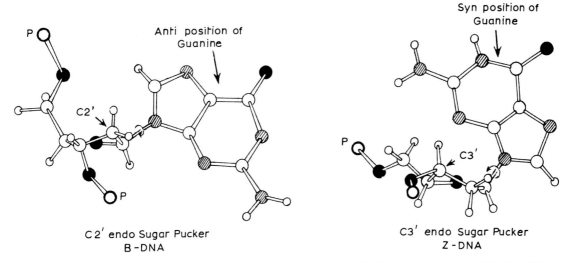

**Fig. 2.16** Conformation of the base–sugar linkage. Guanine is shown in the *anti* conformation linked to C2′ *endo* deoxyribose as in B-DNA and in the *syn* conformation linked to C3′ *endo* deoxyribose as in Z-DNA reproduced with permission, from Rich *et al.* [115] copyright Annual Reviews Inc.

and proteins [36] have been suggested to force the DNA from the B to the A form, and the transition from the B conformation to the C conformation appears to occur in concentrated salt solutions [37] and ethylene glycol [32].

Since the late 1970s there have been several challenges to the B structure of the DNA double helix. It has been suggested that the fibre diffraction data could as readily fit a side-by-side (SBS) model which would have the helix changing from a right- to a left-handed sense at regular intervals [38]. The main attraction of this model is that it would make the unwinding of the duplex in the replication process easier to envisage. X-ray diffraction studies on fibres provide information which relates to the general organization of the sugar phosphate chains but no detail at atomic resolution.

Recently, as a result of the production of crystals of short oligonucleotides, X-ray crystall-

**Table 2.5** The different forms of DNA.

| Form | Pitch (nm) | Residues per turn | Inclination of base pair from horizontal |
|---|---|---|---|
| *A*<br>Na salt<br>75% relative humidity | 2.8 | 11 | 20° |
| *B*<br>Na salt<br>92% relative humidity | 3.4 | 10 | 0° |
| *C*<br>Li salt<br>66% relative humidity | 3.1 | 9.3 | 6° |
| DNA/RNA hybrid | 2.8 | 11 | 20° |

ography has confirmed the right-handed double-helical structure. In contrast to the fibre diffraction patterns analysis of crystals of defined DNA fragments can allow the unequivocal positioning of each atom in the helical frame-work and is a wholly definitive technique. Large oligonucleotides are not readily crystallized so the data currently available relate to small molecules which may well have an atypical structure because the end effects which do not normally make a substantial contribution to polynucleotide conformation may play a domin-ant role.

There have been many experimental attempts to relate the structure of DNA in solution to the structure in crystals or fibres, using techniques such as circular dichroism [37] and low-angle X-ray scattering [39, 40]. One of the best experimental systems for the analysis of secondary structure in solution is circular superhelical DNA (Section 2.8) and experi-mental and theoretical analysis on these molecules suggests that the solution structure is close to, but not entirely compatible with, the B structure of DNA. The molecules are slightly unwound to give 10.4 rather than an integral 10 base pairs per turn [41]. A similar conclusion has been reached by an analysis of the frequency of cleavage by nucleases of DNA adsorbed on to a mica surface [42]. Thus in solution the angle between adjacent base pairs (i.e. helical twist angle) is 34°.

Multistranded structures for DNA have frequently been invoked, especially where the component strands are synthetic polynucleo-tides. One such situation where tetraplexes could occur in chromosomal DNA has been shown to be theoretically feasible if the sequence contains an inverted repeat. Tetraplex DNA has long been suggested as a necessary preliminary structure for genetic recombination [48].

The fact that DNA is a right-handed double helix in solution has been confirmed by Iwamoto and Hsu [43] in an electron microscopy study of small cyclic molecules hydrogen-bonded over a 39 base-pair region only.

Recent studies [108–109] indicate that the DNA need not necessarily assume a straight rod-like shape, but may be bent in a manner dictated by the nucleotide sequence. This may be of great significance when considering the interaction of DNA with proteins, particularly these proteins concerned with packaging the DNA [110, 111], see chapter 3.

### 2.6.2 Variations on the B-form of DNA

The structure considered so far is an average structure for DNA of undefined base sequence. With the availability of crystals of oligonucleo-tides of defined base sequence X-ray crystall-ography has enabled the precise structure of DNA of a specified sequence to be determined. Adjacent bases are related to each other by roll, slide and twist (Fig. 2.17) and the previously defined A, B and C forms are simply structures which exist across a continuum of values. For instance, the bases may slide relative to each other with values of slide ($\sigma$) ranging from $-0.1$ nm to $+0.2$ nm. The A form DNA has a value of $\sigma = 0.15$ nm; the B form has $\sigma = 0$ and in the C form $\sigma = -0.1$ nm. The value of $\sigma$ depends on the actual bases involved and so a stretch of

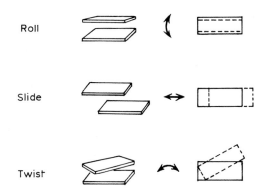

**Fig. 2.17**  Ways in which adjacent bases may move relative to one another (seen from two different positions).

DNA may have regions of A, B and C form DNA. In a similar manner it has become apparent that the helical twist angle is highly dependent on the particular sequence and varies from 27.7° (for ApG) to 40.0° (for GpC). B-DNA is not, therefore, a smooth, regular, double helix but rather the helix shows an irregularity dependent on the base sequence. The alteration of the helical twist angle results, in part, from a rolling of adjacent base pairs over one another along their long axes i.e. along a line from the $C_8$ of the purine across the helix axis to the $C_6$ of the pyrimidine [44–47]. The resulting irregular sequence specific pattern of the sugar phosphate backbone might itself be recognized by proteins interacting with DNA.

When the sequence of bases in DNA shows a regular pattern this may have quite a dramatic effect on the structure of the DNA. When there is a region of DNA containing alternating purines and pyrimidines this can lead to a 'wrinkled' form of DNA [48–50] or in more extreme cases to a complete breakdown of the B form of DNA and conversion to Z-DNA (see below). In wrinkled DNA, although the structure of the purine–pyrimidine dinucleotide (e.g. GpC) is normal, in the next dinucleotide (CpG) the orientation of the linking phosphate is twisted. This has the effect of moving the bases slightly so that the minor groove is made deeper. This could lead to changes in the interactions of the duplex with cations and water molecules. The wrinkled structure may also affect DNA protein interactions and may explain why the *lac* repressor (Section 10.1.1) binds one thousand times more strongly to poly[d(A-T)].poly[d(A-T)] than to DNA of general sequence. In the normal recognition site (the *lac* operator) three-quarters of the bases are arranged as alternating purines and pyrimidines and they may take up a wrinkled configuration.

If, rather than alternating purines and pyrimidines DNA has a purine–pyrimidine dinucleotide followed five bases later by a pyrimidine–purine dinucleotide (i.e. Pu Py a b c Py Pu d e f Pu Py g h i Py Pu) then rather than being wrinkled this would lead to the DNA being bent [51, 52]. Indeed eukaryotic DNA displays just such an anisotropy which may be related to the bending of the DNA around the nucleosome core (see Chapter 3). In a similar manner runs of As, appropriately placed, lead to the formation of a bent DNA molecule [120].

### 2.6.3 Z-DNA

Particularly under conditions of high salt concentration, DNA with an alternating purine–pyrimidine sequence, e.g. poly(dA–dT) poly (dA–dT), tends to form a left-handed double helix known as Z-DNA. This has 12 base pairs per turn in solution. As with wrinkled DNA the repeating unit is the dinucleotide but with the alternating dG.dC polymer the purine sugar linkage is *syn* (Fig. 2.16) placing the $N_7$ and $C_8$ atoms in a very exposed position outside the helix. Because of this, normal base stacking cannot occur but the pyrimidines can stack with pyrimidines on the opposite strands [53, 54, 115]. The pucker of the sugar rings is changed so that the sugar attached to the pyrimidine is now C2' *endo* and that attached to the purine is C3' *endo*. There is one deep groove between the sugar phosphate backbones corresponding to the minor groove in B-DNA. Conversion of B-DNA to Z-DNA brings about a major change in for example the circular dichroism spectrum of DNA (Fig. 2.18).

As well as forming at high salt Z-DNA will form in 1 mM $MgCl_2$ if the carbon 5 of cytosine is substituted with methyl, bromo or iodo groups, e.g. in poly(dG·dm$^5$C) poly(dG·dm$^5$C). It can also form in supercoiled DNA molecules where the region of left-handed DNA serves to relax the tension of the supercoiled molecule (see Section 2.8) [68]. Evidence for the existence of Z-DNA in cells comes from the use of specific antibodies to Z-DNA. It is not certain, however, to what extent Z-DNA forms during the required fixation processes which involve removal of basic proteins and which may induce

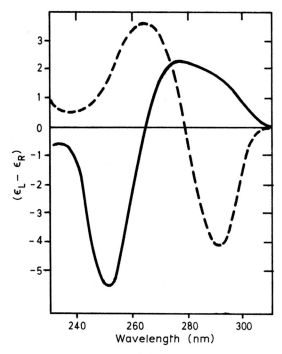

**Fig. 2.18** Circular dichroism spectra of poly(dG-dC) in low-salt 0.2 M NaCl——and high-salt (3.5 M NaCl)–––conditions (reproduced , with permission, from Pohl and Jovin [55] copyright Academic Press, Inc.).

super-coiling and hence stabilize regions of Z-DNA [54, 56, 57].

### 2.6.4 The dynamic structure of DNA

An additional and separate phenomenon is the dynamic secondary structure of DNA. The double helix is not a totally fixed or rigid molecule but undergoes considerable internal deformation in a continuous manner. This aspect of DNA secondary structure is best shown by tritium exchange experiments [112] which indicate that small segments of the double helix can swing apart (breathe) and protrude into the external medium in a manner closely dependent on the environment of the DNA molecule. Whereas X-ray diffraction experiments and circular dichroism show the average conformations that the DNA molecule as a whole can assume,

tritium exchange experiments show the amount of deformation and twisting of these structures that can occur in localized areas. Fluorescence experiments have also confirmed that internal distortions of the DNA molecule occur to a hitherto unsuspected degree [113].

Sequences which breathe most readily have been calculated by Gotoh and Tagashira [75, 76]. The opening of a base pair is determined by the stability of the base pair and its nearest neighbour. 5′-purine-pyrimidine dinucleotide pairs are much more stable than 5′-pyrimidine–purine dinucleotide pairs as a result of the increased overlap (stacking) of the nucleotide rings.

### 2.7 DENATURATION AND RENATURATION

### 2.7.1 DNA denaturation: the helix–coil transition

When double-stranded DNA molecules are subjected to extremes of temperature or pH, the hydrogen bonds of the double helix are ruptured and the two strands are no longer held together. The DNA is said to denature and changes from a double helix to a random coil. When heat is used as the denaturant the DNA is said to melt and the temperature at which the strands separate is the melting or transition temperature $T_m$.

The component bases of a polynucleotide absorb light at 260 nm, but in double-stranded DNA this absorption is partially suppressed. This is because the stacking of the bases, one above the other, leads to coupling between the transition of the neighbouring chromophores, i.e. the $\pi$-electrons of the bases interact with one another.

When duplex DNA melts the hydrogen bonds break and the bases unstack with the consequence that the absorption at 260 nm rises by about 20–30%. This is the *hyperchromic effect* and is used to monitor the melting of DNA (Fig. 2.19).

As well as a change in absorbance of ultra-

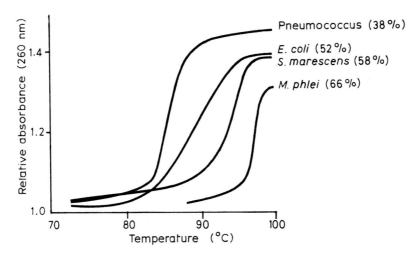

**Fig. 2.19** Denaturation by heat of DNAs isolated from different sources. The figures in brackets indicate the composition of the DNA in G + C (%) (from *Molecular Genetics* by G. S. Stent, W. H. Freeman and Co. Copyright 1971 – after [116]).

violet light the helix coil transition is also accompanied by a change in density of the DNA, the single-stranded molecule being more dense than the corresponding duplex.

The nature of the melting transition is affected by several factors:

(a) *the (G + C) content of the DNA.* There are three hydrogen bonds involved in the G-C base pair and only two in the A-T base pair (Fig. 2.14). Because of this the higher the G + C content of DNA the more stable will the molecule be and the higher will be the melting temperature. Fig. 2.19 shows the melting of DNAs of different G + C content. The equation:

$$\%GC = (T_m - X)2.44$$

[58] expresses the relationship where X is dependent on:

(b) *the nature of the solvent.* In low concentrations of counterion denaturation occurs at relatively low temperatures and over quite a broad range of temperature. At higher concentrations of counterion the $T_m$ is raised and the transition is sharp.

(c) *the nature of the DNA.* Most DNA molecules are mosaics of regions of varying G + C composition [59] and the A + T-rich regions will melt prior to the G + C-rich regions. This can result in the two strands being held together and hence in register by G + C-rich regions. On cooling these double-stranded regions will act as foci to allow rapid reannealing of the two strands of DNA. Only when all the hydrogen bonds are broken does the melting become irreversible on rapid cooling (see below). Short viral DNAs which may be more homogeneous exhibit sharper melting profiles.

**2.7.2 The renaturation of DNA: $C_0t$ value analysis**

When two DNA strands are returned from the extreme conditions which caused them to melt they may reassociate to re-form a double helix. In order to do so correctly they must align themselves perfectly and this is a process dependent on both the concentration of the DNA molecules and the time allowed for reassociation. Very often imperfect matches may be formed which must again dissociate to allow the strands to align correctly. For the dissociation to happen and to encourage the diffusion of the very large molecules involved the temperature must be maintained just below the melting temperature for reannealing to occur. If the solution is quenched to 4° this limits diffusion and prevents the separation of any DNA strands which have become mismatched. Consequently, except for simple DNA molecules which correctly reanneal almost instantaneously, quenching produces solutions of denatured DNA.

If renaturation is allowed to occur under ideal conditions, two DNA samples of identical concentration will take different times to reanneal depending on their complexity. Thus the DNA from a mammal is far more complex and has more different sequences than does the DNA from a small virus. Each sequence is therefore present in a much lower concentration and will take correspondingly longer to find its complementary strand. The complexity of a DNA sample is thus reflected in the time it takes a solution of DNA of given concentration to

reanneal. The $C_0t$ value of a DNA is defined as the initial concentration ($C_0$) in moles nucleotide per litre multiplied by the time ($t$) in seconds it takes for the DNA to reanneal (Fig. 2.20).

The reassociation can be followed spectroscopically or by taking advantage of the fact that a duplex DNA binds more strongly to hydroxyapatite than does single-stranded DNA. This latter method allows the reannealing reactions to be performed over a much broader range of DNA concentration and also allows the isolation of preparations of duplex and single-stranded DNA molecules from a partially reannealed sample. A third method of studying reassociation of DNA molecules makes use of S1 nuclease (see Chapter 4). This enzyme preferentially digests the single-stranded DNA leaving behind the duplex molecules only. For both the hydroxyapatite and the S1 nuclease method it is convenient to have the DNA radioactively labelled thus allowing rapid and accurate quantitation of the proportions of DNA in the reannealed and single-stranded forms.

Further results of renaturation ($C_0t$) analysis will be considered in Chapter 3.

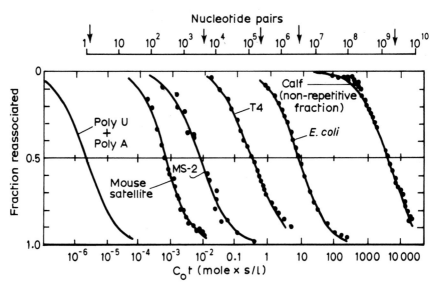

**Fig. 2.20**  The rate of reassociation of double-stranded polynucleotides from various sources showing how the rate decreases with the complexity of the organism and its genome (from [60]).

### 2.7.3 The buoyant density of DNA

The physical properties of DNA are strongly influenced by the percentage of G + C in the molecule, and the buoyant density of DNA in concentrated CsCl solutions is no exception. G + C-rich DNA has a higher buoyant density than A + T-rich DNA [61] and there is a linear relationship between the buoyant densities ($\rho$) of different DNAs and their G + C contents (see Fig. 2.21). This can be expressed by the relationship:

$$\rho = 1.660 + 0.098 \, (GC)$$

where GC is the mole fraction of (G + C). The relative (G + C) content can also be determined from the thermal denaturation temperature discussed above [58] and from the ultraviolet spectrum of the DNA [114].

Several dyes and antibiotics have been shown to have strong binding specificity for either AT or GC base pairs. Thus distamycin and netropsin are AT-specific and actinomycin is GC-specific [62, 63]. This property can be used to improve the separation of DNA molecules which differ in their GC content in CsCl gradients, since dye binding alters the buoyant density of the molecules. In addition, dyes such as the AT-specific

malachite green, or the GC-specific phenyl neutral red, can be immobilized on polyacrylamide columns and used to fractionate DNA molecules of differing base composition [65].

On the basis of such measurements the relative (G + C) contents of the DNAs from a wide variety of sources have been determined and are shown in Table 2.6. While mammalian DNAs show a (G + C) content between 40 and 45%, the range of bacterial DNAs is much wider (30–75%). The significance of these variations in base content has been discussed in relation to the taxonomy of bacteria [67] and protozoa [69] and to the evolution of various organisms [70].

**Table 2.6** The relative (G + C) content of DNAs from various sources [14, 66, 118].

| Source of DNA | Percentage (G + C) |
| --- | --- |
| *Plasmodium falciparum* (malarial parasite) | 19 |
| *Dictyostelium* (slime mould) | 22 |
| *M. pyogenes* | 34 |
| Vaccinia virus | 36 |
| *Bacillus cereus* | 37 |
| *B. megaterium* | 38 |
| *Haemophilus influenzae* | 39 |
| *Saccharomyces cerevisiae* | 39 |
| Calf thymus | 40 |
| Rat liver | 40 |
| Bull sperm | 41 |
| *Diplococcus pneumoniae* | 42 |
| Wheatgerm | 43 |
| Chicken liver | 43 |
| Mouse spleen | 44 |
| Salmon sperm | 44 |
| *B. subtilis* | 44 |
| T1 phage | 46 |
| *E. coli* | 51 |
| T7 phage | 51 |
| T3 phage | 53 |
| *Neurospora crassa* | 54 |
| *Pseudomonas aeruginosa* | 68 |
| *Sarcina lutea* | 72 |
| *Micrococcus luteus* | 72 |
| Herpes simplex virus | 72 |
| *Mycobacterium phlei* | 73 |

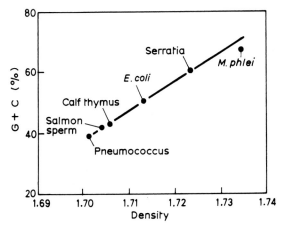

**Fig. 2.21** Relationship of density to content of guanine plus cytosine in DNAs from various sources [64].

## 2.8 SUPERCOILS, CRUCIFORMS AND TRIPLE-STRANDED STRUCTURES

Imagine a covalently closed cyclic DNA molecule (or a linear molecule physically constrained at both ends). Break one of the strands and unwind the double helix by one or two turns and then rejoin the strands. The resulting molecule will try to rewind the two strands back to their normal structure (the most stable form) but will be unable to do so because of the covalent closure. To compensate the duplex will take on a superhelical configuration (Fig. 2.22). The number and nature of the supercoiled turns will depend on the difference between the secondary structure of the DNA when it was sealed and the secondary structure under the conditions of observation.

Covalently closed, cyclic duplex DNA differs very considerably from linear or open cyclic molecules (i.e. molecules with an interruption in the phosphodiester backbone) of the same size and base composition. The two strands of the DNA are unable to separate and thus a high temperature is required to disrupt the structure of the supercoiled DNA. When these DNA molecules eventually do melt, the two strands cannot separate, but the entire molecule col-

lapses into a compact, fast-sedimenting complex of the two interlocked random coils. The secondary structure of such DNA can differ detectably from that of the open circular form because of the tertiary restraints on the molecule [71]. Thus the cyclic duplex DNA molecules of the viruses SV40 and polyoma have been shown to have short regions where the two DNA strands are unwound [72, 73]. The intrinsic viscosity is low, and the sedimentation rate and electrophoretic mobility about 20% faster than in the open circular molecules of the same size. The tertiary structure of the superhelix is most commonly represented by a straight interwound superhelix (Fig. 2.22). Other forms such as toroidal and branched structures have been shown to exist in solution [74].

The linking number ($Lk°$) is the number of Watson–Crick turns present in a DNA molecule. Thus for a molecule of 5000 base pairs $Lk°$ is about 500. If the molecule is partially unwound say by 25 turns there is a change in linking number ($\Delta Lk$) equal to $-25$. The superhelix density ($\sigma$) is defined as $\Delta Lk/Lk°$ ($= -0.05$ in the above example). In other words the superhelix density of a covalently closed cyclic DNA is the number of superhelical turns per 10 base pairs. Superhelix density is determined by 'partial denaturation' of the DNA either with alkali [79] or with an intercalating dye [78]. The binding of the dye molecules to the DNA results in an increase in the number of residues per turn in the duplex and thus a decrease in the number of superhelical turns. The sedimentation rate of the DNA therefore drops to a minimum and then rises as the binding of the further dye molecules causes the DNA to take on superhelical turns of the opposite sense. Frequently, superhelix density is estimated by gel electrophoresis in the presence of increasing amounts of an enzyme which abolishes superhelical turns [77]. The superhelix density of most covalently closed DNA molecules is $-0.06$ in neutral caesium chloride [78, 79]. Superhelix density is affected by both temperature and ionic strength [80].

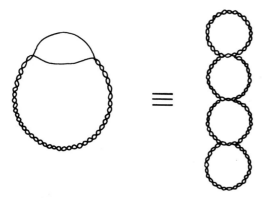

**Fig. 2.22** Supercoiling of DNA. The supercoiled molecule on the right is a low-energy equivalent of the partly underwound molecule.

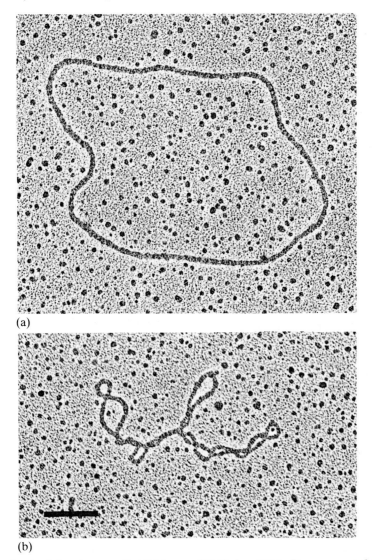

(a)

(b)

**Fig. 2.23** Open circular (a) and supercoiled (b) forms of PM2 virus DNA. Bar represents $0.2\mu$m. (By courtesy of Dr Lesley Coggins.)

Electron micrographs of relaxed and supercoiled molecules are shown in Fig. 2.23.

There is some doubt as to whether supercoiled DNA exists in eukaryotic cells or whether all the tension introduced by underwinding is removed by association of the DNA with proteins. Although this may be true in general it is probable that certain stretches of DNA in the transcriptionally active fraction are in a torsionally stressed state [81, 82] – see Chapter 10.

Another discontinuity in DNA may be produced by palindromic sequences which may fold back on themselves to form cruciform structures (Fig. 2.24). Such structures can be formed *in vitro* by *intramolecular* reassociation of single-stranded DNA but a foldback structure, even

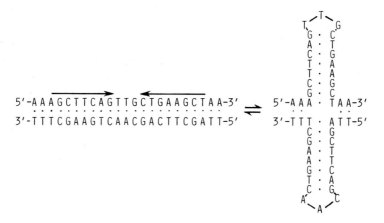

**Fig. 2.24** Palindromes and cruciforms. The arrows indicate the palindromic sequence which can fold back on itself to form the cruciform structure.

when formed from a perfect palindrome, will never be as stable as the linear duplex DNA as there will always be an unpaired region of DNA in the loop region.

As with Z-DNA, cruciforms may be stabilized by specific binding proteins or by their presence in supercoiled molecules as cruciforms help to relieve the supercoil tension. Even in this situation, however, they are unlikely to be present in DNA *in vivo* and form only very slowly *in vitro* [83–85].

Watson–Crick base pairs are not the only ones possible and there is evidence that where regions of DNA contain runs of pyrimidines in one strand these can fold back on themselves to form triple-stranded helices containing one polypurine and two polypyrimidine tracts [86, 87]. One of the latter will run parallel with the polypurine tract (i.e. in the same 5' to 3' orientation) and form with it Hoogsteen base interactions (see Fig. 2.27 and Section 2.9).

## 2.9 THE SECONDARY AND TERTIARY STRUCTURE OF RNA

RNA has a variety of functions within the cell and for each function a specific type of RNA is required. The types of RNA differ in chain length and secondary and tertiary structures. Messenger RNA (mRNA) is involved in carrying the genetic message from the DNA to the site of protein synthesis which takes place on small particles known as ribosomes. The ribosomes are made of protein and RNA and the ribosomal RNA (rRNA) itself consists of several different-sized molecules. A small RNA molecule known as transfer RNA (tRNA) is involved in the transfer of amino acids to the ribosome (see Chapter 11 for a detailed description of protein synthesis and the structure of the ribosome and tRNA). In addition several other small RNA molecules form parts of enzymes or are involved in enzymic transformation of macromolecules (see Chapters 9 and 11).

While RNA molecules do not possess the regular interstrand hydrogen-bonded structure characteristic of DNA, they have the capacity to form double-helical regions. These helices can be formed between two separate RNA chains, but are more frequently found between two segments of the same chain folded back on itself. The secondary structure is similar to the A form of DNA with tilted bases, since the 2' OH hinders B structure formation. The helical

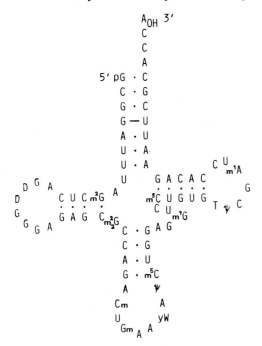

**Fig. 2.25** Yeast tRNA^Phe. The cloverleaf secondary structure is shown with the hydrogen bonds between standard Watson–Crick base pairs denoted by a dot, and the bond between a G-U base pair denoted by a dash. The identity of the bases can be found by consulting the list of abbreviations at the front of the book.

(1) The various base pairs have differing stability and this is modified by the nature of the adjacent base pairs and the presence of interruptions to the duplex region. The most stable base pairs are CG or GC base pairs following a GC base pair. They have a free energy $(\Delta G)$ of $-20$ kJ per mole base pair compared with the very low value of $-1$ kJ per mole base pair for a UG base pair following another UG base pair [88].

(2) Along a duplex region there may be occasional unpaired bases forming bulges or short loops. These have a strong destabilizing effect on the duplex region.

(3) At one end of an intrastrand base-paired stem there is a hairpin loop (Fig. 2.26). A loop must have a minimum of three unpaired nucleotides. Hairpin loops with six unpaired

regions formed in this manner are seldom regular as the segments on the chain brought into opposition do not have entirely complementary sequences so non-bonded residues 'loop out' of the structure (Fig. 2.25). In some RNA molecules in the region of 70% of the bases are involved in secondary structure interactions (Chapter 11).

Such molecules frequently show unusual base pairing in addition to the expected A:U and G:C pairs. For instance G:U pairing is observed. To form a stable duplex requires at least three conventional base pairs and the stability of a duplex region depends on three factors.

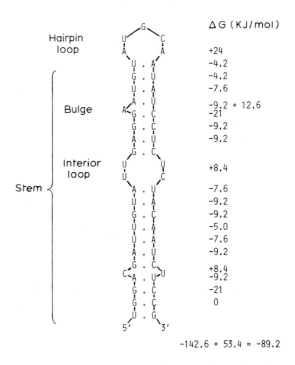

**Fig. 2.26** A stem–loop structure formed by intrastrand hydrogen bonding of a palindromic region of RNA. The predicted stability of the region is calculated from the data of Tinoco [88].

bases are the most stable but even these reduce the stability of the duplex region by 16–24 kJ per mole oligonucleotide. With shorter loops steric hindrance and base stacking interactions destabilize the loop yet the greater the distance between self-complementary regions the less likely it becomes that duplex regions will be formed. Fig. 2.26 indicates the predicted energy of a hairpin region of palindromic RNA. If the stem–loop region is more stable than $\Delta G = -40$ kJ per mole oligonucleotide there is a possibility that such regions will exist *in vivo* but this is only one of the criteria which must be satisfied before confidence can be placed in predictions of secondary structure for RNA [89, 90].

X-ray crystallographic data of many small RNA molecules show that extensive folding of partially duplex arms occurs. The subsequent hydrogen-bonding of those bases not already involved in secondary structure formation is important in stabilizing the folds (see Chapter 11). The types of hydrogen-bonding involved is frequently not that found in the conventional Watson–Crick base pair. In addition, short,

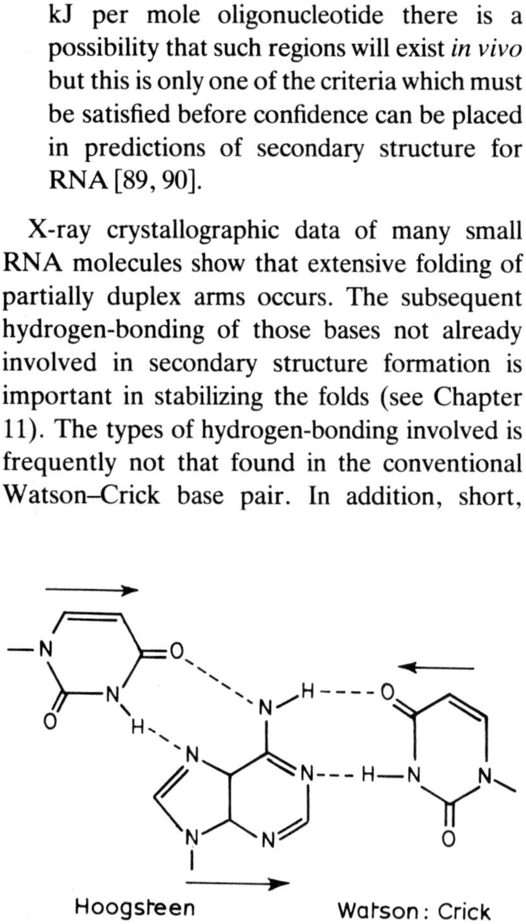

triple-stranded regions can occur in which two of the chains run parallel with one another. Such unusual structures have been studied with model systems where the homopolymer chains poly(A) and poly(U) have been shown to form a triple-stranded structure in which antiparallel poly(A) and poly(U) strands are held together by conventional Watson–Crick base pairs while a second poly(U) strand uses Hoogsteen base pairs to bind in parallel to the poly(A) strand (Fig. 2.27) [87, 91, 92, 103–106].

For further details of the possible secondary and tertiary structures of RNA molecules see Chapter 11.

## 2.10 CHEMICAL REACTIONS OF BASES, NUCLEOTIDES AND POLYNUCLEOTIDES

### 2.10.1 Reactions of ribose and deoxyribose [93, 94]

The sugars are readily acylated and alkylated. In the nucleoside the 5'-OH is most susceptible and reaction with triphenylmethyl chloride gives the 5'-*O*-trityl derivative. In the ribopolynucleotides the 2'-OH is acetylated by acetic anhydride to give a molecule which is now resistant to ribonucleases (see Section 2.10.3 and Chapter 4).

Oxidation with periodate occurs at 2'3'-glycols, i.e. at the terminal nucleotide in ribopolynucleotides. The initial reaction gives a dialdehyde which is lost from the polynucleotide by an amine-catalysed cleavage (Fig. 2.28).

Similarly the 2'3'-glycol can react with aldehydes or ketones under acidic conditions or with boric acid to form crosslinked complexes.

Depurination reactions (see below) lead to the conversion of the sugars into furfural derivatives which give specific colour reactions with orcinol [95] or diphenylamine [96]. These form the basis of quantitative assays for RNA or DNA respectively.

Hoogsteen                    Watson : Crick

**Fig. 2.27** The triple-stranded structure formed by two poly(U) and one poly(A) strand involves a Watson:Crick base paired poly(A)·poly(U) with the second poly(U) strand running in the same direction as the poly(A) strand and bonded to it with Hoogsteen hydrogen bands.

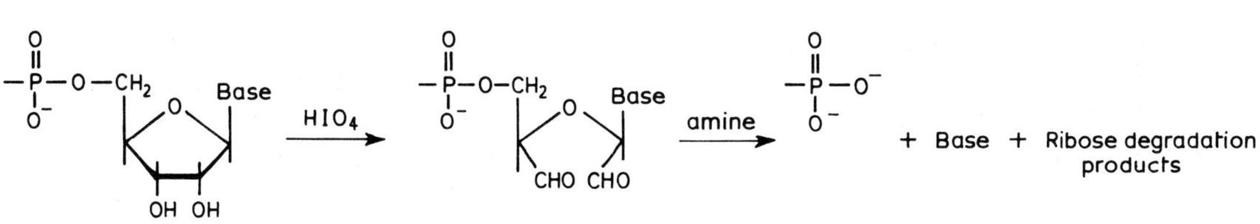

**Fig. 2.28** Interaction of 3′-terminal nucleotide of RNA with periodate.

### 2.10.2 Reactions of the bases

(a) Nitrous acid reacts with amino groups to convert them into hydroxyls (Fig. 2.29). It thus converts:

cytosine → uracil

adenine → hypoxanthine

guanine → xanthine

It is the free acid which is active and hence reaction must take place at a low pH (about 4.25) and native DNA is not very susceptible to deamination. Nitrous acid is used as a mutagenizing reagent as the bases resulting from treatment have altered base-pairing potential (see Chapter 7) [97, 98].

(b) Formaldehyde also reacts with free amino groups to produce methylol derivatives but these can then produce crosslinks between two bases or between nucleic acid and protein [99].

$$R\text{-}NH_2 \rightarrow R\text{-}NH\text{-}CH_2OH \rightarrow$$

$$R\text{-}NH\text{-}CH_2\text{-}NH\text{-}R'$$

As formaldehyde only reacts with single-stranded regions of DNA it can be used to fix such regions for electron microscopy.

(c) Bisulphite reacts with pyrimidines in single-stranded regions of nucleic acids to form an addition product across the 5-6 double bond [100] (Fig. 2.30). The adduct is formed at pH 6 but the reaction proceeds in the reverse direction under mild alkaline conditions. The cytosine adduct is rapidly deaminated at pH 5–6 and this provides a mild way of converting cytosine to uracil.

(d) Bromine and iodine will add to pyrimidines to give for example 5-bromouracil or 5-bromocytosine and with purines to form the 8-bromo derivative.

(e) Hydrazine produces a nucleophilic addition across the 5–6 double bond of pyrimidine bases (e.g. Fig. 2.31).

A slow reaction also occurs with adenine bases. The (deoxy)ribosyl urea breaks down to give apyrimidinic acid with concomitant polynucleotide chain cleavage, a reaction which is used in nucleic acid sequencing (see Section A.5.1).

(f) Hydroxylamine reacts with the amino group of cytosine but it also reacts more slowly with thymine and uracil to bring about pyrimidine ring cleavage, and eventually depyrimidination of nucleic acids.

(g) Strong mineral acids lead to depurination of nucleic acids in an amine-catalysed reaction [93, 96]. The purine-*N*-glycosyl bond is much less stable than the pyrimidine-*N*-glycosyl bond. Purine-*N*-glycosyl bond cleavage occurs much more readily with DNA than with RNA, and concomitant phosphodiester bond cleavage

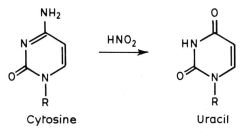

Cytosine          Uracil

**Fig. 2.29** Deamination of cytosine by treatment with nitrous acid.

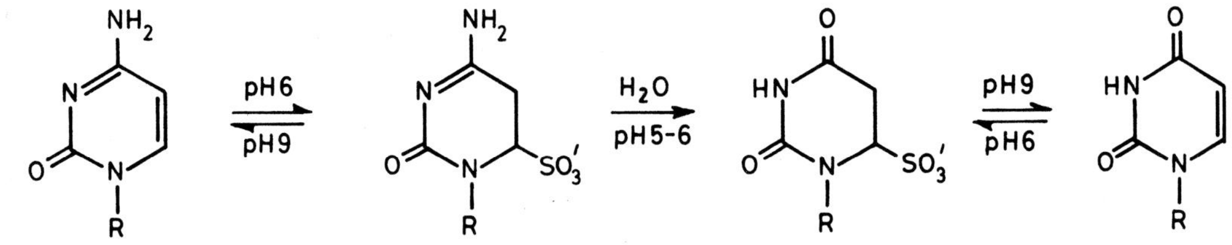

**Fig. 2.30**   Interaction of cytosine with bisulphite.

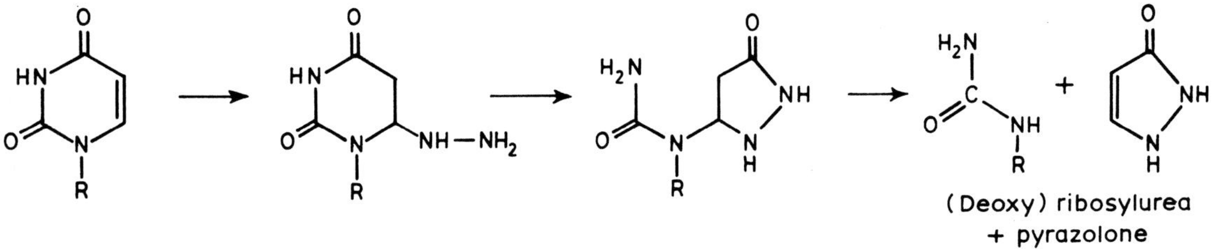

**Fig. 2.31**   The action of hydrazine with pyrimidine bases.

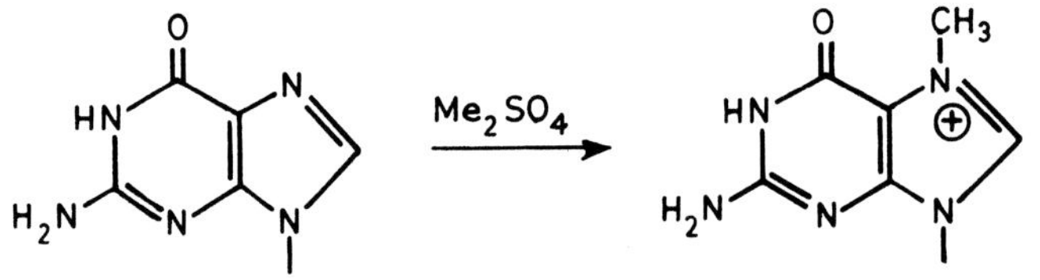

**Fig. 2.32**   Methylation of guanine with dimethyl sulphate.

leads to hydrolysis of the nucleic acid to low-molecular-weight compounds. Such reactions are made use of in the Maxam–Gilbert sequencing technique (see Section A.5.1).

(h) Alkylating agents are often carcinogens and alkylation of nucleic acid bases can occur at several sites (see Chapter 7). The N-7 position of guanine is particularly sensitive (Fig. 2.32).

The positive charge produced at N-7 renders the polynucleotide sensitive to cleavage with piperidine a reaction which is an important step in the Maxam–Gilbert sequencing technique (see Section A.5.1).

### 2.10.3 Phosphodiester bond cleavage

The strong hot acids required to break the phosphodiester backbone of DNA also lead to release of the free bases and breakdown products of deoxyribose. Typical treatments are with 12 M perchloric acid at 100°C or 98% formic acid at 170°C. DNA is not sensitive to mild alkaline hydrolysis.

Alkaline hydrolysis of RNA (0.3 M NaOH at 37°C) yields both the nucleoside 2'- and 3'-phosphates. These isomers are readily interconverted under acidic conditions via the intermediate formation of a nucleoside 2':3'-cyclic

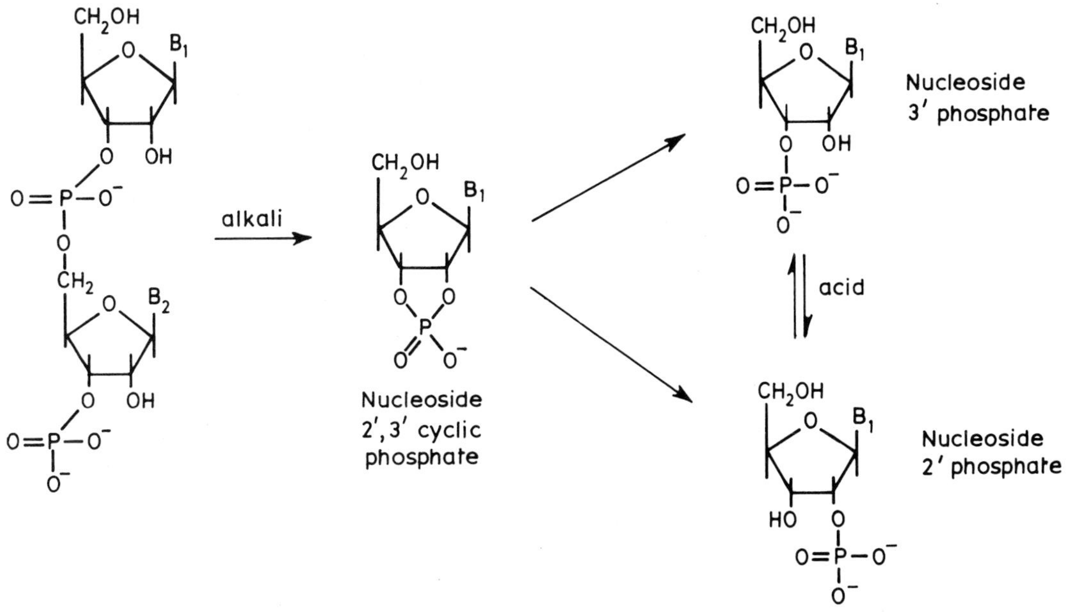

**Fig. 2.33** Hydrolysis of a dinucleotide by alkali. The cyclic phosphate is formed as an intermediate and is hydrolysed to give a mixture of 2′- and 3′-phosphates. These can be interconverted (via the cyclic phosphate) under acidic conditions.

phosphate (Fig. 2.33). Hydrolysis by pancreatic ribonuclease also proceeds via the formation of cyclic phosphate but treatment with venom phosphodiesterase yields the nucleoside 5′-phosphates (Chapter 4).

Alkylation of the 2′-OH group to yield for instance 2′-*O*-methylribonucleosides (which are found in ribosomal and transfer RNA) renders the phosphodiester bond resistant to cleavage by alkali or by pancreatic ribonuclease since the 2′ : 3′-cyclic phosphate cannot be formed.

### 2.10.4 Photochemistry [87, 101, 102]

The bases absorb ultraviolet light but the spectra of the bases are considerably affected by glycosylation and by pH. For example adenine has a p$K$ at 4.1 for the protonation of N-1 and a second p$K$ at 9.8 for the protonation of the amino group both of which affect the spectrum.

**Table 2.7** Wavelength (nm) of maximum absorbance ($\lambda_{max}$) of adenine and adenosine at various pHs.

| pH | Adenine | Adenosine |
|----|---------|-----------|
| 1  | 265.5   | 257.0     |
| 7  | 260.5   | 260.0     |
| 12 | 269.0   | 260.0     |

As described in Section 2.7 absorption is also affected by base stacking interactions.

Ultraviolet irradiation of pyrimidines in solution catalyses the addition of water across the $C_5$–$C_6$ double bond. Adjacent pyrimidines on a polynucleotide chain form cyclobutane-linked dimers (see Chapter 7).

### REFERENCES

1 Adler, M., Weissman, B. and Gutman, A. B. (1958), *J. Biol. Chem.*, **230**, 717.

2 Littlefield, J. W. and Dunn, D. B. (1958), *Biochem. J.*, **70**, 642.

3 Smith, J. D. and Dunn, D. B. (1959), *Biochem. J.*, **72**, 294.

4 Davis, F. F., Carlucci, A. F. and Roubein, I. F. (1959), *J. Biol. Chem.*, **234**, 1525.

5 Nakanishi, K., Blobstein, S., Funamizu, M., Furutachi, N., Van Lear, G., Grunberger, D., Lanks, K. W. and Weinstein, I. B. (1971), *Nature (London) New Biol.*, **234**, 107.

6 Thiebe, R., Zachau, H. G., Baczymskyj, L., Biemann, K. and Sonnenbichler, J. (1971), *Biochim. Biophys. Acta*, **240**, 163.

7 Nichimura, S. (1972), *Prog. Nucleic Acid Res. Mol. Biol.*, **12**, 49.

8 Hutchinson, D. W. (1964), *Nucleosides and Coenzymes*, Methuen, London.

9 Michelson, A. M. (1963), *The Chemistry of Nucleosides and Nucleotides*, Academic Press, New York.

10 Bendich, A. (1957), *Methods Enzymol.*, **3**, 715.

11 Adams, R. L. P., McKay, E. L., Craig, L. M. and Burdon, R. H. (1979), *Biochim. Biophys. Acta*, **563**, 72.

12 Fasman, G. D. (ed.) (1976), *CRC Handbook of Biochemistry and Molecular Biology* (3rd edn), Vol. 2, CRC Press, Boca Raton.

13 Chargaff, E. (1963), *Essay on Nucleic Acids*, Elsevier/North-Holland, Amsterdam.

14 Harpst, J. A., Krasna, A. I. and Zimm, B. H. (1968), *Biopolymers*, **6**, 595.

15 Krasna, A. I., Dawson, J. R. and Harpst, J. A. (1970), *Biopolymers*, **9**, 1017.

16 Krasna, A. I. (1970), *Biopolymers*, **9**, 1029.

17 Schmidt, V. W. and Hearst, J. E. (1969), *J. Mol. Biol.*, **44**, 143.

18 Kavenoff, R. and Zimm, B. H. (1963), *Chromosoma*, **41**, 1.

19 Crothers, D. M. and Zimm, B. H. (1965), *J. Mol. Biol.*, **12**, 525.

20 Lang, D. (1970), *J. Mol. Biol.*, **54**, 557.

21 Southern, C. (1978), *Methods Enzymol.*, **68**, 152.

22 Leighton, S. B. and Rubenstein, I. (1969), *J. Mol. Biol.*, **46**, 313.

23 Freifelder, D. (1970), *J. Mol. Biol.*, **54**, 567.

24 Dublin, S. B., Benedek, G. B., Bancroft, F. C. and Freifelder, D. (1970), *J. Mol. Biol.*, **54**, 547.

25 Astbury, W. T. (1974), *Symp. Soc. Exp. Biol.*, **1**, 66.

26 Franklin, R. and Gosling, R. G. (1953), *Nature (London)*, **171**, 740, **172**, 156.

27 Langridge, R., Wilson, H. R., Hooper, C. W., Wilkins, M. H. F. and Hamilton, L. D. (1960), *J. Mol. Biol.*, **3**, 547.

28 Fuller, W., Wilkins, M. H. F., Wilson, H. R. and Hamilton, L. D. (1965), *J. Mol. Biol.*, **12**, 60.

29 Davies, D. R. (1967), *Annu. Rev. Biochem.*, **36**, 321.

30 Watson, J. D. and Crick, F. H. C. (1953), *Nature (London)*, **171**, 737, 964.

31 Watson, J. D. (1968), *The Double Helix*, Atheneum, New York.

32 Olby, R. (1964), *The Path to the Double Helix*, Macmillan, London.

33 Arnott, S. (1970), *Science*, **167**, 1694.

34 Tunis, M. J. B. and Hearst, J. E. (1958), *Biopolymers*, **6**, 128.

35 Brahms, J. and Mommaerts, W. H. F. M. (1964), *J. Mol. Biol.*, **10**, 73.

36 Shih, T. Y. and Fasman, G. D. (1971), *Biochemistry*, **10**, 1675.

37 Tunis-Schneider, M. J. B. and Maestre, M. F. (1970), *J. Mol. Biol.*, **52**, 521.

38 Rodley, G. A., Scobie, R. S., Bates, R. H. T. and Lewitt, R. M. (1976), *Proc. Natl. Acad. Sci. USA*, **73**, 2959.

39 Bram, S. (1971), *J. Mol. Biol.*, **58**, 277.

40 Bram, S. (1973), *Cold Spring Harbor Symp. Quant. Biol.*, **38**, 83.

41 Wang, J. C. (1979), *Proc. Natl. Acad. Sci. USA*, **76**, 200.

42 Rhodes, D. and Klug, A. (1981), *Nature (London)*, **292**, 378.

43 Iwamoto, S. and Hsu, H-T. (1983), *Nature (London)*, **305**, 70.

44 Trifonov, E. N. (1982), *Cold Spring Harbor Symp. Quant. Biol.*, **47**, 271.

45 Dickerson, R. E., Kopka, M. L. and Pjura, P. (1983), *Proc. Natl. Acad. Sci. USA*, **80**, 7099.

46 Dickerson, R. E. (1983), *Sci. Amer.*, **249(6)**, 87.

47 Dickerson, R. E. (1983), in *Nucleic Acids: The Vectors of Life* (ed. B. Pullman and J. Jortner), D. Reidel, Dordrecht, p. 1.

48 Morgan, A. R. (1979), *Trends Biochem. Sci.*, **4**, N244.

49 Arnott, S., Chandrasekaran, R., Puigjaner, L. C., Walker, J. K., Hall, I. H. and Birdsall, D. L. (1983), *Nucleic Acids Res.*, **11**, 1457.

50 Arnott, S., Chandrasekaran, R., Puigjaner, L. C. and Walker, J. K. (1983), in *Nucleic Acids: The Vectors of Life* (ed. B. Pullman and J. Jortner), D. Reidel, Dordrecht, p. 17.

51 Zurkin, V. B., Lysov, Y. P. and Ivanov, V. I. (1979), *Nucleic Acids Res.*, **6**, 1981.

52 Zurkin, V. B. (1983), *FEBS Lett.*, **158**, 293.

53 Zimmerman, S. B. (1982), *Annu. Rev. Biochem.*, **51**, 359.

54 Rich, A. (1983), *Cold Spring Harbor Symp. Quant. Biol.*, **47**, 1.

55 Pohl, E. M. and Jovin, T. M. (1972), *J. Mol. Biol.*, **67**, 375.

56 Hamada, H. and Kakunaga, T. (1982), *Nature (London)*, **298**, 396.

57 Hill, R. J. and Stollar, B. D. (1983), *Nature (London)*, **305**, 338.

58 Mandel, M. and Marmur, J. (1976), *Methods Enzymol.*, **12**, 195.

59 Adams, R. L. P. and Eason, R. (1984), *Nucleic Acids Res.*, **12**, 5869.

60 Britten, R. J. and Kohne, D. E. (1968), *Science*, **161**, 529.

61 Felsenfeld, G. (1968), *Methods Enzymol.*, **12**, 247.

62 Guttman, T., Votavova, H. and Pivec, C. (1976), *Nucleic Acids Res.*, **3**, 835.

63 Birnsteil, M., Telford, J., Weinberg, G. and Stafford, D. (1974), *Proc. Natl. Acad. Sci. USA*, **71**, 2900.

64 Doty, P. (1961), *Harvey Lect.*, **55**, 103.

65 Bunermann, H. and Muller, W. (1978), *Nucleic Acids Res.*, **5**, 1059.

66 Laskowski, M. (1972), *Prog. Nucleic Acid Res. Mol. Biol.*, **12**, 161.

67 Marmur, J. (1963), *Annu. Rev. Microbiol.*, **17**, 329.

68 Leng, M. (1985), *Biochim. Biophys. Acta*, **825**, 339.

69 Schildkraut, C. L., Mandel, M., Levisohn, S., Smith-Sonneborn, J. E. and Marmur, J. (1962), *Nature (London)*, **196**, 795.

70 Freese, E. (1962), *J. Theor. Biol.*, **3**, 82.

71 Campbell, A. M. and Lochhead, D. S. (1971), *Biochem. J.*, **123**, 661.

72 Beard, P., Morrow, J. F. and Berg, P. (1973), *J. Virol.*, **12**, 1303.

73 Monjardino, J. and James, A. W. (1975), *Nature (London)*, **225**, 249.

74 Campbell, A. M. (1978), *Trends Biochem. Sci.*, **3**, 104.

75 Gotoh, O. and Tagashira, Y. (1981), *Biopolymers*, **20**, 1033.

76 Gotoh, O. and Tagashira, Y. (1981), *Biopolymers*, **20**, 1043.

77 Keller, W. (1975), *Proc. Natl. Acad. Sci. USA*, **72**, 4876.

78 Wang, J. C. (1974), *J. Mol. Biol.*, **89**, 783.

79 Pulleyblank, D. E. and Morgan, A. R. (1975), *J. Mol. Biol.*, **91**, 1.

80 Wang, J. C. (1976), *J. Mol. Biol.*, **43**, 25.

81 Luchnick, A. N., Bakayev, V. V. and Glaser, V. M. (1983), *Cold Spring Harbor Symp. Quant. Biol.*, **47**, 293.

82 Lilley, D. M. J. (1983), *Nature (London)*, **305**, 276.

83 Courey, R. J. and Wang, J. C. (1983), *Cell*, **33**, 817.

84 Sinden, R. R., Carlson, J. O. and Pettijohn, D. E. (1980), *Cell*, **21**, 773.

85 Sinden, R. R., Brogles, S. S. and Pettijohn, D. E. (1983), *Proc. Natl. Acad. Sci. USA*, **80**, 1797.

86 Lee, J. S., Woodsworth, M. L., Latimer, L. J. P. and Morgan, A. R. (1984), *Nucleic Acids Res.*, **12**, 6603.

87 Guschlbauer, W. (1976), *Nucleic Acids Structure*, Springer Verlag, New York.

88 Tinoco, I., Borer, P. N., Dengler, B., Levine, M. D., Uhlenbeck, O. C., Crothers, D. M. and Gralla, J. (1973), *Nature (London) New Biol.*, **246**, 40.

89 Woese, C. R., Magrum, L. J., Gupta, R., Siegel, R. B., Stahl, D. A., Kop, J., Crawford, N., Brosius, J., Guteil, R., Hogan, J. J. and Noller, H. F. (1980), *Nucleic Acids Res.*, **8**, 2275.

90 Atmadja, J., Brimacombe, R. and Maden, B. E. H. (1984), *Nucleic Acids Res.*, **12**, 2649.

91 Massoulie, J. (1968), *Eur. J. Biochem.*, **8**, 423.

92 Massoulie, J. (1968), *Eur. J. Biochem.*, **8**, 439.

93 Kotchetkov, N. K. and Budowsky, E. I. (1969), *Prog. Nucleic Acid Res. Mol. Biol.*, **9**, 403.

94 Brown, D. M. (1974), in *Basic Principles of Nucleic Acid Chemistry* (ed. P. O. P. Ts'O) Academic Press, London, p. 1.

95 Ceriotti, G. (1955), *J. Biol. Chem.*, **214**, 59.

96 Burton, K. (1956), *Biochem. J.*, **62**, 315.

97 Singer, B. and Fraenkal-Conrat, H. (1969), *Prog. Nucleic Acid Res. Mol. Biol.*, **9**, 1.

98 Schuster, H. (1960), *Z. Naturforsch B*, **15**, 298.

99 Feldman, M. Y. (1973), *Prog. Nucleic Acid Res. Mol. Biol.*, **13**, 1.

100 Shapiro, R., Cohen, B. I. and Servis, R. E. (1970), *Nature (London)*, **227**, 1047.

101 Kotchetkov, N. K. and Budowsky, E. I. (1972), *Organic Chemistry of Nucleic Acids*, Plenum Press, New York.

102 Bush, C. A. (1974), in *Basic Principles of Nucleic Acid Chemistry* (ed. P. O. P. Ts'O) Academic Press, London, p. 91.

103 Thrierr, J. C. and Leng, M. (1972), *Biochim. Biophys. Acta*, **272**, 238.

104 Arnott, S. and Bond, P. J. (1973), *Nature (London) New Biol.*, **244**, 99.

105 Arnott, S., Hukins, D. W. L., Dover, S. D., Fuller, W. and Hodgson, A. R. (1973), *J. Mol. Biol.*, **81**, 107.

106 Crick, F. H. C. (1966), *J. Mol. Biol.*, **19**, 548.

107 Travers, A. (1980), *Nature (London)*, **283**, 16.

108 Wu, H-H. and Crothers, D. M. (1984), *Nature (London)*, **308**, 509.

109 Hagerman, P. J. (1984), *Proc. Natl. Acad. Sci. USA*, **81**, 4632.

110 Frederick, C. A., Grable, J., Melia, M., Samudzi, C., Jen-Jacobson, L., Wang, B-C., Greene, P., Boyer, H. W. and Rosenberg, J. M. (1984), *Nature (London)*, **309**, 327.

111 Richmond, T. J., Finch, J. T., Rushton, B., Rhodes, D. and Klug, A. (1984), *Nature (London)*, **311**, 532.

112 McConnell, B. and Von Hippel, P. H. (1970), *J. Mol. Biol.*, **50**, 297.

113 Wahl, P., Paoletti, J. and Pecq, J. B. (1970), *Proc. Natl. Acad. Sci. USA*, **65**, 417.

114 Sober, H. A. (ed.) (1968), *Handbook of Biochemistry*, Chemical Rubber Co., Cleveland, Ohio, pp. H-11, H-30.

115 Rich, A., Nordheim, A. and Wang, A. H. J. (1984), *Annu. Rev. Biochem.*, **53**, 791.

116 Marmur, J. and Doty, P. (1959), *Nature (London)*, **183**, 1427.

117 Heppel, L. A., Ortiz, P. J. and Ochoa, S. (1967), *J. Biol. Chem.*, **229**, 679.

118 Bone, M., Gibson, T., Goman, M., Hyde, J. E., Langsley, G. W., Scaife, J. G., Walliker, D., Yankofsky, N. K. and Zolg, J. W. (1983), *Molecular Biology of Parasites* (eds J. Guardiola, L. Luzzatto and W. Trager) Raven Press, New York, p. 125.

119 Adamiak, R. W. and Gornicki, P. (1985) *Prog. Nucleic Acid Res. Mol. Biol.*, **32**, 27.

200 Koo, H-S., Wu, H-M. and Crothers, D. M. (1986) *Nature (London)*, **320**, 501.

# 3

# Chromosome organization

## 3.1 INTRODUCTION

The function of DNA is to carry the genetic message from generation to generation, and to allow the expression of that message under appropriate conditions. DNA molecules are very large and carry the information for the synthesis of many proteins, not all of which are required at the same time. Problems will obviously arise when it comes to packaging such long lengths of DNA into a cell in such a way as to allow programmed access to the encoded information. These problems are compounded by the fact that the DNA duplex must replicate and the two daughter DNA molecules must segregate to the two daughter cells. In this chapter we shall look at the structures involved in these processes. These structures are called chromosomes and consist not only of the DNA but of a variety of associated proteins.

For the study of the larger chromosomes of eukaryotic cells cytological techniques have been used for many years. Genetical approaches have been most useful for the study of bacterial chromosomes but have been much less successful in the study of the far more complex and slower-growing animal and plant cells. In recent years, the modern techniques of genetic engineering, together with the cloning and sequencing strategies discussed in the Appendix, have enabled rapid advances to be made in our understanding of the organization of the DNA and the genes of all organisms.

The eukaryotes, as typified by the animal cell shown in Fig. 3.1, contain several organelles each bounded by a membrane or series of membranes. In particular the endoplasmic reticulum (er) is a series of membrane-bound vesicles or tubes (40–150 nm in diameter) to some of which (the rough er) are attached ribosomes. Other ribosomes are found free in the cytoplasm (see Chapter 11). Other inclusions in an animal cell are the mitochondria (0.5–5 $\mu$m $\times$ 0.3–0.7 $\mu$m), lysosomes and the nucleus in which most of the DNA is found. The nucleus is bounded by a double membrane which is continuous with the endoplasmic reticulum (Fig. 3.1) and the outer (cytoplasmic) surface is covered in ribosomes. At intervals in the nuclear membrane are pores through which nucleo-cytoplasmic exchange is believed to occur [9]. Each pore is surrounded by a disc of protein molecules known as the nuclear pore complex which form a channel wide enough to allow the passage of small molecules (molecular weight less than 5000). It is believed that special transport systems exist for the nucleocytoplasmic exchange of large RNA and protein molecules [358]. Lying just inside the inner nuclear membrane is the nuclear lamina, the whole complex forming the nuclear envelope [12]. The nuclear lamina is made from three proteins (Lamins A, B and C) of molecular weight 60 000 to 70 000 which are polymerized to form a sort of fibrous skeleton, to which the chromosomes are attached [3, 14, 15]. This

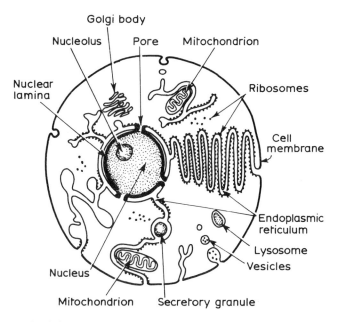

**Fig. 3.1** Schematic representation of a typical animal cell.

attachment of chromosomes to the nuclear lamina may help to align the chromosomes in a manner which is important for their function [149] but as yet little firm evidence is available that this is so. Within the body of the nucleus the chromosomes are associated with what is known as the nuclear matrix, a structure around which condensation is believed to occur at metaphase (see Section 3.2.1). Under these conditions the matrix proteins are believed to form the chromosome scaffold. The role of the matrix and scaffold will be considered further in Sections 3.3.5 and 10.4.4e [1–3, 16–21]. In prokaryotes the DNA is not housed in a membrane-bound organelle but is attached to the plasma membrane which itself is surrounded by a rigid cell wall to protect the cell from osmotic damage. When the cell wall is eliminated by digestion with the enzyme lysozyme the membrane and its contents are released as the osmotically sensitive protoplast.

In addition to the nuclear DNA there is a small amount of DNA present in mitochondria and chloroplasts (see Section 3.4). Prokaryotes also carry small extra pieces of DNA known as extrachromosomal DNA or plasmids (see Section 3.7).

Viruses may also be present in cells where they multiply, eventually leading to the death of the cell. Viruses which grow in bacteria are called bacteriophage or more simply phage.

## 3.2 EUKARYOTE DNA

### 3.2.1 The eukaryote cell cycle

In eukaryote cells, DNA synthesis takes place in a restricted time period during the *cell cycle* (Fig. 3.2). In mitosis the cells divide and form two daughter cells. This is then followed by a gap period designated G1 during which no DNA synthesis occurs. The length of the various gap periods and indeed the length of the entire cell cycle depends very much upon the cell type but Fig. 3.2 shows a typical example of a cell in culture with a replication time of 24 hours. G1 is followed by S phase in which DNA synthesis

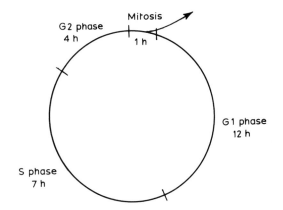

**Fig. 3.2** The eukaryote cell cycle. G1, S and G2 phases are together known as interphase. The actual times involved vary with cell type and growth conditions.

occurs and then by a second shorter gap period known as G2. The cells then go into mitosis again.

A population of growing cells will normally consist of cells at all stages of the cell cycle. A variety of methods may be used to *synchronize* the population so that all are dividing at the same time. In tissue culture mitotic cells may be shaken from the glass or plastic of the tissue culture vessel and hence separated from cells at other stages of the cycle. Alternatively, chemical inhibitors such as thymidine may be used to stop all the cells in S phase from completing the cycle, since excess thymidine inhibits DNA synthesis (see Chapter 5). When this method is used, it is usual to apply the thymidine for two periods each of about 16 hours separated by a period of about 8–10 hours in the absence of thymidine. During the first exposure cells making DNA are stopped throughout S phase while cells in G1, G2 and mitosis continue to grow until they reach the beginning of S phase. On removal of the thymidine all the cells traverse S phase and so during the second exposure they now all accumulate at the G1/S border. Other methods of synchronizing cells are covered in more specialist texts [5, 10].

The combination of G1, S and G2 phases is known as *interphase* and under these conditions the cell nucleus is clearly visible in the light microscope [4, 5]. It is only in mitosis that the chromosomes are visible with the light microscope.

### 3.2.2 Eukaryote chromosomes

Eukaryote DNA is contained in a relatively small number of chromosomes which varies according to the species (Table 3.1). No direct correlation can be made between the amount of DNA in the nucleus and the number of chromosomes in which it is contained [6]. Somatic cells of each species have two copies or homologues of each chromosome with the exception of the sex chromosomes for which the female carries two X chromosomes and the male an X and a Y. Germ cells contain only one copy of each chromosome and it is to these cells that the term *haploid chromosome content* or *haploid DNA content* refers. While most somatic cells are *diploid*, *tetraploid* cells with four and *octaploid* cells with eight copies of each chromosome can also be found, particularly in cells in culture. *Aneuploid* cells have an abnormal chromosome complement which is not necessarily an increase on the diploid condition for each chromosome type so that some may be present in greater numbers than others. The characteristic number and morphology of the chromosomes in any particular cell type is known as the *karyotype* of that cell and is usually determined in metaphase when the chromosomes are highly condensed and readily stained by basic dyes [5, 7]. In interphase, the chromosomes are spread out in the nucleus and cannot be individually distinguished.

The actual process of mitosis can be subdivided into several discrete stages. At the end of the interphase two *poles* are formed in the cell by the *centrioles* (Fig. 3.3). At this time each chromosome consists of two identical chromatids produced as a result of DNA replication

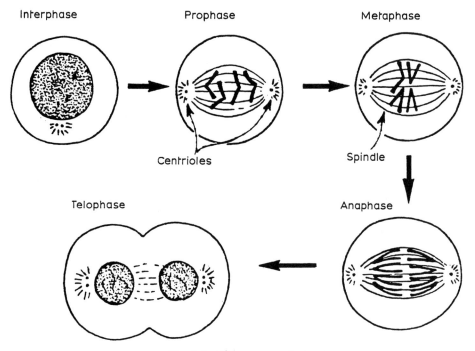

**Fig. 3.3**  The process of mitosis.

(Chapter 6). Each chromatid is a DNA duplex and the two chromatids are joined at the *centromere*. In prophase the chromosomes begin to condense and the nuclear envelope disappears as a result of the phosphorylation of the lamina proteins which depolymerize [8]. Fine fibres of microtubules form between the two poles. In metaphase the highly condensed chromosomes line up in the centre of the cell at the equatorial region to form the *metaphase plate* which is clearly visible using phase contrast microscopy. In anaphase the two daughter chromosomes are pulled apart into the two poles and in telophase the lamina proteins are dephosphorylated and repolymerize causing membrane fragments to associate with individual chromosomes. These fragments then fuse to re-form in the nuclear envelope [9]. Chromosomes with a centromere towards one end are known as *acrocentric* (or telocentric if the short arm is too small to see by light microscopy). Those with a centrally located centromere are *metacentric*, and those defective chromosomes lacking a centromere are *acentric*. Additional secondary constrictions are usually the site of rRNA genes which form the nucleolus in an interphase nucleus and are known as *nucleolar organizers*. These are only found on a few chromosomes. Detailed cytogenetic studies of chromosomes involve the use of fluorescent dyes such as quinacrine mustard or gentle trypsin treatment followed by staining with Giemsa. These treatments give rise to a characteristic band pattern (known as Q-bands or G-bands, respectively) visible on fluorescence or light microscopy [11, 39]. Each chromosome has a banding pattern which allows it to be recognized easily and chromosome deletions or rearrangements may be closely observed. A large variety of banding techniques is now available for cytogenetic studies [5, 12].

In budding yeast (i.e. *Saccharomyces*) chromosome condensation does not occur and the nuclear membrane remains present through-out division. A bud is formed in G1 phase and increases in size during the rest of the cycle. Following nuclear division one of the daughter nuclei migrates into the bud which separates from the parental cell at cell division [13]. Using genetic engineering techniques (see the Appendix) the centromeres of several yeast chromosomes have been isolated and sequenced [22–24]. They contain a 14 bp conserved nucleo-tide sequence separated from an 11 bp conserved nucleotide sequence by an 82–89 bp A + T-rich sequence. The centromere core provides a 220 bp region of the chromosome to which the microtubules can attach at mitosis [173].

Other specific regions of chromosomes which have been isolated are the ends or telomeres. These are considered in more detail in Section 6.10.3. In addition, for a chromosome to replic-ate, it requires an origin of replication. Sequences which confer on chromosomes the ability to replicate and segregate (ARS) have also been cloned and sequenced (see Section 6.9).

By combining these three cloned regions together on a piece of DNA an artificial chromo-some has been constructed. On introduction into yeast cells this chromosome appears to act like a natural chromosome in most respects [25–27].

### 3.2.3 The allocation of specific genes to specific chromosomes

Cells of two different animal species can be induced to fuse in culture to produce hetero-karyons carrying two or more sets of chromo-somes [5, 28]. The redundant chromosomes are readily lost from the hybrid cells. The loss occurs at random but in human:mouse hybrids the human chromosomes are lost preferentially and cell clones can be obtained which retain only one or a few human chromosomes. The clones can then be tested for specific enzymes and their presence related to those chromosomes that remain. For this approach to be successful it is important either to be able to select for a particular characteristic (e.g. thymidine kinase activity) or to be able to distinguish the human enzyme from the corresponding mouse enzyme which will be present in all clones. In this way it has been possible to allocate several hundred enzyme functions to specific chromosomes [29].

The technique of *in situ* hybridization was until recently limited to the detection of repetitive gene families (see Section 3.2.6) but has recently been refined to allow the assignment of single-copy genes to particular segments of specific chromosomes [30]. The method relies on having a radioactive gene probe (see the Appendix) which can be hybridized to a fixed preparation of metaphase or prometaphase chromosomes [31]. Radioautography linked to sensitive chromo-some-staining methods leads to the illumination of the gene on the particular chromosome band.

The two methods can be combined by pre-paring Southern transfers (see the Appendix) of the DNA isolated from the various hybrid clones. These can be treated with the radioactive gene probes to obtain a physical rather than enzymic location of the gene on a particular chromosome [32–36]. This method has to be linked to the use of deletion mutants and the cotransfer of genes on chromosome fragments to obtain more precise assignments to chromosome regions.

Rather than relying on hybrid cells as a mecha-nism of obtaining clones with only one or a few identifiable human chromosomes it is now pos-sible to fractionate a preparation of metaphase chromosomes using a fluorescence-activated cell sorter (FACS). Individual chromosome frac-tions can then be tested to see which hybridize with the radioactive gene probes [37, 38, 91]. Fig. 3.4 shows chromosomes from human cells separated in this manner. Alternatively, DNA molecules of up to 2000 kbp can now be fractionated by pulsed field gradient gel electro-phoresis and this has allowed the production of a molecular karyotype of yeast and trypanosomes [118, 234, 236, 242].

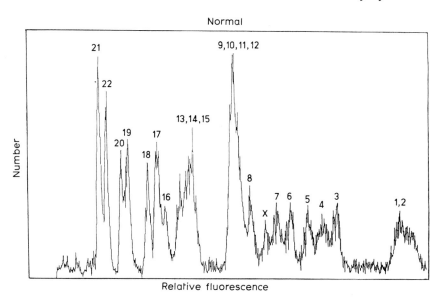

**Fig. 3.4**   Human chromosome identification by relative fluorescence using flow cytometry (reproduced, with permission, from [37] copyright Macmillan Journals Ltd.).

### 3.2.4 Haploid DNA content (C value)

One of the most striking features of eukaryote DNA is the great quantity of it which is present in each cell (Table 3.1).

The amount is at least an order of magnitude in excess of that required for the known gene-coding capacities of the cells and the logical conclusion is that the bulk of the DNA is not expressed. Does this mean that it is redundant (junk) or that it has a function which is as yet imperfectly understood [40–43, 62]? In general terms the minimum size of the genome (i.e. the amount of DNA per cell) increases with the stage of evolutionary development [44] (Fig. 3.5). However, certain amphibia have a C value one hundred-fold in excess of man and even more in excess of other amphibia, and the reason for this is not clear. The phenomenon is known as the *C value paradox* [45]. Logically, the species with the greater amount of DNA should have the advantage of greater coding potential and the disadvantage of the requirement to replicate very large amounts of DNA prior to cell division.

In addition to the C value paradox, different cell types within the same species can vary in their haploid DNA content. Amphibian oocytes have a very large amount of cytoplasm to provide with essential components of protein synthesis such as ribosomes and consequently have greatly amplified the number of genes coding for ribosomal RNA (see Section 7.5). Eukaryote cells in culture tend to become aneuploid with the passage of time, apparently adapting to their culture conditions. In addition, they have the capacity to amplify certain genes several hundred-fold in response to external stimuli. Thus the gene coding for the enzyme dihydrofolate reductase is amplified in cells which are resistant to methotrexate, a drug which interferes with one carbon unit metabolism (see Chapter 5 and Section 7.5).

### 3.2.5 Gene frequency

Eukaryote DNA is, by convention, divided into three frequency classes of unique DNA, moderately repetitive DNA and highly repeated DNA.

**Table 3.1**  Haploid DNA content and chromosome number of a variety of organisms.

| | Haploid DNA content | | Haploid chromosome number |
|---|---|---|---|
| | Picograms | Base pairs | |
| Simian virus 40 | 0.000006 | $5.3 \times 10^3$ | 1 |
| Herpes simplex virus | 0.00017 | $151 \times 10^3$ | 1 |
| *E. coli* | 0.005 | $4.5 \times 10^6$ | 1 |
| *Saccharomyces cerevisiae* (yeast) | 0.025 | $22.5 \times 10^6$ | 17 |
| *Drosophila melanogaster* (fruit fly) | 0.17 | $0.15 \times 10^9$ | 4 |
| Sea-urchin | 0.45 | $0.41 \times 10^9$ | 20 |
| *Gallus domesticus* (chicken) | 0.7 | $0.63 \times 10^9$ | 39 |
| *Mus musculus* (mouse) | 3.0 | $2.7 \times 10^9$ | 20 |
| *Homo sapiens* (human) | 2.7 | $2.4 \times 10^9$ | 23 |
| HeLa cells (human cell culture, aneuploid) | 8.5 | $7.7 \times 10^9$ | 70–164 (Mean 82) |
| *Xenopus laevis* (toad) | 4.2 | $3.8 \times 10^9$ | 18 |
| *Triturus cristatus* (newt) | 35 | $31.5 \times 10^9$ | 12 |
| *Zea mays* (corn) | 7.8 | $7 \times 10^9$ | 10 |

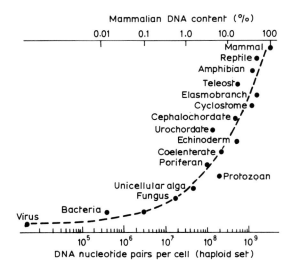

**Fig. 3.5**  Minimum haploid DNA content in species at various levels of organization (reproduced, with permission from [44] copyright the AAS).

There is, in fact, considerable overlap between the three categories which are probably better classified by their $C_0 t_{1/2}$ values (Chapter 2) since this is the manner in which classification is achieved experimentally.

(a) *Unique DNA*

In human cells it is generally classified as that DNA which has a $C_0 t_{1/2}$ value of about 1000 moles of nucleotide seconds litre$^{-1}$. It comprises about half of the total haploid DNA content and is thought to consist of the sequences coding for most enzyme functions for which there is only one or a small number of genes per haploid genome. The genes coding for the various chains of globin (see Section 9.2.2) or the enzyme glucose 6-phosphatase fall into this category. Unique DNA is considered further in Section 3.2.6. *Repetitive DNA* includes all the rest of the DNA. Among the repetitive DNA is a fraction which reanneals with a $C_0t$ value of between 100 and 1000 moles of nucleotide seconds litre$^{-1}$. The sequences represented in this group are generally thought to be those coding for proteins which form major structural components of the cell such as the histones. The genes for rRNA and tRNA also fall into this category.

### (b) Satellite DNA

At the other extreme DNA with a $C_0t_{1/2}$ value as low as $10^{-3}$ moles of nucleotide seconds litre$^{-1}$ consists largely of *satellite DNA*. This represents highly repeated sequences, of which there may be a million or more copies per haploid genome, which are usually quite short and are arranged in tandem arrays. The origin of the name satellite relates to the method of its isolation on caesium chloride buoyant density gradients of sheared DNA where it will sometimes form a satellite band separate from the main DNA band, due to its differing content of adenine and thymine residues. The simplest known satellite DNA is poly[d(A-T)] which occurs in certain crabs. Other satellites can have any number up to several hundred base pairs which are repeated in tandem fashion along the genome.

Human DNA has been shown by density-gradient centrifugation to have four main satellites [46, 47] and by dye binding [45] and restriction endonuclease cleavage [49] to have two additional satellites. The distribution of satellite DNA among chromosomes varies. Some chromosomes have virtually no satellite sequences while others (notably the Y chromosome) are largely composed of satellite sequences [50]. In general, satellite DNA appears to be concentrated near the centromere of the chromosomes in the heterochromatin fraction (Section 3.3). DNA sequence analysis has shown that the basic repeat unit of satellite DNA is itself made up of subrepeats. For example the major mouse satellite has a repeating structure of 234 base pairs made up of four related 58 and 60 bp segments each in turn made up of 28 and 30 bp sequences [48, 52]. The satellites between and within related species are themselves related in an evolutionary sense by cyclical rounds of multiplication and divergence of an initial short sequence [53]. The nature of the multiplication process is not known for certain but probably involves unequal recombination events (see Fig. 3.6 and Section 7.4). The divergence involves single base changes (which generate altered restriction enzyme sites – see Section 4.5.3) and insertions and deletions [54–57].

Despite our detailed knowledge of the sequence and distribution of satellite DNA we have little idea as to the function it has in the cell. Originally, it was thought that satellite DNA was not transcribed since RNA of corresponding sequence was seldom isolated, but occasional cases of satellite transcription have been reported [51, 63]. On the whole, however, its transcriptional inactivity ties in with its localization in heterochromatin. As satellite DNA is often lost in somatic cells (where it always appears to be heavily methylated – see Section 4.6.1) it has been proposed that it may have some function in the germ cells [58–61]. This function may relate to the recombination events which occur during gametogenesis and which may be enhanced by the presence of blocks of similar DNA sequences on several chromosomes.

### (c) Interspersed repetitive DNA

This differs from satellite DNA in that it represents sequences which are not clustered together but are dispersed singly throughout the genome [64–66]. These dispersed repeats fall into two classes: the short interspersed nuclear elements (SINES) consisting of repeats shorter than 500 bp; and the long interspersed nuclear elements (LINES). The SINES are characterized by the *Alu* family found in human cells. This is a family of related sequences each about 300 bp long which is repeated 300 000 times in the human genome. The name *Alu* derives from the presence of a single site for the restriction enzyme *AluI* in the repeat (see Section 4.5.3). The *AluI* sequence is a head-to-tail dimer of a

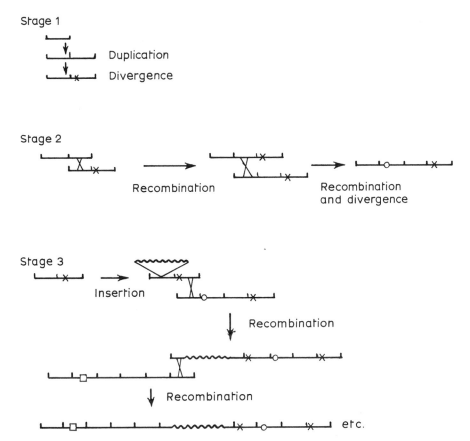

**Fig. 3.6** Possible evolution of satellite DNA involving cyclical duplication and divergence.

sequence which closely resembles the B1 family of mouse DNA. The *Alu* sequence is transcribed and it is believed to have become dispersed throughout the genome by a process involving reverse transcription and reintegration (see Sections 6.4.2.f and 7.7.3). For this reason it has been termed a retroposon [67–70, 73].

One of the best-studied LINES is the L1 (or *Kpn*) family of human DNA. This is a 6–7 kbp sequence which is repeated, at least in part, up to 10 000 times per genome [70–74]. These interspersed repeats are also transcribed and may give rise to proteins. It has been suggested that some of the gene products may be involved in the duplication and dispersal of the family which probably also occurs by a reverse transcription mechanism [74]. The mechanism of dispersion of SINES and LINES is considered in more detail in Section 7.7.3.c.

The interspersion pattern produced in the genome by SINES was first investigated in detail by renaturation kinetics [75]. The conclusion from these studies was that about half the genome consists of alternating regions of 300 bp of repeated DNA and 1000–2000 bp of unique DNA. This pattern of interspersion (which became known as the *Xenopus* pattern) is typical of most eukaryotes. However, some insect genomes lack this short-period interspersion pattern and consist on average of stretches of about 13 000 bp of unique DNA separated by LINES [76–78].

### (d) *Foldback DNA*

A special class of *foldback* or *palindromic* DNA sequences comprising 3–6% of eukaryote DNA has also been characterized. The size range is from 300 to 1200 base pairs and the molecules have a $C_0t_{1/2}$ value of less than $10^{-5}$ moles of nucleotide seconds litre$^{-1}$. This palindromic DNA is represented in all frequency classes and is widely distributed throughout metaphase chromosomes [79–81, 92]. Some palindromic DNA arises when two copies of the *Alu* sequences are present close to one another in opposite orientations and it was as a result of the ability of such DNA to renature instantaneously that the SINES were discovered [64].

### 3.2.6 Eukaryote gene structure

In recent years a fundamental difference between prokaryote and eukaryote genes has been uncovered by experimental analysis. Whereas prokaryote genes occupy a single uninterrupted sequence of DNA, the majority of eukaryote structural genes so far analysed have been shown to have extra sequences inserted into the middle of the gene. These intervening sequences which have also been named *introns* may be small and single as in the case of the gene for tyrosine suppressor tRNA [82] or large and multiple as is the case for the ovalbumin gene which has seven introns so that the entire gene occupies a length of 7.7 kb with only a small fraction of this DNA being actually used for coding purposes [83–85]. These intervening regions are transcribed into RNA and then removed by internal processing events in the cell nucleus to produce the final mRNA product [86–88] (see Section 9.3.4). Intervening sequences appear to have a wider evolutionary freedom for mutation than the coding sequences of the DNA [89]. In general intervening sequences appear to be less frequent in the lower eukaryotes [85]. While the majority of structural genes display this phenomenon, it is not universal even in higher eukaryotes and is not found in the majority of tRNA genes or in the genes coding for the histones [87, 88]. The full functional significance of introns remains improperly understood and has been suggested to reflect the continual process of evolution within the eukaryote chromosome [42, 43, 62, 90]. Introns and their possible origins and functions are further discussed in Section 9.2.1. In addition to these intervening sequences eukaryote genes are preceded by regions of DNA which have a control function. These flanking regions may be up to 1000 nucleotide pairs long. It is clear that at least in part the extra DNA present in higher eukaryotes can be attributed to repeated and intervening and flanking sequences of genes.

### 3.3 CHROMATIN STRUCTURE

Eukaryote chromosomes in metaphase are generally referred to as chromosomes but in interphase the term 'chromatin' is more generally used to describe the nucleoprotein fibres in the cell nucleus. Originally chromatin was loosely defined and subdivided into only two main classes, euchromatin and heterochromatin. Heterochromatin comprised the dense, readily stained areas of the nucleus or chromosome and was thought to represent inactive chromatin which was not undergoing transcription, whereas the more loosely packed euchromatin was thought to represent the transcriptionally active material. These concepts have been largely superseded by recent research in the area but are in occasional use.

Chromatin consists of DNA, RNA and proteins. The actual weight ratio between the three varies greatly with the tissue or cell of origin of the material but in general the amount of protein is equal to or greater than the amount of DNA while the amount of RNA is comparatively small. The protein content of chromatin can be further subdivided into the histone and non-histone proteins. The former group consists of a few types of molecule which are present in very large amounts and the latter of a diverse range of

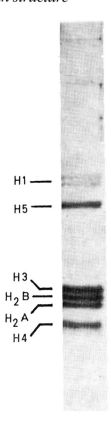

H1 —

H5 —

H3
H₂B
H₂A
H4

**Fig. 3.7** Polyacrylamide gel electrophoresis patterns of histones from chick erythrocytes (reproduced, with permission from Bradbury E. M., Maclean, N. and Matthews, H. R. (1981) *DNA, Chromatin and Chromosomes*, Blackwell Scientific Publications Ltd, Oxford).

protein molecules most of which are present in much smaller quantities. Historically the bulk of the non-histone proteins were referred to as the acidic nuclear proteins but this classification is no longer appropriate.

### 3.3.1 Histones and non-histone proteins

(a) *Histones*
There are five major types of histone molecule in the eukaryote cell nucleus. These are now classified as histones H1, H2A, H2B, H3 and H4 since a *CIBA Foundation Symposium* of 1975

[93]. Prior to this a variety of differing nomenclatures were adopted by different laboratories. The histones are basic proteins of low molecular weight which are readily isolated by salt or acid extraction of chromatin and which can be separately purified on the basis of size or charge [94] (see Fig. 3.7). In 2 M NaCl all the histones are dissociated from DNA. However in 0.5 M NaCl histone H1 alone is dissociated so that the functions of the residual or 'core' histones may be investigated [95, 96]. Each molecule consists of a hydrophobic core region with one or two basic arms. Histone H1 is a *very lysine-rich* protein of about 216 amino acids which shows a high degree of sequence conservation amongst eukaryotes particularly at the central, apolar region. Histones H2A and H2B are even more highly conserved and are known as the *lysine-rich* histones. The most conserved of all are the *arginine-rich* histones H3 and H4. Only three positions in the H4 histone molecule have been found to vary [97–100].

Despite the conservation of amino acid sequence there are multiple genes for each histone in all eukaryotes. This reaches extreme proportions in the sea-urchin and the fruit fly (*Drosophila*) which contain multiple tandem repeats of a region of DNA containing one gene for each histone separated from each other by an A + T-rich spacer region (see Section 9.2.2). Not all these genes produce identical histones. The histone genes present in greatest abundance appear to be active in early development when there is a great demand for new histone synthesis, whereas the minor variants are made at later stages of development and represent the major histones present in somatic cells. The genes for these variant histones may or may not be associated with the tandemly repeated blocks [101–108, 126] (see Section 9.2).

A special type of histone known as histone H5 is found in the nucleated erythrocyte of fish, amphibians and birds (Fig. 3.7). It bears many similarities to histone H1 and is thought to maintain the highly repressed state of the chromatin in these cell types [117]. These are non-dividing cells and in mammals histones H1° and H1e (see below) are present in non-dividing cells while histones H1a and H1b are present in large amounts only in dividing cells [104, 106].

It is not clear what effect the variations in sequence of the histones has on chromatin structure. Neither is it clear what is the effect of post-translational modification. However, the histones may be methylated, phosphorylated, acetylated or ADP-ribosylated, and some of these modifications, by altering the charge on the molecule, may affect the interactions of histones with each other or with DNA [109–114]. For example there are six subtypes of histone H1 (H1a–e and H1°) giving rise to 14 different phosphorylated forms, the prevalence of which varies during the cell cycle. Acetylation and deacetylation of histones H3 and H4 occur rapidly in G2 phase of the slime mould *Physarum* but the importance of this to cell growth or chromatin structure has been questioned [109–110]. As much as 20% of histone H2A is covalently linked to a 76-residue polypeptide called ubiquitin to form a branched-chain protein known as UH2A or A24 [115, 116]. Marked changes in the level of ubiquitination of histones occur during mitosis [359]. The possible role of modified histones in the control of gene expression is further discussed in Section 10.4.4.

In sperm cells, histones are replaced by other small basic proteins known as *protamines* [119, 120].

In the absence of DNA the 'core' histones will associate with each other, the predominant species being a homotypic tetramer in the case of histones H3 and H4 (i.e. a tetramer containing four similar arginine-rich histones) and a dimer in the case of H2A and H2B. At high ionic strength, all four histones form complexes which have variously been described as heterotypic tetramers, with one, or octamers, with two of each type of core histone molecule [121, 122]. Octamers are an integral part of the nucleosomes discussed in Section 3.3.2. They also readily

form long, linear aggregates whose functional significance is not clear.

### (b) *Non-histone proteins*

These are present in chromatin in an amount approximately equal to the histones. On SDS/polyacrylamide gels about 100 different proteins can be seen. Some of these are the enzymes involved in replication and transcription (see Chapters 6 and 8) or form part of the nuclear envelope (see Section 3.1). Others resemble the histones in being of low molecular weight (or high mobility on electrophoresis) and are known as the high-mobility group, or *HMG proteins*. These also resemble histones in being basic proteins and they are present in multiple copies in the chromatin, i.e. they play a structural role. They differ from histones in being only loosely associated with chromatin – they can be extracted with 0.35 M NaCl – and in lacking an apolar centre. They have a basic N-terminal region and an acidic C-terminal region separated by a short region rich in serine, glycine and proline. Those most characterized are HMG1, HMG2, HMG14 and HMG17 [123–125]. They are also considered in Section 10.4.4.

### 3.3.2 The nucleosome

The DNA from a human cell is of the order of 1 m in length and must be condensed into a cell nucleus whose diameter is of the order of 10 $\mu$m. The packing must, however, maintain accessibility and prevent tangling during replication. Eukaryote cells achieve this condensation by a series of packaging mechanisms involving the histones and some other chromosomal proteins. The initial coiling of DNA into nucleosomes and polynucleosomes was the subject of intensive research in the late 1970s and is now substantially elucidated [113, 127–131].

The elucidation of the structure of the nucleosome is a delightful example of how a large number of experimental approaches coalesced to solve the problem. At the same time as the studies on histones were going on, other groups were studying the nucleus using electron microscopy, X-ray and neutron diffraction. Still others employed crosslinking and nuclease studies in which the DNA products were analysed by electrophoresis on agarose or acrylamide gels.

The initial X-ray diffraction patterns of chromatin indicated the presence of a structure repeating every 10 nm but further interpretation was difficult [132]. Electron microscopy of ruptured nuclei showed the presence of a series of spherical particles joined by thin filaments – the so-called beads on a string picture [133] (see Fig. 3.8). The beads have a diameter of 7–10 nm but the length of the filaments is variable. Similar electron micrographs of the SV40 minichromosome present in virus-infected cells indicated the presence of about 21 beads or nucleosomes on each viral DNA molecule [134, 135]. As the contour length of naked SV40 DNA is 1590 nm and that of the minichromosome is only 250 nm it is clear that there has been a six- to sevenfold packing of the DNA into the minichromosome.

Not only is DNA compacted in chromatin it is also rendered partially resistant to nuclease action (see Fig. 3.9). Using micrococcal nuclease which makes double-stranded breaks in DNA it quickly became apparent that at early times of digestion the nuclease was cutting the string (i.e. the linker DNA) which holds the nucleosomes together (see Fig. 3.9). An analysis of the size of the DNA showed that the spacing between successive nucleosomes was about 200 bp. On further digestion the size of the DNA fragments was reduced first to 166 bp and finally to 146 bp [136–139] (see Fig. 3.10).

Meanwhile crosslinking studies using dimethyl suberimidate [140, 141] had shown that in chromatin there was present an octamer of histones of composition (H2A, H2B, H3, H4)$_2$. Analysis of the stoichiometry of histone and DNA suggested that one octamer was present per 200 bp DNA, i.e. per nucleosome.

Another nuclease (DNase I) is able to make nicks all along the length of DNA in chromatin. The nicks occur at ten base intervals which was interpreted to mean that the DNA was wrapped around a core of histones and was accessible to nuclease on its outer surface; a conclusion which was confirmed by neutron diffraction [113, 142–144].

Further studies involving X-ray crystallography [144–146] have shown that the nucleosome is a shallow v-shaped structure around which a 146-bp core of DNA is wrapped making about one-and-three-quarter turns (Fig. 3.10). In order to bend so acutely the DNA double helix makes several sharp bends or kinks [146, 147] which explains how a rigid molecule of persistence length 150 bp [148] can make such tight turns around the histone core. It also explains why certain regions are less sensitive to DNase I than expected.

From DNA–protein crosslinking studies and the X-ray crystallography of nucleosomes it has become clear that an (H3, H4)$_2$ tetramer provides the framework of the nucleosome and an (H2A, H2B) dimer is added to each face of the framework [146, 150, 151]. This is the nucleosome core associated with 146 bp DNA.

Histone H1 holds the two ends of nucleosomal DNA together to form a *chromatosome* of 166 bp and the remaining DNA forms the linker joining nucleosomes together to form oligonucleosomes (Fig. 3.10).

The length of the linker and hence the repeat frequency of nucleosomal DNA is variable both between species and even within tissues. The repeat frequency ranges from 212 bp for chick erythrocytes (an inactive tissue) to 165 bp for yeast (highly active) [120, 152, 153] and although it would be logical to anticipate that active chromatin might be less closely packed than inactive chromatin this is not the case.

Table 3.2 gives the repeat frequencies of chromatin from some eukaryote tissues. It is interesting to note that two cell types which are from the same organ can differ in their repeat distance [154].

Nucleosomes can be reconstituted *in vitro*

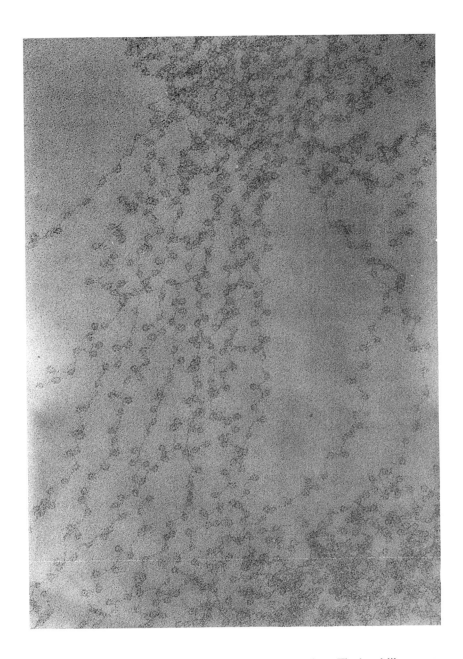

**Fig. 3.8** Chromatin fibres streaming out of a chicken erythrocyte nucleus. The bead-like structures (nucleosomes or $\nu$ bodies) are about 7 nm in diameter. The connecting strands are about 14 nm long. The sample was negatively stained with 5 mM uranyl acetate and the magnification is 285 000. (By courtesy of D. E. Olins and A. L. Olins.)

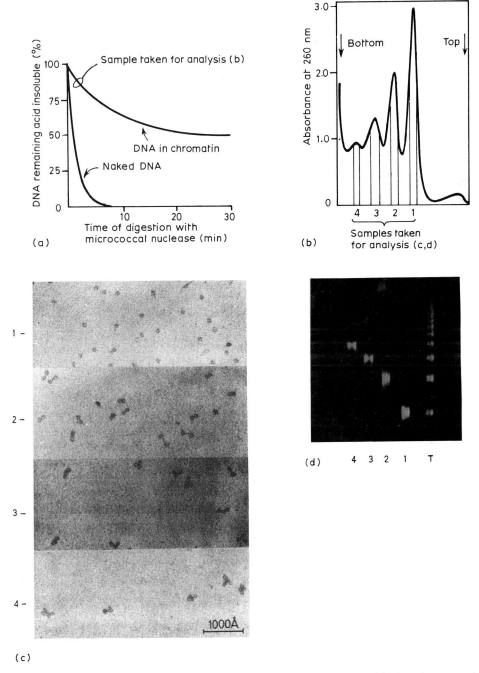

**Fig. 3.9** The time course of digestion of chromatin with micrococcal nuclease (a). Samples were taken at the position shown and fractionated on a sucrose gradient (b); fractions from positions 1, 2, 3 and 4 of the gradient were shown to contain mono-, di-, tri- and tetra-nucleosomes by electron microscopy (c); the DNA in these fractions was shown by electrophoresis (d) to be of approximate length 200 bp, 400 bp, 600 bp, and 800 bp, T shows total DNA before fractionation on the sucrose gradient (reproduced, with permission, from [193]).

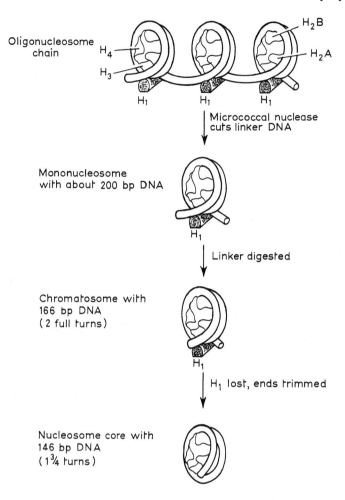

**Fig. 3.10** Diagrammatic representation of the digestion of chromatin with micrococcal nuclease showing the relative positions of the various histones in the nucleosome (from Richmond *et al.* [150]).

**Table 3.2** Repeat frequencies of chromatin from some eukaryote tissues.

| Cell type | Nucleosome repeat frequency |
| --- | --- |
| Rabbit cortical neuron | 162 |
| Yeast | 165 |
| HeLa | 188 |
| Rat foetal liver | 193 |
| Rat liver | 196 |
| Rabbit cerebellar neuron | 200 |
| Chicken erythrocyte | 212 |

from DNA and histones. Reconstitution usually requires the slow dialysis of a mixture of DNA and histones from high- to low-salt conditions [134] but has been achieved under more physiological conditions [155]. *In vivo* proteins are present in the nucleus which are believed to play a part in the assembly of chromatin on newly synthesized DNA (see Section 6.11).

### 3.3.3 Nucleosome phasing

The question arises as to whether the nucleosome cores are present in fixed positions on the

DNA or are (a) free to move or (b) present in random positions so that in a population of cells all the DNA could be found in the linker region. Such nucleosome positioning or phasing may have particular significance with respect to gene expression particularly as certain DNA sequences could be facing into the nucleosome and hence be unavailable for recognition by sequence-recognizing proteins [156] – see Section 10.4.3. Under biological conditions, the nucleosome appears to be remarkably stable and to have small tendency either to travel along a single length of DNA or to move among pre-formed chromosomes [175–177]. The approaches to analysing nucleosome phasing have been summarized by Kornberg [157] but all initially involved the use of micrococcal nuclease (to cleave linker DNA) and a restriction enzyme (see Section 4.5.3) to make a specific cut in a gene which could be identified by Southern blotting (see Appendix, Section A.3). This is illustrated in Fig. 3.11. The limitation with such methods is the preference of micrococcal nuclease for certain regions of DNA rich in sequences such as CATA or CTA [158–160]. This has led to the development of chemical cleavage reagents which show less or no sequence preference [161–164]. The conclusions reached are that micrococcal nuclease and the chemical reagents will cut naked DNA or chromatin in certain well-defined positions relative to the 5′-ends of genes. This can probably be interpreted as showing the presence of a nucleosome-free region of unusual base composition [171]. When a nucleosome-free region exists then the nucleosomes present on either side of this region are very likely to take up particular positions as the nucleosome repeat is about 200 bp [157]. Certain positions may be favoured by particular base compositions (see below) and if these regions recur regularly as in satellite or 5S DNA then the phasing will be reinforced [165–169]. As satellite DNAs are built up of a number of subrepeats (see Section 3.2.5.b) there may be little to choose between

positioning a nucleosome at the beginning of one subrepeat or the next. As these positions may be only 20 or 30 bp apart this results in multiple phases on satellite sequences each related to the other by the subrepeat sequences of the DNA [168, 169]. Phasing may be strongly reinforced on satellite and possibly on other DNA sequences as a result of proteins which bind to DNA so as to hold the DNA into a looped structure. Such a protein has been found for the α-satellite of African green monkey cells [174].

The favoured positions taken up by nucleosomes referred to above may relate to the requirement for DNA to bend or kink as it wraps around the nucleosome core [146, 170] and on reconstitution of chromatin on DNA of defined sequence the nucleosomes have been shown to form at a specific region [172, 252]. Not only may nucleosomes be positioned relative to genes but there is evidence that they take up a precise position relative to the centromere [173].

### 3.3.4 Higher orders of chromatin structure

Histone H1 also plays a role in the association of adjacent nucleosomes and histone H1 molecules can be chemically crosslinked to each other [178, 184]. In the absence of histone H1, nucleosomes and polynucleosomes are soluble over a wide range of ionic strength. However, when histone H1 is present they precipitate above NaCl concentrations of 80 mM and $MgCl_2$ concentrations of 2 mM. This provides a useful mechanism for the separation of nucleosomes with or without histone H1 [179, 180]. In the presence of histone H1 electron microscopy and X-ray diffraction studies show a fibre of 30 nm diameter as well as the 10 nm fibre which is a chain of single nucleosomes. Although different explanations have been proposed for the structure of the 30 nm fibre it now appears that this is a hollow cylinder or *solenoid* into which the nucleosomes are coiled [130, 131, 181–183]. There are about six nucleosomes per turn of the solenoid (Fig. 3.12). The pitch of the solenoid is

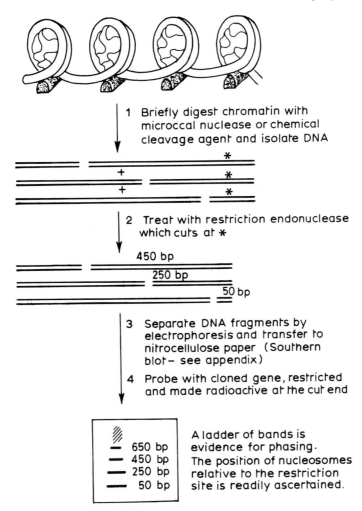

**Fig. 3.11** One method used for studying nucleosome phasing. Other methods involve more extensive digestion with micrococcal nuclease to produce core particle DNA which can be cloned and sequenced or mapped with restriction enzymes [157, 168, 169].

11 nm and the faces of the nucleosomes are approximately parallel to the solenoid axis.

### 3.3.5 Loops, matrix and the chromosome scaffold

When cells, at any stage of the cell cycle, are lysed under certain conditions the DNA can be sedimented intact and displays the dye binding and sedimentation characteristics of supercoiled DNA in loops of 100 kb in size (see Fig. 3.13). The conditions involve the use of non-ionic detergents which disrupt the membranes, high levels of EDTA (ethylenediaminetetra-acetic acid) to chelate the bivalent metal ions which activate nucleases, and high molarities of sodium chloride which dissociate most proteins. These loops are stabilized by proteins rather than RNA since proteases but not ribonucleases abolish the supercoiling [185, 186]. Metaphase chromo-

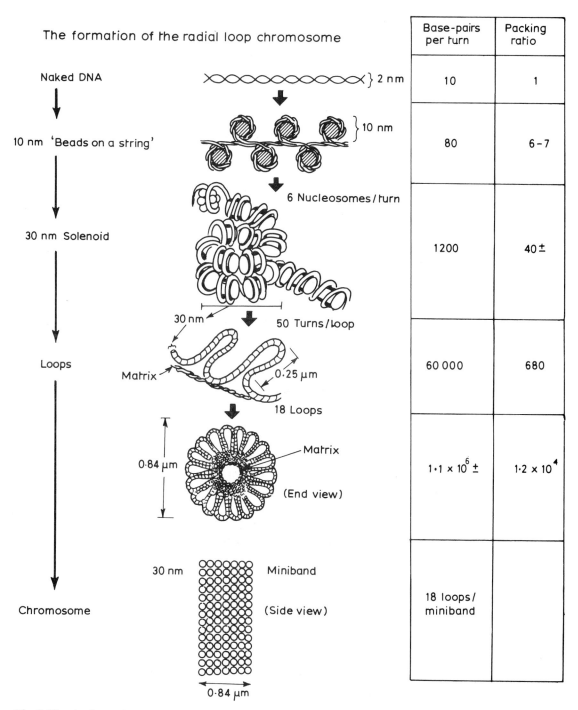

The formation of the radial loop chromosome

| | Base-pairs per turn | Packing ratio |
|---|---|---|
| Naked DNA — } 2 nm | 10 | 1 |
| 10 nm 'Beads on a string' — } 10 nm | 80 | 6–7 |
| 30 nm Solenoid — 6 Nucleosomes/turn, 30 nm, 50 Turns/loop | 1200 | 40± |
| Loops — Matrix, 0·25 μm, 18 Loops | 60 000 | 680 |
| — Matrix (End view), 0·84 μm | $1 \cdot 1 \times 10^{6} \pm$ | $1 \cdot 2 \times 10^{4}$ |
| Chromosome — 30 nm Miniband (Side view), 0·84 μm | 18 loops/ miniband | |

**Fig. 3.12**  A schematic diagram of the higher-order organization of a chromatid (reproduced, with permission from [20] copyright the Company of Biologists Ltd.).

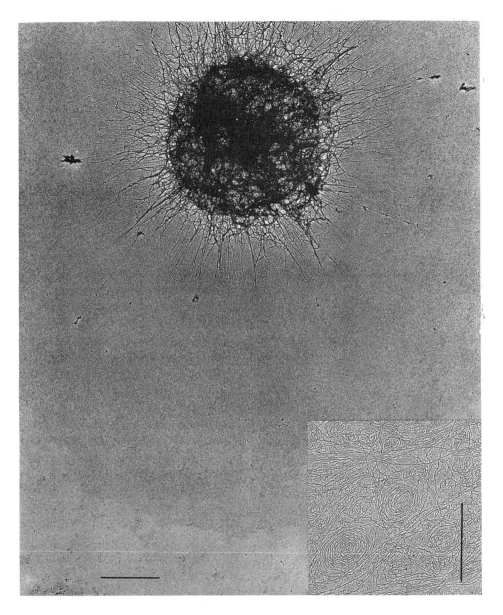

**Fig. 3.13** An electromicrograph of some of the DNA in a human (HeLa) cell. The nuclear DNA has been spread throughout most of the field to form a skirt which surrounds the collapsed skeleton of the nucleus. A tangled network of DNA fibres radiates from the nuclear region. The bar represents 5 $\mu$m.
*Inset*: Only at the very edge of the skirt can individual duplexes be resolved. Most appear as collapsed toroidal or interwound superhelices, indicating that the linear DNA must be unbroken and looped, probably by attachment to the nuclear skeleton. The bar represents 1 $\mu$m. (By courtesy of Dr P. R. Cook and Dr S. J. McReady.)

somes which have been deproteinized are readily visualized in the electron microscope as looped structures with the thread of DNA in each loop entering and leaving the chromosome at the same central point [187, 188]. When the DNA is wrapped around a nucleosome in the cell nucleus its linking number is changed relative to B form DNA (see Section 2.6) and the DNA is essentially relaxed. However, when the histones are removed the DNA tends to return to the B form and as this is prevented by the ends of the loops being linked, the DNA is recovered with superhelical turns. We therefore have the picture of the chromosome consisting of a series of loops of DNA present in a nucleosomal conformation and condensed at least in part into a solenoid (Fig. 3.12). The degree of solenoid formation may be related to the extent of expression of the DNA (see Section 10.4.4). The loops may be attached to a central core from which they radiate [20, 304]. It is possible to digest 99% of the DNA from metaphase chromosome preparations leaving behind a morphologically intact central chromosome 'scaffold'. In a similar manner the material which remains when dehistonated interphase chromatin has most of its DNA removed by nuclease treatment is known as the nuclear 'matrix'. The relationship between the scaffold and matrix is unclear. The scaffold proteins include two high-molecular-weight proteins of molecular weight 170 000 and 135 000 but the matrix, as well as containing an ill-defined fibrous protein network is also associated with the nuclear pore lamina complex [2, 3, 19, 20] to which the chromosomes are attached (see Section 3.1). The chromosomes therefore appear to take up a precise intranuclear location which is retained throughout the cell cycle [189].

The loop model of chromosome structure implies that specific DNA sequences may be associated with the attachment sites of the loops. Many reports exist showing that active, transcribing DNA is associated with the nuclear matrix (see Section 10.4.4.e) and the loop has been associated with the replicon – the unit of replication (Section 6.8.2). There has been some controversy over these results [190]. It seems clear that particular DNA sequences are associated with the nuclear matrix [191, 192] and under certain conditions of isolation these sequences may be actively transcribing. When other, low-salt, isolation conditions are used the matrix-associated DNA may represent sites important in chromosome organization.

### 3.3.6 Lampbrush chromosomes

The diplotene stage of meiosis may last for several months in developing oocytes and during this time the chromosomes are visible under the phase contrast microscope while still undergoing transcription. Such lampbrush chromosomes have been most extensively studied in the oocytes of *Triturus* (newt) and *Xenopus* (toad). At this stage the chromosomes are paired, the two homologues being held together by chiasmata. Each member of the pair consists of two chromatids, i.e. two DNA duplexes. The DNA duplexes run the length of the chromosome, being attached to the axial thread in a series of loops. As there are two chromatids the loops are

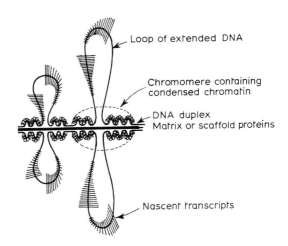

**Fig. 3.14** Diagram of part of a lampbrush chromosome showing the paired loops of extended DNA and the Christmas tree effect produced by the nascent transcripts.

paired (Fig. 3.14) and the structure has the appearance of those brushes which used to be used to clean lamps but which are now used to clean test-tubes and bottles. The axis consists of protein (the matrix or scaffold) and DNA, most of which is compacted (probably in a solenoidal form) to form beads of about 1 $\mu$m in diameter known as chromomeres. These chromomeres contain up to 95% of the total DNA the rest being largely in the 4000 or so loops in each chromosome.

The loops are the site of transcriptional activity and the DNA in the loops may be only partly condensed into nucleosomes. The loops are made more visible by the accumulated RNA. As the RNA polymerase molecules move around a loop the length of the RNA transcript increases, producing the so-called *Christmas tree* effect (Fig. 3.14) (see Section 9.4). *In situ* hybridization has allowed the localization of histone and ribosomal and transfer RNA gene transcripts to particular loops [113, 194–196].

### 3.3.7 Polytene chromosomes

In certain tissues, notably the salivary glands of the fruit fly *Drosophila* and some other species, the DNA molecules do not segregate after replication but remain together during several rounds of the cell cycle. By geometric progression, after ten rounds of replication there are more than·one thousand DNA strands lying alongside each other with their specific sequences and specific bound proteins matched. This forms an excellent experimental system for analysis. Firstly *Drosophila* has a simple chromosome complement of four, the centromere remains under-replicated and a highly characteristic star-like structure with the spokes radiating from the centromere can readily be identified by light microscopy. Unlike metaphase chromosomes, these chromosomes are in the stretched-out interphase condition and sites can be clearly identified where 'puffing' due to transcriptional activity occurs. Puffs can even be

induced in certain sites by the application of specific insect hormones so that the process of transcription can be effectively visualized with respect to time [197–199].

Polytene chromosomes have a highly characteristic banding pattern with dark and light bands of differing densities and dispositions at various points in the arms. In addition, the ease with which *Drosophila* mutants can be isolated means that a lack of a specific gene function can be correlated with the absence of a specific band. While it is not possible to identify unique genes on ordinary metaphase chromosomes because of the low specific activity of the small amount of mRNA or cDNA involved leading to weak radioautography, the large number of binding sites make it possible to perform *in situ* hybridization successfully on polytene chromosomes even in cases where a gene has a low overall frequency with respect to haploid DNA content. As well as their use for the identification of specific RNA hybridization sites, these chromosomes are extensively used in the analysis of the non-histone chromosomal proteins and Z-DNA by immunofluorescence since a specific antibody linked to a fluorescent second antibody can be used to identify the binding site on the DNA of the original antigenic protein or the presence of Z-DNA [200–202].

## 3.4 EXTRANUCLEAR DNA

### 3.4.1 Mitochondrial DNA

Mitochondrial DNA is usually found in cyclic, double-stranded, supercoiled molecules, the exceptions being the linear mitochondrial DNA molecules from *Tetrahymena* and *Paramecium*. Mammalian mitochondrial DNA molecules are not packaged into nucleosomes. They are about 15 kbp long and can therefore code for 15 to 20 proteins. However some yeast ones are considerably larger and *Saccharomyces* mitochondrial DNA is 75 kbp long. Plant mitochondrial DNA is much longer still. Yeast mitochondrial DNA molecules have been widely studied since they are readily amenable to genetic analysis. When yeast cells are grown in the presence of mutagens such as ethidium bromide, 'petite' mutants are formed in which large sections of the mitochondrial genome are deleted with the remaining segments being amplified so that the DNA molecules remain the same size. The mutants are readily propagated by the yeast cells but fail to synthesize mitochondrial proteins. Some petites have been isolated whose mitochondrial DNA contains only adenine and thymine [203] yet these DNA molecules are still able to replicate (see Section 6.9.5). Clearly mitochondrial DNA can only code for a small proportion of mitochondrial proteins yet the organization of the mammalian mitochondrial genome has been described as 'a lesson in economy' [204, 205]. Much of the difference in length between mammalian and yeast mitochondrial DNA is a result of the way the DNA is organized as both appear to contain the same basic set of genes.

Mitochondrial DNA from several mammals, *Xenopus* and *Drosophila*, has been sequenced and the sequence analysed [206–208, 239–241]. There are genes for two ribosomal RNAs, 22 transfer RNAs and for 13 proteins most or all of which are involved in electron transport (Fig. 3.15) [204, 205, 209, 210]. Some of these protein coding regions have still only been identified as unassigned reading frames or URFs (see Section 9.6). None of the mammalian genes contains introns but these are present in some yeast mitochondrial genes.

The only region of mammalian mitochondrial DNA which is non-coding is the D-loop region involved in the initiation of DNA replication (see Section 6.9.5). This is also the region at which transcription of both strands is initiated. Transcription continues uninterrupted around the cyclic molecule and the transcripts are then processed to give individual messenger, ribosomal and transfer RNAs. The transfer RNA forms the punctuation between the various

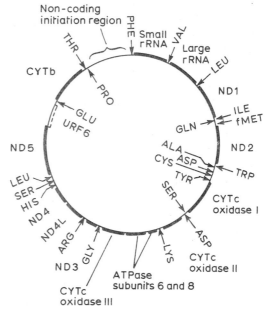

**Fig. 3.15** Map of mammalian mitochondrial DNA (16 569 bp) showing the locations of the genes. Those on the outside are transcribed clockwise from one of the DNA strands; those on the inside anticlockwise from the other. The arrows indicate the positions of tRNA genes. ND, NADH dehydrogenase complex; URF, unassigned reading frame [209, 210].

protein coding regions and provides the signal for the processing enzymes (see Section 9.6). One of the proteins coded for by yeast mitochondrial DNA is a maturase which is involved in the splicing of mitochondrial messenger RNA [211] (see Section 9.6).

Although plant mitochondrial DNA is apparently much more complex than animal mitochondrial DNA it consists of permutations of a basic structure related to each other by recombination [212–214].

One theory for the origin of mitochondria suggests that they arose as symbiotic prokaryotes providing oxidative metabolism to their pre-eukaryotic hosts. Throughout evolution most of the genes were transferred from the symbiont to the nuclear DNA of the host leaving behind only the remnants we find today. The fact that transfer of genes can occur is shown by the finding of one gene (subunit 9 of ATPase) in the mitochondrial DNA of some organisms but in the nucleus in other organisms [215–220]. Since the rate of production of base pair substitutions in mitochondrial DNA is about 10 times that in nuclear DNA it provides a highly sensitive tool for studying short time divergences amongst related species [228].

### 3.4.2 Chloroplast DNA

Chloroplast DNA is in general much larger than mitochondrial DNA, being in the molecular weight range 100 million (150 kbp). In contrast to mitochondria which have from one to ten molecules of DNA per organelle [217] chloroplasts tend to have a very large number of copies of the DNA molecule in each organelle, in some cases greater than one hundred. Like mitochondrial DNA the DNA in chloroplasts carries the coding information for essential membrane components, tRNA and rRNA [217, 221–223]. Mutants similar to the petite mutants may be isolated but show a tendency to revert, possibly because of the multiplicity of copies of the master gene in the chloroplast. All known chloroplast DNA molecules are cyclic and supercoiled.

### 3.4.3 Kinetoplast DNA

The kinetoplast is part of a highly specialized mitochondrion found in certain groups of flagellated protozoa such as the trypanosomes. Its DNA (kDNA) consists of an interlinked series of many thousands of cyclic DNA molecules which vary in size from 0.6 kbp to 2.4 kbp depending on the type of trypanosome from which they are obtained. These components, which are known as minicircles, are further interlinked with a much smaller number of larger circular DNA molecules of above 30 kbp in size known as maxi circles in the majority of systems analysed. While

maxi circles appear to perform the conventional functions of mitochondrial DNA in trypanosomes, minicircles are microheterogeneous in sequence and size and there is no evidence to suggest that they are ever transcribed so that it is possible that they may fulfil some structural rather than coding role [209, 217, 224–226].

### 3.5 BACTERIA

Historically much of the early understanding of gene structure and function has come from the prokaryotes and from the simple bacteriophages which can infect them and grow in them. This has been largely due to the comparative ease with which the prokaryotes can be grown and infected under laboratory conditions and the fast generation time which has made the isolation of genetic mutants a comparatively simple exercise. In addition, the DNA content of prokaryote cells is about one hundredfold less than that of eukaryote cells so that the analytical task is relatively simple.

By far the majority of genetic experiments have been performed in varying strains of the bacterium *Escherichia coli* with a much smaller number being in *Bacillus subtilis*, *Micrococcus luteus* and other bacterial species. There are a wide variety of strains of *E. coli* itself, each possessing its own properties with respect to susceptibility of phage infection and to antibiotic resistance. In addition, many bacteria possess sex pili which are thought to play a major functional role in genetic recombination (Section 3.7). While bacteria normally divide by binary fission, the transfer of genetic information between two types of bacteria is possible. Strains of bacteria which are able to donate their chromosome are known as high-frequency recombinant or HFr strains and have been widely used in the study of bacterial gene function. Mutants are readily isolated by a technique known as *replica plating* whereby colonies of bacteria are grown on a plate of nutrient agar which contains all the essentials for growth. A velvet pad is then used to transfer small portions of these colonies to a second agar plate which contains minimal medium in which certain mutants will not grow. The colonies which are present on the first but not the second plate can then be selected.

Bacteria which possess the full complement of bacterial genes are usually referred to as 'wild type' bacteria or prototrophs and mutants are classified according to their missing functions. Thus a *thr⁻* mutant is a bacterium which requires threonine in its growth medium as it lacks the correct information to make the enzymes involved in threonine synthesis in their fully functional state. Such a nutritional mutant is called an *auxotroph*. The process of mutant selection is greatly enhanced if the cells are first grown in minimal medium in the presence of an inhibitor which will kill all the growing bacteria. Only those mutants survive which are unable to replicate in the deficient medium.

While most bacteria carry all their genetic information on a single, circular chromosome, some possess in addition small extrachromosomal elements known as plasmids which carry genes coding for functions such as drug resistance. These molecules are usually comparatively small and cyclic and can be present in one or several copies per cell (Section 3.7). These plasmids have become widely used as tools in genetic engineering. For a more detailed description of bacterial growth and physiology, there are several specialized texts on the subject which can be consulted [227, 229].

### 3.5.1 The bacterial chromosome

The chromosome of *E. coli* is a single cyclic molecule of 4.5 million base pairs which has an effective circumference of 1 mm but must be contained in a bacterial cell with a diameter in the order of 1 $\mu$m. A complex packaging mechanism is therefore necessary in order to ensure that all the DNA is folded within the bacterial cell in a manner which will not inhibit transcription, nor allow entanglement of the two daughter strands to occur during replication. This is achieved by two main mechanisms. Firstly, the DNA is folded into between 40 and 100 loops, and secondly, each of these quasi circles is itelf supercoiled independently of the others [230, 231]. Each loop is maintained as an independent physical area by DNA:RNA interactions since the loops are abolished by ribonuclease but not, in contrast to eukaryote chromosomes, by proteolytic enzymes. Deoxyribonuclease abolishes the superhelical nature of the loops. Limited treatment with this enzyme will attack only a small number of the loops and therefore leads to the abolition of only some of the supercoiling [232, 233]. The intact folded *E. coli* chromosome or nucleoid can be isolated in the presence of non-ionic detergents in the presence of high molarities of salt in two possible forms, free and membrane-associated. The free form which sediments at 1600 to 1700 S consists of about 60% DNA, 30% RNA and 10% protein, the bulk of which is the enzyme RNA polymerase (Chapter 8). The membrane-bound form which sediments contains an additional 20% of membrane-attached protein. The exact nature of the membrane-attachment sites is not yet known.

### 3.5.2 The bacterial division cycle

As in eukaryotes (see Section 3.2.1) the general rule in bacteria is that chromosome replication (DNA synthesis) and cell division occur alternately. Under slow growth conditions (i.e. doubling times greater than 60 min) there is a gap between completion of a round of DNA synthesis and cell division. Cell division takes place at a fixed time (generally about 60 min) after initiation of a round of DNA synthesis [235]. However, bacteria, under suitable conditions, can grow much more quickly than one division every 60 min and in these situations DNA synthesis becomes continuous, i.e. it no longer occupies a restricted part of the cell cycle [237].

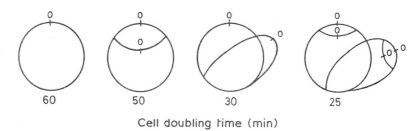

Cell doubling time (min)

**Fig. 3.16** Diagrammatic representation of the structure of one chromosome (of the two present in the cell) immediately prior to cell division in *E. coli* growing at different rates [237]. O is the origin of chromosome replication.

Although cell division still occurs 60 min after initiation of a round of DNA synthesis the cell does not wait until cell division has occurred before initiating a second (or even a third) round of DNA synthesis. This results in some cells, immediately prior to division, having two copies of part of the DNA (the last to be duplicated) and up to eight copies of other regions (those duplicated first) (see Fig. 3.16).

Protein synthesis is required to initiate a round of chromosome replication and hence an amino acid auxotroph, when starved of essential amino acids, will complete ongoing rounds of replication and come to rest with one unreplicated chromosome per cell. In contrast a thymine-requiring mutant when deprived of thymine cannot make DNA but the chromosome is brought into such a condition that when thymine is restored a second round of replication is initiated.

### 3.5.3 Bacterial transformation

Pneumococci (*Diplococcus pneumoniae*) may be classified into a number of different types each characterized by the ability to synthesize a specific serologically and chemically distinct capsular polysaccharide. In 1928 Griffith observed that a particular strain of pneumococci cultivated *in vitro* under specific conditions lost the ability to form the appropriate poly-saccharide and consequently grew on solid media in so-called 'rough' colonies in contrast to the 'smooth' glistening colonies formed by encapsulated cells. If a living culture of such unencapsulated cells was injected into mice together with killed encapsulated pneumococci of type III, the organisms subsequently recovered from the animals were live virulent pneumococci of the encapsulated type III. It appeared therefore that some material present in the dead type III organisms had endowed the unencapsulated pneumococci with the capacity to synthesize the characteristic type III poly-saccharide.

During the next five years it was shown that such pneumococcal transformation could be produced *in vitro*, that a cell-free extract could replace killed cells as the transforming agent, and that organisms which had undergone transformation did not spontaneously revert to their original type.

The chemical nature of the active principle remained obscure until 1944 when Avery, McLeod and McCarty [238] showed that DNA extracted from encapsulated smooth strains of pneumococcus type III could, on addition to the culture medium, transform unencapsulated 'rough' cells into the fully encapsulated smooth type III. The smooth cells so developed could propagate indefinitely in the same form, producing more DNA with the same capabilities. The pneumococcal DNA had therefore initiated its own reduplication as well as inducing the specific inheritable property of capsule synthesis. In other words, it had executed two functions usually associated with the gene.

These observations stimulated further research into bacterial transformation, from which it emerged that the reaction was not limited to pneumococci but could be produced in a wide range of bacteria, e.g. *Haemophilus influenzae*, *E. coli* and the meningococcus. Nor were these transformations limited to changes in serological type since they could also be used to endow bacteria with resistance to specific drugs or antibiotics or the ability to utilize particular nutrients. Transformation can be demonstrated for any characteristic whose acquisition can be measured in the recipient. The two stages required are (a) transfer and (b) utilization of the transferred material.

The proportion of treated cells which may develop a new characteristic after exposure to appropriate DNA is usually small. Among the factors influencing the yield is the capacity of the recipient strain to be transformed, since some strains are much more susceptible than others and *E. coli* requires treatment with $CaCl_2$ in order to act as a recipient at all [243]. However, even in a single population not all cells are competent to take up DNA and this may be a result of variable cell wall permeability. The most appropriate time for transformation to occur is just after cell division. When pneumococci are cooled to a temperature at which growth is arrested and then rewarmed so that they start to divide synchronously, transformations are exceptionally numerous.

The number of transformation events is proportional to the DNA concentration up to a plateau level and hence the frequency of transformation has been used as a measure of gene frequency (see Chapter 6).

Duplex DNA enters the recipient better than does single-stranded DNA and the saturating DNA concentration is reached when about 150 DNA molecules of molecular weight about $10^7$ have been taken up by each bacterium [244].

Just how the transforming DNA enters the cell to be transformed is not fully understood, but it is known that the acquisition of the new characteristic induced by the DNA requires a period up to 1 hour and that after acquiring the new DNA a cell multiplies more slowly for some time than do its unchanged neighbours. The establishment of the mechanism for duplicating the new DNA requires still longer. The mechanism of the transformation reactions has been recently reviewed [245, 246].

Although duplex DNA is better for transformation, on entry into the recipient cell one strand of the duplex is degraded. The major endonuclease of pneumococcus is associated with the cell membrane and it is responsible for degrading one of the DNA strands following action of an enzyme which introduces single-strand breaks every 6000 bases [247]. In *E. coli* the recipient bacterium must be recBC$^-$ otherwise the recBC nuclease degrades the incoming DNA.

For expression, the incoming single strand of DNA finds a complement in the recipient cell DNA and displaces one of the original duplex strands in a recombination-type mechanism (see Chapter 7). This integration event occurs at widely different efficiencies for different markers.

It is possible to transfer more than one inheritable characteristic to susceptible bacteria in a single DNA preparation, e.g. one sample of DNA may carry the three characteristics of resistance to penicillin, resistance to streptomycin, and the ability to form a capsule in pneumococci. Such a sample of DNA might bring about transformation in 5% of the recipient cells. Of the cells transformed, 98% would acquire only one of the three characteristics, 2% would acquire two of the characteristics, and only 0.01% would acquire all three. Clearly, therefore, the DNA preparation cannot convey a complete set of the donor's characteristics to the recipient, although certain characteristics appear to be linked. For example, the

DNA factors responsible for streptomycin resistance tend to be coupled with those responsible for the ability of pneumococci to use mannitol as a source of energy.

Presumably all 'transformations' are essentially processes by which bacteria are endowed with enzyme-synthesizing capacities which they did not previously possess. The first direct proof of the presence of such a new enzyme in an organism after treatment with a DNA transforming factor was made by Marmur and Hotchkiss [248] who demonstrated that the ability to oxidize mannitol could be transferred to a non-utilizing strain of pneumococcus by culturing it in the presence of DNA prepared from a strain which possessed the power of utilizing this sugar. The organisms so transformed differ from the parents in possessing the new enzyme mannitol phosphate dehydrogenase. Bacterial cell transformation plays an essential part in the cloning of DNA as described in the Appendix.

An analogous somatic transformation was first reported by the Szybalskis for mammalian cells [249]. They isolated DNA from cultures of human cells of the strain D98S which contain the

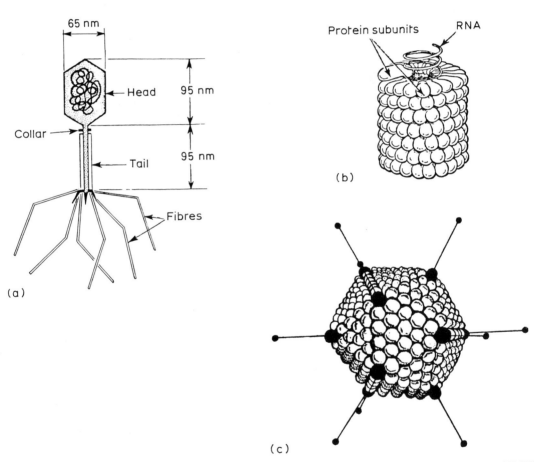

**Fig. 3.17** Structure of virus particles. (a) T-even phage particle; (b) a segment of tobacco mosaic virus (TMV) showing the protein subunits forming a helical array [253]. The RNA lies in a helical groove in the protein subunits, some of which have been omitted to show the top two turns of the RNA helix; (c) a model of adenovirus showing the arrangement of the capsomeres and spikes.

enzyme hypoxanthine phosphoribosyl trans-ferase (HPRT) (EC 2.4.2.8) responsible for the reaction:

$$\text{hypoxanthine} + \text{PRPP} \rightleftharpoons \text{IMP} + \text{PP}_i$$

which is discussed further in Chapter 5. The addition of this DNA to cultures of cells of the strain D98/AH-2 which are deficient in the enzyme resulted in the appearance of HPRT-positive genetically transformed cells detected under highly selective conditions. The trans-forming activity was abolished by deoxyribo-nuclease but not by ribonuclease. Modern methods of introducing foreign DNA into animal cells are discussed in Section A.8.4.

## 3.6 VIRUSES

### 3.6.1 Structure

The general structure of a virus is a nucleic acid core surrounded with a protein coat. The nucleic acid may be single- or double-stranded RNA or DNA. The coat may consist of many identical molecules of protein (capsomeres) or may be a complex structure shown in Fig. 3.17. Viruses infect animals, plants and bacteria and the last are called bacteriophages, or simply phages. As the viruses lack any machinery to catalyse metabolism or protein synthesis they can only multiply within a host cell whose metabolism is subverted to serve the require-ments of the virus [246, 250, 251].

Bacteriophages contain single-stranded RNA (e.g. MS2), single-stranded DNA (e.g. $\phi$X174) or double-stranded DNA (e.g. T-phages). The single-stranded DNA phages are either spherical (e.g. $\phi$X174) or filamentous (e.g. fd or M13) where the circular DNA molecule is wrapped in a protein coat and then formed into a filament of two nucleoprotein strands by bringing the opposite sides of the circle together. All the RNA bacteriophages are spherical. MS2 consists of 180 protein subunits all identical except for one (the maturation protein).

Many of the double-stranded DNA phages have a more intricate structure. The T-even phages (Fig. 3.17) have a head (which contains the DNA), a tail (through which the DNA is injected into the host cell), and a base plate with six tail fibres which recognize, and attach to, sites on the surface of the host cell.

The protein coat protects the nucleic acid from damage and also confers a specific host range on the potential infectivity of the particle.

Most plant viruses contain single-stranded RNA as in TMV (tobacco mosaic virus), but a few contain double-stranded RNA as in wound tumour virus or double-stranded DNA as in cauliflower mosaic virus. The virus particles can be rod-shaped (e.g. TMV) or spherical (e.g. cowpea chlorotic mottle virus).

An extensive study of TMV (for reviews see [253, 254]), which has a particle weight of $4 \times 10^7$ and is rod-shaped measuring $15 \times 300$ nm, has led to a detailed picture of its structure (Fig. 3.17). A helical array of about 2100 identical protein subunits of molecular weight 17 400 surround a single-stranded RNA molecule of molecular weight $2 \times 10^6$. Cowpea chlorotic mottle virus is spherical and is composed of 180 identical subunits of molecular weight about 20 000 arranged on the surface of an icosahedron in 32 morphological units of 20 hexamers (on the faces) and 12 pentamers (on the vertices) [255].

Of the animal viruses adenoviruses are icosa-hedral with 1500 subunits arranged as 240 hexamers on the faces and 12 pentamers at the vertices and a third type of protein forms the fibres which extend from each vertex. Many of the larger animal viruses (e.g. influenza virus) are surrounded by a lipoprotein envelope largely derived from host cell membranes.

### 3.6.2 Virus classification

Viruses can be classified with respect to their structure or their host cell but the Baltimore classification is based on their mode of gene expression and replication [256]. On transcription of DNA only one strand gives rise to messenger RNA (see Chapter 8). The mRNA is given the designation 'plus(+)strand'. The complementary strand is called 'minus(−)strand' RNA.

There are six classes of virus delineated in Table 3.3. Class I contain duplex DNA which may be cyclic or linear. The DNA is replicated in the standard semiconservative manner as described in Chapter 6. Class I viruses infect bacteria, animals and plants and can also bring about cell transformation (see Section 3.6.9).

Class II viruses carry a single strand of DNA which, upon infection, is converted to a duplex replicative form (RF). A final stage in the infectious cycle is formation of single-stranded DNA from the RF (see section 6.7.3). The DNA can be linear or cyclic and the viruses infect bacteria or animals. The DNA is plus strand, i.e. it has the same sequence as the mRNA. No minus-strand DNA viruses are known.

Class III viruses include reovirus of man and plant wound tumour virus. The duplex RNA is 'transcribed' to give + strand mRNA which later serves as a template to re-form duplex RNA. The genomes are segmented, i.e. several different RNA molecules are present in each viral particle (see Section 8.4.2).

Class IV viruses have a + strand RNA genome. They replicate by synthesis of a −RNA strand which in turn serves as a template for viral, +RNA strand synthesis (see Section 8.4).

**Table 3.3**  Baltimore classification of viruses [256].

| Class | Nucleic acid | Replication | Example |
|-------|--------------|-------------|---------|
| I | Duplex DNA | Semiconservative | T4, adeno & herpes viruses |
| II | Single-stranded DNA(+) | Via duplex replicative form (RF) | φX, minute virus of mouse (MVM) |
| III | Duplex RNA | Via + RNA intermediate | Reovirus (produces infantile diarrhoea) |
| IV | + Strand RNA | Via − RNA intermediate | MS2, polio, foot & mouth disease viruses |
| V | − Strand RNA | Via + RNA intermediate | Measles 'flu & rabies viruses |
| VI | + Strand RNA | Via DNA duplex | Leukaemia virus AIDS virus |

A continuous line represents DNA; a dashed line represents RNA.

Class V viruses have a − strand RNA genome which is first copied to give + strand mRNA which in turn give rise to new molecules of − strand RNA (see Section 8.4.2).

Class VI viruses are spectacular in that they have an RNA genome (+ strand) which is converted first to a DNA/RNA hybrid and then to duplex DNA. This is then transcribed to form new + strand RNA. This strategy is used by the leukaemia viruses and the AIDS virus and requires a special enzyme to reverse transcribe the RNA into DNA (see Section 6.4.2 and Section 6.9.10).

### 3.6.3 Life cycle

The replication of a virus can be considered in stages: (1) adsorption of the virus on to the host cell; (2) penetration of the viral nucleic acid into the cell; (3) development of virus-specific functions, alteration of cell functions, replication of the nucleic acid, and synthesis of other virus constituents; (4) assembly of the progeny virus particles; (5) release of virus particles from the cell [246, 250, 251, 257, 258].

The life cycle of animal viruses is similar to that of phage, the main difference being that a cycle takes 20–60 h, rather than 20–60 min.

(1) Viruses will only infect certain specific cells, i.e. they have a limited host range because the coat (or tail) will only recognize and adsorb to specific sites on the appropriate cell walls. The host range of the T-phages is a property of their tail fibres. Some T-even phages are free to adsorb to the bacterial cell wall site only in the presence of tryptophan; in the absence of this amino acid the tail fibres are folded back and attached to the tail sheath [259]. The initial reversible interaction of the tail fibres with the cell wall is followed by the formation of a permanent attachment. The small male-specific phages (e.g. MS2, R17) attach only to the f-pili of male *E. coli* cells [260] (see Section 3.7).

(2) The penetration of the viral nucleic acid into the cell involves a phage mechanism (e.g.

T-phage), host-cell mechanism (e.g. MS2), or possibly the simple removal of the coat once the virus is in the cell.

After attachment to the cell, the lysozyme present in the base of the bacteriophage T4 tail probably hydrolyses part of the cell wall. This allows the tail core to penetrate into the cell as the contractile tail sheath contracts and the small amount of ATP present in the phage tail is hydrolysed to ADP [261]. How the DNA passes from the head through the tail and into the cell (a process equivalent to passing a 10-metre-long piece of string down a straw) is not understood.

The injection of bacteriophage T5 DNA takes place in two stages: 8% of the DNA (the first step transfer DNA) enters the cell and directs mRNA and protein synthesis. One of these proteins is required to complete the injection of the remaining DNA [262].

After attachment of bacteriophage MS2 to the f-pilus of male *E. coli* the RNA leaves the phage and is then transported inside the length of the pilus to the cell. This last step requires cellular energy [263]. An alternative mechanism proposed for entry of M13 into male *E. coli* involves retraction of the pilus with the filamentous bacteriophage attached [264]. Replication to the duplex form is necessary for the bacteriophage DNA to be drawn into the host, and when entry is effected a considerable fraction of the capsid protein is deposited in the inner cell membrane [257].

With animal viruses, uptake mechanisms are simpler [258]. Thus polyoma virus attaches to neuraminidase-sensitive sites on mouse cells [265] and is then taken up entirely by pinocytosis and is carried to the nucleus where the DNA is uncoated.

(3) Once uncoated, isolated viral DNA will not infect cells as well as intact virus. Thus following penetration of viral DNA into a cell the infective titre drops. With bacteriophage it remains low for maybe 20 min (the lag phase) and then suddenly rises up to a hundred-fold, indicating the presence of many new mature

virus particles. During this lag phase after virus infection the metabolic processes of the cell are being modified [262]. Viral mRNA directs the synthesis of specific enzymes, and the rates of host cell DNA, RNA and protein synthesis are altered. The viral nucleic acid replicates and virus constituents are synthesized. These different viral functions are divided into two groups, the early and late functions, which appear to be controlled either at the level of transcription or translation (see Sections 10.1 and 10.2 and 11.10 and 11.11). Early functions include the biosynthesis of enzymes required for the replication of the nucleic acid, and late functions include the formation of the virus coat and other constituents. T-even bacteriophages turn off the synthesis of host cell DNA and redirect the synthesis of DNA precursors to fit their particular requirements (e.g. hydroxy-methylcystosine triphosphate is produced; see Chapter 6).

On the other hand the small tumour viruses, SV40 and polyoma, stimulate the synthesis of host cell DNA [266, 267] particularly if the cells are in a resting state before infection [268].

The replication of virus nucleic acids is described in Chapter 6.

(4) The assembly of virus particles may be either spontaneous (self-assembly) [269] or may involve a series of virus-directed steps [270, 252, 281].

More complex viruses are pieced together step by step. This process has been partially characterized for T4 [251, 271, 272]. Wood and Edgar [270] have shown that about 45 viral genes are involved in T4 assembly, and many of the steps have been characterized by *in vitro* complementation using the partially finished pieces (e.g. heads, tails, fibres) found in cells infected with different mutants under non-permissive conditions. Eight genes have been assigned to the component parts of the head, and a further eight implicated in the assembly of these components. The head, which then appears to be morphologically complete, requires the action of two more gene products before it can interact spontaneously with assembled tails. The base plate components (12 genes) are assembled in two steps, then the core (three genes) and the sheath (one gene) are added by progressive polymerization from the base plate. Two gene products are required to finish the tails before they can be joined to the heads. The assembled heads and tails are modified (one gene) before the tail fibres are added. This last step requires complete tail fibres (two genes for components and three for assembly) and a labile enzyme (L). Phage particles assembled *in vitro* are active and possess characteristics which vary with the source of the parts (e.g. the genotype of the head and the host range of the tail fibres).

*In vitro* systems for the reconstitution of phage lambda particles (packaging) have proved very important for the reintroduction of cloned DNA into cells following its *in vitro* modification [273] (see Section A.7.2).

The reconstitution of the rod-shaped TMV has been studied extensively [251, 274, 275]. When the coat protein and viral RNA are mixed in the correct ionic environment, virus particles are formed which possess up to 80% of the original infectivity. Attempts to reconstitute small spherical viruses have proved more difficult. MS2 RNA and coat protein form morphologically complete particles which are not infective, possibly because they lack the maturation protein [276]. Cowpea chlorotic mottle virus can be partially degraded and reassembled to form particles indistinguishable from the original virus, and separated protein and nucleic acid have been mixed under conditions such that infectious particles are formed which have the same appearance, serological properties and sedimentation coefficient as intact virus [277]. Such experiments are consistent with the suggestion that the nature of the virus coat subunits alone directs the size and shape (spherical or rod) of the completed virus particles [278].

(5) Interference with the normal metabolic processes may lead to the eventual death of the infected cell, followed by natural lysis, but in some instances it has been shown that the virus actively causes cell lysis. Bacteriophage T4 codes for a lysozyme [279] which digests the host cell wall, causing the release of the progeny virus particles. Bacteriophage M13 and related filamentous viruses do not kill the host bacterium but pass out through the cell membrane picking up coat protein on the way [257].

In contrast to production of obvious cytopathic effects (CPE) some viruses disturb cellular metabolism yet do not cause cell death. Thus lymphocytic choriomeningitis virus infections lead to changes in growth and glucose regulation and the infected pituitary glands do not make growth hormone, yet no pathological injury is apparent [280].

### 3.6.4 The Hershey–Chase experiment

The classic experiment of Hershey and Chase [282] showed that on infection of *E. coli* with T2 bacteriophage only the DNA enters the bacterium. Thus only the DNA replicates and carries the information to specify new virus. To demonstrate this they grew T2 phage in the presence of $^{32}$P-labelled phosphate and $^{35}$S-labelled amino acids. This resulted in phage containing DNA labelled with $^{32}$P (there is no sulphur in DNA) and protein labelled with $^{35}$S. The labelled virus was then allowed to infect unlabelled bacteria and the progeny isolated. It was found that much of the $^{32}$P (i.e. the DNA) was present in the progeny virus, but none of the $^{35}$S-labelled protein. Vigorous shaking of the culture within a few minutes of infection was found to dislodge the empty $^{35}$S-labelled protein coat of the virus from the bacteria. On centrifugation this coat remained in the supernatant fraction when the bacteria sedimented. The bacteria carried the $^{32}$P-labelled viral DNA which replicated and produced new phage particles. This shows that the viral DNA carries the genetic material and that the coat protein is not essential once the DNA is inside the bacterium.

### 3.6.5 Virus mutants

The viruses provide a unique opportunity to characterize completely the information content of a functional genome.

A mutation in a viral genome can cause the inactivation of a gene function. If the missing function can be characterized (e.g. a missing enzyme) then the nature of the gene, and the mutant, is defined. However, if the function is essential, the mutation is lethal and the mutant cannot be propagated and studied. The discovery of conditional lethal mutants [284–285], which lack a gene function under one set of conditions but regain it under another set of conditions, has revolutionized the study of virus, and in particular, phage genetics. Stocks of conditional lethal mutants can be grown under permissive conditions and then the mutant gene function studied and identified under non-permissive conditions.

Two classes of conditional lethal mutation are particularly useful. Temperature-sensitive mutants [285] are not viable at the non-permissive temperature (e.g. 42°C) but grow at the permissive temperature (e.g. 30°C). This is due to a single base change in the DNA causing the introduction of the wrong amino acid into the protein, reducing the stability of its configuration at the higher temperature.

Amber mutants [283] of a bacteriophage will only grow in a permissive host (which contains a suppressor). These result from a single base change altering an amino acid-coding triplet to UAG (Section 11.2), which is read as stop under the normal, non-permissive conditions. The protein is terminated at that point. Permissive host bacteria, which suppress the mutation, have a species of tRNA which translates UAG as an amino acid and allows the protein to be completed. Different amino acids are added by different classes of permissive host.

When two virus mutants, with mutations in different genes, infect the same cell under non-permissive conditions, the missing function of each can be supplied by the other, i.e. *complementation* takes place, and progeny viruses are formed. If the two mutants have mutations in the same gene, complementation cannot occur and no progeny are formed. Thus complementation between mutants is used to determine the number of complementation groups, i.e. the number of different genes in which the mutations occur. If sufficient mutants have been isolated for each gene of the virus to be represented, then the total number of essential viral genes can be estimated.

The joint growth of two phage in the same cell can also result in the formation of a few progeny phage, *recombinants*, which carry genetic characters of both parental bacteriophages (see Chapter 7). A study of the frequency of formation of recombinants from parental phage carrying known mutations allows the construction of a genetic map; the frequency of recombinants with both genetic characters is related to the distance between the two mutation points on the genome. Such a map shows the relative positions of the mutations and therefore of the genes in which these mutations occur, and it can be linear or circular [284].

### 3.6.6  Virus nucleic acids (Table 3.4)

Viral DNA varies in molecular weight from a little over $10^6$ to more than $10^8$ (cf. the value of $2.2 \times 10^9$ for *E. coli* DNA) and, unlike the DNA of bacteria and eukaryotes, it can be easily extracted from the virus without degradation. Such intact DNA molecules have revealed a striking variety of tertiary structure. For instance the DNA of bacteriophage $\phi$X174 is single-stranded and in the form of a continuous cyclic chain containing 5386 nucleotides. DNA from

**Table 3.4**  Properties of some viral nucleic acids.

|  | Host cell | $10^{-3} \times$ Base pairs | Single- or double-stranded | Shape |
|---|---|---|---|---|
| **DNA phage** | | | | |
| T2 | *E. coli* | 200 | Double | Linear |
| T5 | *E. coli* | 130 | Double | Linear |
| T7 | *E. coli* | 38 | Double | Linear |
| λ | *E. coli* | 48 | Double | Linear |
| $\phi$X174 | *E. coli* | 5.4 | Single | Cyclic |
| P22 | *Salmonella* | 39 | Double | Linear |
| **RNA phage** | | | | |
| MS2 | *E. coli* (male) | 3.3 | Single | Linear |
| **DNA animal viruses** | | | | |
| Polyoma | Mammals | 5.3 | Double | Cyclic |
| Herpes | Man | 103 | Double | Linear |
| **RNA animal viruses** | | | | |
| Poliovirus | Man | 6.7 | Single | Linear |
| Reovirus | Mammals | 18.2 | Double (segmented) | Linear |
| **Plant virus** | | | | |
| TMV | Tobacco plant | 6.1 | Single | Linear |

bacteriophage G4 and M13 is similar and other variations in viral DNA arrangements are:

### (a) Cohesive (sticky) ends

When DNA extracted from phage λ (molecular weight $30 \times 10^6$) is heated to 65°C and cooled slowly, its sedimentation coefficient is 37S, but when it is quick-cooled its sedimentation coefficient is only 32S [283]. This behaviour is a result of the 5'-ends of the DNA projecting for 12 nucleotides beyond the 3'-ends, the two single-stranded regions being complementary. These cohesive ends can base-pair and convert the DNA into a cyclic molecule which is disrupted at 65°C. The hydrogen-bonded cyclic form can be converted with polynucleotide ligase (Section 6.4.3) into a cyclic form with both strands continuous. *E. coli* DNA polymerase, which adds on nucleotides to the 3'-ends, abolishes the ability to form cyclic molecules, but the ability is regained after treatment with *E. coli* exonuclease III which removes the newly added bases from the 3'-ends. Electron microscopy has shown the 37S form to be cyclic and the 32S form to be linear. Some other lysogenic phage contain DNA with a similar structure (e.g. φ80).

### (b) Terminal repetition

The DNA of phage T7 (molecular weight $25 \times 10^6$) is double-stranded and linear, and the sequence (about 0.7% of the total) at the beginning is repeated at the end of each molecule [286]. Treatment of such molecules with *E. coli* exonuclease III results in the formation of cohesive ends which cause circle formation under suitable conditions. Terminal repetition has been detected in the DNA from several phage (e.g. T2, T4, T3, P22) and is believed to play a role in replication by enabling multiple-length concatamers to be formed (see Section 6.10.2).

The genome of adenoviruses (molecular weight $20$–$25 \times 10^6$) has a sequence of about 100 nucleotide pairs repeated in an inverted fashion at the two ends [287] and the 5'-end of both strands is covalently attached to a protein of molecular weight 55 000 (see Section 6.9.7).

### (c) Circular permutation

If the linear double-stranded DNA of phage T2 (molecular weight $130 \times 10^6$) is denatured and allowed to reanneal slowly, circular molecules are formed which can be detected with the electron microscope [288]. These are formed because the phage DNA molecules do not have a unique sequence, but the population is a collection of molecules with sequences which are

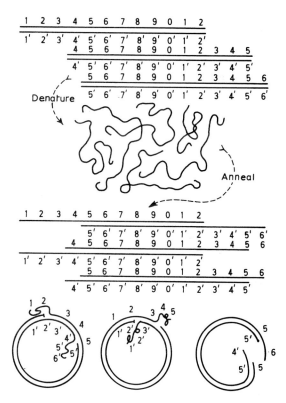

**Fig. 3.18** Formation of cyclic DNA by denaturing and annealing a permuted collection of duplexes [290]. Notice that each permutation is also terminally repetitious. One repetitious terminal from each strand cannot find a complementary partner and is left out of the circular duplex. Their separation depends on the relative permutation of the partner chains. (Reproduced by permission of Academic Press from [290].)

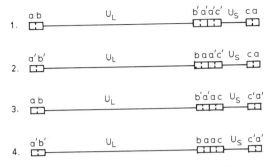

**Fig. 3.19** The four possible arrangements of herpes simplex DNA. $U_L$ and $U_S$ are the long and short unique regions, both of which are bounded by repeated sequences.

duplex at both ends of the DNA while others [e.g. adenoassociated virus (AAV)] have a hairpin at one end only [293].

*(f) Segmented genomes*

Each reovirus particle contains about $12 \times 10^6$ daltons of RNA which, when extracted, is in pieces of three sizes (molecular weight $2.3 \times 10^6$, $1.3 \times 10^6$, and $8 \times 10^5$), plus 50–100 single-stranded oligonucleotides rich in adenine [294].

Similarly the virion of the RNA tumour viruses (retroviruses) contains two identical copies of the 35S viral RNA together with several copies of transfer RNA derived from the host cell (see Section 6.9.10).

circular permutations of each other (Fig. 3.18). The molecules also show terminal repetition of the sequences at the beginning. The DNAs from phages T4 and P22 also exhibit circular permutation and terminal repetition.

The genome of herpes simplex virus (HSV) (molecular weight about $10^8$) is remarkable in being made up of two parts joined in tandem. Each part has a region which is repeated at the two ends in an inverted orientation. As the two parts can be joined in either orientation there are four possible sequences present in the HSV genome [291, 292] (Fig. 3.19).

*(d) Nicks*

Three specific breaks have been found by electron microscopy in one strand of the double-stranded linear DNA extracted from bacteriophage T5 [289]. Accordingly, when the DNA is denatured, five single-stranded pieces are produced (instead of two) from each T5 DNA molecule. Bacteriophages SP8 and SP50 may also contain specific breaks.

*(e) Hairpins*

The smallest animal viruses (the parvoviruses) have a linear single-stranded DNA genome of molecular weight $1.2 \times 10^6$–$2.2 \times 10^6$. Some [e.g. minute virus of mouse(MVM)] have a hairpin

*(g) Modification of viral DNA*

The DNAs of the T-even phage contain 5-hydroxymethylcytosine in place of cytosine [295] and the hydroxyl group of this base can be glucosylated (Table 3.5) [296]. Growth of the phage in a UDP-glucose-deficient host produces phage with DNA which is not glucosylated, and such DNA is degraded when the phage infect the normal host [297]. (This modification is considered in more detail in Section 6.9.2.)

The DNAs of many phage are modified by host-specific mechanisms when grown in certain strains of *E. coli* (*E. coli* K12 and *E. coli* B), but not when grown in other strains (e.g. *E. coli* C). Bacteriophage modified by growth in K12 or B will only grow efficiently in the same strain, K12 or B respectively or in C. Unmodified bacteriophage will only grow efficiently in C.

**Table 3.5** Percentage glucosylation of hydroxymethylcytosine residues in the DNA of T-even bacteriophages [296].

|  | T2 | T4 | T6 |
|---|---|---|---|
| Unglucosylated | 25 | 0 | 25 |
| α-Glucosyl | 70 | 70 | 3 |
| β-Glucosyl | 0 | 30 | 0 |
| β-Glucosyl-α-glucosyl (diglucosyl) | 5 | 0 | 72 |

When the infecting bacteriophage fails to grow (i.e. is restricted), the bacteriophage DNA is degraded. The restriction/modification system is considered in detail in Section 4.5.3.

(h) *Chromatin structure of animal virus DNA*

The small animal tumour viruses SV40 and polyoma (papovaviruses) exist in the cell in the form of nucleoprotein particles where the cyclic duplex DNA molecule (5224 and 5292 nucleotide pairs respectively) is wrapped around histone octamers to form a nucleosomal structure – the minichromosome (see Section 3.3.2). There are 21 nucleosomes associated with the cyclic chromosome which, because of the presence of topoisomerases (see Section 6.4.6), is present in a form of minimum energy. In the cell or when isolated in low-salt conditions the minichromosome is highly compact. When the histones are removed the isolated DNA is present in a supercoiled form (form I) with about 26 superturns per molecule (i.e. there is an average of 1.25 superhelical turns per nucleosome). A single nick in one of the DNA strands will convert the duplex into a relaxed open circular form (form II).

Three forms of double-stranded DNA can be isolated from purified polyoma virus [298, 299]: the supercoiled form (21S, component I), the open cyclic form (16S, component II) and a linear form (14.5S, component III). Viral component III forms a minor fraction of the linear DNA in the virion which is, in fact, mainly cellular DNA that has been encapsulated in virus particles [300]. Electron microscopy (Fig. 2.23) has revealed structures which can be clearly identified with the supercoiled, open cyclic and linear forms and the three forms are readily separated by electrophoresis on agarose gels.

The DNA of the larger animal viruses (e.g. adenovirus or herpes simplex virus) although not naked in the cell does not appear to have a nucleosomal structure similar to that of the host cell DNA. It is for this reason that the smaller SV40 and polyoma virus have been used as models for animal cell chromatin structure.

### 3.6.7 The information content of viral nucleic acids

(a) *DNA phages*

The large phages have many genes. Thus T4 is known to have 135 genes of which 53 are concerned with bacteriophage assembly; 34 of these code for structural proteins. About 60 genes in T4 are non-essential but enable the virus to cope with suboptimal conditions and allow it to extend the host range [257].

The small single-stranded phages, many of whose DNA have been sequenced, possess very few genes. $\phi$X174 and the related phage G4 have eleven genes, six of which are concerned with coat-protein synthesis or assembly, one with cell lysis and the remainder with DNA synthesis (see Section 6.7). The combined molecular weight of these eleven proteins adds up to 262 000 which is greater than can be coded for by the small genome if traditional concepts apply. However in $\phi$X174 the nucleotide sequence [301] shows that gene B is encoded with gene A but in a different reading frame (see Section 11.2) and gene E is encoded within gene D. Gene K is read in a second reading frame and overlaps the gene A/C junction (see Fig. 3.20). In fact gene C starts five nucleotides before the end of gene A so in this region all three reading frames are used. Gene A' produces a protein identical to the carboxy terminus of the gene A protein: in this case there are two sites for initiation of translation. It is clear that the small bacteriophages make full use of their DNA and detailed comparisons of the nucleotide sequence of the genomes of $\phi$X174 and G4 suggest the possible existence of several other genes [302, 303].

(b) *RNA phages*

The small RNA phages (e.g. MS2) contain all their genetic information in a single strand of

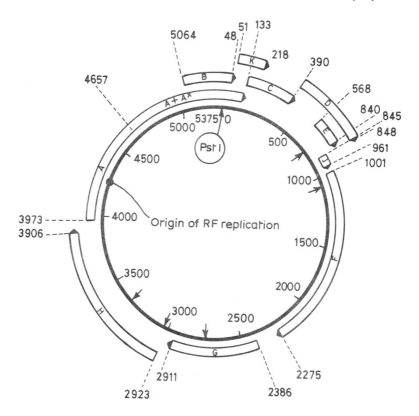

**Fig. 3.20**  Physical map and coding potential of φX174 DNA. The numbers refer to nucleotides relative to the single PstI site. The other arrows in the inner circle show the sites of action of HpaII. The open bars indicate the regions coding for the different proteins (A–K) and bars of different radii demonstrate the use of all three coding frames. The thick arrows indicate the C-terminus of the proteins.

RNA about 3300 nucleotides long (which can code for about 1100 amino acids or about 3–4 proteins of average size). Four viral-coded proteins have been identified (see Fig. 11.27) including an RNA-dependent RNA polymerase or replicase of molecular weight about 50 000. In the infected cell, the replicase is an early protein made about 10 minutes after infection, while the coat and maturation proteins are late proteins made (in appropriate amounts) about 20 minutes later. The mechanism of this control is considered in Chapter 11.

(c) *Animal DNA viruses*

The large animal viruses such as herpes simplex virus (HSV) can code for more than 100 proteins, and, as with the large bacteriophage (e.g. T4) many of these proteins duplicate the functions of host proteins.

In general virus diseases cannot be treated with antibiotics because the virus uses the cell's machinery to replicate itself. Thus any interference with virus metabolism automatically affects the infected animal. However, because the larger viruses substitute their own enzymes (e.g. DNA polymerase and thymidine kinase) for those of the cell it is possible to find some chemicals to which only the viral enzymes are susceptible. Acyclovir (9-(2-hydroxymethyl)-guanine) is one such chemical which can be phosphorylated by the virally induced thymidine kinase (in contrast to the host enzyme). The triphosphate inhibits the virally coded DNA polymerase [306, 307]. Acyclovir is used to treat herpes infections but cannot bring about a complete cure as the virus exhibits a latent phase in which the thymidine kinase is inactive.

The other approach to viral therapy is immunological and many vaccines are now available. Recent advances in this field involve the use of small chemically synthesized viral peptides (fragments of viral coat proteins) [308, 309]. Immunization with these minimizes the risk of side effects.

Unlike the small animal viruses the HSV genome has the expected amount of the dinucleotide CpG which is deficient in the genome of the host and the smaller viruses [277, 278, 310–313].

The papovavirus genome codes for two (SV40) or three (polyoma) early proteins which are known as tumour antigens as they are expressed in transformed cells (see below). Later in infection three virion coat proteins (VP1, VP2 and VP3) are synthesized. The two (or three) tumour antigens are translated from messenger RNA molecules derived by processing a single transcript from the early half of the genome starting near the origin of replication. This region of the DNA is 2700 nucleotide pairs long and in polyoma codes for large T (molecular weight about 100 000), middle T (molecular weight 55 000) and small t (molecular weight 17 000) antigens. In order to achieve this more than one reading frame is used (cf. $\phi$X174) although all three proteins have the same *N*-terminus, showing they are coded for by the same reading frame to start with. This frame has a termination signal after about 600 nucleotides leading to production of small t antigen. RNA splicing (see Sections 9.3.4 and 10.6.2) changes

the reading frame in some of the transcripts to frame 2 (large T antigen) or frame 3 (middle T antigen).

The late region of polyoma virus (coding for the virion proteins) also extends from near the replication origin but in the opposite direction to the early region. It is about 2100 nucleotide pairs long and codes for proteins requiring two reading frames. All three messenger RNAs have the same leader but splicing leads to synthesis of VP2 and VP3 in one reading frame (VP3 is identical to the *C*-terminus of VP2) and VP1 in a different partly overlapping frame (see Fig. 3.21) [310–312].

Thus, like $\phi$X174, polyoma virus and SV40 have more coding information than a co-linear relationship between DNA, RNA and protein would suggest but this is achieved in a somewhat different manner from that with the small bacteriophage.

### 3.6.8 Lysogeny and transduction [248, 252, 271, 314]

When a virulent phage infects a cell, the virus replicates and the cell is killed. With *temperate* phage an alternative outcome is possible. The cell may be *lysogenized* with the viral DNA integrated into the cell genome as a prophage. The genes are transcribed (in a controlled way), replicated and inherited along with the cell genes. Most of the phage functions are repressed (e.g. those involved in the lytic development), but others are not (e.g. those involved in the maintenance of lysogeny). One gene function which is not repressed in the lysogenic cell confers immunity against further infection of the cell by the same phage. This immunity is quite specific.

The survival of virulent phage depends on a continuous supply of susceptible bacteria (e.g. in sewage, a rich source of bacteriophages) while the temperate phage can survive and replicate with a limited population of cells which are protected from further infection. Prophages are

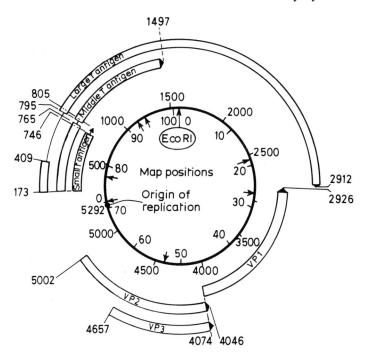

**Fig. 3.21** Physical map and coding potential of polyoma DNA. The numbers inside the inner circle show the polyoma DNA divided into 100 map units starting at the single EcoR1 site. Other arrows show the sites of action of HpaII. All other numbers refer to nucleotides measured from the HpaII 3/5 junction. The open bars indicate the regions coding for the various proteins – the thick arrows indicate the *C*-terminus. The single line joining the open bars for the antigens indicates these proteins are coded for by non-contiguous regions of DNA (Section 9.2). NB Bars of different radii do not imply different reading frames.

either integrated at a specific site on the bacterial chromosome (e.g. λ which normally attaches near the *gal* locus of *E. coli* [315], see Chapter 7) or at random (e.g. phage mu which causes inactivation of a gene into which it integrates).

The prophages in a lysogenic cell can be induced to replicate independently by various agents, most of which damage DNA. This results in the phage DNA being cut out of the cell genome by a recombination event [316] (see Section 7.4.3). The DNA can then replicate and function in a virulent manner, resulting in cell lysis and liberation of a burst of progeny bacteriophage.

Phage lambda is the most studied example of a lysogenic phage. In the integrated state only one prophage gene (*c*I) is expressed. The *c*I gene directs the synthesis of the lambda repressor which turns off all other viral genes while stimulating its own production. Induction inactivates the repressor and switches on the expression of the *cro* gene. The cro protein switches off *c*I transcription and switches on transcription of the other genes involved in lytic growth. Lysogenic phage are induced as a result of treatments (e.g. uv radiation) which damage DNA. As described in Section 7.3.4, damage to DNA leads to activation of Rec A protein and this proteolytically cleaves lambda repressor. Hence repressor synthesis is no longer stimulated and cro begins to accumulate in the cell. For more information on the regulation of transcription in prokaryotes refer to Section 10.1.5.

Occasionally, on induction, mistakes are made in the excision of the prophage DNA and small pieces may be left behind or sections of bacterial DNA may be excised along with the prophage DNA. This bacterial DNA may be replicated along with the phage DNA and packaged into the virus particle. On reinfection of a new bacterium these genes may become integrated into the new genome. This process is called *transduction*. Thus phage λ, which integrates next to the *gal* gene, may pick up this gene on induction to form the composite phage λ *gal*. The *gal* gene is then replicated along with λ DNA and is introduced into a new bacterium. If this recipient is *gal⁻* it could thus acquire the ability to metabolize galactose.

Some prophages (e.g. P1 [317]) are not integrated but remain in a stable extrachromosomal state like plasmids (see Section 3.7). Only very occasionally are they packaged and virions released. This packaging is very non-specific in that any DNA of the right length, including fragments of host chromosome, can be packaged (i.e. a 'headful' mechanism). On infecting a new bacterium genes of the original host are introduced at random and P1 is known as a generalized transducing bacteriophage.

Along with transformation and conjugation, transduction is one of the three ways in which DNA can be transferred between cells. Transfection is the rather inefficient process whereby cells may be infected with naked DNA rather than with intact virus. It involves uptake of the entire viral genome and is described further in Appendix Sections A.7 and A.8.

### 3.6.9 Tumour viruses and animal cell transformation

Some animal viruses (tumour viruses) can alter (transform) the infected cell without killing it, so that it has new properties which are typically neoplastic [318–320]. For example uninfected hamster cells do not form tumours when injected into newborn hamsters, and do not grow when suspended in nutrient agar, but polyoma-transformed hamster cells form tumours and grow in agar [318].

Several DNA-containing tumour viruses have been identified, including polyoma, SV40, rabbit papilloma, human papilloma, adenoviruses 7, 12 and 18 and some herpes viruses.

Some of these viruses can interact with different cells in different ways. Polyoma virus will transform hamster cells but will replicate in, and kill, mouse cells. SV40, on the other hand, transforms mouse cells and kills green monkey cells. A temperature-sensitive mutant of polyoma [321] will transform mouse cells at the non-permissive temperature, 38.5°C, and then replicate when the temperature of the transformed cell is reduced to 31°C [322]. This suggests that at least one virus function necessary for replication is not required for transformation. It is not understood why the different interactions occur with the different cell types.

Several lines of evidence show that the viral genome is present in the transformed cells, in a stable inheritable integrated state (cf. the lysogenic state of some bacteria). mRNA isolated from polyoma transformed cells will hybridize with polyoma virus DNA [323]. SV40-transformed mouse cells can be fused with green monkey cells using inactivated Sendai virus, and the hybrid cells liberate active SV40 particles [320, 324].

Although some SV40-transformed cells appear to contain only one copy of the viral genome per cell [325], others contain up to nine copies [326]. Integration of viral DNA into the host genome suppresses all but the early viral functions. How the T antigens of SV40 and polyoma virus bring about cell transformation is not clear. The large T antigen binds to the origin of replication of the virus and middle T antigen is found on the cell surface where it may mediate cell/cell interactions. The middle T antigen may be the (tumour specific) transplantation antigen (TSTA) which causes rejection of transplantable tumours by animals previously immunized with the virus [329]. The possible roles of T antigen in transformed cells have recently been reviewed [330, 331]. One thing is certain, SV40 large T antigen interacts with and stabilizes a cellular protein known as p53. The p53 gene (which may be a cellular oncogene known as *c-myc*) when introduced into primary cells along with another oncogene (*ras*) is able to transform the cells into immortal cells which show reduced contact inhibition, i.e. into tumour cells [33].

> *An oncogene* is a gene (e.g. *ras* or *myc*) which encodes a protein involved in the control of growth of a cell. Such proteins when present in modified form or in excessive amounts may lead to abnormal cellular growth which is one of the changes associated with cancer [338–340, 361].

The leukaemia viruses and the sarcoma viruses are closely related RNA tumour viruses or retroviruses. The leukaemia viruses are a natural cause of leukaemia and the sarcoma viruses transform cells in tissue culture.

The structure of the genome of these retroviruses together with the mechanism of their conversion to the DNA duplex proviral form is shown in Fig. 6.33. They code for three enzymes only: gag, pol and env; but integration into the host genome is essential for their replication. Integration (see Section 7.7 for the mechanism) is often accompanied by the enhanced expression of adjacent genes which, if these are cellular oncogenes, will lead to cell transformation [332–337].

Some retroviruses (e.g. Rous sarcoma virus of chickens) have incorporated into their RNA genome the transcript of a cellular oncogene (*c-onc*) which is transferred from cell to cell in an infectious manner. This gene is now known as a viral oncogene (*v-onc*) and may be identical to the cellular homologue or it may be a shortened or mutated version of it.

Retroviruses carrying a *v-onc* gene always produce tumours on infection though they are often themselves defective as the *v-onc* gene may replace part of the *env* gene. In this case they have to rely on coinfection with a non-defective virus for transmission to other hosts.

It is by hybridizing *v-onc* sequences to cellular DNA that a series of about 20 *c-onc* genes have been isolated [340] and this is contributing greatly to our increasing understanding of cellular growth control and cancer.

### 3.6.10 Viroids [341, 342, 305]

Viroids are much smaller than viruses and consist of a short strand of RNA and nothing more. They infect higher plants causing diseases such as potato spindle tuber disease (PSTV) and cadang-cadang, a disease of the coconut palm.

The RNA is in the form of a cyclic molecule of about 360 nucleotides which show a high degree of intrastrand base-pairing. The molecule resembles a hairpin with a total of about 110 nucleotides looping out from the duplex stem. There is no evidence that the viroid RNA is translated and the mode of replication is not known.

### 3.6.11 Prions

These are slow viruses producing diseases such as scrapie in sheep and Creutzfeld–Jacob disease and Kuru in man. They may also be the cause of Alzeimer's disease [343]. They are of particular interest as they have not been shown to contain any nucleic acid [344]. Rather they may contain nothing but a single glycoprotein of molecular weight about 30 000 which aggregates into rods. Various intriguing ways in which a protein can be a self-replicating agent are reviewed by Prusiner [344] and Brunori and Talbot [345].

The evidence for the lack of a nucleic acid component in prions has been criticized on several grounds [346, 347, 360] but the problem should be resolved in the near future.

## 3.7 PLASMIDS AND TRANSPOSONS

Plasmids are duplex, supercoiled DNA molecules which range in size from $2.5 \times 10^6$ to $1.5 \times 10^8$ daltons. They are stable elements which exist in bacteria and some eukaryotes in an extrachromosomal state. The large ones are present in only one or two copies per cell while there may be 20 or more copies per cell of the small ones. They enjoy an autonomous, self-replicating status without lowering host viability [246, 257, 348–352].

There is a continuum from the self-replicating plasmid which has no means of transfer from cell to cell, through those plasmids which can be transferred (e.g. the F-factor) and the non-integrative transducing phage (e.g. P1) through to the fully fledged virus which codes for its coat proteins and means of entry into a new host cell.

One type of plasmid is called a sex or F-factor. A bacterium possessing an F-factor ($F^+$) is male and the F factor induces the bacterium to make a tube or pilus by which the male bacterium becomes attached to a female bacterium, i.e. one not carrying the F-factor. When the F-factor replicates (see Chapter 6) one linear single-stranded copy of plasmid DNA may pass through the pilus 5'-end first into the female bacterium. The complementary strand is synthesized in the recipient which thereby becomes male as it now carries the F-factor. Soon all the cells in a population become $F^+$. The presence of an F-factor in a bacterium excludes the entry of a second F-factor, but more than one *type* of plasmid may be stably maintained in a cell if the two are not closely related. Generally $F^+$ cells are rare but each usually contains several F-factors. As well as the F-plasmid, other plasmids and parts of the bacterial chromosome may pass through the pilus and this leads to a low frequency of exchange of genetic markers between bacteria, i.e. a male ($F^+$) prototroph bacterium may transfer the gene required for tyrosine synthesis to a female tyr$^-$ auxotroph. This is known as transfer by *bacterial conjugation*.

In one in 10 000 $F^+$ cells the F-factor becomes integrated into the bacterial chromosome by reciprocal crossing over. A plasmid which can exist either autonomously or in an integrated state is called an episome. For insertion the F-factor breaks at a unique site, but this linear piece of DNA integrates into the host DNA at random with regard to position and orientation (see below). Like the λ-prophage, an integrated plasmid loses its ability for independent replication [353, 354]. However, *E. coli* mutants defective in initiation of chromosomal replication can be rescued by integration of an F-factor which takes over control of initiation [353]. The integrated plasmid still causes the formation of pili and can be transferred by conjugation, but now, between the leading 5'-end of the plasmid DNA and the trailing 3'-end, there is the entire bacterial chromosome which gets dragged along the conjugation tube, into the female cell. Since transfer of the whole chromosome takes about 60 min, breakage of the conjugation tube normally occurs before transfer is complete. When this occurs the second half of the F-factor is not transferred and so the recipient remains female. As the fragments transferred cannot replicate autonomously they are either lost, or are integrated into the recipient's chromosome which thereby shows a high frequency of recombination of genetic markers. Hence cells with an integral F-factor are called *Hfr*. Because of the mode of integration of an F-factor into the bacterial chromosome, different Hfr strains transfer the genes of the bacterial chromosome in a different order and with different polarity. However, closer study shows the gene order to be circularly permuted, providing evidence for the linear arrangement of genes on a circular chromosome [355].

Sometimes an Hfr strain may revert to $F^+$, and when this occurs a section of the bacterial chromosome may also be excised and be found in the plasmid (e.g. Flac contains the *lac* operon integrated into an F-factor which is thereby more than doubled in size). Such plasmids are known as F'-factors. This gene is now transferred to a recipient bacterium at conjugation (cf. transduction) to produce a partial diploid and recombination readily occurs (see Chapter 7). Thus in Flac$^+$ *lac* is transferred infectiously but other markers are only transferred at low frequency or following integration.

R-factors are plasmids which carry genes conferring resistance to one or a number of drugs. Resistance is procured by enzymes which modify or degrade the drug, e.g. chloramphenicol resistance depends on an enzyme which acetylates chloramphenicol; ampicillin resistance by a $\beta$-lactamase. Resistance to tetracycline is acquired by virtue of an inability to take up the drug into the cell. Not all R-factors confer resistance to all drugs; many carry only one drug-resistance marker.

R-factors may or may not carry genes encoding a pilus enabling them to be transferred to other bacteria. For those which can promote their own conjugative transfer it is common to find the antibiotic resistance-determining region is separated from the transfer region. Those R-factors not carrying the transfer (*tra*) genes can be transferred passively from one bacterium to another either by transduction or by relying on mobilization by a conjugative plasmid present in the same cell.

A third group of plasmids synthesize chemicals (*colicins*) which are toxic to bacteria. The larger of these colicinogenic factors can be transferred by conjugation but the smaller ones such as the Col E factors do not code for enzymes to bring about their transmission. Plasmids of *Rhizobium* and *Agrobacterium* carry the genes responsible for nitrogen fixation and crown gall tumour formation in plants [327, 328].

Plasmids, especially the smaller ones existing at high copy numbers, are used as cloning vehicles in genetic engineering (see Section A.7). The presence of drug-resistance markers is also helpful in the selection regimens used. One of the early vectors used was pSC101 derived

from a fragment of a larger R-factor, but derivatives of Col E1 are now in common use. Thus pBR322 has been engineered by removing some regions and inserting others (e.g. the tetracycline-resistance gene from pSC101) to give an autonomous plasmid of molecular weight $2.7 \times 10^6$ [356].

Hybrid plasmids can be constructed containing parts of, for example, pBR322 and a small plasmid found in yeast (the $2\mu$ plasmid) or a segment of an animal virus (e.g. SV40). These plasmids are able to grow both in bacteria and in yeast or animal cells and have been much used in the study of eukaryote gene expression (see Section 10.4).

When first discovered, the location of multiple drug-resistance genes or of genes coding for cognate restriction and modification enzymes on a single plasmid was difficult to explain. However, it transpires that many of these genes are present on transposable genetic elements or *transposons*. These are discrete segments of DNA which have the ability to move around among the chromosomes and extrachromosomal elements, i.e. plasmids or viruses. Thus the ampicillin-resistance gene referred to above rapidly moves from one plasmid to another and from one location to another on the plasmid in the form of a 4800 bp fragment known as transposon 3 (Tn 3). For details on the structure of transposons and mechanism of transposition see Section 7.7. Transposons also occur in eukaryotes and the first circumstantial evidence for their existence was obtained in studies with maize [357].

REFERENCES

1 Franke, W. W., Scheer, U., Krohne, G. and Jarasch, E-D. (1981), *J. Cell Biol.*, **91**, 39S.

2 Lebkowski, J. S. and Laemmli, U. K. (1982), *J. Mol. Biol.*, **156**, 325.

3 Capco, D. G., Wan, K. M. and Penman, S. (1982), *Cell*, **29**, 847.

4 Mitchison, J. M. (1971), *The Biology of the Cell Cycle*, Cambridge University Press, Cambridge.

5 Adams, R. L. P. (1980), *Cell Culture for Biochemists*, Elsevier/North-Holland, Amsterdam.

6 Yunis, J. J. (1976), *Science*, **191**, 1268.

7 Yunis, J. J. (1977), *Molecular Structure of Human Chromosomes*, Academic Press, New York.

8 Gerace, L. and Blobel, G. (1980), *Cell*, **19**, 277.

9 Alberts, B., Bray, D., Lewis, J., Raff, M., Roberts, K. and Watson, J. D. (1983), *The Molecular Biology of the Cell*, Garland Publishing, New York.

10 Lewin, B. (1974), *Gene Expression*, Vol. 2, Wiley, London.

11 Caspersson, T., Hulten, M., Linsten, J. and Zech, L. (1971), *Hereditas*, **67**, 147.

12 Dutrillaux, B. (1977), in *The Molecular Structure of Human Chromosomes* (ed. J. Yunis), Academic Press, New York, p. 233.

13 Nurse, P. (1985), *Trends Genet.*, **1**, 51.

14 Geraci, L., Comeau, C. and Benson, M. (1984), *J. Cell Sci. Suppl.*, **1**, 137.

15 Benavente, R., Krohne, G., Schmidt-Zachmann, M. S., Hugle, B. and Franke, W. W. (1984), *J. Cell Sci. Suppl.*, **1**, 161.

16 Agutter, P. S. and Birchall, K. (1979), *Exp. Cell Res.*, **124**, 453.

17 Detke, S. and Keller, J. M. (1982), *J. Biol. Chem.*, **257**, 3905.

18 Mirkovitch, J., Mirault, M-E. and Laemmli, U.K. (1984), *Cell*, **39**, 223.

19 Lewis, C. D., Lebkowski, J. S., Daly, A. K. and Laemmli, U. K. (1984), *J. Cell Sci. Suppl.*, **1**, 103.

20 Pienta, K. J. and Coffey, D. S. (1984), *J. Cell Sci. Suppl.*, **1**, 123.

21 Jackson, D. A., McCready, S. J. and Cook, P. R. (1984), *J. Cell Sci. Suppl.*, **1**, 59.

22 Carbon, J. (1984), *Cell*, **37**, 351.

23 Blackburn, E. H. and Szostak, J. W. (1984), *Annu. Rev. Biochem.*, **53**, 163.

24 Carbon, J. and Clarke, L. (1984), *J. Cell Sci. Suppl.*, **1**, 43.

25 Blackburn, E. H. (1984), *Cell*, **37**, 7.

26 Murray, A. W. (1985), *Trends Biochem. Sci.*, **10**, 112.

27 Murray, A. W. and Szostak, J. W. (1983), *Nature (London)*, **305**, 189.

28 Ruddle, F. H. (1973), *Nature (London)*, **242**, 165.

29 Ringertz, N. R. and Savage, R. E. (1976), *Cell Hybrids*, Academic Press, London.

30 Gall, J. G. and Pardue, M. L. (1971), *Methods Enzymol.*, **21**, 470.

31 Zabel, B. U., Naylor, S. L., Sakaguchi, A. Y., Bell, G. I. and Shows, T. B. (1983), *Proc. Natl. Acad. Sci. USA*, **80**, 6932.

32 Camerino, G., Grzeschik, K. H., Jaye, M., De La Salle, H., Tolstoshev, P., Lecocq, J. P., Heilig, R. and Mandel, J. L. (1984), *Proc. Natl. Acad. Sci. USA*, **81**, 498.

33 Cheung, P., Kao, F-T., Law, M. L., Jones, C., Puck, T. T. and Chan, L. (1984), *Proc. Natl. Acad. Sci. USA*, **81**, 508.

34 Fisher, J. H., Gusella, M. and Scoggin, C. H. (1984), *Proc. Natl. Acad. Sci. USA*, **81**, 520.

35 Sakaguchi, A. Y., Lalley, P. A., Zabel, B. U., Ellis, R. W., Scolnick, E. M. and Naylor, S. L. (1984), *Proc. Natl. Acad. Sci. USA*, **81**, 525.

36 Koutides, I. A., Barker, P. E., Gurr, J. A., Pravtcheva, D. D. and Ruddle, F. H. (1984), *Proc. Natl. Acad. Sci. USA*, **81**, 517.

37 Davies, K. E., Young, B. D., Elles, R. G., Hill, M. E. and Williamson, R. (1981), *Nature (London)*, **293**, 374.

38 Lebo, R. V., Carrano, A. V., Burkhart-Schulz, K. J., Dozy, A. M., Yu, L-C. and Kan, K. W. (1979), *Proc. Natl. Acad. Sci. USA*, **76**, 5804.

39 Wang, H. C. and Federoff, S. (1973), in *Tissue Culture, Methods and Applications* (ed. P. F. Kruse and M. K. Patterson), Academic Press, London, p. 782.

40 Thomas, C. A. (1971), *Annu. Rev. Genet.*, **5**, 237.

41 Callan, H. G. (1967), *J. Cell Sci.*, **2**, 1.

42 Doolittle, W. F. and Spienza, C. (1980), *Nature (London)*, **284**, 601.

43 Orgel, L. E. and Crick, F. H. C. (1980), *Nature (London)*, **284**, 604.

44 Britten, R. J. and Davidson, E. H. (1968), *Science*, **165**, 349.

45 Ohno, S. (1971), *Nature (London)*, **234**, 134.

46 Jones, K. W. and Corneo, G. (1971), *Nature (London) New Biol.*, **233**, 268.

47 Evans, H. J., Gosden, J. R., Mitchell, A. R. and Buckland, R. A. (1974), *Nature (London)*, **251**, 346.

48 Manuelidis, L. (1978), *Chromosoma*, **66**, 1.

49 Maio, J. J., Brown, F. L. and Musich, P. R. (1977), *J. Mol. Biol.*, **117**, 637.

50 Miklos, G. L. G. and John B. (1979), *Amer. J. Hum. Genet.*, **31**, 264.

51 Varley, J. M., Macgregor, H. C. and Erba, H. P. (1980), *Nature (London)*, **283**, 686.

52 Horz, W. and Altenberger, W. (1981), *Nucleic Acids Res.*, **9**, 683.

53 Southern, E. M. (1975), *J. Mol. Biol.*, **94**, 51.

54 Pech, M., Streeck, R. E. and Zachau, H. G. (1979), *Cell*, **18**, 883.

55 Streeck, R. E. (1981), *Science*, **213**, 443.

56 Taparowsky, E. J. and Gerbi, S. A. (1982), *Nucleic Acids Res.*, **10**, 1271.

57 Taparowsky, E. J. and Gerbi, S. A. (1982), *Nucleic Acids Res.*, **10**, 5503.

58 Gautier, F., Bunemann, H. and Grotjahn, L. (1977), *Eur. J. Biochem.*, **80**, 175.

59 Adams, R. L. P., Burdon, R. H. and Fulton, J. (1983), *Biochem. Biophys. Res. Commun.*, **113**, 695.

60 Sanford, J., Forrester, L., Chapman, V., Chandley, A. and Hastie, N. (1984), *Nucleic Acids Res.*, **12**, 2823.

61 Bostock, C. (1980), *Trends Biochem. Sci.*, **5**, 117.

62 Cavalier-Smith, T. (1980), *Nature (London)*, **285**, 617.

63 Jamrich, M., Warrior, R., Steele, R. and Gall, J. G. (1983), *Proc. Natl. Acad. Sci. USA*, **80**, 3364.

64 Jelinek, W. R., Toomey, T. P., Leinward, L., Duncan, C. H., Biro, P. A., Choudary, P. V., Weissman, S. M., Rubin, C. M., Houck, C. M., Deininger, P. L. and Schmid, C. V. (1980), *Proc. Natl. Acad. Sci. USA*, **77**, 1398.

65 Georgiev, G. P., Kramerov, D. A., Ryskov, A. P., Skryabin, K. G. and Lukanidin, E. M. (1983), *Cold Spring Harbor Symp. Quant. Biol.*, **47**, 1109.

66 Jelinek, W. R. and Haynes, S. R. (1983), *Cold Spring Harbor Symp. Quant. Biol.*, **47**, 1123.

67 Jagadeeswaran, P., Forget, B. G. and Weissman, S. M. (1981), *Cell*, **26**, 141.

68 Sharp, P. A. (1983), *Nature (London)*, **301**, 471.

69 Brown, A. L. (1984), *Nature (London)*, **312**, 106.

70 Hastie, N. (1985), *Trends Genet.*, **1**, 37.

71 Meunier-Rotival, M., Soriano, P., Cuny, G., Strauss, F. and Bernardi, G. (1982), *Proc. Natl. Acad. Sci. USA*, **79**, 355.

72 Sun, L., Paulson, K. E., Schmid, C. W., Kadyk, L. and Leinward, L. (1984), *Nucleic Acids Res.*, **12**, 2669.

73 Rogers, J. (1983), *Nature (London)*, **306**, 113.

74 Singer, M. F. and Skowronski, J. (1985), *Trends Biochem. Sci.*, **10**, 119.

75 Graham, D. E., Neufeld, B. R., Davidson, E. H. and Britten, R. J. (1974), *Cell*, **1**, 127.

76 Crain, W. R., Eden, F. C., Pearson, W. R., Davidson, E. H. and Britten, R. J. (1976), *Chromosoma*, **56**, 309.

77 Wells, R., Royer, H-D. and Hollenberg, C. P. (1976), *Mol. Gen. Genet.*, **147**, 45.

78 Crain, W. R., Davidson, E. H. and Britten, R. J. (1976), *Chromosoma*, **59**, 1.

79 Perlman, S., Phillips, C. and Bishop, J. O. (1976), *Cell*, **8**, 33.

80 Jelinek, W. R. (1978), *Proc. Natl. Acad. Sci. USA*, **75**, 2679.

81 Hardman, N., Bell, A. J. and McLachlan, A. (1979), *Biochim. Biophys. Acta*, **564**, 372.

82 Goodman, H. M., Olson, M. V. and Hall, B. D. (1977), *Proc. Natl. Acad. Sci. USA*, **74**, 5454.

83 Dugaiczyk, A., Woo, S. L., Lai, E. C., Mace, M. L., McReynolds, C. and O'Malley, B. W. (1978), *Nature (London)*, **274**, 328.

84 Mandel, J. L., Breathnach, R., Gerlinger, P., LeMur, M., Gannon, F. and Chambon, P. (1978), *Cell*, **14**, 641.

85 Doel, M. T., Houghton, M., Cook, E. A. and Carey, N. H. (1977), *Nucleic Acids Res.*, **4**, 3701.

86 Abelson, J. (1979), *Annu. Rev. Biochem.*, **48**, 1035.

87 Tilghman, S. M., Curtis, P. J., Tiemcier, D. C., Leder, P. and Weissman, C. (1978), *Proc. Natl. Acad. Sci., USA*, **75**, 1309.

88 Knapp, G., Beckman, J. S., Johnson, P. F., Fuhrman, S. A. and Abelson, J. (1978), *Cell*, **14**, 221.

89 Lomedico, P., Rosenthal, N., Efstratiadis, A., Gilbert, W., Kolodner, R. and Tizard, R. (1979), *Cell*, **18**, 545.

90 Crick, F. H. C. (1979), *Science*, **204**, 264.

91 Watson, J. V. (1984), in *Molecular Biology and Human Disease* (ed. A. Macleod and K. Sikora) Blackwell Scientific, Oxford, p. 66.

92 Schmid, C. W. and Deininger, P. L. (1975), *Cell*, **6**, 345.

93 Fitzsimons, D. W. and Wolstenholme, G. E. W. (eds) (1975), *Ciba Found. Symp. Struct. Funct. Chromatin*, **28**.

94 Von Holt, C. and Brandt, W. F. (1977), *Methods Cell Biol.*, **16**, 205.

95 Frederiq, E. (1971), in *Histones and Nucleohistones* (ed. D. M. P. Philips), Plenum Press, New York, p. 136.

96 Christiansen, G. and Griffith, J. (1977), *Nucleic Acids Res.*, **4**, 1837.

97 Von Holt, C., Strickland, W. N., Brandt, W. F. and Strickland, M. S. (1979), *FEBS Letts.*, **100**, 201.

98 Brandt, W. F., Strickland, W. N., Strickland, M., Carlisk, L., Woods, D. and Von Holt, C. (1979),

*Eur. J. Biochem.*, **94**, 1.

99 Kedes, L. H. (1979), *Annu. Rev. Biochem.*, **48**, 837.

100 Isenberg, I. (1979), *Annu. Rev. Biochem.*, **48**, 159.

101 Kedes, L. H. and Maxson, R. (1981), *Nature (London)*, **294**, 11.

102 Choe, J., Kolodrubetz, D. and Grunstein, M. (1982), *Proc. Natl. Acad. Sci. USA*, **79**, 1484.

103 Torres-Martinez, S. and Ruiz-Carrillo, A. (1982), *Nucleic Acids Res.*, **10**, 2323.

104 Lennox, R. W. and Cohen, L. W. (1983), *J. Biol. Chem.*, **258**, 262.

105 Dashkevich, V. K., Nikolaev, L. G., Zlatanova, J. S., Glotov, B. O. and Severin, E. S. (1983), *FEBS Letts.*, **158**, 276.

106 Sittman, D. B., Graves, R. A. and Marzluff, W. F. (1983), *Proc. Natl. Acad. Sci. USA*, **80**, 1849.

107 Wu, R. S., Tsai, S. and Bonner, W. M. (1982), *Cell*, **31**, 367.

108 Woodland, H. R., Warmington, J. R., Ballantine, J. E. M. and Turner, P. C. (1984), *Nucleic Acids Res.*, **12**, 4939.

109 Waterborg, J. H. and Matthews, H. R. (1983), *Biochemistry*, **22**, 1489.

110 Loidl, P., Loidl, A., Puschendorf, B. and Grobner, P. (1983), *Nature (London)*, **305**, 446.

111 Perry, M. and Chalkley, R. (1982), *J. Biol. Chem.*, **257**, 7336.

112 Song, M. K. H. and Adolph, K. W. (1983), *J. Biol. Chem.*, **258**, 3309.

113 Bradbury, E. M., Maclean, N. and Matthews, H. R. (1981), *DNA, Chromatin and Chromosomes*, Blackwell, Oxford.

114 Smulson, M. (1979), *Trends Biochem. Sci.*, **4**, 225.

115 Levinger, L. and Varshavsky, A. (1982), *Cell*, **28**, 375.

116 Findlay, D., Ciechanover, A. and Varshavsky, S. (1984), *Cell*, **37**, 43.

117 Schiffman, S. and Lee, P. (1974), *Br. J. Haematol.*, **27**, 101.

118 Schwartz, D. C. and Cantor, C. R. (1984), *Cell*, **37**, 67.

119 Somer, J. B. and Castaldi, P. A. (1970), *Br. J. Haematol.*, **18**, 147.

120 Swart, A. C. W. and Hemker, H. C. (1970), *Biochim. Biophys. Acta*, **222**, 692.

121 Campbell, A. M. and Cotter, R. (1976), *FEBS Letts.*, **70**, 209.

122 Thomas, J. O. and Butler, P. J. G. (1977), *J. Mol. Biol.*, **116**, 769.

123 John, E. W. (ed.) (1982), *The HMG Chromosomal Proteins*, Academic Press, London.

124 Cartwright, I. L., Abinger, S. H., Fleischman, G., Lovenhaupt, K., Elgin, S., Keene, M. A. and Howard, G. C. (1982), *Crit. Rev. Biochem.*, **13**, 1.

125 Seale, R. L., Annunziato, A. T. and Smith, R. D. (1983), *Biochemistry*, **22**, 5008.

126 Wu, R. S. and Bonner, W. M. (1984), in *Eukaryotic Gene Expression* (ed. A. Kumar), George Washington University Medical Centre Spring Symposium, Plenum Press, New York, p. 37.

127 Thomas, J. O. (1979), in *Companion to Biochemistry, 2*, (eds R. T. Bull, J. R. Lagnado, J. O. Thomas and K. F. Tipton) Longman, London, p 79.

128 Rill, R. L. (1979), in *Molecular Genetics III* (ed. J. H. Taylor), Academic Press, New York, p. 247.

129 Kornberg, R. D. and Klug, A. (1981), *Sci. Amer.*, **244(2)**, 48.

130 Thomas, J. O. (1984), *J. Cell Sci. Suppl.*, **1**, 1.

131 Butler, P. J. G. (1983), *CRC Crit. Rev. Biochem.*, **15**, 57.

132 Clark, R. J. and Felsenfeld, G. (1971), *Nature (London) New Biol.*, **229**, 101.

133 Olins, A. L. and Olins, D. E. (1974), *Science*, **183**, 330.

134 Germond, J. E., Hirt, B., Oudet, P., Gross-Bellard, M. and Chambon, P. (1975), *Proc. Natl. Acad. Sci. USA*, **72**, 1843.

135 Shure, M., Pulleyblank, D. E. and Vinograd, J. (1977), *Nucleic Acids Res.*, **4**, 1183.

136 Hewish, D. R. and Burgoyne, L. A. (1973), *Biochem. Biophys. Res. Commun.*, **52**, 504.

137 Noll, M. and Kornberg, R. D. (1977), *J. Mol. Biol.*, **109**, 393.

138 Kornberg, R. D. (1974), *Science*, **184**, 868.

139 Oudet, P., Gross-Bellard, M. and Chambon, P. (1975), *Cell*, **4**, 281.

140 Thomas, J. O. and Kornberg, R. D. (1975), *FEBS Letts.*, **58**, 353.

141 Thomas, J. O. and Kornberg, R. D. (1975), *Proc. Natl. Acad. Sci. USA*, **72**, 2626.

142 Noll, M. (1974), *Nucleic Acids Res.*, **1**, 1573.

143 Noll, M. (1977), *J. Mol. Biol.*, **116**, 49.

144 Pardon, J. F., Worcester, D. L., Wooley, J. C., Cotter, R. I., Lilley, D. M. J. and Richards, B. M. (1977), *Nucleic Acids Res.*, **4**, 3199.

145 Finch, J. T., Lutter, L. C., Rhodes, D., Brown, R. S., Rushton, B., Levitt, M. and Klug, A. (1977), *Nature (London)*, **269**, 29.

146 Richmond, T. J., Finch, J. T., Rushton, B.,

Rhodes, D. and Klug, A. (1984), *Nature (London)*, **311**, 532.

147 Crick, F. H. C. and Klug, A. (1975), *Nature (London)*, **255**, 530.

148 Hagerman, P. J. (1981), *Biopolymers*, **20**, 1503.

149 Agard, D. A. and Sedar, J. W. (1983), *Nature (London)*, **302**, 676.

150 Richmond, T. J., Finch, J. T. and Klug, A. (1982), *Cold Spring Harbor Symp. Quant. Biol.*, **47**, 493.

151 Mirzabekov, A. D., Bavykin, S. G., Karpov, V. L., Preobrazhenskayer, O. V., Ebralidze, K. K., Tuneev, V. M., Melnikova, A. F., Goguadze, E. G., Chenchick, A. A. and Beabealashvili, R. S. (1982), *Cold Spring Harbor Symp. Quant. Biol.*, **47**, 503.

152 Morris, N. R. (1976), *Cell*, **9**, 627.

153 Thomas, J. O. and Furber, V. (1976), *FEBS Letts.*, **66**, 274.

154 Thomas, J. O. and Thomson, R. J. (1977), *Cell*, **10**, 633.

155 Ellison, M. J. and Pulleyblank, D. E. (1983), *J. Biol. Chem.*, **258**, 13 307, 13 314 and 13 321.

156 Weintraub, H. (1980), *Nucleic Acids Res.*, **8**, 4745.

157 Kornberg, R. D. (1981), *Nature (London)*, **292**, 579.

158 Horz, W. and Altenberger, W. (1981), *Nucleic Acids Res.*, **9**, 2643.

159 Dingwall, C., Lomonossoff, F. R. and Laskey, R. A. (1981), *Nucleic Acids Res.*, **9**, 2659.

160 Keene, M. H. and Elgin, S. C. R. (1981), *Cell*, **27**, 57.

161 Elgin, S. C. R., Cartwright, I. L., Gleischmann, G., Lowenhaupt, K. and Keene, M. A. (1983), *Cold Spring Harbor Symp. Quant. Biol.*, **47**, 529.

162 Pope, L. E. and Sigman, D. S. (1984), *Proc. Natl. Acad. Sci. USA*, **81**, 3.

163 Cartwright, I. L. and Elgin, S. C. R. (1982), *Nucleic Acids Res.*, **10**, 5835.

164 Cartwright, I. L., Hertzberg, R. P., Dervan, P. B. and Elgin, S. C. R. (1983), *Proc. Natl. Acad. Sci. USA*, **80**, 3213.

165 Gottesfeld, J. M. and Bloomer, L. S. (1980), *Cell*, **21**, 751.

166 Louis, C., Schedl, P., Samal, B. and Worcel, A. (1980), *Cell*, **22**, 387.

167 Zhang, X-Y., Fittler, F. and Horz, W. (1983), *Nucleic Acids Res.*, **11**, 4287.

168 Zhang, X-Y. and Horz, W. (1984), *J. Mol. Biol.*, **176**, 105.

169 Bock, H., Abler, S., Zhang, X-Y., Fritton, H. and Igo-Kemenes, T. (1984), *J. Mol. Biol.*, **176**, 131.

170 Zurkin, V. B. (1983), *FEBS Letts.*, **158**, 293.

171 Samal, B., Worcel, A., Louis, C. and Schedl, P. (1981), *Cell*, **23**, 401.

172 Simpson, R. T. and Stafford, D. W. (1983), *Proc. Natl. Acad. Sci. USA*, **80**, 51.

173 Bloom, K. S., Fitzgerald-Hayes, M. and Carbon, J. (1982), *Cold Spring Harbor Symp. Quant. Biol.*, **47**, 1175.

174 Strauss, F. and Varshavsky, A. (1984), *Cell*, **37**, 889.

175 Beard, P. (1978), *Cell*, **15**, 955.

176 Manser, T., Thacher, T. and Rechsteiner, M. (1980), *Cell*, **19**, 993.

177 Germond, J-E., Bellard, M., Oudet, P. and Chambon, P. (1976), *Nucleic Acids Res.*, **3**, 3173.

178 Ring, D. and Cole, R. D. (1983), *J. Biol. Chem.*, **258**, 15 361.

179 Campbell, A. M. and Cotter, R. I. (1977), *Nucleic Acids Res.*, **4**, 3877.

180 Goodwin, G. H., Mathew, C. G. P., Wright, C. A., Venkov, C. D. and Johns, E. W. (1979), *Nucleic Acids Res.*, **7**, 1815.

181 Finch, J. T. and Klug, A. (1976), *Proc. Natl. Acad. Sci. USA*, **73**, 1897.

182 McGhee, J. D., Nickol, J. M., Felsenfeld, G. and Rau, D. C. (1983), *Cell*, **33**, 831.

183 Labhart, P., Koller, T. and Wenderli, H. (1982), *Cell*, **30**, 115.

184 Thomas, J. O. and Khabaza, A. J. A. (1980), *Eur. J. Biochem.*, **112**, 501.

185 Cook, P. R. and Brazell, I. A. (1978), *Eur. J. Biochem.*, **84**, 465.

186 Benyajati, C. and Worcel, A. (1976), *Cell*, **9**, 393.

187 Adolph, K. W., Cheng, S. M. and Laemmli, U. K. (1977), *Cell*, **12**, 805.

188 Adolph, K. W., Cheng, S. M., Paulson, J. R. and Laemmli, U. K. (1977), *Proc. Natl. Acad. Sci. USA*, **74**, 4937.

189 Agard, D. A. and Sedat, J. W. (1983), *Nature (London)*, **302**, 676.

190 Zakian, V. A. (1985), *Nature (London)*, **314**, 223.

191 Mirkovitch, J., Mirault, M-E. and Laemmli, U. K. (1984), *Cell*, **39**, 223.

192 Jackson, D. A., McCready, S. J. and Cook, P. R. (1984), *J. Cell Sci. Suppl.*, **1**, 59.

193 Finch, J. T., Noll, M. and Kornberg, R. D. (1975), *Proc. Natl. Acad. Sci. USA*, **72**, 3320.

194 Callan, H. G., Gross, K. W. and Old, R. W.

(1977), *J. Cell Sci.*, **27**, 57.

195 Sommerville, J. (1979), *J. Cell Sci.*, **40**, 1.

196 Franke, W. W., Scheer, V., Spring, H., Trendelenburg, M. F. and Krohne, G. (1976), *Exp. Cell Res.*, **100**, 233.

197 Lewis, M., Helmsing, P. J. and Ashburner, M. (1975), *Proc. Natl. Acad. Sci. USA*, **72**, 3604.

198 Korge, G. (1975), *Proc. Natl. Acad. Sci. USA*, **72**, 4550.

199 Lonn, U. (1982), *Trends Biochem. Sci.*, **7**, 24.

200 Silver, L. M. and Elgin, S. C. R. (1977), *Cell*, **11**, 971.

201 Jamrich, M., Greenleaf, A. L. and Bautz, E. K. F. (1977), *Proc. Natl. Acad. Sci. USA*, **74**, 2079.

202 Zarling, D. A., Arndt-Jovin, D. J., Robert-Nicoud, M., McIntosh, L. P., Thomae, R. and Jovin, T. M. (1984), *J. Mol. Biol.*, **176**, 369.

203 Fangman, W. L. and Dujon, B. (1984), *Proc. Natl. Acad. Sci. USA*, **81**, 7156.

204 Attardi, G. (1981), *Trends Biochem. Sci.*, **6**, 86.

205 Attardi, G. (1981), *Trends Biochem. Sci.*, **6**, 100.

206 Van Etten, R. A., Michael, N. L., Bibb, M. J., Brennicke, A. and Clayton, D. A. (1982), in *Mitochondrial Genes* (ed. P. Slonimski, P. Borst and G. Attardi), Cold Spring Harbor, New York, p. 73.

207 Anderson, S., Bankier, A. T., Barrell, B. G., de Bruijn, M. H. L., Coulson, A. R., Drovin, J., Eperon, I. C., Nierlich, D. P., Roe, B. A., Sanger, F., Schreier, P. H., Smith, A. J. H., Staden, R. and Young, I. C. (1982), in *Mitochondrial Genes* (ed. P. Slonimski, P. Borst and G. Attardi), Cold Spring Harbor, New York, p. 5.

208 Anderson, S., Bankier, A. T., Barrell, B. G., de Bruijn, M. H. L., Coulson, A. R., Drovin, J., Eperon, I. C., Nierlich, D. P., Roe, B. A., Sanger, F., Schreier, P. H., Smith, A. J. H., Staden, R. and Young, I. C. (1981), *Nature (London)*, **290**, 457.

209 Grivell, L. A. (1983), *Sci. Amer.*, **248(3)**, 60.

210 Chomyn, A., Mariottini, P., Cleeter, M. W. J., Ragan, C. I., Matsuno-Yagi, A., Hatefi, Y., Doolittle, R. F. and Attardi, G. (1985), *Nature (London)*, **314**, 592.

211 Lazowska, J., Jacq, C. and Slonimski, P. P. (1980), *Cell*, **22**, 333.

212 Palmer, J. D. and Shields, C. R. (1984), *Nature (London)*, **307**, 437.

213 Fox, T. D. (1984), *Nature (London)*, **307**, 415.

214 Levings, C. S. (1983), *Cell*, **32**, 659.

215 Borst, P. and Grivell, L. S. (1978), *Cell*, **15**, 705.

216 Bandlow, W., Schweyen, R. J., Wolf, K. and Kaudewitz, F. (1977), *Genetics and Biogenesis of Mitochondria*, De Gruyter, Berlin.

217 Gillham, N. W. (1978), *Organelle Heredity*, Raven Press, New York.

218 Slonimski, P., Borst, P. and Attardi, G. (eds) (1982), *Mitochondrial Genes*, Cold Spring Harbor Symposium, New York.

219 Tzagoloff, A. and Macino, G. (1979), *Annu. Rev. Biochem.*, **48**, 419.

220 Yaffe, M. and Schatz, G. (1984), *Trends Biochem. Sci.*, **9**, 179.

221 Klein, A. and Bonhoeffer, F. (1972), *Annu. Rev. Biochem.*, **41**, 301.

222 Smith, H. (1975), *Nature (London)*, **254**, 13.

223 Ohta, N., Sager, R. and Inouye, M. (1975), *J. Biol. Chem.*, **250**, 3655.

224 Borst, P. and Hoeijmakers, J. H. J. (1979), *Plasmid*, **2**, 20.

225 Kleisen, C. M., Borst, P. and Weijers, P. J. (1976), *Eur. J. Biochem.*, **64**, 141.

226 Kleisen, C. M., Weislogel, P. O., Fonck, K. and Borst, P. (1976), *Eur. J. Biochem.*, **64**, 153.

227 Mandelstam, J. and McQuillen, K. (1973), *Biochemistry of Bacterial Growth*, Blackwell Scientific, Oxford.

228 Barton, N. and Jones, J. S. (1983), *Nature (London)*, **306**, 317.

229 Dawes, I. W. and Sutherland, I. W. (1976), *Microbiological Physiology*, Blackwell Scientific, Oxford.

230 Stonington, O. and Pettijohn, D. (1971), *Proc. Natl. Acad. Sci. USA*, **68**, 6.

231 Worcel, A. and Burgi, L. (1972), *J. Mol. Biol.*, **71**, 127.

232 Pettijohn, D. and Hecht, R. (1973), *Cold Spring Harbor Symp. Quant. Biol.*, **38**, 31.

233 Worcel, A., Burgi, E., Robinton, J. and Carlson, C. L. (1973), *Cold Spring Harbor Symp. Quant. Biol.*, **38**, 43.

234 Kemp, D. J., Corcoran, L. M., Coppel, R. L., Stahl, H. D., Bianco, A. E., Brown, G. V. and Anders, R. F. (1985), *Nature (London)*, **315**, 347.

235 Cooper, S. (1979), *Nature (London)*, **280**, 17.

236 Carle, G. F. and Olson, M. V. (1984), *Nucleic Acids Res.*, **12**, 5647.

237 Cooper, S. and Helmstetter, C. E. (1968), *J. Mol. Biol.*, **31**, 519.

238 Avery, O. T., McLeod, C. M. and McCarty, M. (1944), *J. Exp. Med.*, **79**, 137.

239 Bibb, M. J., Van Etten, R. A., Wright, C. T., Walberg, M. W. and Clayton, D. A. (1981), *Cell*, **26**, 167.

240 Clary, D. O., Wahleithner, J. A. and Wolsten-holme, D. R. (1984), *Nucleic Acids Res.*, **12**, 3747.

241 De Bruijn, M. H. L. (1983), *Nature (London)*, **304**, 234.

242 Cox, F. E. G. (1985), *Nature (London)*, **315**, 280.

243 Mandel, M. and Higa, A. (1970), *J. Mol. Biol.*, **53**, 159.

244 Hayes, W. (1968), *The Genetics of Bacteria and their Viruses* (2nd edn), Blackwell Scientific, Oxford.

245 Notani, N. K. and Setlow, J. K. (1974), *Prog. Nucleic Acid Res. Mol. Biol.*, **14**, 39.

246 Lewin, B. (1977), *Gene Expression-3*, John Wiley and Sons, New York.

247 Lacks, S. and Neuberger, M. (1975), *J. Bacteriol.*, **124**, 1321.

248 Marmur, J. and Hotchkiss, R. D. (1953), *J. Biol. Chem.*, **214**, 383.

249 Szybalska, E. H. and Szybalski, W. (1962), *Proc. Natl. Acad. Sci., USA*, **48**, 2026.

250 Luria, S. E., Darnell, J. E., Baltimore, D. and Campbell, A. (1978), *General Virology* (3rd edn), John Wiley and Sons, New York.

251 Primrose, S. B. and Dimmock, N. J. (1980), *Introduction to Modern Virology*, Blackwell Scientific, Oxford.

252 Thoma, F. and Simpson, R. T. (1985), *Nature (London)*, **315**, 250.

253 Klug, A. and Caspar, D. L. D. (1960), *Adv. Virus Res.*, **7**, 225.

254 Markham, R. (1963), *Prog. Nucleic Acid Res.*, **2**, 61.

255 Bancroft, J., Hills, G. and Markham, R. (1967), *Virology*, **31**, 354.

256 Baltimore, D. (1971), *Bacteriol. Rev.*, **35**, 235.

257 Kornberg, A. (1980), *DNA Replication*, W. & H. Freeman & Co., San Francisco.

258 Simons, K., Garoff, H. and Helenius, A. (1982), *Sci. Am.*, **246(2)**, 46.

259 Stent, G. S. and Wollman, E. L. (1950), *Biochim. Biophys. Acta*, **6**, 307.

260 Crawford, E. M. and Gesteland, R. F. (1964), *Virology*, **22**, 165.

261 Kozloff, L. M. and Lute, M. (1959), *J. Biol. Chem.*, **234**, 534.

262 McCorquodale, D. J., Oleson, A. E. and Buchanan, J. M. (1967), *The Molecular Biology of Viruses* (ed. J. S. Colter and W. Paranchych), Academic Press, New York, p. 31.

263 Brinton, C. C. and Beer, H. (1967), *The Molecular Biology of Viruses* (ed. J. S. Colter and W. Paranchych), Academic Press, New York, p. 251.

264 Marvin, D. A. and Hohn, B. (1969), *Bacteriol. Rev.*, **33**, 172.

265 Crawford, L. V. (1962), *Virology*, **18**, 177.

266 Dulbecco, R., Hartwell, L. H. and Vogt, M. (1965), *Proc. Natl. Acad. Sci. USA*, **53**, 403.

267 Winocour, E., Kaye, A. N. and Stollar, V. (1965), *Virology*, **27**, 156.

268 Fried, M. and Pitts, J. D. (1968), *Virology*, **34**, 761.

269 Leberman, R. (1968), *Symp. Soc. Gen. Microbiol.*, **18**, 183.

270 Wood, W. B. and Edgar, R. S. (1967), *Sci. Amer.*, **217(1)**, 60.

271 Hendrix, R. W., Roberts, J. W., Stahl, F. W. and Weisberg, R. A. (eds) (1983), *Lambda II*. Cold Spring Harbor, New York.

272 Weigle, J. J. (1966), *Proc. Natl. Acad. Sci. USA*, **55**, 1462.

273 Kaiser, D. and Masuda, T. (1973), *Proc. Natl. Acad. Sci. USA*, **70**, 260.

274 Klug, A. (1980), *The Harvey Lectures*, Academic Press, New York.

275 Butler, P. J. G. and Klug, A. (1978), *Sci. Am.*, **239(5)**, 52.

276 Hohn, T. (1967), *Eur. J. Biochem.*, **2**, 152.

277 Bancroft, J. B. and Hiebert, E. (1967), *Virology*, **32**, 354.

278 Crick, F. H. C. and Watson, J. D. (1956), *Nature (London)*, **177**, 473.

279 Streisinger, G., Mukai, F., Dreyer, W. J., Miller, B. and Horiuchi, S. (1961), *Cold Spring Harbor Symp. Quant. Biol.*, **26**, 25.

280 Oldstone, M. B. A., Rodriguez, M., Daughaday, W. H. and Lampert, P. W. (1984), *Nature (London)*, **306**, 278.

281 Casjens, S. and King, J. (1975), *Annu. Rev. Biochem.*, **44**, 555.

282 Hershey, A. D. and Chase, M. (1952), *J. Gen. Physiol.*, **26**, 36.

283 Hershey, A. D., Burgi, E. and Ingraham, L. (1963), *Proc. Natl. Acad. Sci. USA*, **49**, 748.

284 Edgar, R. S. and Epstein, R. H. (1965), *Sci. Am.*, **212(2)**, 71.

285 Edgar, R. S. and Lielausis, I. (1964), *Genetics*, **49**, 649.

286 Ritchie, D. A., Thomas, C. A., MacHattie, L. A. and Wensinv, P. C. (1967), *J. Mol. Biol.*, **23**, 365.

287 Tolun, A., Alestrom, P. and Pettersson, V.

(1979), *Cell*, **17**, 705.

288 Thomas, C. A. and MacHattie, L. A. (1964), *Proc. Natl. Acad. Sci. USA*, **52**, 1297.

289 Bujard, H. (1969), *Proc. Natl. Acad. Sci. USA*, **62**, 1167.

290 Colter, J. S. and Paranchych, W. (eds) (1967), *The Molecular Biology of Viruses*, Academic Press, New York.

291 Subak-Sharpe, J. H. and Timbury, M. C. (1977), *Comp. Virol.*, **9**, 89.

292 Hayward, G. S., Jacob, R. J., Wadsworth, S. C. and Roizman, B. (1975), *Proc. Natl. Acad. Sci. USA*, **72**, 4243.

293 Astell, G. R., Smith, M., Chow, M. B. and Ward, D. C. (1979), *Cell*, **17**, 691.

294 Bellamy, A. R. and Joklik, W. K. (1967), *Proc. Natl. Acad. Sci. USA*, **58**, 1389.

295 Wyatt, G. R. and Cohen, S. S. (1953), *Biochem. J.*, **55**, 774.

296 Lehman, I. R. and Pratt, E. A. (1960), *J. Biol. Chem.*, **235**, 3254.

297 Hattman, S. and Fukasawa, T. (1963), *Proc. Natl. Acad. Sci. USA*, **50**, 297.

298 Dulbecco, R. and Vogt, M. (1963), *Proc. Natl. Acad. Sci. USA*, **50**, 236.

299 Weil, R. and Vinograd, J. (1963), *Proc. Natl. Acad. Sci. USA*, **50**, 730.

300 Winocour, E. (1967), *The Molecular Biology of Viruses*, (ed. J. S. Colter and W. Paranchych), Academic Press, New York p. 577.

301 Sanger, F., Air, G. M., Barrell, B. G., Brown, N. L., Coulson, A. R., Fiddes, J. C., Hutchinson, C. A., Slocombe, P. M. and Smith, M. (1977), *Nature (London)*, **265**, 687.

302 Godson, G. N., Barrell, B. G., Staden, R. and Fiddes, J. C. (1978), *Nature (London)*, **276**, 236.

303 Fiddes, J. C. and Godson, C. W. (1979), *J. Mol. Biol.*, **133**, 19.

304 Rattner, J. B. and Liu, C. C. (1985), *Cell*, **42**, 291.

305 Diener, T. O. (1984), *Trends Biochem. Sci.*, **9**, 133.

306 Fyfe, J. A., Keller, P. M., Furman, P. A., Miller, R. L. and Elion, G. B. (1978), *J. Biol. Chem.*, **253**, 8721.

307 Derse, D., Bastow, K. F. and Cheng, Y-C. (1984), *J. Biol. Chem.*, **257**, 10 251.

308 Lerner, R. A. (1982), *Nature (London)*, **299**, 592.

309 Dreasman, G. R., Sanchez, Y., Ionescu-Matiu, I., Sparrow, J. T., Six, H. R., Peterson, D. L., Hollinger, F. B. and Melnick, J. L. (1982), *Nature (London)*, **295**, 158.

310 Reddy, V. B., Thimmappaya, B., Dhar, R., Subramanian, K. N., Zain, B. S., Pan, J., Ghosh, P. K., Celma, M. L. and Weissman, S. M. (1978), *Science*, **200**, 494.

311 Fiers, W., Contreras, R., Haegeman, G., Rogiers, R., Van de Vorde, A., van Heuverswyn, H., van Hevreweghe, J., Volckaert, G. and Ysebaert, M. (1978), *Nature (London)*, **273**, 113.

312 Soeda, E., Arrand, J. R., Smoler, N., Walsh, J. E. and Griffin, B. E. (1980), *Nature (London)*, **283**, 445.

313 Adams, R. L. P. and Burdon, R. H. (1985), *Molecular Biology of DNA Methylation*, Springer-Verlag, New York.

314 Ptashne, M., Johnson, A. D. and Pabo, C. O. (1982), *Sci. Am.*, **267(5)**, 106.

315 Lederberg, E. M. and Lederberg, J. (1953), *Genetics*, **38**, 51.

316 Campbell, A. (1962), *Adv. Genet.*, **11**, 101.

317 Jacob, F. and Wollman, E. L. (1957), *The Chemical Basis of Heredity*, (ed. W. D. McElroy and B. Glass), Johns Hopkins, Baltimore, p. 468.

318 Tooze, J. (ed) (1981), *DNA Tumor Viruses* (2nd edn), Cold Spring Harbor Laboratory Monograph 10B.

319 Weiss, R. A., Teich, N. M., Varmus, H. E. and Coffin, J. M. (eds) (1982), *RNA Tumor Viruses* (2nd edn), Cold Spring Harbor Laboratory Monograph 10C.

320 Dulbecco, R. (1967), *Sci. Am.*, **216(4)**, 28.

321 Fried, M. (1965), *Proc. Natl. Acad. Sci. USA*, **53**, 486.

322 Cuzin, F., Vogt, M., Dieckmann, M. and Berg, P. (1970), *J. Mol. Biol.*, **47**, 317.

323 Benjamin, T. L. (1966), *J. Mol. Biol.*, **16**, 359.

324 Watkins, J. F. and Dulbecco, R. (1967), *Proc. Natl. Acad. Sci. USA*, **58**, 1396.

325 Gelb, L. D., Kohne, D. E. and Martin, M. A. (1971), *J. Mol. Biol.*, **57**, 129.

326 Ozanne, B., Vogel, A., Sharp, P., Keller, W. and Sambrook, J. (1973), *Lepetit Colloq. Biol. Med.*, **4**.

327 Nuti, M. P., Lepidi, A. A., Prakash, R. K., Hoogkaas, P. J. J. and Schilperoort, R. A. (1982), in *Molecular Biology of Plant Tumours* (eds G. Khan and J. Schell), Academic Press, New York, p. 561.

328 Braun, A. C. (1982), in *Molecular Biology of Plant Tumours* (eds G. Khan and J. Schell), Academic Press, New York, p. 155.

329 Erikson, R. L., Collett, M. S., Erikson, E. and Purchio, A. F. (1979), *Proc. Natl. Acad. Sci. USA*, **76**, 6260.

330 Crawford, L. V. (1980), *Trends Biochem. Sci.*, **5**, 39.

331 Rigby, P. (1979), *Nature (London)*, **282**, 781.

332 Bishop, J. M. (1982), *Sci. Amer.*, **246(3)**, 68.

333 Weinberg, R. A. (1983), *Sci. Am.*, **249(5)**, 102.

334 Weinberg, R. A. (1982), *Trends Biochem. Sci.*, **7**, 135.

335 Weinberg, R. A. (1982), *Cell*, **30**, 3.

336 Waterfield, M. D., Sarace, G. T., Whittle, N., Stroobaut, P., Johnsson, A., Wasteson, A., Westermark, B., Heldin, C-H., Huang, J. S. and Deuel, T. F. (1983), *Nature (London)*, **304**, 35.

337 Lane, D. P. (1984), *Nature (London)*, **312**, 596.

338 Hunter, T. (1984), *Sci. Am.*, **251(2)**, 60.

339 Berridge, M. J. and Irvine, R. F. (1984), *Nature (London)*, **312**, 315.

340 Verma, I. M. (1984), *Nature (London)*, **308**, 317.

341 Diener, T. O. (1981), *Sci. Am.*, **244(1)**, 58.

342 Haseloff, J., Mohamed, N. A. and Symons, R. H. (1982), *Nature (London)*, **299**, 316.

343 Wurtman, R. J. (1985), *Sci. Am.*, **252(1)**, 48.

344 Prusiner, S. B. (1984), *Sci. Am.*, **251(4)**, 48.

345 Brunori, M. and Talbot, B. (1985), *Nature (London)*, **314**, 676.

346 Masters, C. (1985), *Nature (London)*, **314**, 15.

347 Raulet, D. H. (1985), *Nature (London)*, **314**, 103.

348 Bukhari, A. I., Shapiro, J. A. and Adhya, S. L. (1977), *DNA Insertion Elements, Plasmids and Episomes*, Cold Spring Harbor, New York.

349 Meynell, G. G. (1972), *Bacterial Plasmids*, Macmillan, London.

350 Sherratt, D. J. (1974), *Cell*, **3**, 189.

351 Campbell, A. (1969), *Episomes*, Harper and Row, New York.

352 Novick, R. P. (1980), *Sci. Am.*, **243(6)**, 76.

353 Nishmura, Y., Caro, L., Berg, C. M. and Hirota, Y. (1971), *J. Mol. Biol.*, **55**, 441.

354 Caro, L. G. and Berg, C. M. (1969), *J. Mol. Biol.*, **45**, 325.

355 Jacob, F. and Wollman, E. L. (1961), *Sexuality and the Genetics of Bacteria*, Academic Press, New York.

356 Bolivar, F. (1979), *Life Sci.*, **25**, 807.

357 McClintock, B. (1965), *Brookhaven Symp. Biol.*, **18**, 162.

358 Dingwall, C. (1985), *Trends Biochem. Sci.*, **10**, 65.

359 Mueller, R. D., Yasuda, H., Hatch, C. L., Bonner, W. M. and Bradbury, E. M. (1985), *J. Biol. Chem.*, **260**, 5147.

360 Robertson, H. D., Branch, A. D. and Dahlberg, J. E. (1985), *Cell*, **40**, 725.

361 Mowat, M., Cheng, A., Kimura, N., Bernstein, A. and Beuchimol, S. (1985), *Nature (London)*, **314**, 633.

# 4

# Degradation
# and modification of
# nucleic acids

## 4.1 INTRODUCTION AND CLASSIFICATION OF NUCLEASES

Enzymes which catalyse the breakdown of nucleic acids by hydrolysis of phosphodiester bonds have been found in almost all biological systems [1, 2]. Some, the *ribonucleases*, are quite specific for RNA, others, the *deoxyribonucleases*, act only on DNA, while a third group of non-specific nucleases is active against either nucleic acid.

The *phosphorylases*, polynucleotide phosphorylase and pyrophosphorylase, are also capable of depolymerizing RNA, but their degradative role *in vivo* is uncertain.

The *phosphomonoesterases* act on polynucleotides or oligonucleotides with a terminal phosphate group or on a mononucleotide to liberate inorganic phosphate. Their substrates will often be the products of nuclease action.

Classification schemes for the nucleases have been discussed by Laskowski [1] and by Barnard [2]. Three main features of nuclease action can be used as a basis for classification.

The first of these is substrate specificity, i.e. action on RNA, DNA or both, as discussed above. The second is mode of attack; polynucleotides can be attacked at points within the polymer chain *endolytically* or stepwise from one end of the chain *exolytically*. Thus we may have endonucleases which produce oligonucleotides and cause rapid changes in physical properties (e.g. in viscosity of DNA), and exonucleases which produce mononucleotides but initially with rather less drastic effects on nucleic acid physical properties. A few enzymes appear to act as both endo- and exo-nuclease, e.g. micrococcal nuclease. The third feature is mode of phosphodiester bond cleavage. Most biological polymers like proteins and carbohydrates, can be split in only one way; but, as illustrated in Fig. 4.1, polynucleotides can be cleaved at the phosphodiester bond on either side of the phosphate. Of the ribonucleases, those enzymes that produce 3'-phosphates and 5'-hydroxyl termini are concerned with the degradation of RNA while those enzymes that are involved in the processing of RNA precursor molecules generate 5'-phosphate and 3'-hydroxyl termini.

Additional criteria may be used to define further the action of a nuclease. These include specificity towards secondary structure of substrate, direction of attack by exonuclease (3' → 5' or 5' → 3'), and preferential endonucleolytic bond cleavage, e.g. GpX → Gp, by ribonuclease $T_1$. However, in only a few cases

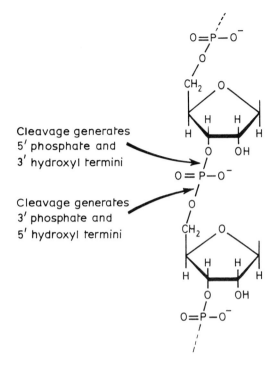

Cleavage generates
5' phosphate and
3' hydroxyl termini

Cleavage generates
3' phosphate and
5' hydroxyl termini

**Fig. 4.1**   The cleavage of phosphodiester bonds.

(e.g. the restriction endonucleases) are base specificities absolute, and relative differences in reaction rates with different bases are more common.

Experimental details for the preparation and handling of several of these enzymes are to be found in *Methods in Enzymology*, volume 65 [3].

### 4.2 NON-SPECIFIC NUCLEASES

#### 4.2.1 Non-specific endonucleases

(a) *Micrococcal nuclease (EC 3.1.31.3)\**
This enzyme is found in cultures of *Staphylococcus* and degrades DNA to a mixture of nucleoside 3'-monophosphates and oligonucleo-

---

*\*These index numbers refer to the enzyme nomenclature recommendations (1984) of the Nomenclature Committee of the International Union of Biochemistry [21].*

tides with 3'-phosphate termini [4]. It attacks RNA and heat-denatured DNA preferentially, though it also actively degrades native DNA when it acts initially as an endonuclease producing double-stranded breaks, and subsequently as an exonuclease. It shows a certain amount of sequence specificity; the primary cutting sites are adenine- and thymine-rich regions of DNA [5, 6]. It requires $Ca^{2+}$ for maximum activity. Its structure and chemical properties have been reviewed [7, 8].

(b) *Neurospora crassa nuclease* [9]
This enzyme has been considerably purified from conidia of *Neurospora* and attacks DNA or RNA to give oligonucleotides with a 5'-phosphate terminus. It exhibits a preference for guanosine or deoxyguanosine residues, but its most interesting property is an absolute requirement for denatured polynucleotide particularly in the presence of high concentrations of NaCl [10]. It is active under a wide range of conditions and requires $Ca^{2+}$ or $Mg^{2+}$.

(c) *Nuclease S1 from Aspergillus oryzae (EC 3.1.30.1)*
This enzyme is very similar to the *N. crassa* nuclease in that it hydrolyses phosphodiester bonds in single-stranded DNA or RNA [10, 11]. It has an acid pH optimum and shows a requirement for $Zn^{2+}$.

(d) *Nuclease $P_1$ from Penicillium citrinum*
Splits 3'5'-phosphodiester bonds in RNA and single-stranded DNA as well as 3'-phospho-monoester bonds in mononucleotides and oligo-nucleotides with limited specificity. It also is a $Zn^{2+}$ enzyme and works well at 70°C [10, 12].

(e) *Mung bean nuclease I*
This is another Zn-dependent single-stranded nuclease which acts on both RNA and heat-denatured DNA. It also has an acid pH optimum but is strongly inhibited by salt. It can dephos-phorylate 3'-mononucleotides [10, 13].

## (f) *BAL 31 nuclease*

An extremely stable enzyme isolated from the marine bacterium *Alteromonas espejiano* (BAL 31). It will degrade both DNA and RNA acting endonucleolytically on single-stranded molecules and exonucleolytically in duplex DNA when it degrades both the 3′ and 5′ termini [10, 14, 17]. It is active at high temperature (60°C), high salt (up to 7 M CsCl) and in the presence of 5% SDS provided that $Ca^{2+}$ and $Mg^{2+}$ ions are present. It has an optimum pH of about 8.0.

The *in vivo* function of these, and many other endonucleases present in a variety of mammalian cells, invertebrates, plants and bacteria is largely unknown. However, they have proved of immense use to molecular biologists as tools for the study of nucleic acids and chromatin [10].

### 4.2.2 Non-specific exonucleases

#### (a) *Venom phosphodiesterase* [15]

The venom of several species of snakes contains a phosphodiesterase which is commonly employed in the preparation of nucleoside 5′-phosphates. The enzyme occurs naturally in association with a high concentration of phosphomonoesterase from which it can be freed by chromatography and acetone fractionation.

Venom diesterase hydrolyses RNA to nucleoside 5′-monophosphates starting at the 3′-hydroxyl end of the chain, and is also active in hydrolysing the oligonucleotides produced by the action of deoxyribonuclease I on DNA (Fig. 4.2). The presence of a 3′-phosphoryl terminal group confers resistance on the substrate.

#### (b) *Spleen phosphodiesterase (EC 3.1.16.1)* [16]

Hydrolyses RNA to nucleoside 3′-monophosphates starting at the 5′-hydroxyl end and also acts on the mixture of oligonucleotides produced from DNA by spleen deoxyribonuclease II (Fig. 4.3). It is inactive with oligonucleotides carrying a 5′-phosphomonoester end group.

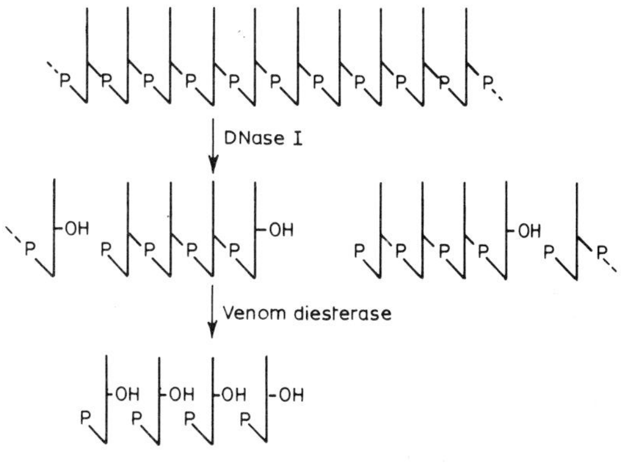

**Fig. 4.2** The digestion of DNA by DNase I followed by venom diesterase to yield deoxyribonucleoside 5′-monophosphates.

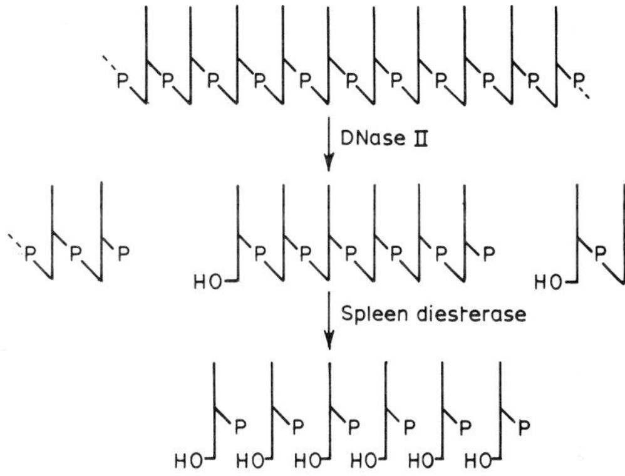

**Fig. 4.3** The digestion of DNA by DNase II followed by spleen diesterase to yield deoxyribonucleoside 3′-monophosphates.

## 4.3 RIBONUCLEASES (RNases)

The following discussion of ribonucleases reviews some of the enzymes which have been purified and studied in detail. They tend to be nucleases associated with RNA degradation and are also often those enzymes most extensively used in the laboratory as a tool to cleave or digest RNA. Many of the ribonucleases involved in the specific processing of RNA precursor species are poorly characterized. As indicated in Table 4.1, those that have been identified are discussed in subsequent sections of the book which deal with the processing events that they catalyse.

**Table 4.1**   Selected ribonucleases.*

| Enzyme | Source* | Type | Product |
|---|---|---|---|
| Pancreatic RNase A | Mammalian pancreas | Endonuclease | 3'-PO$_4$, 5'-OH |
| RNase T1 | *Aspergillus oryzae* | Endonuclease | 3'-PO$_4$, 5'-OH |
| RNase T2 | *Aspergillus oryzae* | Endonuclease | 3'-PO$_4$, 5'-OH |
| | | | |
| RNase U1 | *Ustilago sphaerogena* | Endonuclease | 3'-PO$_4$, 5'-OH |
| RNase U2 | *Ustilago sphaerogena* | Endonuclease | 3'-PO$_4$, 5'-OH |
| RNase L | Mammalian cells | Endonuclease | 3'-PO$_4$, 5'-OH |
| RNase I | *E. coli* | Endonuclease | 3'-PO$_4$, 5'-OH |
| RNase P1 | *Penicillium citrinum* | Endonuclease | 5'-PO$_4$, 3'-OH |
| Mung bean nuclease 1 | Mung bean | Endonuclease | 5'-PO$_4$, 3'-OH |
| Micrococcal nuclease | *Staphylococcus* | Endonuclease | 3'-PO$_4$, 5'-OH |
| | | | |
| S1 nuclease | *Aspergillus* | Endonuclease | 5'-PO$_4$, 3'-OH |
| Endoribonuclease H | Many sources | Endonuclease | 5'-PO$_4$, 3'-OH |
| Exoribonuclease H | Retroviruses | Exonuclease 5' → 3' and 3' → 5' | 5'-PO$_4$, 3'-OH |
| RNase II | *E. coli* | Exonuclease 3' → 5' | 5'-NMP |
| Spleen phosphodiesterase | Bovine spleen | Exonuclease 5' → 3' | 3'-NMP |
| Venom phosphodiesterase | Snake venom | Exonuclease 3' → 5' | 5'-NMP |
| Polynucleotide phosphorylase | Micro-organisms | Exonuclease 3' → 5' | 5'-NDP |
| RNase III | *E. coli* | Endonuclease | 5'-PO$_4$, 3'-OH |
| | | | |
| RNase IV | *E. coli* | Endonuclease | 3'-PO$_4$, 5'-OH |
| | | | |
| RNase P | *E. coli* | Endonuclease | 5'-PO$_4$, 3'-OH |
| RNase E | *E. coli* | Endonuclease | 5'-PO$_4$, 3'-OH |
| RNases M5, M16, M23 | *E. coli* | Endonucleases | Presumed to be 5'-PO$_4$, 3'-OH |
| | | | |
| RNases, F.Q.Y.D.T and BN | *E. coli* | Endo and exonucleases | Presumed to be 5'-PO$_4$, 3'-OH |

* Note that ribonuclease activities related to those described above have been detected and in some cases isolated from many other species. Some of these are described in the text or in the given references. ss and ds refer to single stranded and double stranded respectively.

### 4.3.1 Endonucleases which form 3'-phosphate groups

(a) *Pancreatic ribonuclease (Ribonuclease A) (EC 3.1.27.5) (for reviews see [18, 19])*
In 1920 Jones [20] described a heat-stable enzyme present in the pancreas which was capable of digesting yeast RNA. The enzyme was purified by Dubos and Thompson [22] and was crystallized in 1940 by Kunitz [23] who named it ribonuclease.

Crystalline pancreatic RNase prepared by the method of Kunitz tends to be contaminated with traces of proteolytic enzymes which have on occasion given rise to misleading results. The

**Table 4.1** *(continued)*

| Specificity | Function | Reference |
|---|---|---|
| Cuts 3′ to pyrimidine residues | Degradative | Section 4.3.1.a |
| Cuts 3′ to guanosine residues | Degradative | Section 4.3.1.b |
| Partial specificity to 3′ side of adenosine | Degradative | Section 4.3.1.b |
| Cuts 3′ to guanosine residues | Degradative | Section 4.3.1.b |
| Cuts 3′ to purine residues | Degradative | Section 4.3.1.d |
| Cuts 3′ to uridine residues | Degradative | Section 4.3.1.e |
| Non-specific | Assumed to be degradative | Section 4.3.1.c |
| Single-stranded RNA or DNA | Not known | Section 4.2.1d |
| Single-stranded RNA or DNA | Not known | Section 4.2.1e |
| Double- or single-stranded RNA or DNA | Degradative | Section 4.2.1a |
| Single-stranded RNA or DNA | Degradative | Section 4.2.1c |
| RNA of RNA:DNA duplexes | Degradative | Section 4.3.4 |
| RNA of RNA:DNA duplexes | Degradative | Section 4.3.4 |
| Non-specific but prefers ss RNA | Assumed to be degradative | Section 4.3.3 |
| RNA and DNA, prefers ss substrate | Degradative | Section 4.2.2.b |
| RNA and DNA, prefers ss substrate | Degradative | Section 4.2.2a |
| Reversible reaction, non-specific | Not known | Section 4.4 |
| ds RNA, unknown method of recognition | Processing | Section 9.4.1 |
| Specific sites but recognition method unknown | Not known | Section 4.3.2 |
| Specific cuts to pre-tRNA | Processing | Section 9.5.2.a |
| Specific cuts to pre-5SRNA | Processing | Section 9.4.1.b |
| Maturation of 5S, 23S and 16S precursor | Processing | Section 9.4.1a |
| Maturation of pre-tRNA | Processing | Section 9.5.2.a |

crystallization of pancreatic RNase free from proteolytic contaminants was described by McDonald [24] but more recent purification methods have been much simplified by the use of affinity chromatography [19].

Pancreatic RNase is a very small protein, mol. wt. 13 700, is stable over a wide pH range, and is remarkably resistant to heat in slightly acid solution, although it is readily inactivated by alkali. It has no action on DNA and is strongly antigenic. Its maximum activity is in the range pH 7.0–8.2, with the optimum at pH 7.7. Its optimum temperature is 65°C. The enzyme is commonly referred to as ribonuclease A but glycosylated variants occur which are given different notations. Thus, bovine pancreatic ribonucleases B, C and D contain different carbohydrate side chains attached to the asparagine-34 side chain (the latter derivative should not be confused with *E. coli* RNase D; see Table 4.1 and Section 9.5.2a). A man-made derivative of RNase A, which has featured in many studies of the catalytic activity of the enzyme, is known as RNase S. It is produced by cleavage between residue 20 and 21 by the protease, subtilisin.

Since the initial sequence determination of bovine ribonuclease by Moore and collaborators [25], pancreatic ribonucleases have been isolated and sequenced from many sources. Beintema and colleagues have been particularly active in

this respect and their many sequences [26] together with those of others have been tabulated by Blackburn and Moore [19]. The three-dimensional structure of pancreatic ribonuclease has been resolved by X-ray diffraction both by crystallographic studies [27, 28, 29] and in solution [30] and the nature and mechanism of action of the active site has been determined (reviewed in [18] and [19]). The amino acids

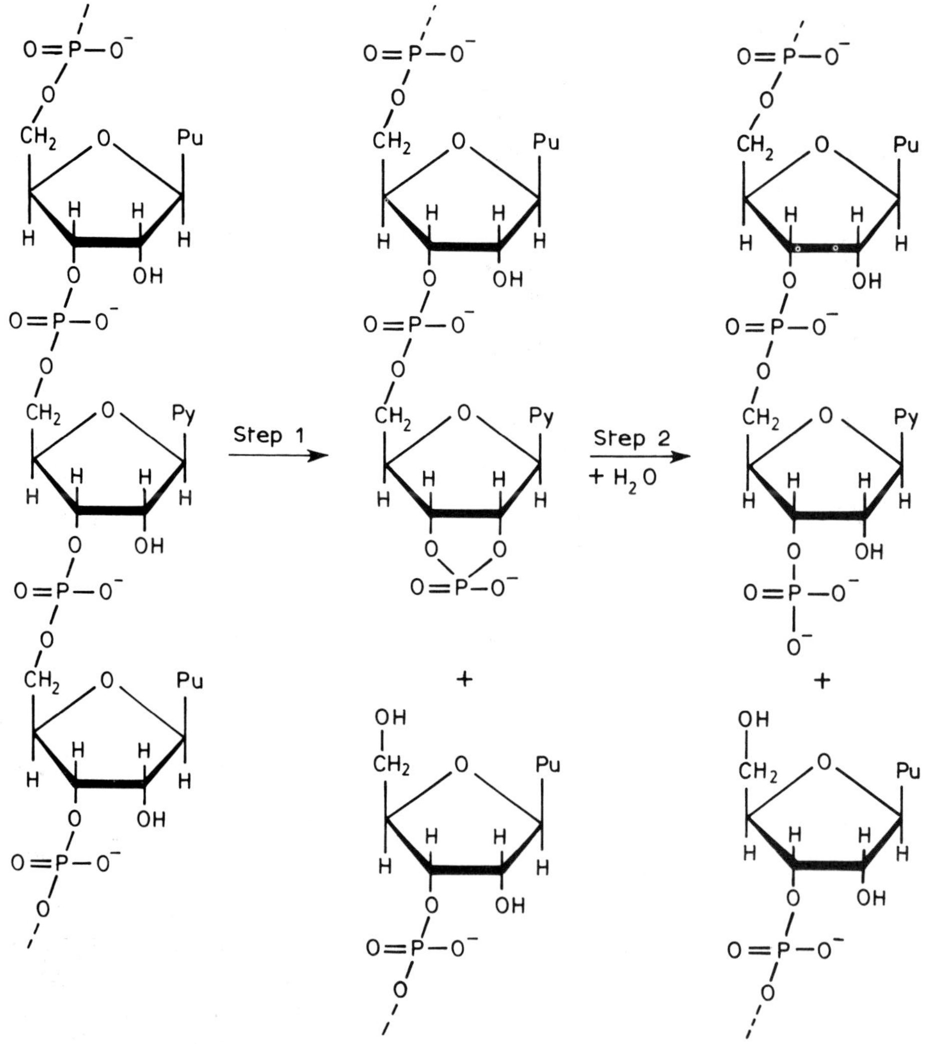

**Fig. 4.4** The action of pancreatic ribonuclease (RNase) on RNA, showing the intermediate formation of cyclic phosphates.

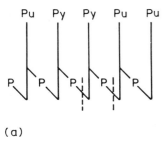

(a)

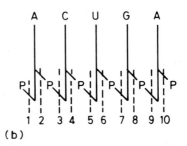

(b)

**Fig. 4.5** (a) The pentanucleotide containing three purine and two pyrimidine nucleotide units is split by pancreatic ribonuclease at the broken lines. (b) A pentanucleotide containing two adenine nucleotide residues and one residue each of cytosine, uracil and guanine nucleotides with monoesterified phosphate residues at each end is split by pancreatic ribonuclease at positions 5 and 7, by ribonuclease $T_1$ at position 9, by 5'-monoesterase at 1, by 3'-monoesterase at 10, by venom diesterase at 2, 4, 6 and 8 and by spleen diesterase at 3, 5, 7 and 9.

involved in catalysis and in maintaining the secondary and tertiary structure are those that are highly conserved. The enzyme has been chemically synthesized by both solid phase and liquid phase methods [31–33].

Pancreatic ribonuclease cleaves the ester bond between the phosphate and the 5'-carbon of the ribose of a pyrimidine nucleotide. It does this in two steps (Fig. 4.4). The ester bond is first transferred to the 2'-hydroxyl group so forming a cyclic ester. This is then specifically hydrolysed to the 3'-monoester. The kinetics and geometry of the two stages have been thoroughly investigated [34–38] but their detailed consideration is outside the scope of this book. Pancreatic RNase and other RNases whose mode of action is similar have the theoretical possibility of forming both the 2'- and 3'-monoester by hydrolysis of the cyclic phosphate (Fig. 4.4); most, if not all, form the 3'-ester exclusively.

The action of pancreatic RNase may be illustrated as follows. The pentanucleotide shown in Fig. 4.5(a), in which Pu and Py represent purine and pyrimidine residues respectively, will be hydrolysed at the points shown by the broken lines, while the ribopolynucleotide chain shown in Fig. 4.5(b), which may also be expressed as pApCpUpGpAp, will be broken at positions 5 and 7 to yield pApCp + Up + GpAp.

To take a slightly more elaborate case, the polynucleotide

ACCCCAGGGUUUAGUCp

would be split by RNase thus:

AC/C/C/C/AGGGU/U/U/AGU/Cp

to yield

ACp + AGGGUp + AGUp + 4Cp + 2Up

Pancreatic RNase also digests certain homopolyribonucleotides. Thus poly(A) and poly(I) are not split by the enzyme whereas poly(C) and poly(U) yield the 3-'mononucleotides. However, the specificity of pancreatic RNase for pyrimidines is not absolute, since Ap diester bonds in a polynucleotide are also attacked, albeit considerably less readily [39].

Modern methods of assaying the activity of ribonuclease continue to employ modifications of the spectrophotometric assay of Kunitz [40], the release from RNA of acid-soluble nucleotides [41] and the hydrolysis of 2',3'-cyclic CMP [42]. The modifications include the use of radioactive or chromogenic substrates, improved precipitations and fluorimetric assays and have been reviewed [19]. Several methods have also been described for the detection of ribonuclease activity after fractionation on SDS/polyacrylamide gels (see for example [43]).

Similar activities to that of pancreatic ribonuclease have been described from many other extracellular and intracellular sources [43, 44]. An activity in bovine semen represents more than 2% of its total protein. It shares considerable sequence homology with the pancreatic enzyme and also preferentially cuts RNA 3' to pyrimidine residues. However, it differs in charge, in normally existing as a dimer and in its ability to hydrolyse double-stranded RNA (reviewed in [19]). Many, probably most, mammalian tissues contain cytoplasmic enzymes that in many respects resemble pancreatic ribonuclease [45–48, 49 and references therein]. They are however present in very small quantities and are hence difficult to study. Furthermore, mammalian tissues also contain a proteinaceous ribonuclease inhibitor [50–52] and much of the cytoplasmic ribonuclease appears to exist as a latent complex with this protein which is often known as RNasin. Barton *et al.* [53] distinguish two groups of mammalian ribonucleases that cleave RNA at pyrimidine residues in two steps via 2', 3'-cyclic phosphate intermediates. The secreted enzymes of the pancreas, serum and urine tend to be heat- and acid-stable and have alkaline pH optima. RNase A typifies this group. The non-secretory RNases typically have pH optima around 7.0 and are relatively heat-labile. The validity of this grouping awaits further purification and characterization.

### (b) *Ribonuclease $T_1$ (EC 3.1.27.3)*

Has been the subject of intensive study, and our knowledge of its chemistry approaches that of pancreatic RNase (for a review see [54]). It is obtained from *Aspergillus oryzae* and at neutral pH specifically hydrolyses the internucleotide bonds of RNA between 3'-GMP and the 5'-OH groups of adjacent nucleotides (Fig. 4.5). Like pancreatic ribonuclease, it does this in a two-stage reaction producing 3'-phosphate termini via 2', 3'-cyclic phosphate intermediates.

Ribonuclease $T_1$ is a small ($M_r$ 11 085) heat-stable and acid-stable extracellular endonuclease. A second enzyme from the same source, *ribonuclease $T_2$*, preferentially attacks Ap residues and will digest tRNA almost totally to 3'-monophosphates [55]. Guanine-specific enzymes similar to ribonuclease $T_1$ are common in fungi and bacteria. These include ribonuclease $U_1$ from *Ustilago sphaerogena* [56], ribonuclease $N_1$ from *Neurospora crassa* [57] and others reviewed by Takahashi and Moore [54].

### (c) *Ribonuclease I from E. coli (for review see [58])*

This endonuclease normally occurs in the periplasmic space of the bacterial cell but on conversion of the cells into spheroplasts it is released into the medium. In broken cell preparations it is found artefactually associated with 30S ribosomal subunits [59]. Its intracellular role is unclear since strains lacking this enzyme are perfectly viable [60]. It has been considered to have a scavenging role as have the latent ribonucleases of mammalian cells. Again, however, rRNA turnover is unchanged in mutants lacking the enzyme [60]. The enzyme is a low-$M_r$ basic protein with a single subunit and a pH optimum of 8.1. It digests single-stranded RNA with no absolute base specificity although it hydrolyses poly U faster than other homopolymers. Like the previous two enzymes, it produces 3'-phosphate termini via 2', 3'-cyclic phosphate intermediates [61].

### (d) *Ribonuclease U2 from Ustilago sphaerogena (EC 3.1.27.4)*

This endonuclease, like those previously considered, hydrolyses phosphodiester bonds in two separable steps; transphosphorylation and hydrolysis [62]. The enzyme is however unique in its specificity in cleaving 3' to purine residues [63] and, under certain conditions, is highly specific for adenyl residues [64]. It consists of a *single polypeptide of $M_r$* (molecular weight) 12 490 [65] and exhibits considerable sequence and catalytic similarity to RNase $T_1$ [62, 66].

(e) *2-5A-dependent endoribonuclease (RNase L)*

Exposure of cells to interferon inhibits viral replication by arresting the synthesis of viral mRNA and protein (reviewed in [67]). Induction of the antiviral state involves the synthesis of a number of new proteins. One of these, produced in response to double-stranded RNA of viral origin, is 2-5A synthetase which synthesizes the unusual oligonucleotide $ppp(A2'p)_n$ where $n = 2-4$. Known as 2-5A, this nucleotide in turn activates a latent endoribonuclease (RNase L) [68, 69] which can degrade mRNA [68, 70] and rRNA [71]. This is discussed in Section 11.11.3.

(f) *Other endonucleases of value in RNA sequence and structural analysis*

Most of the above enzymes have played an important role in the base-specific cleavage of RNA for sequence and structural analysis [72] and continue to be important experimental tools despite the fact that DNA sequencing has largely replaced RNA sequencing. Other enzymes that have been useful in this respect include the pyrimidine-specific endonuclease of *Bacillus cerus* which cleaves RNA after all pyrimidines [73, 74]. Pancreatic nuclease is less reliable in this respect and cleaves between two adjacent pyrimidines with difficulty. Two other useful enzymes PhyI and PhyII have been isolated from *Physarum polycephalum* [75]. PhyI is a 25 000 $M_r$ protein with a pH optimum of 4.3 and cleaves all phosphodiester bonds with an order of susceptibility UpN > ApN > GpN > CpN. It has been useful in distinguishing between C and U in RNA sequences [72]. A relatively unfractionated preparation known as PhyM [76] is commercially available, and under denaturing conditions cleaves specifically at U and A residues. RNase PhyII shows outstanding preference for GpN leaving ApN and Py–py doublets practically unsplit [75]. Another enzyme which has proved useful in distinguishing between C and U in RNA sequence [77] and in structural investigations [78] is RNase CL3 from chicken

liver [79]. It is 61 times more active on cytidylic bonds than uridylic bonds [79].

Other endonucleases that have been detected in the cells and tissues of a wide range of species are not considered here because they are too inadequately characterized for their mode of action, function or their relationship to the above enzymes to be ascertained.

### 4.3.2 Endonucleases which form 5′-phosphate groups

The mechanism by which these nucleases cleave the phosphodiester bond is non-cyclizing. A direct attack of water on a 3′, 5′-phosphodiester is catalysed and thus a 2′-OH group is not required. For this reason, many 5′-monophosphate-forming endonucleases will attack RNA and DNA. Many of the enzymes which catalyse RNA cleavage in this way are involved in the specific processing of RNA precursor species. They include the *E. coli* ribonucleases III, P, E, M5 etc. which, as indicated in Table 4.1, are discussed in the subsequent sections of text on RNA processing. It is certain that many other activities still await discovery, particularly in eukaryotic cells. At least one of these activities *E. coli* RNase P is a ribonucleoprotein and there is evidence that the RNA is the catalytic component of the enzyme (Section 9.5.2.a). In *Tetrahymena*, pre-rRNA functions autocatalytically in its own cleavage (Section 9.4.3.b).

One further, well-characterized, *E. coli* enzyme which generates 5′-phosphates from single-stranded RNA is considered here because its function is unknown.

(a) *Ribonuclease IV from* E. coli *(for a review see [58])*

RNase IV cleaves a number of RNAs at a limited number of specific sites by an unknown mechanism of recognition. The 26S RNA of the bacteriophage R17 is cleaved at five sites in a short section of the RNA so generating 15S and 22S fragments. It seems likely that the enzyme is recognizing some feature of the secondary structure of the substrate and Adams *et al.* [80] have related the cleavage sites to the proposed secondary structure of the phage RNA. The cuts occur in or adjacent to proposed single-stranded regions. Similarly, the single cut made by the enzyme in 5S RNA is in a region which is thought to be single stranded [81]. The specificity of the enzyme indicates a function as a structure-related, site-specific nuclease.

### 4.3.3 RNA exonucleases

(a) *Ribonuclease II from E. coli*

RNase II is a processive exonuclease; that is, it remains bound to its substrate until it has hydrolysed all phosphodiester bonds. This distinguishes it from distributive exonucleases which dissociate after every or many catalytic events, i.e. spleen exonuclease [82]. It degrades RNA in a $3' \rightarrow 5'$ direction producing $5'$-nucleotide monophosphates but cannot degrade DNA or RNAs with secondary structure [83]. It is one of the most active enzymes in lysates of *E. coli* [84] and consists of a single subunit with a $M_r$ of 72 000–75 000 [83, 85]. Donovan and Kushner [86] have presented evidence that RNase II works in conjunction with polynucleotide phosphorylase (see Section 4.3.6) in the degradation of bacterial mRNA. They suggest that the initial steps are endonucleolytic cleavage by polynucleotide phosphorylase followed by exonucleolytic degradation by RNase II.

(b) *Spleen phosphodiesterase (spleen exonuclease) and venom phosphodiesterase (venom exonuclease)*
See Section 4.2.2.

### 4.3.4 Ribonucleases which act on RNA: DNA hybrids (RNase H)

An RNase which specifically digested the RNA of an RNA:DNA hybrid was first described by Stein and Hausen [87]. Since then, RNase H

activity has been demonstrated in a wide range of species and cell types ranging from prokaryotes, lower eukaryotes, higher animals and plants (for a review see [88]). In addition, the reverse transcriptases of retroviruses (see Section 4.5.3) exhibit RNase H activity (for a review of the evidence that the two activities reside within a single polypeptide see [88]). In many cases, a given cell or tissue appears to have more than one species of RNase H. Thus, Büsen [89, 90] has shown that calf thymus contains at least two enzymes and has demonstrated that at least the larger of the two is a nuclear enzyme. Three different RNase H-type activities have been detected in yeast [91, 92, 93].

All characterized RNases H produce $5'$-phosphate and $3'$-hydroxyl termini but some produce oligomers two to nine nucleotides in length while others generate mono- and di-nucleotides. Cellular RNases H are endoribonucleases although some of the fungal enzymes may also possess exonuclease activity. Conversely, the RNase H activity which copurifies with retroviral reverse transcriptase is exonucleolytic [94, 95] and processively degrades the RNA of RNA:DNA duplexes in either a $5' \rightarrow 3'$ or a $3' \rightarrow 5'$ direction [95, 96].

The role of RNase H is unknown but recent experiments with *E. coli* mutants defective in RNase H activity [97, 98] suggest that, in this organism at least, the enzyme plays a role in the initiation of the replication of the bacterial chromosome and in the formation of a primer for DNA polymerase (see Section 6.9.3).

### 4.3.5 Double-stranded RNA-specific ribonucleases

Ribonuclease Vl isolated from the venom of the cobra *Naja naja oxiana* [99] has proved useful in mapping the secondary structure of RNA because its cleavage is primarily restricted to double-stranded regions of RNA [99, 100, 101]. Investigations with various tRNA species [101] reveal that the enzyme does not recognize any specific nucleotide or nucleotide sequence but appears to recognize the stacked nucleotides of RNA duplex. It has been suggested [101] that it interacts with the minor groove of the duplex.

### 4.3.6 Ribonuclease inhibitors

Many nucleotide substrate analogues have been studied with respect to their effect on the catalytic mechanism of ribonucleases [18, 19]. White and co-workers have investigated oligo-nucleotides as competitive inhibitors of RNA hydrolysis. ApUp ($K_i = 0.5$ mM) was the best of those examined at inhibiting pancreatic ribonuclease [102] while 2'-5' GpG ($K_i = 0.165$ mM) was the best at inhibiting RNase $T_1$ [103]. Neither these, nor other derivatives, such as those in which ribose is replaced by arabinose [104], have found much application in the protection of RNA during its isolation and manipulation. The key to such protection is to inhibit endogenous RNase, especially during the early stages of extraction and to avoid exogenous contamination by traces of RNase from glassware, solutions and the hands of investigators. Older methods relied on non-specific adsorbants, such as the inert acidic clay derivatives, bentonite and macaloid [105, 106] and on polyanions which protect RNA from attack as a result of their electrostatic interaction with the cationic enzyme [107]. The latter included polyvinyl sulphate [108] and heparin [109]. Whilst some of these inhibitors are still in use, various investigators have found them less than totally reliable and most modern methods of preparing and working with RNA employ guanidinium thiocyanate [110] and sodium dodecyl sulphate [111] for the disruption of cells and simultaneous inactivation of nuclease. Vanadyl–ribonucleoside complexes [112] and the proteinaceous RNase inhibitor, RNasin (Section 4.3.1.a), are used in many other situations where RNA must be protected. The latter is the inhibitor of choice where RNA must be protected during enzymic reactions [113,

114]. RNase contamination of solutions is normally excluded by the inclusion of 0.1% diethyl pyrocarbonate. This inhibitor destroys ribonuclease [115, 116] but must then itself be destroyed by autoclaving or it can inactivate RNA by carboxymethylation [117].

## 4.4 POLYNUCLEOTIDE PHOSPHORYLASE (PNPase, EC 2.7.7.8)

Polynucleotide phosphorylase was first discovered in *Azotobacter vinelandii* [118] but is now known to be widely distributed in bacteria [119] (for reviews see [119, 120]). It catalyses the reversible reaction

$$n\text{NDP} \rightleftharpoons (\text{NMP})_n + n\text{P}_i$$

and was the first enzyme to be discovered that can catalyse the formation of an RNA. In doing this however it shows little resemblance to RNA polymerase as it uses no template but catalyses the polymerization of a mixture of nucleoside diphosphates into a random polymer. With a single species of diphosphate it will make a homopolymer. It is also able to elongate an oligonucleotide primer which it does without the lag which otherwise characterizes the formation of the first internucleotide bond. The *E. coli* enzyme is composed of three identical subunits of $M_r$ 84 000–95 000 [121] with a pI of 6.1 and a requirement for $\text{Mg}^{2+}$ which can be partially replaced by $\text{Mn}^{2+}$.

Polynucleotide phosphorylase has proved a useful polymerizing tool for the biochemist, and has been used to make homopolymers and heteropolymers [119, 122, 123] as well as RNAs containing modified or radioactive nucleotides [124]. Polymers made in this way were of particular importance in the elucidation of the genetic code (Section 11.2). Nevertheless, the enzyme's function *in vivo* is far from clear and it appears likely that it normally catalyses the reverse reaction, i.e. the phosphorolytic degradation of RNA. As already discussed in Section 4.3.3, evidence has recently been presented that RNaseII and polynucleotide phosphorylase may act co-operatively to degrade bacterial mRNA [86]. Neither enzyme is essential to the bacteria but mutants in which both activities are lost appear not to be viable. The phosphorolytic activity of the enzyme is also used as an analytical tool, particular use being made of the fact that at 0°C it will degrade poly(A) tails without digesting the rest of a mRNA [125].

## 4.5 DEOXYRIBONUCLEASES (DNases)

### 4.5.1 Endonucleases

The two deoxyribonucleases which were the first to be purified and characterized are both endonucleases. The first type exemplified by pancreatic deoxyribonuclease (DNase I) hydrolyses DNA to yield 5'-phosphomonoesters. The second type (DNase II) which is found in spleen and thymus hydrolyses DNA to give 3'-phosphomonoesters (Figs 4.2 and 4.3). As a result of the characterization of these two enzymes it was suggested that there were two classes of endo-deoxyribonuclease (Class I and Class II), but subsequent discoveries have shown that the product of action is only one of the characteristics that distinguish one nuclease from another [1, 126, 153].

The activity of DNA endonucleases is generally measured by estimating the release of acid-soluble products from DNA, whether as ultraviolet-absorbing material or as radioactive label. These methods are, however, useful only in the presence of extensive endonuclease action. Where extreme sensitivity has been required, supercoiled circular viral or plasmid DNA is used as a substrate. Only one phosphodiester bond cleavage is required to alter the physical properties of such molecules and allow separation of intact and cleaved molecules.

(a) *Pancreatic deoxyribonuclease (DNase I) (EC 3.1.21.1) (for reviews see [4, 154])*
This enzyme breaks down DNA into oligonucleotides of average chain length four units with a free hydroxyl group on position 3' and a phosphate group on position 5' (Fig. 4.2). It is produced in the pancreas and used as a digestive enzyme after secretion into the small intestine.

The enzyme has a molecular weight of 31 000, an optimum pH in the range of 6.8–8.2 and an isoelectric point of 4.7. It has two disulphide bonds which are very readily reduced by mercaptoethanol with concomitant inactivation. $Ca^{2+}$ ions prevent this inactivation. The enzyme is activated by magnesium ions (optimum concentration 4 mM) or manganese ions, and the nature of the divalent cation qualitatively affects specificity [127]. Citrate, borate and fluoride inhibit the enzyme by removing the activating magnesium ions.

Pancreatic DNase hydrolyses native DNA more rapidly than denatured DNA. As the early products of the reaction are worse substrates than the initial DNA, the enzyme is auto-retarding. In the early stages of the reaction, single-stranded nicks are produced towards the centre of the DNA molecule [128] but later on the Pu-p-Py bond is preferentially cleaved leading to a final product of di- and oligonucleotides. The biosynthetic polymers poly (dA) · poly(dT), poly(dI) · poly(dC), poly(dG) · poly(dC) are degraded in part by pancreatic DNase. The resistance of the poly(dC) chain in the latter two copolymers to hydrolysis by the enzyme is overcome by adding $Ca^{2+}$ to the $Mg^{2+}$ or by replacing $Mg^{2+}$ by $Mn^{2+}$ [127]. The enzyme can be freed of ribonuclease contamination by electrophoresis or by ion-exchange chromatography.

DNases I have been isolated from a variety of animal (and plant) tissues and exist in multiple forms. In bovine pancreas forms A and B differ from forms C and D by a single amino acid substitution and forms A and C have a high-mannose oligosaccharide linked to them while forms B and D are linked to more complex oligosaccharides [132, 133].

Most tissues, but especially the crop gland of pigeons and calf spleen, contain an inhibitor of DNase I. This inhibitor has been purified and shown to be actin which in the monomeric, globular form associates to form a 1:1 complex with DNase I [4, 154].

### (b) *Deoxyribonuclease II (DNase II)* *(EC 3.1.22.1) (for review see [129])*

A deoxyribonuclease of $M_r$ about 40 000 with a pH optimum in the range 4.5–5.5, and no requirement for magnesium ions, has been isolated from spleen and thymus. DNase II from porcine spleen is a dimeric protein with subunits of 35 000 and 10 000 $M_r$ [268]. Along with other hydrolases with acid pH optima it is found in lysosomes and is believed to be involved in intracellular breakdown of DNA.

Double-stranded DNA is degraded by splenic DNase II, in part by a 'one-hit' process that hydrolyses both strands of the double helix at the same point [128, 130]. This initial phase of the reaction is followed by the slower release of oligonucleotides of chain length from 14 to 100 nucleotides. The final stage produces oligonucleotides of average chain length six units, which have a free 5'-hydroxyl group and a phosphate residue on position 3' (Fig. 4.3).

The properties of the two main animal DNases are summarized in Table 4.2 but many more nucleases have been characterized. These have been tabulated by Linn [131]. In addition, many viruses induce new DNase activities [136, 137].

### (c) *Endodeoxyribonucleases from E. coli*

The unexpected plethora of nucleases is nowhere better illustrated than in *E. coli*. Within this single organism there have been reported nine endonucleases and exonucleases (Tables 4.3 and 4.4) [134, 135] not to mention topoisomerases (Section 6.4.6) and plasmid- and phage-coded enzymes.

(i) *Endonuclease I* [138]. This enzyme is the 12 000-$M_r$ product of the *end* A gene. It is an endonuclease which attacks DNA producing scissions at many points along the DNA chain. At each scission an exonucleolytic activity removes about 400 nucleotides [139] which are released as a mixture of oligonucleotides of average chain length seven units terminated by a 5'-phosphoryl group. It is highly specific for DNA, attacking native DNA seven times more readily than denatured DNA to give random double-stranded breaks [140]. It is found in the periplasmic space of log phase cultures of *E. coli*, often in association with double-stranded RNA which inhibits its action. For this reason it is activated by ribonuclease treatment [134].

(ii) *Endonuclease II*. Some confusion is associated with this enzyme as the activity originally given this name has turned out to be a mixture of enzymes (see Section 7.3.2). The

**Table 4.2** The properties of DNase I and DNase II.

|  | DNase I | DNase II |
|---|---|---|
| Substrate | DNA | DNA |
| pH optimum | 7–8 | 4–5 |
| Activators | $Mg^{2+}$, $Mn^{2+}$, $Ca^{2+}$ | $0.3\,M\,Na^+$ |
| Inhibitors | Citrate, EDTA | $Mg^{2+}$ |
| Initial action | Nicks | Double-strand breaks |
| Product | 5'-Phosphoryl-terminated oligonucleotides | 3'-Phosphoryl-terminated oligonucleotides |

**Table 4.3**   Endodeoxyribonucleases of *E. coli*.

| Name | Gene | $M_r$ | Substrate | Product |
|---|---|---|---|---|
| Endonuclease I | *endA* | 12 000 | Duplex DNA | Oligonucleotides with 5'-phosphoryl group |
| Endonuclease II | | No enzyme known | | |
| Endonuclease III | — | 27 000 | AP sites and u/v-irradiated duplex | Nicks with 3'-deoxyribose and 5'-phosphoryl termini |
| Endonuclease IV | — | 33 000 | AP sites | Nicks with 3'-OH and 5'-deoxyribose termini |
| Endonuclease V | | 27 000 | ss DNA, damaged duplex | Short oligonucleotides |
| Endonuclease VI | *xthA* | Same as exonuclease III | | |
| uvrABC endonuclease | *uvrA* | 114 000 | Damaged DNA | Excises short damaged regions |
| | *uvrB* | 84 000 | | |
| | *uvrC* | 68 000 | | |
| Restriction endonucleases | *hsd R* | 135 000 | Duplex DNA with unmodified recognition sequence | Long duplex fragments |
| | *hsd S* | 60 000 | | |
| | *hsd M* | 55 000 | | |

name was given to the activity which produces single-strand breaks in double-stranded DNA which has been alkylated with monofunctional alkylating agents (e.g. methyl methanesulphonate) [141]. The name is still reserved for such an enzyme but none is known to exist. Rather the activity is the result of the action of an *N*-glycosylase (which removes the alkylated base) and an *AP endonuclease* (an enzyme which cleaves DNA lacking a base, i.e. apurinic or apyrimidinic DNA). The original enzyme preparation contained an AP endonuclease which was given the name endonuclease VI (see below) which has since been shown to be identical to exonuclease III (see Section 4.5.2) [142, 143].

(iii) *Endonuclease III* [144] is a single 27 000-$M_r$ enzyme which shows the combined activities of glycosylase (specific for DNA irradiated with ultraviolet light) and an AP endonuclease (see Section 7.3.2). The glycosylase produces AP-DNA and the endonuclease nicks the duplex DNA 3' to the AP site to give a 3'-terminal sugar (Fig. 4.6). For this reason it is classified as a Class

I AP endonuclease [134]. Similar enzymes have been found in *M. luteus* and phage T4-infected cells [145, 146].

(iv) *Endonuclease IV* [147, 148] is a Class II AP endonuclease of about 33 000 $M_r$. It represents about 10% of the AP endonuclease activity of the cell, the *rest* being due mainly to exonuclease III (endonuclease VI).

(v) *Endonuclease V* [149] is a 27 000-$M_r$ nuclease which shows greatest activity with single-stranded DNA. It also acts on duplex DNA damaged by ultraviolet irradiation and duplex DNA containing uracil which it nicks to give 3'-OH and 5'-phosphoryl termini. Eventually the DNA is degraded to short oligonucleotides.

(vi) *Endonuclease VI* [135] is the product of the *xth*A gene and has been shown to be identical to exonuclease III (see Section 4.5.2) [142, 143]. It is a Class II AP endonuclease which also possesses $3' \rightarrow 5'$ exonuclease and 3'-phosphatase activities which may combine to increase

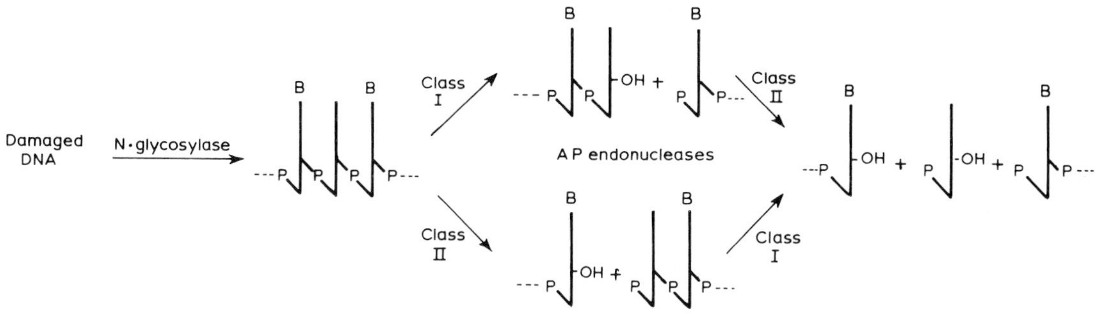

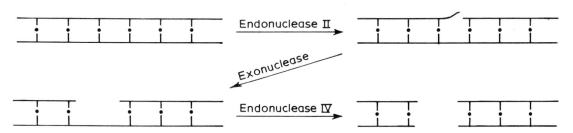

**Fig. 4.6** AP endonucleases. The consecutive action of a class I and class II AP endonuclease (as exemplified by *E.coli* endonucleases II and IV) removes the baseless sugar phosphate from the DNA.

the gap size during the repair of damaged DNA (see Section 7.3.2).

(vii) *uvrABC endonuclease* acts on damaged DNA especially at pyrimidine dimers to produce gaps in DNA. It is considered in detail in Section 7.3.2.

(viii) *Restriction endonucleases* cleave duplex DNA at specific sequences. At least one such enzyme (e.g. EcoB) is present in uninfected *E. coli* cells and many others are carried on plasmids or prophage. These enzymes are considered in Section 4.6.3.

(ix) *ATP-dependent endonucleases.* A number of enzymes are known which can hydrolyse or unwind duplex DNA while hydrolysing ATP. Their roles in DNA replication and recombination are considered in Chapters 6 and 7.

(d) *Phage-induced endonucleases*

Nucleases are induced following infection of bacteria with a variety of bacteriophages. Two of the best studied are endonuclease II and endonuclease IV induced after infection of *E. coli* with phage T4 [150, 151] (see Section 6).

(i) *T4 Endonuclease II* makes single-stranded breaks in double-stranded DNA other than that of T4 to give products (at least from phage λ) of about $10^3$ nucleotides. These have 5'-phosphoryl and 3'-OH termini. It differs from the host cell endonucleases in its inability to attack T4 DNA (glucosylated or not).

(ii) *T4 Endonuclease IV* hydrolyses single-stranded DNA to give considerably smaller products than endonuclease II, again with 5'-phosphoryl termini, but with dCMP exclusively in that position. DNA containing hydroxy-

**Fig. 4.7** Hypothetical scheme for degradation of cell DNA after infection by bacteriophage T4.

methylcytosine (i.e. T4 DNA) is inactive as a substrate.

A mechanism has been suggested [151] whereby these enzymes, with the help of a bacteriophage-induced exonuclease, may be involved in the degradation of host DNA after T4 infection (Fig. 4.7)

(iii) *Other endonucleases* play specific roles in the replication of viral DNA (see Chapter 6). For example that coded for by gene A of φX174 breaks a single phosphodiester bond to initiate phage DNA synthesis [152].

### 4.5.2 Exonucleases

As with the endonucleases a great number of exonucleases have been described from both prokaryotes and eukaryotes [131, 153, 156, 169]. We shall here show the scope of their action by referring to those present in *E. coli* (Table 4.4) [134, 135, 155].

(a) *E. coli exonuclease I (EC3.1.11.1)*
This enzyme hydrolyses single-stranded DNA and has hardly any effect on native, double-stranded DNA. It is an exonuclease hydrolysing the DNA chain stepwise beginning at the 3′-hydroxyl end, and releasing deoxyribonucleoside 5′-monophosphates until only a dinucleotide is left.

The enzyme does not cleave free dinucleotides or the 5′-terminal dinucleotide portion of a polydeoxyribonucleotide chain, but it can degrade bacteriophage DNAs containing glucosylated hydroxymethylcytosine. It has no effect on polyribonucleotides. It is the 72 000-$M_r$ product of the *sbc*B gene, but as mutations in this gene have no deleterious effects on replication, recombination or repair its *in vivo* function is obviously not essential.

(b) *E. coli exonuclease II*
Is the 3′ → 5′ exonuclease activity associated with DNA polymerase I (see below and Section 6.4.2).

**Table 4.4**  Exodeoxyribonucleases of *E. coli*.

| Name | Gene | $M_r$ | Substrate | End group attacked | Product |
|------|------|-------|-----------|--------------------|---------|
| Exonuclease I | *sbc*B | 72 000 | ss DNA | 3′-OH | Mono-and di-nucleoside 5′-phosphates |
| Exonuclease II | *pol*A | 109 000 | ss DNA or nicked duplex | 3′-OH | Nucleoside 5′-phosphates |
| | *pol*B | 120 000 | ss DNA | 3′-OH | Nucleoside 5′-phosphates |
| | *pol*C | 140 000 | ss DNA or nicked duplex | 3′-OH | Nucleoside 5′-phosphates |
| Exonuclease III | *xth*A | 28 000 | Duplex (AP sites) | 3′-OH or 3′-OP | Nucleoside 5′-phosphates and ss DNA |
| Exonuclease IV | — | — | Oligonucleotides | 3′-OH | Nucleoside 5′-phosphates |
| Exonuclease V | *rec*B | 140 000 | Duplex or | 3′-OH or | Oligonucleotides |
| | *rec*C | 130 000 | ss DNA | 5′-OH | (ATP-dependent) |
| Exonuclease VI | *pol*A | 109 000 | Nicked duplex, RNA/DNA | 5′-OH or | Short |
| | *pol*C | 140 000 | hybrid and ss DNA | 5′-OP | oligonucleotides |
| Exonuclease VII | *xse*A | 88 000 | ss DNA | 3′-OH or 5′-OH | Oligonucleotides |
| Exonuclease VIII | *rec*F | (140 000)₂ | Duplex | 5′-OH or 5′-OP | Nucleoside 5′-phosphate |

**(c)** E. coli *exonuclease III (DNA phosphatase-exonuclease)* [154, 157] (EC 3.1.11.2)

This enzyme is found in close association with the DNA polymerase of *E. coli* but can be separated from it by chromatography. Its exonuclease action is very similar to the $3' \rightarrow 5'$ exonuclease activity of DNA polymerase I but, in addition, it acts as a phosphatase highly specific for a phosphate residue esterified to the 3'-hydroxyl terminus of a DNA chain (Fig. 4.8). It does not release inorganic phosphate from monophosphates, from short oligodeoxyribonucleotides or from 3'-phosphoryl-terminated RNA, but it does attack DNA with a phosphoribonucleotide terminus.

As an exonuclease, it carries out a stepwise attack on the 3'-hydroxyl end of the DNA chain releasing mononucleotides but it acts only on double-stranded DNA, degrading it until 35–45% has been digested. If the enzyme begins its attack from both 3'-hydroxyl ends of the double-stranded molecule (Fig. 4.9), when nearly half has been degraded, the residual acid-insoluble DNA will be single-stranded and resistant to further attack although it is still susceptible to the action of exonuclease I.

Exonuclease III will also degrade the RNA strand of a DNA:RNA hybrid (RNase H activity) at more than 10 000 times the rate at which the DNA strand is degraded. More than 85% of the AP endonuclease activity of an *E. coli* cell is present as exonuclease III (also called endonuclease VI – see Section 4.5.1.c). These multitudinous activities can be reconciled in a

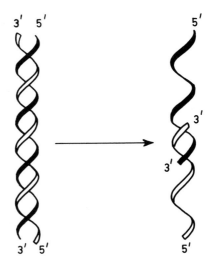

**Fig. 4.9** Mechanism of action of stepwise attack of *E. coli* exonuclease III on native DNA beginning at the 3'-hydroxyl termini.

common-site model [135] which assumes the enzyme has three regions. One region recognizes a deoxyribose on strand A and a second region cleaves 3'-phosphates on strand B of a duplex molecule. However, cleavage only occurs if the third region fails to find a base-paired deoxyribonucleotide, a 3'-terminal phosphate or a 3'-deoxyribonucleotide which because of its terminal position is transiently unpaired.

Exonuclease III is the 28 000-molecular-weight product of the *xth*A gene, mutants in which show little phenotypic change other than an increased level of genetic recombination. Double mutants with *dut* (defective in dUTPase,

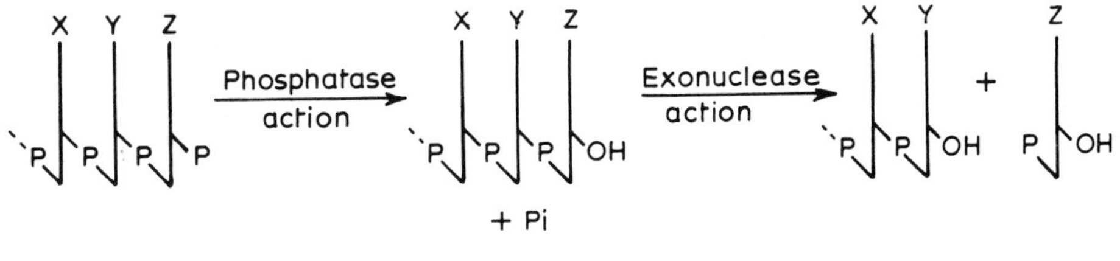

**Fig. 4.8** Sequential action of *E. coli* exonuclease III (DNA phosphatase-exonuclease) on a DNA chain terminated by a nucleotide carrying a 3'-phosphate group.

see Section 6.3.2) are, however, lethal pointing to a role for the enzyme in the repair of AP sites which accumulate in *dut⁻* cells as a result of removal by uracil *N*-glycosylase of the excessive number of uracils from DNA.

Similar enzymes are found in *Diplococcus pneumonia* [158] and *B. subtilis* [159].

### (d) *E. coli exonuclease IV*.

This exonuclease shows little activity towards single- or double-stranded DNA, and exhibits a considerable preference (twentyfold) for DNA predigested with pancreatic deoxyribonuclease. In this sense it could be termed oligonucleotide diesterase. It can be separated by DEAE-cellulose chromatography into two fractions (IVA and IVB) [160]. A similar enzyme is induced following infection of *E. coli* with phage T4. This enzyme which liberates 5'-monophosphates from a 3'-terminus may be involved in the degradation of cellular DNA (see Fig. 4.7) [167, 168].

### (e) *E. coli exonuclease V (recBC nuclease)*

The role of this enzyme in repair and recombination is discussed in Section 7.4.1 and by Muskavitch and Linn [161] who also consider similar enzymes from other sources. The enzyme is made of two subunits which are the products of the *recB* and *recC* genes. It acts exonucleolytically on both double- and single-strand DNA to produce oligonucleotides in a reaction coupled with the hydrolysis of ATP (i.e. it is a DNA-dependent ATPase and an ATP-dependent DNase). Magnesium ions are required for nuclease action but in their absence, or in the presence of calcium ions, the enzyme unwinds duplex DNA. This unwinding action also requires ATP and is stimulated by single-stranded DNA-binding proteins (see Section 6.4.4). The enzyme also shows endonuclease activity on single-stranded DNA. It is the unwinding action to produce single-stranded DNA which is believed to be the important function of this enzyme *in vivo*.

### (f) *Exonuclease associated with E. coli DNA polymerase I*

The terms exonuclease II and VI have been used to define the 3' → 5' and the 5' → 3' DNA exonuclease activities which form part of the protein of *E. coli* polymerase I (see Section 6.4.2.b) [162].

(i) *3'→5' exonuclease activity* [163]. This activity resides in the same protein molecule that possesses DNA polymerase activity. Like exonuclease I it commences attack at the 3'-hydroxyl terminus of a polydeoxyribonucleotide chain, with the stepwise release of deoxyribonucleoside 5'-monophosphates, but unlike exonuclease I it also attacks dinucleotides. It will, for example, hydrolyse single-stranded DNA in preference to native duplex DNA [164] but has no effect on oligonucleotides bearing a 3'-phosphomonoester group or on RNA. Most bacterial and phage DNA polymerases so far investigated (including *E. coli* DNA polymerase II and III but with the possible exception of those from *B. subtilis*) have an associated 3' → 5' exonuclease [162, 164].

Evidence that the enzyme is in fact an exonuclease with these properties comes from several sources. Exhaustive digestion of $^{32}$P-labelled d(A-T) copolymer results in the conversion of 99 per cent of the $^{32}$P into an acid-soluble form which can be accounted for in terms of 5'-monophosphates. Partial digestion results in the release of a proportion of radioactivity which is the same as the proportion of monophosphates formed. When d(A-T) copolymer specifically labelled with $^{32}$P[dTMP] at the 3'-hydroxyl end is used as substrate, 90% of the $^{32}$P-labelled material is made acid-soluble when less than 1% of the unlabelled nucleotides from the interior of the chain have been released. This indicates attack from the 3'-hydroxyl end of the chain.

(ii) *5' → 3' exonuclease activity*. This activity is specific towards native DNA and will function in the presence or absence of a 5'-phosphate group on the substrate. The products are mostly mononucleotides, with 20–25% dinucleotides or longer [165]. The enzyme can act to remove pyrimidine dimers and other regions of damaged DNA and will also remove ribonucleotides from an RNA:DNA hybrid. These properties are important in excision repair (see Section 7.3.2) and in the removal of RNA primers from Okazaki pieces during DNA replication (see Section 6.3). This enzyme is made use of to radiolabel DNA in the process known as nick translation (see Section 6.4.2.b and Fig. A.8).

The DNA polymerase I molecule is susceptible to protease action such that it is specifically split into two fragments. One of these (76 000 mol. wt.) has DNA polymerase I and 3' → 5' exonuclease activity; the other (34 000 mol. wt.) shows 5' → 3' exonuclease activity but only if the cleavage is carried out in the presence of DNA [166]. *B. subtilis* DNA polymerase I contains no 5' → 3' exonuclease activity [159] but a separate enzyme exists with this function. A 5' → 3' exonuclease activity is absent from *E. coli* DNA polymerase II but a modified activity is present in *E. coli* DNA polymerase III. In this case the exonuclease will only act on single-stranded DNA or on duplex DNA with a 5' single-stranded tail to which the enzyme can initially attach.

(g) *Exonuclease VII (EC 3.1.11.6)*
Degrades single-stranded DNA from either end, initially releasing large oligonucleotides which are further degraded to smaller oligonucleotides. Despite this finding the enzyme acts exonucleolytically but cleaves only occasionally as it processes along the DNA strand [135, 155, 170–172]. Exonuclease VII is the 88 000-$M_r$ product of the *xse*A gene, mutants of which show increased frequency of recombination. As it can also act at nicks in duplex DNA and excise thymine dimers it may partly substitute for other enzymes involved in excision repair (see Section 7.3.2). It does not require ATP.

(h) *Exonuclease VIII*
The gene for this enzyme (*recF*) is closely linked to that for exonuclease I (*sbcB*) but it is only expressed in *sbcA⁻* cells when exonuclease VIII can substitute for exonuclease I (see Section 7.4.1). It can also substitute for exonuclease V in *recBC⁻* cells. It is a dimer of 140 000-$M_r$ subunits which degrades linear duplex DNA from the 5'-ends, whether phosphorylated or not, to yield 5'-mononucleotides and residual single-stranded DNA [173, 174], i.e. it is similar but of opposite polarity to exonuclease III. A similar enzyme is found in bacterial cells infected with phage λ and exonuclease VIII can substitute for λ-exonuclease in lambda recombination [137] (see Section 7.4.1).

### 4.5.3 Restriction endonucleases

When a bacteriophage is transferred from growth on one host to a different host, its efficiency is frequently impaired several thousandfold. However, when those phage which do survive and multiply are used to reinfect the second host they now grow normally.

The initial poor growth is caused by the action on the phage DNA of a highly sequence-specific bacterial endonuclease (known as a restriction endonuclease). Following restriction, the invading phage DNA is rapidly degraded to nucleotides by exonuclease action, although certain regions may be rescued by recombination [175].

Host (i.e. bacterial) DNA is not degraded because the nucleotide sequence which is recognized by the restriction endonucleases has been modified by methylation (see Section 4.6.1). The few molecules of phage DNA which survive the initial infection do so because they are themselves modified by the host methylase before the restriction enzyme has time to act. Similarly, methylation of progeny DNA renders it resistant in the second infection [176].

A related phenomenon is the degradation of

cytosine-containing DNA in *E. coli* infected with T-even bacteriophage whose own DNA contains hydroxymethylcytosine (see Section 6.9.2.c).

The *E. coli* chromosome encodes a type I restriction endonuclease known as EcoB (in *E. coli* strain B) or EcoK (in *E. coli* K12). Other restriction enzymes are coded for by plasmids and these are mostly type II enzymes, but a few phage or plasmid-coded type III enzymes are known which are intermediate in properties between type I and type II enzymes [177–183]. Type I and type III enzymes catalyse both endonucleolytic cleavage and methylation of DNA, but these two functions are carried out by separate type II enzymes.

(a) *Type I*

These are complex multifunctional proteins which cleave unmodified DNA in the presence of *S*-adenosyl-L-methionine (AdoMet), ADP and $Mg^{2+}$. They are multisubunit enzymes, which are also DNA methylases (see Section 4.6.1) and ATPases. The subunits are present in different proportions in the EcoB and EcoK enzymes [184–186]. They are coded for by three contiguous genes known as *hsdR* (host specificity for DNA restriction), *hsdM* (modification) and *hsdS* (specificity). The three subunits are respectively responsible for the nuclease action, the methylation and the sequence specificity of the enzyme. Before interacting with DNA the *E. coli* K restriction enzyme binds to AdoMet [187, 188]. Binding to AdoMet takes place rapidly and is followed by a slower allosteric modification of the enzyme to an activated form which then interacts with DNA at a non-specific site. The enzyme moves along the DNA to the recognition site and its subsequent reaction depends on the state of this recognition site [204].

The recognition sites for the EcoK and EcoB type I restriction enzymes are shown in Fig. 4.10. The adenines with an asterisk in the EcoB sequence are methylated by the EcoB enzyme and the corresponding adenines in the EcoK sequences are probably the site of methylation by the EcoK restriction methylase [189–191].

The recognition sequences both consist of a group of three bases and a group of four bases separated by six (EcoK) or eight (EcoB) unspecified bases. Four of the seven specified bases are conserved and the methylated adenines in both cases are separated by eight bases.

If the recognition site is methylated in both strands of the DNA the enzyme does not recognize it. If the site is methylated on one strand only (as would be the case with DNA immediately following synthesis – see Section 4.6.1), the enzyme binds to this site and methylates the second strand in a reaction stimulated by ATP [192]. Methylation can also occur at completely unmethylated sites but only at a rate about 0.2% of that at half-methylated sites. Normally, if the site is unmodified in both strands, the enzyme is triggered into its nuclease (restriction) mode, but restriction does not occur at the binding site.

In the presence of ATP the DNA is made to loop past the enzyme which remains bound to the recognition site. This translocation leads to formation of supercoiled loops which become relaxed on subsequent cleavage [185, 193–195]. Cleavage of the looped-out DNA occurs at non-random sites and in two stages (first one strand, followed some time later by a break some distance away on the opposite strand). At this point the enzyme ceases to be a nuclease (i.e. it performs only one restriction event) and becomes a vigorous ATPase. About $10^5$ ATP molecules are hydrolysed for each restriction event [175, 205].

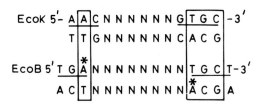

**Fig. 4.10** The recognition sequences for the EcoK and EcoB restriction modification enzyme [191].

(b) *Type II*

These (e.g. *Eco*RI, *Hap*II) are simpler enzymes which require only $Mg^{2+}$ for activity. Most have molecular weights in the range from 30 000 to 40 000 and have two identical subunits. These enzymes, which are neither methylases nor ATPases, appear to translocate along the DNA until they recognize a specific site and, if this is unmodified, cleavage occurs at this site [203]. DNA modified on one strand is not a substrate for restriction, but is a substrate for a complementary methylase which recognizes the same specific nucleotide sequence (see Section 4.6.1). The vast majority of sites are four to six nucleotide pairs long and have twofold rotational symmetry (see Table 4.5) which suggests that the two enzyme subunits may be arranged with twofold symmetry [176, 196]. Cleavage at the site may be staggered by up to five nucleotides (e.g. *Eco*RI) when identical self-complementary, cohesive termini are produced. *Bsu*RI, *Sma*I and *Hind*II produce even breaks without single-stranded termini (see

Table 4.5). The recognition site for the *Hin*fI restriction enzyme has one unspecified base [197] and frequently bases are only partially specified, i.e. they must be a purine or a pyrimidine. Enzymes are known which do not cleave within the recognition sequence; thus enzyme *Hph*I cleaves DNA eight bases before the sequence TCACC [198]. In these cases neither the recognition nor the cleavage sequence is symmetrical. With *Eco*RII the cleavage site is a sequence of five nucleotide pairs and hence not all the ends are mutually cohesive. In this case the complementary methylase modifies two cytosines in different sequences.

At least one restriction endonuclease (*Dpn*I) is specific for methylated DNA but its function is unclear as *Diplococcus pneumoniae* also contains an enzyme of opposite specificity (*Dpn*II) [199]. Hundreds of type II restriction enzymes are now known covering a very broad range of sequence specificities and comprehensive lists can be found in several reviews [200–202].

**Table 4.5** Class II DNA restriction and modification enzymes.

| Enzyme | Restriction and modification site | Bacterial strain |
|---|---|---|
| *Eco*RI | G↓A\*ATTC | *E. coli* RY13 |
| *Eco*RII | ↓C\*C$_T^A$GG | *E. coli* R245 |
| *Ava*I | C↓Y$^0$CGRG | *Anabaena variabilis* |
| *Bam*HI | G↓GATCC | *Bacillus amyloliquifaciens* H |
| *Bgl*II | A↓GATCT | *Bacillus globiggi* |
| *Bsu*RI | GG↓\*CC | *Bacillus subtilis* R |
| *Hha*I | G\*CG↓C | *Haemophilus haemolyticus* |
| *Hae*III | GG↓\*CC | *Haemophilus aegyptius* |
| *Hind*II | GT$_C^T$↓$_G^A$\*AC | *Haemophilus influenzae* Rd |
| *Hind*III | \*A↓AGCTT | *Haemophilus influenzae* Rd |
| *Hin*fI | G↓ANTC | *Haemophilus influenzae* Rf |
| *Hpa*II | C↓\*CGG | *Haemophilus parainfluenzae* |
| *Hga*I | GACGCNNNNN↓ | *Haemophilus gallinarum* |
| *Sma*I | CC\*C↓GGG | *Serratia marcescens* Sb |
| *Dpn*I | G\*A↓TC | *Diplococcus pneumoniae* |

↓shows the site of cleavage and \* shows the site of methylation of the corresponding methylase where known. $^0$shows the site of action of a presumed methylase, i.e. methylation at this site blocks restriction but the methylase has not yet been isolated. More complete lists are found in references [200–202].

A restriction enzyme is found in crown gall tumours [206] and an enzyme which cuts a monkey satellite DNA into discrete size fragments has been isolated from the testes of the African Green monkey, but neither this nor a similar site-specific endonuclease from yeast is a restriction endonuclease as defined by the properties of the type II enzymes herein described [207–209].

### (c) *Type III*

These (e.g. *Eco*P1, *Eco*P15 and *Hinf*III) have two different subunits; a nuclease and a combined methylase and specificity factor. They also differ from type I enzymes in not catalysing extensive ATP hydrolysis following nuclease action [182, 183, 216, 217]. The holoenzyme is essential for restriction and the two activities (endonuclease cleavage and methylation) compete with one another. The recognition sites are non-symmetrical and only one strand of the DNA is modified. The endonuclease makes staggered cuts 20 to 30 bases from the 3'-end of the recognition site.

### (d) *Nomenclature*

The restriction endonucleases are named according to the system described by Smith and Nathans [210]. The first three letters give the name of the bacterium (e.g. *Eco* for *E. coli*) and the fourth indicates the strain (e.g. *Eco*R and *Hin*d for *E. coli* strain R and *Haemophilus influenzae* strain d respectively). Where more than one restriction enzyme is found in a particular strain these are indicated by Roman numerals (e.g. *Hin*aI, *Hin*aII).

Enzymes recognizing identical nucleotide sequences are called isoschizomers. However, isoschizomers do not necessarily cleave in the same position and may respond to methyl groups on different bases in the recognition sequence.

### (e) *Applications*

Since their discovery the class II restriction enzymes have become increasingly used as tools for the biochemist. Because of their ability to make relatively few specific cuts they can be used as a first step in the sequencing of DNA (see Appendix) or in the isolation of specific genes. The use of a series of restriction enzymes allows a physical map to be made of genes or small chromosomes. Thus the small chromosome of the tumour virus SV40 (simian virus 40) is cleaved by each restriction enzyme into a small number of discrete pieces which are readily separated by electrophoresis in gels of polyacrylamide or agarose [179, 211, 212]. The size of the fragments can be determined from their speed of migration in the gel or by direct length measurements using electron microscopy (see Appendix). The order of the fragments in the genome can be determined by a variety of techniques such as analysis of partial digests or successive cleavage by multiple restriction endonucleases [179, 213].

As cleavage by many of these restriction enzymes is staggered, this leads to production of fragments with 'sticky' ends (i.e. termini with overlapping self-complementary sequences) which can be rejoined by DNA ligase (see Section 6.4.3). Moreover, fragments from different genomes can be joined together to form hybrid genomes, and this is the basis for genetic engineering considered in detail in the Appendix.

The use of certain pairs of restriction enzymes has enabled the extent of methylation of sites in individual eukaryotic genes to be investigated. Thus *Hpa*II will only cleave the sequence CCGG if it lacks a methyl group on the internal cytosines of both strands of the duplex DNA whereas *Msp*I will cleave the same sequence irrespective of whether the internal cytosine is methylated [214, 215].

## 4.6 NUCLEIC ACID METHYLATION

### 4.6.1 DNA methylation

Methylated bases occurring in DNA fall into two classes. The first class contains thymine and hydroxymethylcytosine (in T4 DNA) which are incorporated into DNA from dTTP or HMdCTP by DNA polymerase. The second class contains bases arising from methylation of a preformed polydeoxynucleotide, i.e. methylation occurs after DNA synthesis. This second class of methylated bases contains only 5-methylcytosine, first discovered by Hotchkiss in calf thymus DNA [218] and 6-methyladenine and $^4N$-methylcytosine which are largely restricted to the DNA from lower organisms.

The extent of DNA methylation is very variable, ranging from almost nothing in some insect and yeast DNA and in chloroplast DNA, through the bacteria (which may methylate one adenine and/or cytosine in two hundred) and animals (with 3–5% of cytosines methylated) to the higher plants where the nuclear DNA can have a third of the cytosines present as 5-methylcytosine [219–222]. Usually animal virus DNA is not methylated but exceptions are known [223, 224]. Within the genome of eukaryotes there is an uneven distribution of methylcytosine. This is seen to an extreme degree in the sea-urchin whose DNA can be separated into methylated and unmethylated compartments on the basis of its sensitivity to the methyl-sensitive restriction enzyme *Hpa*II [225].

Methylation of DNA takes place at the poly-nucleotide level [226], and the reactions are catalysed by specific enzymes, the DNA methylases (methyltransferases). The source of methyl groups is methionine, in the form of an intermediate with a high free energy of hydrolysis, *S*-adenosyl-L-methionine [221]. The mechanism is illustrated for cytosine methylation in Fig. 4.11.

A number of DNA methylases have been purified from prokaryotes. These enzymes are either part of the type I or type III restriction enzymes or complement the type II restriction enzymes by recognizing and methylating a common sequence and thus protecting it from the action of the cognate nuclease (see Section 4.5.3). In addition, *E. coli* has two other methylases (the products of the *dam* and *dcm* genes) which respectively methylate adenines in the sequence GATC and cytosines in the sequence $CC^A_TGG$. It is not certain what are the functions of such methylations though it is known they play a role in mismatch repair (Section 7.3.3), recombination (Section 7.4) and the initiation of DNA replication (Sections 6.8 and 6.9).

In eukaryotes the DNA sequence recognized and methylated by the methyltransferases is

5′–CG–
–GC–

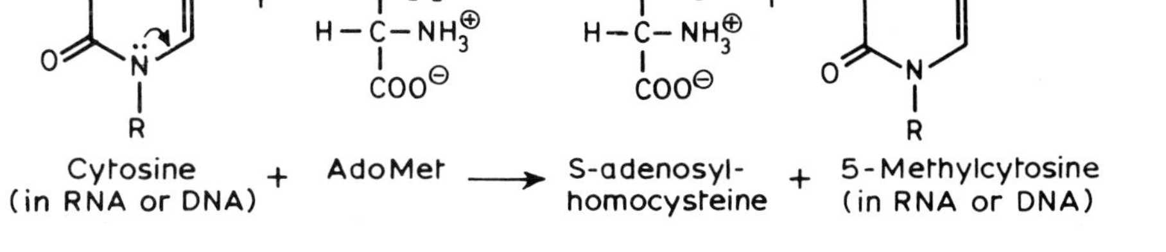

S-adenosylmethionine (AdoMet)
+
base (in DNA or RNA)

Methylase →

S-adenosylhomocysteine
+
methylated base
(in DNA or RNA)

Cytosine (in RNA or DNA) + AdoMet → S-adenosyl-homocysteine + 5-Methylcytosine (in RNA or DNA)

**Fig. 4.11** The nucleic acid methylase reaction (A represents the adenosyl radical).

in animals and

$$5'–CNG–$$
$$–GNC–$$

in plants. As with the prokaryotes these recognition sequences are symmetrical and are methylated on both strands. However, in eukaryotes not all the recognition sites are modified and the pattern of methylation is tissue-specific [219–222]. The recognition of the increased (epi)genetic information with which eukaryotic DNA is endowed by the presence of a fifth base has led to much speculation as to the function of DNA methylation in eukaryotes. A role in the control of gene expression is considered in Section 10.4.4.a, but it is pertinent to point out here that a base change from, say, cytosine to methylcytosine will have a marked effect on DNA–protein interactions as is ex-

emplified by the restriction endonucleases (Section 4.5.3).

As in prokaryotes, methylation of DNA in eukaryotes may have more than one function and the finding that marked changes in the pattern of methylation occur during gametogenesis and in the very early vertebrate embryo indicate that some methylation may be important in determining the potential of a gene for future activity [221, 227, 232]. Such speculation arises from the finding that in somatic cells the DNA methyltransferase serves only to maintain the pattern of methylation in newly synthesized DNA (Fig. 4.12). Thus a pattern of methylation established in, say, the presumptive liver cell DNA in the early embryo can be maintained through the action of the maintenance methylase alone and can be present to affect gene expression in the adult liver.

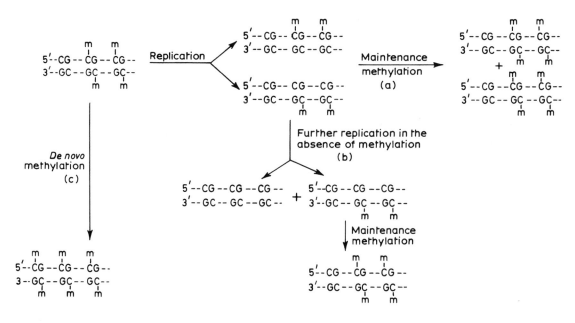

**Fig. 4.12**   The mechanism for the maintenance of the pattern of methylation and ways in which the pattern can be changed. The top line shows how the preferential action on hemimethylated DNA of the vertebrate DNA methyltransferase reproduces, on newly replicated DNA, the pattern of methylation found in parental DNA (a). If, during two rounds of replication, methylation is blocked in some region of DNA (by bound proteins?), that region of DNA will thereafter lack methylcytosine in at least half of the progeny cells (b). *de novo* methylation of unmethylated paired CG dinucleotides takes place to a marked extent only in early embryos (c).

The pattern of methylation can be changed in two ways (Fig. 4.12). Methyl groups can be lost from DNA cytosines by replication in the absence of methylation. Such a block in methylation may arise when the recognition site of the methyltransferase is covered by protein. Methyl groups can be gained by *de novo* methylation of unmethylated -CG- sequences, a process thought to occur to a significant extent only in the early embryo. Very little *de novo* methylation occurs on DNA transfected into animal cells but viral DNA can integrate into cellular DNA to bring about transformation and such a process is accompanied by the inactivation and concomitant *de novo* methylation of much of the viral genome [233, 234]. The packaging of foreign DNA into inactive chromatin and its subsequent methylation may be an effective defence against viral infection. Such an association between DNA methylation and gene and chromosome inactivation was postulated by Sager and Kitchin [235] as a prelude to the permanent loss of DNA in lower eukaryotes and is consistent with the models for the permanent inactivation of one of the X-chromosomes in female mammals [236–238].

Such inactivation can be partly reversed by treatment with 5-azacytidine, an analogue of cytidine which, when incorporated into DNA, leads to inactivation of cellular DNA methyltransferase, an inhibition of methylation and a transient overall demethylation of the genome [238–240].

5-Azacytidine has also been used to bring about phenotypic changes in cultured cells, and in an attempt to induce expression of the foetal globin gene in patients suffering from thalassaemia (a disease caused by a defect in an adult globin gene) [241–242].

All these observations add weight to the theory that changes in methylation are associated with changes in gene expression in eukaryotes.

In *E. coli* certain spontaneous base substitutions occur as a result of deamination of 5-methylcytosine to thymine [266]. Although this reaction was postulated many years ago as a mechanism underlying eukaryotic cell differentiation [267] there is little evidence for such a function, though a role in cell ageing remains possible. Nevertheless, such a deamination occurring over many, many generations is believed to have resulted in the depletion of the -CG- dinucleotide from vertebrate DNA in which it is found at only about one-fifth the expected frequency [228–231].

### 4.6.2 RNA methylation and other RNA nucleotide modifications

Methylation and numerous other forms of nucleotide modification are known for all four of the bases of RNA. They occur most commonly, and in by far the greatest diversity, on tRNA as a result of post-transcriptional modification and are considered under the processing of tRNA in Section 9.5.2.d. Ribosomal RNA is also methylated and contains some pseudouridine (Section 9.4.3.c); mRNA and its precursors are methylated particularly in their 5′ cap structures (Section 9.3.2) and most of the U-series snRNAs also have methylated nucleotides at their 5′-ends (Section 9.3.1). In most cases, the enzymes involved in these modifications are ill-characterized but they have recently been reviewed [243, 244] and some of the tRNA modifying activities that have been examined are tabulated in Section 9.5.2.d (Table 9.2).

### 4.7 NUCLEIC ACID KINASES AND PHOSPHATASES

### 4.7.1 Bacteriophage polynucleotide kinase

Polynucleotide kinases produced by T2-, T4- and T6-infected bacteria are enzymes with a broad specificity which transfer the γ-phosphate of a nucleoside triphosphate to the 5′-OH of DNA, RNA, oligonucleotides and nucleoside 5′-phosphates [for reviews see 245, 246]. The enzyme

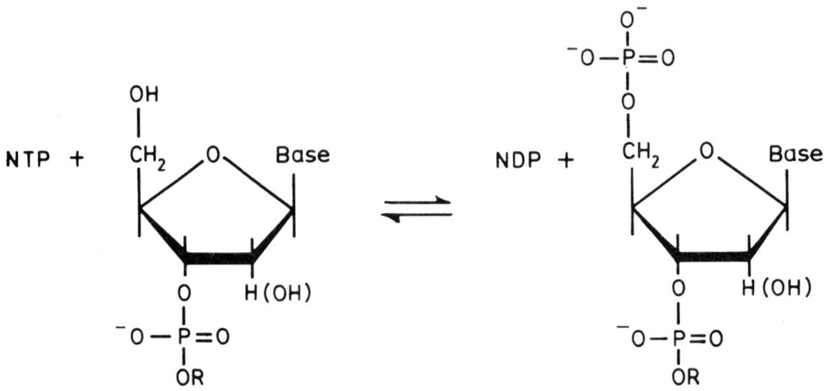

**Fig. 4.13** The phosphorylation reaction of polynucleotide kinase. NTP can be a variety of nucleoside 5'-triphosphates. R can be H, nucleoside, oligonucleotide or polynucleotide.

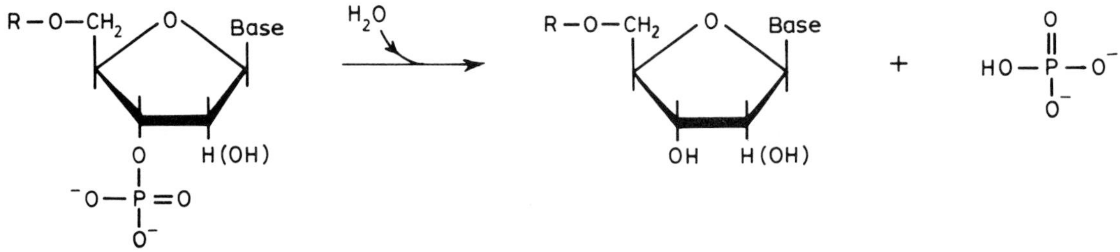

**Fig. 4.14** The phosphatase reaction of polynucleotide kinase. R can be H, a nucleotide or a polynucleotide.

has been purified from T2-infected [247] and T4-infected cells and that of T4 in particular has been extensively characterized [248] and employed in a wide variety of research applications. The phosphorylation reaction (Fig. 4.13) requires magnesium ions and the presence of compounds with reduced -SH groups. Phosphoryl transfer does not involve an enzyme phosphoryl intermediate [249] and the reaction is reversible [250]. Dephosphorylation is not however as efficient as phosphorylation and has a pH optimum of 6.2 compared with 7.6 for phosphorylation. The enzyme also exhibits a totally separate 3'-phosphatase activity which catalyses the hydrolysis of 3'-phosphoryl groups of nucleic acids (Fig. 4.14).

Richardson [246] has reviewed the evidence

that this phosphatase activity is probably identical to the previously purified 3'-phosphatase isolated from T4-infected bacteria despite the fact that early evidence indicated that this latter enzyme was specific for DNA [251]. The phosphatase and kinase activities of T4 polynucleotide kinase appear to involve separate active sites [252]. However there is no evidence for a concerted reaction involving for instance the transfer of phosphate from the 3' to the 5' side of a nick so that it could then be a substrate for DNA ligase [253]. (DNA ligase is considered in Section 6.4.3.) Indeed the role of polynucleotide kinase in bacteriophage infection is not understood.

Polynucleotide kinase has multiple research uses. The reversibility of the phosphorylation reaction in the presence of ADP allows its use to catalyse the exchange of $^{32}$P between the $\gamma$-phosphate of ATP and the 5'-phosphate group of a polynucleotide [250]. Such end-labelled polynucleotides [254], and those end-labelled by the forward reaction after the removal of the 5'-phosphate by alkaline phosphatase [255], are used in the analysis of nucleotide 'finger prints', end group analysis, the mapping of restriction endonuclease-generated fragments and most of all in the sequencing of RNA and DNA (reviewed in [245, 246] see also Sections A.4.2 and A.5.1).

### 4.7.2 Eukaryotic DNA and RNA kinases

DNA kinase activities have been purified from the nuclei of rat liver [256] and bovine thymus [257]. They differ from polynucleotide kinase principally in their inactivity with RNA as a substrate and their lower pH optimum of 5.5 (for review see [258]). Again their function is unknown but they are thought to play a role in DNA repair and are used experimentally to introduce labelled 5'-phosphate groups at single-strand breaks in duplex DNA [259]. The resultant labelled DNA can be used to assay DNA ligase activity [259].

An RNA kinase activity which transfers phosphate to the 5'-OH of RNA much more efficiently than DNA has been isolated from HeLa cell nuclei [260].

### 4.8 BASE EXCHANGE IN RNA AND DNA

The generation of two modified purines, namely queuosine and inosine, which occur in the anticodon wobble position of a number of tRNAs (Section 11.2), occurs by direct base replacement. An enzyme known as tRNA–guanine ribosyl transferase or tRNA guanine transglycosylase (EC 2.4.2.29) catalyses the exchange of guanine in the wobble position of the relevant pre-tRNAs for the highly modified base, queuosine in the mature tRNA [261]. The reaction involves breaking and re-formation of a C–N glycoside bond rather than a 3'–5' phosphodiester bond and it requires no energy (Fig. 4.15). For reviews of queuosine synthesis and insertion into tRNA see [262, 263].

Recently, evidence has been presented that inosine biosynthesis in the wobble position of the anticodon occurs by a similar mechanism in which the base at position 34 of the relevant tRNA precursors (probably adenine) is replaced

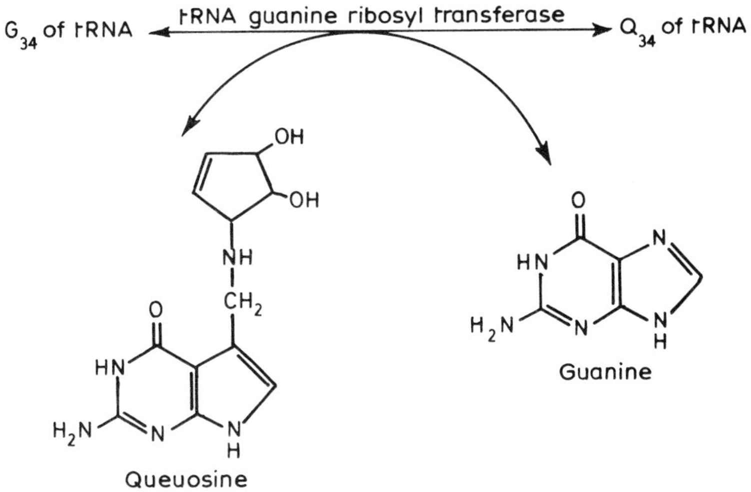

**Fig. 4.15** Insertion of queuosine into the wobble position of tRNA.

by hypoxanthine (inosine is the nucleoside of hypoxanthine) [264]. The hypoxanthine is first derived from adenosine which is converted to inosine by adenosine deaminase and thence to hypoxanthine by purine nucleoside phosphorylase.

Nucleoside deoxyribosyl transferases, which catalyse the exchange of bases in DNA, occur in bacteria [265].

## REFERENCES

1 Laskowski, M., Sr (1967), *Adv. Enzymol.*, **29**, 165.
2 Barnard, E. A. (1965), *Annu. Rev. Biochem.*, **38**, 677.
3 Grossman, L. and Moldave, K. (eds) (1980), *Methods Enzymol.* **65**.
4 Laskowski, M. (1971), *The Enzymes* (3rd edn) (ed. P. D. Boyer), Academic Press, New York, Vol. 4, p. 289.
5 Dingwall, C., Lomonossoff, G. P. and Laskey, R. A. (1981), *Nucleic Acids Res.*, **9**, 2659.
6 Horz, W. and Altenberger, W. (1981), *Nucleic Acids Res.*, **9**, 2643.
7 Cotton, F. A. and Hazen, E. E. (1971), *The Enzymes* (3rd edn) (ed. P. D. Boyer), Academic Press, New York, Vol. 4, p. 153.
8 Anfinsen, C. B., Cuatrecasas, P. and Taniuchi, H. (1971), *The Enzymes* (3rd edn) (ed. P. D. Boyer), Academic Press, New York, Vol. 4, p. 177.
9 Linn, S. and Lehman, I. R. (1965), *J. Biol. Chem.*, **240**, 1287, 1294.
10 Shishido, K. and Ando, T. (1982), in *Nucleases Cold Spring Harbor Monograph*, **14**, p. 155.
11 Shishido, K. and Ando, T. (1972), *Biochim. Biophys. Acta*, **287**, 477.
12 Fujimoto, M., Kuninaka, A. and Yoshina, H. (1974), *Agric. Biol. Chem.*, **38**, 785.
13 Sung, S-C and Laskowski, M. (1962), *J. Biol. Chem.*, **237**, 506.
14 Gray, H. B., Winston, T. P., Hodnett, J. L., Legevski, R. J., Nees, D. W., Wel, C-F. and Robberson, D. L. (1981), in *Gene Amplification and Analysis*, Elsevier/North-Holland, New York, Vol. 2, p. 169.
15 Laskowski, M. (1971), in *The Enzymes* (3rd edn) (ed. P. D. Boyer), Academic Press, New York, Vol. 4, p. 313.
16 Bernardi, R. and Bernardi, G. (1971), in *The Enzymes* (3rd edn) (ed. P. D. Boyer), Academic Press, New York, Vol. 4, p. 319.
17 Wei, C-F., Alianell, G. A., Beneen, G. H. and Gray, H. B. (1983), *J. Biol. Chem.*, **258**, 13 506.
18 Richards, F. M. and Wyckoff, H. W. (1971), in *The Enzymes* (3rd edn) (ed. P. D. Boyer), Academic Press, New York, Vol. 4, p. 647.
19 Blackburn, P. and Moore, S. (1982), in *The Enzymes* (3rd edn) (ed. P. D. Boyer), Academic Press, New York, Vol. 15, p. 317.
20 Jones, W. (1920), *Amer. J. Physiol.*, **52**, 203.
21 Enzyme Nomenclature (1984), *Recommendations of the International Union of Biochemistry*, Academic Press, London.
22 Dubos, R. J. and Thompson, R. H. S. (1938), *J. Biol. Chem.*, **124**, 501.
23 Kunitz, M. (1940), *J. Gen. Physiol.*, **24**, 15.
24 McDonald, M. R. (1955), *Methods Enzymol.*, **2**, 427.
25 Smyth, D. G., Stein, W. H. and Moore, S. (1963), *J. Biol. Chem.*, **238**, 227.
26 Beintema, J-J. and Lenstra, J. A. (1982), in *Macromolecular Sequences in Systematics and Evolutionary Biology* (ed. M. Goodman), Plenum Press, New York.
27 Kartha, G., Bello, J. and Harker, D. (1967), *Nature (London)*, **213**, 862.
28 Wyckoff, H. W., Hardman, K. D., Allewell, N. M., Inagami, T., Johnson, L. N. and Richards, F. M. (1967), *J. Biol. Chem.*, **242**, 3984.
29 Carlisle, C. H., Palmer, R. A., Mazumdar, S. K., Gorinsky, B. A. and Yeates, D. G. R. (1974), *J. Mol. Biol.*, **85**, 1.
30 Timchenko, A. A., Ptitsyn, O. B., Dolgikh, D. A. and Fedorov, B. A. (1978), *FEBS Lett.*, **88**, 105.
31 Gutte, B. and Merrifield, R. B. (1969), *J. Am. Chem. Soc.*, **91**, 501.
32 Hirschmann, R., Nutt, R. F., Veber, D. F., Vitali, R. A., Varga, S. L., Jacob, T. A., Holly, F. W. and Denkewalter, R. G. (1969), *J. Am. Chem. Soc.*, **91**, 507.
33 Bernd, G. and Merrifield, R. B. (1971), *J. Biol. Chem.*, **246**, 1922.
34 Rubsamen, H., Khandker, R. and Witzel, H. (1974), *Hoppe Seyler's Z. Physiol. Chem.*, **355**, 687.
35 Walker, E. J., Ralston, G. B. and Darvey, I. G. (1975), *Biochem. J.*, **147**, 425.
36 Walker, E. J., Ralston, G. B. and Darvey, I. G.

(1978), *Biochem. J.*, **173**, 1.

37 Usher, D. A., Richardson, D. I. and Eckstein, F. (1970), *Nature (London)*, **228**, 663.

38 Usher, D. A., Erenrich, E. S. and Eckstein, F. (1972), *Proc. Natl. Acad. Sci. USA*, **69**, 115.

39 Beers, R. F. (1960), *J. Biol. Chem.*, **235**, 2393.

40 Kunitz, M. (1946), *J. Biol. Chem.*, **164**, 563.

41 Anfinsen, C. B., Redfield, R. R., Choate, W. L., Page, J. and Carroll, W. R. (1954), *J. Biol. Chem.*, **207**, 201.

42 Crook, E. M., Mathias, A. P. and Rabin, B. R. (1960), *Biochem. J.*, **74**, 234.

43 Blank, A. and Dekker, C. A. (1981), *Biochemistry*, **20**, 2261.

44 Maor, D. and Mardiney, M. R. (1979). *CRC Crit. Rev. Clin. Lab. Sci.*, **10**, 89.

45 Aoki, Y., Mizuno, D. and Goto, S. (1981), *J. Biochem. (Tokyo)*, **90**, 737.

46 Little, B. W. and Willingham, L. S. (1981), *Biochim. Biophys. Acta*, **655**, 251.

47 Button, E. E., Guggenheimer, R. and Kull, F. J. (1982), *Arch. Biochem. Biophys.*, **219**, 249.

48 Kumagai, H., Kato, H., Igarashi, K. and Hirose, S. (1983), *J. Biochem. (Tokyo)*, **94**, 71.

49 Niwata, Y., Ohgi, K., Sanda, A., Takizawa, Y. and Irie, M. (1985), *J. Biochem. (Tokyo)*, **97**, 923.

50 Roth, J. S. (1967), *Methods Cancer Res.*, **3**, 151.

51 Blackburn, P. (1979), *J. Biol. Chem.*, **254**, 12 484.

52 McGregor, C. W., Adams, A. and Knowler, J. T. (1981), *J. Steroid Biochem.*, **14**, 415.

53 Barton, A., Sierakowska, H. and Shugar, D. (1976), *Clin. Chim. Acta*, **67**, 231.

54 Takahashi, K. and Moore, S. (1982), in *The Enzymes* (ed. P. D. Boyer), Academic Press, New York, Vol. 15, p. 435.

55 Uchida, T. and Egami, F. (1967), *J. Biochem. (Tokyo)*, **61**, 44.

56 Kenney, W. C. and Dekker, C. A. (1971), *Biochemistry*, **10**, 4962.

57 Kasai, K., Uchida, T., Egami, F., Yoshida, K. and Nomoto, M. (1969), *J. Biochem. (Tokyo)*, **66**, 389.

58 Shen, V. and Schlessinger, D. (1982), in *The Enzymes* (ed. P. D. Boyer), Academic Press, New York, Vol. 15, p. 501.

59 Neu, H. C. and Heppel, L. A. (1964), *J. Biol. Chem.*, **239**, 3893.

60 Gesteland, R. (1966), *J. Mol. Biol.*, **16**, 67.

61 Cohen, L. and Kaplan, R. (1977), *J. Bacteriol.*, **129**, 651.

62 Minato, S. and Hirai, A. (1979), *J. Biochem. (Tokyo)*, **85**, 327.

63 Rushizky, G. W., Mozejko, J. H., Rogerson, D. L. and Sober, H. A. (1970), *Biochemistry*, **9**, 4966.

64 Randerath, K., Gupta, R. C. and Randerath, E. (1980), *Methods Enzymol.* **65**, *Part* **1**, 638.

65 Sato, S. and Uchida, T. (1975), *Biochem. J.*, **145**, 353.

66 Yasuda, T. and Inoue, Y. (1982), *Biochemistry*, **21**, 364.

67 Clemens, M. J. and McNurlan, M. A. (1985), *Biochem. J.*, **226**, 345.

68 Baglioni, C., Minks, M. A. and Maroney, P. A. (1978), *Nature (London)*, **273**, 684.

69 Jacobsen, H., Krause, D., Friedman, R. M. and Silverman, R. H. (1983), *Proc. Natl. Acad. Sci. USA*, **80**, 4954.

70 Clemens, M. J. and Williams, B. R. G. (1978), *Cell*, **13**, 565.

71 Wrescher, D. H., James, T. C., Silverman, R. H. and Kerr, I. H. (1981), *Nucleic Acids Res.*, **9**, 1571.

72 RajBhandary, U. L., Lockard, R. E. and Wurst-Reilly, R. M. (1982), in *Nucleases* (ed. S. M. Linn and R. J. Roberts), Cold Spring Harbor Laboratory, New York, p. 275.

73 Tabor, M. W., Leake, B. H. and MacGee, J. (1976), *Fed. Proc. Abs.*, **35**, 298.

74 Lockard, R. E., Alzner-Deweerd, B., Heckman, J. E., MacGee, J., Tabor, M. W. and RajBhandary, U. L. (1978), *Nucleic Acids Res.*, **5**, 37.

75 Pilly, D., Niemeyer, A., Schmidt, M. and Bargetzi, P. (1978), *J. Biol. Chem.*, **253**, 437.

76 Donis-Keller, H. (1980), *Nucleic Acids Res.*, **8**, 3133.

77 Boguski, M. S., Hieter, P. A. and Levy, C. C. (1980), *J. Biol. Chem.*, **255**, 2160.

78 Forentz, C., Briand, J. P., Romby, P., Hirth, L., Ebel, J. P. and Giege, R. (1982), *EMBO J.*, **1**, 269.

79 Levy, C. C. and Karpetsky, T. P. (1980), *J. Biol. Chem.*, **255**, 2153.

80 Adams, J. M., Cory, S. and Spahr, P. F. (1972), *Eur. J. Biochem.*, **29**, 469.

81 Bellemore, G., Jordan, B. R. and Monier, R. (1972), *J. Mol. Biol.*, **71**, 307.

82 Thomas, K. R. and Olivera, B. M. (1978), *J. Biol. Chem.*, **253**, 424.

83 Gupta, R. S., Kasai, T. and Schlessinger, D. (1977), *J. Biol. Chem.*, **252**, 8945.

84 Singer, R. S. and Tolbert, G. (1965), *Biochemistry*, **4**, 1319.

85 Cudny, H. and Deutscher, M. P. (1980), *Proc. Natl. Acad. Sci. USA*, **77**, 837.

86 Donovan, W. P. and Kushner, S. R. (1983), *Nucleic Acids Res.*, **11**, 265.

87 Stein, H. and Hausen, P. (1970), *Cold Spring Harbor Symp. Quant. Biol.*, **35**, 709.

88 Crouch, R. J. and Dirksen, M.-L. (1982), in *Nucleases* (ed. S. M. Linn and R. J. Roberts), Cold Spring Harbor Laboratory, New York, p. 211.

89 Büsen, W. (1980), in *Biological Implications of Protein–Nucleic Acid Interactions* (ed. J. Augustynaik), Elsevier/North-Holland Biomedical Press, New York.

90 Büsen, W. (1982), *J. Biol. Chem.*, **257**, 7106.

91 Wyers, F., Huet, J., Sentenac, A. and Fromageot, P. (1976), *Eur. J. Biochem.*, **69**, 385.

92 Huet, J., Buhler, M., Sentenac, A. and Fromageot, P. (1977), *J. Biol. Chem.*, **252**, 8848.

93 Iborra, F., Huet, J., Breant, B., Sentenac, A. and Fromageot, P. (1979), *J. Biol. Chem.*, **254**, 10 920.

94 Keller, W. and Crouch, R. (1972), *Proc. Natl. Acad. Sci. USA*, **69**, 3360.

95 Leis, J. P., Burkower, I. and Hurwitz, J. (1973), *Proc. Natl. Acad. Sci. USA*, **70**, 466.

96 Gerard, G. F. (1981), *Biochemistry*, **20**, 256.

97 Ogawa, T., Pickett, G. C., Kogoma, T. and Kornberg, A. (1984), *Proc. Natl. Acad. Sci. USA*, **81**, 1040.

98 Kogoma, T. (1984), *Proc. Natl. Acad. Sci. USA*, **81**, 7845.

99 Lockard, R. and Kumar, A. (1981), *Nucleic Acids Res.*, **9**, 5125.

100 Grinnell, B. W. and Wagner, R. R. (1984), *Cell*, **36**, 533.

101 Auron, P. E., Weber, L. D. and Rich, A. (1982), *Biochemistry*, **21**, 4700.

102 White, M. D., Bauer, S. and Lapidot, Y. (1977), *Nucleic Acids Res.*, **4**, 3029.

103 White, M. D., Rapoport, S. and Lapidot, Y. (1977), *Biochem. Biophys. Res. Commun.*, **77**, 1084.

104 Pollard, D. R. and Nagyvary, J. (1973), *Biochemistry*, **12**, 1063.

105 Singer, B., Fraenkel-Conrat, H. and Tsugita, A. (1961), *Virology*, **14**, 54.

106 Stanley, W. M. and Bock, R. M. (1965), *Biochemistry*, **4**, 1302.

107 Mora, P. T. (1962), *J. Biol. Chem.*, **237**, 3210.

108 Cheng, T., Polmar, S. K. and Kazazian, H. H. (1974), *J. Biol. Chem.*, **249**, 1781.

109 McKnight, G-S. and Schimke, R. T. (1974), *Proc. Natl. Acad. Sci. USA*, **71**, 4327.

110 Chirgwin, J. M., Przybyla, A. E., MacDonald, R. J. and Rutter, W. J. (1979), *Biochemistry*, **18**, 5294.

111 Girard, M. (1967), *Methods Enzymol.*, **12a**, 581.

112 Berger, S. L. and Birkenmeier, C. S. (1979), *Biochemistry*, **81**, 5143.

113 Scheel, G. and Blackburn, P. (1979), *Proc. Natl. Acad. Sci. USA*, **76**, 4898.

114 de Martynoff, G., Pays, E. and Vassart, G. (1980), *Biochem. Biophys. Res. Commun.*, **93**, 645.

115 Solymosy, F., Fedorcsák, I., Gulyás, A., Farkas, G. L. and Ehrenberg, L. (1968), *Eur. J. Biochem.*, **5**, 520.

116 Wiener, S. L., Wiener, R., Urivetzky, M. and Meilman, E. (1972), *Biochim. Biophys. Acta*, **259**, 378.

117 Ehrenberg, L., Fedorcsák, I. and Solymosy, F. (1976), *Prog. Nucleic Acid Res. Mol. Biol.*, **16**, 189.

118 Grunberg-Manago, M., Ortiz, P. J. and Ochoa, S. (1956), *Biochim. Biophys. Acta*, **20**, 269.

119 Grunberg-Manago, M. (1963), *Prog. Nucleic Acid Res.*, **1**, 93.

120 Littauer, U. Z. and Soreq, H. (1982), in *The Enzymes* (ed. P. D. Boyer), Academic Press, New York, Vol. 15, p. 517.

121 Portier, C. (1975), *Eur. J. Biochem.*, **55**, 573.

122 Brenneman, F. N. and Singer, H. F. (1964), *J. Biol. Chem.*, **239**, 893.

123 Leder, P., Singer, M. F. and Brimacombe, R. L. C. (1965), *Biochemistry*, **4**, 1561.

124 Trip, E. M. and Smith, M. (1978), *Nucleic Acids Res.*, **5**, 1539.

125 Soreq, H., Nudel, U., Salomon, R., Revel, M. and Littauer, U. Z. (1974), *J. Mol. Biol.*, **88**, 233.

126 Laskowski, M. (1982), in *Nucleases* (ed. S. M. Linn and R. J. Roberts), Cold Spring Harbor Laboratory, New York, p. 1.

127 Bollum, F. J. (1965), *J. Biol. Chem.*, **240**, 2599.

128 Young, E. T. and Sinsheimer, R. L. (1965), *J. Biol. Chem.*, **240**, 1274.

129 Bernardi, G. (1971), in *The Enzymes* (3rd edn) (ed. P. D. Boyer), Academic Press, New York, Vol. 4, p. 271.

130 Bernardi, G. (1965), *J. Mol. Biol.*, **13**, 603.

131 Linn, S. (1982), in *Nucleases* (ed. S. M. Linn and R. J. Roberts), Cold Spring Harbor Laboratory, New York, p. 341.

132 Salnikow, J. and Murphy, D. (1973), *J. Biol. Chem.*, **248**, 1499.

133 Liao, T-H, (1974), *J. Biol. Chem.*, **249**, 2345.

134 Linn, S. (1982), in *Nucleases* (ed. S. M. Linn and R. J. Roberts), Cold Spring Harbor Laboratory, New York, p. 291.

135 Weiss, B. (1981), in *The Enzymes* (ed. P. D. Boyer), Academic Press, New York, Vol. 14, p. 203.

136 Burlingham, B. T., Doerfler, W., Pettersson, U. and Philipson, L. (1971), *J. Mol. Biol.*, **60**, 45.

137 Pogo, B. G. T. and Dales, S. (1969), *Proc. Natl. Acad. Sci. USA*, **63**, 820.

138 Lehman, I. R. (1971), in *The Enzymes* (3rd edn) (ed. P. D. Boyer), Academic Press, New York, Vol. 4, p. 251.

139 Radloff, R., Bauer, W. and Vinograd, J. (1967), *Proc. Natl. Acad. Sci. USA*, **57**, 1514.

140 Studier, F. W. (1965), *J. Mol. Biol.*, **11**, 373.

141 Friedberg, E. C. and Goldthwaite, D. A. (1969), *Proc. Natl. Acad. Sci. USA*, **62**, 934.

142 Yajko, D. M. and Weiss, B. (1975), *Proc. Natl. Acad. Sci. USA*, **72**, 688.

143 Ljungquist, S., Nyberg, B. and Lindahl, T. (1975), *FEBS Lett.*, **57**, 169.

144 Radman, M. (1976), *J. Biol. Chem.*, **251**, 1438.

145 Grossman, L., Riazuddin, S., Haseltine, W. and Lindan, K. (1978), *Cold Spring Harbor Symp. Quant. Biol.*, **43**, 947.

146 Demple, B. and Linn, S. (1980), *Nature (London)*, **287**, 203.

147 Mosbaugh, D. W. and Linn, S. (1980), *J. Biol. Chem.*, **255**, 11 743.

148 Lindahl, T. (1979), *Prog. Nucleic Acid Res.*, **22**, 135.

149 Demple, B. and Linn, S. (1982), *J. Biol. Chem.*, **257**, 2848.

150 Hurwitz, J., Becker, A., Gefter, M. and Gold, M. (1967), *J. Cell Comp. Physiol.*, *Suppl. 1*, **70**, 181.

151 Sadowski, P. D. and Hurwitz, J. (1969), *J. Biol. Chem.*, **244**, 6182, 6192.

152 Henry, J. J. and Knippers, R. (1974), *Proc. Natl. Acad. Sci. USA*, **71**, 1549.

153 Linn, S. (1981), in *The Enzymes* (ed. P. D. Boyer), Academic Press, New York, Vol. 14, p. 121.

154 Moore, S. (1981), in *The Enzymes* (ed. P. D. Boyer), Academic Press, New York, Vol. 14 p. 281.

155 Chase, J. W. and Richardson, C. C. (1974), *J. Biol. Chem.*, **249**, 4545.

156 Healy, J. W., Stollar, D. and Levine, L. (1963), *Methods Enzymol.*, **6**, 49.

157 Richardson, C. C., Lehman, I. R. and Kornberg, A. (1964), *J. Biol. Chem.*, **239**, 251.

158 Lacks, S. and Greenberg, B. (1967), *J. Biol. Chem.*, **242**, 3108.

159 Okazaki, T. and Kornberg, A. (1964), *J. Biol. Chem.*, **239**, 259.

160 Jorgensen, S. E. and Koerner, J. F. (1966), *J. Biol. Chem.*, **241**, 3090.

161 Muskavitch, T. K. M. and Linn, S. (1981), in *The Enzymes* (ed. P. D. Boyer), Academic Press, New York, Vol. 14, p. 233.

162 Kornberg, A. (1974), *DNA Synthesis*, Freeman, San Francisco.

163 Lehman, I. R. and Richardson, C. C. (1964), *J. Biol. Chem.*, **239**, 233.

164 Cozzarelli, N. R., Kelly, R. B. and Kornberg, A. (1969), *J. Mol. Biol.*, **45**, 513.

165 Kelly, R. B., Atkinson, M. R., Huberman, J. A. and Kornberg, A. (1969), *Nature (London)*, **224**, 495.

166 Klenow, H. and Overgaard-Hansen, K. (1970), *FEBS Lett.*, **6**, 25.

167 Oleson, A. E. and Koerner, J. F. (1964), *J. Biol. Chem.*, **239**, 2935.

168 Short, E. C. Jr and Koerner, J. F. (1969), *J. Biol. Chem.*, **244**, 1487.

169 McAuslan, B. R. (1971), in *Strategy of the Viral Genome*, Ciba Symposium Volume, Churchill-Livingstone, Edinburgh and London, p. 25.

170 Chase, J. W. and Richardson, C. C. (1974), *J. Biol. Chem.*, **249**, 4553.

171 Chase, J. W. and Richardson, C. C. (1977), *J. Bacteriol.*, **129**, 934.

172 Vales, L. D., Chase, J. W. and Richardson, C. C. (1979), *J. Bacteriol.*, **139**, 320.

173 Joseph, J. W. and Kolodner, R. (1983), *J. Biol. Chem.*, **258**, 10 411.

174 Joseph, J. W. and Kolodner, R. (1983), *J. Biol. Chem*, **258**, 10 418.

175 Arber, W. (1974), *Prog. Nucleic Acid Res. Mol. Biol.*, **14**, 1.

176 Murray, K. and Old, R. W. (1974), *Prog. Nucleic Acid Res. Mol. Biol.*, **14**, 117.

177 Boyer, H. (1971), *Annu. Rev. Microbiol.*, **25**, 153.

178 Meselson, M., Yuan, R. and Heywood, J. (1972), *Annu. Rev. Biochem.*, **41**, 447.

179 Nathans, D. and Smith, H. O. (1975), *Annu. Rev. Biochem.*, **44**, 273.

180 Meselson, M. and Yuan, R. (1968), *Nature (London)*, **217**, 1110.

181 Bickle, T. A. (1982) p. 85 and Modrich, P. and Roberts, R. J. (1982), p. 109 in *Nucleases* (ed. S. M. Linn and R. J. Roberts), Cold Spring Harbor Laboratory, New York.

182 Iida, S., Meyer, J., Bachi, B., Stahlhammer-Carlamalin, M., Schrikel, S., Bickle, T. A. and Arber, W. (1983), *J. Mol. Biol.*, **165**, 1.

183 Kauc, L. and Piekarowicz, A. (1978), *Eur. J. Biochem.*, **92**, 417.

184 Lautenberger, J. A. and Linn, S. (1972), *J. Biol. Chem.*, **247**, 6176.

185 Suri, B., Nagaraja, V. and Bickle, T. A. (1984), *Curr. Top. Microbiol. Immunol.*, **108**, 1.

186 Eskin, B. and Linn, S. (1972), *J. Biol. Chem.*, **247**, 6183.

187 Hadi, S. M., Bickle, T. A. and Yuan, R. (1975), *J. Biol. Chem.*, **250**, 4159.

188 Yuan, R., Bickle, T. A., Ebbers, W. and Brack, C. (1975), *Nature (London)*, **256**, 556.

189 Ravetch, J. V., Horiuchi, K. and Zinder, N. D. (1978), *Proc. Natl. Acad. Sci. USA*, **75**, 226.

190 Lautenberger, J. A., Kan, N. C., Lackey, D., Linn, S., Edgell, M. H. and Hutchison, C. A. (1978), *Proc. Natl. Acad. Sci. USA*, **75**, 2271.

191 Kan, N. C., Lautenberger, J. A., Edgell, M. H. and Hutchison, C. A. (1979), *J. Mol. Biol.*, **130**, 191.

192 Vovis, G. F., Horiuchi, K. and Zinder, N. D. (1974), *Proc. Natl. Acad. Sci. USA*, **71**, 3810.

193 Bickle, T. A., Brack, C. and Yuan, R. (1978), *Proc. Natl. Acad. Sci. USA*, **75**, 3099.

194 Yuan, R., Hamilton, D. L. and Burckhardt, J. (1980), *Cell*, **20**, 237.

195 Yuan, R. (1981), *Annu. Rev. Biochem.*, **50**, 285.

196 Frederick, C. A., Grable, J., Melia, M., Samudzi, C., Jen-Jacobsen, L., Wang B-C, Greene, P., Boyer, H. W. and Rosenberg, J. M. (1984), *Nature (London)*, **309**, 327.

197 Subramanian, K. M., Weissman, S. M., Zain, B. S. and Roberts, R. J. (1977), *J. Mol. Biol.*, **110**, 297.

198 Kleid, D., Humayun, Z., Jeffrey, A. and Ptashne, M. (1976), *Proc. Natl. Acad. Sci. USA*, **73**, 293.

199 Lacks, S. and Greenberg, B. (1975), *J. Biol. Chem.*, **250**, 4060.

200 Roberts, R. J. (1976), *CRC Crit. Rev. Biochem.*, **4**, 123.

201 Roberts, R. J. (1982), *Nucleic Acids Res.*, **10**, r117.

202 Roberts, R. J. (1984), *Nucleic Acids Res.*, **12** (*Suppl*), r167.

203 Ehrbrecht, H-J., Pingoud, A., Urbanke, C., Maass, G. and Gualerzi, C. (1985), *J. Biol. Chem.*, **260**, 6160.

204 Endlich, B. and Linn, S. (1985), *J. Biol. Chem.*, **260**, 5720.

205 Endlich, B. and Linn, S. (1985), *J. Biol. Chem.*, **260**, 5729.

206 Le Bon, J. M., Kado, C. I., Rosenthall, L. J. and Chirikjian, J. G. (1978), *Proc. Natl. Acad. Sci. USA*, **75**, 4097.

207 Brown, F. L., Musich, P. R. and Maio, J. J. (1978), *Nucleic Acids Res.*, **5**, 1093.

208 Watabe, H., Iino, T., Kaneko, T., Shibata, T. and Ando, T. (1984), *J. Biol. Chem.*, **259**, 10 499.

209 Shibata, T., Watabe, H., Kaneko, T., Iino, T. and Ando, T. (1984), *J. Biol. Chem.*, **259**, 10 499.

210 Smith, H. O. and Nathans, D. (1973), *J. Mol. Biol.*, **81**, 419.

211 Sharp, P. A., Sugden, B. and Sambrook, J. (1973), *Biochemistry*, **12**, 3055.

212 Sugisaki, H. and Takanami, M. (1973), *Nature (London) New Biol.*, **246**, 138.

213 Danna, K. J., Sack, G. H. Jr and Nathans, D. (1973), *J. Mol. Biol.*, **78**, 363.

214 Singer, J., Roberts-Ems, J. and Riggs, A. D. (1979), *Science*, **203**, 1019.

215 Waalwijk, C. and Flavell, R. A. (1978), *Nucleic Acids Res.*, **5**, 3231.

216 Hadi, S. M., Bachi, B., Iida, S. and Bickle, T. A. (1983), *J. Mol. Biol.*, **165**, 19.

217 Reiser, J. and Yuan, R. (1977), *J. Biol. Chem.*, **252**, 451.

218 Hotchkiss, R. D. (1948), *J. Biol. Chem.*, **175**, 315.

219 Burdon, R. H. and Adams, R. L. P. (1980), *Trends Biochem Sci.*, **5**, 294.

220 Adams, R. L. P. and Burdon, R. H. (1982), *CRC Crit. Rev. Biochem.*, **13**, 349.

221 Adams, R. L. P. and Burdon, R. H. (1985), *Molecular Biology of DNA Methylation*, Springer-Verlag, New York.

222 Razin, A., Cedar, H. and Riggs, A. D. (eds) (1984), *DNA Methylation*, Springer-Verlag, New York.

223 Willis, D. B. and Granoff, A. (1980), *Virology*, **107**, 250.

224 Diala, E. S. and Hoffman, R. M. (1983), *J. Virol.*, **45**, 482.

225 Bird, A. P., Taggart, M. H. and Smith, B. A.

(1979), *Cell*, **17**, 889.

226 Borek, E. and Srinivasan, P. R. (1966), *Annu. Rev. Biochem.*, **35**, 275.

227 Groudine, J. and Conkin, K. F. (1985), *Science*, **228**, 1061.

228 Salzer, W. (1977), *Cold Spring Harbor Symp. Quant. Biol.*, **42**, 985.

229 Bird, A. P. (1980), *Nucleic Acids Res.*, **5**, 1499.

230 Adams, R. L. P. and Eason, R. (1984), *Nucleic Acids Res.*, **12**, 5869.

231 Max, E. E. (1984), *Nature (London)*, **310**, 100.

232 Jaenisch, R. and Jahner, D. (1984), *Biochim. Biophys. Acta*, **782**, 1.

233 Pollack, Y., Stein, L., Razin, A. and Cedar, H. (1980), *Proc. Natl. Acad. Sci. USA*, **77**, 6463.

234 Doerfler, W. (1984), *Curr. Top. Microbiol. Immunol.*, **108**, 79.

235 Sager, R. and Kitchin, R. (1975), *Science*, **189**, 426.

236 Wolf, S. F., Jolly, D. J., Lunnen, K. D., Friedmann, T. and Migeon, B. R. (1984), *Proc. Natl. Acad. Sci. USA*, **81**, 2806.

237 Kratzer, P. G., Chapman, V. M., Lambert, H., Evans, R. E. and Liskay, R. M. (1983), *Cell*, **33**, 37.

238 Shapiro, L. J. and Mohandas, T. (1983), *Cold Spring Harbor Symp. Quant. Biol.*, **47**, 631.

239 Adams, R. L. P., Fulton, J. and Kirk, D. (1982), *Biochim. Biophys. Acta*, **697**, 286.

240 Tanaka, M., Hibasami, H., Nagai, J. and Ikeda, T. (1980), *Austral. J. Exp. Biol. Med. Sci.*, **58**, 391.

241 Taylor, S. M. and Jones, P. A. (1979), *Cell*, **17**, 771.

242 Ley, T. J., Chiang, Y. L., Hiadaris, D., Anagnou, N. P., Wilson, V. L. and Anderson, W. F. (1984), *Proc. Natl. Acad. Sci. USA*, **81**, 6618.

243 Sol, D. and Kline, L. K. (1982), in *The Enzymes* (ed. P. D. Boyer), Academic Press, New York, Vol. 15, p. 557.

244 Kline, L. and Sol, D. (1982), in *The Enzymes*, (ed. P. D. Boyer), Academic Press, New York, Vol. 15, p. 567.

245 Kleppe, K. and Lillehaug, J. R. (1979), *Adv. Enzymol. Relat. Areas Mol. Biol.*, **48**, 245.

246 Richardson, C. C. (1981), in *The Enzymes* (ed. P.

D. Boyer), Academic Press, New York, Vol. 14, p. 299.

247 Novogrodsky, A., Tal, M., Traub, A. and Hurwitz, J. (1966), *J. Biol. Chem.*, **241**, 2933.

248 Lillehaug, J. R. (1977), *Eur. J. Biochem.*, **73**, 499.

249 Jarvest, R. L. and Lowe, G. (1981), *Biochem. J.*, **199**, 273.

250 Van de Sande, J. H., Kleppe, K. and Khorana, H. G. (1973), *Biochemistry*, **12**, 5050.

251 Becker, A. and Hurwitz, J. (1967), *J. Biol. Chem.*, **242**, 936.

252 Soltis, D. A. and Uhlenbeck, O. C. (1982), *J. Biol. Chem.*, **257**, 11 340.

253 Soltis, D. A. and Uhlenbeck, D. C. (1982), *J. Biol. Chem.*, **257**, 11 332.

254 Berkner, K. L. and Folk, W. R. (1977), *J. Biol. Chem.*, **252**, 3176.

255 Maxam, A. M. and Gilbert, W. (1980), *Methods Enzymol.*, **65**, 499.

256 Levin, C. L. and Zimmerman, S. B. (1976), *J. Biol. Chem.*, **251**, 1767.

257 Tamura, S., Teraoka, H. and Tsukada, K. (1981), *Eur. J. Biochem.*, **115**, 449.

258 Zimmerman, S. B. and Pheiffer, B. H. (1981), in *The Enzymes* (ed. P. D. Boyer), Academic Press, New York, Vol. 14, p. 315.

259 Tsukada, K. and Ichimura, M. (1971), *Biochem. Biophys. Res. Commun.*, **42**, 1156.

260 Shuman, S. and Hurwitz, J. (1979), *J. Biol. Chem.*, **254**, 10 396.

261 Okada, N., Noguchi, S., Kasai, H., Shindo-Okada, N., Ohgi, T., Goto, T. and Nishimura, S. (1979), *J. Biol. Chem.*, **254**, 3067.

262 Nishimura, S. (1983), *Prog. Nucleic Acids Res. Mol. Biol.*, **28**, 49.

263 Singhal, R. P. (1983), *Prog. Nucleic Acids Res. Mol. Biol.*, **28**, 75.

264 Elliott, M. S. and Trewyn, R. W. (1984), *J. Biol. Chem.*, **259**, 2407.

265 McNutt, W. S. (1955), *Methods Enzymol.*, **2**, 464.

266 Coulondre, C., Miller, J. H., Farabrough, P. J. and Gilbert, W. (1978), *Nature (London)*, **274**, 775.

267 Scarano, E., Iaccarino, M., Grippo, P. and Parisi, E. (1967), *Proc. Natl. Acad. Sci. USA*, **57**, 1394.

268 Liao, T-H. (1985), *J. Biol. Chem.*, **260**, 10 708.

# 5

# The metabolism of
# nucleotides

## 5.1 ANABOLIC PATHWAYS

A supply of ribonucleoside and deoxyribonucleoside 5'-triphosphates is required for the biosynthesis of nucleic acids (see Chapters 6 and 8). The synthesis of these compounds occurs in two main stages: (1) the formation of the purine and pyrimidine ring systems and their conversion to the parent ribonucleoside monophosphate, inosine 5'-monophosphate (IMP) and uridine 5'-monophosphate (UMP); (2) a series of interconversions involving the reduction of ribonucleotides to deoxyribonucleotides and the phosphorylation of these nucleotides to form a balanced set of 5'-triphosphates (Fig. 5.8).

## 5.2 THE BIOSYNTHESIS
## OF THE PURINES

The pathway of biosynthesis of the purines leads directly to the production of their nucleoside 5'-monophosphates and this subject has been so extensively reviewed [1–3, 83] that only an outline need be given here.

It is known from experiments with isotopes that the sources of the atoms in the purine ring are as shown in Fig. 5.1

The first step in the biosynthetic pathway is the phosphorylation of ribose 5'-phosphate by transfer of the pyrophosphoryl residue of ATP to carbon-1 of the ribose moiety to form 5-phosphoribosyl-1-pyrophosphate (PRPP) (Fig. 5.2) which is involved in reactions other than those of nucleotide biosynthesis.

Synthesis of PRPP is catalysed by *PRPP synthetase* an enzyme subject to control particularly by ADP which acts allosterically as well as competing with ATP [101, 102]. Assembly of the purine ring itself begins with the transfer of the γ-amino group of glutamine to PRPP forming 5-phosphoribosylamine (PRA) (Fig. 5.3) under the influence of the enzyme *phosphoribosyl pyrophosphate amidotransferase (amidophosphoribosyl transferase*, EC 2.4.2.14).

Glycine then reacts with PRA to give glycinamide ribonucleotide (GAR) a nucleotide-like compound in which the amide of glycine takes

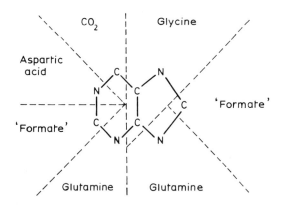

**Fig. 5.1** Sources of the atoms in the purine ring.

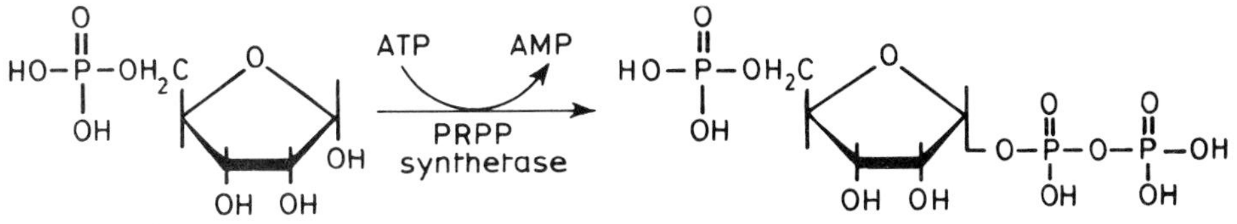

**Fig. 5.2** Formation of 5-phosphoribosyl-1-pyrophosphate (PRPP).

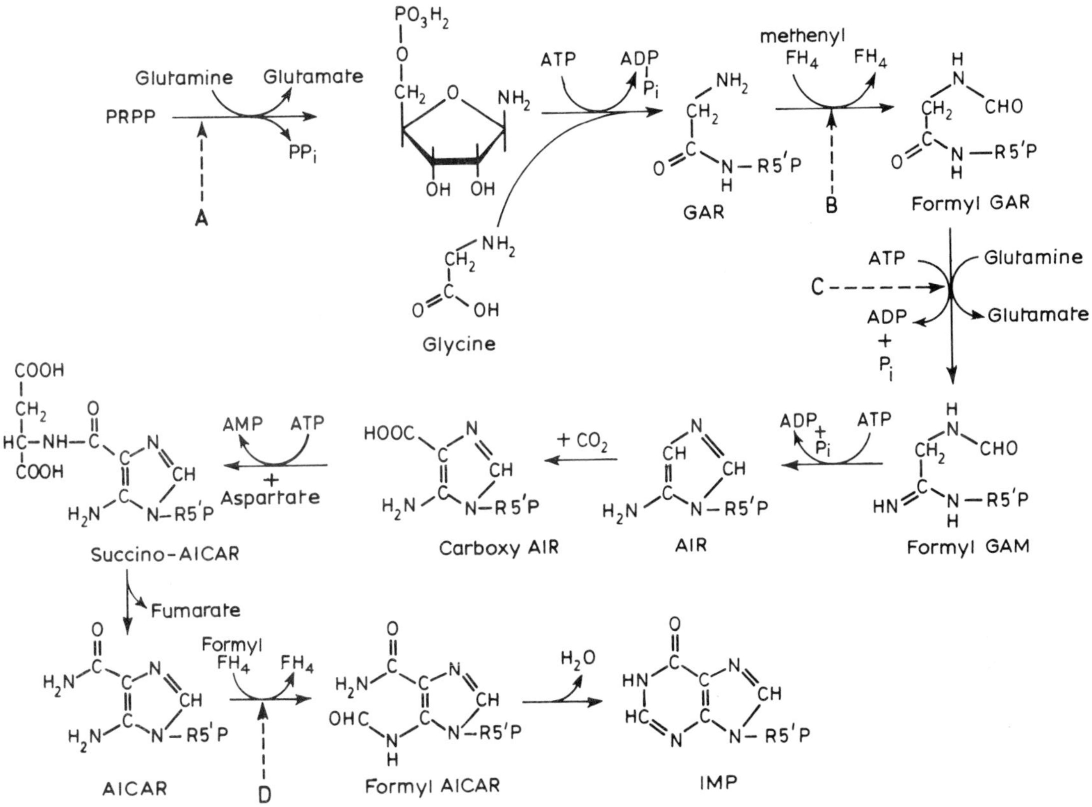

**Fig. 5.3** Purine nucleotide biosynthesis.

the place of the usual purine or pyrimidine base. The reaction sequence continues by formylation from $N^5N^{10}$-methenyltetrahydrofolic acid to give formylglycinamide ribonucleotide (formyl GAR), then amination from glutamine to give formylglycinamidine ribonucleotide (formyl-GAM). Ring closure ensues, producing the imidazole ring compound 5-aminoimidazole ribonucleotide (AIR). Carboxylation of this

compound gives 5-aminoimidazole-4-carboxylic acid ribonucleotide (carboxyl-AIR). The corresponding amide, 5-aminoimidazole-4-carbox-amide ribonucleotide (AICAR) is produced in the subsequent two reactions via an intermediate compound, 5-aminoimidazole-4-succinocarbox-amide ribonucleotide (succino-AICAR). The purine ring system is completed (Fig. 5.3) when $N^{10}$-formyltetrahydrofolic acid donates its formyl group to the 5-amino group of the imidazole carboxamide ribonucleotide. The complete parent ribonucleotide is inosinic acid (inosine 5'-monophosphate, IMP).

By virtue of the role of folic acid derivatives as donors of 1-carbon units at two different stages (B and D) in this sequence of reactions, analogues of folic acid such as aminopterin and amethopterin (methotrexate) are powerful inhibitors of purine biosynthesis. Because of this and because, as will be seen later, they also inhibit the formation of thymine nucleotides, these compounds interfere with the biosynthesis of nucleic acids and are widely used in the treatment of certain forms of cancer and related diseases [4, 5]. In cell biology they are employed in studies on cell hybridization and transformation (see Section 5.7).

Two other important inhibitors of purine nucleotide biosynthesis are azaserine:

$$N\overline{\equiv}N\overset{+}{\equiv}CH-CO-OCH_2CH(NH_2)-COOH$$

and 6-diazo-5-oxonorleucine (DON):

$$N\overline{\equiv}N\overset{+}{\equiv}CH-CO-CH_2-CH_2-CH(NH_2)-COOH$$

both of which are analogues of glutamine:

$$H_2N-CO-CH_2-CH_2-CH(NH_2)-COOH$$

and inhibit the sequence at the amination steps (A and C) from PRPP to PRA and from formyl-GAR to formyl-GAM [6, 7].

Inosine 5'-monophosphate is the common precursor of both adenosine and guanosine 5'-monophosphates. Amination of IMP to AMP proceeds in two stages with the intermediate formation of adenylosuccinic acid (Fig. 5.4).

This reaction, in which the amino group of aspartate is transferred to C-6 of IMP to give AMP, resembles the reaction above in which 5-aminoimidazole-4-carboxamide ribonucleotide is formed from 5-aminoimidazolecarboxylic acid ribonucleotide (Fig. 5.3). One difference, however, is the requirement for GTP as coen-

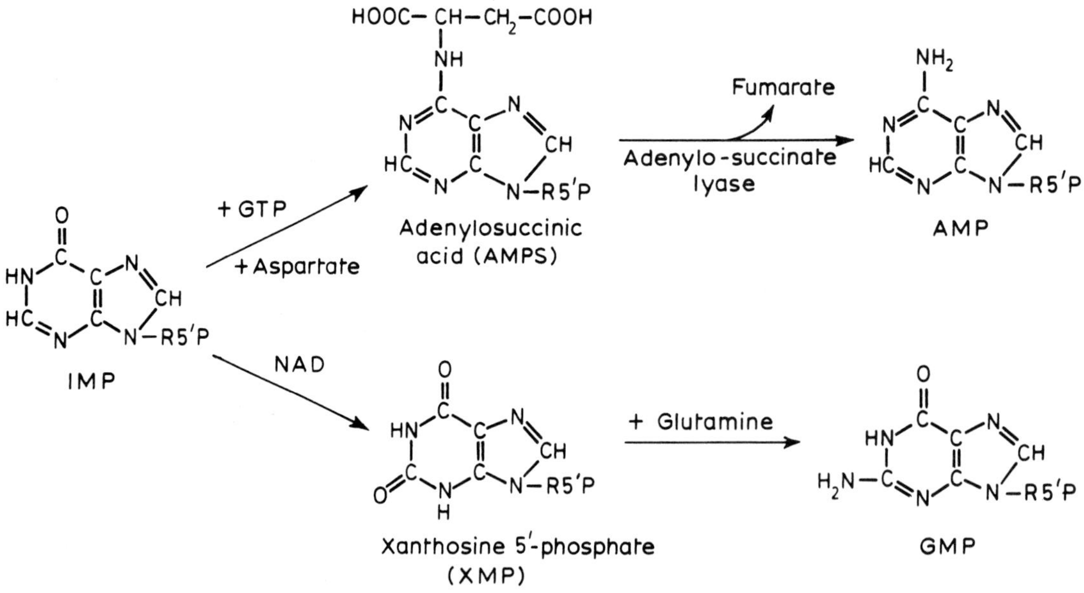

**Fig. 5.4** Formation of AMP and GMP from IMP.

zyme in the reaction forming adenylosuccinic acid from IMP.

The formation of GMP from IMP is also a two-stage reaction in which xanthosine 5'-monophosphate (XMP) is initially formed and then aminated to give GMP (Fig. 5.4). This amination reaction, like the earlier reactions in the sequence that utilize glutamine, is inhibited by azaserine and DON [8].

The two purine mononucleotides AMP and GMP are phosphorylated by kinases through the diphosphate stage to give ATP and GTP (see Section 5.8).

## 5.3 PREFORMED PURINES AS PRECURSORS

In 1947, Kalckar [9] demonstrated the interaction of purine bases and ribose 1-phosphate to yield nucleosides and inorganic phosphate. Such reactions, which are forms of transglycosidation, are reversible and are catalysed by enzymes termed *nucleoside phosphorylases* (EC 2.4.2.1):

Hypoxanthine + ribose 1-phosphate $\rightleftharpoons$ Inosine + $P_i$

5-Aminoimidazole-4-carboxamide can take the place of a purine base in this type of reaction.

Since most nucleosides can be phosphorylated by ATP under the influence of appropriate phosphokinases, a route exists for the biosynthesis of ribonucleotides from preformed purines.

A much more important mechanism for the conversion of bases into nucleotides involves PRPP which is also concerned in the *de novo* pathway described above. Under the influence of enzymes originally termed nucleotide pyrophosphorylases, but more correctly called *phosphoribosyltransferases*, bases can react with PRPP to form nucleotides and pyrophosphate [10].

The reactions catalysed by phosphoribosyl transferases are illustrated in Fig. 5.5.

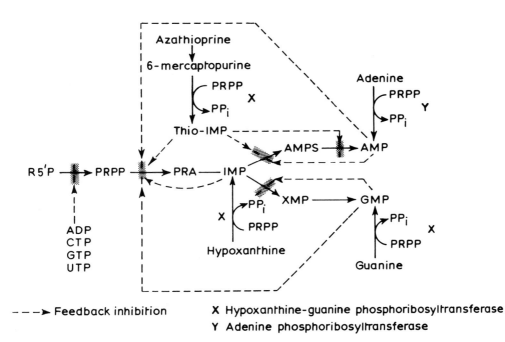

- - -► Feedback inhibition    X Hypoxanthine-guanine phosphoribosyltransferase
Y Adenine phosphoribosyltransferase

**Fig. 5.5** Phosphoribosyl transferases and control of purine nucleotide biosynthesis.

One such enzyme, *adenine phosphoribosyltransferase* (EC 2.4.2.7), (APRT) forms AMP from adenine and PRPP. A second enzyme, *hypoxanthine–guanine phosphoribosyltransferase* (EC 2.4.2.8) (HGPRT) converts hypoxanthine and guanine into IMP and GMP respectively in the presence of PRPP.

Sophisticated control mechanisms govern the pathway of purine nucleotide biosynthesis although there are major variations in these mechanisms from organism to organism [11]. A generalized and consolidated summary of these controls is shown in Fig. 5.5 from which it can be seen that both AMP and GMP exert feedback control on the formation of PRA from PRPP [12,13]. In some systems, AMP also controls the formation of AMP from IMP while in others GMP controls the formation of XMP from IMP. GTP moreover is required as a cofactor in the synthesis of AMP from IMP. In addition to feedback inhibition most of the enzymes of the pathway are repressed by purine bases, nucleosides and nucleotides.

The enzyme HGPRT also catalyses the formation of the corresponding thio-IMP from the drugs 6-mercaptopurine and azathioprine (imuran). This analogue of IMP inhibits the formation of PRA and AMP and so prevents biosynthesis of the normal purine nucleotides [14]. Because such drugs inhibit purine (and therefore nucleic acid) biosynthesis they have been used as immunosuppressive or cancerostatic agents [15]. Azathioprine is also of value in the treatment of gout by inhibiting purine formation.

In the rare condition in children known as the *Lesch-Nyhan syndrome* there is a deficiency of the enzyme hypoxathine–guanine phosphoribosyltransferase [16, 17]. Because of this there is an increase in the concentration of PRPP which in turn leads to increased synthesis of PRA and consequently to overproduction of purine nucleotides and ultimately of uric acid. The condition is associated with excessive uric acid synthesis and is resistant to the action of azathioprine presumably because this compound is only active after conversion to thio-IMP (Fig. 5.5), a reaction that requires the missing enzyme HGPRT.

## 5.4 THE BIOSYNTHESIS OF THE PYRIMIDINES

Whereas purine nucleotide biosynthesis proceeds by growth of the purine ring on PRPP, the formation of the pyrimidine nucleotides involves assembly of a pyrimidine derivative, orotic acid, and the subsequent combination of this moiety with PRPP. The complete series of enzymic reactions giving rise to the parent pyrimidine mononucleotide (uridine 5′-monophosphate, UMP) is shown in Fig. 5.6 [3, 18].

The starting compounds are aspartic acid and carbamoylphosphate which combine under the influence of *aspartate carbamoyltransferase* to form carbamoylaspartate. Formation of the pyrimidine ring is then effected by the action of *dihydro-orotase* giving dihydro-orotic acid, dehydrogenation of which produces the important pyrimidine intermediate orotic acid. A phosphoribosyltransferase reaction then follows in which orotic acid accepts a ribose 5-phosphate group from PRPP. The resulting product is orotidine 5′-monophosphate (OMP) and inorganic pyrophosphate is eliminated.

Decarboxylation of orotidine 5′-monophosphate gives uridine 5′-monophosphate (UMP) which is then converted by kinases through uridine 5′-diphosphate (UDP) into uridine 5′-triphosphate (UTP) and it is at this level of phosphorylation that conversion of uracil into cytosine takes place [19] under the influence of the enzyme *CTP synthetase* (EC 6.3.4.2):

$$UTP + NH_3 + ATP \rightarrow CTP + ADP + P_i$$

In eukaryotes, the first three enzymes in the pyrimidine biosynthetic pathway (carbamoylphosphate synthetase, aspartate carbamoyltransferase and dihydro-orotase) are all part of a single multifunctional protein of molecular

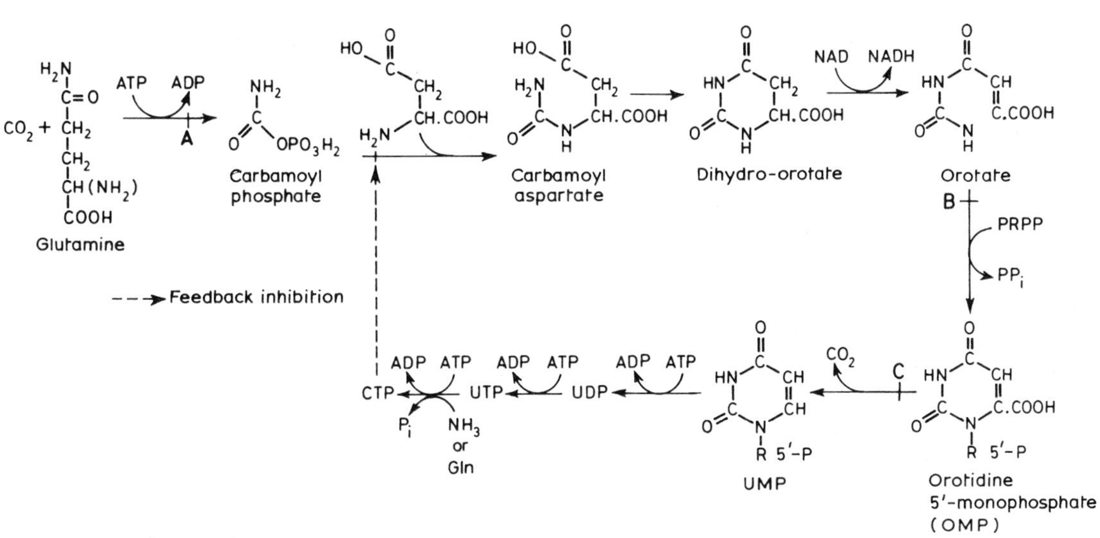

Glutamine — Carbamoyl phosphate — Carbamoyl aspartate — Dihydro-orotate — Orotate

A    B

ATP ADP    NAD NADH

NH₂

Feedback inhibition

PRPP   PP$_i$

CTP ← UTP ← UDP ←   UMP ← Orotidine 5'-monophosphate (OMP)

ADP ATP   ADP ATP   ADP ATP

P$_i$   NH₃ or Gln

CO₂

R 5'-P

**Fig. 5.6**   Biosynthesis of pyrimidine nucleotides.

weight 200 000 [20–23]. The carbamoyltransferase is allosterically controlled being activated by ATP and inhibited by CTP [24, 25]. UMP exerts feedback inhibition on the synthesis of carbamoylphosphate [11, 25] and inhibits the decarboxylation of OMP to UMP [13].

Inhibitors of pyrimidine biosynthesis include azaserine and DON (Fig. 5.6, A) (see Section 5.2) which inhibit carbamoylphosphate synthesis in those systems utilizing glutamine as the amino group donor, 5-azaorotate, which inhibits the formation of OMP from orotic acid (B) and 6-azauridine which blocks the decarboxylation of OMP to UMP (C) [26].

Exposure of cultured cells to *N*-phosphono-acetyl-L-aspartate (a transition state analogue of aspartate carbamoyltransferase) leads to the selection of mutant cells which have amplified the gene coding for the multifunctional enzyme [20–22].

Preformed pyrimidines can be taken up by cells and converted to nucleotides by *salvage pathways*. This may proceed by direct reaction with PRPP or via a phosphorylase and kinase as occurs with purines (Section 5.3) [103]. In bacteria uridine and deoxyuridine are frequently converted to uracil prior to incorporation while cytidine is rapidly deaminated to uridine [96, 97, 103]. The presence of thymidine induces a thymidine phosphorylase in *E. coli* which converts thymidine to thymine [104]. As radioactive thymine and thymidine are often used to label DNA the presence of this inducible enzyme means that thymine should always be used for longer labelling times.

## 5.5 THE BIOSYNTHESIS OF DEOXYRIBONUCLEOTIDES AND ITS CONTROL

The conversion of ribose into deoxyribose takes place at the nucleotide level without breakage of the glycosidic linkage [27] since compounds such as uniformly labelled [¹⁴C]cytidine are incorporated into the dCMP residues of DNA without change in the relative specific radioactivities of sugar and base.

Two related but readily distinguishable systems have been purified for the reduction of the ribosyl moiety of ribonucleotides to the corresponding deoxyribosyl derivative [27, 28]. In *Lactobacillus leichmannii* the reductase uses ribonucleoside triphosphates and requires a cobamide coenzyme. In *E. coli*, mammalian cells and almost all other systems investigated, the reductase uses ribonucleoside diphosphates and requires no coenzyme. The *L. leichmannii* enzyme is monomeric and has a molecular weight of 76 000. The *E. coli* enzyme has two atoms of iron and two pairs of non-identical subunits, i.e. it has the constitution $\alpha\alpha'\beta_2.\alpha$ (the product of the *nrd* A gene), has a molecular weight of 80 000 and has sulphydryl groups at the active site. $\alpha$ differs from $\alpha'$ in the *N*-terminal amino acid. $\beta$ (the product of the *nrd* B gene) has a molecular weight of 39 000 [28]. The *nrd* A and *nrd* B genes are adjacent to each other on the *E. coli* chromosome and are transcribed to form a polycistronic mRNA (see Chapter 8). Control of transcription is dependent on growth conditions [29]. The complete nucleotide sequence of both genes has been determined [30].

The two $\alpha$ chains are tightly associated to form the B1 subunit and the two $\beta$ chains form the B2 subunit which is associated with two iron atoms. The mammalian enzyme has a similar constitution but the two subunits are called M1 and M2 [108]. The genes for the mouse M1 subunit, *E. coli* B1 subunit and the herpes virus-encoded ribonucleotide reductases show considerable homology [110].

Although NADPH is the ultimate hydrogen donor a series of hydrogen carriers are involved in the reduction of ribonucleotides. The mechanism in *E. coli* is known as the *thioredoxin system*. Thioredoxin is a sulphur-containing protein with some 108 amino acid residues [107]. It is reduced by NADPH under the influence of *thioredoxin reductase*, a flavoprotein containing FAD. The reduced thioredoxin in turn reduces ribonucleotides to the corresponding deoxy- derivatives, becoming itself reoxidized to thioredoxin. Recent evidence has suggested that other reducing agents, e.g. glutathione, may be able to substitute for thioredoxin [28].

Ribonucleotide reductase reduces the four nucleotides, ADP, GDP, CDP and UDP to the corresponding deoxyribonucleotides. These are then phosphorylated by kinases to the triphosphates. dATP, dGTP and dCTP are used directly for DNA synthesis but dUTP is rapidly hydrolysed to dUMP by an active dUTPase. This prevents incorporation of dUTP into DNA (see Section 6.3.2). dUMP is converted by a series of steps to dTTP (see below).

There is a single catalytic site on ribonucleotide reductase for all four ribonucleotides. In the enzyme from *Leichmanii* there is also a single allosteric site which can bind any of the four deoxyribonucleoside triphosphates. Binding of an effector increases the affinity of the substrate site for a particular ribonucleotide and is also essential for the binding of the coenzyme [28, 31]. For example reduction of ATP is favoured by the binding of dGTP to the allosteric site. In association with the enzyme trans-*N*-deoxyribosylase [32] this enables a balanced supply of deoxyribonucleoside triphosphates to be produced. The initial step in the reaction is the homolytic cleavage of the cobalt-carbon bond to yield two free radicals one of which removes a dithiol hydrogen to produce HS-E-S˙. This cyclizes on abstraction of the 2′-OH from the ribonucleotide. With retention of the stereochemistry the second hydrogen is added to form the deoxyribose.

The iron-dependent reductases are more complex. Although consisting of four polypeptide chains the enzyme has only one active site which involves the two iron atoms and a single tyrosine-free radical [28, 31]. As this free radical can be detected quantitatively in frozen packed cells by EPR spectroscopy it has been possible to follow enzyme activity during the cell cycle. Moreover, as the EPR signal from deuterated enzyme differs from that of the normal enzyme the synthesis and degradation of the enzyme can be readily followed [33]. As with the cobamide-dependent enzyme the diphosphate to be reduced is determined by the conformation of the enzyme. (Evidence for two enzymes in mammalian systems can be explained on consideration of the allosteric controls [28, 34–36].) Enzyme conformation is altered by nucleotides bound at the allosteric sites. Although most studies have been performed using the enzyme from *E. coli*, the mammalian enzyme responds similarly. One of the allosteric sites (the l or low-affinity site) binds ATP or dATP. When ATP is bound the enzyme is active but dATP inhibits all activity, possibly by causing aggregation [25, 108]. Deoxyadenosine is a potent inhibitor of DNA synthesis as it is rapidly converted to dATP which inhibits ribonucleotide reductase [37]. Binding to the other allosteric site (the h or high-affinity site) is more complex. When ATP (also dATP in *E. coli*) is bound to the h site, reduction of CDP and UDP is enhanced. dGTP in the h site inhibits reduction of GDP, UDP and CDP but stimulates reduction of ADP. dTTP bound to the h site stimulates reduction of GDP but inhibits (possibly via dGTP) reduction of CDP and UDP (see Fig. 5.7).

The fine control exerted over the activity of ribonucleotide reductase is in line with its being a key enzyme in the programmed production of a balanced supply of deoxyribonucleoside triphosphates for synthesis of DNA [28, 38]. In support of this is the finding that the tissue level of ribonucleotide reductase correlates with the growth rate of the tissue and shows variations in activity during the cell cycle [33, 39–41]. This variation in activity has been attributed both to changes resulting from allosteric effects and to variation in the absolute amount of enzyme present. However, direct measurement of the deoxyribonucleoside triphosphate pool sizes in cells [41–45] shows them to be very low in non-dividing cells and to rise during S phase in parallel with increased ribonucleotide reductase activity. Pool sizes reach a maximum in G2 phase and then fall after mitosis. As the amounts of the four deoxyribonucleotides are not the same it is possible that these pools do not represent precursors for DNA synthesis but either excess production or the allosteric pool which acts to regulate production (see below).

Addition to cells of high concentrations of thymidine leads to the formation of a large intracellular pool of dTTP [43, 46–48]. This feeds back (either directly or through stimulating reduction of GDP) to inhibit further reduction of UDP and CDP bringing about a deficiency of dCTP and hence an inhibition of DNA synthesis. Addition of deoxycytidine, which is rapidly phosphorylated to dCTP, bypasses the block once again allowing DNA synthesis to occur [46–49].

By treating cultured cells with normally lethal

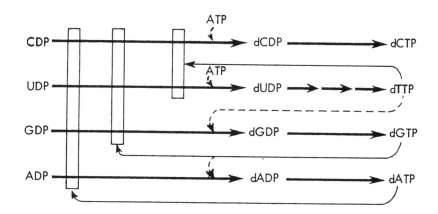

**Fig. 5.7** Allosteric control of ribonucleotide reductase; open bars show inhibition, dashed lines stimulation (based on Thelander and Reichard [28]).

concentrations of deoxyribonucleosides an amplification of the gene coding for ribonucleotide
reductase can be induced [37] (see Section
7.5.2).

Ribonucleotide reductase is inhibited by
hydroxyurea [28, 50] which acts as a free radical
scavenger [51–53]. One of the few differences
between the *E. coli* enzyme and the mammalian
enzyme is that the action of hydroxyurea is
irreversible in *E. coli* but is readily reversed in
mammalian systems allowing its use as a chemo-
therapeutic agent and in cell synchronization
[48, 50].

It has been proposed that ribonucleotide
reductase forms part of a complex of enzymes
which regulates the production and use of
precursors in DNA replication (see Section
6.11). The rate-limiting function of such a
complex could be maintained by the presence of
a constant pool of deoxyribonucleotides acting
allosterically. The concentration of the components of this regulatory pool would be
regulated by balanced synthesis and degradation. When DNA polymerase is inhibited the
excess dNTPs are rapidly degraded but degradation ceases when ribonucleotide reductase is
inhibited by hydroxyurea [43, 54]. A second
source of deoxyribonucleotides in animal cells is
by salvage from circulating deoxyribonucleosides. Indeed in cells such as thymocytes which
have active salvage enzymes this may be the
major source of dCTP and dTTP [82].

## 5.6 THE BIOSYNTHESIS OF THYMINE DERIVATIVES

The essential step in the formation of thymine
nucleotides is the methylation of deoxyuridine
monophosphate (dUMP) to produce thymidine
monophosphate (dTMP) (dUMP → dTMP)
under the influence of thymidylate synthetase.
The process is elaborate and takes place in
several stages. The source of the additional
carbon atom at C-5 is $N^5N^{10}$-methylenetetra-

hydrofolic acid [3, 55]. The reaction is as follows:

$$N^5N^{10}\text{-methylenetetrahydrofolate} +$$
$$\text{dUMP} \rightarrow \text{dihydrofolate} + \text{dTMP}$$

The dihydrofolate is reduced again to tetra-
hydrofolate under the influence of dihydrofolate
reductase:

$$\text{dihydrofolate} + \text{NADPH} + \text{H}^+ \rightarrow$$
$$\text{tetrahydrofolate} + \text{NADP}^+$$

This reaction is powerfully inhibited by the
folic acid analogues aminopterin and amethopterin (methotrexate) which therefore inhibit
the formation of thymine derivatives.

The two enzymes, thymidylate synthetase and
dihydrofolate reductase (DHFR), are coded for
by overlapping DNA sequences in phage T4 [57]
(where the thymidylate synthase gene is the first
intron-containing, prokaryotic gene to be discovered [109]) and are co-ordinately regulated in
*B. subtilis* [58]. This association is taken to
extremes in protozoa where the two activities are
found in the same polypeptide chain [59–61].
The mammalian thymidylate synthetase shows a
high level of amino acid sequence homology with
prokaryote enzymes [62].

Thymidylate synthetase activity is low in non-
growing cells but increases in activity in growing
cells [56]. It is inhibited by 5-fluoro-dUMP. For
this reason 5-fluorodeoxyuridine is an inhibitor
of DNA synthesis.

Extracts of *E. coli* infected with a T-even
phage contain the enzyme deoxycytidylate
hydroxymethylase which brings about the formation of 5-hydroxymethyldeoxycytidylic acid
from formaldehyde and deoxycytidylic acid in
the presence of $N^5N^{10}$-methylenetetrahydrofolic
acid [63] (see Sections 3.6.6g and 6.9.2c).

Another pathway leading to the synthesis of
dTTP involves the deamination of dCMP to
dUMP by the enzyme dCMP deaminase (dCMP
aminohydrolase EC 3.5.4.5). This enzyme is
present in only small amounts in non-dividing
tissues but is active in rapidly growing tissues
such as embryos, spleen and regenerating rat

liver [64–66]. It thus shows a similar response to growth as thymidine kinase, thymidylate synthetase, ribonucleotide reductase and DNA polymerase, enzymes present in the replitase complex (see Chapter 6.11). Although substrate for the enzyme may arise by phosphatase action on dCDP and dCTP the levels of circulating deoxycytidine probably are the major source of cellular dCMP.

dCMP deaminase is under allosteric control [67]. The enzyme is inhibited by dTTP and stimulated by dCTP and hence cellular dCMP is channelled towards the pyrimidine deoxynucleoside triphosphate present in limiting amounts.

## 5.7 AMINOPTERIN IN SELECTIVE MEDIA

Because of the part played by folic acid in purine and thymidylate biosynthesis, analogues of folic acid have become important in cell biology as ingredients in HAT medium (medium containing hypoxanthine, aminopterin and thymidine) [48, 68]. This growth medium is used for selection of animal cells able to bypass the aminopterin-imposed blocks in purine and thymidylate metabolism by incorporating exogenous hypoxanthine and thymidine. Mutant cells lacking the enzymes hypoxanthine phosphoribosyltransferase (HGPRT) or thymidine kinase (TK) cannot grow in HAT medium. However, fusion of a TK⁻ cell with an HGPRT⁻ cell leads to a hybrid cell which can grow in HAT medium [69]. Similarly metabolic co-operation is shown by mixed cultures of TK⁻ and HGPRT⁻ cells which exchange thymidylate and purine nucleotides across gap junctions and hence grow in HAT medium [70].

HAT medium is used extensively to select for transformants of TK⁻ cells treated with DNA containing the thymidine kinase gene from herpes simplex virus, linked by genetic engineering techniques to any other DNA [71] – see Section A.8.4 – and to select for monoclonal-antibody-producing cells arising as a result of

fusion of non-growing spleen cells with HGPRT⁻ myeloma cells [72].

## 5.8 FORMATION OF NUCLEOSIDE TRIPHOSPHATES

*Kinases* convert nucleosides to nucleoside monophosphates, diphosphates and triphosphates by transferring the γ-phosphate from ATP to the substrate. The kinases which act on the pyrimidine nucleosides and deoxyribonucleosides enable the preformed materials to be utilized for nucleic acid synthesis by the so-called 'salvage' pathway, e.g. thymidine → dTMP → dTDP → dTTP [73]. The thymidine kinases are unique in that they are active in dividing tissues but low in non-growing tissues [74, 75] and activity shows regular changes during the cell cycle [75, 76]. Some animal viruses code for an additional thymidine (or deoxypyrimidine nucleoside) kinase [77] (see Section 3.6).

Thymidine kinase is also able to phosphorylate 5-fluorodeoxyuridine but the 5-FdUMP produced is an inhibitor of thymidylate synthetase and because of this fluorodeoxyuridine is an inhibitor of DNA synthesis.

Thymidine kinase is inhibited by dTTP and by dCTP [78–80] but there is also an element of forward promotion [81] which means that on incubating cells with thymidine the size of the dTTP pool produced is proportional to the external thymidine concentration [48, 81].

Nucleotide interconversions are summarized in Fig. 5.8.

## 5.9 GENERAL ASPECTS OF CATABOLISM

It is recognized that cellular DNA tends to be strongly conserved and that the amount of degradation of DNA in normal circumstances is small. Some species of RNA have a relatively short life span and are therefore degraded quite rapidly. There is also evidence that nucleic acids may not undergo total degradation but that some

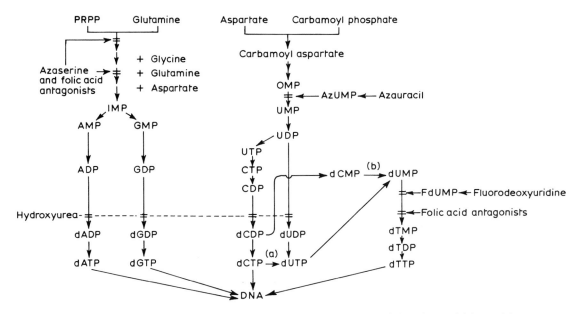

**Fig. 5.8**  Pathways involved in the biosynthesis of deoxyribonucleoside triphosphates: (a) bacterial systems only; (b) eukaryotes. The blocking action of some antimetabolites is indicated.

of the intermediate products such as nucleotides and nucleosides may be reutilized by so-called 'salvage' pathways.

RNA and DNA are hydrolysed by nucleases and diesterases first to oligonucleotides and eventually to mononucleotides and nucleosides, and the glycosidic linkages between the purine or pyrimidine bases and the sugar moieties are cleaved either hydrolytically or phosphorolytically to yield for free purine and pyrimidine bases [83, 84]. The nature and function of nucleases and phosphodiesterases have been considered in Chapter 4, and the pyrophosphatases and phosphatases that attack nucleotides have been reviewed extensively [85–87].

## 5.10  PURINE CATABOLISM

The breakdown of purine nucleotides has been studied over many years. Adenine and its nucleosides and nucleotides can be deaminated hydrolytically under the influence of the enzymes adenine deaminase (present only in

some micro-organisms [105, 106]), adenosine deaminase, and adenylate deaminase to yield hypoxanthine, inosine or inosine monophosphate respectively (Fig. 5.9) and guanine nucleotides are similarly attacked by guanine, guanosine or guanylate deaminases to yield xanthine or its ribose derivatives (Fig. 5.9). The balance between dephosphorylation and deamination at any stage is quite variable and in T lymphoblasts although about half of the GMP is degraded by deamination, little if any deamination of dGMP occurs [88]. This can result in the accumulation of cytotoxic levels of deoxyguanosine in those immunodeficient patients lacking purine nucleoside phosphorylase.

Hypoxanthine and xanthine are oxidized under the influence of xanthine oxidase to yield uric acid (Fig. 5.10). Although the distribution of the enzymes involved is far from uniform in different species, this scheme of purine degradation appears to be of fairly general application, and experiments with $^{15}$N have shown that, as might be expected, the administration of labelled

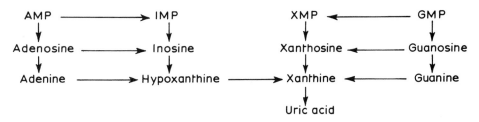

**Fig. 5.9** The degradation of purines at the levels of nucleotides, nucleosides and bases.

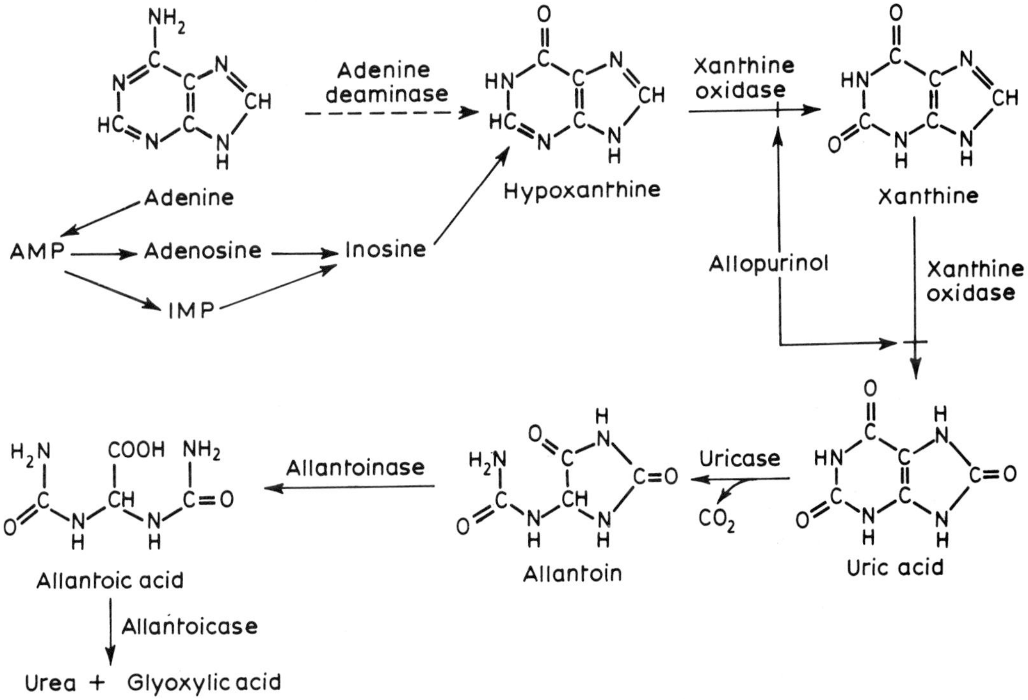

**Fig. 5.10** The catabolism of purines.

purines to animals is followed by the appearance of the isotope in the excreted uric acid or in its further degradation products.

Uric acid itself is excreted by only a few mammals, since most non-uricotelic animals are provided with the enzyme uricase, which oxidizes uric acid to the much more soluble allantoin, and under certain conditions to other end products as well [89]. The conversion of uric acid into allantoin appears to involve a number

of intermediate compounds including the symmetrical compound hydroxyacetylene-diureinecarboxylic acid [83, 90] (Fig. 5.11).

Man and certain higher apes, however, are unable to bring about this step owing to absence of uricase from their tissues, and in them the end product of purine metabolism is uric acid which is excreted in the urine along with very much smaller amounts of xanthine and hypoxanthine [91]. The Dalmatian coach-hound is peculiar

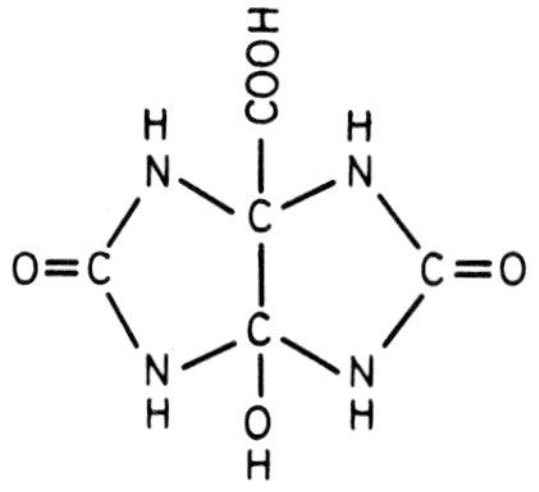

**Fig. 5.11**  Hydroxyacetylenediureinecarboxylic acid.

in that it excretes uric acid in preference to allantoin, owing to lack of tubular reabsorption of uric acid in the kidney [83].

The substance allopurinol, which has a structure very similar to that of hypoxanthine, acts as a competitive inhibitor of xanthine oxidase, and so prevents uric acid formation. It is therefore sometimes used in the treatment of gout, a disease in which uric acid accumulates in the body. Patients treated with allopurinol excrete xanthine and hypoxanthine in place of uric acid [91]. Mention has already been made of the use of azathioprine in the treatment of gout by inhibiting the synthesis of PRA from PRPP, IMP from hypoxanthine and AMP from IMP [26].

In fishes, amphibia and more primitive organisms allantoin is broken down by allantoinase to allantoic acid, and this in turn may be degraded by allantoicase to urea and glyoxylic acid. The main nitrogenous excretory product in the spider is not uric acid but guanine. These aspects of comparative biochemistry are discussed in detail in the books by Baldwin [92], Florkin [93] and Henderson and Paterson [83].

It is in the birds and the uricotelic reptiles that uric acid formation is most pronounced since in them uric acid rather than urea is the main nitrogenous excretory product. In most birds, uric acid production can be shown to take place in the liver, since hepatectomy is followed by cessation of uric acid synthesis and a rise in the blood ammonia level. The obvious inference that in birds and reptiles uric acid is derived ultimately from ammonia is supported by isotopic experiments. Urea does not act as a precursor of uric acid except in so far as it may give rise to ammonia. While the liver of the fowl or goose contains all the enzymes required for uric acid formation, that of the pigeon is lacking in xanthine oxidase. In the pigeon, therefore, hypoxanthine is produced in the liver, and is oxidized to uric acid in the kidney where xanthine oxidase is present.

In the pathological condition known as gout, uric acid is deposited in the joints, particularly in the big toe, and under the skin as nodules called tophi. In this disease the miscible pool of uric acid in the human body is increased to as much as 15 times the normal value of about 1 g [94, 95].

## 5.11 PYRIMIDINE CATABOLISM

The catabolism of pyrimidine nucleotides, like that of purine nucleotides, involves dephosphorylation, deamination and cleavage of glycosidic bonds, and many of the phosphatases that act upon purine nucleotides act also on the corresponding pyrimidine derivatives. As with purine nucleosides, so pyrimidine nucleosides may be hydrolysed to form pyrimidine bases and sugar, or they may be involved in phosphorolytic cleavage [83].

Cytosine can be deaminated by cytosine deaminase; this has been demonstrated in yeasts and other micro-organisms [96], and cytosine nucleosides are broken down to uridine nucleosides by cytidine deaminase which is widespread in animal tissues [97] as well as in bacteria.

The catabolic pathways of uracil [98] and thymine [99, 100] in mammalian tissues involve reduction of the pyrimidines to the dihydro derivatives, ring opening to give the appropriate ureido-acid, and the removal of ammonia and $CO_2$ to give $\beta$-alanine or its methylated derivative (Fig. 5.12). In some bacteria, uracil and thymine undergo oxidative degradation via barbituric acid and methylbarbituric acid to urea and malonic or methylmalonic acids [83].

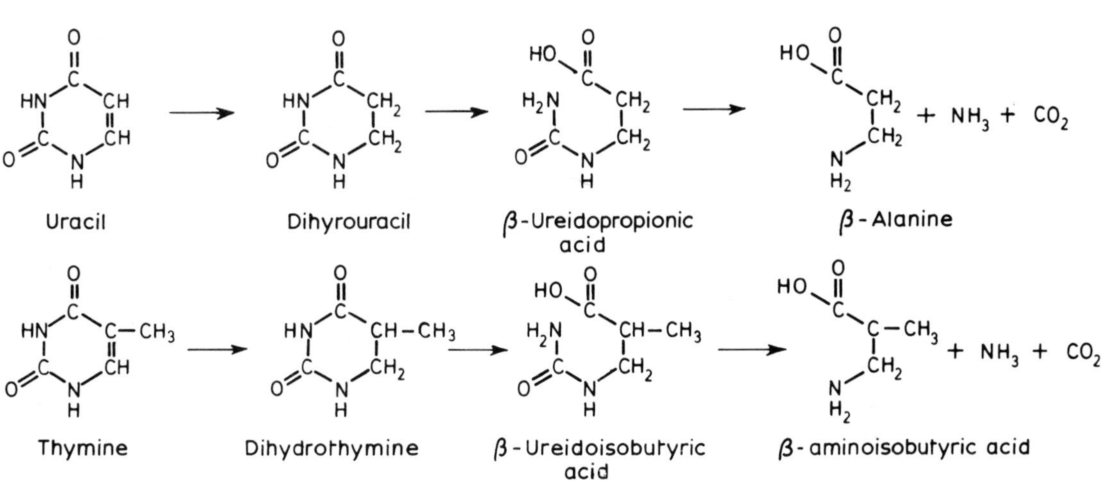

**Fig. 5.12** The catabolism of uracil and thymine.

REFERENCES

1 Colowick, S. P. and Kaplan, N. O. (1978), *Methods Enzymol.* **51**.

2 Muller, M. M., Kaiser, E. and Seegmiller, J. E. (1977), *Advances in Experimental Medicine and Biology. Purine Metabolism in Man – II*, Plenum Press, New York, Vol. 76A and 76B.

3 Kornberg, A. (1980), *DNA Replication*. Freeman, San Francisco.

4 Rhoads, C. P. (ed.) (1955), *Anti-metabolites and Cancer*, American Association for the Advancement of Science, Washington, DC.

5 Bresnick, E. (1974) in *The Molecular Biology of Cancer* (ed. H. Busch), Academic Press, New York, p. 278.

6 Levenberg, B., Melnick, I. and Buchanan, J. M. (1957), *J. Biol. Chem.*, **225**, 163.

7 Hartman, S. C. (1963), *J. Biol. Chem.*, **238**, 3036.

8 Abrams, R. and Bentley, M. (1959), *Arch. Biochem.*, **79**, 91.

9 Kalckar, H. (1947), *Symp. Soc. Exp. Biol.*, **1**, 38.

10 Murray, A. W. (1971), *Annu. Rev. Biochem.*, **40**, 811.

11 Mandelstam, J. and McQuillen, K. (1973), *Biochemistry of Bacterial Growth* (2nd edn), Blackwell Scientific, Oxford, p. 236.

12 Bojarski, T. B. and Hiatt, H. H. (1960), *Nature (London)*, **188**, 1112.

13 Creasey, W. A. and Handschumacher, R. E. (1961), *J. Biol. Chem.*, **236**, 2058.

14 Kelley, W. N., Rosenbloom, F. M., Henderson, J. F. and Seegmiller, J. E. (1967), *Proc. Natl. Acad. Sci. USA*, **57(6)**, 1735.

15 Roitt, I. (1974), *Essential Immunology* (2nd edn), Blackwell Scientific, Oxford.

16 Fujimoto, W. Y., Subak-Sharpe, J. H. and Seegmiller, J. E. (1971), *Proc. Natl. Acad. Sci. USA*, **68**, 1516.

17 Rubin, C. S., Dancis, J., Yip, L. C., Bowinski, R. C. and Balis, M. E. (1971), *Proc. Natl. Acad. Sci. USA*, **68**, 1461.

18 Reichard, P. (1959), *Adv. Enzymol.*, **21**, 263.

19 Lieberman, I. (1956), *J. Biol. Chem.*, **222**, 765.

20 Stark, G. R. (1977), *Trends Biochem. Sci.*, **2**, 64.

21 Kempe, T. D., Swyryd, E. A., Bruist, M. and Stark, G. R. (1976), *Cell*, **9**, 541.

22 Coleman, P. F., Suttle, D. P. and Stark, G. R. (1977), *J. Biol. Chem.*, **252**, 6379.

23 Makoff, A. J., Buxton, F. P. and Radford, A. (1978), *Mol. Gen. Genet.*, **161**, 297.

24 Changeux, J. P. (1965), *Sci. Amer.*, **212(4)**, 36.

25 Kantrowitz, E. R., Pastra-Landis, S. C. and Lipscomb, W. N. (1980), *Trends Biochem. Sci.*, **5**, 124.

26 Roy-Burman, P. (1970), *Analogues of Nucleic Acid Components*, Springer-Verlag, New York, p. 16.

27 Larsson, A. and Reichard, P. (1967), *Prog. Nucleic Acid Res. Mol. Biol.*, **7**, 303.

28 Thelander, L. and Reichard, P. (1979), *Annu. Rev. Biochem.*, **48**, 133.

29 Hanke, P. D. and Fuchs, J. A. (1983), *J. Bacteriol.*, **154**, 1140.

30 Carlson, J., Fuchs, J. A. and Messing, J. (1984), *Proc. Natl. Acad. Sci. USA*, **81**, 4294.

31 Hunting, D. and Henderson, J. F. (1982), *CRC Crit. Rev. Biochem.*, **13**, 325.

32 Witt, L., Yap, T. and Blakley, R. (1978), *Adv. Enzyme Regul.*, **17**, 157.

33 Eriksson, S., Graslund, A., Skog, S., Thelander, L. and Tribukait, B. (1984), *J. Biol. Chem.*, **259**, 11 695.

34 Peterson, D. M. and Moore, E. C. (1976), *Biochim. Biophys. Acta*, **432**, 80.

35 Cory, J. G., Mansell, M. M. and Whitford, T. W. (1976), *Adv. Enzyme Regul.*, **14**, 45.

36 Eriksson, S., Thelander, L. and Akerman, M. (1979), *Biochemistry*, **18**, 2948.

37 Meuth, M. and Green, H. (1974), *Cell*, **3**, 367.

38 Elford, H. L., Freese, M., Passamani, E. and Morris, H. P. (1970), *J. Biol. Chem.*, **245**, 5228.

39 Turner, M. K., Abrams, R. and Lieberman, I. (1968), *J. Biol. Chem.*, **243**, 3725.

40 Murphree, S., Stubblefield, E. and Moore, C. E. (1969), *Exp. Cell Res.*, **58**, 118.

41 Albert, D. A. and Gudes, L. J. (1985), *J. Biol. Chem.*, **260**, 679.

42 Walters, R. A. and Ratliff, R. L. (1975), *Biochim. Biophys. Acta*, **414**, 221.

43 Adams, R. L. P., Berryman, S. and Thomson, A. (1971), *Biochim. Biophys. Acta*, **240**, 455.

44 Skoog, K. L., Nordenskjold, B. A. and Bjursell, K. G. (1973), *Eur. J. Biochem.*, **33**, 428.

45 Skoog, K. L. and Bjursell, G. (1974), *J. Biol. Chem.*, **249**, 6434.

46 Bjursell, G. and Reichard, P. (1973), *J. Biol. Chem.*, **248**, 3904.

47 Morris, N. R. and Fischer, G. A. (1960), *Biochim. Biophys. Acta*, **42**, 183.

48 Adams, R. L. P. (1980), *Cell Culture for Biochemists* (ed. T. E. Work and R. H. Burdon), Elsevier, Amsterdam.

49 Reynolds, E. C., Harris, A. W. and Finch, L. R. (1979), *Biochim. Biophys. Acta*, **561**, 110.

50 Adams, R. L. P. and Lindsay, J. G. (1966), *J. Biol. Chem.*, **242**, 1314.

51 Thelander, L., Larsson, B., Hobbs, J. and Eckstein, F. (1976), *J. Biol. Chem.*, **251**, 1398.

52 Atkin, C. L., Thelander, L., Reichard, P. and Lang, G. (1973), *J. Biol. Chem.*, **248**, 7464.

53 Thelander, M., Gräslund, A. and Thelander, L. (1985), *J. Biol. Chem.*, **260**, 2737.

54 Nicander, B. and Reichard, P. (1985), *J. Biol. Chem.*, **260**, 5376.

55 Friedkin, M. (1963), *Annu. Rev. Biochem.*, **32**, 185.

56 Conrad, R. H. and Ruddle, F. H. (1972), *J. Cell Sci.*, **10**, 471.

57 Purohit, S. and Mathews, C. K. (1984), *J. Biol. Chem.*, **259**, 6261.

58 Myoda, T. T. and Funanage, V. L. (1985), *Biochim. Biophys. Acta*, **824**, 99.

59 Coderre, J. A., Beverley, S. H., Schimke, R. T. and Santi, D. V. (1983), *Proc. Natl. Acad. Sci. USA*, **80**, 2132.

60 Beverley, S. M., Coderre, J. A., Santi, D. V. and Schimke, R. T. (1984), *Cell*, **38**, 431.

61 Meek, T. D., Garvey, E. P. and Santi, D. V. (1985), *Biochemistry*, **24**, 678.

62 Takeishi, K., Kaneda, S., Ayusawa, D., Shimizu, K., Gotoh, O. and Seno, T. (1985), *Nucleic Acids Res.*, **13**, 2035.

63 Flaks, J. G. and Cohen, S. S. (1957), *Biochim. Biophys. Acta*, **25**, 667.

64 Scarano, E., Talarico, M., Bonaduce, L. and de Petrocellis, B. (1960), *Nature (London)*, **196**, 237.

65 Maley, G. F. and Maley, F. (1964), *J. Biol. Chem.*, **239**, 1168.

66 Potter, V. R. (1964), *Cancer Res.*, **24**, 1085.

67 Rossi, M., Dosseva, I., Pierro, M., Cacace, M. G. and Scarano, E. (1971), *Biochemistry*, **10**, 3060.

68 Littlefield, J. W. (1964), *Science*, **145**, 709.

69 Littlefield, J. W. and Goldstein, S. (1970), *In vitro*, **6**, 21.

70 Pitts, J. D. (1971), in *Growth Control in Cell Cultures. Ciba Found. Symp.* (ed. G. E. W. Wolstenholme and J. Knight, p. 89.

71 Mantei, N., Boll, W. and Weissmann, C. (1979), *Nature (London)*, **281**, 35.

72 Campbell, A. M. (1984), *Monoclonal Antibody Technology* (ed. R. H. Burdon and P. H. van-Knippenberg), Elsevier, Amsterdam.

73 Grav, H. J. and Smellie, R. M. S. (1965), *Biochem. J.*, **94**, 518.

74 Weissman, S. M., Smellie, R. M. S. and Paul, J. (1960), *Biochim. Biophys. Acta*, **45**, 101.

75 Bello, L. J. (1974), *Exp. Cell Res.*, **89**, 263.

76 Stubblefield, E. and Dennis, C. M. (1976), *J. Theoret. Biol.*, **61**, 171.

77 Keir, H. M. (1968), *Soc. Gen. Microbiol. Symp.*, **18**, 67.

78 Potter, V. R. (1964), *Metabolic Control Mechanisms in Animal Cells*, National Cancer Institute,

Monograph No. 13, p. 111.

79 Bresnick, E. and Karjala, R. J. (1964), *Cancer Res.*, **24**, 841.

80 Okazaki, R. and Kornberg, A. (1964), *J. Biol. Chem.*, **239**, 275.

81 Ives, D. H., Morse, P. A., Jr. and Potter, V. R. (1963), *J. Biol. Chem.*, **238**, 1467.

82 Cohen, A., Barankiewicz, J., Lederman, H. M. and Gelfand, E. W. (1983), *J. Biol. Chem.*, **258**, 12 334.

83 Henderson, J. F. and Paterson, A. R. P. (1973), *Nucleotide Metabolism*, Academic Press, New York.

84 Smellie, R. M. S. (1955), in *The Nucleic Acids* (ed. E. Chargaff and J. N. Davidson), Academic Press, New York, Vol. II, p. 393.

85 Kielley, W. W. (1961), *The Enzymes* (2nd edn) (ed. P. D. Boyer, H. Hardy and K. Myreback), Academic Press, New York, Vol. 5, p. 149.

86 Morton, R. K. (1965), *Compr. Biochem.*, **16**, 55.

87 Bodansky, O. and Schwartz, M. K. (1968), *Adv. Clin. Chem.* **11**,. 277.

88 Barankiewicz, J. and Cohen, A. (1985), *J. Biol. Chem.*, **260**, 4565.

89 Canellakis, E. S. and Cohen, P. O. (1955), *J. Biol. Chem.*, **213**, 385.

90 Dalgliesh, C. E. and Neuberger, A. (1954), *J. Chem. Soc.*, 3407.

91 Balis, E. W. (1968), *Fed. Proc.*, **27**, 1067.

92 Baldwin, E. (1949), *An Introduction to Comparative Biochemistry*. Cambridge University Press, London.

93 Florkin, M. (1949), *Biochemical Evolution*, Academic Press, New York.

94 Wyngaarden, J. B. (1966), *Adv. Metab. Disorders*, **2**, 1.

95 Bishop, C., Garner, W. and Talbott, J. H. (1951), *J. Clin. Invest.*, **30**, 879.

96 O'Donovan, G. A. and Neuhard, J. (1970), *Bacteriol. Rev.*, **34**, 278.

97 Wisdom, G. B. and Orsi, B. A. (1969), *Eur. J. Biochem.*, **7**, 223.

98 Schulman, M. P. (1954), *Chemical Pathways of Metabolism* (ed. D. M. Greenberg), Academic Press, New York, Vol. 2, p. 223.

99 Canellakis, E. S. (1957), *J. Biol. Chem.*, **227**, 701.

100 Fink, K., Cline, R. E., Henderson, R. B. and Fink, R. M. (1956), *J. Biol. Chem.*, **221**, 425.

101 Jensen, K. F. (1983), in *Metabolism of Nucleotides, Nucleosides and Nucleobases in Microorganisms* (ed. A. Munch-Petersen), Academic Press, London, p. 1.

102 Becker, M. A. and Seegmiller, J. E. (1979), *Adv. Enzymol. Relat. areas Mol. Biol.*, **49**, 281.

103 Neuhard, J. (1983), in *Metabolism of Nucleotides, Nucleosides and Nucleobases in Microorganisms* (ed. A. Munch-Petersen), Academic Press, London, p. 95.

104 Mollgaard, H. and Neuhard, J. (1983), in *Metabolism of Nucleotides, Nucleosides and Nucleobases in Microorganisms* (ed. A. Munch-Petersen), Academic Press, London, p. 149.

105 Nygaard, P. (1983), in *Metabolism of Nucleotides, Nucleosides and Nucleobases in Microorganisms* (ed. A. Munch-Petersen), Academic Press, London, p. 27.

106 Zielke, C. L. and Suelter, C. H. (1971) in *The Enzymes* (3rd edn) (ed. P. D. Boyer), Academic Press, New York, Vol. 4, p. 47.

107 Stryer, L., Holmgren, A. and Reichard, P. (1967), *Biochemistry*, **6**, 1016.

108 Engstrom, Y., Eriksson, S., Thelander, L. and Akerman, M. (1979), *Biochemistry*, **18**, 2941.

109 Belfort, M., Pedersen-Lane, J., West, D., Ehrenman, K., Maley, G., Chu, F. and Maley, F. (1985) *Cell*, **41**, 375.

110 Caras, I. W., Levinson, B. B., Fabry, M., Williams, S. R. and Martin, D. W. (1985), *J. Biol. Chem.*, **260**, 7015.

# 6

# Replication of DNA

## 6.1 INTRODUCTION

Each daughter cell produced on cell division contains an identical copy of the genetic material. Since DNA carries the genetic blueprint of the cell encoded in the sequence of nucleotides, the question as to how DNA is reproduced in the cell has attracted a great deal of attention.

Studies with intact cells have shown that replication is semiconservative (Section 6.2) and that it occurs in a discontinuous manner (Section 6.3), and work with purified enzymes has indicated the detailed mechanism of the polymerization reaction (Section 6.4). However, it was the realization of the complexity of the reaction which is required to replicate DNA faithfully and rapidly that led to the development of several *in vitro* systems and finally to the reconstitution from purified components of complexes capable of limited replication (Sections 6.6 to 6.9).

Modern techniques of gene cloning, *in vitro* mutagenesis and DNA sequencing have facilitated the dissection of several origins of replication (Sections 6.8 and 6.9) and we are now beginning to gain some insight into the controls exerted over the event of initiation of replication which is an important, and possibly the major, site at which cell multiplication is controlled.

## 6.2 SEMICONSERVATIVE REPLICATION

At the end of their paper [1] suggesting the double-stranded structure for DNA with its

complementary base pairs, Watson and Crick wrote: 'It has not escaped our notice that the specific pairing we have postulated suggests a possible copying mechanism for the genetic material'.

Let us suppose that a short length of a DNA double helix has the nucleotide sequence shown in Fig. 6.1 (top), and further that the two strands can be untwisted and separated from one another to form two single chains, as in Fig. 6.1

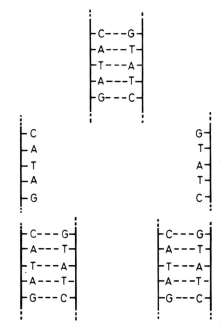

**Fig. 6.1** Schematic illustration of the separation of the two strands of a portion of DNA with the formation of a new strand on each.

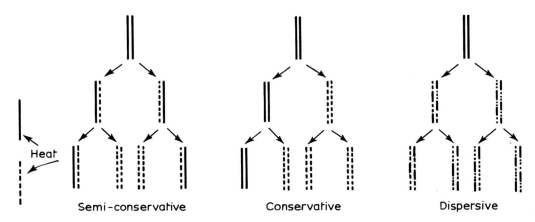

**Fig. 6.2** Possible mechanisms of replication. The dotted lines represent filial strands. For details see text.

(middle), and that each base in the single strands can attach to itself the complementary deoxyribonucleotide by the same hydrogen-bonding which exists in the intact DNA double helix. Finally, if these attached mononucleotides are polymerized to form a polynucleotide chain as in Fig. 6.1 (bottom), the end result will be the formation of two complete DNA double helices identical with each other and with the original molecule. One strand of each daughter molecule will be derived from the original DNA molecule; the other will be the product of the new synthesis.

This mechanism is described as *semiconservative* to distinguish it from the other possible mechanisms. In the *conservative* mechanism the two strands do not come apart but act together as a template to form a completely new double-helical molecule; in this case one daughter molecule would be wholly new and the other totally derived from the parent. In the *dispersive* mechanism the parental molecule is partly degraded and the fragments are incorporated into two new daughter double helices. These possibilities are illustrated in Fig. 6.2.

Convincing evidence for semiconservative replication of DNA was obtained by Meselson and Stahl [2] who grew *E. coli* in a medium containing $^{15}$NH$_4$Cl of 96.5% isotopic purity for fourteen generations so as to label the DNA very heavily with $^{15}$N. The cells were then transferred to a medium containing $^{14}$NH$_4$Cl and samples of bacteria were withdrawn at intervals for several generations. Each sample was lysed by means of sodium dodecyl sulphate and was centrifuged in a concentrated solution of caesium chloride at 140 000$g$ for 20 hours to enable the DNA to attain sedimentation equilibrium. The bands of DNA were found in the CsCl gradient in the region of density 1.71 g cm$^{-3}$ and were well isolated from all other macromolecular components of the bacterial lysate. Ultraviolet absorption photographs taken during the course of the run revealed the position of the DNA bands.

At the start of the experiment the DNA appeared as one single band corresponding to the heavy $^{15}$N-labelled nucleic acid (Fig. 6.3). Macromolecules containing half this level of $^{15}$N then began to appear, and one generation time after the addition of $^{14}$N these hybrid (light–heavy, $^{14}$N–$^{15}$N) molecules alone were present. Subsequently a mixture of light–heavy ($^{14}$N–$^{15}$N) DNA and unlabelled (light–light, $^{14}$N only) DNA was found. When two generation times had elapsed after the addition of $^{14}$N, the half-labelled and unlabelled DNA molecules were present in equal amounts (Fig. 6.3). During

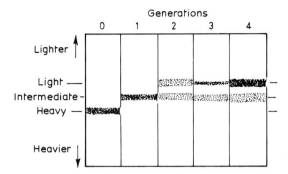

**Fig. 6.3** The pattern of results from the Meselson and Stahl experiment. For explanation see text.

subsequent generations the unlabelled DNA accumulated. Moreover, when the hybrid $^{14}$N–$^{15}$N molecules were heated they separated to give a $^{14}$N strand and a $^{15}$N strand.

Such experiments indicate that in DNA synthesis each existing DNA molecule is split into two subunits (Fig. 6.2), each subunit going to a different daughter molecule. The other subunit of each daughter molecule is the product of new synthesis. These subunits do not undergo any fragmentation but remain intact for many generations.

Experiments at the chromosomal level also support the view that DNA replication is semi-conservative. When the replication of the *E. coli* chromosome is followed by radioautography after labelling of the DNA with tritium, the amount of tritium per unit length of newly synthesized DNA is consistent with the presence of only one newly synthesized strand in a daughter chromosome [3].

In plants to which tritiated thymidine has been given as a specific precursor of DNA during the period of DNA synthesis, both the daughter chromatids are found to be labelled at the time of cell division. However, at the next round of duplication after withdrawal of the [$^3$H]thymidine, these two chromosomes each produce one labelled and one unlabelled chromatid as would be expected [4].

## 6.3 THE REPLICATION FORK

### 6.3.1 Discontinuous synthesis

In the above section we have shown the two parental DNA strands separating from one another before the daughter strands are synthesized. However, in a long DNA molecule replication only occurs over one short stretch at a time and the two parental strands separate only at the point of replication to produce a Y-shaped molecule as the replication fork passes along the DNA (Fig. 6.4).

Replication forks can be seen in electron micrographs when they appear as bubbles in replicating DNA molecules (see Fig. 6.5).

One daughter DNA chain grows in the $5' \rightarrow 3'$ direction and this occurs by the addition of an incoming deoxyribonucleotide to the 3'-hydroxyl at the end of the growing *primer* chain. This is the reaction catalysed by DNA polymerase and considered in more detail in Section 6.4.2.a.

As the two DNA chains are antiparallel (i.e. one is running $5' \rightarrow 3'$ and other $3' \rightarrow 5'$) this poses the problem as to whether the two daughter strands are synthesized by two different mechanisms: one by adding nucleotides to a growing 3'-end and the other to a growing 5'-end. Addition to the 5'-end may involve deoxyribonucleoside 3'-triphosphates or, alternatively, may result in DNA molecules having terminal 5'-triphosphates (see Fig. 6.6). Such

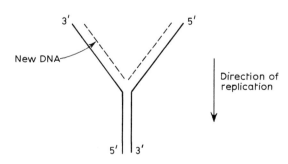

**Fig. 6.4** A replication fork.

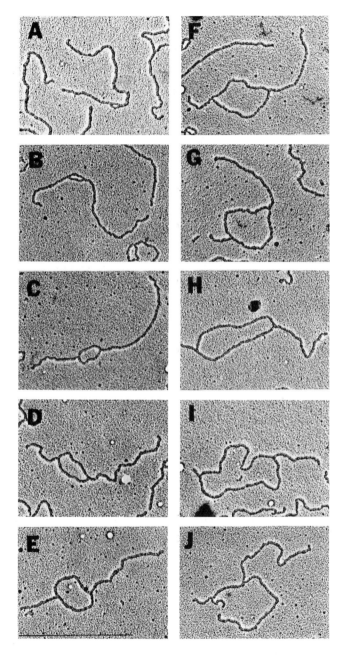

**Fig. 6.5** Replicating SV40 DNA molecules. The molecules have been cut at a unique site with the restriction endonuclease EcoRI and have been arranged in increasing degree of replication (A through J) and oriented with the short branch at the left (reproduced by permission from Fareed, Garon and Salzman [400]).

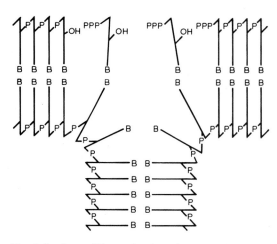

**Fig. 6.6** A possible mechanism of DNA chain growth.

termini have not been found and, moreover, addition of nucleotides to a 5'-triphosphate-terminated primer is incompatible with the proof-reading function of DNA polymerase (see Sections 6.4.2.b and 6.5). This function requires the removal of a mismatched terminal residue. If this residue held the energy for the polymerization, i.e. if it were a triphosphate, then this energy would be dissipated on proof-reading and polymerization would halt.

Other evidence argues against the extension of DNA chains at the 5'-ends.

(1) Many DNA polymerases have been investigated and they all add deoxyribonucleoside 5'-phosphates on to a 3'-OH group on the growing DNA chain, i.e. the primer.
(2) No enzymes have been found which will generate or polymerize deoxyribonucleoside 3'-triphosphates, and neither have these putative substrates been found in cells.
(3) Both daughter strands have been shown to grow in the 5' → 3' direction.

The alternative explanation put forward by Okazaki[5] is that both chains are synthesized by the same mechanism but that one is made 'backwards' in short pieces which are subsequently joined together by DNA ligase (see Fig. 6.7). Careful investigation of electron micrographs (which give a resolution of about 100 nucleotides) shows that on one side of a replication fork the DNA is single-stranded (Fig. 6.8) and this supports the '*discontinuous*' mechanism. In this model the strand of DNA which is made continuously in the direction of fork movement is called the *leading strand* while that

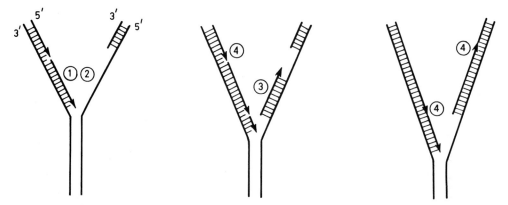

**Fig. 6.7** A working model of DNA synthesis. Synthesis occurs in the 5'→3' direction down the left-hand strand of the parental DNA molecule (1). This synthesis may be continuous or discontinuous and it exposes the right-hand strand of the parental DNA molecule. This single-stranded region (2) is probably stabilized by association with DNA-binding proteins. At some point DNA synthesis starts on the right-hand strand (3) which is copied backwards (i.e. 5'→3') until the growing strand has filled the gap. DNA ligase then joins the newly synthesized pieces (4).

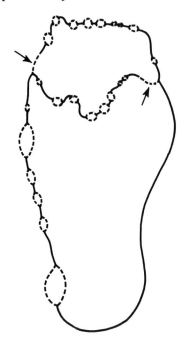

**Fig. 6.8** A partial denaturation map of the lambda replicating chromosome. Broken lines represent single-stranded regions of DNA and the arrows indicate such regions occurring at the replication fork. Drawn from the electron micrographs of Schnös and Inman [6].

made discontinuously in the reverse mode is called the *lagging strand*.

### 6.3.2 Okazaki pieces

The major support for discontinuous synthesis stems from the work of Okazaki with *E. coli* and T4-infected *E. coli*. He showed that the most recently synthesized DNA, i.e. that labelled with a brief pulse of tritiated thymidine, can be isolated *after denaturation* as short pieces now known as Okazaki pieces [5]. In this system the short pieces sediment at 8–10 S on a gradient of alkaline sucrose, representing chain lengths of 1000–2000 nucleotides. In animal cells the fragments are much shorter, being only about 100–200 residues long, i.e. 4–5 S [7, 8].

In order to detect Okazaki pieces it is essential to lower the temperature to slow down the rate of reaction and to give a *very brief* pulse of tritiated thymidine. Under these conditions considerably more than half the radioactivity is found in small pieces which led to the proposal that Okazaki pieces are made on both sides of the replication fork. (There is, however, some controversy over this; see below.) An alternative explanation is that discontinuities are introduced into DNA during repair of regions of DNA containing uracil in place of thymine. It has been calculated that in *E. coli* during DNA replication one uracil will be accidentally incorporated into DNA for every 300 thymines which may lead to a cut in the DNA every 1200 bases; a size not too different from that of the Okazaki pieces [9]. However, the frequency of uracil incorporation in *B. subtilis* is very low and in those mutants which fail to excise it (*ung⁻*) uracil is present less than once per 5000 bases [10].

Incorporation of uracil is accentuated in mutants lacking an active dUTPase (*dut⁻*) [11]. Such cells have elevated levels of dUTP and this competes with and is incorporated into DNA in place of dTTP. A uracil *N*-glycosidase removes the uracil and the damaged DNA is repaired by an excision repair mechanism which involves cutting the DNA chain on the 5′-side of the damage (see Chapter 7). *dut* mutants produce short Okazaki fragments (*sof*) only 100–200 nucleotides long [12]. The discovery of such *sof* mutants was initially misinterpreted as indicating that Okazaki fragments were formed by joining of several much shorter pieces. In double mutants lacking the uracil *N*-glycosidase (*ung⁻*) as well as the dUTPase the DNA contains significant amounts of uracil which is not excised [13]. Normal-sized Okazaki fragments are formed on the lagging strand of *dut ung* double mutants showing that the majority of these fragments are not normally produced as a result of repair but are apparently true intermediates in DNA replication [14, 15]. Moreover Okazaki pieces accumulate in mutant *E. coli* cells deficient in DNA ligase or DNA polymerase I whether or not uracil excision occurs [16, 17] and this implies that they are true intermediates in DNA synthesis and that these two enzymes are involved in their subsequent incorporation into high-molecular-weight DNA.

### 6.3.3 Direction of chain growth

When T4, growing at 8°C to reduce the rate of DNA synthesis, is incubated with [14C]thymidine for 150 seconds, and for the final 6 seconds with [3H]thymidine, Okazaki pieces can be isolated which apparently contain tritium at only the 3'-end [18, 19]. This was shown by degrading the isolated Okazaki pieces with exonuclease I of *E. coli* (which degrades single-stranded DNA from the 3'-end; see Chapter 4) when the 3H label is released before the 14C label. The complementary experiment using a nuclease from *B. subtilis* which acts from the 5'-end causes release of much of the 14C before the 3H is rendered acid-soluble.

These experiments demonstrate that Okazaki pieces are made in the 5' → 3' direction and support the discontinuous mechanism as outlined in Fig. 6.7.

### 6.3.4 Initiation of Okazaki pieces

In wild-type *E. coli* rifampicin (an inhibitor of *E. coli* RNA polymerase) inhibits replication of phage M13 and certain plasmids. Replication is normal in mutants with a rifampicin-resistant RNA polymerase [20]. Moreover, some RNA synthesis has been shown to be essential for the *in vitro* conversion of M13 single strands into the duplex form [21]. Such evidence, considered with the fact that all DNA polymerases require a 3'-hydroxyl priming end, suggested that this end may be provided by an oligoribonucleotide. That rifampicin does not inhibit continued replication of *E. coli* DNA (it does block the initiation of new rounds [22]) or the replication of other

phages such as φX174 was interpreted to mean that in these cases a second RNA polymerase (resistant to the drug) was involved in the synthesis of the priming oligoribonucleotides. Indeed, the product of the *dnaG* gene is believed to be such an enzyme and has been shown *in vitro* to synthesize short lengths of RNA on single-stranded DNA of phage G4 [25] (see Section 6.7.1). Although not required for the initial stages of M13 replication, the *dnaG* gene product is essential for duplication of the replicative form [24]. This and similar enzymes are called primases. The *dnaG* protein has been purified and shown to be a rifampicin-resistant RNA polymerase of molecular weight 64 000 [23, 127]. As well as polymerizing ribonucleotides it can also use deoxyribonucleotides [43, 128] and make a mixed primer though the first nucleotide is always rA.

Okazaki fragments containing oligoribonucleotides at their 5'-end have now been isolated from both prokaryotic and eukaryotic systems. However, the experimental proof of their existence was initially subject to much controversy because non-covalent RNA–DNA interactions gave some spurious results [25, 26]. However, methods (Fig. 6.9) to detect RNA covalently linked to the 5'-end of DNA molecules [25, 27] have shown that, although not all Okazaki pieces have ribonucleotides at their 5'-end, the shortest ones do. Moreover, RNA-linked Okazaki pieces accumulate in mutant *E. coli* cells deficient in either the polymerase or the 5' → 3' exonuclease of DNA polymerase I (see Section 6.4.2.b). Even so, using such a mutant, primers were detected on only about 30% of the Okazaki pieces [30] and in a wild-type *E. coli* only three to six of the total of 20–40 chains in the size range 3000–9000 bases chase into high-molecular-weight DNA [31]. When precautions are taken to eliminate all but covalent attachment of RNA to Okazaki pieces the size of the RNA segments obtained is only one to three nucleotides long and of no particular sequence [32, 33] which contrasts strongly with the 50–100

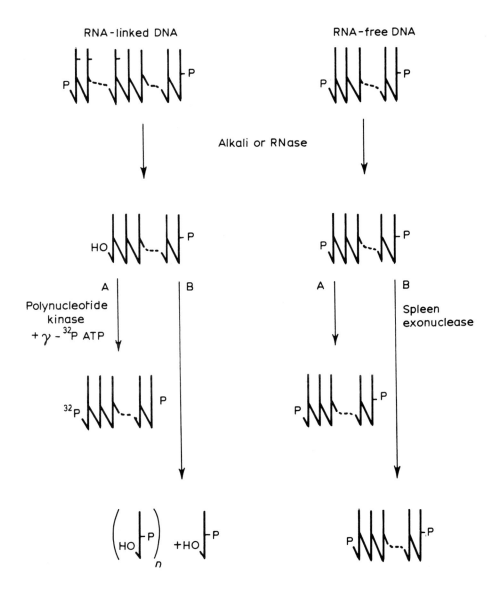

**Fig. 6.9** Methods to detect RNA-linked Okazaki pieces produced *in vivo* [25, 27]. The isolated Okazaki pieces are first treated with polynucleotide kinase and non-radioactive ATP to ensure that all 5′-ends are phosphorylated. Subsequent treatment with alkali or ribonuclease removes any RNA from the 5′ end and leaves a 5′-hydroxyl group. This group can be detected using polynucleotide kinase and [$\gamma$–$^{32}$P]–ATP (Method A). However, this method is not selective for nascent DNA. Method B overcomes this disadvantage by starting with tritiated Okazaki pieces resulting from a pulse labelling experiment. Spleen exonuclease is used under conditions where only 5′-OH terminated DNA (i.e. that initially linked to RNA) is degraded. These methods have been criticized on the grounds that treatment with alkali of DNA cleaved during repair (see chapter 7) may also yield 5′-OH groups [28, 29].

nucleotides erroneously attributed to the RNA primer [27, 34].

In animal cells about 40% of Okazaki pieces have RNA primers [35, 36] and the RNA primers of animal viruses are about 10 nucleotides long. They do not have a specific base sequence [8, 37, 38].

A review of the methods used to detect primers was published in 1976 [39].

The use of toluenized *E. coli* enables $\alpha$-[$^{32}$P]phosphate-labelled *deoxyribo*nucleoside triphosphates to penetrate the cell membrane when transfer of radioactivity to a *ribo*nucleotide occurs [34]. This is definite evidence for a covalent attachment of RNA to DNA (see Fig. 6.10). When DNA synthesis is studied in nuclei isolated from polyoma-infected cells (see Section 6.9.6) the nascent DNA chains are found with a length of about ten ribonucleotides at their 5'-ends [8, 37]. When $\alpha$-$^{32}$P-labelled deoxyribonucleoside triphosphatases are

injected into the slime mould *Physarum polycephalum* [40] there is transfer of the $^{32}$P to ribonucleotides, which is indicative, as described above, of a covalent attachment of RNA and DNA. Failure to detect RNA primers may simply reflect their transient nature. Indeed Okazaki's group have shown that many Okazaki pieces do not have RNA primers [25, 27] and RNA is not found in mature DNA. An exception to this is the DNA of the plasmid Col E1 grown in the presence of chloramphenicol. In this case it is inferred that the drug prevents the normal excision of the RNA primers [41]. Mitochondrial DNA also frequently has a number of ribonucleotides left in the mature duplex [42].

Studies with the primase indicate that it shows a very strong preference to initiate with adenosine followed by guanosine [43] and this suggests that initiation of Okazaki pieces may occur at particular sites on the lagging strand. This idea is supported by evidence from *E. coli*

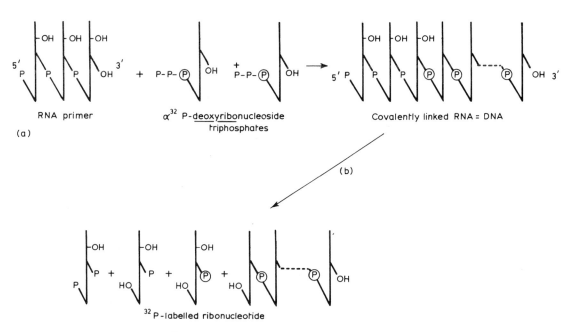

**Fig. 6.10**  The formation of the $^{32}$P-labelled ribonucleotide is indicative of the occurrence of a covalently linked RNA–DNA molecule where the radioactive phosphate forms the bridge between the two nucleic acids. (a) Extension of RNA primer; (b) Alkaline hydrolysis of product; ⓟ = radioactive phosphate.

suggesting that the sites of initiation are at or very close to sites of adenine methylation [44]. However, the small phage P4 which requires only about 20 Okazaki pieces per round of replication shows no preferential initiation sites [45].

### 6.3.5 Continuous synthesis

Discontinuous synthesis of one strand of DNA was postulated in order to overcome certain problems, but no problems appear to exist with the chain growing in the $5' \rightarrow 3'$ direction, and so there is no *a priori* reason why this chain should not be made continuously. Indeed, this may well be the case when DNA synthesis is proceeding rapidly. Under these conditions nascent DNA is found in both large and small pieces. However, when deoxyribonucleotides are limiting and the rate of chain elongation slows, both daughter strands appear to be made discontinuously. One possible explanation is that there is competition between the propagation of the growing chain and the initiation of new chains [46, 47].

Any initiation of an Okazaki piece on the leading strand is unlikely to be far in advance of the growing strand and so ligation will occur very rapidly, most probably before the Okazaki piece has grown to a detectable length. This is in contrast to the synthesis of an Okazaki piece on the lagging strand which (a) cannot be ligated until its synthesis is completed and (b) cannot be initiated until a region of single-stranded DNA of comparable length has been exposed on the lagging strand. Under conditions where primer removal and ligation is restricted (i.e. in the absence of a functional DNA polymerase I) Okazaki pieces are formed on both lagging and leading strands of the phage P2 [48]. However, under non-restrictive conditions discontinuities are not detected on the leading strand [49].

*In vitro* studies using the cellophane disc assay (see Section 6.6) have led to the conclusion that discontinuities on the leading strand are most likely produced as a result of uracil excision [50] and that the normal continuous synthesis of the leading strand can be masked by uracil excision [51].

A similar system for eliminaton of uracil residues from DNA exists in eukaryotes [52].

## 6.4 ENZYMES OF DNA SYNTHESIS

### 6.4.1 Introduction

As we have said DNA synthesis occurs at a replication fork which progresses along the DNA molecule. A variety of proteins is required to bring about this process in an efficient manner. The major proteins are as follows:

(a) A DNA polymerase is required to add dNTPs to the growing leading strand and to growing Okazaki pieces on the lagging strand. In general DNA polymerases act alongside other helper proteins.
(b) A primase is required to help initiate Okazaki pieces.
(c) An exonuclease is required to remove the primers.
(d) A ligase is required to join the Okazaki pieces together.
(e) An unwinding protein is required to help unwind the duplex DNA at the replication fork.
(f) A topoisomerase is required to relax the tension engendered by unwinding duplex DNA.
(g) A binding protein is required to stabilize single-stranded DNA exposed by progression of the leading strand prior to initiation of Okazaki pieces.
(h) A gyrase may be required to help unwind the double helix ahead of the replication fork or for initiation of replication.

The action of these proteins is summarized in Fig. 6.11 and the remainder of this section is devoted to the biochemistry of these proteins.

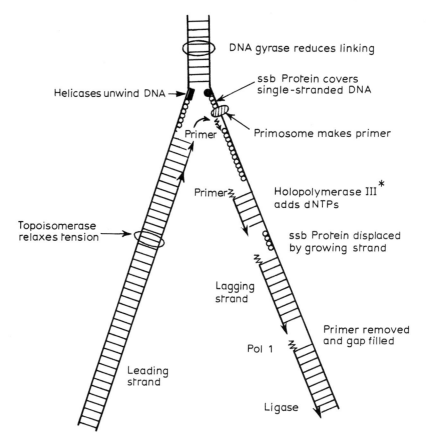

DNA gyrase reduces linking

ssb Protein covers
single-stranded DNA

Helicases unwind DNA →

Primosome makes primer

Primer

Holopolymerase III*
adds dNTPs

Primer₂

Topoisomerase
relaxes tension

ssb Protein displaced
by growing strand

Lagging
strand

Primer removed
and gap filled

Pol 1

Leading
strand

Ligase

**Fig. 6.11**  Site of action of enzymes involved in DNA replication at the growing fork in *E. coli.*

### 6.4.2 DNA polymerases

(a) *Mechanism of action*

DNA polymerase is the name given to an enzyme which catalyses the synthesis of DNA from its deoxyribonucleotide precursors. Such enzymes have been purified from various sources (bacterial, plant and animal) but, apart from some details discussed below, the mechanism of the polymerization reaction is common to all preparations.

DNA polymerase catalyses the formation of a phosphodiester bond between the 3'-OH group at the growing end of a DNA chain (the *primer*) and the 5'-phosphate group of the incoming deoxyribonucleoside triphosphate. Growth is in the 5' → 3' direction, and the order in which the deoxyribonucleotides are added is dictated by base-pairing to a *template* DNA chain. Thus, as well as the four triphosphates and $Mg^{2+}$ ions, the enzyme requires both primer and template DNA. No DNA polymerase has been found which is able to initiate DNA chains.

The simplest case to consider is that in which a single-stranded template has bound to it a growing strand of primer terminating at the growing point in a 3'-hydroxyl group (Fig. 6.12(1)). Such a situation is also illustrated in Fig. 4.9 for a piece of double-stranded DNA partially degraded by exonuclease III.

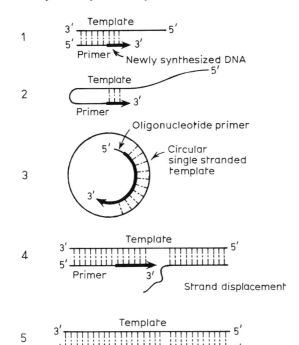

Fig. 6.12  Primer/templates for the replication of DNA.

the enzyme reaches the end of the template strand. *In vivo* the pyrophosphate is hydrolysed by a pyrophosphatase which drives the equilibrium in favour of DNA synthesis.

In the replication of single-stranded DNA the 3'-hydroxyl terminus may loop back on itself and serve as a priming strand as shown in Fig. 6.12(2), or short lengths of oligonucleotide may act as primers by becoming hydrogen-bonded to the template.

When a cyclic single-stranded DNA is used as template, replication can occur only in the presence of short oligonucleotides which become attached by base-pairing to the template and serve as primers (Fig. 6.12(3)). DNA polymerase cannot use double-stranded DNA as a primer/template. However, if a nick or gap is introduced into one strand the exposed 3'-OH

The polymerase binds to the single-stranded template in the region of the 3'-hydroxyl end of the primer (Figs 6.12 and 6.13). An incoming deoxyribonucleoside triphosphate containing a base which can pair with the corresponding base on the template becomes attached to the triphosphate binding site. The polymerase then catalyses a nucleophilic attack by the 3'-hydroxyl group of the primer on the α-phosphate of the deoxyribonucleoside 5'-triphosphate (Fig. 6.14). Inorganic pyrophosphate is released, a phosphodiester bond is formed, and the chain is lengthened by one unit. The enzyme moves along the template by the distance of one unit and the newly added nucleotide with its 3'-hydroxyl group now occupies the primer terminus site. The process is then repeated until

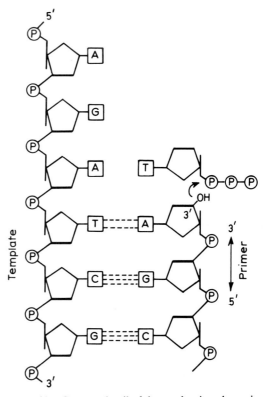

Fig. 6.13  Greater detail of the mechanism shown in Fig. 6.12 (1).

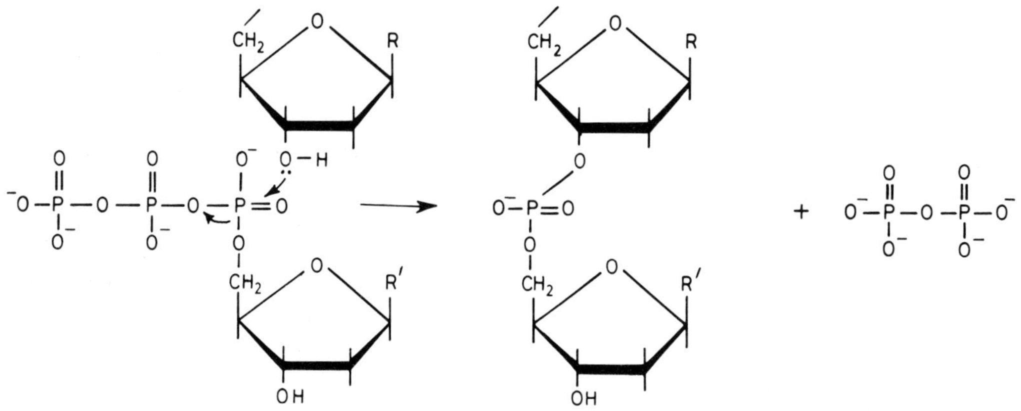

**Fig. 6.14**   The mechanism of the action of DNA polymerase.

group acts as a primer and the new growing strand will fill the gap and then may continue displacing the complementary strand from the 5'-end. In a strand displacement reaction (Fig. 6.12(4)) this can lead to complications since at some point the enzyme may leave the original template strand and begin to copy the complementary strand so that *in vitro* a branched structure is formed. In the presence of a 5' → 3' exonuclease the 3'-OH group at the nick can act as a primer for polymerase action while the exonuclease digests the DNA from the 5'-phosphate at the nick. The nick thus moves along the duplex DNA in a so-called *nick translation* reaction (Fig. 6.12(5)) – see below.

When native DNA is treated with exonuclease III, a partially single-stranded molecule is produced (Fig. 4.9). It can be repaired and restored to the native double-stranded form by the polymerase.

### (b) *E. coli DNA polymerase I (EC 2.7.7.7)*
Since a great amount of effort has been put into the study of DNA polymerase I from *E. coli* (the Kornberg enzyme) this archetypal polymerase will be considered in some detail and then comparisons will be made with other enzymes. Kornberg himself has written two books on the subject [53, 54].

(i) *Historical.* In 1956 Kornberg and his collaborators used cell-free extracts from exponentially growing cultures of *E. coli* to demonstrate that, in the presence of ATP and $Mg^{2+}$, [$^{14}C$]thymidine, a specific precursor for DNA, was incorporated into an acid-precipitable material which was judged to be DNA from the fact that it was no longer acid-precipitable after treatment with deoxyribonuclease. Using this incorporation method for assay purposes, the bacterial extract was fractionated and it soon became evident that crude extracts contained a kinase system which converted thymidine into thymidine triphosphate (dTTP). In subsequent work labelled dTTP was used as substrate and the enzyme responsible for its incorporation into DNA has been purified to homogeneity as judged by sedimentation, chromatography and electrophoresis.

With a partially purified enzyme preparation it was shown that, for good incorporation of dTTP (labelled either with $^{14}C$ in C-2 of the thymine or with $^{32}P$ in the innermost phosphate group), the 5'-triphosphates of all four deoxyribonucleosides which normally occur in DNA had to be present together with $Mg^{2+}$ ions and a denatured DNA template (calf thymus DNA was used in early experiments). Under these conditions excellent incorporation was observed as is

**Table 6.1** Incorporation of [$^{32}$P]dCTP into DNA (Bessman *et al.* [55]).

| Additions | [$^{32}$P]DNA (pmol) |
|---|---|
| dCTP (DNA omitted) | 0.0 |
| dCTP | 2.5 |
| dCTP + dGTP | 5.1 |
| dCTP + dGTP + dTTP | 15.7 |
| dCTP + dGTP + dTTP + dATP | 3300 |

The incubation mixture (0.3 ml) contained 5 nmol of $\alpha$[$^{32}$P]dCTP ($7.2 \times 10^7$ cpm mol$^{-1}$) and 5 nmol of each of the other deoxynucleoside triphosphates as indicated, together with 1.9 $\mu$mol of MgCl$_2$, 20 $\mu$mol of glycine buffer (pH 9.2), 10 $\mu$g of DNA, and 3 $\mu$g of purified 'polymerase' from *E. coli*. The incubation was carried out at 37°C for 30 minutes.

illustrated in Table 6.1. Omission of any one triphosphate or Mg$^{2+}$ or of DNA template, or pretreatment of the template with deoxyribonuclease, reduced the incorporation to a very low level. Deoxyribonucleoside diphosphates could not replace the triphosphates. With the purified enzyme it was also possible by chemical analysis to demonstrate net synthesis of DNA, and increases of DNA by a factor of 10–20 could be obtained indicating that 90–95% of the isolated DNA was derived from the added deoxyribonucleoside triphosphates.

As well as DNA polymerase activity, the purified enzyme shows several other activities each of which is associated with the same enzyme molecule. Thus exonuclease activity (both 3′ → 5′ and 5′ → 3′) is purified along with the polymerase activity in a ratio which is not altered by fractionation.

(ii) *Evidence for the copying of the template.* The DNA polymerase is an enzyme with the unusual property of taking directions from a template and faithfully reproducing the sequence of nucleotides in the product. The evidence for this rests on several experimental observations.

(1) The most significant fact is that the enzyme can faithfully copy the nucleotide sequence of the single-stranded cyclic DNA bacteriophage $\phi$X174. The product can be converted into biologically active material by subsequent cyclization by the enzyme polynucleotide ligase. This experiment was done in 1967 by Kornberg and his colleagues [56] who produced, *in vitro*, infective DNA identical with the natural material. These findings show that DNA polymerase I is an enzyme which has all the requirements necessary for the replication of DNA.

(2) When *E. coli* polymerase is used to prepare DNA under conditions such that only 5% of the sample produced comes from the template, the product has many of the same physical and chemical properties as DNA isolated from natural sources. The product appears to have a hydrogen-bonded structure similar to that of natural DNA and undergoes molecular melting (see Chapter 2) in the same way.

It shows the same equivalence of adenine to thymine and guanine to cytosine that characterizes natural DNA. Moreover the characteristic ratio of A·T pairs to G·C pairs of a given DNA primer is imposed on the product whether the net DNA increase is 1% or 1000%. The base ratios in the product are not distorted when widely differing molar proportions of substrate are used. This is best illustrated by the use of the synthetic template poly(dA-dT) when, from a mixture of all four deoxyribonucleoside triphosphates, only dAMP and dTMP are incorporated into the product.

(3) The nucleotide sequence in the template DNA is reproduced in the product. This has been established by the Kornberg group by using the technique of *nearest-neighbour sequence analysis* [57, 58]. The partially purified *E. coli* enzyme is incubated with a particular template DNA and all four deoxyribonucleoside triphosphates, one of which, say dATP, is labelled with $^{32}$P in the innermost phosphate. During the

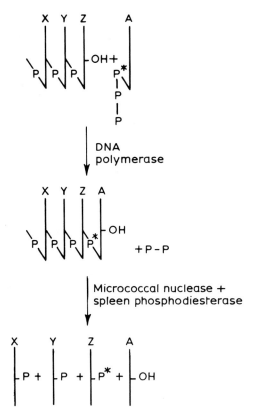

**Fig. 6.15** Illustration of the method of nearest-neighbour sequence analysis.

synthetic reaction this $^{32}$P becomes the bridge between the nucleoside of that labelled triphosphate (A) and the nearest-neighbour nucleotide containing the base Z at the growing end of the

polynucleotide chain (Fig. 6.15a). After the synthetic reaction is complete the DNA is isolated and degraded with micrococcal DNase and spleen phosphodiesterase (Chapter 4) to yield the deoxyribonucleoside 3'-monophosphates. The $^{32}$P is thus transferred to the 3'-carbon of the neighbouring nucleotide in the chain (Z in Fig. 6.15), i.e. the one with which the labelled triphosphate has reacted (Z might be any one of the four bases). The four deoxyribonucleoside 3'-monophosphates are isolated by paper electrophoresis or HPLC[59], and their radioactivities measured to give the relative frequency with which the originally labelled nucleotide locates itself next to other nucleotides in the new chain. This procedure is carried out four times with a different labelled triphosphate each time so as to determine the relative frequencies of all sixteen possible nearest-neighbour (or dinucleotide) sequences [58]. The results of such an experiment with DNA from *Mycobacterium phlei* as primer are shown in Table 6.2.

They illustrate several points:

(1) All sixteen possible nearest-neighbour sequences are present and they occur with widely varying frequencies.

(2) The results show a very striking deviation from the nearest-neighbour frequencies predicted if the arrangement of mononucleotides were completely random. Thus the frequency of TpA in the first row is quite different from that of

**Table 6.2** Nearest-neighbour frequencies of *Mycobacterium phlei* DNA [60].

| Labelled triphosphate | Deoxyribonucleoside 3'-phosphate-isolated | | | |
|---|---|---|---|---|
| | Tp | Ap | Cp | Gp |
| dATP | TpA 0.012 | ApA 0.024 | CpA 0.063 | GpA 0.065 |
| dTTP | TpT 0.026 | ApT 0.031 | CpT 0.045 | GpT 0.060 |
| dGTP | TpG 0.063 | ApG 0.045 | CpG 0.139 | GpG 0.090 |
| dCTP | TpC 0.061 | ApC 0.064 | CpC 0.090 | GpC 0.122 |
| Sums | 0.162 | 0.164 | 0.337 | 0.337 |

ApT in the second row whereas these two frequencies would have to be identical in a random assembly. The nucleotides have therefore been assembled in accordance with a definite pattern.

(3) The sums of the four columns show the equivalence of A to T and of G to C in the product and indicate both the validity of the analytical method and the replication of the overall composition of the primer DNA.

(4) The results indicate that base-pairing occurs in the newly synthesized DNA and that its two strands are of opposite polarity. According to the Watson–Crick model the two strands of the double helix are of opposite polarity, and it is presumed that each can act as a template for the formation of a new chain so as to give precise replication with the formation of two daughter helices identical with each other and with the parent helix. The results of nearest-neighbour sequence analysis support this mechanism. For example, the frequencies of ApA and TpT sequences are equivalent, and so are the frequencies of GpG and CpC. The matching of the other sequences depends upon whether the strands of the double helix are of similar or opposite polarity (Fig. 6.16).

If the strands are of opposite polarity the following matching sequences can be predicted:

CpA and TpG

GpA and TpC

CpT and ApG

GpT and ApC

whereas if the strands are of the same polarity the matching sequences would be:

TpA and ApT

CpA and GpT

GpA and CpT

TpG and ApC

ApG and TpC

CpG and GpC

The results in Table 6.2 favour the helix with strands of opposite polarity. The ApA and TpT,

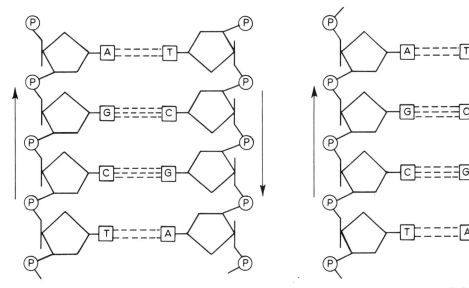

**Fig. 6.16** Possible structures of the DNA molecule showing opposite polarity of strands (left) and the same polarity of strands (right).

and the CpC and GpG sequences match in both models.

(5) The nearest-neighbour frequencies measured by the method described above are those of the newly synthesized DNA. To verify that they are an accurate reflection of those in the original DNA template, an enzymically synthesized sample of calf thymus DNA in which only 5% of the total DNA consisted of the original template was itself used as template in a sequence analysis. The results showed good agreement between the sequence frequencies of the products primed by native DNA and by enzymically produced DNA whereas DNAs from other sources gave quite different results.

It can be concluded therefore that the polymerase yields a DNA product with strands of opposite polarity and that the sequence of bases is faithfully reproduced.

(iii) *The chemical nature of DNA polymerase I.* The purified Kornberg enzyme is a protein of molecular weight 109 000 in the form of a single polypeptide chain [61]. This chain can be unfolded in guanidine–HCl–mercaptoethanol so as to denature the protein. When the reagent is diluted out renaturation occurs with restoration of activity.

The protein migrates as a single band on SDS/acrylamide gel electrophoresis; it contains only one sulphydryl group and one disulphide group; the residue at the N-terminus is methionine. One *E. coli* cell contains about 400 molecules of enzyme, each spherical and of diameter 6.5 nm. It can form dimeric forms which can be visualized in electron micrographs [62].

(iv) *The active centre of the enzyme.* On the basis of binding experiments Kornberg has concluded that the enzyme is able to catalyse the following operations [54, 63]:

(1) Extension of a DNA chain in the $5' \rightarrow 3'$ direction by the addition to the 3'-hydroxyl terminus of mononucleotides from deoxy-ribonucleoside triphosphates at the rate of 1000 nucleotides per minute.

(2) Hydrolysis of a DNA chain from the 3'-hydroxyl end in the $3' \rightarrow 5'$ direction to yield 5'-monophosphates (the exonuclease II action referred to in Chapter 4).

(3) Hydrolysis of DNA chain from the 5'-phosphate (or 5'-hydroxyl) terminus in the $5' \rightarrow 3'$ direction to yield mainly 5'-monophosphates.

(4) Pyrophosphorolysis of a DNA chain from the 3'-end; this is essentially the reversal of the polymerization reaction.

(5) Exchange of inorganic pyrophosphate with the terminal pyrophosphate group of a deoxyribonucleoside triphosphate as a result of alternating polymerization and pyrophosphorolysis. These last two reactions are of doubtful *in vivo* significance since they require a high concentration of inorganic pyrophosphate.

Treatment of the enzyme with the protease *subtilisin* from *B. subtilis* breaks it into two fragments, a larger fragment known as the Klenow fragment of molecular weight 76 000 which retains polymerase activity and $3' \rightarrow 5'$ nuclease activity but not the $5' \rightarrow 3'$ nuclease activity, and a smaller fragment of molecular weight 34 000 which retains nuclease $5' \rightarrow 3'$ activity in the presence of DNA.

Kornberg envisages the active centre of the enzyme as a specially adapted polypeptide surface comprising at least five major sites.

(1) A site for the binding of the template chain in the region where base pairs are formed and for a few nucleotides on each side of this.

(2) A site for the growing primer chain which is, of course, base-paired to the template.

(3) A site for the special recognition of the terminal 3'-hydroxyl group of the primer. This point is the start of the $3' \rightarrow 5'$ hydrolytic cleavage.

(4) A triphosphate binding site for which all four triphosphates compete.

(5) A site which allows for the $5' \rightarrow 3'$ cleavage of a 5'-phosphoryl-terminated chain. It is presumably this area that is broken off by subtilisin.

These sites determine the nature of the DNA which can bind to the enzyme. For example, linear single-stranded DNA binds readily on site (1) whereas an intact linear duplex such as the DNA of bacteriophage T7 does not bind if it has been prepared with great care so as to avoid internal breaks. An intact circular duplex such as plasmid DNA or $\phi$X174 replicative form DNA does not bind to the enzyme until a 'nick' has been introduced in one of the strands by an appropriate nuclease yielding 3'-hydroxyl and 5'-phosphate termini. Such nicks are active points for replication whereas nicks introduced by micrococcal nuclease with 5'-hydroxyl and 3'-phosphoryl termini are not replication points although they bind the enzyme. One molecule of enzyme is bound at each nick in either case.

Recently evidence has been obtained that there are two distinct sites which recognize the 3'-OH group of the primer [64]. When the terminal nucleotide is correctly matched to the template the 3'-OH group resides in the polymerase site. However, if the terminal nucleotide is not correctly matched to the template the 3'-OH group is displaced into a site where it is susceptible to the $3' \rightarrow 5'$ exonuclease (see Section 6.5).

The gene coding for *E. coli* DNA polymerase I has been sequenced and the sequence has been used to predict the primary and secondary structure of the protein [275, 276]. X-ray crystallography has shown the enzyme to have a deep crevice capable of binding duplex DNA [668].

(v) *Poly d(A–T) and poly(dG). poly (dC).* When DNA polymerase I is incubated without primer in the presence of dATP, dTTP and $Mg^{2+}$ an interesting polymer is formed containing adenine and thymine nucleotides [54, 65].

Polymer formation occurs only after a lag period of several hours and then takes place rapidly until 60–80% of the triphosphates have been utilized.

The product of the reaction contains equal amounts of adenine and thymine. Nearest-neighbour frequency analysis shows that the frequencies of ApT and TpA are each 0.500 whereas the sequences ApA and TpT are undetectable. The polymer therefore contains alternating residues of A and T.

The molecular weight calculated from the sedimentation value and reduced viscosity is between $2 \times 10^6$ and $8 \times 10^6$. The polymer melts sharply at 71°C with an increase of 37% in the absorbance at 260 nm. The process is completely reversible on cooling. Such physical data, including the X-ray diffraction pattern, suggest that the molecule is a long fibrous double-stranded structure with the strands joined by hydrogen bonds between adenine and thymine bases.

A somewhat similar polymer [54, 57] is formed when *E. coli* DNA polymerase I is incubated with high concentration of dGTP and dCTP in the presence of $Mg^{2+}$. Again, a lag period of several hours is found. This product contains guanine and cytosine, not necessarily in equal amounts, and nearest-neighbour sequence analysis shows that the frequencies of GpG and CpC are each 0.500. Mild acid hydrolysis releases all the dGMP but none of the dCMP. Sedimentation and viscosity measurements yield values similar to those found with the poly[d(A–T)] but the $T_m$ value is much higher (83°C). These observations are consistent with the view that the molecule consists of two homopolymers, one containing only guanine and the other only cytosine, hydrogen-bonded throughout their lengths.

(vi) *The $3' \rightarrow 5'$ exonuclease and proof-reading.* Before it was realized that this was an integral function of the polymerase, the $3' \rightarrow 5'$ exonuclease activity was ascribed to an exonuclease

II. It appears that this exonuclease functions to recognize and eliminate a non-base-paired terminus on the primer DNA [66]. When *E. coli* DNA polymerase I is provided with four triphosphates and a primer/template with a mismatched end, the non-matching terminus is removed by the $3' \rightarrow 5'$ exonuclease before polymerization begins. (In the absence of the triphosphates the exonuclease continues to remove nucleotides from the frayed ends of the DNA molecules [67].)

Such a nuclease by correcting errors occurring on polymerization may be expected to increase dramatically the fidelity of the base-pairing mechanism [53] (see Section 6.5).

This proof-reading mechanism provides a justification for involving 5'-deoxyribonucleoside triphosphates in the $5' \rightarrow 3'$ extension of a 3'-OH on the growing DNA chain. Addition of nucleotides to a growing chain terminating in a triphosphate (see Fig. 8.6) would result, during proof-reading, in the removal of the energy required for further extension.

However, there are arguments for and against the significance of this proposed proof-reading mechanism. Certain *E. coli* mutants show an increased or decreased rate of mutation which can be correlated with a change in the relative activities of the polymerase and $3' \rightarrow 5'$ activities of DNA polymerase I. However, as we shall see below, DNA polymerase I is believed to have a role not as the major polymerase involved in replication but as an enzyme involved in the repair of DNA. Furthermore, the rate of generation of dNMP by exonuclease action could account for only a minor increase in fidelity [68].

The *error rates* in eukaryote polymerases (see below) which do not have associated $3' \rightarrow 5'$ exonucleases are comparable with that for *E. coli* DNA polymerases I [68–70].

On the other hand in *E. coli* infected with phage T4 (which codes for its own DNA polymerase) there are *mutator* and *antimutator polymerases* which show (a) decreased or increased rates of the pyrophosphate exchange reaction and (b) increased or decreased rates of incorporation of non-complementary nucleotides, both of which have been correlated with changes in $3' \rightarrow 5'$ exonuclease activities [71–73]. This is considered in more detail in Section 6.5.

(vii) *The $5' \rightarrow 3'$ exonuclease.* This activity, which is present in the smaller fragment released from DNA polymerase I by *subtilisin* treatment, cleaves base-paired regions of DNA, releasing oligonucleotides from 5'-ends. Because of its ability to jump several bases at a time, this nuclease can act on DNA molecules containing mismatched bases or distortions which render them unsuitable as substrates for polymerase [74]. It may thus serve a function, for instance, in the elimination of thymine dimers from DNA exposed to ultraviolet radiation (see Chapter 7).

When *E. coli* polymerase I binds to a nick on double-stranded DNA two reactions occur simultaneously. Polymerization extends the 3'-OH end and $5' \rightarrow 3'$ exonuclease degrades the 5'-phosphate terminus. This results in *nick translation*. If DNA polymerase I is incubated with a nicked duplex DNA molecule and radioactive deoxyribonucleoside triphosphates then the nick translation reaction results in replacement of long sections of the DNA molecule with a radioactive stretch of identical sequence, a reaction of great importance to modern molecular biology (see Section A.4.1). Nick translation may only end when the enzyme reaches the end of the DNA molecule [75]. Alternatively nick translation may end if a long 5'-terminus is released from the nick thereby rendering it no longer susceptible to the exonuclease. The branched product typical of reaction with native DNA primer/templates would then be formed. Normally DNA polymerase I is highly processive. For each DNA/enzyme association event hundreds of nucleotides are polymerized before dissociation occurs [76]. However, a mutant of DNA polymerase I

(PolA5) shows all the normal properties of the wild-type enzyme but has a reduced rate of polymerization and nick translation caused by a lowered affinity of the enzyme for the DNA primer/template [77].

(viii) *The role of the Kornberg enzyme in vivo*. Although the DNA polymerase I is exceedingly effective in the copying of a single-stranded DNA template when provided with a primer, it is much less effective with double-stranded DNA. This observation and other considerations led to doubts as to the role of the Kornberg enzyme (DNA polymerase I) in the replication of DNA *in vivo* and to the suggestion that it is concerned merely in maintenance and repair of DNA (see Chapter 7). These considerations may be summarized as follows:

(1) The purified enzyme cannot replicate intact double-stranded DNA semiconservatively to yield a biologically active product.
(2) The purified enzyme catalyses the incorporation of 1000 nucleotides per minute per molecule of enzyme whereas the estimated rate of incorporation *in vivo* is 100 times faster.
(3) Mutants of *E. coli* have been isolated which contain apparently normal Kornberg enzyme but are defective in DNA duplication. This demonstrates that other enzymes (which *may* include other polymerases) are required for *in vivo* DNA synthesis, and this may help to explain the deficiencies enumerated under (1) and (2) where only the purified Kornberg polymerase was present.

The evidence against the Kornberg enzyme being *essential* for replication *in vivo* is based on the properties of a mutant of *E. coli* (PolA1 or PolA$^-$) isolated by de Lucia and Cairns [78, 79]. This mutant and several others discovered later multiply normally but contain 1% or less of the Kornberg enzyme activity present in wild-type cells. Such mutants, however, show a reduced

ability to join Okazaki fragments [80] and an increased sensitivity to ultraviolet [78] and ionizing radiation [81] and to alkylating reagents [78, 82]. This implies a role for the enzyme in 'gap' filling, and perhaps also in excision of RNA primers and mismatched base pairs. However, these functions are not completely lacking in mutants lacking DNA polymerase I and it is suggested that they may also be performed to a limited extent by other enzymes, e.g. polymerases II and III and the RecB and C nuclease (see Chapter 7). Double mutants of PolA and RecB are non-viable [83], suggesting an essential function which can be carried out by more than one enzyme.

### (c) *E. coli DNA polymerase II*

The discovery of the PolA$^-$ mutant of *E. coli* which grows well and replicates its DNA in the usual manner in spite of the absence of the Kornberg enzyme gave an impetus to the search for another enzyme apparatus which can synthesize DNA. Two further polymerases, designated DNA polymerase II and III, were found in extracts of PolA$^-$ cells. These enzymes had not been detected previously because, in extracts of wild-type cells, they show little activity relative to DNA polymerase I when single-stranded or nicked DNA is used as template. DNA polymerase II and III show significant activity only with a 'gapped' DNA template (Table 6.3). The enzymes can be separated from one another by chromatography on phosphocellulose, DEAE-cellulose, or DNA–agarose [54].

Purified DNA polymerase II has a molecular weight of 90 000–120 000 and is homogeneous as judged by SDS/polyacrylamide gel electrophoresis. It synthesizes DNA in the $5' \rightarrow 3'$ direction and for maximal activity it requires all four triphosphates, $Mg^{2+}$, $NH_4^+$, and a native DNA template containing single-stranded gaps 50–200 bases long. The rate of reaction falls off with longer gaps, but may be restored by addition of *E. coli* DNA-binding protein (see

**Table 6.3**  DNA polymerases of *E. coli*.

| Polymerase (gene) | Molecular weight | Molecules per cell | Nucleotides polymerized/s (a) per enzyme molecule (b) per bacterial cell | Direction of (a) polymerization (b) exonuclease action | Template (all require 3'-OH primer) |
|---|---|---|---|---|---|
| I (*polA*) | 109 000 | 400 | (a) 16–20 (b) 8000 | (a) 5' → 3' (b) 3' → 5' and 5' → 3' | Denatured Nicked Gapped |
| II (*polB*) | 120 000 | 17–100 | (a) 2–5 (b) 500 | (a) 5' → 3' (b) 3' → 5' | Gapped |
| III (*polC*) | 180 000 | 10 | (a) 250–1000 (b) 10 000 | (a) 5' → 3' (b) 3' → 5' and 5' → 3' | Gapped |

Section 6.4.4). The enzyme also requires a 3'-OH primer. It is sensitive to sulphydryl reagents and is not affected by antiserum to DNA polymerase I. The purified enzyme, however, like polymerase I, only synthesizes DNA at rates a fraction of those found *in vivo*. The enzyme also possesses 3' → 5' exonuclease activity, but no 5' → 3' exonuclease activity [84].

The function of *E. coli* DNA polymerase II *in vivo* is unknown and mutants lacking the enzyme appear normal in all respects. However, double mutants lacking both polymerase I and polymerase II join Okazaki fragments even more slowly than do mutants lacking polymerase I alone [85].

### (d)  *E. coli DNA polymerase III*

*E. coli* with a temperature-sensitive mutation in the gene for DNA polymerase III (*dnaE* or *polC*) are not viable at the restrictive temperature [86] and lysates prepared from them are defective in DNA synthesis [84, 87] (see Section 6.5.2). Complementation of such lysates with DNA polymerase III purified from normal cells restores their DNA synthetic ability. This is strong evidence that, unlike DNA polymerases I and II, polymerase III is *essential* for DNA synthesis.

DNA polymerase III has been purified about 20 000 fold and shown to be highly sensitive to salt in dilute solutions (i.e. normal assay conditions) but not when assayed by complementation (i.e. at high protein concentration) [53, 84, 88]. Although there appear to be only about 10 molecules per bacterial cell, its high rate of polymerization of nucleotides (Table 6.3) shows it to be capable of its proposed role in DNA synthesis.

The best template for DNA polymerase III is double-stranded DNA with many small gaps containing 3'-OH priming ends.

DNA polymerase III exists in a complex holoenzyme form in the cell and it proved difficult to determine the molecular weight and subunit composition of the enzyme. The $\alpha$ subunit of the holoenzyme (that coded for by the *dnaE* or *polC* gene) has a molecular weight of 140 000 [54, 89–93] and in DNA polymerase III core enzyme it is associated with two small subunits: $\epsilon$ (molecular weight 25 000) and $\theta$ (molecular weight 10 000) [54, 94] (see Table 6.4). The core enzyme has both 3' → 5' exonuclease (which could be involved in proofreading) and 5' → 3' exonuclease activities though the latter is only manifest *in vitro* on duplex DNA with a single-stranded 5'-tail [89, 91]. For this reason the core enzyme cannot use a

**Table 6.4** Subunit composition of *E. coli* DNA polymerase III [54, 94, 238, 480].

| Subunit | Gene | Molecular weight | Name |
|---|---|---|---|
| $\alpha$ | *polC* | 140 000 | Polymerase III |
| $\epsilon$ | *dnaQ* | 25 000 | $3' \rightarrow 5'$ exonuclease |
| $\theta$ |  | 10 000 | |
| $\tau$ |  | 83 000 | |
| $\gamma$ | *dnaZ* | 52 000 | EFII |
| $\delta$ | *dnaX* | 32 000 | EFIII |
| $\beta$ |  | 440 000 | EFI, copol III* |

Groupings (braces at right): $\alpha$, $\epsilon$, $\theta$ = Polymerase III core enzyme; core enzyme + $\tau$ = pol III′; pol III′ + $\gamma$, $\delta$ = pol III*; pol III* + $\beta$ = Polymerase III holoenzyme.

nicked primer/template but requires a gap. The $3' \rightarrow 5'$ exonuclease activity resides in the $\epsilon$ subunit and its role in proof-reading will be considered in Section 6.5.

DNA polymerase III core enzyme cannot use long single-stranded DNA molecules as a template even when provided with a primer. In the presence of the tau ($\tau$) subunit it dimerizes to form polymerase III′ – $(\alpha\epsilon\theta)\tau_2(\alpha\epsilon\theta)$. Polymerase III′ associates with subunits $\gamma$ and $\delta$ to form polymerase III*. In the presence of the $\beta$ subunit (also known as copolymerase III* or factor I) the holoenzyme is formed which, with lipids and ATP, is able to extend an RNA or DNA primer hydrogen-bonded to circular, single-stranded M13, G4, or $\phi$X174 DNA [53, 54, 89, 93, 95, 96].

In the presence of ATP (which interacts with the $\alpha$, $\tau$, $\gamma$ and $\delta$ subunits) the holopolymerase III forms an initiation complex with a primed template which results in hydrolysis of two ATP molecules and a rearrangement of the holo-polymerase such that the $\beta$-subunit becomes inaccessible to antibody and the enzyme is unaffected by a concentration of KCl (150 mM) which inhibits the non-activated enzyme [97–100]. The ATP-activated holoenzyme shows a much enhanced processivity, the one complex being capable of complete replication of a molecule of primed circular phage DNA. On completing replication the holoenzyme dissociates and, in the presence of ATP, will re-form on a new primed template [480].

*B. subtilis*, a bacterium widely divergent from *E. coli*, also has three DNA polymerases which closely resemble those of *E. coli*. They differ in that polymerases I and II and also possibly III are devoid of nuclease activity and, in addition, DNA polymerase III is sensitive to 6-(*p*-hydroxyphenylazo)uracil (HPUra), an antibiotic active against Gram-positive bacteria. DNA synthesis in lysates of *B. subtilis* is also inhibited by HPUra [101].

**(e) *DNA polymerases in eukaryotes***
While the DNA polymerases of micro-organisms have been studied more intensively, similar enzymes are present in eukaryotic cells. In general they resemble the bacterial enzymes in their requirement for a template, a 3′-OH primer and four deoxyribonucleoside triphosphates [102–104].

In higher eukaryotes there are three DNA polymerases known as $\alpha$, $\beta$ and $\gamma$. In Yeast, DNA polymerase I corresponds to DNA polymerase $\alpha$ [141, 142, 149]. On cell fractionation the bulk of DNA polymerase $\beta$ is found in the nuclear fraction and there is considerable evidence that the *in vivo* location of DNA polymerase $\alpha$ is intranuclear despite its recovery in high-speed supernatant fractions on cell disruption [105, 106]. In those cells making DNA there is evidence for association of DNA polymerase $\alpha$ with the nuclear matrix [107, 108].

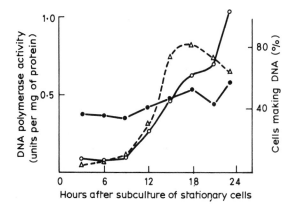

**Fig. 6.17** Variation in the activities of nuclear ($\beta$) and supernatant ($\alpha$) DNA polymerase with growth phase of cultured mouse L cells: ●–●, nuclear; ○–○ supernatant; △–△ % cells making DNA.

Indeed DNA polymerase $\alpha$ activity varies dramatically with the rate of cell division, being undetectable in non-growing cells and tissues [109–113] (Fig. 6.17).

DNA polymerase $\alpha$ has a pH optimum in the range 6.5–8.0 and is highly susceptible to inhibition by thiol-active reagents (i.e. *N*-ethyl-maleimide and *p*-chloromercuribenzoate). It is also inhibited strongly by araCTP and by aphidicolin (a tetracycline diterpene tetraol) [114, 688]. In both instances inhibition is competitive with dCTP. Strangely it is resistant to inhibition by 2′,3′-dideoxythymidine triphosphate (ddTTP) [115, 116]. Use of these inhibitors together with consideration of the high enzymic activity found in growing cells has led to the conclusion that this enzyme plays the major role in DNA replication in eukaryotes. A temperature-sensitive mutant cell line which is defective in DNA synthesis at the non-permissive temperature has a heat-labile DNA polymerase $\alpha$ [119].

DNA polymerase $\alpha$ shows optimal activity with a gapped DNA template but shows a remarkable ability to use single-stranded DNA by forming transient hairpins [117]. It will use synthetic single-stranded DNA templates pro-

vided with either oligoribo- or deoxyribo-nucleotide primers, e.g. poly(dT)·oligo(dA) or poly(dT)·oligo(A) [118, 119]. It will not bind to duplex DNA. As normally isolated, polymerase $\alpha$ is only poorly processive, i.e. it adds about 11 nucleotides before it dissociates from the DNA [120, 121]. The processivity can be dramatically increased in the presence of other proteins and ATP [130]. In contrast to the enzymes from *E. coli* no nuclease activity normally is associated with the eukaryotic enzymes. Exceptions include a high molecular weight form, DNA polymerase $\delta$, which has 3′ → 5′ exonuclease activity [122], and an enzyme from herpes-virus-infected cells [123].

Because of its extreme lability to proteolysis during purification and its apparent heterogeneity, DNA polymerase $\alpha$ has only recently been purified to the same extent as *E. coli* DNA polymerase I [124] despite the fact that a 200-fold purification of the calf thymus enzyme was achieved in 1965 when it was separated from a terminal transferase (see below).

DNA polymerase $\alpha$ sediments in sucrose gradients as a broad peak in the 6–11S region and on gel filtration activity is found in a number of regions [117, 125, 126, 129]. It is probable, but by no means certain, that the native, undegraded enzyme consists of two subunits: a 185 000-molecular-weight polymerase together with a 60–70 000-molecular-weight primase [697, 698, 134, 137]. However, on isolation the bulk of the polymerase is recovered as an active 155 000-molecular-weight protein together with several smaller active proteins which all share some primary sequence homology [145–148]. Of particular interest is the presence of the associated *primase* activity which allows the enzyme to initiate replication on unprimed single-stranded cyclic DNAs [130–144, 699]. The primase is a novel RNA polymerase (i.e. not RNA polymerase I, II or III – see Section 8.1.2) which synthesizes a short (8–15) oligoribo-nucleotide which, in the presence of deoxy-ribonucleotides, is extended by the DNA polymerase $\alpha$ activity. Primase does not act at random but primers are nearly always initiated with a GTP two to ten nucleotides 5' to a hexanucleotide resembling the sequence $C_2A_2C_2$ [701]. The association of primase with DNA polymerase $\alpha$ is restricted to the DNA-synthetic phase of the cell cycle and the complex forms part of the nuclear DNA replitase or replicase (see Section 6.11).

DNA polymerase $\beta$ is recovered in the nuclear fraction but shows little correlation between activity and the growth rate of the cells. This enzyme, which has been purified to homogeneity, has a pH optimum of around 8.6–9.0 [150] and a molecular weight of 43 000 [151]. However, in the final purification step used for the enzyme from Novikoff hepatoma or guinea-pig liver a polypeptide is removed from the enzyme whose molecular weight falls to 32 000. Re-addition of the polypeptide greatly stimulates enzymic activity [152]. DNA polymerase $\beta$ shows optimal activity with native DNA activated by limited treatment with DNase I (to produce single-stranded nicks and short gaps bearing 3'-OH priming termini) [153, 154] and shows negligible activity with denatured DNA. DNA polymerase $\beta$ is relatively insensitive to thiol-active reagents, araCTP and aphidicolin but is strongly inhibited by ddTTP. The fact that DNA polymerase $\beta$ never processes, i.e. it leaves the DNA primer after addition of a single nucleotide [76], may account for its insensitivity to araCTP. It resembles DNA polymerase $\alpha$ in being devoid of nuclease activities. Thus neither enzyme has the 'proof-reading' capacity of the bacterial enzymes, and in fact both will add on to a mismatched primer terminus [151, 155].

DNA polymerase $\beta$ cannot use an oligoribonucleotide primer but, unlike polymerase $\alpha$, given an oligodeoxyribonucleotide primer it is able to copy a ribonucleotide template [e.g. poly(A)·oligo(dT)] [109, 119]. DNA polymerase $\beta$ is believed to play a role in repair of DNA and studies *in vitro* using nuclei from irradiated neuronal cells [112, 156] or from non-dividing mouse L929 cells treated with nuclease [157] show the repair synthesis to be sensitive to ddTTP. However, damage by ultra-violet light or alkylating agents, especially when it is severe, involves DNA polymerase $\alpha$ in the repair reaction as well as DNA polymerase $\beta$ [158–160].

The use of an antibody to DNA polymerase $\beta$ from calf thymus has demonstrated cross-reacting molecules of similar size in eukaryotes from protozoa to man indicating a lasting and essential role for this enzyme [179].

Animal cells also have a small amount of DNA polymerase $\gamma$ [119, 161, 162]. This enzyme, probably identical to the previously identified mitochondrial DNA polymerase [112, 161–165], shows greatest activity using an oligodeoxyribonucleotide primer and a ribonucleotide template (e.g. poly(A)·oligo(dT)). Other than a presumed role in replication of mitochondrial DNA, polymerase $\gamma$ has been implicated in the replication of adenovirus DNA [166–168]. DNA polymerase $\gamma$ from chick embryos is a tetramer having four identical subunits each of molecular weight 47 000 [169]. It is highly processive in contrast to DNA polymerases $\alpha$ and $\beta$ [170] and this may be relevant to its function (see Section 6.9.5).

Chloroplasts also possess the enzymic mechanisms for synthesizing their own DNA [171, 172] and new enzymes are induced following viral infection [173–176]. The DNA polymerase coded for by herpes simplex virus is very sensitive to inhibition by acyclovir and phosphonoacetic acid which makes these drugs suitable for use in the treatment of herpes virus infections [201, 202] (see Section 3.6.7.c).

In addition to replicative DNA polymerase activities, which require the presence of all four deoxyribonucleoside triphosphates, animal cell extracts contain a separable enzyme responsible for the addition of nucleotidyl units to the *ends* of polynucleotide chains. This second activity, which was originally described by Krakow in 1962 [177], does not require a template strand but catalyses the incorporation of nucleotide units from single triphosphates into terminal positions in the DNA primer molecule. It is not further stimulated by the addition of the other three triphosphates but it is stimulated by cysteine. It has been called 'terminal transferase enzyme' and may be used in the biosynthesis of homopolymers of deoxyribonucleotides [178]. It is possible to distinguish between the two types of enzyme by the use of actinomycin D [60].

Unlike the replicative enzymes, which are stimulated by low levels of EDTA, the terminal enzyme is completely inhibited by micromolar concentrations of EDTA which is believed to exert its effect by binding $Zn^{2+}$.

Terminal transferase has a molecular weight of 58 000–62 000 [180, 181] but the purified enzyme shows bands at 26 500 and 8 000 molecular weight on electrophoresis in sodium dodecyl sulphate. This is a result of proteolysis during purification [181, 182].

The activity of the terminal transferase is very low in all tissues except thymus [183] and acute leukaemic lymphoblasts, i.e. lymphoid progenitor cells [184, 185]. To what extent this reflects the peculiar physiological functions of such cells is unclear but the generation of short insertions at recombinational junctions formed during the rearrangements of immunoglobulin heavy chain genes is positively correlated with the presence of the enzyme [186].

### (f) *Reverse transcriptase or RNA-dependent DNA polymerase*

The enzymes so far discussed copy DNA strands in the synthesis of DNA, but great excitement was created in 1970 by the announcement of the existence in certain RNA viruses of RNA-dependent DNA polymerases which use RNA as a template for the synthesis of DNA. These enzymes were discovered simultaneously by Temin [188] in the virus particles (virions) of Rous sarcoma virus (RSV) and by Baltimore [189] in Rauscher mouse leukaemia virus (R-MLV). The observation was confirmed for more

than half a dozen RNA viruses by Spiegelman [190], and was seen as an important breakthrough in cancer research since the RNA viruses involved were oncogenic, i.e. capable of bringing about malignant change (see Chapter 3).

On treating the viral 70S RNA genome with agents which disrupt hydrogen bonds it dissociates into two genetically identical RNA molecules [191], and to several molecules of tRNA [192]. The unique RNA-directed DNA polymerase is also present in the virion. The enzyme uses one of the tRNA molecules (tRNA$^{Trp}$) [193] as a primer to synthesize DNA on the template 35S single-stranded RNA (Section 6.9.10), and this distinguishes it from any cellular enzyme. Ribonucleoside triphosphates are without effect as substrates and the process is not sensitive to actinomycin D.

The immediate product of the reaction is a double-stranded RNA–DNA hybrid which is the result of the synthesis of a complementary strand of DNA on the single-stranded viral RNA as template.

The enzyme also shows ribonuclease H and DNA-dependent DNA polymerase activities which convert the hybrid into a duplex DNA. The polymerase then proceeds to replicate the duplex DNA so as to provide more copies which may be integrated into the genome of the host cell.

Reverse transcriptase may be used to make DNA copies of eukaryotic messenger RNA molecules. A thymidylate oligomer will anneal to the 3'-poly(A) tail of the mRNA and serve as a primer in a reaction which is of great importance in the production of complementary DNA (cDNA) molecules for cloning (see Section A.7.2).

Reverse transcriptase is the product of the *pol* gene of retroviruses. It is synthesized as a gag-pol polyprotein as the result of differential splicing of the transcript from the integrated proviral DNA [395] or by translational suppression [194]. The polyprotein is processed by a protease, present in the *N*-terminal region of the polymerase, to give the β subunit (92 000 molecular weight) and then the α subunit (58 000 molecular weight) which in the leukosis virus (ALV) combine to give the most active (αβ) form of the enzyme [194–197]. The enzyme from other retroviruses is a single polypeptide [187].

The enzyme shows only low fidelity and lacks associated exonuclease activity though it is capable of specifically hydrolysing the ribo strand of an RNA–DNA duplex [198–200]. The 3'-end of the pol gene also encodes an endonuclease important for replication and integration [187, 194] which altogether makes this a remarkable enzyme (see Section 7.7.3).

### 6.4.3 DNA ligases

The ligase or joining enzymes catalyse the repair of a single-stranded phosphodiester bond cleavage of the type introduced by endonuclease. They were first described in *E. coli* [203] and have since been described in both animal and plant cells.

DNA ligases catalyse the formation of a phosphodiester bond between the free 5'-phosphate end of an oligo- or poly-nucleotide and the 3'-OH group of a second oligo- or poly-nucleotide positioned next to it (Fig. 6.18). A ligase–AMP complex seems to be an obligatory intermediate and is formed by reaction with NAD in the case of *E. coli* and *B. subtilis* [204] and with ATP in mammalian and phage-infected cells [205–207] (Fig. 6.18). The adenyl group is then transferred from the enzyme to the 5'-phosphoryl terminus of the DNA. The activated phosphoryl group is then attacked by the 3'-hydroxyl terminus of the DNA to form a phosphodiester bond.

DNA ligase will close single-strand breaks in double-stranded DNA or in either strand of a polyribonucleotide–polydeoxyribonucleotide hybrid polymer; double-stranded ribopolymers are not substrates. Breakage of a single phosphodiester bond without removal of a nucleotide to give 5'-phosphoryl and 3'-OH termini is essential for repair by ligase activity (see Fig. 6.19). Reaction is independent of the base composition around the cleavage point.

The enzymes from *E. coli* or bacteriophage T4

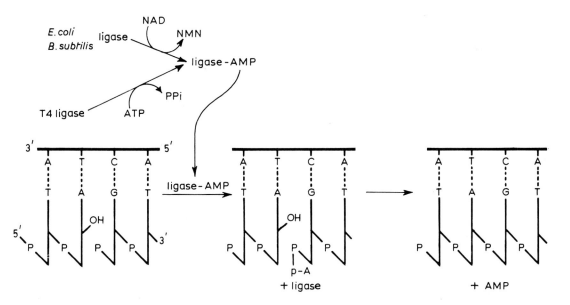

**Fig. 6.18** Action of DNA ligases.

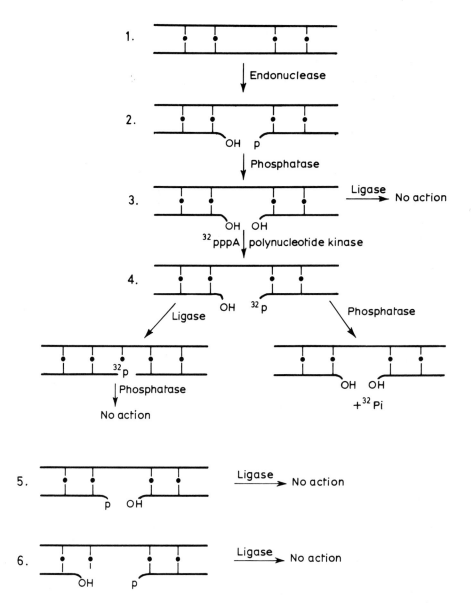

**Fig. 6.19** An assay for DNA ligase. The substrate (4) is formed by treating double-stranded DNA (1) with DNase I to give single-stranded breaks (2), removing the 5'-phosphate residues with phosphatase (3) and replacing them with [$^{32}$P]phosphate groups with the aid of polynucleotide kinase and γ-labelled [$^{32}$P]ATP. Forms (3), (5) and (6) are inactive as substrates for the ligase.

will join short oligodeoxynucleotides in the presence of a long complementary strand, e.g. d(T-G)$_3$, d(T-G)$_4$ or d(T-G)$_5$ can be joined in the presence of poly(dC-dA) and blunt end ligation (i.e. the joining of two duplex DNA molecules in the absence of any complementary overlapping sequence) can occur in the presence of excess ATP (see Section A.7.2).

### (a) *Assay of DNA ligase*

DNA ligase has been assayed in a variety of systems, including the formation of covalently closed circles of double-stranded DNA, restoration of transforming activity of nicked DNA (e.g. [204]) and the formation of phosphatase-resistant radioactive phosphate [208] (see Fig. 6.19).

### (b) *Role of DNA ligase*

The nature of the reaction catalysed by DNA ligase makes it an important enzyme involved in DNA synthesis, DNA repair and genetic recombination. In all of these cases, its postulated role is in re-establishing continuity by joining a stretch of newly synthesized DNA to pre-existing DNA by the formation of a phosphodiester bond.

It has, however, proved difficult to define a situation in which DNA ligase could be shown to be absolutely necessary for any one of these functions. Nevertheless, since *E. coli* ligase-deficient mutants selected in different ways exhibit abnormal uv sensitivity [209], it appears that ligase has some function in the repair process. Temperature-sensitive ligase mutants also accumulate Okazaki pieces at the restrictive temperature. That these pieces are eventually joined is probably a reflection of the presence of a few molecules of ligase still remaining in the mutant cells. There are normally about 300 molecules of ligase per cell, and it has been calculated that these are capable of sealing 7500 breaks per minute at 30°C [210]. Since only about 200 breaks per cell are formed on

replication each minute there is a vast excess of enzyme.

Infection of *E. coli* by T4 leads to the synthesis of a T4-specified ligase which, in addition to joining two DNA chains, will also join RNA to DNA in an RNA–DNA hybrid [210–212]. Some ligase is essential for T4 development [209], and phage with a temperature-sensitive ligase show increased susceptibility to uv radiation [213] (see Chapter 7).

There are two DNA ligases present in cells of higher eukaryotes. DNA ligase I (molecular weight 165 000–200 000) resembles DNA polymerase $\alpha$ in being recovered partly in supernatant fractions on cell homogenization and in being present in much greater amounts in dividing than in resting tissues. The smaller, nuclear DNA ligase II does not change in activity with the growth rate of the cells [673–675].

### 6.4.4 Helix-destabilizing proteins (HD) or single-stranded DNA-binding proteins (ssb)

These are a group of proteins brought together under this heading because of a common property. They have been isolated from many sources and they do not necessarily all perform the same function *in vivo*.

The first of these to be characterized was that coded for by gene 32 of bacteriophage T4 [214]. This and similar proteins isolated from T7 and uninfected *E. coli* [215] have the ability to convert double-stranded DNA into single-stranded form at a temperature 40°C below the normal melting temperature ($T_m$). The gene 32 protein has a molecular weight of 35 000 and it binds co-operatively to single-stranded regions of DNA, each protein covering some 10 nucleotides. Thus if a limiting amount of gene 32 protein is mixed with excess single-stranded DNA some DNA molecules will become covered with protein and others remain protein free. There are from 300 to 800 molecules of binding protein per uninfected *E. coli* cell, which is enough to cover about 1600 nucleotides at each replication fork [215, 216]. A similar length of T4 DNA can be covered by the 10 000 molecules of gene 32 protein present in an infected cell. (There are about 60 T4 replication forks per infected cell [215].)

Purified polymerases are stimulated in a very specific manner by the presence of their complementary DNA-binding proteins. Thus the T4 DNA-binding protein will stimulate only the T4 polymerase and the T7 DNA-binding protein will stimulate the T7 polymerase [216, 217], the *E. coli* DNA-binding protein stimulates *E. coli* polymerase II and holopolymerase III but does not stimulate polymerases I and III [218]. It is suggested that the polymerase and DNA-binding protein form a specific complex that interacts with DNA and other proteins during replication [219].

The HD proteins from *E. coli* and T4-infected cells are believed to play a role in replication, recombination and repair [177, 178, 220] and an *E. coli* mutant (ssb-1) has been isolated showing a defective DNA-binding protein and extracts fail to convert phage G4 single strands into duplex form [221] (see Section 6.7.1). In contrast an HD protein coded for by gene 5 of phage M13 plays an essential role in preventing further replication by stabilizing the single-stranded progeny viral DNA [222].

DNA-binding proteins have also been isolated from eukaryotic cells.

Thus helix-destabilizing proteins serve a passive role. Although *in vitro* they can bring about denaturation of DNA their role *in vivo* may simply be to stabilize single-stranded DNA.

### 6.4.5 DNA unwinding proteins or DNA helicases (DNA-dependent ATPases)

DNA helicases have the property of being able to unwind a DNA duplex obtaining the necessary energy from the hydrolysis of ATP. In *E. coli* there are several helicases which have been partially characterized by Hoffman

Berling's group [223–227]. Helicase I is present only in F[+] strains of *E. coli* (i.e. strains carrying the F sex factor – see Section 3.7) and is the product of one of the transfer genes (*tra*I) of the F factor [228]. Thus a role for helicase I in bacterial conjugation is implied and it plays no role in cell DNA replication [229]. It is a fibrous protein of molecular weight 180 000, similar to myosin. About 80 molecules bind to a single-stranded region of DNA near a replication fork and travel down the single strand in a 5′ → 3′ direction towards the fork unwinding the DNA (Fig. 6.20) [230]. There are about 500 molecules per cell.

There are about 3000 molecules of helicase II (molecular weight 75 000) per cell. It also moves towards a replication fork in a 5′ → 3′ direction but remains bound to the single-stranded DNA generated (Fig. 6.20). It interacts specifically with holopolymerase III to promote unwinding of DNA at the replication fork in the presence of *E. coli* ssb [239]. It is the product of the *E. coli* uvrD gene, mutants in which show increased rates of mutation and recombination [231, 267]. Immunological studies show that helicase II is involved in the replication of *E. coli*, phage lambda and the plasmid Col E1 [229]. Rep is so called because it is essential for the replication of small single-stranded phages (e.g. phage G4 and φX174 – see Section 6.7.3) as well as *E. coli*. It

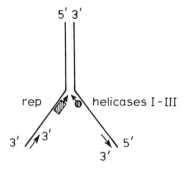

**Fig. 6.20** DNA helicases acting at the replication fork.

differs from helicase I, II and III as it binds to a single-stranded region and translocates in a $3' \rightarrow 5'$ direction until it reaches the replication fork (i.e. along the leading strand). For its unwinding action it requires a single-stranded binding protein to bind to the two arms of single-stranded DNA produced (Fig. 6.20) [215, 232].

It is clear that rep acting on the leading side of the fork and helicase II acting on the lagging side would bring about efficient unwinding of the DNA duplex. Two molecules of ATP are hydrolysed per nucleotide unwound [233]. This is a high price to pay for unwinding DNA as it involves the release of 60 kJ of energy to break H-bonds whose energy varies from 5 kJ (A≡T) to 21 kJ (G≡C).

Helicase III (molecular weight $2 \times 20\ 000$) is similar to rep but moves along the bound strand of DNA in a $5' \rightarrow 3'$ direction [234, 235].

It is clear that these helicases move along the DNA in combination with a DNA-polymerizing complex and similar enzymes are induced by T7 and T4 phage.

The gene 4 product of bacteriophage T7 is similar to rep except that it forms a complex with the T7-coded DNA polymerase (the gene 5 product). Together they are able to extend a $3'$-OH at a gap in the duplex, unwinding the DNA ahead of the complex and hydrolysing two ATP molecules for every dNMP incorporated [236, 447]. The gene 4 protein, in addition to acting as an unwindase, is also able to synthesize a specific primer for DNA synthesis (see below). A similar function is catalysed by the T4 gene 41/61 complex [393], and a second T4-coded helicase – the product of the non-essential *dda* gene – greatly speeds up the rate of replication fork movement in *in vitro* systems [240, 241]. The *dda* protein overcomes the inhibitory action exerted by RNA polymerase molecules which, when bound to the DNA template, can completely block fork movement [242]. Unwinding enzymes are also found in eukaryotes [243].

Other DNA-dependent ATPases are present in cells but in these cases the ATP hydrolysis is probably required to translocate an enzyme along the DNA chain (this may also be required for the unwinding enzymes). Some of these are considered in Section 6.7.1.

### 6.4.6 Topoisomerases

When two DNA strands unwind as the replication fork moves forwards either the whole DNA molecule has to rotate at 10 000 rev./min (an impossibility for closed cyclic DNA molecules) or some sort of swivel must be present. Cairns first proposed the idea of a swivel [3, 244] which might act by alternating action of an endonuclease and a ligase. However, the two functions are present in a single enzyme, examples of which have been isolated from both prokaryotes and eukaryotes [245–247]. The activity was initially called a nicking closing enzyme but is now known as a topoisomerase as it catalyses the interconversion of different topological isomers of DNA (see Chapter 2). The difference between a topoisomerase and the earlier suggested mechanism is that with the topoisomerase the cut ends are not released from the enzyme although rotation of the two strands about one another is effectively achieved.

There are two types of topoisomerase. Topoisomerase I cuts only one strand of duplex DNA whereas both strands are cut by topoisomerase II. The cuts usually produce a $5'$-OH group (which can be phosphorylated by polynucleotide kinase) and a $3'$-phosphate which is covalently linked to a tyrosine in the enzyme [248, 249, 252, 317].

*In vitro*, topoisomerases are studied by following the result of their action on closed cyclic supercoiled DNA molecules such as SV40 DNA or the plasmid Col E1 (Fig. 6.21). The topoisomerases will bring about a step-by-step relaxation of such supercoiled molecules by changing the linking number. Type I topoisomerases change the linking number in single steps indicating that they produce an enzyme-

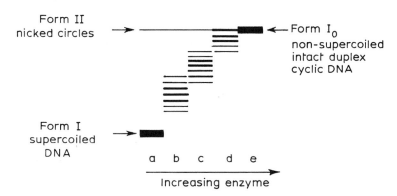

**Fig. 6.21** Topoisomerase action. Diagram of agarose gel electrophoresis of SV40 DNA incubated with increasing amounts of topoisomerase I. Whereas an endonuclease would convert form I to form II directly, the topoisomerase produces DNA molecules of intermediate linking number.

linked cut in one strand only of the DNA. The other strand is then passed through the cut which is resealed [253, 316]. This is equivalent to free rotation of the bond opposite to the nick.

Type II topoisomerases produce an enzyme-bridged break in both strands of DNA. In an ATP-requiring action, another region of duplex DNA passes through the gap thus changing the linking number two steps at a time (Fig. 6.22). In most instances both types of topoisomerase will act until all the supercoils are removed and the DNA is released as a fully relaxed closed cyclic DNA molecule [254]. If the two regions of duplex DNA involved in topoisomerase II action are part of two different DNA molecules then these two molecules will become interlocked (Fig. 6.22). Because of this the enzyme is capable of producing and resolving catenated and knotted molecules [255, 256]. In a similar manner topoisomerase I can act opposite to a nick in a cyclic molecule and bring about the catenation and knotting of several supercoiled molecules to the nicked cyclic molecule [253].

Both types of topoisomerase are found in prokaryotes and eukaryotes [251, 252, 255–264, 317, 700]. The activity of topoisomerase II is higher in rapidly dividing eukaryotic cells and it is thought to be involved in DNA replication and the segregation of intertwined duplex DNA

molecules (see Section 6.10.1) [262, 703]. In replicating cells it exists as a complex in association with other enzymes involved in DNA synthesis [317]. Bacterial mutants defective in topoisomerase II are not viable unless compensatory mutations in topoisomerase I are also present [265, 266].

There are at least two types of topoisomerase II. The eukaryotic enzymes and the enzyme from T4 are only capable of relaxing supercoiled DNA. That is, they catalyse the conversion of supercoiled DNA into relaxed cyclic duplex. They are homodimers having two subunits each of approximate molecular weight 170 000 [251, 264, 268, 269, 317].

Bacterial type II topoisomerases are known as DNA gyrases. *E. coli* DNA gyrase is a 400 000-mol.wt. tetramer, the subunits of which are the products of two genes *gyrA* and *gyrB* (also known as *nal* and *cou*).

The *gyrB* product is a DNA-dependent ATPase (molecular weight 95 000) which is sensitive to the antibiotics coumermycin and novobiocin.

The *gyrA* product is a 100 000–110 000 molecular-weight 'nicking-closing' enzyme sensitive to nalidixic acid and oxolinic acid. In the presence of oxolinic acid or strong detergent these subunits remain bound to the 5'-ends of the

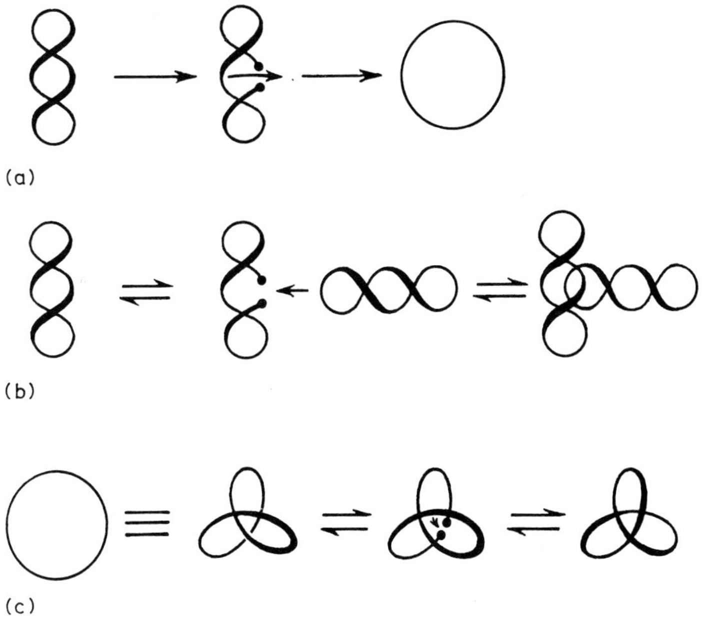

**Fig. 6.22** Reactions catalysed by type II topoisomerases: (a) relaxation; (b) catenation and decatenation; (c) knotting and unknotting.

cut [277, 278] and hence the tetrameric enzyme holds the cut ends together. DNA gyrase, as well as relaxing supercoiled molecules can also reduce the linking number of cyclic duplex DNA molecules. Thus when incubated with ATP and DNA gyrase a relaxed duplex ring can be converted to a negatively supercoiled molecule with a linking number as much as 10% lower than that of the most stable B-DNA structure [248, 250, 271–274]. Although DNA gyrase is the only enzyme that can catalyse this reaction *in vitro* it has been suggested that it is the normal reaction catalysed *in vivo* by all type II topoisomerases [316].

A third type of topoisomerase II has been reported in archaebacteria [270]. This *reverse gyrase* is able to convert a relaxed cyclic DNA duplex to a *positively* supercoiled molecule.

DNA gyrase cleaves duplex DNA to give a four-base-pair stagger, but there is controversy over whether or not it shows sequence speci-ficity. Topoisomerase I may also show site selectivity [257, 280]. A model for the action of DNA gyrase has been constructed [279]. The DNA first wraps itself around the protein to form a structure similar to a nucleosome (see Chapter 3) [281, 282]. ATP then binds and causes a conformational change which drives the trans-location of the DNA relative to the enzyme, leading to the formation of a positive super-helical loop on the DNA [283]. A double-stranded breaking rejoining event removes the positive superhelical turn [284] and on hydrolysis of ATP the enzyme returns to its original conformation leaving the DNA with a negative superhelical twist, i.e. partly unwound. The wrapping of the DNA around the gyrase pro-duces overwound and underwound regions of duplex DNA such that when the DNA is left with an underwound region the overwound regions are selectively relaxed.

DNA gyrase may play a role either by

removing positive superhelical turns induced into DNA during replication or alternatively or additionally by putting negative supertwists into the DNA ahead of the replication fork thus making unwinding easier. As treatment of *E. coli* with either coumermycin or nalidixic acid immediately stops DNA synthesis, the contribution of DNA gyrase is essential [285, 286]. DNA gyrase is also required for replication of phage T7 DNA [287] and phage T4 codes for a corresponding activity [268, 288]. The products of T4 genes 39, 52 and 60 form a complex with topoisomerase activity which is essential for initiation of replication of T4 DNA.

Decatenation (also catalysed by gyrase) may be an essential step in the physical separation of cyclic DNA molecules after replication (see Section 6.10.1). Single-stranded regions of DNA are required not only for initiation of DNA synthesis but also for recombination and transcription and DNA gyrase is believed to be required in these reactions [289–292].

It is possible that topoisomerase activity is markedly affected by enzyme phosphorylation [293] or poly(ADP)ribosylation [294].

## 6.5 FIDELITY OF REPLICATION

To produce a sequence of bases in the DNA of a daughter cell which is *identical* to that in the parent the process of DNA replication would need to be completely faithful. This is obviously not the case as mutants do arise spontaneously. This is not altogether a bad thing as it is on this varied counterpane that natural selection is able to act.

There are also mechanisms which come into play to repair damage to DNA post-replicatively (see Chapter 7) but even these mechanisms rely on the faithful production of DNA complementary to the template strand.

Fidelity may depend on a number of steps:

(a) The polymerase must select the correct deoxyribonucleoside triphosphate.

(b) A checking mechanism may exist which would allow rejection of incorrect triphosphates. Such enhanced discrimination may result from template-induced changes in enzyme conformation dictating the selection of the correct substrate or increased binding of the enzyme to the template in the presence of the complementary nucleotide.

(c) A kinetic proof-reading model proposed by Hopfield [295] predicts that a high-energy [enzyme:template:dNMP] complex is produced with release of pyrophosphate. The polymerase can check the complex to see whether or not the correct base has been incorporated. The nucleotide (dNMP) may be released at this time but the chance of an incorrect nucleotide being released is much higher than for the correct nucleotide.

(d) Following incorporation, an exonucleolytic proof-reading mechanism could remove incorrect 3'-terminal nucleotides which are mismatched with the template strand prior to elongation (see Section 6.4.2.b). These nucleotides are also released as monophosphates.

Both the proof-reading mechanisms rely on the reduced rate of reaction with a mismatched nucleotide. In the kinetic mechanism it is the joining of the bound nucleotide which is delayed whereas with exonucleolytic proof-reading it is the addition of the next nucleotide which is delayed allowing a greater opportunity for exonuclease action.

It is only if all these mechanisms fail that post-replicative repair is required but as explained above this is also subject to the demands that it produce a faithful complementary DNA strand.

One of the reasons for increased interest in fidelity of replication arises from the fact that, unlike their prokaryote counterparts, eukaryote DNA polymerases lack the 3' → 5' proof-reading exonuclease activity yet still exhibit error rates much less than those predicted by a

**Table 6.5**  Nucleotide substitutions possible at the am16 codon of the φX174 genome.

| Gene A | | Phe  Arg | | |
|---|---|---|---|---|
| Mutant DNA sequence | | 5'-A G A T T T A G G C T G G 3' | | |
| Gene B | | Ile   am16   Ala | | |
| Position at amber codon | Mispaired nucleotide | Gene A | Gene B | |
| T | G | Phe-Arg | Glu | |
| | T | Leu-Arg | Lys | |
| | C (wt) | Leu-Arg | Gln | |
| A | G | Phe-Arg | Ser | |
| | A | Phe-Trp | Leu | |
| | C | Phe-Gly | Trp | |
| G | G | Phe-Thr | Tyr | |
| | T | Phe-Lys | Ochre | |
| | A | Phe-Met | Tyr | |

consideration of base-pairing alone [54, 296]. An assay that has proved particularly useful in assessing the fidelity of *in vitro* replication involves the copying of the single-stranded DNA of the phage φX174 [297, 298]. An amber mutant (φX174 am16 – see Table 6.5) is unable to grow on wild-type *E. coli* (but can grow on the suppressor strain *E. coli* C62). Faithful copying by polymerase of the mutant φX174 DNA will reproduce mutant progeny, but a single base substitution at any one of the three bases of the amber codon will result in progeny viral DNA which can grow on wild-type *E. coli*. Using purified DNA polymerase α from calf thymus Grosse *et al.* showed a reversion frequency of about $1 \times 10^{-4}$ [298]. Of course all the base changes would not result in substitution with the original wild-type base and, as this region of the φX174 genome codes for two different proteins (see Section 3.6.7.a), a variety of progeny phage can result. For instance, insertion of a guanine

opposite to the guanine of the amber codon (TAG) results in a change of an arginine into a threonine (AGG → ACG) in gene A and insertion of a tyrosine into gene B. (See Chapter 11 for a discussion of the genetic code.) This produces small phage plaques which show reduced growth at temperatures above 35°C.

From a study of the plaque morphology and related sequence data it can be estimated that the frequency of mispairing dATP with a guanine in the template (1 in 20 700) is an order of magnitude more common than the reciprocal mispairing of dGTP with an adenine in the template. Nonetheless, the major mutations are purine:purine mismatches (transversions) and GT transitions [299].

An alternative assay system involves comparing the competition between the correct and incorrect nucleotide for incorporation in place of a dideoxynucleotide in the DNA sequencing reaction (see Section A.5.1). This system has the advantage that the effect of the environment of the substituted nucleotide can be studied [311].

Having established an assay system, the effect on mispairing of altering the relative concentrations of precursors can be studied. For instance, when the concentration of dATP is kept low and that of dCTP increased the chances of mispairing with the first base of the amber triplet is increased. However, other examples of an effect of pool bias were not found using polymerase $\alpha$.

The fidelity of the various eukaryote DNA polymerases is very variable and it is only the high-molecular-weight 9S form of polymerase $\alpha$ which shows a fidelity comparable to that of the prokaryote polymerases, both making only 1 error in $10^5$ nucleotides incorporated [143, 300, 301, 303]. The eukaryotic enzymes are working in the absence of a proof-reading exonuclease and the results suggest that a domain of polymerase $\alpha$ which is lost on proteolysis is vital for maintaining fidelity in a parallel but perhaps a different way from that in which the proof-reading exonuclease of E. coli polymerases is used.

In the absence of exonucleolytic removal of a mismatched primer terminus, DNA polymerase $\alpha$ shows an elongation rate of only 5% that with a fully matched terminus [302]. It is not only the first base after the mismatch which is added more slowly, for the effect extends to additions five bases following the mismatch.

In contrast, the prokaryotic DNA polymerases have associated proof-reading exonucleases and in the $\phi$X174 am16 assay pool effects are more marked. Thus the frequency of a TAG $\rightarrow$ TGG transversion depends not only on the ratio of incorrect to correct triphosphate but also on the concentration of the next triphosphate, i.e. $[dGTP]^2/[dATP]$ [303]. By using deoxycytidine [1-thio]triphosphates which, once incorporated, cannot be removed by exonuclease action, Kunkel [304] calculated that proof-reading can increase the fidelity of E. coli

DNA polymerase I or phage T4 polymerase by up to 20-fold. The efficiency of proof-reading depends on the adjacent nucleotide which is stacked together with the mismatched base [316] and the fidelity of these enzymes and E. coli DNA polymerase III holoenzyme is dependent on the divalent ions [305] and accessory proteins present [100, 306–310].

In an assay of fidelity which measures the production of dNMP, Loeb's group have failed to find significant evidence for proof-reading of either the kinetic or exonuclease variety [312]. They favour a direct action of the polymerase in enhancing the specificity of base-pairing and show that the affinity of E. coli DNA polymerase I for a complementary nucleotide exceeds that for non-complementary nucleotides by one to three orders of magnitude. Similar results have been obtained for the T4 DNA polymerase [313, 314].

It is difficult to reconcile the failure to find evidence for the importance of proof-reading by $3' \rightarrow 5'$ exonuclease with the occurrence of mutator and antimutator mutations in prokaryotes (see Section 6.4.2.b). A way out is to envisage excision repair reactions occurring following an endonucleolytic cut in DNA on the $3'$-side of a region of damaged DNA. The $3' \rightarrow 5'$ exonuclease would then act to remove the defective region in a way that the $5' \rightarrow 3'$ exonuclease can act if the endonucleolytic cut is $5'$ to the lesion (see Chapter 7). A defective $3' \rightarrow 5'$ exonuclease would fail to excise the mismatch leading to an increase in mutation frequency [68].

It is the post-replicative mismatch correction mechanisms which contribute to the increase in fidelity (up to one error per $10^9$–$10^{11}$ nucleotides incorporated) found in vivo (see Chapter 7).

The RNA-dependent DNA polymerases of retroviruses (reverse transcriptase) are very error-prone showing a fidelity similar to that of RNA polymerases, i.e. one error per $10^3$–$10^4$ nucleotides incorporated [198–200, 315].

## 6.6 *IN VITRO* SYSTEMS FOR STUDYING DNA REPLICATION

### 6.6.1 dna mutants

Increased understanding of DNA replication has come about as a result of the intensive use of bacterial mutants, especially temperature-sensitive mutants, i.e. bacteria which grow normally at low temperatures but cease to grow at a higher, restrictive temperature.

The study of such conditional lethal mutants, defective in DNA synthesis, has demonstrated that a number of proteins are essential for DNA replication in *E. coli*. Some of these proteins together with the genes which code for them are listed in Table 6.6. Most of these proteins have now been purified and used with some success in attempts to reconstruct a system which will replicate DNA *in vitro*.

The mutant cells, when transferred to the restrictive temperature, may cease immediately to make DNA or they may continue to grow for some time. The former mutants are defective at the stage of the elongation of the replicating DNA molecule whereas the latter are defective in the process of initiation of new rounds of DNA replication.

The genetic approach thus indicates the involvement of a number of gene products in DNA replication but is able to say little about the detailed function of these proteins. The biochemist seeks to purify the proteins involved and to characterize their individual functions. A first step in this process involves the circumvention of the barrier of the cell membrane. This has been done in a number of ways [318, 319].

### 6.6.2 Permeable cells

Treatment of *E. coli* with lipid solvents (ether [320] or toluene [321]) whilst inhibiting cell division appears to leave the cells and constituent enzymes intact yet renders the membrane permeable to small molecules including nucleotide precursors. Such cells when

**Table 6.6** *E. coli* genes whose products play a role in DNA synthesis.

| Gene | Enzyme | Effect of mutation |
|---|---|---|
| *dnaA* | | No initiation |
| *dnaB* | | No synthesis |
| *dnaC* | | No initiation |
| *dnaE* (*polC*) | DNA polymerase III ($\alpha$) | No synthesis |
| *dnaF* (*nrd*) | Ribonucleotide reductase | No synthesis |
| *dnaG* | Primase | No synthesis |
| *dnaQ* (*mutD*) | DNA polymerase III ($\epsilon$) | Increased mutation rate |
| *dnaS* (*dut*) | dUTPase | Short Okazaki pieces |
| *dnaX* | DNA polymerase III ($\delta$) | |
| *dnaZ* | DNA polymerase III ($\gamma$) | |
| *polA* | DNA polymerase I | Slow joining of Okazaki pieces. uv-sensitive |
| *polB* | DNA polymerase II | Normal |
| *lig* | DNA ligase | Slow joining of Okazaki pieces |
| *ssb* | DNA-binding protein | |
| *gyrA* *gyrB* | DNA gyrase | |
| *top* | DNA topoisomerase I | |
| *uvrD* | Helicase II | Increased mutation rate |

incubated with the four deoxyribonucleoside triphosphates, ATP and $Mg^{2+}$ continue semi-conservative replication. The products of such replication are the 10S Okazaki pieces which are only joined into high-molecular-weight DNA in the presence of NAD, the cofactor for *E. coli* DNA ligase. It was with toluenized cells and $\alpha$-$^{32}$P-labelled deoxyribonucleotides that Okazaki was able to show the covalent linkage of RNA and DNA (see Section 6.32).

### 6.6.3 Cell lysates

The lysate formed on disintegration of the cell membrane is capable of maintaining DNA synthesis under certain conditions [322, 323]. It is very important not to dilute the macromolecular constituents of the cell but to maintain them near their *in vivo* concentration. One way in which this has been achieved is by lysis of cells on a cellophane disc held on an agar surface. Small molecules are then allowed to diffuse through the disc into the concentrated lysate. Semiconservative DNA synthesis occurs in this and similar systems if the lysate is prepared from wild-type *E. coli*, but when it is prepared from *E. coli* cells defective in DNA synthesis then the lysate reflects the *in vivo* capabilities of the cells. Thus lysates prepared from mutants temperature-sensitive in the genes *dnaB*, *dnaD* or *dnaG* (fail to extend growing DNA chains at the non-permissive temperature; Table 6.6) only synthesize DNA at low temperatures. In lysates prepared from mutants temperature-sensitive in the genes *dnaA* or *dnaC* (fail to initiate DNA synthesis at non-permissive temperatures) DNA synthesis fails to occur only if the cells are maintained at the non-permissive temperature for one generation time prior to lysis [324].

This system has proved very useful in the purification of the products of these *dna* genes [318, 319]. Fractions containing the *dnaG* gene product from wild-type cells, when added to a lysate from a mutant temperature-sensitive in *dnaG*, are able to restore DNA synthesis at the

non-permissive temperature. Although the products of the *dna* genes have all been at least partially purified by this *in vitro* complementation assay, functions for all these proteins have not yet been established. However, they have been used with some success in reconstruction experiments using fully defined components (see below).

Permealysed cells [325] and lysates from eukaryotic cells [326–328] also show a limited ability to continue DNA synthesis when provided with the four deoxyribonucleoside triphosphates, ATP and $Mg^{2+}$. Eukaryotic cells can be rendered permeable to deoxyribonucleotides by treatment with hypo-osmotic buffers. DNA synthesis continues in these cells for about 30 minutes before slowing down. This pattern of synthesis is probably due to completion of ongoing replicons coupled with an inability to initiate new replicons [329] (see Section 6.7.2).

Cell lysates show similar characteristics [330] but isolated nuclei are much more limited in their capabilities [331–334] unless supplemented with cytosol fractions [335]. In general, isolated nuclei appear capable of extending Okazaki pieces initiated *in vivo* but have only a limited capacity for ligating or initiating Okazaki pieces. DNA synthesis in isolated nuclei is sensitive to inhibitors of DNA polymerase $\alpha$ but not to inhibitors of DNA polymerase $\beta$ or $\gamma$ indicating a role for DNA polymerase $\alpha$ in this reaction [115, 336]. However, removal of all *soluble* DNA polymerase $\alpha$ activity from nuclei does not markedly reduce their DNA-synthetic capacity, possibly indicating the presence of an enzyme complex [337, 338] (see Section 6.11).

Cell lysates do initiate the synthesis of Okazaki pieces and these have been shown to have RNA primers 8–11 nucleotides in length [339]. The sequence of these oligoribonucleotides is diverse.

The disadvantage with lysates of eukaryotic cells is that very little is known about the nature of the DNA template or product. As with prokaryotes (see below), model systems have been used. Thus lysates of cells infected with SV40 or polyoma virus synthesize predominantly viral DNA *in vitro* [326, 327], and soluble replication systems have also been prepared from adenovirus-infected cells and lysates made from cells infected with herpes simplex virus synthesize predominantly herpes simplex DNA *in vitro*.

### 6.6.4 Soluble extracts

Although useful in the identification and purification of some of the factors necessary for DNA synthesis, the cellophane-disc assay is of limited value in defining the function of the various gene products and moreover it relies almost entirely on possession of temperature-sensitive mutants for every protein involved.

A simplification involves the use of soluble extracts from *E. coli* free of the bacterial DNA. When cells are lysed with lysozyme the bacterial chromosome can be pelleted by centrifugation at 200 000 *g*. The soluble preparation remaining is capable of performing specific stages of DNA synthesis when supplemented with simple DNA substrates, e.g. single-stranded cyclic DNA from the small phages $\phi$X174 or M13. Once again possession of mutants is an invaluable aid but the advantage of this system over the crude lysate is that specific steps in DNA replication may be studied and hence information on the function of particular proteins may be obtained.

### 6.6.5 Reconstruction experiments

The ultimate aim of these experiments is to achieve the ordered replication of a complex DNA molecule such as the *E. coli* chromosome using totally defined components. Initially simpler systems have been studied. The requirements for the conversion of single-stranded phage DNA into the double-stranded replicative form (RF) and the replication of RF have been delineated and the reactions carried out using purified components. Progress was not all plain

sailing for in addition to the components identified with the aid of mutants and purified by the complementation assay (see above) addition of soluble extract was required. From this extract several further components have been purified for which, as yet, no mutants are known.

The detailed results of these reconstruction experiments will now be considered. The simplest situation to investigate is the elongation of a primer hydrogen-bonded to a template DNA strand. When the components of such a system are known the synthesis of the primer can be investigated. Synthesis of a primer and its elongation as DNA is the reaction which occurs on the lagging strand of the replication fork, i.e. Okazaki piece synthesis (Section 6.3). In order to ensure that a primer is being synthesized it is important to use a cyclic, single-stranded template molecule which has no free ends which can bend back and hybridize to the template strand (see Fig. 6.11). Such templates are naturally provided by the DNA of the small bacteriophage fd (M13), G4 and $\phi$X174. The replication of these phage in *E. coli* takes place in three stages [53, 54].

(i) The original infecting cyclic DNA molecules (known as the viral or plus strand) is used as a template on which a complementary minus strand is synthesized (SS → RF).

(ii) The resulting replicative form (RF) undergoes several cycles of replication by the so-called rolling circle mechanism (RF → RF).

(iii) The RF molecules are used as templates from which progeny plus strands are synthesized. These are subsequently packaged into the viral coat as the mature virus leaves the cell (RF → SS).

Although of similar size the replication of the three small phage mentioned above differs in a number of significant steps which can be explained as a result of differences in DNA sequence.

## 6.7 MOLECULAR BIOLOGY OF THE REPLICATION FORK

### 6.7.1 Lagging-strand synthesis

(a) *Prokaryotes*

In the cell single-stranded DNA is always complexed with a DNA-binding protein and such a complex is formed as soon as the DNA from the small single-stranded phages (fd, M13, G4, φX174) enters the host cell or is incubated with a host cell lysate. In fd there remains exposed a hairpin loop which is able to bind *E. coli* RNA polymerase [676–678]. This enzyme synthesizes a short RNA which then acts as a primer for minus strand synthesis [92, 679, 680].

In G4 a similar hairpin loop is exposed but this has a binding site for the *dnaG* protein – the primase [128, 681]. This enzyme is responsible for synthesizing the RNA primer in most prokaryotes including phage G4. The sequence of DNA around the origin of phage G4 and the sequence of the RNA primer synthesized by the primer is shown in Fig. 6.23 [23, 92, 681–683]. Studies with phage φK have shown that the primase-binding site is more complex than shown in the figure and in fact two molecules of

primase bind to a region of complex secondary structure extending over 100 bases having three hairpin loops [684, 685]. Although a primer of 28 nucleotides is shown in Fig. 6.23, in the presence of DNA polymerase as few as two ribonucleotides may be incorporated by primase before DNA polymerase takes over [686, 687].

In φX174 no binding site for either RNA polymerase or primase is exposed. Comparison of the nucleotide sequences of the DNA of the closely related bacteriophages G4 and φX174 reveals a deletion in the latter covering the primase-binding site [681, 340]. In φX174 a complex is formed involving a series of proteins identified as the products of the genes *dnaB*, *dnaC* and proteins i, n, n′ and n″ (called X, Y and Z by Hurwitz) [54, 341–344].

This complex, called the preprimosome, binds to a 55-nucleotide-long sequence which occurs once in φX174 DNA and contains a hairpin loop between bases 2308 and 2350. It also occurs once on each strand of the duplex Col E1 plasmid DNA as well as on *E. coli* DNA [345, 346, 361] and this sequence when introduced into fd DNA renders this DNA capable of assembling the preprimosome. Protein n′ is a 76 000-molecular-weight protein responsible for site selection

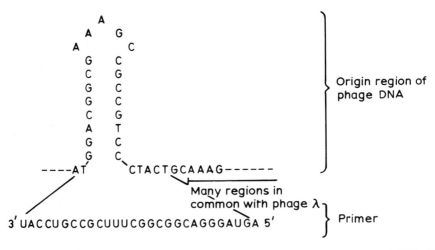

**Fig. 6.23**  Location of primer in phage G4 (this untranslated region is absent from φX174) [682, 686, 687].

[347–349]. A complex of proteins n, n' and n'' interacts with DNA and this then reacts with protein i and a *dnaB–dnaC* protein complex which contains six molecules of each protein [354–357]. This preprimosome complex of six different proteins interacts with primase to form the primosome (Fig. 6.24) and an RNA primer is synthesized. As the primase moves off in the $3' \rightarrow 5'$ direction along the template strand the preprimosome translocates in the opposite direction ($5' \rightarrow 3'$) in a reaction activated by the DNA-dependent ATP hydrolysis catalysed by protein n' [341, 342, 344, 350, 351]. At various sites the *dnaB* protein is able to affect the structure of the DNA template thereby causing another molecule of primase to interact with the preprimosome and initiate synthesis of another primer [342, 350, 352, 353].

In all, six to eight primers are made per round of replication on $\phi$X174 DNA [677, 342]. They range in size from one to nine nucleotides long and although they initiate with pppA-Pu their internal sequence is random [351]. Their length is in part dependent on the availability of deoxyribonucleoside triphosphates and the DNA-polymerizing enzymes.

In all cases the RNA primers formed are extended by DNA polymerase III holoenzyme which catalyses the addition of deoxyribonucleotides until the growing strand reaches the next RNA primer (Fig. 6.24). Because of the limited availability of DNA polymerase III holoenzyme several primers may be present awaiting extension, and rapid transfer of polymerase occurs from a completed Okazaki piece to the adjacent primer [362]. DNA polymerase I removes the RNA primer and completes the DNA chain in a nick-translation type of reaction (see Section 6.4.2.b). Finally the chains are joined by DNA ligase to form the cyclic duplex molecule (RF). These reactions have been recently reviewed [709].

These reactions, as well as describing the initial (SS $\rightarrow$ RF) stages of replication of the small single-stranded phages, also represent the

reaction occurring on the lagging side of the replication fork (Fig. 6.24 and Fig. 6.11).

A similar set of reactions is catalysed by enzymes induced following phage T4 infection of *E. coli*. Seven T4-coded proteins will catalyse initiation of lagging-strand synthesis on single-stranded cyclic templates and will catalyse displacement-strand synthesis on nicked duplex DNA (see Section 6.7.2) [385–390].

### (b) *Eukaryotes*

In an attempt to characterize the enzymes involved in lagging-strand synthesis in eukaryotes similar closed cyclic single-stranded DNA molecules have been incubated with extracts of *Xenopus* eggs or *Drosophila* embryos [358–360]. Such extracts were initially chosen as these tissues have the potential for very high rates of *in vivo* replication. These extracts, and extracts of cultured cells, catalyse the synthesis of an RNA primer which is extended to form a DNA chain of about 1000 nucleotides in length complementary to the single-stranded template. The sequence of the RNA/DNA junction is random but the primers initiate preferentially with the sequence pppA. On fractionation of the extracts it appears that single-stranded DNA-binding protein and DNA polymerase $\alpha$ with its associated primase activity are the only proteins required (see Section 6.4.2.e) [130–144].

### 6.7.2 Leading-strand synthesis

This reaction requires the extension of the growing daughter DNA strand and the concomitant unwinding of the duplex ahead of the point of polynucleotide synthesis. A suitable substrate to study this reaction is a nicked duplex molecule and this had been used in particular by Albert's group to investigate the proteins required for phage T4 DNA replication.

On a nicked DNA duplex a complex of six enzymes is able to displace the free 5'-end at the nick and extend the 3'-OH. The proteins required include a DNA polymerase (product of

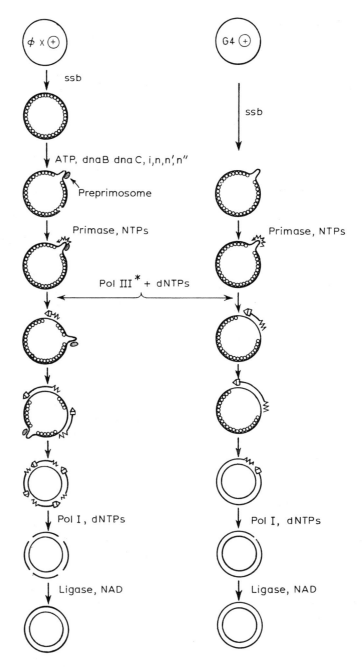

**Fig. 6.24** Conversion of $\phi$X174 and G4 single-stranded DNA into duplex form (M13 resembles G4 except *E. coli* RNA polymerase is used in place of primase): ssb single stranded DNA binding protein; ᴐ, DNA holopolymerase III; ᴡ, primer. After Kornberg [53, 54, 357, 676].

T4 gene 43); the archetypal single-stranded DNA-binding protein (gene 32 protein); polymerase accessory proteins (gene 44, 62 and 45 proteins) which increase the rate and processivity of polymerase action, and a DNA-unwinding protein (gene 41) which is required to unwind the displaced strand from the template [306, 307, 385–393]. (Another helicase is required to allow the replicating complex to pass a bound RNA polymerase molecule [394].)

In conjunction with gene-61 protein the gene-41 protein also has a primase function and the complex will synthesize and extend pentanucleotide primers on the displaced (lagging) strand [393–396]. This mixture of seven enzymes is thus able to catalyse the reaction going on at the replication fork. The Okazaki pieces synthesized on the lagging side are, however, some 10 000 nucleotides long as the complex lacks the capacity to remove RNA primers and join Okazaki pieces.

### 6.7.3 RF replication

Nicked cyclic duplexes are produced as the first step in the replication of the RFs of the small single-stranded DNA phage and it is convenient now to consider this reaction [709].

To initiate replication of the RF requires the intervention of a specific bacteriophage-coded protein. This protein makes an endonucleolytic cut in the plus (viral) strand of the RF DNA (Fig. 6.25). The site of the cut is specific and occurs between nucleotides 4305 and 4306 of $\phi$X174 RF DNA, within the region coding for the endonuclease. There is evidence that the endonuclease acts preferentially on the phage DNA molecule by which it was coded. The enzyme from phage $\phi$X174 is called the *cisA* protein and it recognizes a site of two-fold symmetry which may be present as a hairpin structure [689, 384].

The unique site of action of the gene II nuclease from bacteriophage fd is between bases 5781 and 5782 which is adjacent to (24 bases 3′ on the plus strand) the unique binding site of RNA

polymerase (f in Fig. 6.25). In bacteriophage G4 the *cisA* protein acts on the opposite side of the DNA molecule (between nucleotides 506 and 507) to the primase-binding site (nucleotides 3972–3992) (g in Fig. 6.25). The sequences recognized by the *cisA* protein of G4 and $\phi$X174 are identical and can be inserted into a plasmid. This makes plasmid replication dependent on the A protein and allows packaging of single-stranded plasmid DNA into preformed phage coats (see below) [361, 362].

Once the A protein has cut the plus strand it remains attached to a dAMP residue at the 5′-terminus by means of a phosphotyrosine bond [379]. The 3′-end can act as a primer for plus strand synthesis by the DNA polymerase III holoenzyme which together with the primosome has remained attached to the DNA following RF synthesis [362–365] (Fig. 6.25). Unwinding of the DNA is accomplished by combined action of an *E. coli*-coded ATP-dependent DNA-unwinding protein, the *cisA* protein, and the single-stranded DNA-binding protein [232, 364–371]. DNA gyrase may also help to unwind the DNA duplex ahead of the replicating fork but it is not required *in vitro* [364, 371].

With phage fd the gene II protein does not remain bound to the 5′-end but rather there is an additional requirement for the host *dnaA* protein [373, 376, 377].

As plus strand synthesis proceeds in the 5′ → 3′ direction the 5′-end of the plus strand is peeled off the rolling minus strand and it associates with single-stranded DNA-binding protein (Fig. 6.25). The single plus strand with its associated binding protein thus resembles the initial bacteriophage DNA. In phage fd it thus remains unaltered until plus strand synthesis has gone almost a complete cycle when the RNA polymerase-binding site is exposed and minus strand synthesis is initiated [372] (Fig. 6.25). In phage G4 the primase-binding site is exposed when plus strand synthesis has gone half way round the molecule [374] (Fig. 6.25). With $\phi$X174 it is not certain whether or not an intact

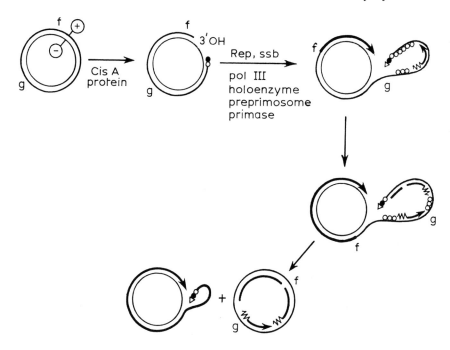

**Fig. 6.25**  RF replication of φX174 DNA (for explanation see text). g and f correspond to the primase and RNA polymerase-binding sites on single-stranded DNA from bacteriophage G4 and fd respectively, and hence indicate the position of initiation of –strand synthesis with these bacteriophage DNAs; ssb, DNA-binding protein (based on the work of Kornberg and others [53, 54, 342, 366, 371–375, 676, 683]).

single-stranded plus strand is produced before the preprimosome reinitiates lagging-strand synthesis. However, as the preprimosome is retained on completion of SS → RF synthesis it appears likely that minus strand synthesis is initiated at several non-random sites leading to the production of Okazaki pieces (Fig. 6.25) [366, 375].

When one round of plus strand synthesis is complete the *cisA* protein at the 5′-end of the plus strand is again brought adjacent to its site of action (Fig. 6.25). It now recognizes the left half of the origin sequence, nicks it, and circularizes the displaced plus strand, i.e. the origin sequence is both an initiation and a termination sequence [371, 375, 378]. A second round of plus strand synthesis is immediately reinitiated using the same complex of *cisA* protein and rep [380].

As the φX174 infectious cycle proceeds viral capsid proteins accumulate in the cell in the form of a prohead. This involves the products of viral genes F (major capsid protein), G (spike protein), H (spike-tip protein), B and D (non-structural proteins).

The prohead interacts with the replicating DNA–gene A protein complex in a reaction requiring phage gene C and J proteins, and this prevents initiation of minus strand synthesis. Thus the single-stranded tail of the rolling circle is no longer available for conversion to duplex but rather is packaged into the phage prohead. Synthesis of mature phage particles by a reconstituted mixture of purified proteins has been achieved [53, 54, 381–383].

In the case of phage M13 the single-stranded tail becomes associated with gene V protein which is exchanged for coat proteins when the viral DNA leaves the host cell.

## 6.8 INITIATION OF REPLICATION – GENERAL

The results of the Meselson–Stahl experiment suggested that DNA replication would be found to be a sequential process and that successive rounds of replication would not begin at random positions on the chromosome. It is now clear that chromosomes from most viruses and prokaryotes, whether containing linear or cyclic DNA, initiate their replication at one specific site. For small DNA molecules, or for fragments of larger molecules, these sites can be visualized using the electron microscope when they appear as double-stranded 'bubbles' (Figs 6.5 and 6.8) [396–400].

Cairns was able to visualize the whole of the replicating *E. coli* chromosome autoradiographically by growing the bacteria for several generations in the presence of tritiated thymidine (Fig. 6.26), and he showed that the chromosome exists as a continuous piece of double-stranded cyclic DNA [3].

### 6.8.1 Methods of locating the origin and direction of replication

That the initiation of the replication bubble occurs at a unique site has been shown by a number of different techniques for viral, plasmid, bacterial and mitochondrial DNA. Electron microscopy studies have shown that initiation of replication of the linear T7 phage chromosome occurs 17% from one end [398] but cyclic chromosomes appear featureless in electron micrographs and so it is not immediately possible to define the position of replicating bubbles. However, Schnös and Inman [6] were able to show that initiation of replication occurred at a particular site with reference to the partial denaturation map of phage λ and in the plasmid Col E1 [396] and in the simian virus 40

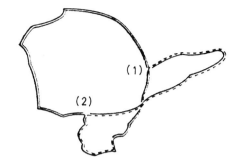

**Fig. 6.26** Diagrammatic representation of an autoradiograph of the chromosome of *E. coli* labelled with tritiated thymidine for two generations [4]. The chromosome has been two-thirds duplicated and the two replication forks are indicated as (1) and (2). (By courtesy of Dr John Cairns.)

[400] chromosome replication is always initiated at a fixed distance from the site of action of the *Eco* RI restriction endonuclease (Fig. 6.5, see Chapter 4).

In most cases (Col E1 is an exception) the midpoint of the replication bubble stays fixed with respect to the ends of the molecule. As the bubble gets larger it is thus clear that replication is occurring in both directions from the origin, i.e. replication is bidirectional.

As explained in Chapter 3 those genes located near the origin of replication of the *E. coli* or *B. subtilis* chromosome are present in more than one copy per cell in growing cells (see Fig. 3.16). Knowing the genetic map it is possible to measure gene copy number in growing cells relative to non-growing cells (where all genes are present in one copy per cell) and hence locate the origin of replication [401, 402]. Early experiments to measure gene copy number relied on measuring transforming or transducing activity and required a large panel of mutant strains. More recent experiments measure the copy number or specific radioactivity of identifiable restriction fragments, an approach which has shown that replication of adenovirus DNA is initiated at both ends of the linear duplex molecule [403].

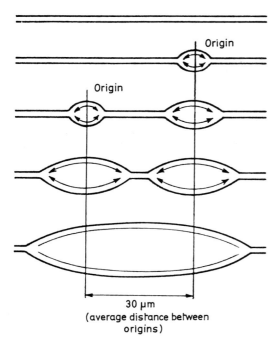

**Fig. 6.27**  Replication of mammalian DNA (two replicons are shown) (after Huberman and Riggs [405]).

Replicating DNA from eukaryotes shows a large number of replication bubbles in electron micrographs [399, 404], i.e. there is not a single origin of replication for each chromosome.

### 6.8.2  Replicons

Fibre radioautography studies suggest that DNA from their origins (Fig. 6.27). Fibre radioauto-tandemly joined sections [405, 407] or 'replicons' which are separately replicated bidirectionally from their origin (Fig. 6.27). Fibre radioauto-graphy involves labelling cells with tritiated thymidine and then lysing the cells in such a way that the DNA is released from the nucleus as a long fibre which attaches to a slide. On radio-autography the position of the DNA is indicated by lines of grains produced by the $\beta$-particles

released on tritium decay. What is seen are two parallel lines of grains whose length is dependent on the period for which the cells were exposed to the tritiated thymidine. The spacing between neighbouring lines gives the distance between replicons. The size of replicons varies between 15 and 60 $\mu$m (40–200 kbp) in cultured mammalian cells but is considerably shorter in the very rapidly dividing cells of amphibian and insect early embryos, where S phase may last for only 20 minutes, and is much longer in slowly dividing spermatocytes [406, 408].

When the specific radioactivity of the tritiated thymidine is reduced after a short initial period of labelling at high specific radioactivity the density of the radioautographic grains is seen to decrease at both ends of the line of grains indicating that replication is bidirectional (Fig. 6.27). It is important in these experiments to start labelling the cells at the time of initiation of replicon synthesis otherwise gaps will be present in the middle of the lines of grains which may lead to misinterpretation of the results.

With an average size of 30 $\mu$m the replicons of eukaryotes are very much smaller than the single replicon which is the *E. coli* chromosome (1300 $\mu$m). They are similar in size and may be physically identical with the loops of DNA seen in electron micrographs of dehistonized chromatin (see Chapter 3). In a mammalian cell there are typically about 60 000 replicons.

### 6.8.3  Rate of replication

The rate of replication of DNA is dependent on the temperature and the supply of nutrients, particularly deoxyribonucleotides. The mini-mum time required to replicate the *E. coli* chromosome is about 40 minutes, implying a rate of synthesis of about 1700 base pairs per second [244]. As replication is bidirectional the rate at each fork is about 850 base pairs per second or 14 $\mu$m per minute. (Compare this with the rate of transcription which proceeds at 35–40 nucleo-tides per second in *E. coli*.) In mammalian cells the rate of fork movement is only 0.5–1.2 $\mu$m per minute or about 60 base pairs per second [409]. However, in eukaryotes this slower rate is compensated for by the smaller size of the replicon which is completely replicated in about 15 minutes. It is clear that only about 5% of replicons in a eukaryote cell are simultaneously active [410] and some control must be exercised over the order in which replicons initiate DNA synthesis.

### 6.8.4  Origin strategies

Implicit in the fact that initiation of DNA synthesis occurs at a unique site on the chromosome is the suggestion that this site has a distinctive structure which allows the binding of the replicative machinery and the unwinding of the double helix.

It is considered probable that DNA synthesis is only initiated on a region of the DNA which is rendered single stranded [127] perhaps by reason of its being involved in transcription, and specific RNA transcripts known as origin RNA (oriRNA) have been implicated. Such RNA molecules may also serve as primers for subsequent DNA synthesis.

Many origin regions have been sequenced and whether present in regions coding for protein or not they contain unusual sequences, e.g. direct repeats, palindromes and true palindromes which may enable these regions (or the RNA derived therefrom) to assume a hairpin configuration. The DNA may take up a hairpin or cruciform conformation in response to initiation proteins [411] which thereby allow an RNA polymerase to synthesize a primer on which DNA synthesis may be initiated. Alternatively, as with the single-stranded bacteriophages a hairpin may be the only region of the DNA not covered by binding proteins and as such may be the only region accessible to a primase (see Section 6.7.1). Normally an RNA molecule, once synthesized, is released from the DNA template which returns to its double-stranded

form. If the non-transcribed strand of the DNA were able to stabilize itself by forming a hairpin this would have the effect of retaining the RNA strand at the replication origin to serve as a primer for DNA synthesis. It should also be borne in mind that rather than assuming a cruciform structure, four-stranded regions of DNA may be formed which could act as structures recognized by specific initiation proteins [412, 413]. Section 6.9 illustrates the ways some of these strategies are used but it may be helpful at this stage to point out the ways in which replication occurs in different replicons (Fig. 6.28).

(a) Duplex circles may replicate via a theta ($\theta$) or Cairns structure in which the parental strands remain intact. Initiation involves production of a primer and usually relies on some mechanism to open up the duplex. The primer is extended as DNA to produce the leading strand. Lagging-strand synthesis may start almost immediately or may be delayed for a short time to produce a D-loop – (Fig. 6.28c) – or until the replication fork has traversed the whole replicon in which case synthesis leads to displacement of a single strand (displacement synthesis).

If lagging-strand synthesis proceeds through the origin region, or if separate initiation mechanisms exist for clockwise and anticlockwise replication, then replication will be bidirectional.

(b) Linear duplex molecules may initiate internally by similar mechanisms (Fig. 6.28b).

(c) Duplex circles often replicate by the rolling circle mechanism when initiation requires the action of a sequence-specific nuclease (Fig. 6.28d).

(d) Some linear duplex molecules initiate replication at the ends leading to displacement synthesis (Fig. 6.28e). Initiation may occur at both ends or a linear single-stranded molecule may be produced.

(e) Linear single-stranded DNA molecules can initiate by formation of a hairpin loop at the end which supplies a 3'-OH primer.

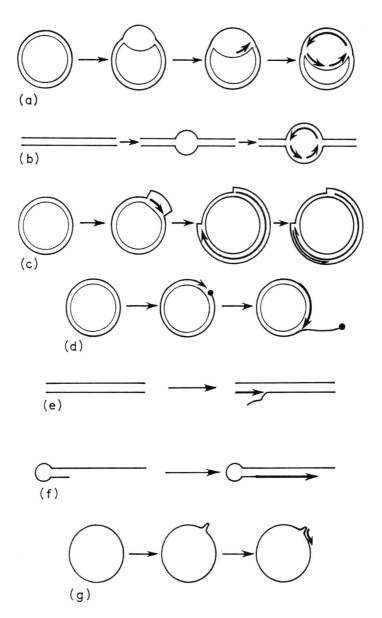

**Fig. 6.28** Different ways in which initiation of DNA synthesis can occur. (a) and (b) represent Cairns (Θ) structures formed by cyclic and linear duplex molecules respectively; (c) is a D-loop and (d) a rolling circle; (e) represents an alternative method of initiation of replication in linear duplex DNA molecules and (f) illustrates the mechanism of initiation in linear single-stranded DNA molecules. The initiation of replication of single-stranded cyclic molecules is shown in (g).

(f) Single-stranded cyclic DNA is converted to duplex by producing an RNA primer. This requires specific sequences on the DNA which can be recognized by the primase or primosome complex.

### 6.8.5 Positive or negative control of initiation

DNA replication and, in many cases, cell division is initiated by a single event which occurs at the origin of replication of each replicon. This led Jacob, Brenner and Cuzin to propose a theory of positive control of replication similar to the induction/repression model for the control of protein synthesis in prokaryotes (see Chapter 10) [414]. This model proposes that a product of a replicon is a (protein) molecule which interacts with the origin to induce replication. The inducer molecule is *trans* acting and its concentration dictates whether or not replication will be initiated.

An alternative model of Pritchard, Barth and Collins [415] proposes negative control over initiation. They suggest that there is an inhibitor of replication present in cells. The inhibitor is synthesized by the replicon, and only when its concentration is diluted (by growth of the cells) will initiation occur.

Experiments of Rao and Johnson [416] where animal cells in the G1 phase of the cell cycle were fused with S-phase cells strongly supported the idea of a trans acting factor exerting control over initiation of DNA synthesis but did not distinguish between positive and negative control. However, experiments involving fused plasmids strongly supported a negative control mechanism. Cabello *et al.* [417] fused Col E1 with the related plasmid pSC101 to form plasmid pSC134 which used only the Col E1 origin and maintained the copy number of Col E1. However, in *polA* mutants Col E1 cannot grow (see Section 6.9.3) and under these conditions pSC134 used the pSC101 origin and assumed the copy number of pSC101. These results are what would be expected of incompatible plasmids responding to

identical repressors where the pSC101 origin is more sensitive to inhibition and hence is only induced in large cells in which the repressor has been diluted. Thus normally the Col E1 origin is activated first and maintains the level of repressor. Only when the Col E1 origin is inactive does the level of repressor fall sufficiently for the pSC101 origin to be activated.

Related plasmids are incompatible if they respond to the same inhibitor molecule. Thus, as a result of changes in base sequence around the target for the inhibitor, the plasmids pMB9 and pBR322 are less sensitive than Col E1 and rapidly exclude the latter [413].

As we shall see in the next section there may be elements of both positive and negative control over the initiation of replication.

## 6.9 INITIATION OF REPLICATION – SPECIFIC EXAMPLES

This section considers several examples of what we know about initiation of replication. Every system has its own peculiarities and this section can in no way be comprehensive. However, it gives an idea from which some generalities can be obtained. A comprehensive review of prokaryote DNA replication has been published recently [702].

### 6.9.1 Small single-stranded phage

In Section 6.7.2 it was described how replication of the duplex RF forms of the small single-stranded DNA virus is initiated. A viral-coded protein nicks one strand of the duplex molecule at a unique site and initiates replication by the so-called rolling circle mechanism.

### 6.9.2 Double-stranded phage

Three phage (λ, T7 and T4) have been extensively studied. No common rule can be applied so each will be briefly considered.

(a) *Lambda*

The λ DNA molecule is linear but has sticky ends (see Section 3.6.6a) and cyclizes immediately on entry into the cell [418, 419]. Replication then is initiated at a unique origin and proceeds bi-directionally to give theta ($\theta$) forms (cf. *E. coli* replication). However, at later stages of infection rolling circles are found.

In addition to host functions the products of λ genes O and P are required for initiation of DNA replication [344, 420, 421]. The P gene product appears to take over the function of *E. coli dnaC* gene product. It interacts with *E. coli dnaB* protein and the complex interacts with lambda gene O product and the host cell *dnaJ* and *dnaK* proteins to form a preprimosome at a specific site (ori) rendered single-stranded by transcription by host RNA polymerase. This is called tran-scriptional activation but a similar preprimo-some can form on any single-stranded DNA in a sequence-independent manner [704]. Eight molecules of the O protein interact with four 19 base pair repeats [422] within a 65-bp origin region which itself is within the gene coding for the O protein and adjacent to that coding for the P protein. This region which is necessary for phage lambda replication has been excised and propagated as a plasmid called λdv which is transcribed from the $P_R$ promoter. This whole region has been sequenced [421, 423, 424].

It has been proposed that *dnaG* primase synthesizes a 78-base RNA from a region to the left of ori. This *oop* RNA [420, 421] may prime DNA polymerase III-catalysed DNA synthesis.

Alternatively *oop* RNA may be extended by RNA polymerase leading to synthesis of λ repressor and the state of lysogeny results where the genome becomes integrated into the host genome and the two replicate under host cell controls (see Sections 3.6.8 and 10.1.5) [420, 425, 426].

*In vitro* systems have been developed which will replicate λdv plasmid using purified proteins [427–429]. They indicate that the transfer from RNA to DNA synthesis takes place on both sides

of the *ori* sequence in such a way that DNA synthesis from the two sides converges [470].

(b) *Phage T7*

Initiation of T7 DNA synthesis occurs 17% from the so-called left end of the molecule. A repli-cation bubble forms which expands bidirection-ally. When the short side is completely syn-thesized a Y-shaped replicating chromosome is formed (Fig. 6.29).

Concatamers, i.e. double length molecules, are found later in infection [430–432].

Replication requires phage genes 1,2,3,4,5 and 6 [433, 434]. Gene 1 product is an RNA polymerase essential for the transcription of genes 2–6 and gene 2 protein inactivates the host RNA polymerase [435]. Gene 3 protein is an endonuclease and gene 6 protein an exonuclease involved among other things in the breakdown of host DNA.

There are two strong promoters for the T7 RNA polymerase in the origin region, closely followed by an AT-rich region (Fig. 6.30) [436–438]. Transcription causes melting of the origin region which can be detected by treatment with a single-stranded endonuclease [439]. Within the AT-rich region there is a change-over from RNA synthesis to DNA synthesis which proceeds unidirectionally towards the right.

In the presence of T7 gene 5 protein and the host protein thioredoxin (which together form a DNA polymerase [440–444]) the T7 gene 4 protein catalyes an ATP-dependent unwinding of duplex DNA thereby promoting the DNA polymerase action [441, 445–447]. This complex is thus able to catalyse leading-strand synthesis and DNA-binding protein and gyrase further promote this action. Gyrase either relieves the positive supertwist generated by replication or it may actively help unwinding by imposing a negative supercoiling ahead of the fork.

Gene 4 protein has a second function in addition to its action as a DNA helicase. It also acts as a primase for lagging strand synthesis which is initiated by the synthesis of tetra-

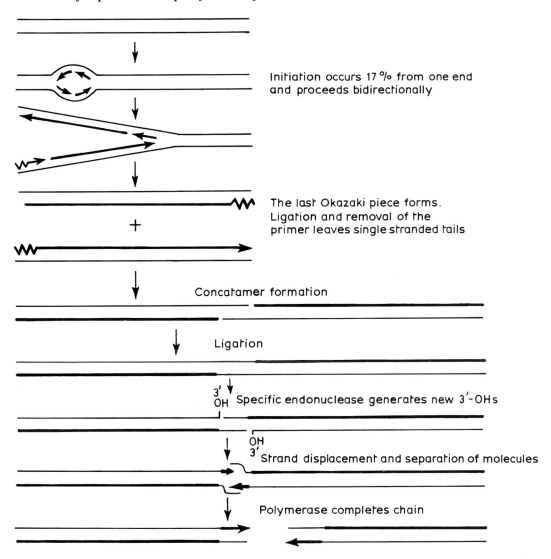

Initiation occurs 17 % from one end and proceeds bidirectionally

The last Okazaki piece forms. Ligation and removal of the primer leaves single stranded tails

Concatamer formation

Ligation

Specific endonuclease generates new 3′-OHs

Strand displacement and separation of molecules

Polymerase completes chain

**Fig. 6.29**  Replication of T7 DNA. After Watson [430].

ribonucleotides of sequence pppACCC and pppACCA. Thus when a primase site is exposed by the rightward movement of the leading strand, synthesis of the lagging strand is initiated using the T7 gene 4 protein. The tetraribonucleotide primer is extended by T7 DNA polymerase. DNA synthesis now proceeds leftwards on the lagging strand through the origin region. With the help of the helicase action of

gene 4 protein and single-stranded DNA-binding protein a fork is now established moving leftwards (Fig. 6.30) and replication is now bidirectional.

Thus as well as catalysing leading-strand synthesis the gene 4–gene 5 protein complex can also catalyse initiation and elongation of Okazaki pieces on the lagging strand [33, 434, 445, 448–455].

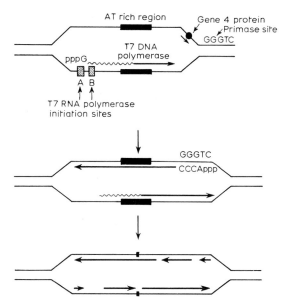

**Fig. 6.30**   Initiation of T7 DNA replication [438].

Gene 6 exonuclease or host DNA polymerase I 5′ → 3′ exonuclease can catalyse removal of RNA primers and the Okazaki pieces can be joined using either the host ligase or a ligase coded for by T7 gene 1.3.

*(c) Phage T4*
Replication of T4 DNA is far more complex and initiation of DNA synthesis appears to occur at several points along the long, circularly permuted, linear molecule [456, 457]. It has been proposed [458, 459] that, as with T7, primary initiation occurs following synthesis of an RNA primer but in this case it is supposed that only the host RNA polymerase can act in this manner. Mutations in T4 genes 39, 52 and 60 lead to a delay in the onset of T4 replication and the DNA synthesis that does occur is dependent on the host cell gyrase [460]. These three genes code for an ATP-dependent topoisomerase [460–462] which may help to open up the duplex molecule. However, shortly after infection the host RNA polymerase is modified by interaction with the products of viral gene 33 and gene 55 and this

affects promoter recognition rendering it unable to initiate DNA replication. Initiation now becomes dependent on genes 46 and 47 which code for enzymes involved in recombination. The model which has been proposed for these secondary initiations involves recombination between partially replicated molecules which contain single-stranded regions and the circularly permuted duplex molecules (see Chapter 7). This model explains the finding of complex branched molecules later in the T4 infectious cycle.

When cells of *E. coli* are infected with the T-even bacteriophages, the economy of the cells is completely altered so as to lead to the production of new phage DNA which differs from the host DNA in containing hydroxy-methylcytosine in place of cytosine. These changes result in the production in the infected cell of a series of new and interesting enzymes [57, 153, 463, 464].

(i) Within a few minutes of infection a hydroxymethylase (the product of phage T4 gene 42) appears which brings about the conversion of dCMP into hydroxymethyldeoxy-cytidine monophosphate (dHMCMP).

(ii) At about the same time a *kinase* is produced which phosphorylates dHMCMP to the corresponding triphosphate dHMCTP.

Neither of these new enzymes is found in cells infected with phage T5 which does not contain HMC.

The kinases for dTMP and dGMP are also greatly increased but not that for dAMP. This increase is due to the production of new enzymes which can be distinguished from the kinases present in the host cell prior to infection.

(iii) The formation of host DNA is prevented by the appearance of a pyrophosphatase (the product of phage T4 gene 56) which converts dCTP into pyrophosphate and dCMP which then acts as a substrate for the hydroxymethylase.

(iv) Five distinct glucosyltransferases are known to be induced after infection with T-even phages for the purpose of transferring glucose residues from uridine diphosphate glucose to the HMC of phage DNA in the proportions shown in Table 3.5. The glucosylase found in T2 phage-infected cells transfers a glucose residue to HMC in the $\alpha$-configuration. Two glucosylating enzymes are produced after T4 infection; one adds a glycosyl group in $\alpha$-linkage to HMC while the second also adds a glucose group but in $\beta$-configuration. After T6 infection two gluco-syltransferases are also produced. One adds a monoglucosyl residue to HMC in $\alpha$-linkage while the other reacts with the monoglucosylated groups on HMC to add a second glucose residue, the linkage between the residues being of the $\beta$-configuration.

(v) A nuclease is produced which degrades DNA containing cytosine. T4 DNA which contains hydroxymethylcytosine is resistant.

*In vivo* the enzymes synthesizing the precursors for DNA synthesis are closely associated with the DNA polymerase [465–468] and *in vitro* the complex of enzymes preferentially uses deoxynucleoside monophosphates to make T4 DNA [469] (see Section 6.11).

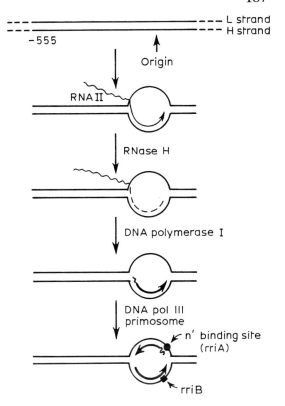

**Fig. 6.31**  Initiation of Col E1 replication.

### 6.9.3  Plasmids

#### (a)  *Col E1*
This small double-stranded cyclic DNA molecule replicates from a unique origin but in one direction only [677]. Theta ($\theta$) structures are formed and initiation does not involve permanent nicking of either strand [471, 472]. Up to three proteins bind near the origin and bring about supercoiling of the cyclic molecule [472] thereby exposing single-stranded regions to which *E. coli* RNA polymerase binds [474]. An RNA transcript (RNA II) is initiated 555 base pairs upstream from the origin ($-555$). This transcript is not released from the origin region but a D-loop is formed with a DNA/RNA hybrid and single-stranded DNA [475, 476, 490] (see Fig. 6.31). This hybrid is sensitive to RNase H (see Chapter 4) which fragments the RNA

producing several primers which can be extended by *E. coli* DNA polymerase I. This enzyme synthesizes a 6S piece of DNA (400 nucleotides) joined to a 20-long ribonucleotide primer, the whole remaining associated in the D-loop configuration [475–479, 497]. The DNA polymerase III replicating complex now takes over to extend the fork [329, 472, 475]. On the lagging side of the fork (on the L strand) there is an n' binding site (*rriA*) (see Section 6.7.1) at which lagging-strand synthesis can be initiated [329]. However, lagging-strand synthesis cannot pass the origin region and Col E1 replication remains unidirectional.

There are two complications to this story. As well as RNA II a second RNA (RNA I) is transcribed from the region $-455$ to $-555$ from the opposite strand of the DNA. This RNA, which can assume a complex secondary structure, interacts with RNA II and either prevents its forming a stable hybrid at the origin or interferes with RNase H action in some other way [473, 482–486, 708]. RNA I thus acts in trans to limit initiation and hence copy number of Col E1 and other compatible plasmids (see Section 6.8.5).

A second trans acting inhibitory element exists downstream from the origin (184 to 806 bp). The product of this region is the *rop* or *rom* protein (63 amino acids long) which enhances the binding of RNA I to RNA II [487–490]. This leads either to premature termination of transcription of RNA II or to inhibition of processing. This protein appers to play no role in incompatibility but may regulate the balance between replication of duplex plasmid molecules (as described here) and the replication involved in the single-strand transfer reaction of conjugation (see Section 3.7).

In conjugal transfer a nick is introduced into one strand (the H-strand) and the endonuclease remains bound to the 5′-side of the nick. Rolling circle replication now leads to the synthesis of a new duplex cyclic molecule and the parental H-strand can be passed to another bacterium through a conjugation tube in the form of a single strand. Cyclization follows transfer and conversion to a duplex molecule is initiated by the binding of n′ protein to a preprimosome recognition site (*rriB*) [481, 491].

Thus in conjugal transfer, replication of Col E1 closely resembles replication of φX174 except that the order of reactions is

$$RF \rightarrow SS \rightarrow RF$$

### (b) *Other plasmids*

The drug-resistance plasmids (e.g. R1, R6 and R100) differ from Col E1 in that RNA II may act as a primer for initiation of replication but, in addition, it can be extended to form a messenger RNA from which a protein (π) is translated. π protein acts as a cis-acting initiation protein [492–497].

The origin region of pSC101 shows a similarity to that of *E. coli* and replication of this plasmid is dependent on the *dnaA* protein (see Section 6.9.4). In addition a plasmid-coded protein which binds to the origin region is also required [498–500]. These plasmid-coded replication initiator proteins also regulate their own transcription by acting as autorepressors [496, 499].

### 6.9.4 Bacteria

*E. coli* temperature-sensitive mutants *dnaA* and *dnaC* fail to initiate new rounds of DNA replication when placed at restrictive temperature but complete those rounds already in progress. On cooling such mutants to the permissive temperature in the presence of tritiated thymidine the first region to become radioactive is a 1.3 kilobase fragment formed by treatment of the DNA with the restriction endonuclease *Hind*III [501]. The amount of incorporation is less in the case of *dnaA* mutants than *dnaC* mutants indicating that the *dnaA* gene product is involved several minutes before the *dnaC* gene product in the initiation process.

Initiation of replication in *E. coli* is sensitive to rifampicin and chloramphenicol and a protein located in the membrane which is normally made about 15 min before initiation is not made in *dnaA* mutants [502]. Ori RNA of *E. coli* has been isolated linked to high-molecular-weight DNA. It is not made in *dnaA* mutants and is not linked to DNA in *dnaC* mutants suggesting a role for these gene products in initiating ori RNA synthesis (*dnaA*) and in the change-over to DNA synthesis (*dnaC*) [503].

The origin of replication of the *E. coli* chromosome (*oriC*) has been located on a 245 bp sequence which when cloned into small Col E1-type plasmids allows them to replicate in the absence of DNA polymerase I. Such replication is bidirectional and dependent on *dnaA* protein. The *oriC* region binds with high affinity to membrane proteins [512].

The *oriC* sequence is conserved in a number of

Gram-negative species though it has been difficult closely to define the essential features [690–692]. In this region some base substitutions are deleterious; others not; but deletions are harmful suggesting some of the sequences may serve as spacers. The origin region probably provides multiple sites of interaction with replication initiation factors and must be present in a precise structure for activity.

It appears, however, that initiation does not take place within the 245 bp region but just outside [504]. In addition to the proteins required for leading- and lagging-strand synthesis, initiation shows a requirement for the products of the *dnaA*, *dnaI* and *dnaP* genes (and possibly also the *dnaJ*, *dnaK* and *dnaL* genes), RNA polymerase and a histone-like DNA-binding protein called HU [505–507].

The *dnaA* requirement is highly characteristic and multiple copies of this protein bind to four 9-bp regions in *oriC* of sequence [508]: TTATCCACA.

Initiation is severalfold more efficient in *dam*+ cells indicating that DNA methylation may play a role here [509] (see Section 4.6).

A system has been reconstructed from purified proteins which has enabled initiation to be separated into two stages [510, 511]. In the absence of DNA polymerase III holoenzyme an RNA primer is made but not extended. If an inhibitor of *E. coli* RNA polymerase is now added along with DNA polymerase III holoenzyme, no further primers are made but those already present can now be extended. Topoisomerase I, RNase H and HU protein are required to prevent initiation at sites other than *oriC*. In addition to RNA polymerase the production of a specific primer requires DNA gyrase and the products of the *dnaA*, *dnaB* and *dnaC* genes. There are two RNA polymerase promoter sequences in *oriC* which can lead to initiation of transcription in both directions [513]. However, there is evidence that initiation occurs first on one strand to produce a fork which moves in a counter-clockwise direction [710].

Primase (the *dnaG* gene product) is not essential but when present it will interact with the preprimosome to synthesize a primer, i.e. the RNA polymerase functions to catalyse transcriptional activation [705]. The second stage requires gyrase, single-stranded-DNA-binding protein, and the products of the *dnaB* and *dnaC* genes in addition to the polymerase III holoenzyme.

### 6.9.5 Mitochondria

In animal cells mitochondria contain a closed circular 16-kbp duplex DNA molecule which replicates unidirectionally from a unique origin [514–517]. Prior to initiation the duplex undergoes partial unwinding by removal of about 48 Watson–Crick turns in the presence of a nicking closing enzyme [518, 519]. Initiation of DNA synthesis then occurs with synthesis of a 450-base length of DNA (the leading or heavy strand) to produce a D-looped structure [163, 515–517, 519, 520]. This small piece of newly synthesized 7S DNA is unstable and may turn over several times [518, 521, 522] before it is extended. Extension occurs asymmetrically and unidirectionally until 60% of the heavy strand (99% in *Drosophila* mitochondria [523]) has been synthesized [519, 520], when light (lagging)-strand synthesis is initiated. Thus although θ forms are seen in the electron microscope one loop is a single-stranded parental H-strand displaced by synthesis of daughter H-strand.

The short primer RNA found at the 5′-end of human mitochondrial 7S D-loop DNA does not have a unique 5′-end probably as a result of rapid processing [525, 526]. This RNA, which may be synthesized by a unique RNA polymerase [527], is initiated in the region of a palindrome which could form a stable hairpin structure when the DNA helix is unwound [517, 526]. The 3′-end of 7S DNA is just downstream from a 13-bp sequence which occurs four times in mouse mitochondrial DNA but only once in human mitochondrial DNA.

D-loop DNA is extended and when H-strand synthesis has proceeded two-thirds of the way round the genome a sequence is exposed which can form a stable hairpin structure. An RNA primer is synthesized here which leads to initiation of L-strand synthesis [524, 528, 529]. Both strands are now extended and the two daughter molecules separate with one being still almost half single stranded.

The available evidence suggests that DNA synthesis in mitochondria is catalysed by DNA polymerase $\gamma$ [530].

In sea-urchin oocytes mitochondrial DNA synthesis is similar to that found in mouse mitochondria but duplex synthesis occurs early

with multiple initiations [531].

Yeast mitochondrial DNA contains several origins of replication. In petite strains in which a short random stretch of DNA is excised and amplified, surrogate origins are used. In one petite strain the mitochondrial DNA contains only adenine and thymine nucleotides [532–534].

Mitochondrial DNA in *Tetrahymena* and *Paramecium* is not cyclic but is a linear duplex [535–539]. In the former, replication is initiated near the centre of the molecule and 'eye'-shaped replication intermediates are seen in the electron microscope. Replication is bidirectional. In *Paramecium* replication proceeds unidirectionally from a crosslinked terminus [538, 539].

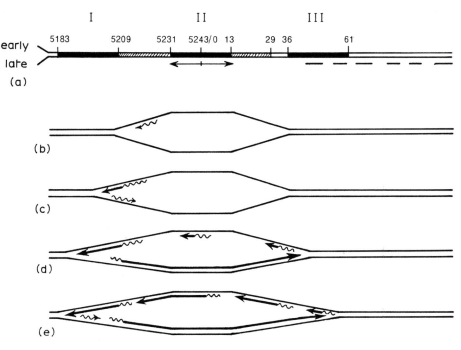

**Fig. 6.32** Initiation of SV40 DNA replication: (a) The origin region showing the three T antigen binding sites (filled boxes), the minimal origin region (hatched box), the six 8 base pair repeats (lines under the diagram) and the palindromic region (arrows); (b) T antigen binding denatures the origin region and allows primase to synthesize a short primer (∿→) at one of several sites; (c) The primer is extended as DNA synthesis exposes primase binding sites for retrograde synthesis; (d) The retrograde strand is extended through the origin region to become the leading strand for late strand synthesis; (e) Other primers are formed on the early strand to initiate lagging-strand synthesis. Binding of T antigen to site III may also allow late strand synthesis to be initiated at the 8 bp repeats (based on [543–545]).

### 6.9.6 Double-stranded cyclic DNA viruses (SV40 and polyoma)

These similar, small, animal tumour viruses contain a cyclic DNA molecule which replicates as a $\theta$ form [540], the parental strands remaining intact. They have been studied, not only for their own sake but as models for animal cell DNA replication. This is only poorly justified for, although the DNA is present in the nucleus complexed with histones (i.e. as a minichromosome) replication is not restricted to one round per cell division.

Initiation occurs at a unique site and proceeds bidirectionally. The origin region is also the site of initiation of transcription which also occurs in both directions (Fig. 3.20 and Fig. 6.32). One of the early SV40 gene products (T antigen) is essential for initiation of replication [541–544]. There are three sites in and around the origin region to which T antigen can bind. However, only one of the T antigen-binding sites (site II) falls within the 65-bp essential origin region, and the central 26-bp region of site II is palindromic. Just outside the 65-bp origin regions are six repeats of the sequence GGGCGGRR (Fig. 6.32).

Isolation of nascent DNA from the origin region showed initiation to occur at many sites, each time with a short (7–8 base) ribonucleotide primer. Some 70% of primers start with adenosine, the remainder with guanosine [543–546]. On the early strand (i.e. the template for early mRNA synthesis) DNA primase will initiate at four major sites around nucleotide 5215 between T antigen-binding sites I and II, and transfer to continuous (leading strand) DNA synthesis occurs at about nucleotide 5210 (i.e. at the beginning of T antigen-binding site I). There are no initiations for late-strand synthesis in the origin region and it is supposed that only when leading-strand synthesis exposes primase-binding sites does initiation of late-strand synthesis occur [543, 544]. *In vitro* results suggest that late-strand synthesis may also be initiated in

any of the six repeated sequences, removal of which reduces the efficiency of DNA replication *in vivo* [545, 552]. *In vitro* systems have shown that DNA polymerase $\alpha$ and its associated primase are involved in both initiation and continued SV40 DNA replication [116, 545, 547, 548].

The origin region of polyoma virus DNA is similar to SV40 but for initiation, in addition to large T antigen, there is a requirement for a *cis*-acting enhancer element [544, 548–551] (See Section 10.4).

### 6.9.7 Adenoviruses

These are animal viruses containing a linear duplex DNA molecule of about 36 000 bp. This replicates by a strand-displacement mechanism, initiation occurring at either end of the molecule (Fig. 6.33). The adenovirus DNA contains an inverted terminal repetition extending to the very ends of the genome. In different adenoviruses the length of the terminal repeat varies from 103 to 162 bp [553–556]. Two regions in this terminal repeat are required for initiation. An AT-rich stretch from positions 9–18 is essential for binding of a 'terminal protein' (see below) and the region up to position 48 is required for binding of a host cell nuclear protein [557–561]. Covalently bound to both 5'-termini is a 55 000-molecular-weight protein. Binding is by a phosphodiester link to the $\beta$-OH of a serine residue [553, 554, 562, 563].

Initiation requires host cell DNA topoisomerase I [564] and at least three virus-coded proteins: a single-stranded DNA-binding protein which may also bind to the ends of duplex molecules [553, 565]; a DNA polymerase [173, 174, 560]; and an 80 000-molecular-weight 'terminal protein' [553, 526, 566–573]. The terminal protein forms a complex with the DNA polymerase and in the presence of adenovirus DNA it interacts with dCTP so that a dCMP residue becomes covalently linked to the terminal protein. This dCMP then serves as a primer to

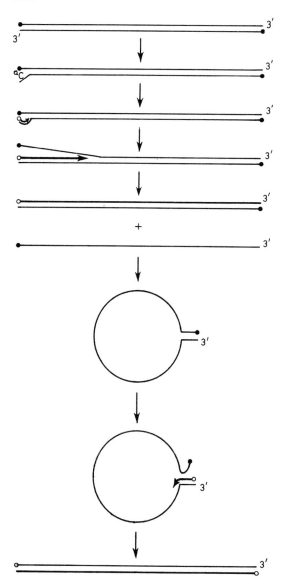

**Fig. 6.33** Replication of adenovirus DNA. The terminal protein is represented by a circle (●, 55 K; ○, 80 K).

pulse-labelling with [³H]thymidine and analysis of completed molecules it has been shown that termination occurs at both ends of adenovirus DNA showing that initiation can occur at either end and some molecules are seen in the electron microscope which are replicating from both ends [553, 555]. However, if a single strand of DNA is produced, the terminal repeat will allow it to form a panhandle structure to initiate conversion to a duplex DNA molecule [553, 556]. At some time during infection the 80 000-molecular-weight terminal protein is reduced in size to 55 000 molecular weight but this processing does not appear to be essential [574].

Adenovirus is not unique in adopting this strategy for initiation of replication. Thus phage φ29 initiates using a dAMP covalently linked to a 30 000-molecular-weight terminal protein [575, 576] and Pearson *et al.* [577] have pointed out the similarity between this strategy and that adopted by φX174 which initiates replication of duplex circles by nicking one strand and covalently attaching a protein to the exposed end (see Section 6.7.2).

### 6.9.8 Yeast

Yeast cells contain a small 2 *μ* plasmid which replicates autonomously. For this to occur the plasmid requires two proteins for which it codes and a short origin-containing region at which bidirectional replication is initiated [578–580]. Other plasmids have been constructed which can also replicate in yeast provided they contain a particular region from the yeast chromosomal DNA. Such regions are known as autonomous replicating sequences (ARS) as they confer this property on the plasmid [581–586]. This system has allowed the isolation of several ARS elements which all contain a very AT-rich core region which has the consensus sequence [583, 586]: A/TTTTATA/GTTTT/A. Similar sequences obtained from non-yeast cells will also confer autonomous replication upon yeast plasmids but we have no evidence yet

initiate displacement synthesis (Fig. 6.33). Electron microscopy has been used to show that initiation usually occurs at only one end of the molecule so that one round of replication results in the formation of a duplex and a linear single-stranded molecule [553]. However, by that they act as origins of replication in their normal environment.

### 6.9.9 Higher eukaryotes

As we have said earlier there are multiple origins of replication in eukaryotic cells each of which initiates once per cell cycle; yet not all replicons are simultaneously active (see Section 6.8.2).

In the situation where genes occur in long tandem repeats it has been possible to identify an origin of replication in each repeat [587–590]. For example, electron microscopy has been used to localize the centre of replication bubbles of sea-urchin ribosomal DNA to the non-transcribed spacer, and at the beginning of S phase one particular restriction fragment of the amplified dihydrofolate reductase gene in hamster cells is labelled first.

In contrast, although initially subject to some controversy, it now appears clear that any DNA sequence when injected into *Xenopus* eggs will replicate under the strict cell cycle control pertaining [591–595]. Although even bacterial DNA replicates in *Xenopus* eggs, there is some evidence that sequences containing copies of the short interspersed repeated sequences (e.g. *Alu* sequences – see Section 3.2.5.c) may be replicated more efficiently [596]. Nevertheless, it is difficult to reconcile observations of specific origin sequences with the changes in size of replicons found in rapidly dividing embryos and very slowly growing cells [406, 408], observations which imply that a particular origin may be used in some situations, but not in others.

Relevant to this finding is the observation that origins of replication and/or replicating forks are found associated with the nuclear matrix, that residual nucleoprotein complex which remains following treatment of nuclei with high concentrations of NaCl followed by deoxyribonuclease digestion [597–601] (see Section 3.1). Thus initiation may require the binding of short interspersed repeated sequences to the nuclear matrix. Certain sequences may be more susceptible to binding than others and hence may initiate earlier in S phase. Other sequences may fail to initiate at all in slowly growing cells leading to an increased replicon size. This situation is not radically different from that found in *E. coli* where the *oriC* region is required to bind to the membrane-associated *dnaA* protein for initiation to occur.

### 6.9.10 Retroviruses

These pose a unique problem, for, as described in Chapter 3, they replicate by first converting their RNA genome into DNA using RNA-directed DNA polymerase (reverse transcriptase – Section 6.4.2.f). Models of reverse transcriptase action have been proposed [220, 621–624] to explain certain unexpected observations, i.e. the duplex DNA produced is longer at both ends than the viral RNA [625, 626] and during the reaction short molecules of both plus and minus strand DNA ('strong stop DNAs') are formed [627]; as well as two genetically identical copies of the viral RNA the virion contains several molecules of tRNA [191, 192]; reverse transcriptase also exhibits ribonuclease H activity (which will hydrolyse the RNA strand of an RNA:DNA hybrid) and endonuclease activity.

Avian reverse transcriptases are initiated with tRNA [Trp] [193, 628–630] but the reverse transcription from Moloney murine leukaemia virus starts by using a tRNA[Pro] which attaches to the viral RNA at about 150 nucleotides from the 5′-end. This tRNA[Pro] serves as a primer for synthesis of minus strand DNA. Synthesis stops when the DNA strand reaches the 5′-end of the template, i.e. after 100–150 nucleotides have been added [631]. This is minus strand strong stop DNA (Fig. 6.34b).

The RNA genome is terminally redundant and digestion of the 5′-end by ribonuclease H allows the molecule to cyclize (Fig. 6.34d) and this provides a template for the further extension of the DNA to form a copy of the whole of the viral

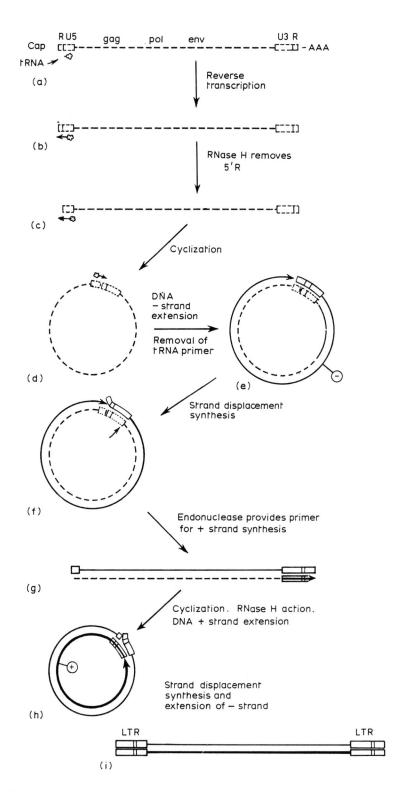

**Fig. 6.34** Postulated mechanism of conversion of retroviral RNA into duplex DNA. See text for details.

strand. Some strand-displacement synthesis causes the minus strand to duplicate part of the long terminal repeat sequences (LTR – Fig. 6.34f). It is speculated that specific endonuclease action on the viral RNA by the reverse transcriptase could now provide a primer for plus strand DNA synthesis. The plus strand strong stop DNA is extended to produce the complete duplex DNA which is usually integrated into the host cell chromosome before transcription leads to production of new viral RNA. The terminal nucleotides of the linear DNA duplex are required for integration into the host chromosome and are lost in the process [624] (see Section 7.7.3).

Similar conversion of RNA into duplex DNA followed by integration is believed to occur in the production of pseudogenes and in the movement of certain transposable genetic elements (see Chapter 7). The replication of the plant DNA virus cauliflower mosaic virus is also believed to involve reverse transcription from a 35S cyclic RNA transcript [669–672], and a similar enzyme has been implicated in hepatitis B virus replication [194].

## 6.10 TERMINATION OF REPLICATION

### 6.10.1 Cyclic chromosomes

A complication arises when the two replication forks moving in opposite directions around a cyclic chromosome approach one another. Unwinding takes place ahead of the fork for a distance of about 100 bp and when the unwound regions meet the result is two gapped duplex molecules interlinked (i.e. catenated) about 20 times (Fig. 6.35). Replication of one of the daughter molecules appears to go to completion before the other, and a DNA topoisomerase II is required to decatenate the dimer [602].

### 6.10.2 Small linear chromosomes

These pose a different problem. A typical situation arises with T7 DNA replication (Fig. 6.29). The last Okazaki piece to be made on the lagging strand might start at the extreme 3'-end of the template, but when the primer is removed this will leave a short single-stranded tail at each end of the linear molecule, a tail which cannot be replicated by any mechanism considered so far.

James Watson [430] postulated a mechanism whereby the problem might be solved. T7 DNA has an identical sequence of about 260 nucleotides at each end of the molecule, i.e. the molecule is terminally redundant. Thus the unreplicated single-stranded tail of one molecule could H-bond to the similar region of another molecule giving rise to a concatamer of almost two unit lengths (see Fig. 6.29). The two pieces can be joined together by ligase action and two new nicks introduced several nucleotides 5' to the original nick. The gene 3 endonuclease is a candidate for the enzyme capable of making these specific cuts. Nick translation (see Section 6.4.2.b) from the newly introduced 3'-OHs will allow the concatamer to separate into two halves and the replication to be completed.

Other small linear genomes cope differently with the problem of end replication. We have seen (Section 6.9.7) that adenovirus DNA uses an unusual mechanism for initiation at the very ends of the DNA duplex which removes the extra stages required for termination by phage T7 DNA. Phage lambda DNA has sticky ends and forms a cyclic molecule on entering its host cell (see Section 6.9.1).

### 6.10.3 Telomeres

Eukaryote chromosomes are faced with the same problem as is T7 but the ends of these chromosomes are not like the ends of the small linear chromosomes. Recombination does not readily take place between the normal ends of chromosomes showing them to be different from new ends produced by chromosome breakage (which very rapidly recombine with another piece of duplex DNA). The telomere is the terminus of a linear cellular DNA molecule

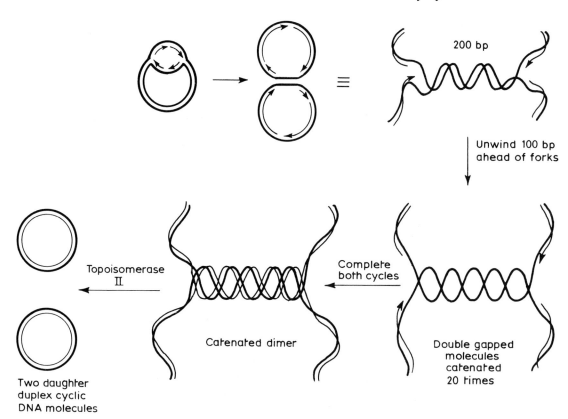

**Fig. 6.35**    Termination of replication of cyclic molecules.

which is able to support the complete replication and segregation of the two daughter molecules [603]. Telomere structure has been studied in a number of simpler systems for example the ribosomal DNA of *Tetrahymena*, the termini of trypanosome chromosomes and yeast chromosomes and Pox virus DNA [603–607].

In all cases it has been concluded that telomeres have certain characteristic properties. They contain multiple repeats of short sequences such as $(CCCCAA)_n$. The ends of the two DNA strands are covalently linked (Fig. 6.36). Frequently there is a subterminal nick in one or both strands. The size of the telomeric region is variable. Several models have been put forward to explain these findings and one of these is presented in Fig. 6.36 [608, 609].

Replication round the corner of the covalently joined telomere leads to the structure shown in Fig. 6.36b. The covalently joined duplexes can be resolved by nicking (c) and partial unpairing (d) so that when the gaps are filled in by $5' \rightarrow 3'$ extension (e) and the nicks ligated (f) the products are two daughter telomeres each somewhat longer than the parental molecule. Obviously, from time to time, a section of the telomere must be removed to restore it to the original length. Other models have been proposed by Van der Ploeg *et al.* and by Walmsley *et al.* [706, 707].

Similar mechanisms are believed to be important in the replication of the small linear single-stranded DNA of the parvoviruses [610].

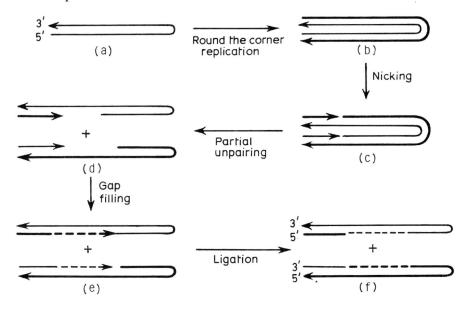

**Fig. 6.36** Replication of telomeres. See text for details.

## 6.11 REPLICATION COMPLEXES

We have seen (Section 6.7.1) how at the point of replication a number of enzymes may be involved together in catalysing initiation and elongation of the daughter DNA molecules. There is evidence from both prokaryotes and eukaryotes that all the enzymes involved in synthesizing the deoxyribonucleotide precursors and modifying the resulting DNA may be assembled into a huge replicating complex.

During phage T4 replication a complex of 10 or more enzymes catalyses the conversion of UDP to dTTP and 5-hydroxymethyl dCTP. This complex is associated with the phage-coded DNA polymerase, topoisomerase, gene 32 DNA-binding protein etc. so that ribonucleotides are efficiently channelled into DNA [611–613].

Evidence for a similar complex in Chinese hamster cells and human lymphoblastoid cells has also been obtained [614–619] although some of this evidence has been questioned [620]. The complex catalyses the conversion of ribonucleo-side diphosphates into DNA, a reaction which is inhibited by hydroxyurea and aphidicolin illustrating the involvement of ribonucleotide reductase and DNA polymerase $\alpha$.

During development of the amphibian oocytes large amounts of so-called 'DNA enzymes' are stockpiled. They are available along with histones and deoxyribonucleotides to catalyse the fantastic rates of DNA synthesis which occur following fertilization [237].

## 6.12 CHROMATIN REPLICATION

Once the basic structure of chromatin in eukaryotic cells was known (see Chapter 3) two questions were asked: (1) what happens to the nucleosomal structures at the replication fork and, (2) to what extent is the structure of chromatin responsible for the mechanism of DNA synthesis which involves (a) Okazaki pieces (Section 6.3.2) and (b) replicons (Section 6.8.2)?

That nucleosomes quickly become associated

with newly synthesized DNA was shown by the cleavage of nascent chromatin by micrococcal nuclease into approximately 200-bp fragments [693–696]. However, the rate and extent of cleavage of nascent chromatin is greater than that of mature chromatin indicating a somewhat different structure which requires 15 minutes to mature, i.e. a time sufficient to synthesize 22 000–54 000 bp of DNA or approximately one replicon [405, 629, 632, 633].

Histones, as well as DNA, are made predominantly in S phase and inhibition of DNA synthesis inhibits synthesis of histones and causes pre-existing histone mRNA to break down with a much reduced half-life [634–636]. In yeast, co-ordinate control of histone gene transcription and replication appears to involve the location of the promoter for the H2A and H2B genes adjacent to an ARS element [644, 645]. Some histone synthesis is not dependent on DNA synthesis, however, and for instance histone H3.3 is synthesized in quiescent cells and during G1 and G2 phase together with four H2A variants and H1A and H1° [637–643].

How histones associate with DNA is not clear but DNA injected into *Xenopus* oocytes rapidly assumes a nucleosomal conformation, and *in vitro* studies have implicated a heat-stable protein, nucleoplasmin, in the process [646–648].

As the replicating fork progresses along a region of nucleosome-associated DNA a number of different situations might result, some of which are illustrated in Fig. 6.37. Approaches used to distinguish between the possibilities include: (a) electron microscópic studies; (b) isolation and analysis of replicating chromatin; (c) crosslinking studies usually associated with attempts to separate the crosslinked species [651–656]; (d) nuclease digestion experiments [649, 650, 693]; and (e) the above approaches carried out with cells exposed to inhibitors of protein synthesis in an attempt to visualize the histone-free nascent DNA [649, 657, 658, 696]. The conclusion from these experiments is that

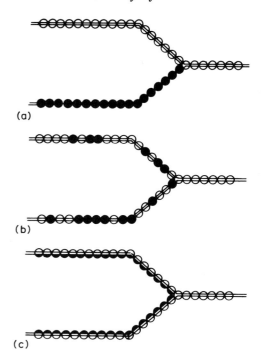

**Fig. 6.37** Three possible mechanisms are shown for the distribution of nucleosomes at a replication fork (after Weintraub [649]). ○, Parental nucleosome; ●, newly synthesized nucleosome. (a) All the old nucleosomes remain with one of the two parental strands when they separate at the replication fork–co-operative alignment; (b) The old nucleosomes are dispersed at random between the two strands of DNA; (c) The old nucleosomes dissociate into half nucleosomes, one half going to each side of the fork.

new histones are associated with newly synthesized DNA and that the nucleosomes form by initial addition of histones H3 and H4 followed by histones H2A and H2B. Histone H1 is added last and turns over quite readily. Newly synthesized nucleosomes consist predominantly of eight newly synthesized histones rather than a mixture of new and old histones. The major controversy concerns the question whether newly synthesized nucleosomes are found adjacent to one another (as required for conservative segregation) or whether they are dispersed amongst old nucleosomes. Although some conservative segregation definitely appears to occur, the evidence is not yet strong enough to conclude that it represents the only mechanism [654, 657–661].

If segregation of nucleosomes is conservative the evidence favours the continued association of the old nucleosomes with the leading strand at the replication fork, with the newly synthesized nucleosomes being added to the lagging side [658].

Thus DNA fork movement may be considered to occur as a series of steps caused by relocation of nucleosomes on the leading side of the fork which is always made up of duplex DNA. At each relocation a region of about 200 bases of DNA on the lagging side of the fork is rendered single stranded providing a target for random initiation of RNA primers. For this reason the Okazaki pieces resulting have a mean size somewhat shorter than the 200 bases of nucleosomal DNA [335].

Replicons have a length approximately equal to that of chromatin loops [662] or the equivalent of 400 nucleosomes or about 70 turns of a solenoid [663]. As adjacent replicons tend to replicate at the same time in the cell cycle [405–407, 664–666], it is possible that a region of a chromosome becomes partially expanded and single-stranded DNA is exposed at the ends of the loops. Bidirectional replication would be initiated at such regions and hence a replicon and a loop would be the same structure. Alternatively initiation of replicon synthesis may occur on the matrix to which the loops are attached in which case each loop would be part of two adjacent replicons. As replicon size varies dramatically in developing amphibia and insects from 5 $\mu$m long in early cleavage to up to 350 $\mu$m long during spermatogenesis [406] chromatin structure would be expected to show parallel changes and this is indeed the case [633].

The altered structure associated with transcribing chromatin (see Section 10.4.4) requires between 1 and 3 minutes to re-form following replication. During this time, from 18 to 54 nucleosomes will have been added to the newly synthesized DNA [667].

## REFERENCES

1 Watson, J. D. and Crick, F. H. C. (1953), *Nature (London)*, **171**, 737.
2 Meselson, M. and Stahl, F. W. (1958), *Proc. Natl. Acad. Sci. USA*, **66**, 671.
3 Cairns, J. (1966), *Sci. Am.*, **214(1)**, 36.
4 Taylor, J. H. (1963), in *Molecular Genetics* (ed. J. H. Taylor), Academic Press, New York, Part I, p. 65.
5 Okazaki, R., Okazaki, T., Sakabe, K., Sugimoto, K., Kainuma, R., Sugino, A. and Iwatsuki, N. (1968), *Cold Spring Harbor Symp. Quant. Biol.*, **33**, 129.
6 Schnös, M. and Inman, R. B. (1970), *J. Mol. Biol.*, **51**, 61.
7 Gautschi, J. R. and Clarkson, J. M. (1975), *Eur. J. Biochem.*, **50**, 403.
8 Magnussen, G., Pigiet, V., Winnacker, E. L., Abrams, R. and Reichard, P. (1973), *Proc. Natl. Acad. Sci. USA*, **70**, 412.
9 Schlomai, J. and Kornberg, A. (1978), *J. Biol. Chem.*, **253**, 3305.
10 Tamanoi, F., Machida, Y. and Okazaki, T. (1978), *Cold Spring Harbor Symp. Quant. Biol.*, **43**, 239.
11 Tye, B-K., Nyman, P. O., Lehman, I. R., Hochhauser, S. and Weiss, B. (1977), *Proc. Natl. Acad. Sci. USA*, **74**, 154.
12 Tye, B-K. and Lehman, I. R. (1977), *J. Mol. Biol.*, **117**, 293.
13 Warner, H. R. and Duncan, B. K. (1978), *Nature (London)*, **272**, 32.
14 Olivera, B. M., Manlapaz-Ramos, P., Warner, H. R. and Duncan, B. K. (1979), *J. Mol. Biol.*, **128**, 265.
15 Tye, B-K., Chien, J., Lehman, I. R., Duncan, B. K. and Warner, H. R. (1978), *Proc. Natl. Acad. Sci. USA*, **75**, 233.
16 Okazaki, R., Sugimoto, K., Okazaki, T., Imac, Y. and Sugino, A. (1970), *Nature (London)*, **228**, 223.
17 Lehman, I. R., Tye, B-K. and Nyman, P. O. (1978), *Cold Spring Harbor Symp. Quant. Biol.*, **43**, 221.

18 Okazaki, T. and Okazaki, R. (1969), *Proc. Natl. Acad. Sci. USA*, **64**, 1242.

19 Sugino, A. and Okazaki, R. (1972), *J. Mol. Biol.*, **64**, 61.

20 Brutlag, D., Schekman, R. and Kornberg, A. (1971), *Proc. Natl. Acad. Sci. USA*, **68**, 2826.

21 Wickner, W., Brutlag, D., Schekman, R. and Kornberg, A. (1972), *Proc. Natl. Acad. Sci. USA*, **69**, 965.

22 Bagdasarian, M. M., Izakawska, M. and Bagdasarian, M. (1977), *J. Bacteriol.*, **130**, 577.

23 Bouché, J. P., Zeckel, K. and Kornberg, A. (1975), *J. Biol. Chem.*, **250**, 5995.

24 Ray, D. S., Duebner, J. and Suggs, S. (1975), *J. Virol.*, **16**, 348.

25 Kurosawa, Y., Ogawa, T., Hirose, S., Okazaki, T. and Okazaki, R. (1975), *J. Mol. Biol.*, **96**, 653.

26 Sugino, A., Hirose, S. and Okazaki, R. (1972), *Proc. Natl. Acad. Sci. USA*, **69**, 1863.

27 Okazaki, R., Okazaki, T., Hirose, S., Sugino, A., Ogawa, T., Kurosawa, Y., Shinozaki, K., Tamanoi, F., Seki, T., Machida, Y., Fujiyama, A. and Kohara, Y. (1975), *DNA Synthesis and its Regulation* (ed. M. Goulian and P. Hanawalt), Vol. III (Series ed. F. Fox), ICN-UCLA Symposium on Molecular and Cellular Biology, Benjamin, California.

28 Seidel, H. (1967), *Biochim. Biophys. Acta*, **138**, 98.

29 Thomas, K. R., Manlapaz-Ramos, P., Lundquist, R. and Olivera, B. M. (1978), *Cold Spring Harbor Symp. Quant. Biol.*, **43**, 231.

30 Miyamoto, C. and Denhardt, D. T. (1977), *J. Mol. Biol.*, **116**, 681.

31 Anderson, M. L. M. (1978), *J. Mol. Biol.*, **118**, 277.

32 Ogawa, T., Hirose, S., Okazaki, T. and Okazaki, R. (1977), *J. Mol. Biol.*, **112**, 121.

33 Okazaki, T., Kurasawa, Y., Ogawa, T., Seki, T., Shinozaki, K., Hirose, S., Fujiyama, A. A., Kohara, Y., Machida, Y., Tamanoi, F. and Hozumi, T. (1978), *Cold Spring Harbor Symp. Quant. Biol.*, **43**, 203.

34 Sugino, A. and Okazaki, R. (1973), *Proc. Natl. Acad. Sci. USA*, **70**, 88.

35 Kowalski, J. and Denhardt, D. T. (1979), *Nature (London)*, **281**, 704.

36 Tseng, B. Y., Erickson, J. M. and Goulian, M. (1979), *J. Mol. Biol.*, **129**, 531.

37 Reichard, P., Eliasson, R. and Soderman, G. (1974), *Proc. Natl. Acad. Sci. USA*, **71**, 4901.

38 Pigiet, V., Eliasson, R. and Reichard, P. (1974), *J. Mol. Biol.*, **84**, 197.

39 Geider, K. (1976), *Curr. Top. Microbiol. Immunol.*, **74**, 58.

40 Huberman, J. A. and Horwitz, H. (1973), *Cold Spring Harbor Symp. Quant. Biol.*, **38**, 233.

41 Blair, D. G., Sherratt, D. J., Clewel, D. B. and Helinski, D. R. (1972), *Proc. Natl. Acad. Sci. USA*, **69**, 2518.

42 Martens, P. A. and Clayton, D. A. (1979), *J. Mol. Biol.*, **135**, 327.

43 Rowen, L. and Kornberg, A. (1978), *J. Biol. Chem.*, **253**, 770.

44 Gomez-Eichelmann, M. C. and Lark, K. G. (1977), *J. Mol. Biol.*, **117**, 621.

45 Kahn, M. and Hanawalt, P. (1979), *J. Mol. Biol.*, **128**, 501.

46 Olivera, B. M. and Bonhoeffer, F. (1972), *Nature (London) New Biol.*, **240**, 233.

47 Olivera, B. M., Lark, K. G., Herrmann, R. and Bonhoeffer, F. (1973), in *DNA Synthesis In Vitro* (ed. R. D. Wells and R. B. Inman), University Park Press, Baltimore, p. 124.

48 Kurosawa, Y. and Okazaki, R. (1975), *J. Mol. Biol.*, **94**, 229.

49 Kuroda, R. K. and Okazaki, R. (1975), *J. Mol. Biol.*, **94**, 213.

50 Olivera, B. M. (1978), *Proc. Natl. Acad. Sci. USA*, **75**, 238.

51 Thomas, K. R., Manlapaz-Ramos, P., Lundquist, R. and Olivera, B. M. (1978), *Cold Spring Harbor Symp. Quant. Biol.*, **43**, 231.

52 Brynolf, K., Eliasson, R. and Reichard, P. (1974), *Cell*, **13**, 573.

53 Kornberg, A. (1974), *DNA Synthesis*, Freeman, San Francisco.

54 Kornberg, A. (1980), *DNA Replication*, Freeman, San Francisco, and supplement.

55 Bessman, M. J., Lehmann, I. R., Simms, E. S. and Kornberg, A. (1958), *J. Biol. Chem.*, **233**, 171.

56 Kornberg, A. (1968), *Sci. Am.*, **21(4)**, 64.

57 Kornberg, A. (1961), *The Enzymatic Synthesis of DNA*, Wiley, London.

58 Josse, J., Kaiser, A. D. and Kornberg, A. (1961), *J. Biol. Chem.*, **236**, 864.

59 Eick, D., Fritz, H-J. and Doerfler, W. (1983), *Anal. Biochem.*, **135**, 165.

60 Becker, A. and Hurwitz, J. (1971), *Prog. Nucleic Acid Res. Mol. Biol.*, **11**, 423.

61 Jovin, T. M., Englund, P. T. and Bertsch, L. L.

(1969), *J. Biol. Chem.*, **244**, 2996.

62 Griffith, J., Huberman, J. A. and Kornberg, A. (1971), *Proc. Natl. Acad. Sci. USA*, **55**, 209.

63 Kornberg, A. (1969), *Science*, **163**, 1410.

64 Que, B. G., Downey, K. M. and So, A. G. (1979), *Biochemistry*, **18**, 2064.

65 Schachman, H. K., Adler, J., Radding, C. M., Lehman, I. R. and Kornberg, A. (1960), *J. Biol. Chem.*, **235**, 3242.

66 Brutlag, D. and Kornberg, A. (1972), *J. Biol. Chem.*, **247**, 241.

67 Galas, D. J. and Branscomb, E. W. (1978), *J. Mol. Biol.*, **124**, 653.

68 Loeb, L. A., Weymouth, L. A., Kunkel, T. A., Gopinathan, K. P., Beckman, R. A. and Dube, D. K. (1978), *Cold Spring Harbor Symp. Quant. Biol.*, **43**, 921.

69 Seal, G., Shearman, C. W. and Loeb, L. A. (1979), *J. Biol. Chem.*, **254**, 5229.

70 Agarwal, S. S., Dube, D. K. and Loeb, L. A. (1979), *J. Biol. Chem.*, **254**, 101.

71 Reha-Kranz, L. J. and Bessman, M. J. (1977), *J. Mol. Biol.*, **116**, 99.

72 Lo, K. Y. and Bessman, M. J. (1976), *J. Biol. Chem.*, **251**, 2475.

73 Lo, K. Y. and Bessman, M. J. (1976), *J. Biol. Chem.*, **251**, 2480.

74 Cozzarelli, N. R., Kelly, R. B. and Kornberg, A. (1969), *J. Mol. Biol.*, **45**, 513.

75 Kelly, R. B., Cozzarelli, N. R., Deutscher, M. P., Lehman, I. R. and Kornberg, A. (1970), *J. Biol. Chem.*, **245**, 39.

76 Bambara, R. A., Vyemura, D. and Choi, T. (1978), *J. Biol. Chem.*, **253**, 413.

77 Matson, S. W., Capaldo-Kimball, F. N. and Bambara, R. A. (1978), *J. Biol. Chem.*, **253**, 7851.

78 de Lucia, P. and Cairns, J. (1969), *Nature (London)*, **224**, 1164.

79 Gross, J. and Gross, M. (1969), *Nature (London)*, **224**, 1166.

80 Okazaki, R., Arisawa, M. and Sugino, A. (1971), *Proc. Natl. Acad. Sci. USA*, **68**, 2954.

81 Kato, T. and Konda, S. (1970), *J. Bacteriol.*, **104**, 871.

82 Smirnov, G. B., Favorskaya, Y. N. and Skavronskaya, A. G. (1971), *Mol. Gen. Genet.*, **111**, 357.

83 Emmerson, P. T. and Strike, P. (1974), *Mechanism and Regulation of DNA Replication* (eds A. R. Kolber and M. Kohiyama), Plenum Press, New York, p. 47.

84 Gefter, M. L. (1974), *Prog. Nucleic Acid Res. Mol. Biol.*, **14**, 101.

85 Tait, R. C. and Smith, D. W. (1974), *Nature (London)*, **249**, 116.

86 Gefter, M. L., Hirota, Y., Kornberg, T., Wechsler, S. A. and Barnoux, C. (1971), *Proc. Natl. Acad. Sci. USA*, **68**, 3150.

87 Klein, A., Nusslein, V., Otto, B. and Powling, A. (1973), in *DNA Synthesis In Vitro* (ed. R. D. Wells and R. B. Inman), University Park Press, Baltimore, p. 185.

88 Otto, B. (1973), *Biochem. Soc. Trans.*, **1**, 629.

89 Livingstone, D. M., Hinckle, D. C. and Richardson, C. C. (1975), *J. Biol. Chem.*, **250**, 461.

90 Kornberg, T. and Kornberg, A. (1974), in *The Enzymes* (ed. P. D. Boyer), Academic Press, New York, Vol. 10, p. 119.

91 Lehman, I. R. (1974), in *The Enzymes* (ed. P. D. Boyer), Academic Press, New York, Vol. 10, p. 237.

92 Kornberg, A. (1977), *Trans. Biochem. Soc.*, **5**, 359.

93 Wickner, W., Schekman, T., Geider, K. and Kornberg, A. (1973), *Proc. Natl. Acad. Sci. USA*, **70**, 1764.

94 McHenry, C. S. and Crown, W. (1979), *J. Biol. Chem.*, **254**, 1748.

95 Wickner, W. and Kornberg, A. (1974), *J. Biol. Chem.*, **249**, 6244.

96 Hurwitz, J. and Wickner, S. (1974), *Proc. Natl. Acad. Sci. USA*, **71**, 6.

97 Burgers, P. M. J. and Kornberg, A. (1982), *J. Biol. Chem.*, **257**, 11 474.

98 Fay, P. J., Johanson, K. O., McHenry, C. S. and Bambara, R. A. (1982), *J. Biol. Chem.*, **257**, 5692.

99 Crute, J. J., La Duca, R. J., Johanson, K. O., McHenry, C. S. and Bambara, R. A. (1983), *J. Biol. Chem.*, **258**, 11 344.

100 Biswas, S. B. and Kornberg, A. (1984), *J. Biol. Chem.*, **259**, 7990.

101 Gass, K. B. and Cozzarelli, N. R. (1974), *Methods Enzymol.*, **29**, 17.

102 Bollum, F. J. (1974), in *The Enzymes* (ed. P. D. Boyer), Academic Press, New York, Vol. 10, p. 145.

103 Loeb, L. A. (1974), in *The Enzymes* (ed. P. D. Boyer), Academic Press, New York, Vol. 10, p. 174.

104  Weissbach, A. (1975), *Cell*, **5**, 101.

105  Nakamura, H., Morita, T., Masaki, S. and Yoshida, S. (1984), *Exp. Cell Res.*, **151**, 123.

106  Bensch, K. G., Tanaka, S., Hu, S-Z., Wang, T. S-F. and Korn, D. (1982), *J. Biol. Chem.*, **257**, 8391.

107  Smith, H. C. and Berezney, R. (1983), *Biochemistry*, **22**, 3042.

108  Jones, C. and Su, R. T. (1982), *Nucleic Acids Res.*, **10**, 5517.

109  Chang, L. M. S. and Bollum, F. J. (1972), *J. Biol. Chem.*, **247**, 7948.

110  Ove, P., Jenkins, M. D. and Lazlo, J. (1970), *Cancer Res.*, **30**, 535.

111  Lindsay, J. G., Berryman, S. and Adams, R. L. P. (1970), *Biochem. J.*, **119**, 839.

112  Hubscher, U., Kuenzle, C. C., Limacher, W. K., Sherrer, P. and Spadari, S. (1978), *Cold Spring Harbor Symp. Quant. Biol.*, **43**, 625.

113  Wang, H. F. and Popenoe, E. A. (1977), *Biochim. Biophys. Acta*, **474**, 98.

114  Ogura, M., Suzuki-Hari, C., Nagano, H., Mano, Y. and Ikegami, S. (1979), *Eur. J. Biochem.*, **97**, 603.

115  Waqar, M. A., Evans, M. J. and Huberman, J. A. (1978), *Nucleic Acids Res.*, **5**, 1933.

116  Edenberg, H. T., Anderson, S. and de Pamphilis, M. L. (1978), *J. Biol. Chem.*, **253**, 3273.

117  Mechali, M., Abadiedebat, J. and de Recondo, A. M. (1980), *J. Biol. Chem.*, **255**, 2114.

118  Bollum, F. J. (1975), *Prog. Nucleic Acid Res. Mol. Biol.*, **15**, 109.

119  Murakami, Y., Yasuda, H., Miyazawa, H., Hanaoka, F. and Yamada, M. (1985), *Proc. Natl. Acad. Sci. USA*, **82**, 1761.

120  Korn, D., Fisher, P. A., Battey, J. and Wang, T.S-F. (1978), *Cold Spring Harbor Symp. Quant. Biol.*, **43**, 613.

121  Fisher, P. A., Wang, T.S-F. and Korn, D. (1979), *J. Biol. Chem.*, **254**, 6128.

122  Crute, J. J., Whal, A. F. and Bambara, R. A. (1986) *Biochemistry*, **25**, 26.

123  Knopf, K. W. (1979), *Eur. J. Biochem.*, **98**, 231.

124  Fichot, P., Pascal, M., Mechali, M. and de Recondo, A. M. (1979), *Biochim. Biophys. Acta*, **561**, 29.

125  Holmes, A. M., Hesselwood, I. P. and Johnston, I. R. (1974), *Eur. J. Biochem.*, **43**, 487.

126  Hesselwood, I. P., Holmes, A. M., Wakeling, W. F. and Johnston, I. R. (1978), *Eur. J. Biochem.*, **84**, 123.

127  Rowen, L. and Kornberg, A. (1978), *J. Biol. Chem.*, **253**, 758.

128  Wickner, S. (1977), *Proc. Natl. Acad. Sci. USA*, **74**, 2815.

129  Pritchard, C. G. and De Pamphilis, M. L. (1983), *J. Biol. Chem.*, **258**, 9801.

130  Konig, H., Riedel, H. D. and Knippers, R. (1983), *Eur. J. Biochem.*, **135**, 435.

131  Kozu, T., Yagura, T. and Seno, T. (1982), *Nature (London)*, **298**, 180.

132  Yagura, T., Kozu, T. and Seno, T. (1982), *J. Biol. Chem.*, **257**, 11 121.

133  Shioda, M., Nelson, E. M., Bayne, M. L. and Benbow, R. M. (1982), *Proc. Natl. Acad. Sci. USA*, **79**, 7209.

134  Kaguni, L. S., Rossinol, J-M., Conaway, R. C. and Lehman, I. R. (1983), *Proc. Natl. Acad. Sci. USA*, **80**, 2221.

135  Yagura, T., Tanaka, S., Kozu, T., Seno, T. and Korn, D. (1983), *J. Biol. Chem.*, **258**, 6698.

136  Kagani, L. S., Rossingol, J-M., Conaway, R. C., Banks, G. R. and Lehman, I. R. (1983), *J. Biol. Chem.*, **258**, 9037.

137  Tseng, B. Y. and Ahlem, C. N. (1983), *J. Biol. Chem.*, **258**, 9845.

138  Yagura, T., Kozu, T., Seno, T., Saneyoshi, N., Hiraga, S. and Nagano, H. (1983), *J. Biol. Chem.*, **258**, 13 070.

139  Wang, T.S.-F., Hu, S-Z. and Korn, D. (1984), *J. Biol. Chem.*, **259**, 1854.

140  Hu, S-Z., Wang, T. S-F. and Korn, D. (1984), *J. Biol. Chem.*, **259**, 2602.

141  Plevani, P., Badaracco, G., Augl, C. and Chang, L. M. S. (1984), *J. Biol. Chem.*, **259**, 7532.

142  Singh, H. and Dumas, L. B. (1984), *J. Biol. Chem.*, **259**, 7936.

143  Kaguni, L. S., DiFrancesco, R. A. and Lehman, I. R. (1984), *J. Biol. Chem.*, **259**, 9314.

144  Gronostajski, R. M., Field, J. and Hurwitz, J. (1984), *J. Biol. Chem.*, **259**, 9479.

145  Albert, W., Grummt, F., Hubscher, U. and Wilson, S. H. (1982), *Nucleic Acids Res.*, **10**, 935.

146  Masaki, S., Tanabe, K. and Yoshida, S. (1984), *Nucleic Acids Res.*, **12**, 4455.

147  Goscin, L. P. and Byrnes, J. J. (1982), *Nucleic Acids Res.*, **10**, 6023.

148  Ottiger, K-P. and Hubscher, U. (1984), *Proc. Natl. Acad. Sci. USA*, **81**, 3993.

149  Tsuchiya, E., Kimura, K., Miyakawa, T. and Fukui, S. (1984), *Nucleic Acids Res.*, **12**, 3143.

150  Chang, L. M. S. (1975), *Methods Enzymol.*, **29**, 81.

151 Wang, T.S.-F., Sedwik, W. D. and Korn, D. (1974), *J. Biol. Chem.*, **249**, 841.

152 Kunkel, T. A., Tcheng, J. E. and Meyer, R. R. (1978), *Biochim. Biophys. Acta*, **520**, 302.

153 Aposhian, H. V. and Kornberg, A. (1962), *J. Biol. Chem.*, **237**, 519.

154 Loeb, L. A. (1969), *J. Biol. Chem.*, **244**, 1672.

155 Chang, L. M. S. (1973), *J. Biol. Chem.*, **248**, 6983.

156 Waser, J., Hubscher, V., Kuenzle, C. C. and Spadari, S. (1979), *Eur. J. Biochem.*, **97**, 361.

157 Adams, R. L. P. and Kirk, D., unpublished results.

158 Miller, M. R. and Chinault, D. N. (1982), *J. Biol. Chem.*, **257**, 10 204.

159 Dressler, S. L. and Lieberman, M. W. (1983), *J. Biol. Chem.*, **258**, 9990.

160 Mosbaugh, D. W. and Linn, S. (1984), *J. Biol. Chem.*, **259**, 10 247.

161 Spadari, S. and Weissbach, A. (1974), *J. Mol. Biol.*, **86**, 11.

162 Spadari, S. and Weissbach, A. (1974), *J. Biol. Chem.*, **249**, 5809.

163 Kasamatsu, H., Robberson, D. L. and Vinograd, H. (1971), *Proc. Natl. Acad. Sci. USA*, **68**, 2252.

164 Meyer, R. R. and Simpson, M. V. (1970), *J. Biol. Chem.*, **245**, 3426.

165 Hubscher, U., Kuenzle, C. C. and Spadari, S. (1979), *Proc. Natl. Acad. Sci. USA*, **76**, 2316.

166 Abboud, M. M. and Horwitz, M. S. (1979), *Nucleic Acids Res.*, **6**, 1025.

167 Ito, K., Arens, M. and Green, M. (1976), *Biochim. Biophys. Acta*, **447**, 340.

168 Krokan, H., Schaffer, P. and de Pamphilis, M. L. (1979), *Biochemistry*, **18**, 4431.

169 Yamaguchi, M., Matsukage, A. and Takahasi, T. (1980), *J. Biol. Chem.*, **255**, 7002.

170 Yamaguchi, M., Matsukage, A. and Takahashi, T. (1980), *Nature (London)*, **285**, 45.

171 Tewari, K. K. and Wildman, S. G. (1967), *Proc. Natl. Acad. Sci. USA*, **58**, 689.

172 Spencer, D. and Whitfield, P. R. (1967), *Biochem. Biophys. Res. Commun.*, **28**, 538.

173 Stillman, B. W., Tamanoi, F. and Mathews, M. B. (1982), *Cell*, **31**, 613.

174 Field, J., Gronostajaski, R. M. and Hurwitz, J. (1984), *J. Biol. Chem.*, **259**, 9487.

175 Weissbach, A. (1979), *Arch. Biochem. Biophys.*, **198**, 386.

176 Knopf, K. (1979), *Eur. J. Biochem.*, **98**, 231.

177 Krakow, J. S., Coutsogeorgopoulos, C. and Canellakis, E. S. (1962), *Biochim. Biophys. Acta*, **55**, 639.

178 Bollum, F. (1965), *J. Biol. Chem.*, **240**, 2599.

179 Chang, L. M. S., Plevani, P. and Bollum, F. J. (1982), *Proc. Natl. Acad. Sci. USA*, **79**, 758.

180 Deibel, M. R. and Coleman, M. S. (1979), *J. Biol. Chem.*, **254**, 8634.

181 Chang, L. M. S., Plevani, P. and Bollum, F. J. (1982), *J. Biol. Chem.*, **257**, 5700.

182 Chang, L. M. S. and Bollum, F. J. (1971), *J. Biol. Chem.*, **246**, 909.

183 Chang, L. M. S. (1971), *Biochem. Biophys. Res. Commun.*, **44**, 124.

184 Srivastava, B. I. S. (1974), *Cancer Res.*, **34**, 1015.

185 Srivastava, B. I. S. (1975), *Res. Commun. Chem. Pathol. Pharm.*, **10**, 715.

186 Desiderio, S. V., Yancopoulos, G. D., Paskind, M., Thomas, E., Boss, M. A., Landau, N., Alt., F. W. and Baltimore, D. (1984), *Nature (London)*, **311**, 752.

187 Panganiban, A. T. and Temin, H. M. (1984), *Proc. Natl. Acad. Sci. USA*, **81**, 7885.

188 Temin, H. M. and Mizutani, S. (1970), *Nature (London)*, **226**, 1211.

189 Baltimore, D. (1970), *Nature (London)*, **226**, 1209.

190 Spiegelman, S., Burny, A., Das, M., Keydar, J., Schlom, J., Travnicek, M. and Watson, K. (1970), *Nature (London)*, **228**, 430.

191 Wang, R. M. (1978), *Annu. Rev. Microbiol.*, **32**, 561.

192 Erikson, E. and Erikson, R. L. (1971), *J. Virol.*, **8**, 254.

193 Harada, F., Sawyer, R. C. and Dahlberg, J. E. (1975), *J. Biol. Chem.*, **250**, 3487.

194 Varmus, H. E. (1985), *Nature (London)*, **314**, 583.

195 Papas, T. S., Pry, T. W. and Marciani, D. J. (1977), *J. Biol. Chem.*, **252**, 1425.

196 Hizi, A. and Joklik, W. K. (1977), *J. Biol. Chem.*, **252**, 2281.

197 Hizi, A., Leis, J. P. and Joklik, W. K. (1977), *J. Biol. Chem.*, **252**, 2290.

198 Chang, L. M. S. and Bollum, F. J. (1971), *Biochemistry*, **10**, 536.

199 Seal, G. and Loeb, L. A. (1976), *J. Biol. Chem.*, **251**, 975.

200 Battula, N., Dube, D. K. and Loeb, L. A. (1975), *J. Biol. Chem.*, **250**, 8404.

201 Mao, J. C-H. and Robishaw, E. E. (1975), *Biochemistry*, **14**, 5475.

202 Furman, P. A., St. Clair, M. H., Fyfe, J. A., Rideout, J. L., Keller, P. M. and Elion, G. B. (1979), *J. Virol.*, **32**, 72.

203 Gellert, M. (1967), *Proc. Natl. Acad. Sci. USA*, **57**, 48.

204 Laipis, P. J., Oliver, B. M. and Ganesan, A. T. (1969), *Proc. Natl. Acad. Sci. USA*, **62**, 289.

205 Weiss, B. and Richardson, C. C. (1967), *Proc. Natl. Acad. Sci. USA*, **57**, 1021.

206 Lindahl, T. and Edelman, G. M. (1968), *Proc. Natl. Acad. Sci. USA*, **61**, 680.

207 Tsukada, K. and Ichimura, M. (1971), *Biochem. Biophys. Res. Commun.*, **42**, 1156.

208 Gefter, M. L., Becker, A. and Hurwitz, J. (1967), *Proc. Natl. Acad. Sci. USA*, **58**, 240.

209 Gellert, M. and Bullock, M. L. (1970), *Proc. Natl. Acad. Sci. USA*, **67**, 1580.

210 Lehman, I. R. (1974), in *The Enzymes* (ed. P. D. Boyer), Academic Press, New York, Vol. 10, p. 237.

211 Mate, K. and Hurwitz, J. (1974), *J. Biol. Chem.*, **249**, 3650.

212 Sano, H. and Feix, G. (1974), *Biochemistry*, **13**, 5110.

213 Baldy, M. W. (1969), *Cold Spring Harbor Symp. Quant. Biol.*, **33**, 333.

214 Alberts, B. M. and Frey, L. (1970), *Nature (London)*, **227**, 1313.

215 Weiner, J. H., Bertsch, L. L. and Kornberg, A. (1975), *J. Biol. Chem.*, **250**, 1972.

216 Sigal, N., Delius, H., Kornberg, T., Gefter, M. L. and Alberts, B. (1972), *Proc. Natl. Acad. Sci. USA*, **69**, 3537.

217 Reuben, R. C. and Gefter, M. L. (1973), *Proc. Natl. Acad. Sci. USA*, **70**, 1846.

218 Geider, K. and Kornberg, A. (1974), *J. Biol. Chem.*, **249**, 3999.

219 Barry, J. and Alberts, B. M. (1972), *Proc. Natl. Acad. Sci. USA*, **69**, 2717.

220 Gilboa, E., Mitra, S. W., Goff, S. and Baltimore, D. (1979), *Cell*, **18**, 93.

221 Meyer, R. R., Glassberg, J. and Kornberg, A. (1979), *Proc. Natl. Acad. Sci. USA*, **76**, 1702.

222 Van Darp, B., Schneck, P. K. and Staudenbauer, W. L. (1979), *Eur. J. Biochem.*, **94**, 445.

223 Abdel-Monem, M., Chanel, M. C. and Hoffmann-Berling, H. (1977), *Eur. J. Biochem.*, **79**, 33.

224 Abdel-Monem, M., Chanal, M. C. and Hoffmann-Berling, H. (1977), *Eur. J. Biochem.*, **79**, 39.

225 Kuhn, B., Abdel-Monem, M. and Hoffmann-Berling, H. (1978), *Cold Spring Harbor Symp. Quant. Biol.*, **43**, 63.

226 Kuhn, B., Abdel-Monem, M., Krell, H. and Hoffmann-Berling, H. (1978), *J. Biol. Chem.*, **254**, 11 343.

227 Abdel-Monem, M. and Hoffmann-Berling, H. (1980), *Trends Biochem. Sci.*, **5**, 128.

228 Abdel-Monem, M., Taucher-Scholz, G. and Klinkert, M-Q. (1983), *Proc. Natl. Acad. Sci. USA*, **80**, 4659.

229 Klinkert, M-Q, Klein, A. and Abdel-Monem, M. (1980), *J. Biol. Chem.*, **255**, 9746.

230 Abdel-Monem, M., Lauppe, H. F., Kartenbeck, J., Durwald, H. and Hoffmann-Berling, H. (1977), *J. Mol. Biol.*, **110**, 667.

231 Oeda, K., Horiuchi, T. and Sakiguchi, M. (1982), *Nature (London)*, **298**, 98.

232 Yarranton, G. T. and Gefter, M. L. (1979), *Proc. Natl. Acad. Sci. USA*, **76**, 1658.

233 Alberts, B. and Sternglanz, R. (1977), *Nature (London)*, **269**, 655.

234 Yarranton, G. T., Das, R. H. and Gefter, M. L. (1979), *J. Biol. Chem.*, **254**, 11 997.

235 Yarranton, G. T., Das, R. H. and Gefter, M. L. (1979), *J. Biol. Chem.*, **254**, 12 002.

236 Kolodner, R., Masamune, Y., Le Clere, J. E. and Richardson, C. C. (1978), *J. Biol. Chem.*, **253**, 566.

237 Zierler, M. K., Marini, N. J., Stowers, D. J. and Benbow, R. M. (1985), *J. Biol. Chem.*, **260**, 974.

238 Scheuermann, R. H. and Echolo, H. (1984), *Proc. Natl. Acad. Sci. USA*, **81**, 7747.

239 Kohn, B. and Abdel-Monem, M. (1982), *Eur. J. Biochem.*, **125**, 63.

240 Jongeneel, C. V., Formosa, T. and Alberts, B. M. (1984), *J. Biol. Chem.*, **259**, 12 925.

241 Jongeneel, C. V., Bedinger, P. and Alberts, B. M. (1984), *J. Biol. Chem.*, **259**, 12 933.

242 Bedinger, P., Hochstrasser, M., Jongeneel, C. V. and Alberts, B. M. (1983), *Cell*, **34**, 115.

243 Graw, J., Schlaeger, E-J. and Knippers, R. (1981), *J. Biol. Chem.*, **256**, 13 207.

244 Cairns, J. (1963), *J. Mol. Biol.*, **6**, 208.

245 Wang, J. C. (1982), *Sci. Am.*, **247**, 84.

246 Wang, J. C. (1971), *J. Mol. Biol.*, **55**, 523.

247 Champoux, J. J. and Dulbecco, R. (1972), *Proc. Natl. Acad. Sci. USA*, **69**, 143.

248 Champoux, J. J. (1978), *J. Mol. Biol.*, **118**, 441.

249 Champoux, J. J. (1981), *J. Biol. Chem.*, **256**, 4805.

250 Fisher, L. M. (1981), *Nature (London)*, **294**, 607.

251 Sander, M. and Hsieh, T-S. (1983), *J. Biol. Chem.*, **258**, 8421.

252 Liu, L. F., Rowe, T. C., Yang, L., Tewey, K. M. and Chen, G. L. (1983), *J. Biol. Chem.*, **258**, 15 365.

253 Brown, P. O. and Cozzarelli, N. R. (1981), *Proc. Natl. Acad. Sci. USA*, **78**, 843.

254 Shure, M. and Vinograd, J. (1976), *Cell*, **8**, 215.

255 Liu, L. F., Liu, C-C. and Alberts, B. M. (1980), *Cell*, **19**, 697.

256 Hsieh, T-S. (1983), *J. Biol. Chem.*, **258**, 8413.

257 Edwards, K. A., Halligan, B. D., Davis, J. L., Nivera, N. L. and Liu, L. F. (1982), *Nucleic Acids Res.*, **10**, 2565.

258 Ishii, K., Hasegawa, T., Fujisawa, K. and Andoh, T. (1983), *J. Biol. Chem.*, **258**, 12 728.

259 Dynan, W. S., Jendrisak, J. J., Hagan, D. A. and Burgess, R. R. (1981), *J. Biol. Chem.*, **256**, 5860.

260 Attardi, D. G., De Padis, A. and Tocchini-Valentini, G. P. (1981), *J. Biol. Chem.*, **256**, 3654.

261 Miller, K. G., Liu, L. F. and Englund, P. T. (1981), *J. Biol. Chem.*, **256**, 9334.

262 Dugnet, M., Lavenot, C., Harper, F., Mirambeau, G. and de Recondo, A-M. (1983), *Nucleic Acids Res.*, **11**, 1059.

263 Goto, T., Laipis, P. and Wang, J. C. (1984), *J. Biol. Chem.*, **259**, 10 422.

264 Shelton, E. R., Osheroff, N. and Brutlag, D. L. (1983), *J. Biol. Chem.*, **258**, 9530.

265 Pruss, G. J., Manes, S. H. and Drlica, K. (1982), *Cell*, **31**, 35.

266 DiNardo, S., Voelkel, K. A., Sternglanz, R., Reynolds, A. E. and Wright, A. (1982), *Cell*, **31**, 43.

267 Kumura, K. and Sekiguchi, M. (1984), *J. Biol. Chem.*, **259**, 1560.

268 Liu, L. F., Liu, C-C. and Alberts, B. M. (1979), *Nature (London)*, **281**, 456.

269 Wang, J. C. (1984), *Nature (London)*, **309**, 669.

270 Kikuchi, A. and Asai, K. (1984), *Nature (London)*, **209**, 677.

271 Sugino, A., Peebles, C. L., Kreuzer, K. N. and Cozzarelli, N. R. (1977), *Proc. Natl. Acad. Sci. USA*, **74**, 4767.

272 Mizuuchi, K., O'Dea, M. H. and Gellert, M. (1978), *Proc. Natl. Acad. Sci. USA*, **75**, 5960.

273 Gellert, M., Mizuuchi, K., O'Dea, M. H., Itoh, T. and Tomizawa, J. I. (1977), *Proc. Natl. Acad. Sci. USA*, **74**, 4772.

274 Peebles, C. L., Higgins, N. P., Kreuzer, K. N., Morrison, A., Brown, P. O., Sugino, A. and Cozzarelli, N. R. (1978), *Cold Spring Harbor Symp. Quant. Biol.*, **43**, 41.

275 Joyce, C. M., Kelley, W. S. and Grindley, N. D. F. (1982), *J. Biol. Chem.*, **257**, 1958.

276 Brown, W. E., Stump, K. H. and Kelley, W. S. (1982), *J. Biol. Chem.*, **257**, 1965.

277 Morrison, A. and Cozzarelli, N. R. (1979), *Cell*, **17**, 175.

278 Sugino, A., Higgins, N. P. and Cozzarelli, N. R. (1980), *Nucleic Acids Res.*, **8**, 3865.

279 Denhardt, D. T. (1979), *Nature (London)*, **280**, 196.

280 Been, M. D., Burgess, R. R. and Champoux, J. J. (1984), *Nucleic Acids Res.*, **12**, 3097.

281 Liu, L. F. and Wang, J. C. (1978), *Cell*, **15**, 979.

282 Kirkegaard, K. and Wang, J. C. (1981), *Cell*, **23**, 721.

283 Sugino, A., Higgins, N. P., Brown, P. O., Peebles, C. L. and Cozzarelli, N. R. (1978), *Proc. Natl. Acad. Sci. USA*, **75**, 4838.

284 Gellert, M., Mizuuchi, K., O'Dea, M. H., Ohmari, H. and Tomizawa, J. (1978), *Cold Spring Harbor Symp. Quant. Biol.*, **43**, 35.

285 Gellert, M., O'Dea, M. H., Itoh, T. and Tomizawa, J. (1976), *Proc. Natl. Acad. Sci. USA*, **73**, 4474.

286 Sumida-Yasumoto, C., Yudelevich, A. and Hurwitz, J. (1976), *Proc. Natl. Acad. Sci. USA*, **73**, 1887.

287 De Wyngaert, M. A. and Hinckle, D. C. (1979), *J. Virol.*, **29**, 529.

288 McCarthy, D. (1970), *J. Mol. Biol.*, **127**, 265.

289 Beattie, K. L., Wiegand, R. C. and Radding, C. M. (1977), *J. Mol. Biol.*, **116**, 783.

290 Smith, C. L., Kubo, M. and Imamoto, F. (1978), *Nature (London)*, **275**, 420.

291 Mizuuchi, K., Gellert, M. and Nash, H. A. (1978), *J. Mol. Biol.*, **121**, 375.

292 Marians, K. J., Ikeda, J-E., Schlagman, S. and Hurwitz, H. (1977), *Proc. Natl. Acad. Sci. USA*, **74**, 1965.

293 Tse-Dinh, Y-C., Wong, T. W. and Goldberg, A. R. (1984), *Nature (London)*, **312**, 785.

294 Ferro, A. M., Higgins, N. P. and Olivera, B. M. (1983), *J. Biol. Chem.*, **258**, 6000.

295 Hopfield, J. J. (1974), *Proc. Natl. Acad. Sci. USA*, **71**, 4135.

296 Domingo, E., Sabo, D., Taniguchi, T. and Weisman, C. (1978), *Cell*, **13**, 735.

297 Kunkel, T. A., Shaper, R. M., Beckman, R. A. and Loeb, L. A. (1981), *J. Biol. Chem.*, **256**, 9883.

298 Grosse, F., Krauss, G., Knill-Jones, J. W. and Fersht, A. R. (1983), *EMBO J.*, **2**, 1515.

299 Fersht, A. R. and Knill-Jones, J. W. (1983), *J. Mol. Biol.*, **165**, 633.

300 Seal, G., Shearman, C. W. and Loeb, L. A. (1979), *J. Biol. Chem.*, **254**, 5229.

301 Brosius, S., Grosse, F. and Krauss, G. (1983), *Nucleic Acids Res.*, **11**, 193.

302 Rechmann, B., Grosse, F. and Krauss, G. (1983), *Nucleic Acids Res.*, **11**, 7251.

303 Fersht, A. R. (1979), *Proc. Natl. Acad. Sci. USA*, **76**, 4946.

304 Kunkel, T. A. (1981), *Proc. Natl. Acad. Sci. USA*, **78**, 6734.

305 Hillebrand, G. G. and Beattie, K. L. (1984), *Nucleic Acids Res.*, **12**, 3173.

306 Bedinger, P. and Alberts, B. M. (1983), *J. Biol. Chem.*, **258**, 9649.

307 Topal, M. D. and Sinha, N. K. (1983), *J. Biol. Chem.*, **258**, 12 274.

308 Kunkel, T. A., Loeb, L. A. and Goodman, M. F. (1984), *J. Biol. Chem.*, **259**, 1539.

309 Echols, H., Lu, C. and Burgers, P. M. J. (1983), *Proc. Natl. Acad. Sci. USA*, **80**, 2189.

310 Schenermann, T., Tam, S., Burgers, P. M. J., Lu, C. and Echols, H. (1983), *Proc. Natl. Acad. Sci. USA*, **80**, 7085.

311 Lasken, R. S. and Goodman, M. F. (1985), *Proc. Natl. Acad. Sci. USA*, **82**, 1301.

312 Loeb, L. A., Dube, D. K., Beckmann, R. A., Koplitz, M. and Gopinathan, K. P. (1981), *J. Biol. Chem.*, **256**, 3978.

313 Travaglini, E. C., Mildvan, A. S. and Loeb, L. A. (1975), *J. Biol. Chem.*, **250**, 8647.

314 Gillin, F. D. and Nossal, N. G. (1975), *Biochem. Biophys. Res. Commun.*, **64**, 457.

315 Reanney, D. (1984), *Nature (London)*, **307**, 318.

316 Mhaskar, D. N. and Goodman, M. F. (1984), *J. Biol. Chem.*, **259**, 11 713.

317 Liu, L. F. (1984), *CRC Crit. Rev. Biochem.*, **15**, 1.

318 Wells, R. D. and Inman, R. B. (1973), *DNA Synthesis In Vitro*, University Park Press, Baltimore.

319 Wickner, R. B. (1974) in *Methods in Molecular Biology*, Vol. 7 (ed. R. B. Wickner), DNA Replication, Dekker, New York.

320 Vosberg, H. P. and Hoffmann-Berling, H. (1971), *J. Mol. Biol.*, **58**, 739.

321 Moses, R. E. and Richardson, C. C. (1970), *Proc. Natl. Acad. Sci. USA*, **67**, 674.

322 Schaller, H., Otto, B., Nusslein, V., Huf, J., Herrmann, R. and Bonhoeffer, F. (1972), *J. Mol. Biol.*, **63**, 183.

323 Schekman, R., Wickner, W., Westergaard, P., Brutlag, D., Geider, K., Bertsch, L. L. and Kornberg, A. (1972), *Proc. Natl. Acad. Sci. USA*, **69**, 2691.

324 Nusslein, V. and Klein, A. (1974), in *Methods in Molecular Biology*, Vol. 7 (ed. R. B. Wickner), Dekker, New York.

325 Burgoyne, L. A. (1972), *Biochem. J.*, **130**, 959.

326 Hunter, T. and Francke, B. (1974), *J. Virol.*, **13**, 125.

327 Winnacker, E. L., Magnussen, G. and Reichard, P. (1972), *J. Mol. Biol.*, **72**, 523.

328 De Pamphilis, M. L., Beard, P. and Berg, P. (1975), *J. Biol. Chem.*, **250**, 4340.

329 Berger, N. A., Petzold, S. J. and Johnson, E. S. (1977), *Biochim. Biophys. Acta*, **478**, 44.

330 Fraser, J. M. K. and Huberman, J. A. (1977), *J. Mol. Biol.*, **117**, 249.

331 Wist, E., Krokan, H. and Prydz, H. (1976), *Biochemistry*, **15**, 3647.

332 Tseng, B. Y. and Goulian, M. (1975), *J. Mol. Biol.*, **99**, 317.

333 Krokan, H., Wist, E. and Prydz, H. (1977), *Biochim. Biophys. Acta*, **475**, 553.

334 Hershey, H. V. and Taylor, J. H. (1974), *Exp. Cell Res.*, **85**, 78.

335 De Pamphilis, M. L., Anderson, S., Bar-Shavit, R., Collins, E., Edenberg, H., Herman, T., Karas, B., Kaufmann, G., Krokan, H., Shelton, E., Su, R., Tapper, D. and Wassarman, P. M. (1978), *Cold Spring Harbor Symp. Quant. Biol.*, **43**, 679.

336 Wist, E. and Prydz, H. (1979), *Nucleic Acids Res.*, **6**, 1583.

337 Butt, T. R., Wood, W. M., McKay, E. L. and Adams, R. L. P. (1978), *Biochem. J.*, **173**, 309.

338 Elford, H. L. (1974), *Arch. Biochem. Biophys.*, **163**, 537.

339 Tseng, B. Y. and Goulian, M. (1977), *Cell*, **12**, 483.

340 Godson, G. N., Barrell, B. G., Staden, R. and Fiddes, J. C. (1978), *Nature (London)*, **276**, 236.

341 Sumida-Yasamoto, C., Ikeda, J-E., Benz, E., Marians, K. J., Vicuna, R., Sugrue, S., Zipurzky, S. L. and Hurwitz, J. (1978), *Cold Spring Harbor Symp. Quant. Biol.*, **43**, 311.

342 Meyer, R. R., Shlomai, J., Kobon, J., Bates, D. L., Rowen, L., McMacken, R., Ueda, K. and Kornberg, A. (1978), *Cold Spring Harbor Symp. Quant. Biol.*, **43**, 289.

343 Wickner, S., and Hurwitz, J. (1976), *Proc. Natl. Acad. Sci. USA*, **73**, 1053.

344 Wickner, S. H. (1975), *Cold Spring Harbor Symp. Quant. Biol.*, **43**, 303.

345 Schlomai, J. and Kornberg, A. (1980), *Proc. Natl. Acad. Sci. USA*, **77**, 799.

346 Nomura, N., Low, R. L. and Ray, D. S. (1982), *Proc. Natl. Acad. Sci. USA*, **79**, 3153.

347 Shlomai, J. and Kornberg, A. (1981), *J. Biol. Chem.*, **256**, 6789.

348 Shlomai, J. and Kornberg, A. (1981), *J. Biol. Chem.*, **256**, 6794.

349 Arai, K-I. and Kornberg, A. (1981), *Proc. Natl. Acad. Sci. USA*, **78**, 69.

350 Arai, K-I., Low, R. L. and Kornberg, A. (1981), *Proc. Natl. Acad. Sci. USA*, **78**, 707.

351 Ogawa, T., Arai, K-I. and Okazaki, T. (1983), *J. Biol. Chem.*, **258**, 13 353.

352 McMacken, R. and Kornberg, A. (1978), *J. Biol. Chem.*, **253**, 3313.

353 Arai, K-I. and Kornberg, A. (1979), *Proc. Natl. Acad. Sci. USA*, **76**, 4308.

354 Arai, K-I., McMacken, R., Yasuda, S. and Kornberg, A. (1981), *J. Biol. Chem.*, **256**, 5281.

355 Kobori, J. A. and Kornberg, A. (1982), *J. Biol. Chem.*, **257**, 13 757.

356 Kobori, J. A. and Kornberg, A. (1982), *J. Biol. Chem.*, **257**, 13 762.

357 Kobori, J. A. and Kornberg, A. (1982), *J. Biol. Chem.*, **257**, 13 770.

358 Reidel, H-D, Konig, H., Stahl, G. and Knippers, R. (1982), *Nucleic Acids Res.*, **10**, 5621.

359 Mechali, M. and Harland, R. M. (1982), *Cell*, **30**, 93.

360 Yoda, K-Y. and Okazaki, T. (1983), *Nucleic Acids Res.*, **11**, 3433.

361 van der Ende, A., Teertstra, R., van der Avoort, H. G. A. M. and Weisbeek, P. J. (1983), *Nucleic Acids Res.*, **11**, 4957.

362 Burgers, P. M. J. and Kornberg, A. (1983), *J. Biol. Chem.*, **258**, 7669.

363 Aoyama, A. and Hayashi, M. (1982), *Nature (London)*, **297**, 707.

364 Shlomai, J., Polder, L., Arai, K. and Kornberg, A. (1981), *J. Biol. Chem.*, **256**, 5233.

365 Arai, N., Polder, L., Arai, K. and Kornberg, A. (1981), *J. Biol. Chem.*, **256**, 5239.

366 Eisenberg, S., Griffith, J. and Kornberg, A. (1977), *Proc. Natl. Acad. Sci. USA*, **74**, 3198.

367 Koths, K. and Dressler, D. (1978), *Proc. Natl. Acad. Sci. USA*, **75**, 605.

368 Duget, M., Yarranton, G. and Gefter, M. (1978), *Cold Spring Harbor Symp. Quant. Biol.*, **43**, 335.

369 Ikeda, J. E., Yudelevich, A. and Hurwitz, J. (1976), *Proc. Natl. Acad. Sci. USA*, **73**, 2669.

370 Eisenberg, S. and Kornberg, A. (1979), *J. Biol. Chem.*, **254**, 5328.

371 Eisenberg, S., Scott, J. F. and Kornberg, A. (1978), *Cold Spring Harbor Symp. Quant. Biol.*, **43**, 295.

372 Meyer, T. F., Geider, K., Kurz, C. and Schaller, H. (1979), *Nature (London)*, **278**, 365.

373 Geider, K. and Meyer, T. F. (1978), *Cold Spring Harbor Symp. Quant. Biol.*, **43**, 59.

374 Godson, G. N. (1977), *J. Mol. Biol.*, **117**, 353.

375 Denhardt, D. T. (1975), *J. Mol. Biol.*, **99**, 107.

376 Meyer, T. F. and Geider, K. (1970), *J. Biol. Chem.*, **254**, 12 642.

377 Mitra, S. and Stallions, D. R. (1976), *Eur. J. Biochem.*, **67**, 37.

378 Ende, A. van der, Langeveld, S. A., Teertstra, R., Arkel, G. A. van and Weisbeek, P. J. (1981), *Nucleic Acids Res.*, **9**, 2037.

379 Roth, M. J., Brown, D. R. and Hurwitz, J. (1984), *J. Biol. Chem.*, **259**, 10 556.

380 Brown, D. R., Roth, M. J., Reinberg, D. and Hurwitz, J. (1984), *J. Biol. Chem.*, **259**, 10 545.

381 Aoyama, A., Hamatake, R. K. and Hayashi, M. (1983), *Proc. Natl. Acad. Sci. USA*, **80**, 4195.

382 Koths, K. and Dressler, D. (1980), *J. Biol. Chem.*, **255**, 4328.

383 Aoyama, A., Hamatake, R. K. and Hayashi, M. (1981), *Proc. Natl. Acad. Sci. USA*, **78**, 7285.

384 Brown, D. R., Reinberg, D., Schmidt-Glenewinkel, T., Roth, M., Zipursky, S. L. and Hurwitz, J. (1983), *Cold Spring Harbor Symp. Quant. Biol.*, **47**, 701.

385 Morris, C. F., Sinha, N. K. and Alberts, B. M. (1975), *Proc. Natl. Acad. Sci. USA*, **72**, 4800.

386 Piperno, J. R. and Alberts, B. M. (1978), *J. Biol. Chem.*, **253**, 5174.

387 Piperno, J. R., Kallen, R. G. and Alberts, B. M. (1978), *J. Biol. Chem.*, **253**, 5180.

388 Liu, C. C., Burke, R. L., Hibner, U., Barry, J. and Alberts, B. (1978), *Cold Spring Harbor Symp. Quant. Biol.*, **43**, 469.

389 Nossal, N. G. (1979), *J. Biol. Chem.*, **254**, 6026.

390 Nossal, N. G. and Peterlin, B. M. (1979), *J. Biol.*

*Chem.*, **254**, 6032.

391 Sinha, N. K., Morris, C. F. and Alberts, B. M. (1980), *J. Biol. Chem.*, **255**, 4290.

392 Hibner, U. and Alberts, B. M. (1980), *Nature (London)*, **285**, 300.

393 Alberts, B. M., Barry, J., Bedinger, P., Formosa, T., Jongeneel, C. V. and Krenzer, K. N. (1983), *Cold Spring Harbor Symp. Quant. Biol.*, **47**, 655.

394 Bedinger, P., Hochstrasser, M., Jongeneel, C. V. and Alberts, B. M. (1983), *Cell*, **34**, 115.

395 Muesing, M. A., Smith, D. H., Cabradilla, C. D., Benton, C. V., Lasky, L. A. and Capon, D. J. (1985), *Nature (London)*, **313**, 450.

396 Lovett, M. A., Katz, L. and Helinski, D. R. (1974), *Nature (London)*, **251**, 337.

397 Hirt, B. (1969), *J. Mol. Biol.*, **40**, 141.

398 Wolfson, J. and Dressler, D. (1972), *Proc. Natl. Acad. Sci. USA*, **69**, 2682.

399 Kriegstein, H. J. and Hogness, D. S. (1974), *Proc. Natl. Acad. Sci. USA*, **71**, 135.

400 Fareed, G. C., Garon, C. F. and Salzman, N. P. (1972), *J. Virol.*, **10**, 484.

401 Masters, M. and Broda, P. (1971), *Nature (London) New Biol.*, **232**, 137.

402 Hara, H. and Yoshikawa, H. (1973), *Nature (London) New Biol.*, **244**, 200.

403 Weingartner, B., Winnacker, E. L., Tolun, A. and Pettersson, U. (1976), *Cell*, **9**, 259.

404 Zakian, V. A. (1976), *J. Mol. Biol.*, **108**, 305.

405 Huberman, J. A. and Riggs, A. D. (1968), *J. Mol. Biol.*, **32**, 327.

406 Callan, H. G. (1972), *Proc. R. Soc. London Ser. B*, **181**, 19.

407 Hand, R. (1978), *Cell*, **15**, 317.

408 Woodland, H. R. and Pestell, R. Q. W. (1972), *Biochem. J.*, **127**, 597.

409 Housman, D. and Huberman, J. A. (1975), *J. Mol. Biol.*, **94**, 173.

410 Lewin, B. (1974), *Gene Expression*, Vol. 2, John Wiley & Son, New York.

411 Tolun, A., Alestrom, P. and Pettersson, U. (1979), *Cell*, **17**, 705.

412 Soeda, E., Arrand, J. R., Smoler, N. and Griffin, B. E. (1979), *Cell*, **17**, 357.

413 Lim, V. I. and Mazanov, A. L. (1978), *FEBS Lett.*, **88**, 118.

414 Jacob, F., Brenner, S. and Cuzin, F. (1963), *Cold Spring Harbor Symp. Quant. Biol.*, **28**, 329.

415 Pritchard, R. H., Barth, P. T. and Collins, J. (1969), *Symp. Soc. Gen. Microbiol.*, **19**, 293.

416 Rao, P. N. and Johnson, R. T. (1970), *Nature (London)*, **225**, 159.

417 Cabello, F., Timmis, K. and Cohen, S. N. (1976), *Nature (London)*, **259**, 285.

418 Takahashi, S. (1975), *J. Mol. Biol.*, **94**, 385.

419 Carter, B. J., Shaw, B. D. and Smith, M. G. (1969), *Biochim. Biophys. Acta*, **195**, 494.

420 Hayes, S. and Szybakski, W. (1975), in *DNA Synthesis and its Regulation* (ed. M. Goulian, P. Hanawalt and C. F. Fox), W. A. Benjamin, Menlo Park, CA, p. 486.

421 Tsurimoto, T. and Matsubara, K. (1983), *Cold Spring Harbor Symp. Quant. Biol.*, **47**, 681.

422 Tsuromoto, T. and Matsubara, K. (1981), *Nucleic Acids Res.*, **9**, 1789.

423 Furth, M. E. and Yates, J. L. (1978), *J. Mol. Biol.*, **126**, 227.

424 Sherer, G. (1978), *Nucleic Acids Res.*, **5**, 3141.

425 Honigman, A., Hu. S-L., Chase, R. and Szybalski, W. (1976), *Nature (London)*, **262**, 112.

426 Walz, A., Pirrotta, V. and Ineichen, K. (1976), *Nature (London)*, **262**, 665.

427 Wold, M. S., Mallary, J. B., Roberts, J. D., Le Bowitz, I. H. and McMacken, R. (1982), *Proc. Natl. Acad. Sci. USA*, **79**, 6176.

428 Tsurimoto, T. and Matsubara, F. (1982), *Proc. Natl. Acad. Sci. USA*, **79**, 7639.

429 Le Bowitz, J. H. and McMacken, R. (1984), *Nucleic Acids Res.*, **12**, 3069.

430 Watson, J. D. (1972), *Nature (London) New Biol.*, **239**, 197.

431 Kolodner, R. and Richardson, C. C. (1977), *Proc. Natl. Acad. Sci. USA*, **74**, 1525.

432 Kolodner, R. and Richardson, C. C. (1978), *J. Biol. Chem.*, **253**, 574.

433 Masker, W. E. and Richardson, C. C. (1976), *J. Mol. Biol.*, **100**, 543.

434 Richardson, C. C., Romano, L. J., Kolodner, R., Le Clerc, J. E., Tamanoi, R., Engler, M. J., Dean, F. B. and Richardson, D. S. (1978), *Cold Spring Harbor Symp. Quant. Biol.*, **43**, 427.

435 De Wyngaert, M. A. and Hinkle, D. C. (1979), *J. Biol. Chem.*, **254**, 11 247.

436 Romano, L. J., Tamanoi, T. and Richardson, C. C. (1981), *Proc. Natl. Acad. Sci. USA*, **78**, 4107.

437 Fuller, C. W., Beauchamp, B. B., Engler, M. J., Lechner, R. L., Matson, S. W., Tabor, S., White, J. H. and Richardson, C. C. (1983), *Cold Spring Harbor Symp. Quant. Biol.*, **47**, 669.

438 Richardson, C. C. (1983), *Cell*, **33**, 315.

439 Strothkamp, R. E., Oakley, J. L. and Coleman, J. E. (1980), *Biochemistry*, **19**, 1074.

440 Mark, D. F. and Richardson, C. C. (1976), *Proc. Natl. Acad. Sci. USA*, **73**, 780.

441 Hori, K., Mark, D. F. and Richardson, C. C. (1979), *J. Biol. Chem.*, **254**, 11 591.

442 Nordstran, B., Randahl, H., Slaby, I. and Holmgren, A. (1981), *J. Biol. Chem.*, **256**, 3112.

443 Engler, M. J., Lechner, R. L. and Richardson, C. C. (1983), *J. Biol. Chem.*, **258**, 11 165.

444 Lechner, R. L., Engler, M. J. and Richardson, C. C. (1983), *J. Biol. Chem.*, **258**, 11 174.

445 Scherzinger, E., Lanka, E., Morelli, G., Seiffert, D. and Yuki, A. (1977), *Eur. J. Biochem.*, **72**, 543.

446 Matson, S. W. and Richardson, C. C. (1983), *J. Biol. Chem.*, **258**, 14 009.

447 Matson, S. W., Tabor, S. and Richardson, C. C. (1983), *J. Biol. Chem.*, **258**, 14 017.

448 Scherzinger, E., Lanka, E. and Hillenbrand, G. (1977), *Nucleic Acids Res.*, **4**, 4151.

449 Hillenbrand, G., Morelli, G., Lanka, E. and Scherzinger, E. (1978), *Cold Spring Harbor Symp. Quant. Biol.*, **43**, 449.

450 Romano, L. J. and Richardson, C. C. (1979), *J. Biol. Chem.*, **254**, 10 476.

451 Romano, L. J. and Richardson, C. C. (1979), *J. Biol. Chem.*, **254**, 10 482.

452 Engler, M. J. and Richardson, C. C. (1983), *J. Biol. Chem.*, **258**, 11 197.

453 Fischer, H. and Hinkle, D. C. (1980), *J. Biol. Chem.*, **255**, 7956.

454 Wever, G. H., Fischer, H. and Hinkle, D. C. (1980), *J. Biol. Chem.*, **255**, 7965.

455 Fujiyama, A., Kohara, Y. and Okazaki, T. (1981), *Proc. Natl. Acad. Sci. USA*, **78**, 903.

456 Alberts, B., Morris, C. F., Mace, D., Sinha, N., Bittner, M. and Moran, L. (1975), in *DNA Synthesis and its Regulation* (ed. M. Goulian, P. Hanawalt and C. F. Fox), W. A. Benjamin, Menlo Park, CA, p. 241.

457 Delius, H., House, C. and Lozinski, A. W. (1971), *Proc. Natl. Acad. Sci. USA*, **68**, 3049.

458 Mosig, G., Luder, A., Rowen, L., Macdonald, P. and Bock, S. (1981), *ICN-UCLA Symposium on Molecular and Cellular Biology*, (ed. D. S. Ray), W. A. Benjamin, Vol. 22, p. 277.

459 Luder, A. and Mosig, G. (1982), *Proc. Natl. Acad. Sci. USA*, **79**, 1101.

460 Liu, C. F., Liu, C. C. and Alberts, B. M. (1979), *Nature (London)*, **281**, 456.

461 Stetler, G. L., King, C. J. and Huang, W. M. (1979), *Proc. Natl. Acad. Sci. USA*, **76**, 3737.

462 Liu, L. F., Liu, C-C. and Alberts, B. M. (1980), *Cell*, **19**, 697.

463 Somerville, R., Ebisuyaki, K. and Greenberg, G. R. (1959), *Proc. Natl. Acad. Sci. USA*, **45**, 1240.

464 Prashad, N. and Hosoda, J. (1972), *J. Mol. Biol.*, **70**, 617.

465 Chiu, C-S., Tomack, P. K. and Greenberg, G. R. (1976), *Proc. Natl. Acad. Sci. USA*, **73**, 757.

466 Chao, J., Leach, M. and Karam, J. (1977), *J. Virol.*, **24**, 557.

467 Reddy, G. P. V., Singh, A., Stafford, M. E. and Mathews, C. K. (1977), *Proc. Natl. Acad. Sci. USA*, **74**, 3152.

468 Wirak, D. O. and Greenberg, G. R. (1980), *J. Biol. Chem.*, **255**, 1896.

469 Reddy, G. P. V. and Mathews, C. K. (1978), *J. Biol. Chem.*, **253**, 3461.

470 Tsurimoto, T. and Matsubara, K. (1984), *Proc. Natl. Acad. Sci. USA*, **81**, 7402.

471 Reichard, P., Rowen, L., Eliasson, R., Hobbs, J. and Eckstein, F. (1978), *J. Biol. Chem.*, **253**, 7011.

472 Blair, D. G. and Helinsky, D. R. (1975), *J. Biol. Chem.*, **250**, 8785, and following papers.

473 Hashimoto-Gotoh, T. and Timmis, K. N. (1981), *Cell*, **23**, 229.

474 Backman, K., Betlach, M., Boyer, H. W. and Yanofsky, S. (1978), *Cold Spring Harbor Symp. Quant. Biol.*, **43**, 69.

475 Itoh, T. and Tomizawa, J. (1978), *Cold Spring Harbor Symp. Quant. Biol.*, **43**, 409 and (1980), *Proc. Natl. Acad. Sci. USA*, **77**, 2450.

476 Bastia, D. (1977), *Nucleic Acids Res.*, **4**, 3123.

477 Tomizawa, J., Itoh, T., Selzer, G. and Som, T. (1981), *Proc. Natl. Acad. Sci. USA*, **78**, 1421.

478 Stuitje, A. R., Spelt, C. E., Veltkamp, E. and Nijkamp, H. J. J. (1981), *Nature (London)*, **290**, 264.

479 Itoh, T. and Tomizawa, J. (1982), *Nucleic Acids Res.*, **10**, 5949.

480 Johanson, K. G. and McHenry, C. S. (1984), *J. Biol. Chem.*, **259**, 4589.

481 Zipursky, S. L. and Marians, K. J. (1981), *Proc. Natl. Acad. Sci. USA*, **78**, 6111.

482 Davison, J. (1984), *Gene*, **28**, 1.

483 Tomizawa, J. (1984), *Cell*, **38**, 861.

484 Tomizawa, J. and Itoh, T. (1982), *Cell*, **31**, 575.

485 Tamm, J. and Polisky, B. (1985), *Proc. Natl. Acad. Sci. USA*, **82**, 2257.

486 Lacatena, R. M. and Cesareni, G. (1981), *Nature (London)*, **294**, 623.

487 Tomizawa, J. and Som, T. (1984), *Cell*, **38**, 871.

488 Cesareni, G., Muesing, M. A. and Polisky, B. (1982), *Proc. Natl. Acad. Sci. USA*, **79**, 6313.

489 Lacatena, R. M., Banner, D. W., Castagnoli, L. and Cesareni, G. (1984), *Cell*, **37**, 1009.

490 Masukata, H. and Tomizawa, J. (1984), *Cell*, **36**, 513.

491 Nomura, N., Low, R. L. and Ray, D. S. (1982), *Proc. Natl. Acad. Sci. USA*, **79**, 3153.

492 Kolter, R., Inuzuka, M. and Helinski, D. R. (1978), *Cell*, **15**, 1199.

493 Danbara, H., Brady, G., Timmis, J. K. and Timmis, K. N. (1981), *Proc. Natl. Acad. Sci. USA*, **78**, 4699.

494 Stalker, D. M., Shafferman, A., Tolun, A., Kolter, R., Yand, S. and Helinski, D. R. (1981), *The Initiation of DNA Replication*, (ed. D. S. Ray), W. A. Benjamin, Menlo Park, California, p. 113.

495 Masai, H., Kazio, Y. and Arai, K. (1983), *Proc. Natl. Acad. Sci. USA*, **80**, 6814.

496 Germino, J. and Bastia, D. (1982), *Proc. Natl. Acad. Sci. USA*, **79**, 5475 and Kelly, W. and Bastia, D. (1985), *Proc. Natl. Acad. Sci. USA*, **82**, 2574.

497 Rosen, J., Ryder, T., Ohtsubo, H. and Ohtsubo, E. (1981), *Nature (London)*, **290**, 794.

498 Churchward, G., Linder, P. and Caro, L. (1983), *Nucleic Acids Res.*, **11**, 5643.

499 Vocke, C. and Bastia, D. (1983), *Cell*, **35**, 495, and (1985), *Proc. Natl. Acad. Sci. USA*, **82**, 2252.

500 Armstrong, K. A., Acosta, R., Ledner, E., Machida, Y., Pancotto, M., McCormick, M., Ohtsubo, H. and Ohtsubo, E. (1984), *J. Mol. Biol.*, **175**, 331.

501 Marsh, R. C. and Worcel, A. (1977), *Proc. Natl. Acad. Sci. USA*, **74**, 270.

502 Gudas, L. J., James, R. and Pardee, A. B. (1976), *J. Biol. Chem.*, **251**, 3470.

503 Messer, W., Dankworth, L., Tippe-Schindler, R., Womack, J. E. and Zahn, G. (1975), *DNA Synthesis and its Regulation* (eds M. Goulian and P. Hanawalt) (Series ed. F. Fox), ICN-UCLA Symposium on Molecular and Cellular Biology, Benjamin, California.

504 Tabata, S., Oka, A., Sugimoto, K., Takanami, M., Yasuda, S. and Hirota, Y. (1983), *Nucleic Acids Res.*, **11**, 2617.

505 Dixon, N. E. and Kornberg, A. (1984), *Proc. Natl. Acad. Sci. USA*, **81**, 424.

506 Fuller, R. S. and Kornberg, A. (1983), *Proc.*

*Natl. Acad. Sci. USA*, **80**, 5817.

507 Kornberg, A. (1983), *Eur. J. Biochem.*, **137**, 3377.

508 Fuller, R. S., Funnell, B. E. and Kornberg, A. (1984), *Cell*, **38**, 889.

509 Smith, D. W., Garland, A. M., Herman, G., Enns, R. E., Baker, T. A. and Zyskind, J. W. (1985), *EMBO J*, **4**, 1319.

510 Kaguni, J. M. and Kornberg, A. (1984), *J. Biol. Chem.*, **259**, 8578.

511 Kaguni, J. M. and Kornberg, A. (1984), *Cell*, **38**, 183.

512 Hendrickson, W. G., Kusano, T., Yamaki, H., Balakrishnan, R., King, M., Murchie, J. and Schaechter, M. (1982), *Cell*, **30**, 915.

513 Lotter, J. and Messer, W. (1981), *Nature (London)*, **294**, 376.

514 Robberson, D. L., Clayton, D. A. and Morrow, J. F. (1974), *Proc. Natl. Acad. Sci. USA*, **71**, 4447.

515 Gillum, A. M. and Clayton, D. A. (1978), *Proc. Natl. Acad. Sci. USA*, **75**, 677.

516 van Etten, R. A., Michael, N. L., Bibb, M. J., Brennicke, A. and Clayton, D. A. (1982), *Cold Spring Harbor Symp. on Mitochondrial Genes* (ed. P. Slominski, P. Borst and G. Attardi), p. 73.

517 Clayton, D. A. (1982), *Cell*, **28**, 693.

518 Berk, A. J. and Clayton, D. A. (1974), *J. Mol. Biol.*, **86**, 801.

519 Berk, A. J. and Clayton, D. A. (1976), *J. Mol. Biol.*, **100**, 85.

520 Robberson, D. L., Kasamasu, H. and Vinograd, J. (1972), *Proc. Natl. Acad. Sci. USA*, **69**, 736.

521 Bogenhagen, D. and Clayton, D. A. (1978), *J. Mol. Biol.*, **119**, 49.

522 Bogenhagen, D. and Clayton, D. A. (1978), *J. Mol. Biol.*, **119**, 69.

523 Goddard, J. M. and Wolstenholme, D. R. (1978), *Proc. Natl. Acad. Sci. USA*, **75**, 3886.

524 Wong, T. W. and Clayton, D. A. (1985), *Cell*, **42**, 951.

525 Tapper, D. P. and Clayton, D. A. (1981), *J. Biol. Chem.*, **256**, 5109.

526 Crews, S., Ojala, D., Posakeny, J., Nishiguchi, J. and Attardi, G. (1979), *Nature (London)*, **277**, 192.

527 Yaginuma, K., Kobayashi, M., Taira, M. and Koike, K. (1982), *Nucleic Acids Res.*, **10**, 7531.

528 Martens, P. A. and Clayton, D. A. (1979), *J. Mol. Biol.*, **135**, 327.

529 Tapper, D. P. and Clayton, D. A. (1982), *J. Mol. Biol.*, **162**, 1.

530 Zimmerman, W., Chen, S. M., Bolden, A. and Weissbach, A. (1980), *J. Biol. Chem.*, **255**, 11 847.

531 Matsumoto, L., Kasamatsu, H., Piko, L. and Vinograd, J. (1977), *J. Cell Biol.*, **63**, 146.

532 Bernardi, G. (1982), *Trends Biochem. Sci.*, **7**, 404.

533 Baldacci, G. and Bernardi, G. (1982), *EMBO J.*, **1**, 987.

534 Fangman, W. L. and Dujon, B. (1984), *Proc. Natl. Acad. Sci. USA*, **81**, 7156.

535 Arnberg, A. C., Van Bruggen, E. F. J., Clegg, R. A., Upholt, W. B. and Borst, P. (1974), *Biochim. Biophys. Acta*, **361**, 266.

536 Clegg, R. A., Borst, P. and Weijers, P. J. (1974), *Biochim. Biophys. Acta*, **361**, 277.

537 Arnberg, A. C., Van Bruggen, E. F. J., Borst, P., Clegg, R. A., Schutgens, R. B. H., Weigers, P. J. and Goldbach, R. W. (1974), *Biochim. Biophys. Acta*, **383**, 359.

538 Goddard, J. M. and Cummings, D. M. (1977), *J. Mol. Biol.*, **109**, 327.

539 Pritchard, A. E., Laping, J. L., Seilhamer, J. J. and Cumings, D. J. (1983), *J. Mol. Biol.*, **164**, 1.

540 Jaenisch, R., Mayer, A. and Levine, A. (1971), *Nature (London) New Biol.*, **233**, 72.

541 Varshavsky, A. J., Sundin, O. and Bohn, M. (1979), *Cell*, **16**, 453.

542 Tjian, T. (1978), *Cell*, **13**, 165.

543 Hay, R. T. and De Pamphilis, M. L. (1982), *Cell*, **28**, 767.

544 De Pamphilis, M. L. and Wassarman, P. M. (1982), in *Organisation and Replication of Viral DNA* (ed. A. S. Kaplan), CRC press, Cleveland, p. 37.

545 Tseng, B. Y. and Ahlem, C. N. (1984), *Proc. Natl. Acad. Sci. USA*, **81**, 2342.

546 Hay, R. T., Hendrickson, E. A. and De Pamphilis, M. L. (1984), *J. Mol. Biol.*, **175**, 131.

547 Li, J. J. and Kelly, T. J. (1984), *Proc. Natl. Acad. Sci. USA*, **81**, 6973.

548 Ariga, H. and Sugano, S. (1983), *J. Virol.*, **48**, 481.

549 Soedo, E., Arrand, J. R., Smoler, N. and Griffin, B. E. (1979), *Cell*, **17**, 357.

550 Friedman, T., Esty, A., Laporte, P. and Deininger, P. (1979), *Cell*, **17**, 715.

551 De Villiers, J., Schaffner, W., Tyndall, C., Lupton, S. and Kamen, R. (1984), *Nature (London)*, **312**, 242.

552 Bergsma, D. J., Olive, D. M., Hartzell, S. W. and Subramanian, K. N. (1982), *Proc. Natl. Acad. Sci. USA*, **79**, 381.

553 Kelly, T. J. (1982) in *Organisation and Replication of Viral DNA* (ed. A. S. Kaplan), CRC press, Cleveland, p. 115.

554 Winnacker, E-L. (1978), *Cell*, **14**, 761.

555 Tolun, A., Alestrom, P. and Pettersson, U. (1979), *Cell*, **17**, 705.

556 Stillman, B. W. (1983), *Cell*, **35**, 7.

557 Stillman, B. W. and Tamanoi, F. (1983), *Cold Spring Harbor Symp. Quant. Biol.*, **47**, 741.

558 Tamanoi, F. and Stillman, B. W. (1983), *Proc. Natl. Acad. Sci. USA*, **80**, 6446.

559 Challberg, M. D. and Rawlings, D. R. (1984), *Proc. Natl. Acad. Sci. USA*, **81**, 100.

560 Guggenheimer, R. A., Stillman, B. W., Nagata, K., Tamanoi, F. and Hurwitz, J. (1984), *Proc. Natl. Acad. Sci. USA*, **81**, 3069.

561 Hay, R. T. (1985), *EMBO J.*, **4**, 421.

562 Rekosh, D. M. K., Russell, W. C. and Bellet, A. J. D. (1977), *Cell*, **11**, 283.

563 Nagata, K., Guggenheimer, R. A. and Hurwitz, J. (1983), *Proc. Natl. Acad. Sci. USA*, **80**, 6177.

564 Nagata, K., Guggenheimer, R. A. and Hurwitz, J. (1983), *Proc. Natl. Acad. Sci. USA*, **80**, 4266.

565 Kaplan, L. M., Ariga, H., Hurwitz, J. and Horwitz, M. S. (1979), *Proc. Natl. Acad. Sci. USA*, **76**, 5534.

566 Rijnders, A. W. M., van Bergen, B. G. M., van der Vliet, P. S. and Sussenbach, J. S. (1983), *Nucleic Acids Res.*, **11**, 8777.

567 Ikeda, J-E., Enomoto, T. and Hurwitz, J. (1981), *Proc. Natl. Acad. Sci. USA*, **78**, 884.

568 Ikeda, J-E., Enomoto, T. and Hurwitz, J. (1982), *Proc. Natl. Acad. Sci. USA*, **79**, 2442.

569 Lichy, J. H., Nagata, K., Friefeld, B. R., Enomoto, T., Field, J., Guggenheimer, R. A., Ikeda, J-E., Horwitz, M. S. and Hurwitz, J. (1983), *Cold Spring Harbor Symp. Quant. Biol.*, **47**, 731.

570 Lichy, J. H., Horwitz, M. S. and Hurwitz, J. (1981), *Proc. Natl. Acad. Sci. USA*, **78**, 2678.

571 Enomoto, T., Lichy, J. H., Ikeda, J-E. and Hurwitz, J. (1981), *Proc. Natl. Acad. Sci. USA*, **78**, 6779.

572 Lichy, J. H., Field, J., Horwitz, M. S. and Hurwitz, J. (1982), *Proc. Natl. Acad. Sci. USA*, **79**, 5223.

573 Nagata, K., Guggenheimer, R. A., Enomoto, T., Lichy, J. H. and Hurwitz, J. (1982), *Proc. Natl. Acad. Sci. USA*, **79**, 6438.

574 Challberg, M. D. and Kelly, T. J. (1981), *J. Virol.*, **38**, 272.

575 Watabe, K., Shih, M. F., Sugino, A. and Ito, J. (1982), *Proc. Natl. Acad. Sci. USA*, **79**, 5245.

576 Wimmer, E. (1982), *Cell*, **28**, 199.

577 Pearson, G. D., Chow, K-C., Corden, J. L. and Harpst, J. A. (1981), in *The Initiation of DNA Replication* (ed. D. S. Ray), W. A. Benjamin, Menlo Park, California, p. 581.

578 Broach, J. R. (1982), *Cell*, **28**, 203.

579 Kojo, H., Greenberg, B. D. and Sugino, A. (1981), *Proc. Natl. Acad. Sci. USA*, **78**, 7261.

580 Kikuchi, Y. (1983), *Cell*, **35**, 487.

581 Tschumper, G. and Carbon, J. (1982), *J. Mol. Biol.*, **156**, 293.

582 Celniker, S. E. and Campbell, J. L. (1982), *Cell*, **31**, 201.

583 Broach, J. R., Li, Y-Y., Feldman, J., Jayaram, M., Abraham, J., Nasmyth, K. A. and Hicks, J. B. (1983), *Cold Spring Harbor Symp. Quant. Biol.*, **47**, 1165.

584 Fangman, W. L., Hice, R. H. and Chlebowicz-Sledziewska, E. (1983), *Cell*, **32**, 831.

585 Chan, C. S. M. and Tye, B-K. (1983), *Cell*, **33**, 563.

586 Kearsey, S. (1984), *Cell*, **37**, 299.

587 Vogt, V. M. and Braun, R. (1977), *Eur. J. Biochem.*, **86**, 557.

588 Bozzoni, I., Baldari, C. T., Armaldi, F. and Buongiorno-Nardelli, M. (1981), *Eur. J. Biochem.*, **118**, 585.

589 Botchan, P. M. and Dayton, A. I. (1982), *Nature (London)*, **299**, 453.

590 Heintz, N-H., Millbrandt, J. D., Greisen, K. S. and Hamlin, J. L. (1983), *Nature (London)*, **302**, 439.

591 Harland, R. M. and Laskey, R. A. (1980), *Cell*, **21**, 761.

592 Harland, R. M. (1981), *Trends Biochem. Sci.*, **6**, 71.

593 Laskey, R. A. and Harland, R. M. (1981), *Cell*, **24**, 283.

594 Hines, P. J. and Benbow, R. M. (1982), *Cell*, **30**, 459.

595 Mechali, M. and Kearsey, S. (1984), *Cell*, **38**, 55.

596 Chambers, J. C., Watanabe, S. and Taylor, J. H. (1982), *Proc. Natl. Acad. Sci. USA*, **79**, 5572.

597 McCready, S. J., Godwin, J., Mason, D. W., Brazell, I. A. and Cook, P. R. (1980), *J. Cell Sci.*, **46**, 365.

598 Pardoll, D. M., Vogelstein, B. and Coffey, D. S. (1980), *Cell*, **19**, 527.

599 Valenzuela, M. S., Mueller, G. C. and Dasgupta, S. (1983), *Nucleic Acids Res.*, **11**, 2155.

600 Aalen, J. M. A., Opstelten, R. J. G. and Waika, F. (1983), *Nucleic Acids Res.*, **11**, 1181.

601 Cook, P. R. and Lang, J. (1984), *Nucleic Acids Res.*, **12**, 1069.

602 Sundin, O. and Varshavsky, A. (1981), *Cell*, **25**, 659.

603 Johnson, E. M., Bergold, P. and Campbell, G. R. (1984), in *Recombinant DNA and Cell Proliferation* (ed. G. Stein and J. Stein), Academic Press, New York, p. 303.

604 Blackburn, E. H. and Szostak, J. W. (1984), *Annu. Rev. Biochem.*, **53**, 163.

605 Rogers, J. (1983), *Nature (London)*, **305**, 101.

606 Blackburn, E. H. (1984), *Cell*, **37**, 7.

607 Baroudy, B. M., Venkatesan, S. and Moss, B. (1983), *Cold Spring Harbor Symp. Quant. Biol.*, **47**, 723.

608 Bateman, A. J. (1975), *Nature (London)*, **253**, 379.

609 Bernards, A., Michels, P. A. M., Lincke, C. R. and Borst, P. (1983), *Nature (London)*, **303**, 592.

610 Berns, K. I. and Hauswirth, W. W. (1982), in *Organisation and Replication of Viral DNA* (ed. A. S. Kaplan), CRC press, Cleveland, p. 3.

611 Mathews, C. K. and Sinha, N. K. (1982), *Proc. Natl. Acad. Sci. USA*, **79**, 302.

612 Chiu, C-S., Cook, K. S. and Greenberg, G. R. (1982), *J. Biol. Chem.*, **257**, 15 087.

613 Allen, J. R., Lasser, G. W., Goldman, D. A., Booth, J. W. and Mathews, C. K. (1983), *J. Biol. Chem.*, **258**, 5746.

614 Reddy, G. P. and Pardee, A. B. (1980), *Proc. Natl. Acad. Sci. USA*, **77**, 3312.

615 Wickremasinghe, R. G., Yaxley, J. C. and Hoffbrand, A. V. (1982), *Eur. J. Biochem.*, **126**, 589.

616 Reddy, G. P. and Pardee, A. B. (1982), *J. Biol. Chem.*, **257**, 12 526.

617 Reddy, G. P. and Pardee, A. B. (1983), *Nature (London)*, **304**, 86.

618 Noguchi, H., Reddy, G. P. and Pardee, A. B. (1983), *Cell*, **32**, 443.

619 Wickremasinghe, R. G. and Hoffbrand, A. V. (1983), *FEBS Lett.*, **159**, 175.

620 Spyron, G. and Reichard, P. (1983), *Biochem. Biophys. Res. Commun.*, **115**, 1022.

621 Baltimore, D., Gilboa, E., Rottenberg, E. and Yoshimura, F. (1978), *Cold Spring Harbor Symp.*

*Quant. Biol.*, **43**, 869.

622 Swanstram, R., Varmus, H. E. and Bishop, J. M. (1981), *J. Biol. Chem.*, **256**, 1115.

623 Even, J., Anderson, S. J., Hampe, A., Galibert, F., Lowry, D., Khoury, G. and Sherr, C. J. (1983), *J. Virol.*, **65**, 1004.

624 Panganiban, A. T. and Temin, H. M. (1983), *Nature (London)*, **306**, 155.

625 Hsu, T. W., Sabran, J. L., Mark, G. E., Guntaka, R. V. and Taylor, J. M. (1978), *J. Virol.*, **28**, 810.

626 Varmus, H. E., Shank, P. R., Hughes, S. E., Kung, H-J., Heasley, S., Majors, J., Vogt, P. K. and Bishop, J. M. (1978), *Cold Spring Harbor Symp. Quant. Biol.*, **43**, 851.

627 Varmus, H. E., Heasley, S., Kung, H-J., Oppermann, H., Smith, V. C., Bishop, J. M. and Shank, P. R. (1978), *J. Mol. Biol.*, **120**, 55.

628 Flint, J. (1976), *Cell*, **8**, 151.

629 Haseltine, W., Kleid, D. G., Panet, A., Rothenberg, E. and Baltimore, D. (1976), *J. Mol. Biol.*, **106**, 109.

630 Coffin, J. M. and Haseltine, W. A. (1977), *J. Mol. Biol.*, **117**, 805.

631 Bishop, J. M. (1978), *Annu. Rev. Biochem.*, **47**, 35.

632 Burgoyne, L. A., Mobbs, J. D. and Marshall, A. J. (1976), *Nucleic Acids Res.*, **3**, 3293.

633 Buongiorno-Nardelli, M., Michele, G., Carri, M. T. and Marilley, M. (1982), *Nature (London)*, **298**, 100.

634 Stahl, H. and Gallwitz, D. (1977), *Eur. J. Biochem.*, **72**, 385.

635 Huchhauser, S. J., Stein, J. L. and Stein, G. S. (1981), *Int. Rev. Cytol.*, **71**, 95.

636 Stein, G. S., Plumb, M. A., Stein, G. L., Marashi, F. F., Sierra, L. F. and Baumbach, L. L. (1984), in *Recombinant DNA and Cell Proliferation* (ed. G. S. Stein and G. L. Stein), Academic Press, New York, p. 107.

637 Melli, M., Spinelli, G. and Arnold, E. (1977), *Cell*, **12**, 167.

638 Plumb, M., Stein, J. and Stein, G. (1983), *Nucleic Acids Res.*, **11**, 2391.

639 Plumb, M., Stein, J. and Stein, G. (1983), *Nucleic Acids Res.*, **11**, 7927.

640 Sittman, D. B., Graves, R. A. and Marzluff, W. F. (1983), *Proc. Natl. Acad. Sci, USA*, **80**, 1849.

641 Nurse, P. (1983), *Nature (London)*, **302**, 378.

642 Wu, R. S., Tsai, S. and Bonner, W. M. (1982), *Cell*, **31**, 367.

643 Djondjurov, L. P., Yancheva, N. Y. and Ivanova, E. C. (1983), *Biochemistry*, **22**, 4095.

644 Hereford, L., Bromley, S. and Osley, M. (1982), *Cell*, **30**, 305.

645 Osley, M. and Hereford, L. (1982), *Proc. Natl. Acad. Sci. USA*, **79**, 7689.

646 Laskey, R. A. and Earnshaw, W. C. (1980), *Nature (London)*, **286**, 763.

647 Ryoji, M. and Worcel, A. (1984), *Cell*, **37**, 21.

648 Glikin, G. C., Ruberti, I. and Worcel, A. (1984), *Cell*, **37**, 33.

649 Weintraub, H. (1976), *Cell*, **9**, 419.

650 Annunziato, A. T. and Seale, R. L. (1983), *J. Biol. Chem.*, **258**, 12 675.

651 Cremisi, C., Chestier, A. and Yaniv, M. (1977), *Cell*, **12**, 947.

652 Worcel, A., Han, S. and Wong, M. L. (1978), *Cell*, **15**, 969.

653 Leffak, I. M., Grainger, K. R. and Weintraub, H. (1977), *Cell*, **12**, 837.

654 Russev, G. and Hancock, R. (1982), *Proc. Natl. Acad. Sci. USA*, **79**, 3143.

655 Jackson, V. and Chalkley, R. (1981), *Cell*, **23**, 121.

656 Russev, G. and Tsanev, R. (1979), *Eur. J. Biochem.*, **93**, 123.

657 Riley, D. and Weintraub, H. (1979), *Proc. Natl. Acad Sci. USA*, **76**, 328.

658 Seidman, M. M., Levine, A. J. and Weintraub, H. (1979), *Cell*, **18**, 439.

659 Pospelov, V., Russev, G., Vassilev, L. and Tsanev, R. (1982), *J. Mol. Biol.*, **156**, 79.

660 Tack, L. C., Wassarman, P. M. and De Pamphilis, M. L. (1981), *J. Biol. Chem.*, **256**, 8821.

661 Leffak, I. M. (1984), *Nature (London)*, **307**, 82.

662 Marsden, M. P. F. and Laemmli, U. K. (1979), *Cell*, **17**, 849.

663 Finch, J. T. and Klug, A. (1976), *Proc. Natl. Acad. Sci. USA*, **73**, 1897.

664 Kowalski, J. and Cheevers, W. P. (1976), *J. Mol. Biol.*, **104**, 603.

665 Edenberg, H. J. and Huberman, J. A. (1975), *Annu. Rev. Biochem.*, **44**, 245.

666 Sheinin, R. and Humbert, J. (1978), *Annu. Rev. Biochem.*, **47**, 277.

667 Weintraub, H. (1979), *Nucleic Acids Res.*, **7**, 781.

668 Ollis, D. L., Brick, P., Hamlin, R., Xuong, N. G. and Steitz, T. A. (1985), *Nature (London)*, **313**, 762.

669 Hull, R. and Covey, S. N. (1983), *Trends*

*Biochem. Sci.*, **8**, 119.

670 Pfeiffer, P. and Hohn, T. (1983), *Cell*, **33**, 781 and Hohn, T., Hohn, B. and Pfeiffer, P. (1985), *Trends Biochem. Sci.*, **10**, 205.

671 Varmus, H. (1983), *Nature (London)*, **304**, 116.

672 Marco, Y. and Howell, S. H. (1984), *Nucleic Acids Res.*, **12**, 1517.

673 Soderhall, S. (1976), *Nature (London)*, **260**, 640.

674 David, J. C. and Vinson, D. (1979), *Exp. Cell Res.*, **119**, 69.

675 Mezzina, M., Sarasin, A. Politi, N. and Bertazzoni, U. (1984), *Nucleic Acids Res.*, **12**, 5109.

676 Kornberg, A. (1978), *Cold Spring Harbor Symp. Quant. Biol.*, **43**, 1.

677 Wickner, S. (1978), *Annu. Rev. Biochem.*, **47**, 1163.

678 Schaller, H., Uhlmann, A. and Geider, K. (1976), *Proc. Natl. Acad. Sci. USA*, **73**, 49.

679 Gray, C. P., Sommer, R., Polke, C., Beck, E. and Schaller, H. (1978), *Proc. Natl. Acad. Sci. USA*, **75**, 50.

680 Beck, E., Sommer, R., Averswald, E. A., Kurz, C., Zinc, B., Osterburg, G., Schaller, H., Sugimoto, K., Sugisaki, H., Okamoto, T. and Takanami, M. (1978), *Nucleic Acids Res.*, **5**, 4495.

681 Sims, J., Capon, D. and Dressler, D. (1979), *J. Biol. Chem.*, **254**, 12 615.

682 Fiddes, J. C., Barrell, B. G. and Godson, G. N. (1978), *Proc. Natl. Acad. Sci. USA*, **75**, 1081.

683 Martin, D. M. and Godson, G. N. (1977), *J. Mol. Biol.*, **117**, 321.

684 Sims, J. and Benz, E. W. (1980), *Proc. Natl. Acad. Sci. USA*, **77**, 900.

685 Stayton, M. M. and Kornberg, A. (1983), *J. Biol. Chem.*, **258**, 13 205.

686 Godson, G. N. (1978), *Cold Spring Harbor Symp. Quant. Biol.*, **43**, 367.

687 Bouché, J. P., Rowen, L. and Kornberg, A. (1978), *J. Biol. Chem.*, **253**, 756.

688 Spadari, S., Sala, F. and Pedrali-Noy, G. (1982), *Trends Biochem. Sci.*, **7**, 29.

689 Sumida-Yasumoto, C., Yudelevich, A. and Hurwitz, J. (1976), *Proc. Natl. Acad. Sci. USA*, **73**, 1887.

690 Zyskind, J. W., Harding, N., Clearly, M., Takeda, Y. and Smith, D. W. (1982), *Proc. Natl. Acad. Sci. USA*, **80**, 1164.

691 Asada, K., Sugimoto, K., Oka, A., Takanami, M. and Hirota, Y. (1982), *Nucleic Acids Res.*, **10**, 3745.

692 Oka, A., Sasaki, H., Sugimoto, K. and Takanami, M. (1984), *J. Mol. Biol.*, **176**, 443.

693 Seale, R. L. (1975), *Nature (London)*, **9**, 247.

694 Seale, R. L. (1976), *Cell*, **9**, 423.

695 Annunziato, A. T., Schindler, R. K., Thomas, C. A. and Seale, R. L. (1981), *J. Biol. Chem.*, **266**, 11 880.

696 Annunziato, A. T. and Seale, R. L. (1984), *Nucleic Acids Res.*, **12**, 6179.

697 Chang, L. M. S., Rafter, E., Augl, C. and Bollum, F. J. (1984), *J. Biol. Chem.*, **259**, 14 679.

698 Karawya, E., Swack, J., Albert, W., Fedorko, J., Minna, J. D. and Wilson, S. H. (1984), *Proc. Natl. Acad. Sci. USA*, **81**, 7777.

699 Grosse, F. and Krauss, G. (1985), *J. Biol. Chem.*, **260**, 1881.

700 Halligan, B. D., Edwards, K. A. and Liu, L. F. (1985), *J. Biol. Chem.*, **260**, 2475.

701 Faust, E. A., Nagy, R. and Davey, S. K. (1985), *Proc. Natl. Acad. Sci. USA*, **82**, 4023.

702 Marians, K. J. (1985), *CRC Crit. Rev. Biochem.*, **17**, 153.

703 Holm, C., Goto, T., Wang, J. C. and Botstein, D. (1985), *Cell*, **41**, 553.

704 Le Bowitz, J. H., Zylicz, M., Georgopoulos, C. and McMacken, R. (1985), *Proc. Natl. Acad. Sci. USA*, **82**, 3988.

705 van der Ende, A., Baker, T. A., Ogawa, T. and Kornberg, A. (1985), *Proc. Natl. Acad. Sci. USA*, **82**, 3954.

706 Van der Ploeg, L. H. T., Lui, A. Y. C. and Borst, P. (1984), *Cell*, **36**, 459.

707 Walmsley, R. W., Chan, C. S. M., Tye, B. K. and Petes, T. D. (1984), *Nature (London)*, **310**, 157.

708 Wong, E. M. and Polisky, B. (1985), *Cell*, **42**, 959.

709 Baas, P. D. (1985), *Biochim. Biophys. Acta*, **825**, 111.

710 Kohara, Y., Tohdoh, N., Jiang, X-W. and Okazaki, T. (1985), *Nucleic Acids Res.*, **13**, 6847.

# 7

# Repair, recombination and DNA rearrangement

## 7.1 INTRODUCTION

This chapter is concerned with alterations in the structure of DNA, and considers several different facets of this topic. On the one hand it is important to protect the genome from inadvertent damage, and cellular mechanisms exist to repair such damage. On the other hand genetic variation is the *sine qua non* for evolution, and recombination together with transposition mechanisms allow an extent of genetic flux. Then there are examples of precisely controlled changes in DNA, involving amplifications or rearrangement, that are part of the developmental programme of particular cells or organisms.

## 7.2 MUTATIONS AND MUTAGENS

Damage to DNA resulting in the insertion of incorrect bases or distortion of the normal double-helical structure must be corrected if the cell is to survive. This chapter starts with a consideration of some of the ways in which damage may arise and the various strategies the cell may adopt to rectify the damage. If the damage is quickly rectified little harm is done, but if the DNA is replicated before repair is complete alternative repair mechanisms must be called into play and these usually require recombination events.

Alterations in the base pattern of DNA may arise in various ways. For example, existing bases may be replaced by others, or they may be deleted, or new bases may be inserted in the DNA chain. Occasional mistakes in the normal duplication of DNA give rise to spontaneous mutations but such mistakes are surprisingly rare [1, 2] (see Section 6.5). Misincorporation may occur as a result of spontaneous tautomerization leading to mispairing of iminocytosine with adenine, iminoadenine with cytosine, enolguanine with thymine or enolthymine with guanine. The frequency of mutations depends on conditions of temperature, pH, composition of growth medium, and the like, but it can be greatly increased by exposure of cells to ultraviolet and ionizing radiations (Sections 7.2.4 and 7.2.5) or to certain types of chemical which are known collectively as mutagens. Such substances include base analogues, some dyes of the acridine series, alkylating agents, certain antibiotics, urethane, hydroxylamine and nitrous acid. This last substance has been used very effectively in studying mutations in certain viruses such as tobacco mosaic virus. Nitrous acid may arise in the body from nitrates which are being used in increasing amounts in fertilizers. Bacterial action in the mouth, stomach and bladder leads to reduction of nitrates to nitrites (see Section 2.10.2) [5].

Mutagenic substances are the subject of an extensive literature [1, 3–11] which reflects the considerable effort now being devoted to attempts to inhibit cell division, especially in neoplastic tissues, by the use of compounds which might be expected to inhibit nucleic acid biosynthesis. Research in this field has been stimulated by the hope of finding a basis for an improved therapy for cancer. Reference has already been made (Chapter 5) to the use of such compounds as azaserine and the folic acid antagonists in preventing the synthesis of the purine and pyrimidine nucleotides. Some of the other substances which have been used to prevent nucleic acid biosynthesis and to bring about mutations artificially are discussed below.

### 7.2.1 Base and nucleoside analogues

Some of the artificially produced base analogues are incorporated into RNA and DNA and may have powerful mutagenic effects [1, 9–11]. Among the most important analogues are the halogenated pyrimidines, and those bases where nitrogen has been substituted for a –CH= group (see Fig. 7.1).

The action of these unnatural bases seems, at least in some cases, to be twofold:
(1)   They generally block some stage in the biosynthesis of the normal purine and pyrimidine nucleotides. Thus 8-azaguanine inhibits the biosynthesis of GMP and 6-mercaptopurine blocks the conversion of IMP into AMP [12]. In general, these inhibitions are brought about only after the inhibitor itself has been converted into its nucleotide. Thus 6-azauracil is converted first into its nucleoside (Aza-U) then into its nucleotide (Aza-UMP) which inhibits the action of orotidine 5′-phosphate decarboxylase (Fig. 5.6) and so prevents pyrimidine biosynthesis [9, 10, 13]. 5-Fluorouracil, which has proved to be a potent inhibitor of the growth of certain tumours, is converted first into its ribonucleotide (F-UMP) and then into its deoxyribonucleotide (F-dUMP) which exerts its main effect by inhibiting conversion of dUMP into dTMP [9, 10,

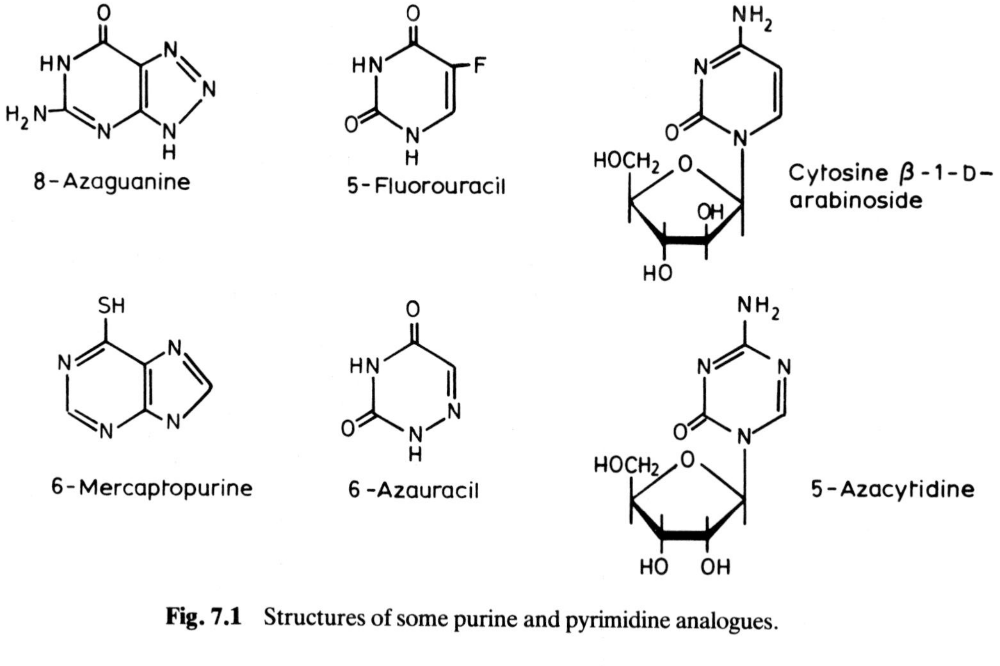

**Fig. 7.1**   Structures of some purine and pyrimidine analogues.

14] and hence inhibiting DNA synthesis (Fig. 5.8).

When 5-fluorodeoxyuridine is added to cells in culture it is converted into 5F-dUMP which blocks DNA synthesis. As no other processes are affected, cells progress around the cycle and accumulate at the beginning of the S phase (see Section 3.2.1). The inhibition can be overcome by the addition of thymidine with the result that a population of cells in synchronized growth results [9, 15].

(2)   They are themselves, after conversion into nucleotides, incorporated to varying degrees into RNA and/or DNA although the incorporation may take an abnormal form. Thus 8-azaguanine can be incorporated at the expense of guanine into the RNA of TMV [16] and, to a much larger extent, into the RNA of *B. cereus* [17]. Only very small amounts are incorporated into the DNA.

5-Azacytidine is incorporated largely into RNA and this rapidly interferes with protein synthesis [11, 18]. Some is, however, incorporated into DNA and like 5-azadeoxy-cytidine it brings about changes in gene expression probably as a result of its inhibiting DNA methylation (see Chapter 4) [19, 20]. 5-Bromouracil can replace thymine in DNA where it normally base-pairs with adenine. However, in its rare enol state (which it assumes more readily than does thymine) it may pair with guanine instead of adenine so bringing about the base-pair transition A–T into G–C. DNA containing 5-bromouracil instead of thymine is very susceptible to breakage at light-induced bromo-uracil dimers [21] (see below).

The main mutagenic effects of bromodeoxy-uridine, however, appear not to arise from its incorporation into DNA in place of thymidine but rather from its effect on ribonucleotide reductase (see Section 5.5). This leads to an imbalance in the deoxyribonucleoside triphosphate pools [22] which may result in base misincorporation. In particular the pool size of dCTP is reduced and some substitution of deoxy-cytidine with bromodeoxyuridine (in the enol form) occurs leading to GC → AT transitions as the misincorporated bromodeoxyuridine reverts to its more stable ketoform. Addition of thymidine accentuates the mutagenic effect of bromodeoxyuridine by further decreasing the dCTP pool size whereas deoxycytidine antagonizes the mutagenic effects.

2-Aminopurine can also base-pair with either cytosine or thymine and, because it inhibits adenosine deaminase, it also increases the pool size of dATP which in turn increases the size of the dCTP pool relative to the dTTP pool leading to misincorporation of cytosine opposite adenine or 2-aminopurine [23].

The D-arabinosyl nucleosides are effectively analogues of deoxyribonucleosides (e.g. cyto-sine $\beta$-D-arabinoside is incorporated in place of deoxycytidine into DNA where it causes chain termination or a marked reduction in the rate of further chain extension [9–11, 21, 24]).

### 7.2.2 Alkylating agents

The alkylating agents exert a variety of biological effects including mutagenesis, carcinogenesis and tumour growth inhibition [1, 4, 25–27]. They all carry one, two, or more alkyl groups in reactive form and include the well-known compounds sulphur mustard or di-(2-chloroethyl) sulphide and nitrogen mustard or methyl-di-(2-chloroethyl)amine.

The action of the alkylating agents on DNA is complex. They are electrophilic reagents and react with nucleophilic centres, particularly with guanine at the N-7 atom. The order of reactivity of different nucleophilic centres is $N^7$–$G >> N^3$–$A > N^1$–$A = N^3$–$G = O^6$–$G$   [3]. The bifunctional alkylating agents (i.e. those with two reactive alkyl groups) may bring about crosslinking between the opposing strands in the DNA molecule. Alkylation of purines in position 7 also gives rise to unstable quaternary nitrogens so that the alkylated purine may separate from the deoxyribose leaving a gap

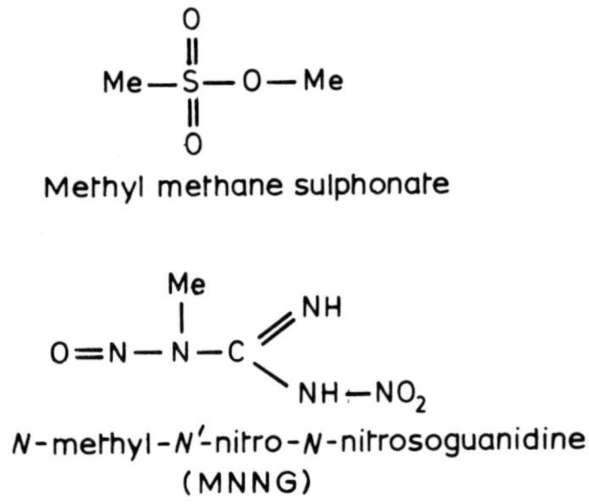

$$\text{Me} - \underset{\underset{\text{O}}{\|}}{\overset{\overset{\text{O}}{\|}}{\text{S}}} - \text{O} - \text{Me}$$

Methyl methane sulphonate

N-methyl-N'-nitro-N-nitrosoguanidine
(MNNG)

**Fig. 7.2** Alkylating agents.

which might interfere with DNA replication or cause the incorporation of the wrong base [1, 25]. The phosphate groups may also be alkylated. The phosphate triester so formed is unstable and may hydrolyse between the sugar and the phosphate so that the DNA chain is broken.

The relationship between DNA alkylation and carcinogenesis (tumour induction) is tenuous though certain alkylations do cause miscoding (i.e. are mutagenic [7]). This is not so for 7-methylguanine but applies to the presence of 3-methylcytosine in the RNA of tobacco mosaic virus [4]. Good alkylating agents, e.g. dimethyl sulphate and methylmethanesulphonate, are poor carcinogens but the reverse is true for N-methyl-N'-nitro-N-nitrosoguanidine (MNNG) (Fig. 7.2). 2-O-Methylated bases may cause miscoding as may also 6-O-methylguanine and 3-N-methyladenine [26, 28, 29].

Mitomycin C inhibits DNA synthesis by causing covalent crosslinking of the complementary DNA strands [10, 30]. The drug is reduced in the cell to produce an active bifunctional alkylating agent which crosslinks guanine residues [31]. It is used to produce a feeder layer of dead cells on which sensitive mammalian cells will grow.

Phleomycin attaches covalently to thymine residues in DNA [10, 32] to inhibit DNA

polymerase action [33]. As with bleomycin, which binds in a similar way, this results in single-strand breaks in the DNA [10, 34].

### 7.2.3 Intercalating agents

The acridine dyes (e.g. proflavine – Fig. 7.3), ethidium bromide and propidium di-iodide are flat molecules which are able to insert themselves between the stacked bases of the DNA double helix. This brings about some unwinding of the DNA duplex which interferes with transcription [36–38]. The unwinding is resisted in closed circular DNA molecules (see Section 2.8) which therefore bind less dye and are therefore considerably denser than nicked cyclic molecules. This forms the basis of the separation of supercoiled molecules from non-supercoiled molecules on gradients of caesium chloride containing the dye [38] (Section A.2.1).

The anthracenes and benzpyrenes (Fig. 7.4) present in soot react with the 2-amino group of guanine prior to intercalating into the DNA duplex. Aflatoxins, which are potent liver carcinogens produced by moulds (*Aspergillus* sp.), also preferentially intercalate next to purines while actinomycins react with deoxyguanosine and intercalate between dG-dC nucleotide pairs. With actinomycin D the cyclic peptide (Fig. 7.4) lies in the groove in the double helix, hydrogen-bonded to the guanine [8]. Actinomycin D inhibits both DNA and RNA polymerases though the former is much less sensitive [39]. At

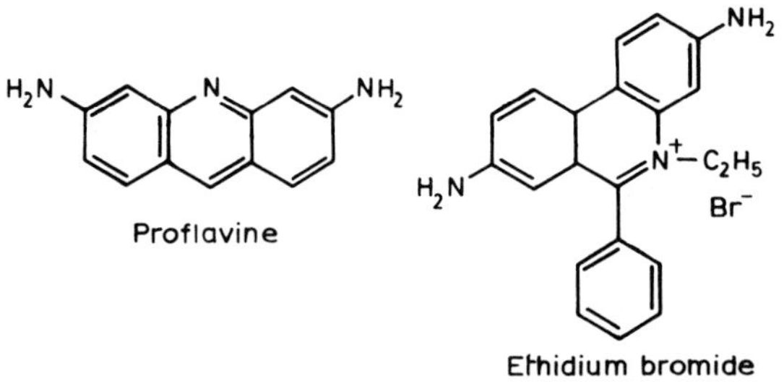

**Fig. 7.3** Structures of proflavine and ethidium bromide

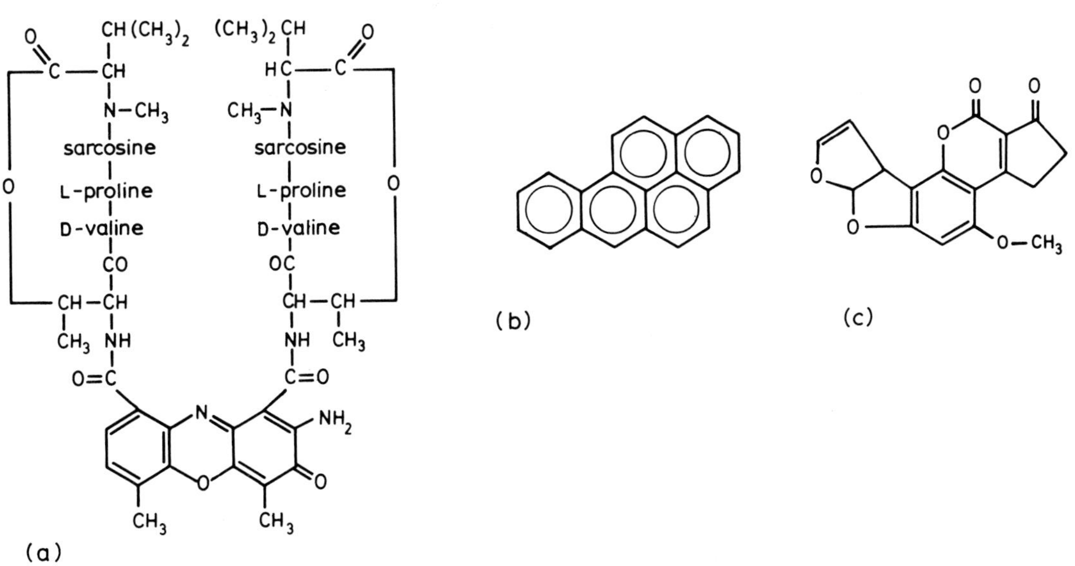

**Fig. 7.4** Structure of (a) actinomycin D; (b) benzo(a)pyrene; (c) aflatoxin B1.

a concentration of actinomycin of 1.0 $\mu$M, for example, the DNA-dependent RNA polymerase is almost completely inhibited whereas the DNA polymerase is only slightly affected [40]. At lower concentrations the synthesis of the different sorts of RNA is affected to different extents, ribosomal RNA being particularly sensitive [41].

Intercalating agents by partially unwinding the DNA double helix can lead to the production of frameshift mutations in which at replication, extra bases are inserted or deleted, opposite to the intercalated molecule. Frameshift mutations can also arise in repeated sequences as a result of misalignment of one of the repeated units with the complement of another unit [42]. They can also arise in palindromic sequences where imperfect alignment occurs [43] (see Section 3.2.5.d).

### 7.2.4 The effects of ionizing radiation

In growing animal and plant cells there is an association between the onset of DNA synthesis and resistance to X-irradiation. Thereafter survival rates remain approximately constant throughout this period and then decline in $G_2$ [44, 45]. Although DNA synthesis is an important factor in X-radiation sensitivity, it is not the only one as is shown by changes in sensitivity which occur following the addition in $G_1$ phase of inhibitors of DNA synthesis [46]. Similar results found with bone marrow cells by Lajtha and his colleagues led them to postulate the existence of a 'system connected with but not identical to DNA synthesis which is more radiosensitive than the process of DNA synthesis' [47], and studies with regenerating rat liver have given strong support for this opinion. The most sensitive period in the cell cycle is the period immediately prior to mitosis. Cells irradiated at this time with as little as 9 rads show a delay in entering mitosis [48].

What may be of importance is the distinction between rapidly growing cells which appear to possess throughout the cell cycle all the enzymes required for DNA synthesis, and non-growing ($G_0$) cells which must synthesize these enzymes when stimulated to grow. Regenerating liver and most primary cell cultures fall into this second category and the initiation of DNA synthesis is very sensitive to X-radiation in such systems.

Single-strand breaks are a major lesion induced into DNA by ionizing radiation and the efficiency of repair depends in part on the nature of the termini produced, e.g. those breaks bounded by a 3'-OH and a 5'-phosphate can readily be rejoined by DNA ligase [49]. A review of the effect on DNA of ionizing radiation has recently appeared [346].

### 7.2.5 Ultraviolet radiation

Large doses of ultraviolet radiation can damage living cells by causing the formation of chemical bonds between adjacent pyrimidine nucleotides in the DNA. Two pyrimidine bases joined in this way in one strand form what is known as a dimer, and of the three possible types of pyrimidine dimer, the thymine dimer is formed most readily (Fig. 7.5). The presence of such dimers

blocks the action of the DNA polymerase and so prevents replication [50]. The major product of ultraviolet irradiation is the cyclobutane: thymine dimer shown in the figure in which adjacent thymine bases are linked at the 5 and 6 atoms. However, it has been shown recently that the mutagenic effect of ultraviolet irradiation may result from a minor product in which the 5'-pyrimidine is linked between its 6 position and the 4 position of the 3'-pyrimidine (the 6:4 photoproduct – Fig. 7.5) [51].

### 7.3 REPAIR MECHANISMS

When DNA is damaged or when mistakes occur in its synthesis, it is important that it is repaired rapidly. Some damage can be directly reversed whereas other types of damage require that the faulty stretch of DNA is removed and replaced. If DNA replication intervenes between damage and repair the consequences are often more extreme.

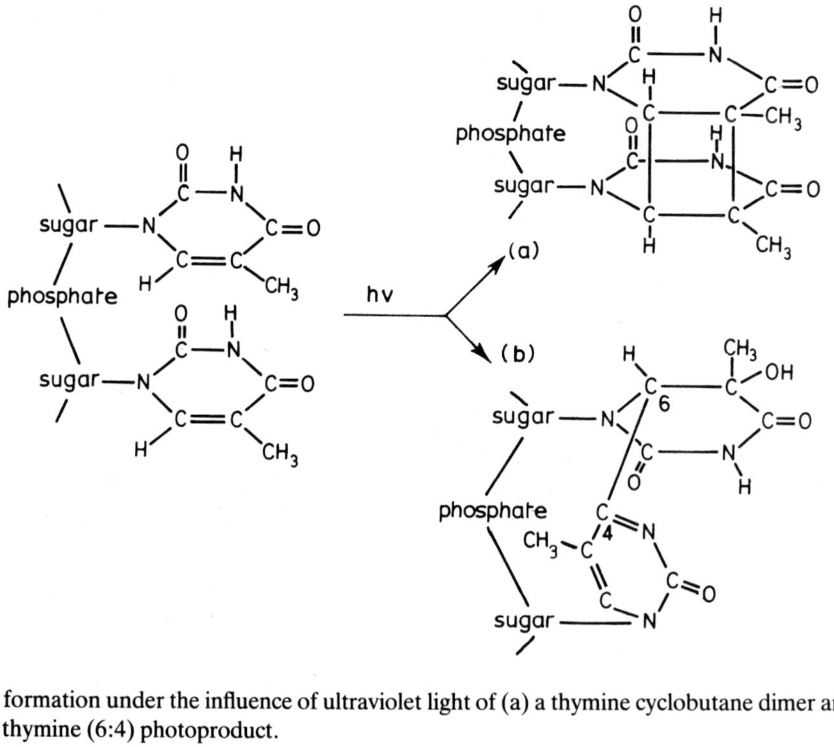

**Fig. 7.5** The formation under the influence of ultraviolet light of (a) a thymine cyclobutane dimer and (b) a thymine:thymine (6:4) photoproduct.

### 7.3.1 Reversal of damage

#### (a) *Photoreactivation*

When bacteria damaged by ultraviolet light are exposed to an intense source of visible light (wavelengths between 320 and 370 nm) a large proportion of the damaged cells recover. This process is known as photoreactivation and is due to the activation by visible light of an enzyme which cleaves the pyrimidine dimers and restores the two bases to their original form. This photoreactivating enzyme (photolyase) has been obtained in pure form [3, 52, 55] and in *E. coli* it is the product of the *phr* gene. It cleaves only the cyclobutane dimers and not the 6:4 adduct [55]. A similar enzyme occurs in eukaryotes [53, 54] and may be deficient in people who suffer from the disease xeroderma pigmentosum characterized by extreme sensitivity to sunlight.

#### (b) *Removal of methyl groups*
An adaptive response to methylation of DNA with MNNG is the appearance of a DNA methyltransferase which transfers methyl groups from $O^6$-methylguanine onto cysteine residues in the protein [3, 28, 56–58, 153]. Each molecule of enzyme acts only once and is consumed in the reaction. The 18 000-molecular-weight protein (the product of the *E. coli ada* gene) also removes other alkyl groups but shows greatest activity with methyl groups. A similar enzyme has been found in mammalian cells [155].

#### (c) *Joining of single-strand breaks*
Although this may occur by action of DNA ligase more complex mechanisms are required following γ-irradiation as 3′-hydroxyls are not available [49].

#### (d) *Purine insertion*
An activity has been found which will insert purines into apurinic sites. The enzyme does not work at random but re-forms the correct base-pairing [69, 70, 155]

### 7.3.2 Excision repair

This more complex repair mechanism acts by excising the damaged section of DNA and then repairing the resultant gaps. The reaction takes place in a number of separate stages of which the first is unique to *repair* of DNA (Fig. 7.6). (i) An endonuclease recognizes the local distortion and breaks the adjoining phosphodiester bond so as to introduce a nick, usually on the 5′-side of the damage with a 3′-hydroxyl terminus at the nick. The nature of the endonuclease will be considered in more detail below. (ii) A second enzyme excises a short stretch of the DNA strand including the damaged region. (iii) DNA polymerase uses the intact complementary strand as template to synthesize a piece of DNA to fill the gap. (iv) The repair is completed by ligase action [50, 59–61].

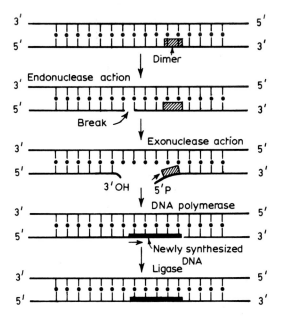

**Fig. 7.6** The repair of DNA damaged by ultraviolet light. The action of uv has been to produce a dimer which is excised by the sequential action of an endonuclease and exonuclease, leaving a short gap which is filled in by the action of DNA polymerase. The final join is effected by polynucleotide ligase.

In *E. coli* DNA polymerase I may be the enzyme which first excises the portion of the affected strand (including, on average, 30 nucleotides [62]) by virtue of its 5′ → 3′ nuclease action (Section 6.4.2.b) and then fills the gap [63, 64]. *polA*⁻ mutants show increased uv sensitivity [65, 79] and, although they do manage to repair lesions, the patch size is considerably bigger than in the wild-type [65–71]. This may be because DNA polymerases II and III take over the role of polymerase I or that repair is occurring by a recombination mechanism (see below) [68].

The endonucleolytic step may be performed by one of a number of different enzymes depending on the nature of the damage. Most commonly the damaged base is first recognized by a glycosylase which removes the base from the sugar phosphate backbone [71–78, 153–155]. Glycosylases are a family of small proteins (less than 30 000 molecular weight), each one specific for a different unusual base. Glycosylases have been isolated which will remove uracil,

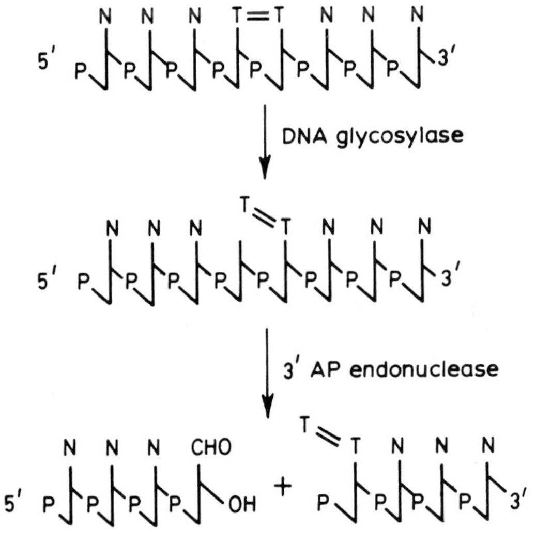

**Fig. 7.7** The combined glycosylase and AP endonuclease action of the uv endonuclease of *M. luteus* and phage T4.

hypoxanthine, 3-methyladenine, 7-methylguanine, 3-methylguanine, formamidopyrimidine, 5:6 hydrated thymine, urea or pyrimidine dimers. The role of *N*-glycosylases has been reviewed by Lindahl [86]. Following glycosylase action the DNA is cleaved by an AP endonuclease (apurinic/apyrimidinic endonuclease). Although cleavage usually occurs 5′ to the AP site there are AP endonucleases which can cleave 3′ to the AP site and the combined action of the AP endonuclease and subsequent 3′ → 5′ or 5′ → 3′ exonuclease step may determine the size of the patch produced – about 20–30 nucleotides long [3]. Some glycosylases appear to have intrinsic endonuclease activity. The *M. luteus* and the phage T4 uv endonucleases first cleave the *N*-glycosyl bond between the 5′-pyrimidine of a pyrimidine dimer and its sugar and then cleave the phosphodiester bond 3′ to the AP site leaving an aldehyde sugar [51] (Fig. 7.7). Endonuclease III of *E. coli* is a similar enzyme which recognizes reduced thymine rings in DNA and removes them at the same time as cleaving the phosphodiester backbone [80, 81].

The major AP endonuclease of *E. coli* is that encoded in the *uvrA*, *uvrB* and *uvrC* genes, which have been cloned and sequenced and their products purified [3, 82–84]. The *uvrA* protein (molecular weight 114 000) binds to damaged DNA and seeds the binding of the *uvrB* protein (molecular weight 84 000) and the *uvrC* protein (molecular weight 70 000). Nicks are made on *both* sides of the damaged region 12 to 13 nucleotides apart [85]. A 3′-hydroxyl is produced 5′ to the damaged region and can serve as a primer for DNA polymerase which produces a patch only 12 to 13 nucleotides long. Helicase II (the product of the *uvrD* gene) is required to unwind the damaged region of DNA and release the *uvrC* protein [188, 342]. This mechanism may be particularly appropriate for bulky adducts as the cleavage sites are a short distance away from the distortion. *E. coli* AP endonucleases are listed in Table 4.3.

The mechanism of repair synthesis is defective

in the skin fibroblasts of some patients suffering from the condition known as xeroderma pigmentosum. One of the repair endonucleases is missing and such people are therefore abnormally sensitive to exposure to sunlight and tend readily to develop skin cancer [3, 53, 87–90, 155].

### 7.3.3 Mismatch repair

There are many *E. coli* mutants known which show an increased frequency of mutations. Some of these have already been considered under proof-reading (Sections 6.4.2.b and 6.5). Others are involved in mismatch correction. If a section of DNA duplex contains a mismatch but both bases are normal (e.g. a T opposite a C) some signal must exist if the cell is to be able to correct the damage. Most mismatches arise as a result of the failure of the proof-reading mechanisms and hence the base in the daughter strand is the incorrect one. It has been suggested that one of the functions of DNA methylation (see Section 4.6.1) is to define which is the parental strand. Methylation occurs shortly after synthesis of DNA but for a finite time the daughter strand is unmethylated and could be subject to mismatch repair. *Dam* mutants which have a much reduced level of DNA adenine methylation show increased rates of mutation [19, 91–94] and the repair involves the products of the *uvrD*, *uvrE*, *mutH*, *mutL*, *mutR* and *mutS* genes. In *dam⁻* cells neither strand is methylated and the repair system cannot discriminate between parental and daughter strands and so repair occurs at random. However, in for example *mut⁻* cells the repair system is absent and mismatches are not removed which again leads to a high rate of spontaneous mutagenesis.

Mismatches can also arise by incorrect alignment of the DNA strands where several repeating units are arranged in tandem or when imperfect palindromic sequences form cruciform structures. These often result in repair reactions leading to deletion or insertion of one or two nucleotides [43].

### 7.3.4 Post-replication repair

If repair is quickly and accurately effected before the damaged DNA is replicated the integrity of the genetic message is maintained. Such excision repair is very effective, of high fidelity and leads to the insertion of only small repair patches. If excision repair is slow or defective, as in *uvr⁻* cells or in patients with xeroderma pigmentosum or actinic keratosis [154] then the cell tries to replicate the damaged region of DNA. When the advancing polymerase reaches a lesion such as a thymine dimer it stops and a gap is left in the daughter strand opposite the lesion. The gap may be filled by one of two mechanisms: error-prone repair or recombination.

#### (a) *Error-prone repair*

In *E. coli* the presence of single-stranded regions produced in DNA as a result of damage induces the so-called SOS functions [54, 96, 97]. The *recA* protein binds to the single-stranded regions and is activated to form a proteolytic enzyme which causes the cleavage of *lexA* protein, the common repressor of at least 17 genes. Only following repair of the damage is the *recA* protease inactivated and the level of *lexA* protein allowed to build up to again repress this pleiotropic effect. Among the proteins under *recA*/*lexA* control is the phage lambda repressor (explaining the induction of prophage by agents which damage DNA – see Section 3.6.8) and the *uvrA*, *uvrB* and *uvrC* genes. In addition the *mutD* gene (*dnaQ* gene – see Section 6.4.2d) product is inactivated. This is the $\epsilon$ subunit of DNA polymerase III responsible for the $3' \rightarrow 5'$ proof-reading exonuclease activity. As a consequence the polymerase shows decreased fidelity allowing nucleotide incorporation across the gaps opposite to damaged regions (translesion replication). A modified error-prone form of DNA polymerase I is also formed in cells showing the SOS response [96]. This decreased fidelity is not, however, restricted to the regions of DNA actually damaged but occurs generally

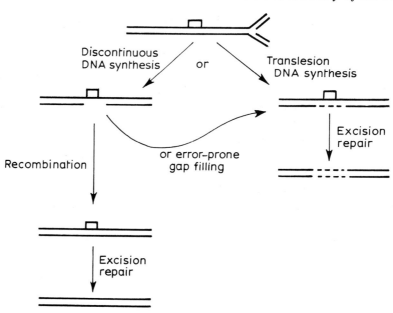

**Fig. 7.8**   Different repair processes occurring following replication of a region of damaged DNA.

within the cell [3, 53, 54, 91] (Fig. 7.8). This may be the origin of many of the mutations arising as a result of damage to DNA.

### (b) *Repair by recombination*

The gap produced by discontinuous replication (Fig. 7.8) may also be filled by *recA* protein-mediated recombination. The details of recombination are considered in the next section but in this type of repair a preformed patch supplied by another copy of the DNA is used to replace the defective segment [3, 98].

Other mechanisms for DNA repair have also been postulated, e.g. repair by strand displacement and branch migration at the replication fork so that the intact daughter strand may act as a template to bridge the gap opposite the damage [99].

### 7.4 RECOMBINATION

Recombination is the production of new DNA molecule(s) from two parental DNA molecules,

such that the new DNA molecules carry genetic information derived from both parental DNA molecules. Recombination involves a physical rearrangement of the parental DNA and may occur during (1) a mixed infection of two viruses or plasmids carrying different genetic markers, (2) crossing over at meiosis in eukaryotic cells, (3) integration of bacteriophage, viral or plasmid DNA into a host cell chromosome, (4) transformation, (5) gene conversion, (6) postreplication repair or (7) a variety of other processes considered later in this chapter.

In higher eukaryotes recombination between chromosomes leads to chiasma formation at meiosis in which two, non-sister chromatids have become linked together at corresponding sequences. Each chromatid is a duplex DNA molecule (with associated protein) and following replication a chromosome consists of two sister chromatids. At chromosome pairing in meiosis the homologous chromosomes become associated to form a synaptonemal complex and it is within this complex that crossing over occurs.

However, there is a conceptual gap between what is observed to occur between these huge nucleoprotein complexes and what is presumed to occur at the DNA level.

Usually recombination occurs between any regions of DNA which are wholly or largely complementary in sequence. Occasionally so-called illegitimate recombination occurs where very little complementarity exists between the incoming DNA and the recipient genome. In site-specific recombination (see Section 7.4.2) complementarity exists over only a specific limited region but the reaction is catalysed by enzymes which recognize this complementary site.

### 7.4.1 *E. coli rec* **system and single-strand invasion**

General recombination [100] can take place anywhere along the length of two complementary DNA molecules. It has been studied in *E. coli* infected with two bacteriophages carrying different genetic markers when multiple recombination events may occur. In *E. coli* general recombination is dependent on the host *rec* system as well as other enzymes and proteins involved in DNA replication.

*Rec BC* nuclease (exonuclease V, the product of genes *recB* and *recC*) has multiple enzymic activities. Commencing at an end of a duplex DNA molecule it can travel along the DNA at about 300 bp per second unwinding the DNA. Rewinding occurs behind the enzyme but single-stranded loops form transiently and these are sensitive to exo- and endo-nuclease cleavage (101–107). Endonucleolytic cleavage is greatly enhanced when the travelling *rec BC* nuclease approaches a Chi site [114]. The Chi sites are therefore recombination hotspots. The Chi site has the sequence 5'GCTGGTGG3' and the *recBC* nuclease, as it approaches from the right, cleaves the single-stranded loop of DNA (produced by its unwinding action) four to six nucleotides on the 3'-side of Chi [105–107].

This leads to the production of a single-stranded tail of DNA.

The *recA* protein, among other functions, is able to catalyse the annealing of single-stranded DNA to a homologous duplex at the expense of ATP hydrolysis [108–113]. The single strands may arise as a result of *rec BC* nuclease action or by a variety of other mechanisms. Thus *recA*-mediated recombination is involved in the integration of the single-stranded DNA arising from conjugal transfer (see Section 3.7) or bacterial transformation (see Section 3.5.3). Alternatively single strands may be produced during replication of damaged DNA or by a strand-displacement mechanism acting from a nick in duplex DNA (see Fig. 7.9). Double-stranded breaks may produce DNA molecules with overlapping (sticky) ends [116–118] and sticky ends may be important as sites of secondary initiation of replication of phage T4 DNA (see Section 6.9.2.c). High rates of recombination are found in mutants defective in ligase, DNA polymerase I or dUTPase [115], situations which lead to an increased frequency of nicks or gaps in DNA.

*RecA* protein is thought to initiate its action by binding co-operatively to duplex DNA, in the presence of ATP, leading to the production of underwound DNA with about 18.6 bp per turn of the helix [110, 113, 119–121].

*RecA* protein also binds very firmly to single-stranded DNA, one molecule binding per five nucleotides [112, 123, 124]. These binding events promote aggression by the single-stranded DNA leading to strand displacement from the duplex to form a D-loop [100–122] (see Fig. 7.9). The initial complex (pre-D-loop) does not involve hydrogen-bonding of DNA strands nor ATP hydrolysis [112]. However, ATP hydrolysis is required for the branch migration required to align regions of complementarity [128].

Using a cyclic single-stranded DNA molecule and linear duplex DNA, recombination produces a region of cyclic duplex by removing the complementary strand from the linear molecule.

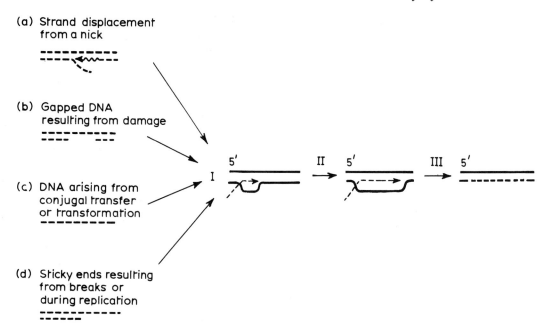

**Fig. 7.9**   Recombination initiated by single stranded regions of DNA: I, *Rec A*-mediated single-strand aggression leading to formation of D loops; II, Branch migration over region of homology; III, Nuclease and ligase action completing formation of recombinant DNA molecule. This series of reactions is believed to take place within a sheath of *recA* protein filaments [121].

This occurs much more readily starting at the end of the duplex as topological problems are not involved. Annealing is initiated with the 3'-end of the complementary strand, i.e. the single strand invades with 5' → 3' polarity [125–127]. Strand transfer is blocked when a region of non-homology arises and three mismatches in a row are sufficient [129].

*Rec A* protein is also able to catalyse reciprocal exchange of DNA strands between two duplex molecules [130, 131] and this gives substance to the models devised by Holliday [132] and Meselson and Radding [133] to explain reciprocal recombination (see below).

The *red* genes of bacteriophage λ code for a 5' → 3' exonuclease (*redα*) and the β-protein (*redβ*) of unknown function [134]. The exonuclease can act on redundant single strands but

acts preferentially on 5'-phosphate ends of duplex DNA to produce molecules with 3' single-stranded tails. Annealing of homologous single-stranded regions followed by exonuclease action leads to formation of a hybrid molecule by a non-reciprocal method, i.e. only one complete DNA molecule results (Fig. 7.10). If the tails are not removed by exonuclease [135] but endonuclease and polymerase intervene (as in bacteriophage T7 replication, see Section 6.10.2) a reciprocal recombination can occur [136].

### 7.4.2   Reciprocal recombination between duplex DNA molecules

The original Holliday model [132] was proposed to explain genetic results obtained in fungi but

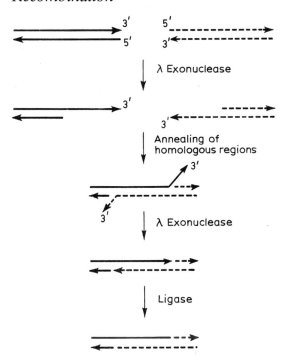

**Fig. 7.10** Action of phage lambda exonuclease to produce 'lap-jointed' molecules.

applies equally well to *rec*-mediated recombination in *E. coli*. Indeed it is this latter system which has provided the experimental evidence to support the model.

Holliday proposed that recombination between homologous duplex molecules is initiated by endonuclease nicking at corresponding regions of homologous strands of the paired duplexes (Fig. 7.11b). Reciprocal strand invasion (Fig. 7.11c) produces a joint molecule in which each duplex has a heteroduplex region derived from one strand from each parental duplex. Ligase action generates stable intact duplexes (Fig. 7.11d). Branch migration (which requires topoisomerase action) leads to the formation of extensive heteroduplex regions (Fig. 7.11e). The joint molecule is known as a Holliday or Chi($\chi$) structure. Its resolution requires two nicks which may occur in one of two ways leading to either a short length of heteroduplex DNA in the other-

wise undisturbed parental molecules (Fig. 7.11g) or the production of duplex molecules derived from different halves of the two parental molecules (Fig. 7.11h).

A variation on the Holliday model was proposed by Meselson and Radding [133] to explain formation of heterozygous regions which are not perfectly reciprocal. They suggested that recombination is initiated by a nick in only one of the parental duplexes followed later by the second nick and ligase action to generate Chi structures. Potter and Dressler [131, 137] have suggested that no nicking is required if strand invasions can occur at transiently single-stranded regions. More recently as a result of studies with the RAD 52 mutant in yeast, Szostak has proposed that recombination is initiated with a double-stranded break in one of the aligned duplexes [117, 118]. Nucleases extend the break to form a gap with 3'-extensions which can invade the homologous region on the other parental duplex. Repair synthesis leads to production of two Holliday junctions which can be resolved in two different ways.

Chi structures can be visualized in the electron microscope and so provide proof of the existence of joint molecules [131]. When the two parental molecules are cyclic duplexes then the joint molecule is a figure 8 which, on resolution, will give either two cyclic molecules each with a heteroduplex region or alternatively a double-length cyclic molecule (the cointegrate) made up from both parental molecules. Both the figure 8 and the double-length cyclic molecules can be seen in electron micrographs [138, 139].

Phage-encoded endonucleases (e.g. T4 endonuclease VII) have been isolated which will resolve Chi structures *in vitro* [140] and they will also cleave cruciform structures which are formally analogous [141]. T4 endonuclease VII acts exclusively at the junction point.

The 1984 *Cold Spring Harbor Symposium on Quantitative Biology* was devoted to papers on recombination and gives a far more detailed picture than can be summarized here.

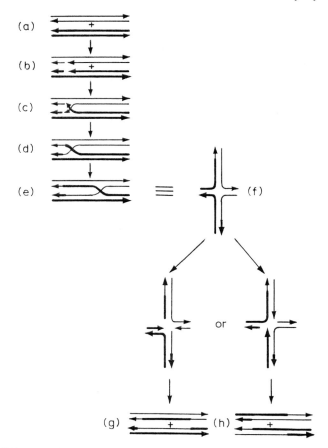

**Fig. 7.11** The Holliday model of recombination. See text for details. Structures e and f are equivalent.

### 7.4.3 Site-specific recombination

Whereas general recombination can occur between any homologous sequences, the enzymes which catalyse site-specific recombination act only at specific short sequences which must be present in both parental duplex DNA molecules. Site-specific recombination has been studied primarily in the lambdoid phages but it also occurs in other systems, both prokaryotic and eukaryotic [142, 143, 340].

In the lysogenic state phage lambda is integrated at a specific site in the bacterial chromosome (see Section 3.6.8). Integration does not require the *rec* or *red* systems of general recombination but does require specific sequences on both the phage and bacterial DNA (the *att* sites) as well as phage-coded proteins. Phage $\phi$80 integrates into the *E. coli* chromosome at the site *att*80 (near *lac*) and phage λ at the site *att*λ (between *gal* and *bio*). The site on the phage DNA and the site on the bacterial DNA are known as POP' (*attP*) and BOB' (*attB*) respectively. Both are made of three parts. The O (or core) region is a sequence of 15 bases which is identical in the two regions. During recombination staggered cuts are made in the two core regions to give a seven-base 3' extension. These then cross-hybridize so that recombination results in formation of BOP' (*attL*) and POB' (*attR*) (140, 145–147) (Fig. 7.12 and 7.13).

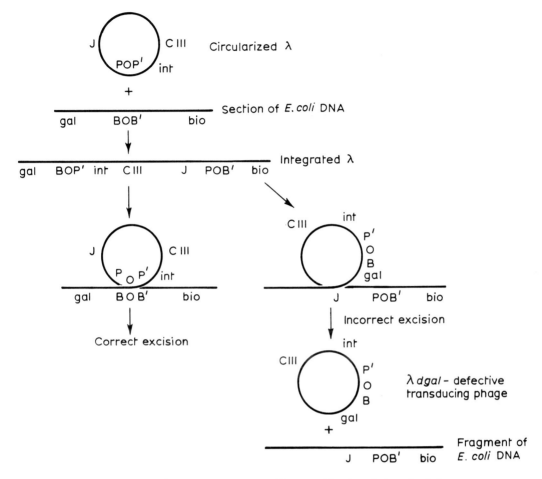

**Fig. 7.12** Campbell model for site specific recombination of phage lambda.

-P- G C T T T T T T A T A C T A A -P'-
      C G A A A A A A T A T G A T T

-B- G C T T T T T T A T A C T A A -B'-
      C G A A A A A A T A T G A T T

↓

-B- G C T T T T T T A T A C T A A -P'- λ -P- G C T T T T T T A T A C T A A -B'-
      C G A A A A A A T A T G A T T           C G A A A A A A T A T G A T T

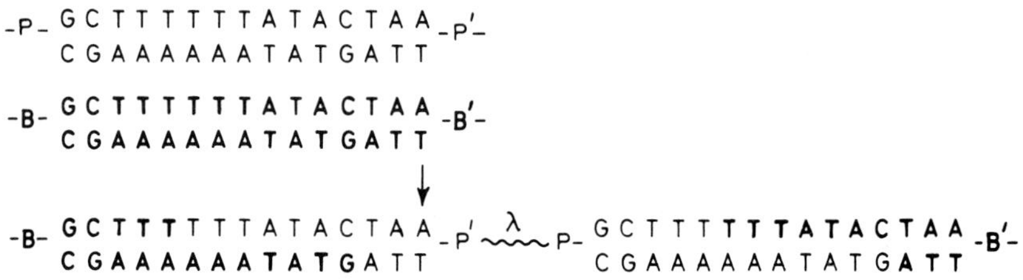

**Fig. 7.13** Details of the cleavage and rejoining of the core regions which occur on site specific recombination of phage lambda.

structure requires the phage-coded Int protein and the host-coded IHF (integration host factor). The Int protein is a 42 000-molecular-weight topoisomerase I which binds to four sites in POP'; two in the 152-bp P region; one in O; and one in the P' region [148–152]. It also binds to the core region on *attB*. *In vitro* Int cleaves the DNA at the positions indicated on Fig. 7.13 and becomes covalently bound to the 3'-phosphate of the DNA at the site of breakage. IHF is a DNA-binding protein which is required to form the Holliday-type intermediates but not for their resolution.

Excision of phage DNA requires a second phage-coded protein (Xis) which serves as a directional switch by inhibiting the integration and allowing only the excisional recombination event [144].

Incorrect excision leads to defective bacteriophage carrying genetic markers adjacent to *att*, i.e. for bacteriophage λ either *gal* or *bio*. Such defective transducing bacteriophage, e.g. λ *dgal* can integrate into *rec*$^+$ cells at a *gal* site and have proved very useful to viral geneticists.

## 7.5 GENE AMPLIFICATION [161]

The selective replication of specific sequences of DNA is known as amplification. It can occur as part of the developmental programme of an organism or in response to treatment with a drug or similar selection agent.

### 7.5.1 Developmental amplification

During oogenesis in *Xenopus* the ribosomal RNA genes are amplified many thousandfold such that the ribosomal DNA content of an oocyte may be as much as 2500 times the normal haploid value [156]. Amplification occurs in two phases. The first phase results in the 40-fold amplification of *selected* ribosomal repeat units and occurs in both sexes when only 9 to 16 primordial germ cells are present [157]. In the male these amplified sequences are lost at

meiosis but in the female oocyte there is a further amplification presumably to satisfy the special demand of the oocyte for large amounts of ribosomal RNA [156–158]. The amplified ribosomal DNA is produced in the form of extrachromosomal circles each containing several copies of the ribosomal DNA repeat and each capable of autonomous replication [159, 160].

Ribosomal DNA is also amplified in *Tetrahymena* but in this case the amplified unit consists of two copies of rDNA joined head to head to form an inverted repeat bounded by from 20 to 70 copies of the telomeric sequence 5'CCCCAA3' [167, 168] (see Section 6.10.3). Ciliates contain two nuclei; a macronucleus which contains a small selection of genes which are transcriptionally active, and a micronucleus or germinal nucleus which gives rise to both nuclei in the next generation. The diploid micronucleus contains a single integrated rDNA gene per haploid genome. For the formation of the macronucleus, chromosome breakage occurs at specific sites on each side of the rDNA to produce a linear duplex containing a short inverted repeat at the 5'-end of the molecule. By intramolecular rearrangement this repeat could act as a primer to produce the double-length molecule to the ends of which are added multiple copies of the CCCCAA sequence [167]. These minichromosomes resemble others found in the macronucleus of ciliates which contains about 1000 copies of each of 24 000 different linear molecules bounded by telomeric repeats [169].

A similar selective amplification occurs to the chorion genes during *Drosophila* development. These genes which code for egg-shell proteins are amplified 16-fold in follicle cells. However, the amplified genes are not present as extrachromosomal circles but arise by multiple initiation of replication of specific origin regions within the gene [161–165]. The actual region amplified (80–100 kbp) extends well beyond the centrally located chorion genes but it is the central region which shows the highest degree of amplification.

A short segment of the amplified sequence acts in *cis* to control amplification [164]. This process of chorion gene amplification may be typical of a number of insect genes which can be seen to 'puff' in the polytene chromosomes of larval salivary glands under the influence of the steroid hormone ecdysone [165, 166].

### 7.5.2 Amplification by chemical selection

When undergoing selection for drug resistance, animal cells in culture may respond by over-producing a gene product. For example, cells treated with methotrexate, an inhibitor of di-hydrofolate reductase (DHFR) (see Section 5.6), may respond by overproducing the in-hibited enzyme. This is often a result of selective amplification of the gene coding for dihydro-folate reductase (*dhfr*) [161, 170–175]. Corresponding results are found with other drugs but it is not always the gene for the target enzyme which is amplified and chromosomal location markedly affects the frequency of amplification (183–185). Similar amplification probably occurs in whole animals and has also been reported in *Leishmania* and bacteria [176, 177]. The mechanisms which give rise to the amplifications are active in the absence of the drug which acts as a selecting agent. Thus amplification of chromosomal segments is occurring spontaneously (see below) and does not simply involve one gene but rather includes a region up to 1000 kbp in size. Under selective pressure this whole region may be amplified as many as a thousandfold when it can exist in one of two forms: (a) as a homogeneously staining region (hsr) integrated into a chromosome when it remains stable for more than 12 months in the absence of the drug; or (b) as small acentric chromosome fragments (known as double min-utes) which are rapidly lost (within two weeks) following removal of the drug. In the first case multiple rearrangements and translocations occur and the hsr is found at variable chromo-somal locations.

### 7.5.3 Mechanism of amplification

The frequency of spontaneous two-fold ampli-fication of a gene has been estimated at $1 \times 10^{-3}$ per cell generation [178] and this rate can be increased by a number of factors. For instance the *dhfr* gene is replicated early in S phase but replication can be reinitiated if cells are treated with hydroxyurea (see Section 5.5) after 2 h of S phase [171, 179, 180]. This suggests that amplification normally results from rereplication of variable regions of the genome. If selection pressure is applied the amplification becomes apparent.

Repeated initiation of replication from an origin – so-called 'onion skin' replication – is also believed to be the mechanism whereby copies of integrated viral DNA are produced linked to adjacent host sequences when transformed cells are induced [181, 182] (Fig. 7.14). Develop-mentally programmed reinitiation may be the first stage in the amplification of chorion and ribosomal RNA genes.

Once produced, extra copies of DNA regions undergo rearrangement and recombination events *via* short repeated sequences in a manner similar to DNA transfected into cells [161, 162, 171, 185–187] (see Section A.8.4). They may form multiple-length structures and also cyclic forms which may replicate autonomously pos-sibly by a rolling-circle mechanism [159]. Reintegration into the chromosome followed by further recombination events will lead to homo-geneously staining regions containing multiple tandemly arranged copies of the gene region. Similar duplicative recombination events are believed to have been involved in the production of satellite DNA (see Section 3.2.5) and in the production of the minisatellites which are a major source of the heterozygosity found in human DNA [35]. These minisatellites are tandem repeats of short (e.g. 33 bp) sequences each one of which contains a core region of consensus sequence 5'GGGCAGGARG3' which is believed to be analogous to the

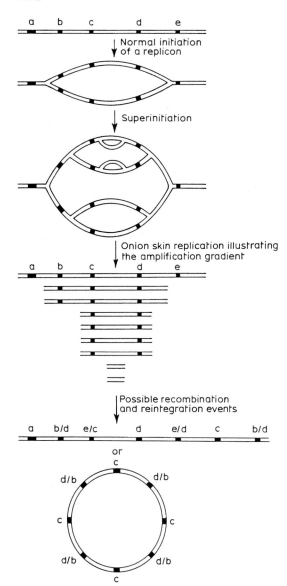

**Fig. 7.14** Amplification by means of onion skin replication and recombination.

Chi sequence of *E. coli*; i.e. it is the recognition signal for a 'recombinase'. Frequent recombinations result in different people having differing numbers of copies of the minisatellite in a repeat and this is the source of a heterozygosity which can be used for pedigree analysis and paternity testing [95].

## 7.6 GENE DUPLICATION AND PSEUDOGENES

### 7.6.1 Multiple related copies of eukaryotic genes

The amplification systems discussed in Section 7.5 above are examples of the rapid, sizeable and, often transient, increase in number of copies of a particular gene in cells during development or under selective pressure. However, hybridization experiments have shown that for many eukaryotic protein-encoding genes there exist a more limited number of relatively stable related copies; and in the case of the genes encoding the stable RNAs a large number of non-transient related copies occur (see Sections 9.4 and 9.5). The mechanism by which some of these related sequences have arisen involves a retroviral-like transposition, which results in the non-functional 'processed pseudogenes' described in Section 7.7.3 below. In this section, however, we are concerned with related sequences that are thought to have arisen by the tandem duplication of an original sequence. Such gene duplication is important in evolutionary terms, as it provides the potential for one copy to diverge while the other still retains its original function. Nevertheless, in some cases the products of such duplication have been maintained as identical or near-identical copies (e.g. members of the histone gene cluster – see Section 9.2.2), whereas in others they have diverged to produce different isoforms of the same protein (e.g. the tissue-specific actin and myosin isoforms [189]). In some instances much greater divergence occurs, and this may result in the relationship between the genes being undetectable by DNA hybridization. Such distant relationships have been revealed by computer-assisted comparison of nucleotide or protein sequences, an extreme example being the genes coding for epidermal growth factor and its receptor [190]. Divergence does not, however, always result in new functional genes, but can instead give rise to pseudogenes (see Section 7.6.3 below).

### 7.6.2 Mechanism of tandem gene duplication

A gene family which provides clear evidence of repeated duplication events is the human globin gene family, which is described in greater detail in Section 9.2.2. Extensive nucleotide sequence analysis, together with a knowledge of the number and organization of globin genes in other species, indicates the evolutionary relationships shown in Fig. 7.15 [191, 192]. It is thought that an initial duplication gave rise to the prototypes of two families, the alpha and beta families. Although in many species these are now on separate chromosomes, the result, it is presumed, of a translocation event, their common origin is indicated by the fact that in amphibians they are still linked on the same chromosome [193]. The human $\alpha$-globin family (on chromosome 16) contains $\alpha$ and $\zeta$ genes, and the $\beta$-globin family (on chromosome 11) contains $\beta$, $\gamma$, $\delta$ and $\epsilon$ genes. There are two similar functional forms of the $\alpha$- and $\gamma$-genes (see Fig. 7.15) and, in addition, certain non-functional pseudogenes, not shown in Fig. 7.15 (but see Fig. 9.7). These latter will be discussed in Section 7.6.3 below.

The best understood mechanism in which tandem copies of a gene can be generated is homologous reciprocal recombination at meiosis (see Section 7.4.2), but involving *unequal crossing-over*. This is most easily envisaged in a situation where two similar tandem copies of the gene already exist, in which case one of the recombinant chromosomes produced will have three copies of the gene (Fig. 7.16). Evidence that such unequal crossing-over can occur among globin genes is provided by the analysis of certain haemoglobinopathies [194]. In haemoglobin anti-Lepore an extra gene has arisen between the $\delta$ and $\beta$ genes and this has a $\beta/\delta$ fusion structure, as would be predicted for such a crossing-over mechanism (cf. Fig. 7.16 where the genes A and A′ would be formally equivalent to $\delta$ and $\beta$ respectively). The other predicted possible result of unequal crossing-over, contraction of gene number, is illustrated in haemoglobin Lepore, where the $\beta$ and $\delta$ genes have been replaced by a $\delta/\beta$ fusion gene. Another, more ancient, example of such contraction is the fusion of the $\psi\beta$ and $\delta$ genes (see Fig. 9.7) in lemur [195].

The model for gene duplication by unequal crossing-over presented in Fig. 7.16 cannot, as such, be applied to the duplication of a single gene (e.g. the original ancestral globin gene in Fig. 7.15). One possibility is that a single gene might undergo 'incidental' duplication when it becomes flanked by similar copies of repeated DNA, which then undergo recombination with unequal crossing-over. Among the candidates for repeated DNA that might arise by chance in such a flanking position are the highly repetitive eukaryotic LINES and SINES (see Sections 3.2.5 and 7.7.3). This is because their mode of duplication and dispersal is independent of recombination, being thought to involve retroviral-like reintegration of DNA copies of RNA transcripts (see Section 7.7.3). (Such a mechanism is excluded for the duplication of the protein-coding genes themselves as it leads to the formation of the processed pseudogenes described in Section 7.7.3.)

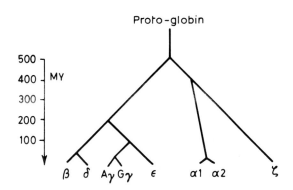

**Fig. 7.15** An evolutionary tree of human globins. MY = million years. After [210], with permission.

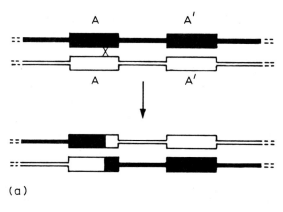

(a)

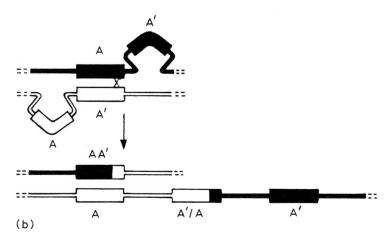

(b)

**Fig. 7.16**   Increase in gene number by homologous recombination involving unequal crossing over. The diagram shows a representation of two tandem copies, A and A', of a gene (or region of DNA) on two recombining non-sister chromatids (distinguished by light and dark shading) at meiosis. Reciprocal crossing-over on copy A of properly aligned chromatids (a) does not affect gene number, whereas unequal crossing over between A and A' on misaligned chromatids (b) generates different products containing one and three copies of the gene, respectively.

The arrangement and copy number of the predicted products of homologous recombination – the three copies of the target DNA and the two copies of the unique DNA between the original double-copy target – should allow one to determine whether such events were responsible for a given gene duplication. Except in the case of relatively recent events, however, the accumulation of mutations in such non-functional spacer DNA, and the occurrence of gene correction or conversion events (see Section 7.6.4), generally obscure the picture. Nevertheless it is still of interest to examine the structures of the more recently duplicated globin genes. In the case of the human $^G\gamma/^A\gamma$, $\delta/\beta$, $\psi\zeta_1/\zeta_2$ and $\alpha_1/\alpha_2$ pairs (Fig. 7.15 and Section 9.2.2), duplicated regions of approx. 5, 7, 12 and 4 kb DNA are still evident [192, 196–198]. These compare with approx. 2 kb for the length of the globin gene including introns and the 5' and 3' untranslated regions. In the case of the $^G\gamma/^A\gamma$ pair, Maniatis and coworkers found a triply repeated sequence of 110–170 bp that could theoretically have flanked a single proto-$\gamma$ gene to produce the duplication [197]. Nevertheless these workers were not prepared to conclude that this repeated sequence was the cause of the $\gamma$-globin duplication, as they were able to construct alternative models of gene duplication via non-homologous recombination that could generate such triply repeated DNA sequences from an original single sequence [197]. In the case of other duplicated genes, such as the human $\alpha_1/\alpha_2$ globin pair [199] or the human haptoglobin pair [200], no such triply repeated sequences are apparent.

Although this section is concerned with duplication events involving several kb of DNA, it should be mentioned in passing that the duplication of much smaller stretches of DNA can occur giving rise to internal repeated units within genes, and to simple tandemly repeated DNA such as satellite DNA [201] (see Fig. 3.6).

### 7.6.3 Pseudogenes

Whereas certain mutations might lead to new and useful functions in a duplicated gene, it is clear that others could be so deleterious as to render the gene functionally useless. Examples of such genes are termed *pseudogenes* [202], and they occur in several instances in the globin gene family [203] where they are designated by the greek letter $\psi$ preceding the active counterpart (Fig. 9.6). For example, the human $\alpha$-globin pseudogene, $\psi\alpha_1$, contains both a mutation in the initiation codon and frameshift deletions that render it incapable of producing a globin product [199].

The evolutionary history of globin genes can be deduced in two different ways, both of which provide evidence that contemporary pseudogenes have been active for part of their lives. One method involves comparison of mutations in the coding region of the gene at silent sites (i.e.

those that would not produce a change in the encoded amino acid) and at replacement sites (those that would produce such a change). For an inactive gene the frequency of mutations at these sites should be equal, whereas in an active gene mutations at replacement sites are selected against. By these criteria $\psi\alpha_1$ is judged to have been active for part of its history [199]. The other method involves phylogenic comparisons. For example, the goat equivalent of the human pseudogene $\psi\beta_1$ (now more generally referred to as $\psi\eta$) is still functional [195], implying that the human $\psi\beta_1$ is the descendant of a relatively ancient active gene. Conversely, although the $\delta$ globin gene is still active in man, it has become a pseudogene in the Old World monkeys [198]. It is interesting that the relatively recent silencing of this gene has allowed the mutation responsible to be identified as involving the region of DNA 5' to the site for initiation of transcription (see Chapter 10).

Finally it should be pointed out that such pseudogenes are not only found as counterparts to genes that encode protein products, but in fact were originally discovered for ribosomal 5S RNA genes in *X. laevis* [204] (see Fig. 9.22).

### 7.6.4 Concerted evolution of duplicated genes

After gene duplication has occurred it might be expected that the two copies would accumulate mutations independently and thus diverge. Although this has, of course, happened in many instances, in others (e.g. the repeated histone gene cluster and the repeated 5S RNA genes of *X. laevis*) the different copies have been maintained in a remarkably similar state. In order to explain such concerted evolution it is necessary to postulate that a mechanism or mechanisms exist to correct divergences as they occur. Once again the intensely studied globin gene family provides an opportunity to explore this problem. Thus there is clear evidence for the operation of such a correction mechanism in the case of the human $\alpha_1/\alpha_2$ [205] and $^G\gamma/^A\gamma$ [197]

globin pairs. This evidence is the discrepancy between the time of the gene duplication based on phylogenic comparison and that deduced from the apparent rate of accumulation of mutations in amino acid or nucleotide.

One proposal is that a series of unequal homologous crossing-over events of the type depicted in Fig. 7.16 can lead to homogenization of the structure of the members of a pair [196, 205]. What is envisaged is a series of successive expansions and contractions involving the interchange of genetic material. That such expansions and contractions are continually occurring is indicated by polymorphisms in $\alpha$-globin number in human populations [205]. Furthermore the human $\delta$-globin gene has a structure that indicates it may be derived from the rescue of a silenced member of the $\beta$-globin pair by its active counterpart, and that this could have occurred through unequal crossing-over [198].

Although there is good evidence that gene correction by unequal crossing-over operates in the case of rRNA genes in *Drosophila* [206] and yeast [207], it is now clear that another mechanism can operate on multigene families. This mechanism, 'gene conversion', involves the replacement of one of the genes by a copy of the other in a non-reciprocal process [208]. It is considered in more detail in Section 7.8 in relation to situations where it is employed to control specific gene expression. It has been suggested that this mechanism has been operative in the case of the human $^{G}\gamma/^{A}\gamma$ pair because of a greater similarity between the $^{A}\gamma$ and $^{G}\gamma$ genes on one chromosome than the $^{A}\gamma$ alleles on different chromosomes [209]. Although the mechanism of such conversion in this case is only a matter of speculation, it was suggested that certain simple sequences might act as recognition signals [197]. Such sequences do not, however, appear to be a universal feature of genes undergoing correction.

Any mechanism for gene correction must also explain how genes can escape from such mechanisms to undergo divergence. In the case of a mechanism involving recombination, it has been proposed that the more rapid accumulation of extensive mutations (e.g. involving deletions and duplications) in the introns (see Section 9.2.1) might allow escape. The larger size of the introns in the $\beta$-globin family compared with those in the $\alpha$-globin family has been suggested as the reason these appear to have become more diverged [205]. Another way in which the duplicated regions could diverge rapidly is if one member acquired an insertion of transposed mobile DNA. An example of this might be the *Alu* sequences (see Section 7.7.3) in the flanking regions of the human $\alpha$-globin genes [269]. Finally, an extreme mechanism of severely decreasing (although not entirely eliminating) the probability of gene correction through homologous crossing-over is separation of the pair through chromosome translocation.

## 7.7 TRANSPOSITION OF DNA

### 7.7.1 Transposable elements

The dogma that the genome is stable arose from the observation of stable phenotypes in individual cells and organisms, and the success of genetic analysis based on the concept of fixed positions for genetic loci within the genome. It was the observation by Barbara McClintock of a non-stable phenotype in maize (see [211]) that led to the first proposal of the existence of the mobile genetic elements. She named the elements in maize 'controlling elements', but such mobile elements are now more generally termed 'transposable elements', or transposons. Transposable elements were later detected in bacteria by the mutations they produce on moving to positions within a functional gene [212, 213]. Because of the wealth of knowledge regarding the genome of *E. coli*, and the power of classical bacterial genetics, it was natural that the bacterial transposable elements should have come under most intense study, and are currently the most completely understood. The tools of molecular genetics have not only added structural information on bacterial transposable elements but have facilitated the study of eukaryotic transposable elements, especially in organisms such as *Drosophila* where classical genetic analysis could also contribute significantly.

Transposable elements may be defined as segments of DNA that are bounded by specific terminal DNA sequences and that have the ability to move to new DNA sites with little or no specificity for the latter. This limited target site specificity distinguishes transposition from site-specific recombination (e.g. of bacteriophage lambda – see Section 7.4.3). In many, although not in all, cases transposition may be further distinguished from site-specific recombination by the fact that it involves DNA replication. In these cases transposition involves the duplication of the element, one copy being retained at the original site and the second being transferred to the target. Because this replication phase can involve a fundamentally different mechanism in eukaryotic transposable elements, the elements of prokaryotes and eukaryotes will be considered separately. However, one common

feature of more general significance should be emphasized at this juncture. That is the ability of transposons to cause restructuring of the genome. On the one hand they can cause deletion or inversion of regions of DNA lying between them (see Fig. 7.17); and on the other hand, at least in some cases, they have the ability to acquire and transpose other genomic DNA. The recent evolutionary consequences of this latter event can be seen in the spread of genes specifying antibiotic resistance to pathogenic organisms. It is in many ways an attractive idea that analogous transposition might have had a major role throughout evolution [214, 215].

### 7.7.2 Transposition in prokaryotes [216–218]

Bacterial transposable elements can generally be assigned to one of three classes on the basis of structure and coding potential (Fig. 7.18). Class I contains elements that primarily code for an enzyme termed a *transposase*, and, even where they also code for other functions, unlike Class II elements, they do not encode a *resolvase* (see below). The simplest members of Class I are the insertion or IS elements (termed IS1, IS5 etc.)

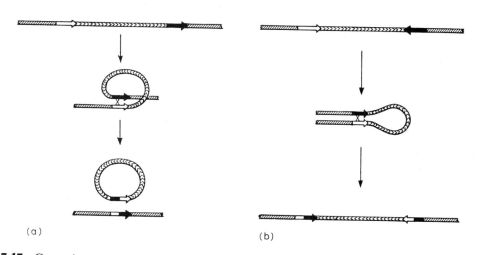

(a)                                           (b)

**Fig. 7.17** Genomic rearrangements promoted by homologous recombination between adjacent copies of mobile elements, represented by light and dark arrows: (a) similarly oriented elements: deletion; (b) oppositely oriented elements: inversion.

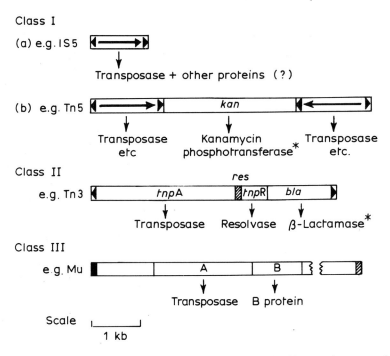

**Fig. 7.18** The structure of bacterial transposable elements. The terminal inverted repeats of Class I and II elements (represented by solid arrowheads) and the termini of the 40–kb bacteriophage Mu (solid and hatched areas) are not to scale. *Products of 'passenger' genes, not necessary for transposition.

[219], the termini of which consist of *inverted repeat* sequences, the size and structure of which characterize a particular IS element. The lengths of the terminal inverted repeats of different IS elements usually lie in the range 15–25 bp. Class I also contains what are termed *composite transposons* (e.g. Tn5, Tn9, Tn10) [220] which comprise two IS or IS-like elements (in inverted or direct relative orientation) flanking a piece of 'passenger' DNA, which is generally a gene specifying antibiotic resistance. In composite transposons both IS or IS-like elements need not be functional, and the extreme terminal repeat contributed by each member of the pair may be regarded as defining the limits of the transposon. Nevertheless individual functional modules can transpose independently.

Class II elements (e.g. Tn3, γδ) [221, 222] also carry various 'passenger' genes (e.g. the *bla*

gene, encoding β-lactamase), but have only single terminal repeats (of approx. 38 bp) at their extremities. In addition to a gene coding for a transposase (*tnpA*), they are typified by containing a gene (*tnpR*) coding for a site-specific recombinase, the resolvase, and an internal site (*res*) which the latter protein recognizes.

A separate class, Class III, has been created to accommodate the bacteriophages such as Mu [223], which replicate via a transpositional mechanism. In addition to encoding a considerable number of bacteriophage functions (coat proteins etc.), Mu possesses two genes, A and B, involved in transposition. It does not, however, encode a resolvase. Gene A is thought to be the transposase and gene B greatly enhances the efficiency of transposition. It is, in fact, the high efficiency of transposition of Mu (*c.* 100 transpositions per lytic infection) compared with that

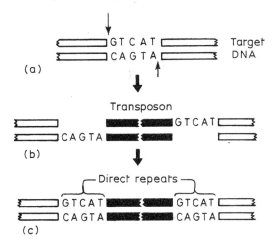

**Fig. 7.19** Model to explain the generation of direct repeats of the target site of transposon integration: (a) endonucleolytic cleavage of the target DNA generates a staggered break; (b) the transposable element is ligated to the protruding ends of the cleaved target site; (c) cellular enzymes repair the single-stranded gaps, generating flanking direct repeats.

of other transposons ($10^{-7}$ to $10^{-5}$ per generation) that has made it particularly suitable for studying the mechanism of transposition. Mu differs from other bacterial transposable elements in that its termini are not inverted repeat sequences. Nevertheless its termini possess precise structural features that are recognized during transposition, and there is every reason to believe that Mu transposes by a mechanism similar to other transposable elements.

A diagnostic feature of transposition is the presence of short direct repeats flanking the element at its site of integration (target site). These direct repeats are not part of the element but are generated from the original target site, and it was proposed [224] (and is generally believed) that they are generated by the formation of a staggered endonucleolytic break in the DNA to the ends of which the transposed DNA is attached. The filling of the remaining single-stranded gaps on either side generates the

direct repeat (Fig. 7.19), the size of which (generally in the range 3–13 bp) is characteristic of a given transposable element.

To appreciate the current model for the mechanism of bacterial transposition it is necessary to consider the different possible products of transposition. Experiments in which the donor transposable element is on a separate circular genome from the acceptor DNA have yielded two types of product (Fig. 7.20). One is termed 'simple insertion', and involves the acceptor DNA acquiring a copy of the element (Fig. 7.20b); the other involves the formation of a so-called cointegrate (Fig. 7.20c), a fusion of donor and acceptor DNA with duplication of the transposon. The relative proportions of these two types of product vary between different

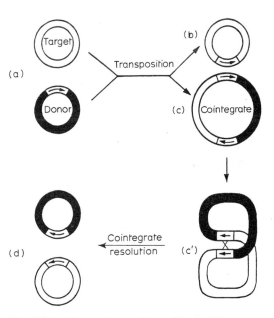

**Fig. 7.20** Alternative products of bacterial transposition. The interaction of a donor replicon containing a transposable element with a target replicon (a) may result in a simple insertion with loss of donor (b) or a cointegrate (c). Recombination between the two copies of the element in the cointegrate (c') can resolve this into both donor and target replicons, each carrying a copy of the element (d).

transposable elements, and in the case of Mu varies according to whether it is performing lytic infection (cointegrates predominating) or lysogeny (where simple insertions occur). Although in some cases (e.g. Tn3) is is clear that simple insertions are generated through the 'resolution' of cointegrates by homologous recombination between the two copies of the transposable element (see Fig. 7.20d and cf. Fig. 7.17) [221], in other cases (e.g. IS10) the evidence argues against such an intermediate (see [218]). The generation of simple insertions in the latter case is thought to be non-replicative, and to involve loss of the element from the donor replicon, which is itself destroyed.

For all classes of transposable element the initial integrative phase of transposition has an absolute requirement for expression of the transposase enzyme. In the case of the IS elements this is not well characterized (although the size of IS5, for example, would preclude a protein of more than about 40 000 daltons), but the A gene product of Mu, and the *tnpA* gene product of Tn3 have molecular weights of 70 000 and 120 000 respectively. It is clear that each transposase specifically recognizes the termini of the transposable element which encodes it, and, in the case of the Tn3 transposase, the binding has been shown to require ATP. It is presumed that the transposase brings the element to its site of integration, and, although there is no direct evidence for this, that it catalyses the staggered endonucleolytic cleavage at the target site and the precise nicking at the ends of the element. One reason for thinking this is the fact that the size of the target direct repeat is characteristic of transposon rather than host.

The model of transposition shown in Fig. 7.21 [218] derives from that of Shapiro [225], modified as proposed by Ohtsubo *et al.* [226] to allow for non-replicative formation of simple insertions, and owes much to the original ideas of Grindley and Sherratt [224]. Its key feature is the formation of an intermediate, evidence for the existence of which has been obtained by

Mizuuchi and coworkers [227, 228] using bacteriophage Mu. The initial step involves the staggered endonucleolytic break in the target DNA that produces 5′ protruding ends, and a single nick at each end of the transposable element leaving a free 3′-OH on the sides adjacent to the element (Fig. 7.21(a)). The 5′-ends of the target are then ligated to the 3′-OH groups on the element, resulting in a structure which resembles two replication forks (Fig. 7.21(b)). In replicative transposition it is assumed that host enzymes attach to the forks of this intermediate and replicate through the transposable element, without necessarily replicating any other DNA (Fig. 7.21(c)). The product of this (Fig. 7.21(d)) can be seen to be equivalent to the cointegrate of Fig. 7.20 if the donor and target were originally on separate circular replicons (i.e. A linked to B, C linked to D). In Class II transposons the resolvase encoded by the *tnpR* gene (Fig. 7.18) binds to the *res* site to promote the homologous recombination required to regenerate host and target replicons or to resolve the scrambled structure (termed inversion–inversion) formed in intramolecular transposition if, respectively, A and C, and B and D were originally joined (Fig. 7.21(e). (If A and D, and B and C were originally joined, two hybrid circular products will be generated.) In other cases the host recombination system may promote this.

The structure in Fig. 7.21(b) has also been proposed as an intermediate in the formation of non-replicative simple insertions. It is assumed that some feature of the forks in these cases prevents replication occurring from them, and allows nucleolytic attack on the donor replicon (Fig. 7.21 (f)). Gap repair then yields the simple insertion of Fig. 7.21 (g). It has been suggested that the 5′-exonuclease activity of DNA polymerase I might act on the intermediate in the absence of replication to effect such destruction of the donor DNA [218].

In focusing on the mechanism of bacterial transposition, other aspects of the process such

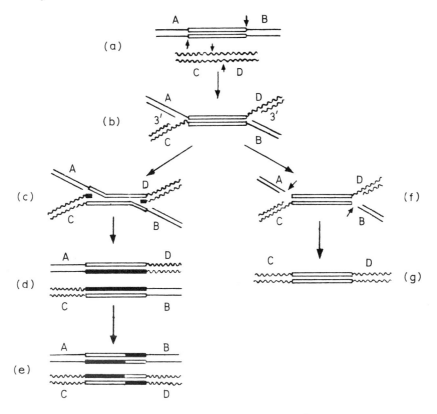

**Fig. 7.21** Model for transposition. The sites of nucleolytic cleavage are indicated by arrows and the original and newly synthesized copies of the transposable element by open and solid areas respectively. For other details see text. Adapted from [218], with permission.

as regulation (see Section 10.5.1) and immunity have been omitted, as has the question of the function of the second protein encoded by most IS elements. The reader interested in these topics is directed elsewhere [217].

### 7.7.3 Transposition in eukaryotes [217, 229]

#### (a) *Retroviruses* [230]

Retroviruses have already been considered (Section 6.9.10) from the standpoint of the replication of a single-stranded viral RNA genome. However, it is possible to start from the standpoint of the double-stranded proviral DNA, and to regard the single-stranded viral RNA as an intermediate in the genomic trans-

position of this to other sites within the same or a different genome. This is no mere academic exercise for it is becoming increasingly apparent that in many cases transposition in eukaryotes operates via reverse transcription, and that the proretroviral DNA is a paradigm for certain eukaryotic transposable elements [231].

The structure of the proviral DNA is presented in Fig. 7.22 in a manner that allows comparison to be made with bacterial (Fig. 7.18) and eukaryotic (Fig. 7.23) transposable elements. One point of similarity, omitted from the figure for increased clarity, is the target-site direct repeat sequences (usually 5 bp in length) flanking the proviral DNA. The most striking feature of the proviral DNA is that it is flanked

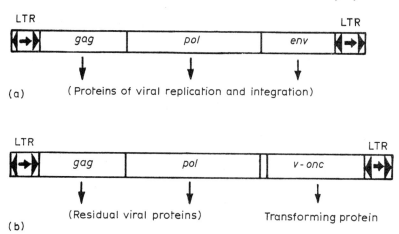

**Fig. 7.22** Structure of proretroviral DNAs: (a) retrovirus competent to perform replication; (b) defective oncogenic retrovirus (e.g. Moloney murine sarcoma virus) with transforming ability because of acquisition of an oncogene. The position of the oncogene and the extent of deletion (if any) of viral functions varies between different oncogenic retroviruses. The internal inverted repeats of the long terminal repeats (LTRs) are represented by arrowheads (not to scale).

by direct repeats (LTR) of the large U3, R, U5 unit (250–1400 bp), each of which includes a short inverted repeat sequence (5–13 bp long) at its end. The similarity of this organization to that of the composite bacterial transposons such as Tn9 (which has IS units in a directly repeated orientation, rather than the inverted orientation of Tn5 – Fig. 7.18) has been remarked. It must be stressed, however, that, unlike Tn9, the protein(s) required for transposition are encoded in the DNA between the LTRs of retroviruses and not in the LTRs themselves.

The three most immediately apparent viral genes, *gag*, *pol* and *env* (cf. Fig. 6.34), disguise a greater coding capacity, aspects of which will be stressed in view of their relevance to transposition. Because of the functional monocistronic nature of eukaryotic mRNAs (see Section 11.6.1) the *env* gene product is translated from a subgenomic mRNA and the main product of translation of the complete mRNA is the *gag* gene product which is subsequently subject to proteolytic cleavage to four smaller species (cf. Sinbis virus – Fig. 11.16). The reverse transcriptase encoded by the *pol* gene is generated from a 180 kilodalton (kDa) polyprotein translated from the *gag* initiation codon, and, at least in the case of Moloney murine leukaemia virus [232], this involves readthrough suppression of an amber termination codon (see Section 11.13.2). (In Rous sarcoma virus a frameshift would be required to produce this.) Other proteins produced from the 180 kDa polyprotein are the 13 kDa viral protease, which is located at the 5′-end of the *pol* gene, and an endonuclease which is located at the 3′-end of the *pol* gene.

What is known about the mechanism of integration of the viral DNA into the host genome? It proceeds via reverse transcription of the viral RNA to generate the linear double-stranded DNA molecule (i) of Fig. 6.34. Two covalently cyclized forms of this DNA are also found, one containing only a single copy of the LTR, and the other containing two tandem copies. It is this latter form that engages in transposition [233], and the short inverted repeats (two of which are now merged at the cyclization point) are indispensable for integration [234]. They must be cleaved by a site-specific endonuclease, which causes the loss of the last two base pairs of these butted inverted repeats, and ligated to a staggered break produced in a random site in the target DNA. Gap-filling produces direct repeats of the target DNA flanking the insertion. As integration has been shown to be prevented by mutation of the 3'-end of the *pol* gene [235], it is likely that the endonuclease encoded in this region is involved. However, as with bacterial transposable elements, it is not clear whether a single endonuclease cleaves both circular viral and target DNA (the reverse transcriptase itself also has endonuclease activity), and whether a separate ligase is involved. Although the involvement of reverse transcription is a point of striking difference between certain types of eukaryotic transposition (see also below) and prokaryotic transposition, it should be pointed out that the circular double-stranded DNA precursor could theoretically be integrated into the target DNA by mechanisms similar to those proposed for bacterial transposable elements (Fig. 7.21).

Another similarity between retroviruses and, at least in prokaryotes, transposable elements is their ability to acquire and transpose 'passenger' host DNA. The most well-known example of this is provided by the oncogenic retroviruses which have acquired cellular genes or parts thereof, the inappropriate expression of which leads to cell transformation and tumour formation. It is not known whether non-oncogenic retroviruses containing 'neutral' host DNA exist, but it seems likely that the oncogenic retroviruses, like bacterial transposons carrying antibiotic resistance genes, have a selective advantage that facilitates their survival. Most (but not all) oncogenic retroviruses are defective, having lost part of their own genes (see Fig. 7.22), and require the assistance of 'helper' retroviruses to replicate. The oncogenic retroviruses are presumed to have arisen in a two-step mutational process [230]. The first step is thought to be deletion of the right-hand LTR and flanking

DNA, bringing the oncogene under the transcriptional control of the promoter in the left-hand LTR. The second step would appear to require recombination between chance homologies in the RNA transcript from this mutated unit and a normal viral RNA in order to regenerate the right-hand LTR. Oncogenic retroviruses are currently the focus of intensive research and have recently been reviewed [236].

The consideration above of the proretrovirus as a transposable element was made in order to bring out the similarities with eukaryotic transposable elements (see below). In most cases the provirus stage is an intermediate in the horizontal transmission of the virus via somatic cells. However, in certain cases there is evidence for the vertical transmission of *endogenous retroviruses* via non-somatic cells, in which cases they are, in fact, acting as transposable elements [237]. One such endogenous retrovirus is responsible for the dilute coat colour mutation in mouse [238]. In addition there are defective elements related to retroviral genes but lacking the possibility of an extracellular phase. These are the genes that encode the *intercisternal A-particles* [270].

(b) *Eukaryotic transposable elements*
We now turn to eukaryotic transposable elements, the mobility of which is testified by the production of mutations at the sites to which they have moved. The most extensively characterized of these are the *copia*-like elements of *Drosophila* [239] and Ty elements of yeast [240], and in both these examples there is persuasive evidence for transposition by a mechanism closely resembling that of the retroviruses, described above. (Class II yeast mitochondrial introns appear similar – see Chapter 9.)

With *copia* the structural similarity to retroviruses is more immediately apparent in the relatively large (*c.* 300 bp) terminal direct repeats, each of which is bounded by a short (17 bp) inverted repeat (Fig. 7.23). The size of the

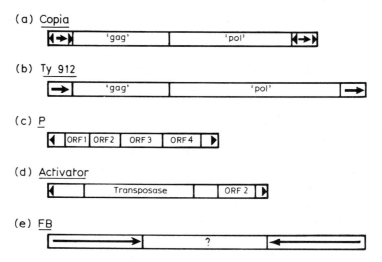

**Fig. 7.23** Structures of some eukaryotic transposable elements. The relative sizes of these are roughly to scale but the sizes of the smaller terminal repeats have been exaggerated for clarity. For actual sizes see text. ORF denotes open reading frame.

target direct repeat is 5 bp. The determination of the complete sequence of the *copia*-like element 17.6 [241] shows open reading frames with similarities to the *gag*, *pol* and *env* genes, although in the shorter (5.1 kbp) *copia* itself [242], where there is a single extensive open reading frame, the homologue of the *env* gene is absent. There is evidence that *copia* is transcribed, and RNA sequences related to *copia* have been found associated with reverse transcriptase in virus-like particles [243]. Circular extrachromosomal copies of *copia* have also been found containing one or two LTRs [244], although there is as yet no evidence that these are generated by reverse transcription, and only indirect evidence that they may be intermediates in transposition [245].

In the case of the (5.9 kbp) Ty elements, although there are long (*c*. 300 bp) terminal direct repeats, these do not contain terminal inverted repeats. (There are, of course, target-site direct repeats, and these are 5 bp long.) Nevertheless there are again open reading frames with similarities to *gag* and *pol* [246, 247] and evidence for the synthesis of a *gag-pol* fusion

protein which, as in the case of Rous sarcoma virus, would appear to require translational frame-shifting [246, 248] and to undergo post-translational proteolysis [248]. Although putative transposition intermediates have not yet been found for Ty, elegant experiments in which an intron was inserted into a Ty element under the artificial control of a yeast GAL 1 promoter showed that transposition was induced by galactose (i.e. inducing transcription), and that the intron had been correctly spliced out of the transposed copy [249]. It is thus beyond doubt that RNA is an intermediate in the transposition of Ty.

There are, however, other eukaryotic transposable elements of spotless pedigree and having coding potential that appear in no way to resemble retroviruses. The best characterized of these are the *P factors* of *Drosophila melanogaster* and the 'controlling element' *Activator* of maize, which are bounded by short (31 bp and 11 bp respectively) inverted repeat sequences (Fig. 7.23). In this respect, at least superficially, they resemble the Tn3 family of bacterial transposons.

P factors (and defective derivatives called P elements) are involved in a complex genetic phenomenon called hybrid dysgenesis [250] in which they have been shown to be responsible for mutagenesis of the progeny of certain inter-strain crosses by transpositional insertion in which 8-bp target-site direct repeats are generated. The complete sequence of a 2.9-kbp P factor [251] shows the presence of four open reading frames (none of which is homologous to a reverse transcriptase) and the genetic evidence suggests that one of these must encode a transposase (i.e. a protein the activity of which is required for transposition) and another must encode a repressor that regulates expression of the transposase. (It is the derepression in hybrids that results in the mutative transposition.)

The *Activator* element of maize and the defective element *Dissociation* (now known to be derived from *Activator* by deletion [252, 253]) were the first transposable elements studied by Barbara McClintock. Recent nucleotide sequence analysis [252, 253] indicates that the 4.6-kbp element, which generates 8-bp target-site direct repeats, has two open reading frames (again showing no similarity to retroviral sequences), one of which contains the deletion in *Dissociation*, and must encode a transposase.

Although there is as yet no biochemical characterization of the transposases in the P *factor* and in *Activator*, the structural and genetic similarities to Tn3 transposons (the resolvase of Tn3 also acts as a repressor) have led workers in these fields to favour an analogous mechanism of transposition. In view of the emphasis given above and in Section 7.7.3.c to reverse transcription in eukaryotic transposition, it is important not to ignore the possibility of other mechanisms of transposition in cases such as these.

We conclude this brief treatment of eukaryotic transposable elements with a mention of the FB (foldback) elements of *Drosophila* which represent a third structural variety (Fig. 7.23). These are bounded by long and variable (200–1200 bp) inverted repeat sequences made up of simple tandem direct repeats, except for a unique conserved 31-bp sequence at their extreme ends [254, 255]. The transposition of these elements is the basis of certain eye-colour mutations in *Drosophila* [256]. The size of the central region in FB elements varies, and some elements actually lack any central region. Furthermore different sequences have been found in the central regions of different FB elements, some of which are found elsewhere in the genome not associated with the inverted repeats [257]. It appears likely that many of the FB elements are defective versions of a parent element (perhaps FBw$^C$ [257]), but whether this has the capacity to encode the transposase involved in its transposition is as yet unknown.

(c) *Processed pseudogenes and other retroposons* [258]

Although the best-characterized eukaryotic transposable elements are those of plants and invertebrates, described above, there are signs (e.g. [259]) that analogous elements (other than the endogenous retroviruses) will be found in vertebrates. At present, however, the best evidence for transpositional events involves so-called *retroposons*. This latter term is used to describe certain sequences that have been inserted into pre-existing sites in the genome by a transposition mechanism (they are characteristically flanked by target-site direct repeats), and that have structural features that clearly indicate that they were generated from the reverse transcription of RNA. They differ from transposable elements in being derived from a parent that lacked the ability to code for the machinery of its own transposition and, in most cases, themselves lacking the potential for further transposition.

Processed pseudogenes [260] differ (at least in the ideal case – there are exceptions) from the type of tandem duplicate pseudogene already discussed (Section 7.6.3) in the following characteristics: (i) extending only from the cap site to the site of polyadenylation, (ii) possessing a poly(A) tail (which is absent from the functional gene) in the expected position following a polyadenylation/processing signal (see Section 9.3.3), (iii) cleanly lacking all introns present in the functional gene, (iv) being flanked by direct-repeat sequences immediately preceding the transcriptional start and immediately following the poly(A) tail, and (v) residing at chromosomal locations distant from the functional gene. These characteristics clearly indicate that these pseudogenes originated from spliced polyadenylated mRNAs, DNA copies of which were inserted at sites in the chromosome created by a staggered endonucleolytic break (typically 11–15 bp) of the type discussed above. Most of such processed pseudogenes have been inserted at sites at which the lack of a promoter renders transcription impossible, and most have accumulated mutations that would render any transcript functionless. However, in the case of a chicken calmodulin processed pseudogene lacking frameshift or nonsense mutations there is evidence for expression [261]. A similar situation obtains for the functional rat insulin I gene, which, because of its location on the same chromosome as the rat insulin II gene and the fact that it contains one of the three introns of the latter, had been thought to have been derived from gene II by a tandem duplication. Recent analysis [262] indicates that the gene I has a poly(A) tail, is flanked by direct repeats and is approx. 100 000 kb from gene II. It appears that the insulin I gene arose from the reverse transcription of a partially processed transcript of gene II that had initiated upstream from the cap site and included the promoter.

Vertical transmission of processed pseudogenes can only occur if they arise in germ cells, and the most numerous processed pseudogenes correspond to mRNAs that one would expect to be abundant in undifferentiated egg or sperm cells. These include multiple processed pseudogenes corresponding to glycolytic enzymes and to cytoskeletal proteins such as actin. In the latter case it is mRNAs corresponding to the cytoplasmic ($\beta$ and $\gamma$) isotypes of actin, rather than the muscle-specific isotypes, that have given rise to processed pseudogenes [263]. Nevertheless processed pseudogenes corresponding to $\alpha$-globin [264] and immunoglobulin $\lambda$ [265], normally the products of differentiated cells, have been described. These, like the insulin I gene, mentioned above, are atypical in corresponding to transcripts extending upstream beyond the normal 5' cap site. The possibility that such mRNAs are expressed (albeit transiently and at a low level) in the germ line (rather than being somehow transmitted horizontally from differentiated cells) is interesting in relation to the gene conversion thought to operate on globin (Section 7.6.4) and immunoglobulin genes [208]. It has been suggested [231] that this conversion, which would, of course, have to occur in the germ line, might possibly involve reverse transcripts of such transiently expressed mRNAs.

Two fundamental questions arise regarding the reverse transcription of the mRNAs that give rise to processed pseudogenes. The first is the source of the reverse transcriptase (and any other components of the transpositional machinery that may be required); the second is the nature of the primer for reverse transcription. The source of the reverse transcriptase is difficult to decide as no such enzyme is detectable in normal cells. One possibility is that it derives from retroviral infection, and it may be more than a coincidence that invertebrates, which are not subject to retroviral infection, generally lack processed pseudogenes. An alternative possibility is that it might derive from the reverse transcriptase encoded by endogenous transposable elements, e.g. endogenous retroviruses or intercisternal A-particles. (It should be pointed out that the characteristic size of the direct repeats flanking processed pseudogenes is greater than those flanking retroviral sequences, which might be regarded as an argument against the same enzyme being responsible for the generation of these.) The mechanism of priming is also uncertain because, in the absence of specific binding sites, it is unlikely to involve a tRNA species as in the case of retrovirus minus-strand synthesis (see Section 6.9.10), and possibly *copia* and Ty. Several models have been proposed for priming of reverse transcription in processed pseudogenes [258, 260, 266]. One attractive model [260] rests on the observation [266] that the target sites of transposition frequently have a dA-rich segment at their 5'-ends. A staggered break at such a point would provide an oligo(dT) attachment site for the poly(A) tail of the mRNA, and this oligo(dT) segment would act as the primer for reverse-transcribing the mRNA (Fig. 7.24). However, this mechanism, despite solving the problem of priming for which there is other circumstantial evidence of the involvement of poly(A), is so different from that for retroviral integration that it is difficult to conceive that it could employ the same enzymes.

Types of RNA other than mRNA can gener-

ate pseudogenes by mechanisms which must involve reverse transcription. The small nuclear RNAs (U1 to U6) give rise to two structural classes of pseudogenes flanked by target-site direct repeats of similar size range to those flanking the processed pseudogenes described above [267]. One class of these are full length (or almost so) but terminate with poly(A) tails, absent from the parent RNA. This suggests a requirement for aberrant polyadenylation for retroposition, and would be consistent with the mechanism of priming suggested in Fig. 7.24. The other class are truncated to various extents at their 3'-ends. In the case of the severely truncated, non-polyadenylated U3 pseudogenes this clearly relates to the potential for self-hybridization at the 3'-end that could allow the self-priming of reverse transcripts of the size observed [268]. It seems also likely that self-priming is involved in the generation of less severely 3' truncated non-polyadenylated pseudogenes of these small nuclear RNAs.

The mobility of certain families of highly

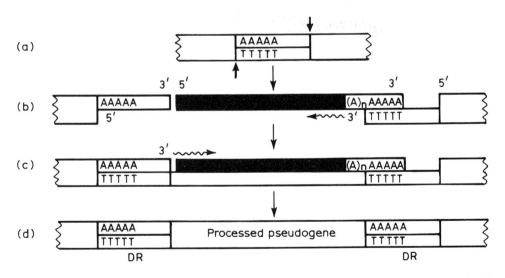

**Fig. 7.24** Possible mechanism for priming of reverse transcription of a polyadenylated RNA to give rise to a processed pseudogene: (a) staggered nucleolytic cut at A/T-rich target site; (b) annealing of mRNA (solid shading) to 3' protruding oligo(dT) end. This serves as priming site for reverse transcription (indicated by wavy arrow), which gives (c), from which the RNA is displaced or degraded for the synthesis of the second DNA strand to give (d) a processed pseudogene flanked by target site direct repeats (DR).

repeated vertebrate DNA, the members of which lack any defined function, appears to depend on retroposition by mechanisms similar to those thought to operate in the case of processed pseudogenes. The families involved are short and long interspersed repetitive elements (SINES and LINES respectively), previously mentioned in Section 3.2.5. Each of these families has particular features distinct from the retroposons already described, and thus merit separate discussion.

The SINES are typified by the *Alu* family in man and the B1 family in rodents [271]. These SINES have structures that are related to the 7S RNA component of the signal recognition particle (see Section 11.7.1), but are flanked by (target site) direct repeats, lack the central section of the latter and have acquired a 3'-oligo(dA) tail (see Fig. 7.25). In the case of *Alu*, the structure is an imperfect head-to-tail dimer. Although it has been suggested that they may have originated from something equivalent to processed pseudogenes of 7S RNA [272], it is important to emphasize that the processed pseudogenes of 7S RNA that have been described are of the 3' truncated variety, assumed to have been generated by self-priming facilitated by a suitable 3' secondary structure [273]. The SINES (like the 7S RNA genes) are transcribed by RNA polymerase III and thus are likely to terminate at oligo(dT) stretches in flanking 3'

DNA (see Section 8.3.4). Transcripts would thus have the potential to form secondary structure interactions between the corresponding 3'-oligo(T) sequence and the intrinsic oligo(A) sequence, which could act as a priming site for reverse transcription, leading to a cDNA copy that could be reintegrated into the genome [274]. In contrast to retroposed processed pseudogenes that (normally) lack their (RNA polymerase II) promoters and hence the potential for further mobility, transposed *Alu* sequences will still contain their (RNA polymerase III) internal promoters and thus have the potential for further transposition. This may explain the extreme abundance of *Alu* sequences (*c.* $10^5$ copies in the human genome). The accumulation of mutations may, of course, inactivate the promoters of some *Alu* sequences. Other SINES with internal RNA polymerase III promoters appear to be derived from tRNA genes [341]. Although SINE sequences may also be transcribed as part of other transcription units (they are found in introns and 3' flanking regions of genes that code for proteins), there is good evidence that independent copies reach the cytoplasm [271]. Furthermore closed circular double-stranded DNAs containing *Alu* sequences with 5' and 3' ends joined have been described [275], although there is no evidence that these are intermediates in transposition.

The LINES [276] are typified by the *Kpn*

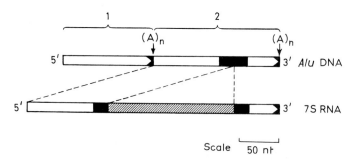

**Fig. 7.25** Structural features of *Alu* DNA and relationship to 7S RNA. The two imperfectly repeated copies in the *Alu* dimer are numbered 1 and 2, the extra DNA in the latter being indicated by solid shading. The hatched area represents the central sequences of 7S RNA absent from either *Alu* repeat and from other SINES.

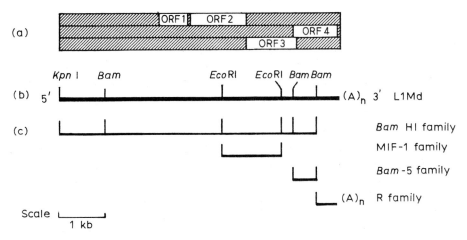

**Fig. 7.26** Structural features of vertebrate LINES: (a) apparent open reading frames (ORFs) common to different vertebrate LINES: (b) restriction endonuclease map of mouse LINE, L1Md; (c) its relationship to various repeated subfamilies reported for the mouse genome.

family in man, which is a member of what is now referred to more generally as the vertebrate L1 family. These are about 6–7 kbp long, may contain four open reading frames, and terminate in a polyadenylation/processing signal followed by a poly(dA) tail (Fig. 7.26). It appears that they are transcribed by RNA polymerase II [277]. Transposed copies of the L1 family have the peculiarity that they lack varying amounts of the 5'-end of the full-length version, consistent with their generation by varying extents of reverse transcription. Because of this, and the fact that they were initially studied as repeated DNA generated by digestion with particular restriction endonucleases, the impression arose of several different families (e.g. the *Bam*HI, MIF-1, *Bam*-5 and R families of the mouse LINE, L1Md). These are now recognized to be parts of a single family (Fig. 7.26). It is assumed that the many defective LINES arose from functional parental copies; however, these latter are difficult to identify. It would clearly be of great interest to obtain these and study their protein products. As well as their own intrinsic interest, if it should transpire that LINES encode the proteins required for their own transposition,

this might also explain the origin of the machinery of transposition of other retroposons.

## 7.8 GENE CONVERSION

As described in Section 7.6.4 this is a process which can maintain homogeneity amongst dispersed copies of similar genes. This may involve exchange or transfer of quite short sequences by a strand invasion mechanism. In this section we use gene conversion to refer to the phenomenon where a gene is replaced in entirety by a *copy* of a very similar, homologous gene. In this case conversion is always non-reciprocal.

Examples of such gene conversion events are known in prokaryotes [278, 280] and in eukaryotes where gene activation is associated with the conversion event. The process involves the transfer of a gene from one position where it is inactive into an expression-linked site.

The prokaryotic examples involve switching of genes coding for surface antigens of a number of pathogenic bacteria and also variations in the type of pilus produced. We shall consider two eukaryote examples.

### 7.8.1 Yeast mating-type locus

In *Saccharomyces* there are two mating types: a and $\alpha$. Fusion of an a and an $\alpha$ haploid cell leads to formation of a diploid cell which can sporulate to form haploid spores. However, in a clone derived from a single haploid cell (a or $\alpha$) some cells will switch their mating type [281].

Mating type is conferred by a specific region (Ya or Y$\alpha$) in the mating-type locus (*MAT*). Inactive copies of Ya and Y$\alpha$ also occur at other loci to the right (*HMRa*) and left (*HML$\alpha$*) respectively of *MAT* (Fig. 7.27). The inactive copies are repressed by the products of two other genes: *MAR* and *SIR*. However, one of the inactive genes may duplicate itself and the copy can replace the gene in the mating-type locus and this may bring about a change from an a-type to an $\alpha$-type cell, or vice versa [282].

Ya (642 bp) is non-homologous to Y$\alpha$ (747 bp) but in all three sites the Y sequence is flanked by homologous regions within which recombination is thought to occur, e.g. XYaZ or XY$\alpha$Z (Fig. 7.27). The mating-type switch is initiated when a double-stranded cleavage occurs in a loop region to the 3'-side of the gene in the Z region at the *MAT* locus. This cleavage is brought about by an endonuclease which is the product of the *HO* gene. Although similar sequences are also present at the YZ junction at the *HMR* and *HML* loci they are not cleaved, perhaps as a result of their being in a different chromatin conformation. Yeast strains lacking an active *HO* gene switch mating types at a frequency of only $10^{-6}$ whereas $HO^+$ cells may switch every cell division.

A nuclease has been isolated which generates the required site-specific double-stranded break with four-base 3' extensions [283, 284]. Further biochemical details of the mechanisms for the mating-type switch are not yet available.

### 7.8.2 Variant surface glycoprotein (VSG) genes in trypanosomes

Trypanosomes are protozoan parasites responsible for sleeping-sickness and nagana [285–288]. The immune response engendered by infection is directed against the surface glycoprotein of the protozoan. However, the parasite is able to change the surface glycoprotein and so avoid the host's defence mechanisms.

There are about a thousand VSG genes in each trypanosome making up about 10% of the genome, but only one – the expression-linked copy (ELC) – is active at any one time. Most of the genes (the basic copy or BC genes) are arranged in tandem, separated by several copies of incompletely homologous 70 bp repeats. All the genes have a common sequence of 6–8 bp at their 3'-ends.

As well as the BC genes there are telomeric VSG genes. In addition to the normal chromosomes trypanosomes have about 100 mini-chromosomes each one of which has a VSG gene at each end. These telomeric VSG genes have copies of the 70-bp repeats in their upstream region and towards the end of the chromosomes they have tandem repeats of the sequence 5'CCCTAA3' (see Section 6.10.3) [289–291]. It is one of the telomeric genes which is the

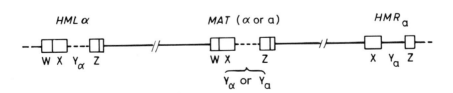

**Fig. 7.27**   Yeast mating type genes.

Gene rearrangements

251

expression-linked copy and it appears to differ from the other genes by the presence of a 5′-exon of 35 bp. This miniexon, which is absent from all silent copies of the gene, is present in 200 or more copies as part of a tandem repeat which may form part of a complex promoter (see Section 10.4) [288].

A change in surface glycoprotein results from the transposition of an inactive VSG gene to the expression-linked site when it replaces the resident gene. For transposition a basic copy gene is duplicated (as yet we know no details of this duplication) and recombines with the ELC. Recombination involves the 70-bp repeats on the 5′-side of the gene and the common 6–8-bp homologous region at the 3′-ends of the genes [292, 34]. The 5′-miniexons are not lost from the ELC.

Transposition of a telomeric gene into the EL site may involve recombination only at the 5′-end of the gene so that the whole of the end of the chromosome is exchanged. Alternatively activation of a telomeric VSG may occur by a non-duplicative event involving reciprocal recombination between the ends of two chromosomes – chromosome end exchange [288, 293].

## 7.9 GENE REARRANGEMENTS

We have considered how genes may be corrected in an apparently non-directed process leading to sequence homogenization (Section 7.6.4) and how gene conversion can lead to the replacement of a gene by a copy of a similar gene (Section 7.8). In this section we look at some examples where genes are constructed by rearrangement of the genome physically to bring together different DNA sequences. Such processes, in which the intervening DNA is excised, have been studied primarily in the lymphocytes (white cells) present in mammalian and avian blood and lymph.

B lymphocytes have the ability to recognize a wide range of foreign antigens in solution and bind them by means of a cell surface *antibody*

(*immunoglobulin*) molecule. Later, these lymphocytes secrete soluble immunoglobulins. In contrast T lymphocytes interact with cellular antigens and this is achieved by the *T-cell receptor* only recognizing the antigen when it is presented along with a major histocompatibility (MHC) protein [294–296, 331].

For the interaction of the antigen with the antibody or T-cell receptor to be specific requires millions of different antibodies and T-cell receptors to react with the millions of different antigens which may possibly be encountered in the life of an animal.

### 7.9.1 Immunoglobulin genes

The ability of animals to generate a multiplicity of antibodies to novel antigenic stimuli posed a problem to scientists for many years. Dreyer and Bennett [297] proposed in 1965 that a single polypeptide might be synthesized using combined information present in two genes and since then numerous experiments have elucidated the details showing that this theory for the generation of antibody diversity is broadly correct [298, 299].

Antibodies (or immunoglobulins) are made up of four polypeptide chains: two light and two heavy chains. Each chain has a region (the constant region) which is of similar sequence for all antibodies of a given class and a region (the variable region) which can have a wide range of amino acid sequences [296]. The constant and variable regions are encoded separately in germ-line DNA but the gene segments are brought together in antibody-producing cells by a somatic recombination event [298, 300–302, 331].

#### (a) *Light chains*

In the mouse and man (two species extensively investigated) there are two classes of light chains (κ and λ) each with a characteristic constant region gene (Cκ and Cλ). A short distance upstream from the C region is a series of joining

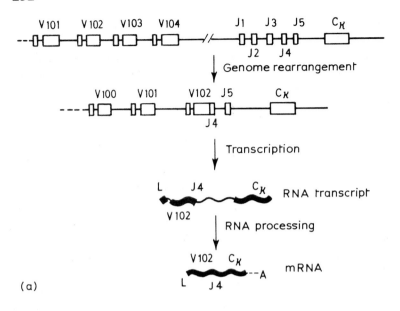

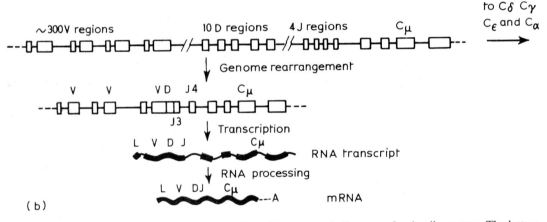

**Fig. 7.28**  Structure, rearrangement and expression of immunoglobulin genes, for details see text. The letters refer to the leader (L), variable (V), joining (J), diversity (D) and constant (C) regions of the kappa light chain genes (a) or the heavy chain genes (b).

(J) sequences each of about 30 nucleotides in length. Much further upstream are a large number of V sequences (Fig. 7.28a). The somatic cell gene rearrangement involves the joining of a particular V region to one of the J regions to form a VJ region with elimination of the intervening DNA. In different cells this recombination event probably occurs at random to produce a battery of lymphocytes each with a different VJ combination. Transcription leads to the synthesis of a premessenger RNA molecule which is processed to eliminate the intervening sequences within the V region and between the selected J region and the C region and thus produce the mRNA for the immunoglobulin light chain (see Section 9.3 and Fig. 7.28a).

A mouse has two alleles each of the $\kappa$ and $\lambda$ light chain constant regions but only one functional rearranged light chain gene is produced [302]. The other genes may not have rearranged or may have undergone a non-productive rearrangement to produce an aberrant gene. This process is called allelic exclusion and implies a feedback regulation such that only when a functional rearrangement has occurred are further rearrangements inhibited [302, 303].

### (b) *Heavy chains*

The heavy-chain genes are also made up of variable (V), joining (J) and constant (C) regions but there are several different C regions (e.g. $C\mu$, $C\gamma$, $C\alpha$) expressed in turn during the life of a single lymphocyte. Additionally each C region is made up of several exons representing the different domains of the resulting heavy chains [304, 338]. Initially the $C\mu$ gene is active to produce IgM antibodies in immature lymphocytes. Later IgG (from $C\gamma$) and IgA (from $C\alpha$) are produced by subsequent DNA rearrangements. This is known as class switching (see Section 7.9.1d below). The diversity of heavy-chain sequences is further increased by the presence of another group of gene segments (the diversity or D regions) which lie between the V and J regions (Fig. 7.28b).

Thus the construction of light-chain genes from V, J and C regions and heavy-chain genes from V, J, D and C regions allows the con-struction of a different immunoglobulin in each of many millions of lymphocytes. This diversity is further increased as a result of (a) variability in the joining reactions and (b) somatic mutations.

### (c) *The joining reactions*

The details of the joining reactions are not yet known but each of the rearrangements leading to a functional immunoglobulin gene probably occurs by the same mechanism of site-specific recombination [298, 305]. In the germ line each segment is bounded by a palindromic heptameric oligonucleotide of consensus sequence 5'CACAGTG3' followed by a spacer of 12 bp or 23 bp and a nanomeric oligonucleotide of consensus sequence ACAAAAACC. The nanomer always has the orientation in which the terminal CC is directed away from the gene segment (Fig. 7.29). There are several possible ways to bring together the two regions which are to undergo recombination. Intrastrand recombination could occur by looping out the intervening sequences and aligning the short consensus sequences (see Fig. 7.30). Or unequal interstrand recombination could occur at mitosis between two chromatids to produce one daughter cell with the required rearrangement. A third possibility which could bring together V and J regions of opposite orientation would involve the inversion of a region of DNA.

Whichever mechanism is the correct one (and intrastrand recombination is favoured) the two

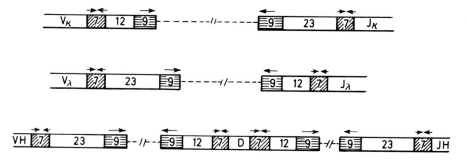

**Fig. 7.29** Details of the arrangement of the heptamer and nonamer sequences flanking the V, D and J regions of immunoglobulin genes.

regions which undergo the recombination always consist of one with a 12-bp spacer and the other with a 23-bp spacer (one and two turns of the double helix respectively). This is known as the 12/23 rule. Vκ regions all end with a 12-bp spacer and Vλ regions with a 23-bp spacer while the corresponding J regions begin with a 23-bp spacer (κ) or a 12-bp spacer (λ). This thus ensures that a V region always joins to a J region and not, for instance, to another V region. Similar rules apply to the rearrangements of the heavy-chain genes [298].

Although the two regions to be joined may align themselves at the heptameric and nanomeric sequences the actual site of the recombination is not fixed but can be at one of several different positions. This leads to a series of different junctions for every combination of V and J segments and hence a further increase in sequence diversity. If the recombination event leads to a change in reading frame (see Section 11.2) then J regions with entirely new amino sequences could be generated but when this occurs the joining is non-productive.

In the D/J joining reaction of heavy chain genes there appears to be some deletion of nucleotides at the joining site followed by insertion of nucleotides in an apparently random fashion. This insertion which may be catalysed by terminal transferase (see Section 6.4.2.e and Fig. 7.30) obviously leads to even greater diversity of nucleotide sequence [306]. In order to bring about the observed recombination, Alt and Baltimore [306] point out that the reaction must be initiated by breaks involving all four strands when they are aligned at the signal sequences (Fig. 7.30a). Joining of the signal sequences leads to their excision as an extrachromosomal circle (Fig. 7.30b). Double-stranded exonuclease and terminal transferase leads to the production of a short region of random nucleotide sequence (the N region – Fig. 7.30d) prior to ligation of the D and J regions [306]. Formation of the N region is restricted to D/J joining, a reaction which occurs at a time

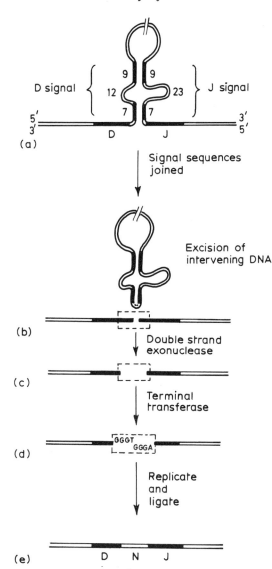

**Fig. 7.30**  A model for the D–$J_H$ recombination involved in formation of heavy-chain genes illustrating the involvement of terminal transferase in the generation of the hypervariable N (for nucleotides) region (based on the model of Alt and Baltimore [306], reproduced with permission).

when terminal transferase is active, prior to joining of the V regions in immature lymphocytes.

## (d) *The mechanism of class switching*

Because the $C\mu$ gene is closest to the $J_H$ segments it is this constant region which is initially expressed following the rearrangements of the variable region discussed above (see Fig. 7.28b). When class switching occurs the $C\mu$ gene region is replaced by another $C_H$ region but no change occurs in the variable region at this time, i.e. a lymphocyte only expresses one variable region but during its lifetime the constant region changes. Switching is brought about by recombination reactions which lead to the deletion of stretches of DNA. Recombination occurs at or near certain short, S sequences (e.g. TGAGC and TGGGG) which occur in the 5'-flanking region of each $C_H$ gene except the $C\delta$ gene [307–310]. These sequences resemble the Chi sequences which are recombination hotspots in *E. coli* (see Section 7.4.2).

Recombination between S sequences may occur on the same chromosome by a looping out mechanism or by sister chromatid exchange either of which mechanism leads to the deletion of the intervening DNA [302, 310, 311].

The change to expression of the $C\delta$ gene is believed to occur not by DNA rearrangement (there is no S sequence) but by read-through transcription and differential processing of the transcript (see Section 9.3) at a stage in development before class switching occurs [312, 313]. Such differential splicing may also lead to the transient coexpression of $C\mu$ and $C\gamma$ genes (and even $C\epsilon$ or $C\alpha$ genes) at a stage prior to DNA rearrangement [309, 314].

## (e) *Somatic mutation*

A comparison with germ-line sequences shows the presence of multiple single-base mutations in the V, J and D segments of both light- and heavy-chain genes [298, 315]. It has been proposed that these mutations arise in combination with the joining and switching rearrangements discussed above. The mechanism envisaged involves error-prone repair of gaps generated by nicking of the DNA followed

by exonuclease action (see Section 7.3.4). When such a mutation generates an antibody with increased affinity for the antigen it will be selected during the clonal expansion phase (see below).

## (f) *Clonal selection*

It is the variable region of the antibody molecule which represents the antigen-combining site. A particular antigen will interact with only a small number of lymphocytes each carrying on their surface an IgM with affinity for the antigen. When an antigen interacts with a particular lymphocyte that lymphocyte is induced to divide. This leads to the production of a clone of lymphocytes each having the particular DNA rearrangement coding for the active immunoglobulin (which may now be secreted as an IgG). This clonal selection was initially postulated to occur by MacFarlane Burnet [316] to explain how the immune system works, not by inducing change, but by selecting for and amplifying those lymphocytes carrying the most appropriate antibody. It has taken 20 years to elucidate the mechanisms used to generate the diversity of antibodies and lymphocytes on which clonal selection may act and in so doing a whole new insight has been obtained into the plasticity of DNA.

### 7.9.2 T-cell receptor genes

T-cell receptors are dimeric molecules containing an $\alpha$ and a $\beta$ chain both of which are divided into a variable (V) and a constant (C) region. The gene for the $\beta$ chain resembles the gene for the immunoglobulin H chain in that it is made up of $V\beta$, $D\beta$ (diversity), $J\beta$ (joining) and $C\beta$ segments that rearrange during T-cell differentiation. There are two adjacent $C\beta$ genes each associated with seven $J\beta$ segments and one $D\beta$ segment. In the mouse there are eleven $V\beta$ genes but some of these are closely related to each other (subfamilies) [317, 318]. The $\alpha$ chain genes consist of V, J and C regions only. There are at least ten subfamilies of Vα genes each containing from one to ten members. There is only one Cα gene associated with 19 Jα segments [319–322].

Rearrangement of the T cell receptor genes is required for expression and takes place in a manner strictly analogous to the rearrangement of the immunoglobulin genes in that the 12/23 rule is obeyed and there is the same scope for generating diversity [338, 339]. There is evidence in this case, however, that some rearrangements occur by chromosomal inversion [343, 344] (see below).

In addition, the receptor on cytotoxic T cells contains, in place of the α chain, a γ chain whose gene contains V, J and C regions. In this case diversity seems to be restricted to the V–J join but there are three Cγ segments each with its associated J segment. There is only limited expression of γ chain genes in the early stages of development of helper T cells [323, 338, 339].

### 7.9.3 Other gene rearrangements

The rearrangement of the immunoglobulin and T-cell receptor genes is an irreversible event occurring in terminally differentiating cells. It leads to the loss of the excised DNA. In this it resembles the excision of a lysogenic phage (Section 7.4.3) where a site-specific recombination joins together two previously distant regions of the genome with the loss of the intervening (phage) DNA.

A similar loss of DNA occurs during the development of the cyanobacterium *Anabaena* from a vegetative cell to a heterocyst when there is a rearrangement of the genes required for nitrogen fixation (*nif*). In the vegetative cell the *nifK* gene is separated by 11 kbp from the *nifD* and *nifH* genes, but during heterocyst formation the intervening DNA is excised to form a circular molecule which remains in the heterocyst. The rearrangement allows the three genes to function as a single transcriptional unit [324]. Heterocysts develop at regular intervals along a filament of

vegetative cells and, as with the antibody-producing cells, such differentiation is irreversible. Here we have an example of a colony of bacteria in which some cells sacrifice their potential for multiplication in order to supply fixed nitrogen compounds to the remaining cells.

Such rearrangements differ from the situation where the inversion of an intervening stretch of DNA can bring together two previously separated regions of DNA (e.g. flagellar phase variation in *Salmonella* [325] and *E. coli* [332], and genome isomerization in herpes simplex virus [334]) which is a potentially reversible reaction in which no DNA is lost. The molecular mechanisms of segment inversion and elimination events may not, however, be fundamentally different. Rather they may depend only on the relative orientation of the two sequences undergoing recombination [333, 344] (Fig. 7.11 and Fig. 7.17).

These examples illustrate the communality between transposition, lysogeny, the rearrangements of the antibody genes and the interchange of genetic material which occurs for instance during the exchange of the VSG genes of trypanosomes.

### 7.10 CHROMOSOMAL TRANSLOCATIONS

A chromosomal translocation is the reciprocal exchange of material between two chromosomes and although translocations have been observed for many years they have recently attracted greater interest as a result of their prevalence in certain cancers. Thus in chronic myeloid leukaemia (CML) there has been (in at least 90% of cases) an exchange between the end of the long arms of human chromosomes 9 and 22 such that chromosome 22 is shortened to form the characteristic Philadelphia chromosome. Other translocations associated with cancers are between chromosomes 8 and 21 (acute myeloid leukaemia: AML); 8 and 2, 8 and 14 or 8 and 22 (Burkitt's lymphoma); 15 and 17 (acute non-lymphocytic leukaemia), 11 and 14 (acute

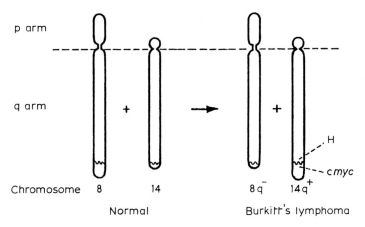

p arm

q arm

Chromosome    8      14         $8q^-$    $14q^+$

H

cmyc

Normal             Burkitt's lymphoma

**Fig. 7.31** A chromosomal translocation involved in the initiation of Burkitt's lymphoma brings the c-*myc* oncogene from chromosome 8 under the control of the immunoglobulin heavy chain promoter on chromosome 14.

lymphoblastic and chronic lymphocytic leukaemia); and 18 and 14 (follicular lymphoma) (see Fig. 7.31).

These diseases are all leukaemias, i.e. occur in the B or T lymphocytes, and many of the translocations occur within or close to the immunoglobulin genes on chromosomes 14 (heavy-chain genes), 22 (lambda light-chain genes) and 2 (kappa light-chain genes). Others occur near to the T cell receptor genes (e.g. the $\alpha$ chain gene on chromosome 14) [326–328].

It would therefore appear that these translocations occur as aberrations of events normally involved in the rearrangement of the immunoglobulin genes. The aberrant rearrangements bring together an oncogene, e.g. c-*myc* on chromosome 8 (see Section 3.6.9) and part of the antibody gene leading to the production of an altered oncogene product or enhanced oncogene transcription [329, 330].

Not all translocations involve aberrant immunoglobulin gene rearrangements. For instance, the small number of females who suffer from Duchenne muscular dystrophy have an X:autosome translocation [345].

The site of the initial pairing at meiosis between the X and Y chromosomes is a common sequence near the telomeres. This is also the site at which obligatory recombination occurs between one chromatid of the X and one chromatid of the Y chromosomes. The effect of this recombination is to make it appear that genes distal to the site of recombination are not sex-linked, i.e. they are pseudoautosomal [335–337].

REFERENCES

1 Singer, B. and Kusmievek, J. T. (1982), *Annu. Rev. Biochem.*, **51**, 655.
2 Pauling, L. (1964), *Bull. N.Y. Acad. Med.*, **40**, 334.
3 Friedberg, E. C. (1985), *DNA Repair*, W. H. Freeman and Co., San Francisco.
4 Singer, B. (1975), *Prog. Nucleic Acid Res. Mol. Biol.*, **15**, 219.
5 Tannenbaum, S. R., Weisman, M. and Fett, D. (1976), *Food Cosmet. Toxicol.*, **114**, 549.
6 Stock, J. A. (1975), *Biology of Cancer* (2nd edn) (ed. E. J. Ambrose and F. S. C. Roe), Ellis Horwood, Chichester, p. 279.
7 Zarbl, H., Sukumar, S., Arthur, A. V., Martin-Zanka, D. and Barbacid, M. (1985), *Nature (London)*, **315**, 382.
8 Sobell, H. M. (1973), *Prog. Nucleic Acid Res. Mol. Biol.*, **13**, 153.

9  Walker, R. T., De Clercq, E. and Eckstein, F. (1979), *Nucleoside Analogues*, Plenum Press, New York.

10 Kersten, H. and Kersten, W. (1974), *Inhibitors of Nucleic Acid Synthesis*, Chapman and Hall, London.

11 Suhadolnik, R. J. (1979), *Prog. Nucleic Acid Res. Mol. Biol.*, **22**, 193.

12 Lasnitski, J., Matthews, R. E. F. and Smith, J. D. (1954), *Nature (London)*, **173**, 346.

13 Lipmann, F. (1963), *Prog. Nucleic Acid Res.*, **1**, 135.

14 Loeb, L. A., Fansler, B., Williams, R. and Mazia, D. (1969), *Exp. Cell Res.*, **57**, 298.

15 Priest, J. H., Heady, J. E. and Priest, R. E. (1967), *J. Cell Biol.*, **35**, 483.

16 Matthews, R. E. F. (1953), *Nature (London)*, **171**, 1065.

17 Hilmoe, R. J. and Heppell, L. A. (1957), *J. Am. Chem. Soc.*, **79**, 4810.

18 Zain, B. S., Adams, R. L. P. and Imrie, R. C. (1973), *Cancer Res.*, **33**, 40.

19 Adams, R. L. P. and Burdon, R. H. (1985), *The Molecular Biology of DNA Methylation*, Springer Verlag, New York.

20 Taylor, S. M. and Jones, P. A. (1979), *Cell*, **17**, 771.

21 Roy-Bowman, P. (1970), *Analogues of Nucleic Acid Components*, Springer Verlag, New York.

22 Kaufman, E. R. and Davidson, R. L. (1978), *Proc. Natl. Acad. Sci. USA*, **75**, 4982.

23 Hopkins, R. L. and Goodman, M. F. (1980), *Proc. Natl. Acad. Sci. USA*, **77**, 1801.

24 Hunter, T. and Francke, B. (1975), *J. Virol.*, **15**, 759.

25 Lawley, P. D. and Brookes, P. (1961), *Nature (London)*, **192**, 1081.

26 Lijinsky, W. (1976), *Prog. Nucleic Acid Res. Mol. Biol.*, **17**, 247.

27 Roberts, J. J. (1975), *Biology of Cancer* (ed. E. J. Ambrose and F. J. C. Roe), Ellis Horwood, Chichester.

28 Lindahl, T. (1981), *Chromosome Damage and Repair* (ed. E. Seeberg and K. Kleppe), Plenum Press, New York, p. 207.

29 Eadie, J. S., Conrad, M., Toorchen, D. and Topal, M. D. (1984), *Nature (London)*, **308**, 201.

30 Iyer, V. N. and Szybalski, W. (1963), *Proc. Natl. Acad. Sci. USA*, **50**, 355.

31 Goldberg, I. H. and Friedman, P. A. (1971), *Annu. Rev. Biochem.*, **40**, 775.

32 Pietsch, P. and Garrett, H. (1968), *Nature (London)*, **219**, 488.

33 Falaschi, A. and Kornberg, A. (1964), *Fed. Proc.*, **23**, 940.

34 Stern, R., Rose, J. A. and Friedman, R. M. (1974), *Biochemistry*, **13**, 307.

35 Jeffreys, A. J., Wilson, V. and Thein, S. L. (1985), *Nature (London)*, **314**, 67.

36 Goldberg, I. H., Reich, E. and Rabinowitz, M. (1963), *Nature (London)*, **199**, 44.

37 Neidle, S. and Abraham, Z. (1985), *CRC Crit. Rev. Biochem.*, **17**, 73.

38 Radloff, R., Bawer, W. and Vinograd, J. (1967), *Proc. Natl. Acad. Sci. USA*, **57**, 1514.

39 Reich, E. and Goldberg, I. H. (1964), *Prog. Nucleic Acid Res.*, **3**, 184.

40 Hurwitz, J. and August, J. T. (1963), *Prog. Nucleic Acid Res.*, **1**, 59.

41 Penman, S., Vesco, C. and Penman, M. (1968), *J. Mol. Biol.*, **34**, 49.

42 Okada, Y., Shreisinger, G., Owen, J. E., Newton, J., Tsugita, A. and Inouye, M. (1972), *Nature (London)*, **236**, 338.

43 deBoer, J. G. and Ripley, L. S. (1984), *Proc. Natl. Acad. Sci. USA*, **81**, 5528.

44 Pelc, S. R. and Howard, A. (1955), *Radiat. Res.*, **3**, 135.

45 Terasima, T. and Tolmach, L. J. (1963), *Science*, **140**, 490.

46 Sinclair, W. K. (1967), *Proc. Natl. Acad. Sci. USA*, **58**, 115.

47 Lajtha, L. G. (1960), in *The Nucleic Acids* (ed. E. Chargaff and J. N. Davidson), Academic Press, New York, Vol. 3, p. 527.

48 Puck, T. T. and Steffen, J. (1963), *Biophys. J.*, **3**, 379.

49 Henner, W. D., Grunberg, S. M. and Haseltine, W. A. (1983), *J. Biol. Chem.*, **258**, 15 198.

50 Hanawalt, P. C. (1972), *Endeavour*, **31**, 83.

51 Haseltine, W. A. (1983), *Cell*, **33**, 13.

52 Sutherland, B. M., Chamberlin, M. J. and Sutherland, J. C. (1973), *J. Biol. Chem.*, **248**, 4200.

53 Lehman, A. R. and Bridges, B. A. (1977), *Essays Biochem.*, **13**, 71.

54 Hanawalt, P. C., Cooper, P. K., Ganesan, A. K. and Smith, C. A. (1979), *Annu. Rev. Biochem.*, **48**, 783.

55 Sancar, A., Franklin, K. A. and Sancar, G. B. (1984), *Proc. Natl. Acad. Sci. USA*, **81**, 7397.

56 Schendel, P. and Robins, P. (1978), *Proc. Natl.*

*Acad. Sci. USA*, **75**, 6077.

57 Karran, P., Lindahl, T. and Griffin, B. E. (1979), *Nature (London)*, **280**, 76.

58 Olsson, M. and Lindahl, T. (1980), *J. Biol. Chem.*, **255**, 10 569.

59 Hornsey, S. and Howard, A. (1956), *Ann. N.Y. Acad. Sci.*, **63**, 915.

60 Kushner, S. R., Kaplan, J. C., Ono, H. and Grossman, L. (1971), *Biochemistry*, **10**, 3325.

61 Kaplan, J. C., Kushner, S. R. and Grossman, L. (1971), *Biochemistry*, **10**, 3315.

62 Setlow, R. B. and Carrier, W. L. (1964), *Proc. Natl. Acad. Sci. USA*, **51**, 226.

63 Monk, M., Peacey, M. and Gross, J. D. (1971), *J. Mol. Biol.*, **58**, 623.

64 Kato, T. and Kondo, S. (1970), *J. Bacteriol.*, **104**, 871.

65 Cooper, P. K. and Hanawalt, P. C. (1972), *J. Mol. Biol.*, **67**, 1.

66 Glickman, B. W. (1974), *Biochim. Biophys. Acta*, **335**, 115.

67 Dorson, J. W., Deutsch, W. A. and Moses, R. E. (1978), *J. Biol. Chem.*, **253**, 660.

68 Grossman, L. (1974), *Adv. Radiat. Biol.*, **4**, 77.

69 Livneh, Z., Elad, D. and Sperling, J. (1979), *Proc. Natl. Acad. Sci. USA*, **76**, 1089.

70 Deutsch, W. A. and Linn, S. (1979), *J. Biol. Chem.*, **254**, 12 099.

71 Gates, F. T. and Linn, S. (1977), *J. Biol. Chem.*, **252**, 1647.

72 Talpaert-Barle, M., Clerici, L. and Campagnari, F. (1979), *J. Biol. Chem.*, **254**, 6387.

73 Wist, E., Uhjem, O. and Krokan, H. (1978), *Biochim. Biophys. Acta*, **520**, 253.

74 Tamanoi, F. and Okazaki, T. (1978), *Proc. Natl. Acad. Sci. USA*, **75**, 2195.

75 Karran, P. and Lindahl, T. (1978), *J. Biol. Chem.*, **253**, 5877.

76 Laval, J. (1977), *Nature (London)*, **269**, 829.

77 Kirtikar, D. M. and Goldthwait, D. A. (1974), *Proc. Natl. Acad. Sci. USA*, **71**, 2022.

78 Kirtikar, D. M., Dipple, A. and Goldthwait, D. A. (1975), *Biochemistry*, **14**, 5548.

79 Cooper, P. (1977), *Mol. Gen. Genet.*, **150**, 1.

80 Demple, B. and Linn, S. (1980), *Nature (London)*, **287**, 203.

81 Breimer, L. H. and Lindahl, T. (1984), *J. Biol. Chem.*, **259**, 5543.

82 Seeberg, E. (1981), *Prog. Nucleic Acid Res. Mol. Biol.*, **26**, 217.

83 Sancar, A. and Rupp, W. D. (1983), *Cell*, **33**, 249.

84 Seeberg, E. and Steinum, A-L. (1983), in *Cellular Responses to DNA Damage* (ed. E. D. Friedberg and B. A. Bridges), Alan R. Liss, New York, p. 39.

85 Yeung, A. T., Mattes, W. B., Oh, E. Y. and Grossman, L. (1983), *Proc. Natl. Acad. Sci. USA*, **80**, 6157.

86 Lindahl, T. (1979), *Prog. Nucleic Acid Res. Mol. Biol.*, **22**, 135.

87 Tanaka, K., Sekiguchi, M. and Okada, Y. (1975), *Proc. Natl. Acad. Sci. USA*, **72**, 4071.

88 Smith, C. A. and Hanawalt, P. C. (1978), *Proc. Natl. Acad. Sci. USA*, **75**, 2598.

89 Marx, J. L. (1978), *Science*, **200**, 518.

90 Park, S. D. and Cleaver, J. E. (1979), *Nucleic Acids Res.*, **7**, 1151.

91 Radman, M., Villani, G., Boiteux, S., Kinsella, A. R., Glickman, B. W. and Spadari, S. (1978), *Cold Spring Harbor Symp. Quant. Biol.*, **43**, 937.

92 Marinus, M. G. and Morris, N. R. (1973), *J. Bacteriol.*, **114**, 1143.

93 Glickman, B. W. and Radman, M. (1980), *Proc. Natl. Acad. Sci. USA*, **77**, 1063.

94 Lu, A-L., Clark, S. and Modrich, P. (1983), *Proc. Natl. Acad. Sci. USA*, **80**, 4639.

95 Jeffreys, A. J., Brookfield, J. F. Y. and Semeonoff, R. (1985), *Nature (London)*, **317**, 818.

96 Lackey, D., Krauss, S. W. and Linn, S. (1985), *J. Biol. Chem.*, **260**, 3178.

97 Witkin, E. M. and Kogoma, T. (1984), *Proc. Natl. Acad. Sci. USA*, **81**, 7539.

98 Cole, R. S. (1973), *Proc. Natl. Acad. Sci. USA*, **70**, 1064.

99 Higgins, N. P., Kato, K. and Strauss, B. (1976), *J. Mol. Biol.*, **101**, 417.

100 Radding, C. M. (1978), *Annu. Rev. Biochem.*, **47**, 847.

101 McKay, V. and Linn, S. (1976), *J. Biol. Chem.*, **251**, 3716.

102 Rosamond, J., Telander, K. M. and Linn, S. (1979), *J. Biol. Chem.*, **254**, 8646.

103 Telander-Muskavitch and Linn, S. (1981), in *The Enzymes* (ed. P. D. Boyer), Academic Press, New York, Vol. 14, p. 233.

104 Taylor, F. and Smith, G. R. (1980), *Cell*, **22**, 447.

105 Chaudhury, A. M. and Smith, G. R. (1984), *Proc. Natl. Acad. Sci. USA*, **81**, 7850.

106 Ponticelli, A. S., Schultz, D. W., Taylor, A. F. and Smith, G. R. (1985), *Cell*, **41**, 145.

107 Taylor, A. F., Schultz, D. W., Ponticelli, A. S.

and Smith, G. R. (1985), *Cell*, **41**, 153.

108 Bridges, B. A. (1979), *Nature (London)*, **277**, 514.

109 Shibata, T., Das-Gupta, C., Cunningham, R. P. and Radding, C. M. (1979), *Proc. Natl. Acad. Sci. USA*, **76**, 1638.

110 Cunningham, R. P., Shibata, T., Das-Gupta, C. and Radding, C. M. (1979), *Nature (London)*, **281**, 191.

111 Weinstock, G. M., McEntee, K. and Lehman, I. R. (1979), *Proc. Natl. Acad. Sci. USA*, **76**, 126.

112 McEntee, K., Weinstock, G. M. and Lehman, I. R. (1980), *Proc. Natl. Acad. Sci. USA*, **77**, 857.

113 Chrysogelos, S., Register, J. C. and Griffith, J. (1983), *J. Biol. Chem.*, **258**, 12 624.

114 Smith, G. R. (1983), *Cell*, **34**, 709.

115 Konrad, E. B. (1977), *J. Bacteriol.*, **130**, 167.

116 Bianchi, M. E. and Radding, C. M. (1983), *Cell*, **35**, 511.

117 Szostak, J. W., Orr-Weaver, T. L. and Rothstein, R. J. (1983), *Cell*, **33**, 25.

118 Whitehouse, H. L. K. (1983), *Nature (London)*, **306**, 645.

119 West, S. C., Cassuto, E., Mursalim, J. and Howard Flanders, P. (1980), *Proc. Natl. Acad. Sci. USA*, **77**, 2569.

120 DiCapua, E., Engel, E., Stasiak, A. and Koller, T. (1982), *J. Mol. Biol.*, **157**, 87.

121 Howard Flanders, P., West, S. C. and Stasiak, A. (1984), *Nature (London)*, **309**, 215.

122 Potter, H. and Dressler, D. (1978), *Proc. Natl. Acad. Sci. USA*, **75**, 3698.

123 Shibata, T., Cunningham, R. F., Das-Gupta, C. and Radding, C. M. (1979), *Proc. Natl. Acad. Sci. USA*, **76**, 5100.

124 Cassuto, E., West, S. C., Mursalim, J., Conlon, S. and Howard Flanders, P. (1980), *Proc. Natl. Acad. Sci. USA*, **77**, 3962.

125 West, S. C., Cassuto, E. and Howard Flanders, P. (1981), *Proc. Natl. Acad. Sci. USA*, **78**, 6149.

126 Kahn, R., Cunningham, R. P., Das-Gupta, C. and Radding, C. M. (1981), *Proc. Natl. Acad. Sci. USA*, **78**, 4786.

127 Cox, M. M. and Lehman, I. R. (1981), *Proc. Natl. Acad. Sci. USA*, **78**, 6018.

128 Das-Gupta, C. and Radding, C. M. (1981), *Proc. Natl. Acad. Sci. USA*, **78**, 4786.

129 Cox, M. M. and Lehman, I. R. (1981), *Proc. Natl. Acad. Sci. USA*, **78**, 3433.

130 West, S. C., Cassuto, E. and Howard Flanders, P. (1981), *Proc. Natl. Acad. Sci. USA*, **78**, 2100.

131 Dressler, D. and Potter, H. (1982), *Annu. Rev. Biochem.*, **51**, 727.

132 Holliday, R. (1964), *Genet. Res.*, **5**, 282.

133 Meselson, M. S. and Radding, C. M. (1975), *Proc. Natl. Acad. Sci. USA*, **72**, 358.

134 Manly, K. F., Signer, E. R. and Radding, C. M. (1969), *Virology*, **37**, 177.

135 Cassuto, E. and Radding, C. (1971), *Nature (London) New Biol.*, **229**, 13.

136 Boon, T. and Zinder, N. D. (1971), *J. Mol. Biol.*, **58**, 133.

137 Potter, H. and Dressler, D. (1979), *Cold Spring Harbor Symp. Quant. Biol.*, **43**, 969.

138 Potter, H. and Dressler, D. (1976), *Proc. Natl. Acad. Sci. USA*, **73**, 3000.

139 Thompson, B. J., Camien, M. N. and Warner, R. C. (1976), *Proc. Natl. Acad. Sci. USA*, **73**, 2299.

140 Craig, N. L. (1984), *Nature (London)*, **311**, 706.

141 Lilley, D. M. J. and Kemper, B. (1984), *Cell*, **36**, 413.

142 Broda, P. (1967), *Genet. Res.*, **9**, 35.

143 Vetter, D., Andrews, B. J., Roberts-Beatty, L. and Sadowsky, P. D. (1983), *Proc. Natl. Acad. Sci. USA*, **80**, 7284.

144 Yin, S., Bushman, W. and Landy, A. (1985), *Proc. Natl. Acad. Sci. USA*, **82**, 1040.

145 Campbell, A. (1969), *Episomes*, Harper and Row, New York.

146 Hsu, P-L, Ross, W. and Landy, A. (1980), *Nature (London)*, **285**, 85.

147 Mizuuchi, K. (1981), *Cold Spring Harbor Symp. Quant. Biol.*, **45**, 429.

148 Kotewicz, M., Chung, S., Takeda, Y. and Echols, H. (1977), *Proc. Natl. Acad. Sci. USA*, **74**, 1511.

149 Ross, W. and Landy, A. (1983), *Cell*, **33**, 261.

150 Ross, W. and Landy, A. (1982), *Proc. Natl. Acad. Sci. USA*, **79**, 7724.

151 Hsu, P. L. and Landy, A. (1984), *Nature (London)*, **311**, 721.

152 Craig, N. L. and Nash, H. A. (1983), *Cell*, **35**, 795.

153 Warner, H. (1983), in *Enzymes of Nucleic Acid Synthesis and Modification* (ed. S. T. Jacob), CRC Press, Cleveland, Vol. 1, p. 145.

154 Abo-Darub, J. M., Mackie, R. and Pitts, J. D. (1978), *Bull. Cancer*, **63**, 357.

155 Pegg, A. E. and Bennett, R. A. (1983), in *Enzymes of Nucleic Acid Synthesis and Modification* (ed. S. T. Jacob), CRC Press, Cleveland, Vol. 1, p. 79.

156 Gall, J. G. (1969), *Genetics*, **61**, Suppl. 121.

157 Bird, A. P. (1978), *Cold Spring Harbor Symp. Quant. Biol.*, **42**, 1179.

158 MacGregor, H. C. (1972), *Biol. Rev.*, **47**, 173.

159 Hourcade, D., Dressler, D. and Wolfson, J. (1974), *Cold Spring Harbor Symp. Quant. Biol.*, **38**, 537.

160 Rochaix, J-D. and Bird, A. P. (1975), *Chromosoma*, **52**, 317.

161 Stark, G. R. and Wahl, C. M. (1984), *Annu. Rev. Biochem.*, **53**, 447.

162 Osheim, Y. N. and Miller, O. L. (1983), *Cell*, **33**, 543.

163 Spradling, A. C. and Mahawold, A. P. (1981), *Cell*, **27**, 203.

164 deCicco, D. V. and Spradling, A. C. (1984), *Cell*, **38**, 45.

165 Ish-Horowitz, D. (1982), *Nature (London)*, **296**, 806.

166 Glover, D. M., Zaha, A., Stocker, A. J., Santelli, R. V., Pueyo, M. T., deToledo, S. H. and Lara, F. J. S. (1982), *Proc. Natl. Acad. Sci. USA*, **79**, 2947.

167 Yao, M-C., Zhu, S-G. and Yao, C-H. (1985), *Mol. Cell Biol.*, **5**, 1260.

168 Karrer, K. and Gall, J. G. (1976), *J. Mol. Biol.*, **104**, 421.

169 Swanton, M. T., Heumann, J. M. and Prescott, D. M. (1980), *Chromosoma*, **77**, 217.

170 Bostock, C. J. (1984), *J. Embryol. Exp. Morphol.*, **83**, Supplement 7.

171 Schimke, R. T. (1984), *Cell*, **37**, 705.

172 Schimke, R. T., Kaufman, R. J., Alt, F. W. and Kellems, R. F. (1978), *Science*, **202**, 1051.

173 Hunt, S. W. and Hoffee, P. A. (1983), *J. Biol. Chem.*, **258**, 13 185.

174 Yeung, C-Y., Frayne, E. G., Al-Ubaidi, M. R., Hook, A. G., Ingolia, D. E., Wright, D. A. and Kellems, R. E. (1983), *J. Biol. Chem.*, **258** 15 179.

175 Coleman, P. F., Suttle, D. P. and Stark, G. R. (1977), *J. Biol. Chem.*, **252**, 6379.

176 Beverley, S. M., Coderre, J. A., Santi, D. V. and Schimke, R. T. (1984), *Cell*, **38**, 431.

177 Anderson, R. P. and Roth, J. R. (1977), *Annu. Rev. Microbiol.*, **31**, 473.

178 Johnston, R. N., Beverley, S. M. and Schimke, R. T. (1983), *Proc. Natl. Acad. Sci. USA*, **80**, 3711.

179 Mariani, B. D. and Schimke, R. T. (1984), *J. Biol. Chem.*, **259**, 1901.

180 Denhardt, D. T. (1984), *Nature (London)*, **309**, 575.

181 Botchan, M. R., Topp, W. C. and Sambrook, J. (1979), *Cold Spring Harbor Symp. Quant. Biol.*, **45**, 709.

182 Baran, N., Neer, A. and Manor, H. (1983), *Proc. Natl. Acad. Sci. USA*, **80**, 105.

183 Wahl, G. M., deSaint Vincent, B. R. and DeRose, M. L. (1984), *Nature (London)*, **307**, 516.

184 Roninson, I. B., Abelson, H. T., Housman, D. E., Howell, N. and Varshavsky, A. (1984), *Nature (London)*, **309**, 626.

185 Borst, P. (1984), *Nature (London)*, **309**, 580.

186 Roberts, J. H., Buck, L. B. and Axel, R. (1983), *Cell*, **33**, 53.

187 Bostock, C. I. and Tyler-Smith, C. (1981), *J. Mol. Biol.*, **153**, 219.

188 Caron, P. R., Kushner, S. R. and Grossman, L. (1985), *Proc. Natl. Acad. Sci. USA*, **82**, 4925.

189 Buckingham, M. E. (1985), *Essays Biochem.*, **20**, 77.

190 Pfeffer, S. and Ullrich, A. (1985), *Nature (London)*, **313**, 184.

191 Efstratiadis, A., Posakony, J. W., Maniatis, T., Lawn, R. M., O'Connell, C., Spritz, R. A., DeRiel, J. K., Forget, B. G., Weissman, S. M., Slightom, J. L., Blechl, A. E., Smithers, O., Baralle, F. E., Shoulders, C. C. and Proudfoot, N. (1981), *Cell*, **26**, 653.

192 Proudfoot, N. J., Gil, A. and Maniatis, T. (1982), *Cell*, **31**, 553.

193 Jeffreys, A. J., Wilson, V., Wood, D., Simons, J. P., Kay, R. M. and Williams, J. G. (1980), *Cell*, **21**, 555.

194 Lang, A. and Lorkin, P. A. (1976), *Brit. Med. Bull.*, **32**, 239.

195 Goodman, M., Koop, B. F., Czelusniak, J., Weiss, M. L. and Slightom, J. L. (1984), *J. Mol. Biol.*, **180**, 803.

196 Lauer, J., Shen, C-K. S. and Maniatis, T. (1980), *Cell*, **20**, 119.

197 Shen, S., Slightom, J. L. and Smithies, O. (1981), *Cell*, **26**, 191.

198 Martin, S. L., Vincent, K. A. and Wilson, A. C. (1983), *J. Mol. Biol.*, **164**, 513.

199 Proudfoot, N. J. and Maniatis, T. (1980), *Cell*, **21**, 537.

200 Maeda, N. (1985), *J. Biol. Chem.*, **260**, 6698.

201 Smith, G. P. (1976), *Science*, **191**, 528.

202 Jeffreys, A. J. and Harris, S. (1984), *BioEssays*, **1**, 253.

203 Little, P. F. R. (1982), *Cell*, **28**, 683.
204 Ford, P. (1978), *Nature (London)*, **271**, 205.
205 Zimmer, E. A., Martin, S. L., Beverley, S. M., Kan, Y. and Wilson, A. C. (1980), *Proc. Natl. Acad. Sci. USA*, **77**, 2158.
206 Tartof, K. D. (1974), *Proc. Natl. Acad. Sci. USA*, **71**, 1272.
207 Petes, T. D. (1980), *Cell*, **19**, 765.
208 Baltimore, D. (1981), *Gene*, **24**, 592.
209 Slightom, J. L., Blechl, A. E. and Smithies, O. (1980), *Cell*, **21**, 627.
210 Jeffries, A. J., Harris, S., Barrie, P. A., Wood, D., Blanchetot, A. and Adams, S. M. (1983), in *Evolution from Molecules to Man* (ed. D. S. Bendall), Cambridge University Press, p. 175.
211 McClintock, B. (1984), *Science*, **226**, 792.
212 Jordan, E., Saedler, H. and Starlinger, P. (1968), *Mol. Gen. Genet.*, **102**, 353.
213 Shapiro, J. A. (1969), *J. Mol. Biol.*, **40**, 93.
214 Reznikoff, W. S. (1983), in *Gene Function in Prokaryotes* (eds J. Beckwith, J. Davies and J. A. Gallant), Cold Spring Harbor Monograph Series, p. 229.
215 Syvanen, M. (1984), *Annu. Rev. Genet.*, **18**, 271.
216 Calos, M. P. and Miller, J. H. (1980), *Cell*, **20**, 579.
217 Shapiro, J. A. (1983), *Mobile Genetic Elements*, Academic Press, New York.
218 Grindley, N. D. F. and Reed, R. R. (1985), *Annu. Rev. Biochem.*, **54**, 863.
219 Iida, S., Meyer, J. and Arber, W. (1983), in *Mobile Genetic Elements* (ed. J. A. Shapiro), Academic Press, New York, p. 159.
220 Kleckner, N. (1983), in *Mobile Genetic Elements* (ed. J. A. Shapiro), Academic Press, New York, p. 261.
221 Heffron, F. (1983), in *Mobile Genetic Elements* (ed. J. A. Shapiro), Academic Press, New York, p. 223.
222 Grindley, N. G. F. (1985), *Cell*, **32**, 3.
223 Toussaint, A. and Resibois, A. (1983), in *Mobile Genetic Elements* (ed. J. A. Shapiro), Academic Press, New York, p. 105.
224 Grindley, N. G. F. and Sherratt, D. J. (1979), *Cold Spring Harbor Symp. Quant. Biol.*, **49**, 1257.
225 Shapiro, J. A. (1979), *Proc. Natl. Acad. Sci. USA*, **76**, 1933.
226 Ohtsubo, E., Zenilman, M., Ohtsubo, H., McCormick, M., Machida, C. and Machida, Y. (1981), *Cold Spring Harbor Symp. Quant. Biol.*, **45**, 283.
227 Mizuuchi, K. (1984), *Cell*, **39**, 395.
228 Craigie, R. and Mizuuchi, K. (1985), *Cell*, **41**, 867.
229 Finegan, D. J. (1985), *Int. Rev. Cytol.*, **93**, 281.
230 Varmus, H. E. (1983), in *Mobile Genetic Elements* (ed. J. A. Shapiro), Academic Press, New York, p. 411.
231 Baltimore, D. (1985), *Cell*, **40**, 481.
232 Yoshinaka, Y., Katoh, I., Copeland, T. D. and Oroszlan, S. (1985), *Proc. Natl. Acad. Sci. USA*, **82**, 1618.
233 Panganiban, A. T. and Temin, H. M. (1984), *Cell*, **36**, 673.
234 Panganiban, A. T. and Temin, H. M. (1983), *Nature (London)*, **306**, 155.
235 Panganiban, A. T. and Temin, H. M. (1984), *Proc. Natl. Acad. Sci. USA*, **81**, 7885.
236 Moelling, K. (1985), *Adv. Cancer Res.*, **43**, 205.
237 Jaenisch, R. (1983), *Cell*, **32**, 5.
238 Jenkins, N. A., Copeland, N. G., Taylor, B. A. and Lee, B. K. (1981), *Nature (London)*, **293**, 370.
239 Rubin, G. M. (1983), in *Mobile Genetic Elements* (ed. J. A. Shapiro), Academic Press, New York, p. 329.
240 Roeder, G. S. and Fink, G. R. (1983), in *Mobile Genetic Elements* (ed. J. A. Shapiro), Academic Press, New York, p. 299.
241 Saigo, K., Kugimiya, W., Matsuo, Y., Inouye, S., Yoshioka, K. and Yuki, S. (1984), *Nature (London)*, **312**, 659.
242 Mount, S. M. and Rubin, G. M. (1985), *Mol. Cell. Biol.*, **5**, 1630.
243 Shiba, T. and Saigo, K. (1983), *Nature (London)*, **302**, 119.
244 Flavell, A. J. and Ish-Horowicz, D. (1981), *Nature (London)*, **292**, 591.
245 Flavell, A. J. and Ish-Horowicz, D. (1983), *Cell*, **34**, 415.
246 Clare, J. and Farabaugh, P. (1985), *Proc. Natl. Acad. Sci. USA*, **82**, 2829.
247 Hauber, J., Nelböck-Hochstetter, P. and Feldmann, H. (1985), *Nucleic Acids Res.*, **13**, 2745.
248 Mellor, J., Fulton, A. M., Dobson, M. J., Roberts, N. A., Wilson, W., Kingsman, A. and Kingsman, S. (1985), *Nucleic Acids Res.*, **13**, 6249.
249 Boeke, J. D., Garfinkel, D. J., Styles, C. A. and Fink, G. R. (1985), *Cell*, **40**, 491.
250 Engels, W. R. (1983), *Annu. Rev. Genet.*, **17**, 315.

251 O'Hare, K. and Rubin, G. M. (1983), *Cell*, **34**, 25.

252 Döring, H. P., Tillmann, E. and Starlinger, P. (1984), *Nature (London)*, **307**, 127.

253 Pohlman, R. F., Fedoroff, N. V. and Messing, J. (1984), *Cell*, **37**, 635.

254 Truett, M. A., Jones, R. S. and Potter, S. S. (1981), *Cell*, **24**, 753.

255 Potter, S. S. (1982), *Nature (London)*, **297**, 201.

256 Collins, M. and Rubin, G. M. (1984), *Nature (London)*, **308**, 323.

257 Brierley, H. L. and Potter, S. S. (1985), *Nucleic Acids Res.*, **13**, 485.

258 Rogers, J. H. (1985), *Int. Rev. Cytol.*, **93**, 187.

259 Wichman, H. A., Potter, S. S. and Pine, D. S. (1985), *Nature (London)*, **317**, 77.

260 Vanin, E. F. (1984), *Biochim. Biophys. Acta*, **782**, 231.

261 Stein, J. P., Munjaal, R. P., Lagace, L., Lai, E. C., O'Malley, B. W. and Means, A. R. (1983), *Proc. Natl. Acad. Sci. USA*, **80**, 6485.

262 Soares, M. B., Schon, E., Henderson, A., Karathanasis, S. K., Cate, R., Zeitlin, S., Chirgwin, J. and Efstratiadis, A. (1985), *Mol. Cell. Biol.*, **5**, 2090.

263 Ponte, P., Gunning, P., Blau, H. and Kedes, L. (1983), *Mol. Cell. Biol.*, **3**, 1783.

264 Vanin, E. F., Goldberg, G. I., Tucker, P. W. and Smithies, O. (1980), *Nature (London)*, **286**, 222.

265 Hollis, G. F., Heiter, P. A., McBride, O. W., Swan, D. and Leder, P. (1982), *Nature (London)*, **296**, 321.

266 Moss, M. and Gallwitz, D. (1983), *EMBO J.*, **2**, 757.

267 Van Ardsall, S. W., Denison, R. A., Bernstein, L. B., Weiner, A. M., Manser, T. and Gesteland, R. F. (1981), *Cell*, **26**, 11.

268 Bernstein, L. B., Mount, S. M. and Weiner, A. M. (1983), *Cell*, **32**, 461.

269 Hess, J. F., Fox, M., Schmid, C. and Shen, C-K. J. (1983), *Proc. Natl. Acad. Sci. USA*, **80**, 5970.

270 Ono, M., Toh, H., Miyata, T. and Awaya, T. (1985), *J. Virol.*, **55**, 387.

271 Jelinek, W. R. and Schmid, C. W. (1982), *Annu. Rev. Biochem.*, **51**, 813.

272 Ullu, E. and Tschudi, C. (1984), *Nature (London)*, **312**, 171.

273 Ullu, E. and Weiner, A. M. (1984), *EMBO J.*, **3**, 3303.

274 Jagadeeswaran, P., Forget, B. G. and Weissman, S. M. (1981), *Cell*, **26**, 141.

275 Krolewski, J. J. and Rush, M. G. (1984), *J. Mol. Biol.*, **174**, 31.

276 Singer, M. F. and Skowronski, J. (1985), *Trends Biochem. Sci.*, **10**, 119.

277 Shafit-Zagardo, B., Brown, F. L., Zavodny, P. J. and Maio, J. J. (1983), *Nature (London)*, **304**, 277.

278 Saunders, J. R. (1985), *Nature (London)*, **315**, 100.

279 Meier, J. T., Simon, M. I. and Barbour, A. G. (1985), *Cell*, **41**, 403.

280 Segal, E., Billyard, E., So, M., Starzbach, S. and Meyer, T. F. (1985), *Cell*, **40**, 293.

281 Nasmyth, K. A. (1983), *Nature (London)*, **302**, 670.

282 Nasmyth, K. A. (1982), *Annu. Rev. Genet.*, **16**, 439.

283 Strathern, J. N., Klar, A. J. S., Hicks, J. B., Abraham, J. A., Ivy, J. M., Nasmyth, K. A. and McGill, C. (1982), *Cell*, **31**, 183.

284 Kostriken, R., Strathern, J. N., Klar, A. J. S., Hicks, J. B. and Heffron, F. (1983), *Cell*, **35**, 167.

285 Donelson, J. E. and Rice-Ficht, A. C. (1985), *Microbiol. Rev.*, **49**, 107.

286 Bernards, A. (1984), *Biochim. Biophys. Acta*, **824**, 1.

287 Donelson, J. E. and Turner, M. J. (1985), *Sci. Am.*, **252(2)**, 32.

288 Borst, P., Bernards, A., Van der Ploeg, L. H. T., Michels, P. A. M., Lin, A. Y. C., De Lange, T. and Kooter, J. M. (1983), *Eur. J. Biochem.*, **137**, 383.

289 Blackburn, E. H. and Challoner, P. B. (1984), *Cell*, **36**, 459.

290 Van der Ploeg, L. H. T., Lin, A. Y. C. and Borst, P. (1984), *Cell*, **36**, 459.

291 Michels, P. A. M., Lin, A. Y. C., Bernards, A., Sloff, P., Van der Bijl, M. M. W., Schinkel, A-H., Menke, H. H., Borst, P., Veeneman, G. H., Tromp, M. M. and Van Broom, J. H. (1983), *J. Mol. Biol.*, **166**, 537.

292 Borst, P. and Cross, G. A. M. (1982), *Cell*, **29**, 291.

293 Pays, E., Guyaux, M., Aerts, D., Van Meirvenne, N. and Steinert, M. (1985), *Nature (London)*, **316**, 562.

294 Alberts, B., Bray, D., Lewis, J., Raff, M., Roberts, K. and Watson, J. D. (1983), *The Molecular Biology of the Cell*, Garland, New York.

295 Cushley, W. and Williamson, A. R. (1982), *Essays Biochem.*, **18**, 1.

296 Roit, I. (1980), *Essential Immunology* (4th edn), Blackwell, Oxford.

297 Dreyer, W. J. and Bennett, J. C. (1965), *Proc.*

*Natl. Acad. Sci. USA*, **54**, 864.

298 Tonegawa, S. (1983), *Nature (London)*, **302**, 575.

299 Robertson, R. (1982), *Nature (London)*, **297**, 184.

300 Tonegawa, S., Brack, C., Hozumi, N. and Schuller, R. (1977), *Proc. Natl. Acad. Sci. USA*, **74**, 3518.

301 Rabbitts, T. H. and Forster, A. (1978), *Cell*, **13**, 319.

302 Seidman, J. G. and Leder, P. (1978), *Nature (London)*, **276**, 790.

303 Early, P. and Hood, L. (1981), *Cell*, **24**, 1.

304 Sakano, H., Rogers, J. H., Huppi, K., Brack, C., Trannecker, A., Maki, R., Wall, R. and Tonegawa, S. (1979), *Nature (London)*, **277**, 627.

305 Seidman, J. G., Max, E. E. and Leder, P. (1979), *Nature (London)*, **280**, 370.

306 Alt, F. and Baltimore, D. (1982), *Proc. Natl. Acad. Sci. USA*, **79**, 4118.

307 Sakano, H., Maki, R., Kurosawa, Y., Roeder, W. and Tonegawa, S. (1980), *Nature (London)*, **282**, 676.

308 Kataoka, T., Miyata, T. and Honjo, T. (1981), *Cell*, **23**, 357.

309 Shimizu, A. and Honjo, T. (1984), *Cell*, **36**, 801.

310 Honjo, T. and Kataoka, T. (1978), *Proc. Natl. Acad. Sci. USA*, **75**, 2140.

311 Obata, M., Kataoka, T., Nakai, S., Yamagishi, H., Takahashi, N. and Yamawaki-Kataoka, Y. (1981), *Proc. Natl. Acad. Sci. USA*, **78**, 2437.

312 Liu, C-P, Tucker, P. W., Mushinski, J. F. and Blattner, F. R. (1980), *Science*, **209**, 1348.

313 Knapp, M. R., Liu, C-P, Newell, N., Ward, R. B., Tucker, P. W., Stroker, S. and Blattner, F. (1982), *Proc. Natl. Acad. Sci. USA*, **79**, 2996.

314 Yaoita, Y., Kumagai, Y., Okumura, K. and Honjo, T. (1982), *Nature (London)*, **297**, 697.

315 Baltimore, D. (1981), *Cell*, **26**, 295.

316 Burnet, F. M. (1959), *The Clonal Selection Theory of Acquired Immunity*, Vanderbilt University Press, Nashville.

317 Gascoigne, N. R. J., Chien, Y.-H., Becker, D. M., Kavaler, J. and Davis, M. M. (1984), *Nature (London)*, **310**, 387.

318 Sims, J. C., Tunnacliffe, A., Smith, W. J. and Rabbitts, T. H. (1984), *Nature (London)*, **312**, 541.

319 Arden, N., Klotz, K. L., Siu, G. and Hood, L. E. (1985), *Nature (London)*, **316**, 783.

320 Yoshikai, Y., Clark, S. P., Taylor, S., Sohn, U., Wilson, B. I., Minden, M. D. and Mak, T. W.

(1985), *Nature (London)*, **316**, 837.

321 Winoto, A., Mjolsness, S. and Hood, L. (1985), *Nature (London)*, **316**, 832.

322 Hayday, A. C., Diamond, D. J., Tanigawa, G., Heilig, J. S., Folsan, V., Saito, H. and Tonegawa, S. (1985), *Nature (London)*, **316**, 828.

323 Heilig, J. S., Glimcher, L. H., Kranz, D. M., Clayton, L. K., Greenstein, J. L., Saito, H., Mokam, A. M., Burakoff, S. J., Eisen, H. N. and Tonegawa, S. (1985), *Nature (London)*, **317**, 68.

324 Golden, J. W., Robinson, S. J. and Haselkorn, R. (1985), *Nature (London)*, **314**, 419.

325 Silverman, M. and Simon, M. (1983), in *Mobile Genetic Elements* (ed. J. A. Shapiro), Academic Press, New York, p. 622.

326 Adams, J. M. (1985), *Nature (London)*, **315**, 542.

327 Graham, M., Adams, J. M. and Cory, S. (1985), *Nature (London)*, **314**, 740.

328 Lewis, W. H., Michalopoulos, E. E., Williams, D. L., Minden, M. D. and Mak, T. W. (1985), *Nature (London)*, **317**, 544.

329 Rabbitts, J. H. (1984), in *Molecular Biology and Human Disease* (ed. A. Macleod and K. Sikora), Blackwell, Oxford, p. 164.

330 Croce, C. M. and Klein, G. (1985), *Sci. Am.*, **252(3)**, 44.

331 Tonegawa, S. (1985), *Sci. Am.*, **253(4)**, 104.

332 Abraham, J. M., Freitag, C. S., Clements, J. R. and Eisenstein, B. I. (1985), *Proc. Natl. Acad. Sci. USA*, **82**, 5724.

333 Craig, N. L. (1985), *Cell*, **41**, 649.

334 Chou, J. and Roizman, B. (1985), *Cell*, **41**, 803.

335 Cooke, H. J., Brown, W. R. A. and Rappold, G. A. (1985), *Nature (London)*, **317**, 687.

336 Simmler, M. C., Rouyer, F., Vergnaud, G., Nystrom-Lahti, M., Ngo, K. Y., de la Chapelle, A. and Weissenbach, J. (1985), *Nature (London)*, **317**, 692.

337 Buckle, V., Mandello, C., Darling, S., Craig, I. W. and Goodfellow, P. N. (1985), *Nature (London)*, **317**, 739.

338 Cushley, W. (1986), in *Multidomain Proteins – Structure and Function* (ed. D. G. Hardie and J. R. Coggins), Elsevier, Amsterdam.

339 Robertson, M. (1985), *Nature (London)*, **317**, 768.

340 Gronostajski, R. M. and Sadowski, P. D. (1985), *J. Biol. Chem.*, **22**, 12 320.

341 Daniels, G. R. and Deininger, P. L. (1985), *Nature (London)*, **317**, 819.

342 Husain, I., van Houten, B., Thomas, D. C.,

Abdel-Monem, M. and Sancar, A. (1985), *Proc. Natl. Acad. Sci., USA*, **82**, 6774.

343 Malissen, M., McCoy, C., Blanc, D., Trucy, J., Devaux, C., Schmitt-Verhulst, A-M., Fitch, F., Hood, L. and Malissen, B. (1986), *Nature (London)*, **319**, 28.

344 Baltimore, D. (1986), *Nature (London)*, **319**, 12.

345 Ray, P. N., Belfall, B., Duff, C., Logan, C., Kean, V., Thompson, M. W., Sylvester, J. E., Gorski, J. L., Schmikel, R. D. and Worton, R. G. (1985), *Nature (London)*, **318**, 672.

346 Hutchinson, F. (1985), *Prog. Nucleic Acid Res. Mol. Biol.*, **32**, 115.

347 Pays, E. (1985), *Prog. Nucleic Acid Res. Mol. Biol.*, **32**, 1.

# 8

# RNA biosynthesis

Gene expression involves a sequential flow of information. In the first stage of this process, known as transcription, one strand of double-stranded DNA is copied into an RNA. In contrast to DNA replication the genome is not copied in its entirety. Rather, defined units of genetic information are copied into RNA molecules which may function as messenger (mRNA), form a component of ribosomes (rRNA) or have an adapter function (tRNA). All three of these RNA products take part in the subsequent stage of gene expression, known as translation, in which mRNAs are used as coded messages for the synthesis of protein. Translation is discussed in Chapter 11.

$$DNA \xrightarrow{\text{transcription}} RNA \xrightarrow{\text{translation}} Protein$$

---

*Conventions and terms associated with the transcription unit*

A *transcription unit* is a segment of DNA (gene or genes) that is transcribed into RNA. Associated with this segment are various non-transcribed DNA elements such as *promoters*, *enhancers* and *operators* that may influence the frequency at which the unit is transcribed. These elements may have a *consensus sequence*; a sequence of nucleotides from which any similar sequence deviates only marginally. There may also be associated elements, such as *regulator genes* that are independently transcribed and cause the formation of proteins that influence the frequency of transcription. Segments of DNA that influence the activity of genes with which they are contiguous are described as *cis-acting*. This differentiates them from *trans-acting* elements that give rise to diffusible products that can act at many sites whether on the same or a different chromosome.

RNA is synthesized in a 5' to 3' direction from the complementary strand of duplex DNA. However, when describing the nucleotide sequence of genes and of elements controlling their function, it is conventional to describe the coding strand as that having the same sequence as the RNA transcript (except that uridines are replaced by thymidines). For the same reason, sequences to the left of the transcription unit are described as being 5' to the start site. Such elements are also described as being *upstream* of the initiation site while those on the 3'-side are described as *downstream*. Individual nucleotides are numbered away from the start site and are given positive values when downstream and negative values when upstream. Thus, a nucleotide 20 base pairs before the start site is −20 and that ten base pairs after the start site is +10.

## 8.1 DNA-DEPENDENT RNA POLYMERASES

The enzymes which catalyse the synthesis of RNA from a DNA template are known as DNA-dependent RNA polymerases. These enzymes require ribonucleoside triphosphates as substrates and transfer nucleoside monophosphates onto the 3'-OH terminus of the growing RNA chain. Polymerization is thus in a 5' to 3' direction and produces an RNA chain the sequence of which is determined by Watson and Crick base-pairing with the DNA template.

### 8.1.1 Bacterial DNA-dependent RNA polymerase

RNA polymerase activity was first detected in rat liver nuclei by Weiss and Gladstone [1] but it is the bacterial enzymes, particularly that of *E. coli*, that are most fully characterized (for reviews see [2, 3, 4]). The complete *E. coli* enzyme or holoenzyme contains five polypeptide chains: two $\alpha$ chains of $M_r$ 36 512; one $\beta$ chain of $M_r$ 150 619; one $\beta'$ chain of $M_r$ 155 162; one $\sigma$ chain of $M_r$ 70 236.

The $\beta$ and $\beta'$ subunits each contain an atom of $Zn^{2+}$ [5] which is essential for catalytic activity. The enzyme requires all four ribonucleoside triphosphates and a divalent metal ion which *in vivo* is $Mg^{2+}$ but *in vitro* can be $Mg^{2+}$ or $Mn^{2+}$. When isolated, the enzyme normally consists of a mixture of the holoenzyme and the core enzyme from which the $\sigma$ chain, commonly known as the sigma factor, is missing. The core enzyme is the catalytic component of the enzyme. It has a general ability to bind to DNA and, if the DNA is nicked, can catalyse the synthesis of RNA from the nicks. It is not, however, able to catalyse specific initiation. The addition of a sigma factor to the core enzyme reconstitutes a fully active holoenzyme [6] that now has reduced affinity for non-specific DNA sequences but a considerably increased affinity for specific recognition sites near the beginning

of the sequence to be transcribed. These sequences are known as *promoters*. Their nature and role in the initiation of transcription by RNA polymerase is further discussed in Section 8.2.

*E. coli* appears to have only one sigma factor but other prokaryotes have more. Thus *Bacillus subtilis* uses different sigma factors at different points in its life cycle. Furthermore, the bacterial virus SP01 redirects the transcriptional processes of its host towards transcribing viral DNA by substituting viral sigma factors into bacterial RNA polymerase. These transcriptional control processes are further discussed in Chapter 10 (Section 10.3).

Rather little is known of the function and mode of action of the subunits within the core enzyme. Numerous studies employing inhibitors, affinity labelling, reconstitution and genetic methods [7, 8, 9] confirm that $\alpha$, $\beta$ and $\beta'$ are all necessary components of the core enzyme. They also show that the $\beta$ subunit is involved in binding nucleotide substrates and is the site of action of antibiotics such as the streptolydigins which inhibit chain elongation. The $\beta$ subunit also binds rifamycins which inhibit initiation by preventing the formation of the first phosphodiester bond. The $\beta'$ subunit has been implicated as the component of RNA polymerase which binds the template DNA. It is basic as would be expected of a protein with such a function and it binds the polyanion heparin that inhibits transcription *in vitro*. There are approximately 7000 RNA polymerase molecules in an *E. coli* cell and depending on the rate of growth about 2000–5000 of these will be actively transcribing RNA at any given time.

The RNA polymerases of other bacteria as well as those of blue–green algae appear to be similar to those of *E. coli* and contain subunits that are homologous to the $\alpha$, $\beta$, $\beta'$ and $\sigma$ chains. Indeed, in many though not all cases, the homology is such that reconstituted enzymes containing heterologous mixtures of subunits derived from different species are still catalytically active [3].

### 8.1.2 Eukaryotic DNA-dependent RNA polymerases

#### (a) *Nuclear polymerases*

Transcription in eukaryotic nuclei is performed by three separate enzymes (for reviews see [10, 11, 12]. RNA polymerase I is located in the nucleoli and transcribes the genes for rRNA (50–70% of total RNA transcription). RNA polymerase II occurs in the nucleoplasm and synthesizes mRNA precursors and the U-series of small nuclear RNAs (SnRNAs) (20–40% of total RNA transcription). RNA polymerase III is also located in the nucleoplasm and transcribes tRNA, 5SRNA and other small RNA species such as 7SRNA (approximately 10% of total transcribed RNA).

All three enzymes catalyse RNA synthesis on a DNA template as previously described for the *E. coli* enzyme and require a divalent metal ion which can be $Mg^{2+}$ or $Mn^{2+}$. When assayed *in vitro*, polymerase II is much more active with $Mn^{2+}$ and polymerase III is slightly so. However, concentrations of $Mn^{2+}$ *in vivo* are low and there is evidence that its use *in vitro* may alter the binding properties of the enzyme [13]. The enzymes from most species also differ in their sensitivity to the fungal amatoxins ($\alpha$-, $\beta$-, $\gamma$-amanitin, amanin and amanullin) of which the most commonly used is $\alpha$-amanitin. RNA polymerase II is inhibited by $\alpha$-amanitin at 50 ng/ml. RNA polymerase III is much less sensitive but is inhibited at concentrations of 5 $\mu$g/ml while RNA polymerase II is insensitive.

Sedimentation analyses show that all three enzymes are very large with molecular masses of 500 000–600 000 and comprising up to 14 subunits. Tabulation of the components of the three polymerases from various sources [10, 11, 12] reveals substantial variation in the $M_r$ of the subunits between species. In general terms, however, each enzyme possesses two non-identical, large subunits with $M_r$ of 120 000–220 000 and up to 12 smaller subunits which with the exception of an 80 000–90 000 subunit in

polymerase III, have $M_r$ of less than 50 000. Lewis and Burgess [11] have discussed the difficulty of ascertaining the precise number of subunits in the active enzymes. Some subunits are only present in the purified enzyme at a molar ratio of less than one molecule per enzyme molecule. Several explanations are possible for this finding. The polypeptide concerned could be a contaminant or it could be a genuine subunit that is differentially lost during enzyme purification. Since it is not yet possible to reconstitute active enzyme from its subunits, the functional significance of any component is hard to ascertain. If a subunit functioned as a factor, this could also explain low abundance. A factor like sigma in *E. coli*, only required at certain points in the transcription cycle, would not necessarily occur in stoichiometric amounts. Evidence is growing that eukaryotic polymerases employ a range of factors for the transcription of specific genes. The best characterized of these are required for the specific transcription of 5S RNA genes by RNA polymerase III. These, and other factors which control eukaryotic transcription are not, however, components of RNA polymerase and they are discussed in Sections 8.3 and 10.4.3.

#### (b) *Mitochondrial and chloroplast RNA polymerases*

Mitochondria and chloroplasts appear to have RNA polymerase activities that are distinct from the nuclear enzymes. The mitochondrial enzyme is difficult to solubilize and may be membrane-bound. It appears to be a single polypeptide ranging in $M_r$ from 45 000 in *Xenopus* [14] and yeast [15] to 66 000 in rat liver [16]. The active enzyme may well be a multimer of these subunit molecular weights. The chloroplast enzyme on the other hand has a subunit structure which in higher plants resembles the nuclear polymerase II with two large subunits and a series of smaller ones (for brief reviews of both enzymes see [11]). However, peptide mapping shows the two enzymes to be distinct [17]. Furthermore, there is evidence from *Euglena* and higher plants [18, 19] that chloroplasts may contain more than one RNA polymerase; one catalysing rRNA gene expression and the other tRNA biosynthesis.

## 8.2 PROKARYOTIC RNA SYNTHESIS

### 8.2.1 Prokaryotic initiation of transcription

The mechanism by which RNA polymerase initiates transcription has been most extensively studied with the enzyme from *E. coli* and has been shown to occur in three stages. Firstly, the holoenzyme recognizes and binds to specific promoters. Secondly, this binding leads to the formation of the so-called '*open complex*' in which a portion of the double helix is unwound. Finally RNA synthesis is initiated. For a recent review see [20].

(a) *Binding of RNA polymerase to prokaryotic promoters*

The binding of RNA polymerase to transcriptional promoters appears to require an initial association with non-specific binding sites on the DNA. There are $4 \times 10^6$ of these non-specific binding sites in the *E. coli* genome. RNA polymerase binds to them relatively loosely with a binding constant of approximately $2 \times 10^{11}$ M$^{-1}$ and the DNA remains in the double-stranded (*closed*) form. The precise role of these sites and the mechanism by which the enzyme moves from non-specific sites to specific promoters is not fully understood. Von Hippel and coworkers [21, 22] have shown that 3D diffusion processes are not fast enough to account for the transfer. Furthermore, they have reviewed evidence that suggests that non-specific binding is electrostatic and that, in this mode, facilitated transfer may

occur by sliding along the DNA [23, 24] or by intersegment transfer or by a combination of both processes [25].

Transfer from non-specific binding to a tight binding complex with a specific promoter depends on the presence of sigma factor. The binding constant for the association varies from $10^{12}$ to $10^{14}$ M$^{-1}$ and the strength of the complex, referred to as *promoter strength*, is strongly related to the efficiency with which the associated gene is transcribed.

*E. coli* RNA polymerase holoenzyme will protect promoter regions from nuclease attack. The length of the protected segment varies with the nuclease and digestion conditions used but it is clear that the enzyme binds asymmetrically to the start site of transcription such that it covers up to 50 nucleotides on the 5'-side of the gene and up to 20 nucleotides into the gene [23]. Within the 5'-side of this protected area there are two segments, centred around nucleotides $-10$ and $-35$, the sequences of which are strongly conserved. It is these segments that are the promoters to which the holoenzyme binds. The $-10$ sequence, often known as the Pribnow Box, has the consensus sequence TATAAT and the $-35$ region has a consensus sequence TTGACA [26, 27]. It should be emphasized that not all promoters have these sequences. However, any given promoter will vary from the consensus at no more than a few positions and this variation is more likely to occur in some nucleotides than in others. Thus, in a recent compilation of 112 well-defined *E. coli* promoter regions [28], a consensus sequence for the region $-50$ to $+10$ was derived which is illustrated in Fig. 8.1 where upper case letters indicate strongly conserved nucleotides, lower case letters represent weakly conserved nucleotides

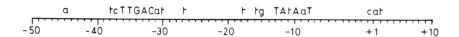

**Fig. 8.1** Consensus sequence of the *E. coli* promoter.

and no letters are inserted where there is no evidence for conservation.

The importance of the consensus sequences has also been investigated in promoter mutants [28, 29] which are described as *down or up mutants* depending on whether they decrease or increase transcription from the promoter. In an analysis of 98 *E. coli* mutants, nearly all followed a general rule that down mutants showed decreased homology and up mutants had increased homology with the consensus sequence [29]. The length of DNA between the −10 and −35 regions is also fairly stable; 100 of the 112 promoter sequences compiled by Hawley and McClure [28] had 17 ± 1 intervening nucleotides, however sequences ranging from 15 to 20 nucleotides are known. Similarly the distance between the −10 region and the transcription start site can vary from 5 to 9 nucleotides.

Contact points between RNA polymerase and its promoter have been studied by examining the extent to which the enzyme could protect specific nucleotides from base modifying reagents. Alternatively, mutations and base-specific reagents can be tested for their effect prior to enzyme binding. These experiments show that RNA polymerase binds to one face of the DNA molecule. Most of its contact points are on the non-coding strand and are heavily concentrated in the −35 and −10 consensus regions [27]. Contact points are also clustered in the −16 area, however, and the contact nucleotides are not always those with strongly conserved bases. Thus, it would appear that features additional to base sequence, such as three-dimensional structure and hydrogen-bond pattern [22], are important in enzyme recognition. That three-dimensional structure must also be important is indicated by the finding that the length of the DNA segment between the −10 and −35 regions has considerable influence on promoter strength. In the B form of DNA, the −35, −10 and −16 regions all lie on the face to which RNA polymerase binds; a change in the length of the intervening segment would skew this arrangement (Fig. 8.2). The arrangement might also be susceptible to different degrees of supercoiling [30].

The sites on the enzyme that interact with the promoter are little understood but photochemical crosslinking experiments have demonstrated contacts between the σ subunit and the base of the nucleotide at position −3 and between the β subunit and the base at position +3 [31]. Both are on the non-coding strand of the DNA.

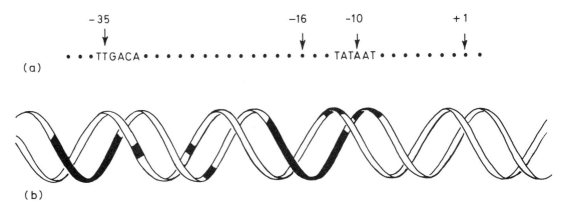

**Fig. 8.2** Contact points for *E. coli*. RNA polymerase on the promoter of the β-galactosidase gene: (a) sequence of the promoter coding strand; (b) distribution of contact points on one face of the DNA indicated by darkened areas on the double helix.

Variant prokaryotic RNA polymerase promoters associated with the expression of complex gene sets or with bacteriophage use of the host enzyme are discussed in Chapter 10 (Section 10.3).

### (b) *Formation of an open promoter complex and the initiation of RNA synthesis*

With the tight binding of RNA polymerase to a promoter site, there is a rapid transition to the so-called *open-complex* in which the DNA is partially unwound (Fig. 8.3). Unwound regions of DNA are accessible to base-modifying reagents. Siebenlist *et al.* [27] exploited this and used dimethyl sulphate to methylate the N-1 and N-3 positions of adenines and cytosines respectively in the unwound segments of the open complex. They showed that a 12-bp region of the promoter is unwound and stretches from the middle of the −10 region to just beyond the RNA start site (specifically from −9 to +3). The fact that this region overlaps with a contact point between one strand of the DNA and sigma factor [31] indicates that the essential role of this subunit in correct initiation may be associated with the formation of open complexes.

Very recently, McClure, Buc and colleagues [32, 33] have suggested that the formation of the open complex occurs via an intermediate step, i.e.:

$$R + P \xrightarrow{\text{Fast}} RP_c \xrightarrow{\text{Rate-limiting step}} RP_i \xrightarrow{\text{Major induced-fit}} RP_o$$

R is RNA polymerase and P is the promoter. $RP_c$ is the closed enzyme promoter complex in which the DNA is still double-stranded and the −10 and −35 regions are out of register. $RP_i$ is seen as an intermediate stage in which the DNA is still double stranded but the −10 and −35 regions are in register and the enzyme is strictly positioned with respect to the DNA backbone. Finally, the open complex, $RP_o$, is formed in which the DNA strands separate.

Transcription in prokaryotes is most commonly initiated with a purine nucleotide at a site 6 or 7 bp downstream from the −10 region. Of 88 promoter start sites examined by Hawley and McClure [28], A and G occurred at position +1 of the coding strand 45 and 37 times respectively and pyrimidines occurred in only six sequences. They also showed a preference for pyrimidines at positions −1 and +2 with CAT as the overall consensus sequence for the triplet −1, +1, +2. Thus a consensus transcript would begin pppApU. The binding of the initiation nucleotide to RNA polymerase is an order of magnitude stronger than that of succeeding nucleotides and it has been suggested that it occupies a separate initiating site on the enzyme [34].

At an early point in the incorporation of nucleoside phosphates into new transcripts, two further changes occur in the initiation complex. In the first of these, the open complex is transformed into a ternary complex which is much more resistant to dissociation by high salt concentrations, is resistant to inhibition by rifampicin and in which RNA polymerase is more resistant to denaturation and to proteolytic enzymes. The point at which these changes occur appears to depend upon the source of the DNA and the method by which they are analysed and may require the nascent RNA chain to be anything from two to 16 residues in length [22].

Initiation finishes with the release of sigma factor from the ternary complex. This appears to occur when transcripts are eight to nine nucleotides in length [35]. The released subunit is recycled and continued elongation of the transcript is catalysed by the core enzyme (Fig. 8.3).

### 8.2.2 Elongation of RNA transcripts

Elongation proceeds by the successive addition of ribonucleoside monophosphates from substrate triphosphates on to the 3′-OH terminus of the growing RNA chain (Fig. 8.4). The nature of the incoming nucleotide is governed by Watson–Crick base-pairing rules and bond formation is accompanied by the release of pyrophosphate.

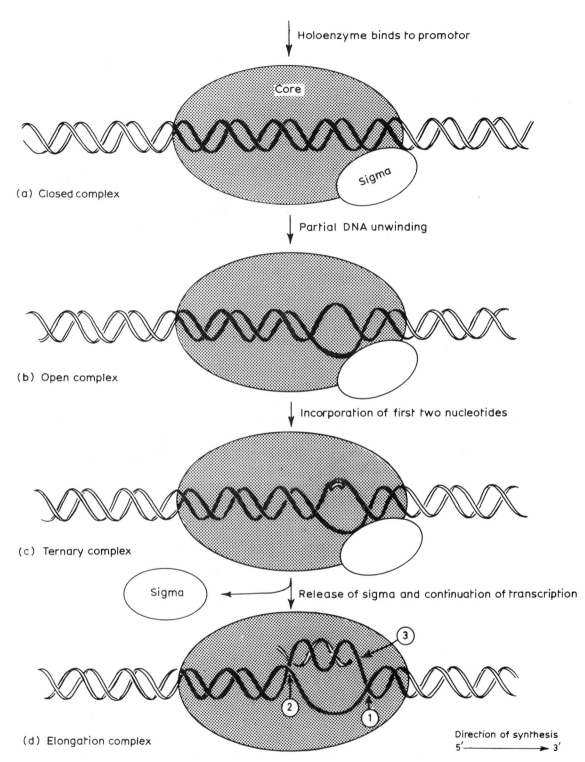

**Fig. 8.3** Diagrammatic representation of the stages in the initiation of transcription. 1, 2 and 3 on the elongation complex represent the active centres for unwindase, windase and catalytic activity respectively (although there may be two catalytic centres).

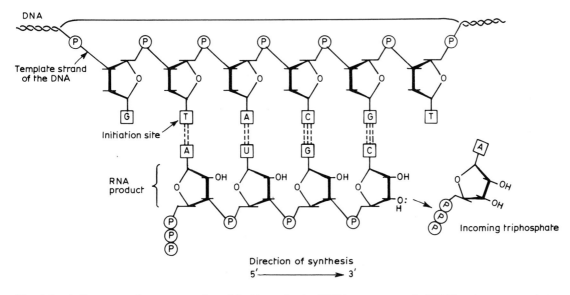

**Fig. 8.4** A diagrammatic representation of the biosynthesis of RNA on one strand of DNA acting as a template.

The reaction can thus be represented as:

$$pppXpY \xrightarrow{nZTP} pppXpY(pZ)_n + nPP_i$$

X, Y and Z can be any of the four ribonucleotides although evidence has already been presented for the preferred insertion of purines as the 5′-nucleotide.

The elongation complex is diagrammatically illustrated in Fig. 8.3d. The catalytic core of the enzyme can be thought of as having a number of active centres including those for the winding and unwinding of DNA as well as catalytic sites for the incoming nucleotides. There is direct evidence from the use of photoaffinity probes that there are two catalytic sites that are used alternatively for the synthesis of successive phosphodiester bonds [36]. Both sites are on the β subunit. As already discussed, the catalytic sites may also be distinct from the initiation site at which the first nucleotide is bound. During elongation, the unwindase activity opens the DNA helix while the windase rejoins it and displaces the nascent RNA. Thus the enzyme moves down the DNA like a nut on a bolt [37]

using alternate catalytic sites and progressing by rotational translocation [36]. The windase and unwindase activities have been estimated to be 17 bp apart [37] implying an opening for the open promoter complex rather larger than the 12 bp estimated by base modification experiments [27].

RNA polymerases from a wide variety of bacteria catalyse the synthesis of RNA from a natural template at 12–19 nucleotides per second [38]. This is slower than the theoretical rate of up to 60 nucleotides per second because transcription from natural templates is not uniform. Many groups have shown that elongation is a discontinuous process in which there are a number of pause sites with rapid elongation between these sites (see, for example, [39]). Pausing occurs at GC- rich regions and at points 16–20 nucleotides downstream of regions of diad symmetry. At these latter regions the delay is associated with the formation, by the nascent RNA, of stem–loop structures either because it base pairs with itself or with the coding strand of the DNA [37, 40]. Pausing is a necessary preliminary to termination (Section 8.2.3) and premature termination can occur at internal pausing sites (see Section 10.2.1). Viral RNA polymerases are apparently not delayed by their host's DNA and have been reported to elongate at up to 200 nucleotides per second [41].

A number of inhibitors have proved useful for the study of elongation [38–41]. The antibiotics rifampicin and streptovaricin bind to the $\beta$ subunit and block initiation while streptolydigin also binds to the $\beta$ subunit but interferes with the elongation steps. Another agent which prevents elongation is actinomycin D. However, it does so, not by binding to the enzyme, but by complexing with deoxyguanosine residues in the DNA template and thus preventing movement of the core enzyme along the template.

### 8.2.3 Termination of transcription in prokaryotes

Three events are required when the transcription of a gene is terminated; elongation ceases, the transcript is released and RNA polymerase is released. Studies with prokaryotes have demonstrated that these events occur by at least two mechanisms that are known as independent termination and factor-dependent termination. Both of these have been the subject of recent reviews [22, 42, 43, 44]. Before discussing each in turn it should be stressed that the precise identification of termination sites is not easy. The 5'-end of a prokaryote transcript can be unambiguously identified by its triphosphate group. No such marker defines the 3'-end and an observed 3'-end could therefore have been subjected to post-transcriptional cleavage. Termination has been largely studied *in vitro* using methods which involve the synthesis, by purified RNA polymerase, of specific transcripts from bacteriophage or defined bacterial genes. Commonly used systems include the bacterial *trp* operon (genes encoding the enzymes that synthesize tryptophan), the DNA coliphages T7, T3, $\phi$X174 and lambda. Termination occurs at

defined sites in these systems and, in a number of them, concern that the sites might be artefacts of *in vitro* incubation conditions has been dispelled by confirmation of their use *in vivo*.

#### (a) *Independent termination*
Sites on the DNA at which independent termination occurs have a characteristic structure that comprises a GC-rich inverted repeat (inverted repeats are defined in Section 2.8) followed, on the template strand, by a run of adenylate residues. The former of these regions results in the formation of GC-rich regions in the transcript, which are able to base-pair into a 'hairpin' or stem-loop structure (Fig. 8.5). Such loops, which typically contain seven to ten G:C base pairs, have been shown to cause RNA polymerase to pause [45]. Mutations that lengthen the region of dyad symmetry over a minimum of 6 bp strengthen the efficiency of termination as does the incorporation into the transcript of nucleotide analogues, such as bromo- or iodo-CMP, that stabilize G:C pairs. Conversely, weakening of G:C pairs, such as occurs when inosine is substituted for guanosine, decreases termination efficiency [42].

Shortly after the GC-rich region, a run of four to eight adenylates in the template will dictate that a string of uridylate residues are transcribed into the nascent RNA. Decreasing the number of these residues has been shown to reduce the efficiency of termination [46] and it appears that the importance of the sequence resides in the very unstable nature of the rU:dA hybrid [47]. Thus three features, the disruption of the RNA:DNA hybrid caused by G:C base-pairing, the pausing of RNA polymerase and instability of the rU:dA region, all combine to facilitate the release of the transcript from the template. O'Hare and Hayward [48], working with a termination site, T1, of coliphage T7, have shown that *in vitro* this recognition of the stop signal and release of RNA occurs comparatively rapidly with a $t_{1/2}$ of 3 min or less. Dissociation of RNA polymerase from the DNA is slower with a

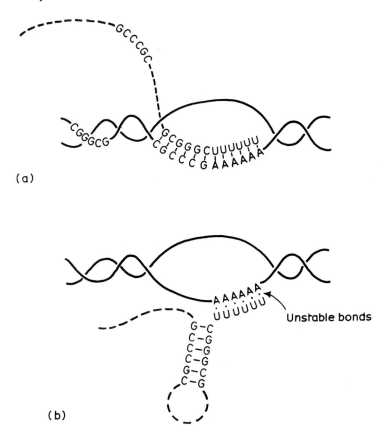

(a)

(b)

**Fig. 8.5** Diagrammatic representation of independent termination. RNA polymerase is omitted for reasons of clarity (adapted from Platt (1981) [43]: (a) Point at which elongation stops; (b) formation of hairpin followed by termination.

$t_{1/2}$ of approximately 12 min and is probably the rate-limiting step in the recycling of the enzyme. Platt [43] has suggested that RNA polymerase undergoes a conformational change to a termination form at some time in this process and this is supported by the isolation of polymerase mutants that are modified in termination efficiency.

*(b) Factor-dependent termination*

Roberts [49] discovered a protein, which he called rho ($\rho$), that when added to an *E. coli in vitro* transcription system causes the generation of transcripts with discrete 3'-ends. The protein

has a $M_r$ of 46 000, has been sequenced from its gene [50] and, in its functional form, is probably a hexamer [51]. It possesses NTPase activity and the hydrolysis of nucleotide triphosphate is necessary for its role in termination [52]. The factor also requires single-stranded RNA both for its NTPase activity and for termination [53, 54], and various physical analyses and RNase protection studies indicate that it binds to approximately 72–84 nucleotides of nascent mRNA [22].

As with independent termination, the activity of rho requires RNA polymerase to pause in its elongation [55]. However, rho-dependent sites do not appear to have the conserved sequence features associated with pausing at rho-independent sites. Although many can be drawn as step-loop structures, they tend to be relatively unstable and largely involve A:U base-pairing. Indeed, a considerable body of evidence has been accumulated showing that rho-dependent termination is enhanced by decreased secondary structure in the RNA transcript [53, 55, 56] and that rho binds to nascent mRNA that is relatively free of secondary structure. Models for rho-dependent termination [42, 46, 57, 58] propose that the factor first binds to nascent RNA that lacks secondary structure. It may then move along the transcript until it finds a paused RNA polymerase and this movement could require the hydrolysis of nucleoside triphosphate [42]. Alternatively, the hydrolysis of NTP may result from a conformational change in rho brought about by the interaction of the nucleotide with the ternary complex and resulting in the release of the nascent RNA chain [22].

It seems likely that there are a number of factors that bring about termination in prokaryotes. *Nus A* protein, which is considered in Chapter 10 under the control of termination (Section 10.2.2), is a termination factor that, in bacteriophage lambda, interacts with an anti-termination factor N.

## 8.3 EUKARYOTIC RNA SYNTHESIS

The study of eukaryotic RNA polymerase activity has been much enhanced by the development of systems in which the enzymes correctly transcribe specific genes *in vitro*. The first of these was a transcription system for polymerase III [59] and led to the use of extracts of *Xenopus* oocytes as a major system to support transcription by this enzyme [60, 61]. Similar systems, employing extracts of tissue culture cells, were also described for the transcription of viral genes and cloned eukaryotic genes by both polymerase II and III [62, 63]. Most recently, a cell-extract system has been developed that correctly initiates the transcription of 45S pre-

rRNA from cloned rDNA [64]. Such polymerase I-dependent systems are species-specific [65].

Transcription is also studied by the micro-injection of cloned eukaryotic genes into *Xenopus* oocytes [66], by the stable introduction of genes into mammalian cells in culture [67] and by the insertion of cellular genes into an SV40 viral vector [68] or a plasmid containing the SV40 or other viral replication origins [69].

The use of such systems has shown that RNA synthesis by eukaryotic polymerases has much in common with that of the prokaryotic enzyme. Since there are three nuclear enzymes, however, one may expect differences in promoter recognition, termination and control of transcription. Promoters and termination are discussed here while control mechanisms are discussed in Chapter 10.

### 8.3.1  Initiation by RNA polymerase II

The promoter for RNA polymerase II consists of a number of sequence elements required for accurate and efficient initiation of transcription (for a recent review see [70]). Two of these show similarities with the conserved sequences of prokaryotes.

The first conserved element identified when eukaryote gene sequences were compared was the Goldberg/Hogness, TATA or ATA box [71] AT-rich region with the consensus sequence

$$5'\text{TATA}^{A}_{T}{}^{A}\text{A}^{A}_{T}3'$$

It occurs approximately 30 bp upstream from the start site (Fig. 8.6) and its function is to fix the location of the start site. Mutations in the consensus sequence can profoundly affect correct initiation. Conversely, nucleotide deletions in the start site region result in initiation at a new site still approximately 30 bp downstream from the TATA box. While this element is obviously very similar to the prokaryotic Pribnow box two differences are immediately apparent. Firstly, the Pribnow box is 10 bp (one DNA helical turn) from the start

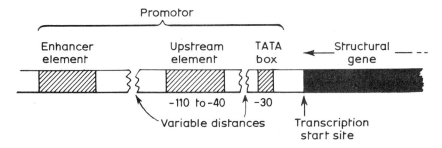

**Fig. 8.6** Diagram of the promoter region for RNA polymerase II.

site, while the TATA box is three DNA turns away. Secondly, there is a degree of sequence conservation in the prokaryotic start site but much greater flexibility in eukaryotes. Transcriptional analysis and S1 nuclease mapping have shown that a functional TATA box precedes most genes transcribed by RNA polymerase II. There are, however, genes, such as the late SV40 genes and U1 snRNA genes, that contain equivalent sequences that can at best only be described as TATA-like [72, 73].

A second less strongly conserved element is located a variable distance ($-40$ to $-110$) upstream of the start site (Fig. 8.6). It appears to be important in promoting efficient initiation but plays no role in determining its accuracy. One variant of this element, which shows similarity with the $-35$ region of prokaryotes, is the so-called CAAT or CCAAT box which has the consensus sequence

$$5'GG{{}^{T}_{C}}CAA{{}^{T}_{A}}CT3'$$

and occurs 70—90 bp upstream from the start site. A further variant is a GC-rich sequence with the consensus

$$5'CCGCCC$$

or its complement

$$GGGCGG$$

which can occur in one or more copies [74, 75] and may be present in addition to a CAAT box as the so-called – 100 element [76].

A third class of promoter element, the en-hancers (Fig. 8.6), was first detected in viruses but have since been identified in mammalian genes. They, and the recently reported down-stream elements, appear to be specific modulators of initiation and are considered in greater detail in Chapter 10 on the control of gene expression (Section 10.4.2).

There is growing evidence that initiation by RNA polymerase II requires the involvement of trans-acting protein factors. *In vitro* transcription systems cannot accurately initiate transcription unless supplemented with crude cell extracts. Chomatography of these generates multiple fractions required for accurate transcription [77, 78] and at least one of these binds to the region of the template that contains the TATA box [79]. The binding occurs in the absence of RNA polymerase. Similarly, a factor has been described which binds to the GC-rich element of SV40 and is required for efficient transcription from that promoter [80]. Chambon *et al.* [70] have suggested that enhancer elements might be bidirectional entry sites for transcription factors which then move along the DNA until they find the promoter elements which they activate.

As in prokaryotes, transcription by RNA polymerase II appears to start with the first nucleotide of the mature mRNA. Eukaryote messages and their precursors are capped (see Section 9.3.2) and the *in vivo* capping site corresponds to the 5'-terminus of the product of *in vitro* transcription systems. Furthermore, capping is very rapid *in vivo* with no evidence for preliminary 5' processing and the capping enzymes require a 5'-di- or tri-phosphate. There are indications that some genes have multiple initiation sites and give rise to a heterogeneous mRNA population [81]. Paul and coworkers [82] have detected a minority of globin gene transcripts that originate from points upstream of the normal initiation site. Some, but not all, of these are associated with alternative TATA-like elements but their significance is unknown.

### 8.3.2 Initiation by RNA polymerase III

The promoters for the genes transcribed by RNA polymerase III do not lie in their 5' flanking sequences but within the coding region of the gene itself (for reviews see [83, 84, 85]). This totally unexpected finding resulted from the work of Brown and colleagues on the expression of *Xenopus* 5S RNA genes in oocytes and in *in vitro* transcription systems. They found that the entire 5' flanking sequence of the gene could be removed without the loss of 5S RNA transcription. Furthermore, when they deleted the 5'-end of the coding sequence, they found that 5S-sized RNAs were still made and not until 50 nucleotides had been deleted was there a drop in the efficiency of transcription [86]. Subsequent experiments in which the 3'-end of the gene was deleted [87], or in which extra nucleotide sequences were inserted into the gene [86], showed that the synthesis of 5S RNA (or a 5S-sized RNA) was controlled by a promoter located between residues 50 and 83 of the *Xenopus* gene. (The promoter is known as the *internal control region or ICR*.) The sequence is required for and sufficient for transcription. It is not, however, recognized by RNA polymerase III [88] but by a transcription factor, known as TF IIIA or 40 kd factor [89, 90], which specifically interacts with two factor-binding domains at the boundaries of the promoter region [91]. This binding to the ICR is followed by the sequential binding of TF IIIC and TF IIIB [92]. It is the complete complex that is recognized by RNA polymerase [92] and it is

assumed that the recognition event positions the catalytic site of the enzyme at the transcription start site.

The binding of TF IIIA to DNA differs from that of the prokaryotic repressor and catabolite activator proteins described in section 10.1.6. In the prokaryotic proteins, the conserved structural motif used in DNA recognition is an alpha helix - beta turn - alpha helix configuration. Eukaryotic proteins including TF IIIA, the glucocorticoid hormone receptor protein and a *Drosophila* protein involved in segmentation share repeated metal binding domains that have been called fingers. It is suggested that DNA either interacts with or lays between the domains [140, 141, 142].

Similar studies with tRNA genes have shown that they too have an internal promoter but in this case it is split into two blocks termed A and B (also termed D and T or 5' and 3' block respectively) [84]. Block A extends from nucleotides 8 to 19 and includes the portion of the gene that encodes the tRNA D loop and four invariant nucleotides. The B block runs from nucleotides 52 to 62 including those encoding the T loop and five invariant nucleotides (Fig. 8.7a). Thus the invariant nature of some tRNA nucleotides appears to be important both in tRNA tertiary structure (see Section 11.1) and in the initiation of tRNA synthesis. Because of the length heterogeneity of the tRNA variable loop and the presence in some tRNA genes of inserts, the distance between the A and B blocks can vary from 31 to 74 bp. It has been suggested that the A and B regions of the gene form stem-loop structures (Fig. 8.7b) that are important for their role as a promoter [84] but the evidence for this is contradictory [93, 94]. Sequences in the 5' flanking sequence of tRNA genes are also required for maximum transcriptional activity of RNA polymerase III [95, 96] and, as in the 5S gene, transcription factors appear to be involved in the formation of the initiation complex [97, 98].

The promoter region of other genes that are

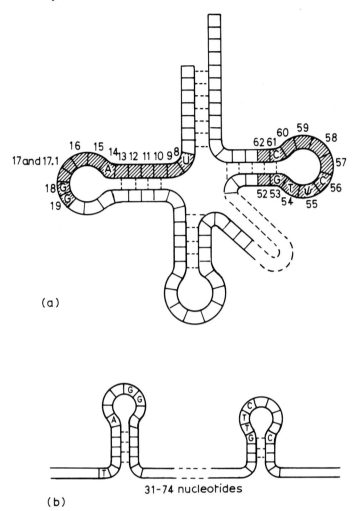

**Fig. 8.7** The split promoter of tRNA genes: (a) the tRNA clover leaf structure showing the portions of the molecule (shaded) and the invariant nucleotides that are transcribed from the internal promoter; (b) a model for the arrangement of the coding strand of the promoter for tRNA genes [84].

transcribed by RNA polymerase III, such as those of 7S RNA, adenovirus VA RNA and Epstein–Barr virus, appear to be very similar to that of tRNA. Furthermore, within the 5S promoter, the first 11 nucleotides bind factor TF IIIA weakly [99] and appear to be the functional equivalent of the tRNA A block. They can be exchanged with it in 5S tRNA hybrid genes [100]. The 3′-end of the 5S promoter binds strongly to factor TF IIIA [99] and seems to have a similar function to the tRNA B block which also associates with specific transcription factors [92, 101, 102].

The role which factor TF IIIA may play in the control of 5S RNA gene expression is discussed in Sections 10.4.3 and 10.4.4.c.

### 8.3.3 Initiation by RNA polymerase I

The relatively late development of *in vitro* and *in vivo* transcription systems for rRNA genes has delayed the identification of sequences required for initiation by RNA polymerase I. Moss, however, micro-injected *Xenopus* rDNA into oocytes and demonstrated the correct initiation of rRNA precursor [103]. By using rDNA mutants, he showed that the promoter for rRNA synthesis extends from approximately −142 to +6 and thus lies largely in the so-called non-transcribed spacer that separates each tandemly repeated transcript unit from its neighbour [103] (ribosomal gene spacers are considered in detail in Section 9.4.2). Further upstream into the spacer there are two to seven imperfect copies of this promoter region (the so-called *Bam* islands) and these are separated from each other by 60 bp and 81 bp repeating units which appear to function as enhancers (for a review see [104]).

When the promoter region of different *Xenopus* species is examined, conserved sequences are found from −9 to +4 and from −24 to −33. When, however, the sequences of more distantly related species are examined, they show little resemblance to the *Xenopus* sequences although they do contain other groups of conserved nucleotides [105]. Thus, those mammals examined show strong homology from −1 to +18, −16 to −20 and in a hexanucleotide occurring between −33 and −43. This and the analysis of mutant sequences has led to the suggestion that the promoter in mammals consists of a limited number of dispersed nucleotides including G residues at −7, −16 and −23, a TTT triplet around −34 and the first few transcribed nucleotides [105, 106]. The sequences for *Drosophila*, slime moulds and yeasts have little in common with those of *Xenopus* or mammals but again exhibit their own conserved elements. Thus, there appears to be considerable variation in the promoters for RNA polymerase I which may explain the species-specific nature of *in vitro* transcription systems and may indicate the existence of species-specific transcription factors. Mishima *et al.* have presented evidence for the existence of such factors [107].

### 8.3.4 Eukaryotic termination

Little is known about the process by which transcription is terminated in eukaryotic cells. Often it is difficult even to be sure of the relationship between observed 3′-termini of newly made RNA molecules and the real point at which transcription terminates. This is especially so with the transcripts of RNA polymerase II, the majority of which are post-transcriptionally modified by 3′-polyadenylation. It has recently become apparent, however, that both poly-adenylated [108] and non-polyadenylated mRNAs [109, 110] are transcribed past their apparent 3′-termini and are then subject to post-transcriptional cleavage (polyadenylation and post-transcriptional cleavage are further considered under the maturation of mRNA in Chapter 9). β-Globin transcripts terminate about 1 kbp past the end of the gene [108, 111] while termination of amylase gene transcripts occurs at multiple downstream sites [112]. Little is known of the sequences that bring about termination except that the sequence, AAUAAA, found in polyadenylated messages is involved in polyadenylation and/or processing, and apparently plays no role in termination. A further conserved sequence

YGTGTTYY

located 30 bp downstream of the AATAAA signal in many viral and mammalian genes has been proposed to have a role in the formation of the 3′-terminus but it is not clear whether it is a termination or a processing signal [113]. It has also been suggested that a repeated 8 bp sequence TTTTTATA controls termination in some yeast genes but that termination actually occurs at a sequence

$$CAA_{G}^{T}CTTG$$

[114]. Similar sequences to the latter have been implicated as premature termination sites in adenovirus [115]. However, the above sequences do not occur at the 3′-ends of all yeast genes neither are they generally found at the end of the genes of higher eukaryotes [114].

Termination by RNA polymerase III of 5S RNA transcripts occurs in a single consensus sequence that consists of a run of four or more T residues in the non-coding strand surrounded by GC-rich sequences [116]. Similar terminator sequences occur immediately or soon after the coding regions of virtually all known tRNA genes. In contrast to prokaryotes (Section 8.2.3), the terminator sequence does not require significant dyad symmetry. Furthermore, it is recognized by the enzyme alone without the need for additional factors [88].

Termination of rRNA transcripts by RNA polymerase I also appears to involve a cluster of four Ts at the 3′-end of the gene [117]. Termination still occurs when there are only three Ts but is severely affected by reduction to two. Because clusters of Ts occur at other points in the gene, it is assumed that other sequences, perhaps those adjacent to the Ts, are also important. Once again, however, it is difficult to be sure that a string of Ts is a termination site rather than a signal at which the rapid processing of longer transcripts stops. Grummt *et al.* [118] have recently presented evidence that mouse rRNA transcripts do not terminate at the end of the 28S gene but continue 300 bp into the spacer.

### 8.3.5 Transcription of mitochondrial and chloroplast genes

Promoters for chloroplast genes appear to be homologous to the prokaryotic −10 and −35 sequences [119]. Similarly, the nonanucleotide

ATATAAGTA

which forms the initiation site (nucleotides −8 to +1) of yeast mitochondrial genes [120] has obvious sequence homologies with both pro-

karyotic and RNA polymerase II promoters; though its position relative to the transcription start site is obviously different. The base composition of the promoter region of human mitochondrial DNA, however, is very different and is unusual in being enriched in A and C residues. It lies adjacent to the D loop replication origin (See Fig. 3.15) and contains a repetitive sequence

AAACCCC

[121, 137]. The transcription of mitochondrial genes has been reviewed [138, 139] and the arrangement and expression of both chloroplast and mitochondrial genes is further discussed in Section 9.6.

### 8.4 RNA POLYMERASES AND RNA SYNTHESIS IN DNA VIRUSES

Some bacteriophage redirect their hosts' RNA polymerase to transcribe the phage genome (Sections 10.1.5, 10.2.2, 10.3.1, 10.3.2). Others, at least in part, employ their own enzymes (for a general review of these enzymes see [122]).

The RNA polymerase gene of bacteriophage T7 and its relatives (T3, SP6, gh-1) is one of the so-called early genes and is transcribed by host RNA polymerase early in the infective cycle (for a review see [123]). The transcript is translated by the host's protein-synthesizing system and the enzyme is then responsible for the transcription of the late viral genes. The T7 enzyme consists of a single polypeptide of $M_r$ 98 000. It transcribes exclusively from promoters on the r stand of the viral DNA at a very fast rate of 200 nucleotides per s. The promoters share a strongly conserved 23 bp consensus sequence for which T7 polymerase has a stringent specificity [124]. It will selectively transcribe DNA linked to the promoter. All transcripts start with the sequence pppGGG [122].

The bacteriophage N4 does not rely on the host enzyme for early gene transcription but carries its own, very large ($M_r$ 350 000) single polypeptide, RNA polymerase within the bacteriophage particle.

Some eukaryotic viruses, such as SV40 [125] and Epstein–Barr virus (EBV) [126], are entirely dependent on host RNA polymerase. The genome of EBV has been totally sequenced and 24 promoters for host RNA polymerase II have been identified, as well as AATAAA poly-adenylation signals [126]. RNA polymerase III is also involved in EBV transcription and transcribes two early (EBER) genes [127].

Other eukaryotic viruses encode their own enzyme and, of these, vaccinia virus RNA polymerase was the first to be purified and characterized. It has a $M_r$ of 425 000, comprises seven subunits, is $\alpha$-amanitin-resistant and dependent on $Mn^{2+}$ ions for activity [128].

## 8.5 THE REPLICATION OF RNA VIRUSES BY RNA-DEPENDENT RNA POLYMERASE (REPLICASE)

### 8.5.1 RNA bacteriophage

The RNA bacteriophage R17, Q$\beta$, MS2 and f2 have a 3600–4500 nucleotide, single-stranded RNA genomes which function both as a mRNA from which viral proteins are translated, and as a template for a RNA-dependent RNA poly-merase (replicase). Part of the replicase is translated from the infective (+) strand of the viral RNA which it then copies into (−) strands. These are then also copied to produce more (+) strands for packaging into progeny infective bacteriophage.

Replicase transcribes RNA from its 3′-end, initiating a new RNA chain with a 5′-GTP and continuing its synthesis in a 5′ to 3′ direction. It therefore moves along the viral genome in the opposite direction to the ribosomes which are reading it as a mRNA. The control of these two opposing events is discussed in Chapter 11 (Section 11.10.1). In *in vitro* systems, a RNA primer can replace GTP in initiating synthesis. The purification and properties of Q$\beta$ replicase have been reviewed by Blumenthal [129]. The enzyme consists of four subunits of which sub-unit II is encoded by the viral genome while the

others are normally part of the host's protein-synthesizing machinery. These are the ribosomal protein S1 and the elongation factors EF-Tu and EF-Ts (the normal function of these proteins is discussed in Chapter 11). A hexameric protein called host factor is also required for Q$\beta$ replication. The role that proteins normally involved in protein synthesis could play in RNA transcription is of considerable interest but is incompletely understood. EF-Tu and EF-Ts are required for replication but are only loosely bound to the complex of S1 and II. They appear to have a stabilizing role while S1 and host factor probably function in the binding of replicase to Q$\beta$ RNA.

### 8.5.2 Eukaryotic RNA viruses

#### (a) *Picornaviruses (class IV)*

These have single-stranded RNA genomes which can function as mRNAs (+ strands). The most thoroughly studied example is poliovirus, the 7.5 kb genome of which is translated by host ribosomes into a giant polypeptide. This is then cleaved by proteases (see Section 11.11.3) into a number of proteins including the four viral capsid proteins, part of the replicase and a number of other products including a 22 amino acid peptide, VPg. The complete replicase consists of the viral 53 000 $M_r$ polypeptide and an unidentified host cell protein of 63 000 $M_r$. Baltimore [130] has reviewed and produced a model for poliovirus that VPg, linked to two uridine monophosphates in the form VPg-pUpU, primes RNA synthesis. He suggests that it hybridizes to the 3′-poly(A) tail of the (+) strand RNA and is elongated by the replicase. The (−) strand so produced is then copied by the replicase after VPg-pUpU has hybridized to its two 3′A residues.

The VPg of another picornavirus, encephalo-myocarditis virus, has been characterized by Golini *et al.* [131].

#### (b) *Rhabdoviruses (class V)*

Rhabdoviruses have single-stranded RNA

genomes that cannot function as mRNAs (−strands) and must be copied into (+) strands before they can be expressed. To achieve this, the infective virus contains a replicase (also known as a transcriptase). The most thoroughly studied example is vesicular stomatitis virus the replicase of which transcribes the viral RNA into five mRNAs which are translated by host cell ribosomes into the five viral proteins. The same replicase is also responsible for the replication of the virus by first producing complete (+) strands and then copying them into (−) strands (for a review see [132]).

### (c) *Myxovirus (class V)*

These viruses have a segmented single-stranded RNA genome in seven or more distinct non-overlapping pieces and with negative polarity. The most-studied example is influenza. The virus encodes and packages its own replicating system. However, viral mRNA synthesis also requires cellular mRNA, the 5′-terminal portions of which are used as primers. The 5′-terminal cap ($m^7$ GpppXm – see Section 9.3.2) of cellular mRNA is cleaved by a cap-dependent viral endonuclease at a purine residue 10–13 nucleotides in from the end [133]. The resulting fragments are weakly bonded to the viral RNA segments, it is presumed by a single base pair between a 3′-terminal A of the primer and a 3′-U of the viral RNA (Fig. 8.8). They then function as a primer for elongation by the viral transcriptase (for reviews see [134, 135]).

### (d) *Reoviruses (class III)*

These are double-stranded RNA viruses of which human reovirus type 3 is a typical example (for a review see [136]). The genome consists of ten segments of double-stranded RNA, each specifying a single polypeptide. Each is transcribed asymmetrically and conservatively by a virion-associated RNA polymerase into (+) RNA strands. These mRNAs are capped but contain no poly(A). However, they are functional mRNAs and are translated by host cell ribosomes. In viral replication the ten (+) strand mRNA molecules complex with some of the viral proteins. The (−) strand RNAs are synthesized on this complex.

### (d) *Retroviruses (class VI)*

The single-stranded RNA genomes of retroviruses are copied by reverse transcriptase into DNA which serves as a template for the synthesis of mRNA. Reverse transcriptase is discussed in Chapter 6 (Sections 6.4.2 and 6.9.10).

REFERENCES

1 Weiss, S. and Gladstone, L. (1959), *J. Amer. Chem. Soc.*, **81**, 4118.

2 Chamberlin, M. J. (1976), in *RNA Polymerase* (ed. R. Losick and M. Chamberlin), Cold Spring Harbor Laboratory, New York, p. 159.

3 Chamberlin, M. J. (1982), in *The Enzymes* (ed. P. Boyer), Academic Press, New York, Vol. 15, p. 61.

4 von Hippel, P. H., Bear, D. G., Morgan, W. D. and McSwiggen, J. A. (1984), *Annu. Rev. Biochem.*, **53**, 389.

5 Miller, J. A., Serio, G. F., Howard, R. A., Bear, J. L., Evans, J. E. and Kimball, A. P. (1979), *Biochim. Biophys. Acta*, **579**, 291.

6 Burgess, R. R., Travers, A. A., Dunn, J. J. and Bautz, E. K. (1969), *Nature (London)*, **221**, 43.

7 Zillig, W., Palm, P. and Heil, A. (1976), in *RNA Polymerase* (ed. R. Losick and M. Chamberlin), Cold Spring Harbor Laboratory, New York, p. 101.

8 Krakow, J. S., Rhodes, G. and Jovin, T. M. (1976), in *RNA Polymerase* (ed. R. Losick and M. Chamberlin), Cold Spring Harbor Laboratory, New York, p. 127.

9 Scaife, J. (1976), in *RNA Polymerase* (ed. R. Losick and M. Chamberlin), Cold Spring Harbor Laboratory, New York, p. 207.

10 Paule, M. (1981), *Trends Biochem. Sci.*, **6**, 128.

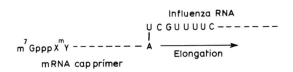

**Fig. 8.8** The priming of influenza RNA synthesis.

11 Lewis, M. K. and Burgess, R. R. (1982), in *The Enzymes* (ed. P. Boyer), Academic Press, New York, Vol. 15, p. 109.

12 Sentenac, A. (1985), *Crit. Rev. Biochem.*, **18**, 31.

13 Anderson, J. A., Juntz, G. P. P., Evans, H. H. and Swift, T. J. (1971), *Biochemistry*, **10**, 4368.

14 Wu, G. J. and Dawid, I. B. (1972), *Biochemistry*, **11**, 3589.

15 Levens, D., Lustig, A. and Rabinowitz, M. (1981), *J. Biol. Chem.*, **256**, 1474.

16 Mukerjee, H. and Goldfeder, A. (1973), *Biochemistry*, **12**, 5096.

17 Kidd, G. H. and Bogorad, C. (1979), *Proc. Natl. Acad. Sci. USA*, **76**, 4890.

18 Greenberg, B. M., Narita, J. O., DeLuca-Flaherty, C., Gruissem, W., Rushlow, K. A. and Hallick, R. B. (1984), *J. Biol. Chem.*, **259**, 14 880.

19 Gruissem, W., Greenberg, B. M., Zurawski, G., Prescott, D. M. and Hallick, R. B. (1983), *Cell*, **35**, 815.

20 McClure, W. R. (1985), *Annu. Rev. Biochem.*, **54**, 171.

21 Berg, O. G., Winter, R. B. and von Hippel, P. H. (1981), *Biochemistry*, **20**, 6929.

22 von Hippel, P. H., Bear, D. G., Morgan, W. D. and McSwiggen, J. A. (1984), *Annu. Rev. Biochem.*, **53**, 389.

23 Winter, R. B., Berg, O. G. and von Hippel, P. H. (1981), *Biochemistry*, **20**, 6961.

24 Park, C. S., Wu, F. Y-H. and Wu, C-W. (1982), *J. Biol. Chem.*, **257**, 6950.

25 Berg, O. G., Winter, R. B. and von Hippel, P. H. (1982), *Trends Biochem. Sci.*, **7**, 52.

26 Rosenberg, H. and Court, D. (1979), *Annu. Rev. Genet.*, **13**, 319.

27 Siebenlist, U., Simpson, R. B. and Gilbert, W. (1980), *Cell*, **20**, 269.

28 Hawley, D. K. and McClure, W. R. (1983), *Nucleic Acids Res.*, **11**, 2237.

29 Youderian, P., Bouvier, S. and Susskind, M. M. (1982), *Cell*, **30**, 843.

30 Smith, G. R. (1981), *Cell*, **24**, 599.

31 Simpson, R. B. (1979), *Cell*, **18**, 277.

32 Buc, H. and McClure, W. R. (1985), *Biochemistry*, **24**, 2712.

33 Spassky, A., Kirkegaard, K. and Buc, H. (1985), *Biochemistry*, **24**, 2723.

34 Krakow, J. S., Rhodes, G. and Jovin, T. M. (1976), in *RNA Polymerase* (ed. R. Losick and M. Chamberlin), Cold Spring Harbor Laboratory, New York, p. 127.

35 Hansen, U. M. and McClure, W. R. (1980), *J. Biol. Chem.*, **255**, 9564.

36 Panka, D. and Dennis, D. (1985), *J. Biol. Chem.*, **260**, 1427.

37 Gamper, H. B. and Hearst, J. E. (1982), *Cell*, **29**, 81.

38 Chamberlin, M. J., Nierman, W. C., Wiggs, J. and Neff, N. (1979), *J. Biol. Chem.*, **254**, 10 061.

39 Kassavetis, G. A. and Chamberlin, M. J. (1981), *J. Biol. Chem.*, **256**, 2777.

40 Morgan, W. D., Bear, D. G. and von Hippel, P. H. (1983), *J. Biol. Chem.*, **258**, 9565.

41 Chamberlin, M., Kingston, R., Gilman, M., Wiggs, J. and de Vera, A. (1983), *Methods Enzymol.*, **101**, 540.

42 Adhya, S. and Gottesman, M. (1978), *Annu. Rev. Biochem.*, **47**, 967.

43 Platt, T. (1981), *Cell*, **24**, 10.

44 Holmes, W. M., Platt, T. and Rosenberg, M. (1983), *Cell*, **32**, 1029.

45 Farnham, P. J. and Platt, T. (1981), *Nucleic Acids Res.*, **9**, 563.

46 Bertrand, K., Korn, L., Lee, F. and Yanofsky, C. (1977), *J. Mol. Biol.*, **117**, 227.

47 Martin, F. and Tinoco, I. (1980), *Nucleic Acids Res.*, **8**, 2295.

48 O'Hare, K. M. O. and Hayward, R. S. (1981), *Nucleic Acids Res.*, **9**, 4689.

49 Roberts, J. W. (1969), *Nature (London)*, **224**, 1168.

50 Pinkham, J. and Platt, T. (1983), *Nucleic Acids Res.*, **11**, 3531.

51 Finger, L. R. and Richardson, J. P. (1982), *J. Mol. Biol.*, **156**, 203.

52 Richardson, J. P. and Conway, R. (1980), *Biochemistry*, **19**, 4293.

53 Richardson, J. P. and Macy, M. R. (1981), *Biochemistry*, **20**, 1133.

54 Sharp, J. A. and Platt, T. (1984), *J. Biol. Chem.*, **259**, 2268.

55 Morgan, W. D., Bear, D. G. and von Hippel, P. H. (1983), *J. Biol. Chem.*, **258**, 9565.

56 Adhya, S., Sarkar, P., Valenzuela, D. and Maitra, U. (1979), *Proc. Natl. Acad. Sci. USA*, **76**, 1613.

57 Morgan, W. D., Bear, D. G. and von Hippel, P. H. (1984), *J. Biol. Chem.*, **259**, 8664.

58 Morgan, W. D., Bear, D. G., Litchman, B. L. and von Hippel, P. H. (1985), *Nucleic Acids Res.*, **13**, 3739.

59 Wu, G-J. (1978), *Proc. Natl. Acad. Sci. USA*, **75**, 2175.

60 Weil, P. A., Segall, J., Harris, B., Ng, S. Y. and Roeder, R. G. (1979), *J. Biol. Chem.*, **254**, 6163.

61 Birkenmeier, E. H., Brown, D. D. and Jordan, E. (1978), *Cell*, **15**, 1077.

62 Weil, P. A., Luse, D. S., Segall, J. and Roeder, R. G. (1979), *Cell*, **18**, 469.

63 Wasylyk, B., Kedinger, C., Corden, J., Brison, O. and Chambon, P. (1980), *Nature (London)*, **285**, 367.

64 Grummt, I. (1981), *Proc. Natl. Acad. Sci. USA*, **78**, 727.

65 Grummt, I., Roth, E. and Paule, M. R. (1982), *Nature (London)*, **296**, 173.

66 Brown, D. P. and Gurdon, J. B. (1977), *Proc. Natl. Acad. Sci. USA*, **74**, 2064.

67 Pellicer, A., Robins, D., Wold, B., Sweet, R., Jackson, J., Lowry, I., Roberts, J. M., Sim, G-K., Silverstein, S. and Axel, R. (1980), *Science*, **209**, 1414.

68 Hamer, D. and Leder, P. (1979), *Cell*, **21**, 697.

69 Mellon, P., Parker, V., Gluzman, Y. and Manniatis, T. (1981), *Cell*, **27**, 279.

70 Chambon, P., Dierich, A., Gaub, M-P., Jakowlev, S., Johnstra, J., Krust, A., LePennec, J-P., Oudet, P. and Reudelhuber, T. (1984), *Recent Prog. Hormone Res.*, **40**, 1.

71 Goldberg, M. (1979), Ph.D. thesis, Stanford University, California.

72 Brady, J., Radonovich, M., Vodkin, M., Natarajan, V., Thoren, M., Das, G., Janik, J. and Salzman, N. P. (1982), *Cell*, **31**, 625.

73 Skuzeski, J. M., Lund, E., Murphy, J. T., Steinberg, T. H., Bargess, R. R. and Dahlberg, J. E. (1984), *J. Biol. Chem.*, **259**, 8345.

74 McKnight, S. L. (1982), *Cell*, **31**, 355.

75 Everett, R. D., Baly, D. and Chambon, P. (1983), *Nucleic Acids Res.*, **11**, 2447.

76 Dierks, P., van Ooyen, A., Cochran, M. D., Dobkin, C., Reiser, J. and Weissmann, C. (1983), *Cell*, **32**, 695.

77 Matsui, T., Segall, J., Weill, P. A. and Roeder, R. G. (1980), *J. Biol. Chem.*, **255**, 11 992.

78 Horikoshi, M., Sekimizu, K., Hirashima, S., Mitsuhashi, Y. and Natori, S. (1985), *J. Biol. Chem.*, **260**, 5739.

79 Davidson, B. L., Egly, J-M., Mulvihill, E. R. and Chambon, P. (1983), *Nature (London)*, **301**, 680.

80 Dynan, W. S. and Tjian, R. (1983), *Cell*, **35**, 79.

81 Grez, M., Land, H., Giesecke, K. and Schütz, G. (1981), *Cell*, **25**, 743.

82 Grindlay, G. J., Lanyon, W. G., Allan, M. and Paul, J. (1984), *Nucleic Acids Res.*, **12**, 1811.

83 Korn, L. J. (1982), *Nature (London)*, **295**, 101.

84 Hall, B. D., Clarkson, S. G. and Toccini-Valentini (1982), *Cell*, **29**, 3.

85 Brown, D. D. (1984), *Cell*, **37**, 359.

86 Sakonja, S., Bogenhagen, D. F. and Brown, D. D. (1980), *Cell*, **19**, 13.

87 Bogenhagen, D. F., Sakonju, S. and Brown, D. D. (1980), *Cell*, **19**, 27.

88 Cozzarelli, N. R., Gerrard, S. P., Schlissel, M., Brown, D. D. and Bogenhagen, D. F. (1983), *Cell*, **34**, 829.

89 Engelke, D. R., Ng, S-Y., Shastry, B. S. and Roeder, R. G. (1980), *Cell*, **19**, 717.

90 Bieker, J. J. and Roeder, R. G. (1984), *J. Biol. Chem.*, **259**, 6158.

91 Bogenhagen, D. F. (1985), *J. Biol. Chem.*, **260**, 6466.

92 Bieker, J. J., Martin, P. L. and Roeder, R. G. (1985), *Cell*, **40**, 119.

93 Folk, W. R. and Hofstetter, H. (1983), *Cell*, **33**, 585.

94 Allison, D. S., Goh, S. H. and Hall, B. D. (1983), *Cell*, **34**, 655.

95 Sprague, K. U., Larson, D. and Morton, D. (1980), *Cell*, **22**, 171.

96 Sharp, S., Dingermann, T. and Soll, D. (1982), *Nucleic Acids Res.*, **10**, 5393.

97 Stillman, D. J., Caspers, P. and Geiduschek, E. P. (1985), *Cell*, **40**, 311.

98 Camier, S., Gabrielsen, O., Baker, R. and Sentenac, A. (1985), *EMBO J.*, **4**, 491.

99 Smith, D. R., Jackson, I. J. and Brown, D. D. (1984), *Cell*, **37**, 645.

100 Ciliberto, G., Raugei, G., Costanzo, F., Dente, L. and Cortese, R. (1983), *Cell*, **32**, 725.

101 Klemenz, R., Stillman, D. J. and Geiduschek, E. P. (1982), *Proc. Natl. Acad. Sci. USA*, **79**, 6191.

102 Ruet, A., Camier, S., Smagowicz, W., Sentenac, A. and Fromageot, P. (1984), *EMBO J.*, **3**, 343.

103 Moss, T. (1982), *Cell*, **30**, 835.

104 Reeder, R. H. (1984), *Cell*, **38**, 349.

105 Sommerville, J. (1984), *Nature (London)*, **310**, 189.

106 Kishimoto, T., Nagamine, M., Sasaki, T., Takakusa, N., Miwa, T., Kominami, R. and Muramatsu, M. (1985), *Nucleic Acids Res.*, **13**, 3515.

107 Mishima, Y., Financsek, I., Kominami, R. and

Muramatsu, M. (1982), *Nucleic Acids Res.*, **10**, 6659.

108 Hofer, E. and Darnell, J. E. (1982), *Cell*, **29**, 887.

109 Krieg, P. A. and Melton, D. A. (1984), *Nature (London)*, **308**, 203.

110 Price, D. H. and Parker, C. S. (1984), *Cell*, **38**, 423.

111 Salditt-Georgieff, M. and Darnell, J. E. (1983), *Proc. Natl. Acad. Sci. USA*, **80**, 4694.

112 Hagenbüchle, D., Wellauer, P. K., Cribbs, D. L. and Schibler, U. (1984), *Cell*, **38**, 737.

113 McLauchlan, J., Gaffney, D., Whitton, J. L. and Clements, J. B. (1985), *Nucleic Acids Res.*, **13**, 1347.

114 Henikoff, S., Kelly, J. D. and Cohen, E. H. (1983), *Cell*, **33**, 607.

115 Maderious, A. and Chen-Kiang, S. (1984), *Proc. Natl. Acad. Sci. USA*, **81**, 5931.

116 Bogenhagen, D. F. and Brown, D. D. (1981), *Cell*, **24**, 261.

117 Bakken, A., Morgan, G., Sollner-Webb, B., Roan, J., Busby, S. and Reeder, R. H. (1982), *Proc. Natl. Acad. Sci. USA*, **79**, 56.

118 Grummt, I., Sorbaz, H., Hofmann, A. and Roth, E. (1985), *Nucleic Acids Res.*, **13**, 2293.

119 Kung, S. D. and Lin, C. M. (1985), *Nucleic Acids Res.*, **13**, 7543.

120 Christianson, T. and Rabinowitz, M. (1983), *J. Biol. Chem.*, **258**, 14 025.

121 Bogenhagen, D. F., Appelgate, E. F. and Yoza, B. K. (1984), *Cell*, **36**, 1105.

122 Chamberlin, M. and Ryan, T. (1982), in *The Enzymes* (ed. P. Boyer), Academic Press, New York, Vol. 15, p. 87.

123 Studier, F. W. and Dunn, J. J. (1983), *Cold Spring Harbor Symp. Quant. Biol.*, **47**, 999.

124 Davanloo, P., Rosenberg, A. H., Dunn, J. J. and Studier, F. W. (1984), *Proc. Natl. Acad. Sci. USA*, **81**, 2035.

125 Das, G. C. and Niyogi, S. K. (1981), *Prog. Nucleic Acid Res. Mol. Biol.*, **25**, 187.

126 Buer, R., Bankier, A. T., Biggin, M. D., Deininger, P. L., Farrell, P. J., Gibson, T. J., Hatfull, G., Hudson, G. S., Satchwell, S. C., Séguin, C., Tuffnell, P. S. and Barrell, B. G. (1984), *Nature (London)*, **310**, 207.

127 Arrand, J. R. and Rymo, L. (1982), *J. Virol.*, **41**, 376.

128 Nevins, J. R. and Joklik, W. K. (1977), *J. Biol. Chem.*, **252**, 6930.

129 Blumenthal, T. (1982), in *The Enzymes* (ed. P. D. Boyer), Academic Press, New York, Vol. 15, p. 267.

130 Baltimore, D. (1984), in *The Microbe 1984: Society for General Biology Symposium*, **36**, *part 1* (ed. B. W. J. Mahy and J. R. Pattison), Cambridge University Press, Cambridge, p. 109.

131 Golini, F., Nomoto, A. and Wimmer, E. (1978), *Virology*, **89**, 112.

132 Shatkin, A. J. and Both, G. W. (1976), *Cell*, **7**, 305.

133 Plotch, S., Bouloy, M., Ulmanen, I. and Krug, R. M. (1981), *Cell*, **23**, 847.

134 Krug, R. M. (1981), *Curr. Top. Microbiol. Immunol.*, **93**, 125.

135 Krug, R. M. (1985), *Cell*, **41**, 651.

136 Wagner, R. R. (1975), in *Comprehensive Virology* (ed. H. Fraenkel-Conrat and R. R. Wagner), Plenum Press, New York, Vol. 4, p. 1.

137 Chang, D. D. and Clayton, D. A. (1984), *Cell*, **36**, 635.

138 Tubak, H. F., Grivell, L. A. and Borst, P. (1983), *CRC Crit. Rev. Biochem.*, **14**, 297.

139 Clayton, D. A. (1984), *Annu. Rev. Biochem.*, **53**, 573.

140 Miller, J. L., McLachlan, A. D. and Klug, A. (1985), *EMBO J.*, **4**, 1609–14.

141 Weinberger, C., Hollenberg, S. M., Rosenfeld, M. G. and Evans, R. M. (1985), *Nature (London)*, **318**, 670–2.

142 Rosenberg, U. B., Schröder, C., Preiss, A., Kienlin, A., Côté, S., Riede, I. and Jäckle, H. (1986), *Nature (London)*, **319**, 336–9.

# 9

# The arrangement of genes, their transcription and processing

## 9.1 TRANSCRIPTION AND PROCESSING OF PROKARYOTIC AND BACTERIOPHAGE mRNA

Messenger RNA was discovered in bacteria. In 1953 Hershey [1] noticed that the appearance of metabolically active RNA followed the infection of *E. coli* with the bacteriophage T2. Subsequent analysis [2, 3] showed that the RNA was rapidly synthesized and had been formed from and would hybridize to the viral DNA. By pre-labelling the host bacterial cells with $^{13}$C and $^{15}$N-labelled growth medium and then infecting them in the presence of $^{12}$C and $^{14}$N media, Brenner, Jacob and Meselson were able to show that the newly made (isotopically light) viral mRNA was translated on pre-existing (isotopically heavy) cellular ribosomes. Furthermore, when they employed radioactive amino acids, they demonstrated that the translation product was viral protein [4]. Rapidly labelled RNA with similar properties could also be detected in uninfected cells [5] but two other bacteriophage, $\alpha$ and SP8, were of importance in the demonstration that only one strand of the DNA serves as a template for mRNA synthesis. The two strands of these viruses are sufficiently different in density to allow their separation on caesium chloride density gradients, and the mRNA formed in infected bacterial cells was shown to hybridize to only one DNA strand [6, 7]. In the case of phage $\phi$X174, however, it was shown that both strands of the replicative form of the DNA were used as a template [8] thus establishing the principle that genes are not necessarily localized on one strand of the genome.

In prokaryotes, transcription and translation are not physically separated by the barrier of a nuclear membrane and the two processes are so tightly coupled that ribosomes will attach to the 5′-end of the mRNA and commence translation before its 3′ synthesis is complete [9]. Since the rates of the two processes are similar [10], the unit can be regarded as a transcription–translation complex in which RNA polymerase transcribing the gene is followed by a string of ribosomes translating the message (Fig. 9.1). Indeed the electron micrographs of Miller and colleagues allowed the visualization of complexes [9].

Many transcripts are of considerable length because prokaryote genes are often clumped in functional units known as *operons* in which, for instance, the genes encoding the enzymes of a particular metabolic pathway might occur consecutively on the DNA. Thus, the lactose or *lac*

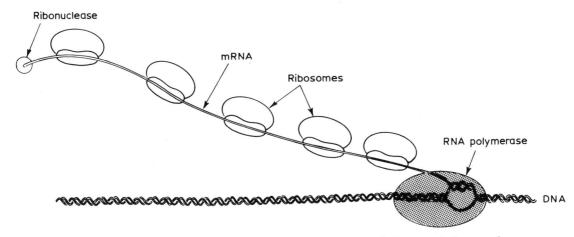

**Fig. 9.1**  Diagrammatic representation of a prokaryotic transcription–translation complex.

operon encodes the three enzymes of lactose metabolism and the *trp* operon contains the genes for the five polypeptides which make up the three enzymes that catalyse tryptophan synthesis. For a discussion of the value of this system in the control of gene expression see Section 10.1. Typically, an operon has a single 5′-promoter and the genes within it are transcribed by RNA polymerase into a single *polycistronic mRNA*. Such messages may encode up to 20 proteins.

Another feature of prokaryote mRNA is a very short half-life, which is typically about 2 min [11]. This, combined with its lack of the 3′-poly(A) tail that provides such a convenient 'handle' for the purification of the more stable eukaryote mRNA, makes it very difficult to study. Some bacterial mRNAs, such as those for *E. coli* lipoprotein, *lac* operon and *gal* operon, have been isolated but for the most part they can only be detected by their translation products in cell-free protein-synthesizing systems or by their hybridization to a radioactive complementary DNA (cDNA). Indeed, much of the control of bacterial gene expression rests on the fine tuning that mRNA instability allows. With long transcripts the degradation of the mRNA may also be initiated before synthesis is complete;

ribonuclease in effect following the last ribosome along the message [12, 13]. The transcription–translation complex may thus be imagined as illustrated in Fig. 9.1. The mRNAs of bacteriophage, particularly the messenger-genome of the RNA phage R17, Qβ, MS2 and f2, are more stable than those of their host, in part at least because they exhibit extensive secondary structure (see Section 11.10.1). This has made them an important source of prokaryote mRNA.

As is the case with many other prokaryote and eukaryote gene transcripts, bacterial and bacteriophage mRNAs are synthesized with additional non-coding sequences at their 5′- and 3′-ends. These sequences do not encode protein but the 5′ or leader sequence does include the Shine-Dalgarno sequence which is involved in the binding of mRNA to the ribosome (see Section 11.5.1). Additional, intercistronic sequences of variable length (from 1 to over 100 nucleotides in some bacteriophage) also occur between the coding regions of polycistronic mRNAs. Usually, the translating ribosomes initiate protein synthesis separately at each coding region on the message. There is some evidence, however, that polycistronic messages may occasionally be processed into their individual protein-encoding elements [see 14,

15, 16 for reviews]. The best evidence for this comes from studies with the polycistronic mRNA that encodes that five early bacteriophage T7 proteins. RNase III may process the transcript to monocistronic mRNAs [17]. It does not appear, however, that such processing is necessary for the expression of this [18] or any other polycistronic mRNA although mutations in the genes of the nucleases that are known to process bacterial rRNA and tRNA precursors do affect the efficiency of production of some proteins [19].

## 9.2 THE ORGANIZATION OF EUKARYOTE PROTEIN-ENCODING GENES

### 9.2.1 Genes are often discontinuous

The availability of a specific purified mRNA species (see Section 9.3), the use of reverse transcriptase to make a radioactive probe by copying the messenger into a complementary DNA (cDNA) and the availability of restriction enzymes that cleave DNA at specific sites led to the discovery that the genes encoding eukaryotic proteins are often discontinuous. Leder and colleagues [20, 21] digested mouse DNA with restriction enzymes, separated the fragments on agarose gels and transferred them to cellulose nitrate by Southern blotting (see Section A.3). They then searched the blotted DNA for fragments that contained globin-encoding sequences by hybridization with a [32]P-labelled $\beta$-globin cDNA. Their results showed that the $\beta$-globin gene did not exist as a contiguous stretch of DNA but as three blocks of coding sequence separated by sections of non-coding DNA which have since become known as *introns, inserts or intervening sequences*. The sections of coding sequence have become known as *exons*. Leder and colleagues [21] further showed that the largest of the two globin introns could be visualized in the electron microscope. Hybrids between the mRNA and a cloned globin

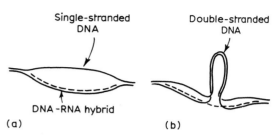

**Fig. 9.2** Diagrammatic representation of R loops: (a) pattern of hybridization which would be expected between an mRNA and a gene containing no intron; (b) pattern of hybridization which would be expected between (i) an mRNA and a gene containing a single large intron, (ii) a messenger cDNA and a pre-mRNA containing a transcript of a single large intron.

gene, formed under conditions in which RNA:DNA hybrids were more stable than the DNA:DNA double helix, had an unusual appearance. Instead of the expected single continuous R loops of displaced DNA illustrated in Fig. 9.2a, two smaller loops of displaced single-stranded DNA were separated by a loop of double-stranded DNA (Fig. 9.2b). This latter loop was caused by the larger intron, having no complement in the mRNA and thus continuing to exist as a DNA duplex. Small introns cannot be detected by this method.

It is now known that most protein-encoding genes of eukaryotes and of their smaller DNA viruses are discontinuous. Known exceptions include most histone genes, interferon, most herpes and Pox virus genes and adenovirus polypeptide IX [22, 23, 24, 25]. It is also known that the initial transcript of genes includes introns and exons and that the former are spliced out of the pre-mRNA. This processing and its possible mechanism is discussed in Section 9.3.4. The arrangement of the coding and non-coding blocks of some of the first genes to be characterized is illustrated in Fig. 9.3. In many genes, the combined length of the introns exceeds that of the exons. Thus, the chicken ovalbumin gene contains approximately 7564

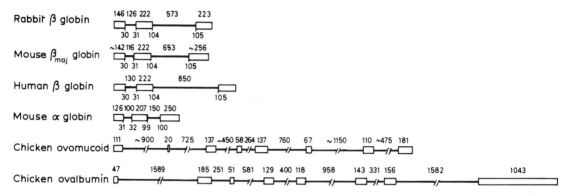

Rabbit β globin

Mouse β_maj globin

Human β globin

Mouse α globin

Chicken ovomucoid

Chicken ovalbumin

**Fig. 9.3**   The organization of introns and exons in some of the first genes characterized [26–31]. The boxes represent exons and the intervening lines are introns. The numbers above the diagrams indicate the numbers of nucleotides in the segments. In some cases they are approximate. The numbers below the globin diagrams represent the amino acid codons that occur on either side of the introns.

base pairs while the mRNA is only 1872 nucleotides long. The difference of 5692 nucleotides represents non-coding sequences arranged in seven introns [31]. The total length of the avian vitellogenin gene is approximately 23 000 base pairs; 25 large and 6–10 small introns are interspersed among the 6700 base pairs that encode the mRNA [32]. The chicken pro-α-2(1)-collagen gene contains about 50 introns [33]. Figure 9.4 is an electron micrograph of a hybrid molecule formed between a cloned ovomucoid DNA and the ovomucoid mRNA. It shows the R loops formed by the seven introns of the gene [34].

The existence of introns provokes an obvious question: What possible evolutionary or structural function might they possess? Why for instance do all known histone genes lack introns except for two genes in chicken, two in the fungus *Neurospora crassa* [35] and one in man [438]? Several authors have suggested that introns have no function; that they are an evolutionary remnant in eukaryotes which prokaryotes have been able to discard because of their rapid rate of evolution [36, 37]. That this might be occurring, though at a slow rate in eukaryotes, is supported by the finding that the simple rapidly dividing eukaryote, yeast, tends

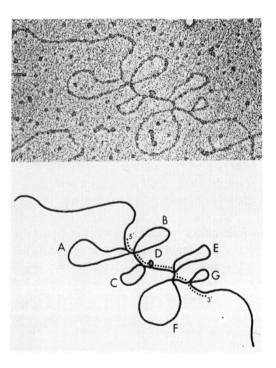

**Fig. 9.4**   Electron micrograph and line drawing of a hybrid molecule formed between a cloned avian ovomucoid gene and its mRNA. The R loops labelled A–F are formed by the seven introns in the DNA. Reproduced from [34] with permission of the authors and publishers.

to have smaller introns than those of higher cells and many of its genes have no introns. Furthermore, there is evidence that introns can be edited out in higher eukaryotes. All known insulin genes have two introns except for one of the two rat genes which has one. Sequence homology studies indicate that the two rat genes have resulted from a gene duplication and the inference is, therefore, that an intron was lost from one copy during the duplication [38] (see Section 7.7.3c). In fact introns are not totally absent in prokaryotes but have been found in the thymidylate synthase gene of bacteriophage T4 [417] and in the tRNA genes of archaebacteria; a group of micro-organisms that are as different from the remaining eubacteria as eubacteria are from eukaryotes [39].

An extension of these ideas, that introns represent an early stage of gene evolution, is that split genes allow the rapid evolution of new genes [36]. An exon might simply represent a function-encoding protein domain able, for instance, to bind a particular coenzyme, span a membrane or locate in the main groove of DNA. Genetic recombinations might then bring together new combinations of exons and generate a protein with a new combination of functions. By this theory introns are seen as functionless sequences trapped between reorganized exons. Without them, however, and the cells' ability to remove their transcripts (see Section 9.3.4), recombination would have exactly to join functional domains so as to preserve both their identity and their reading frames. Such precise reorganization would be very rare and gene evolution would be much slower.

Support for these concepts has come from a number of studies. Tonegawa and coworkers [40] have demonstrated a functional relationship between the domains of the heavy chain of immunoglobulins and the exons of its gene. Each of the six functional polypeptide units, including the 14-amino acid hinge region, is encoded in a separate exon. Similarly, Stein *et al.* [30] have shown that the intervening sequences of chicken

ovomucoid gene separate coding portions that correspond to the functional domains of the protein, and other recent examples include pyruvate kinase [41] and the low-density lipoprotein (LDL) receptor [42]. The latter example contains 18 exons most of which correlate with functional domains in the LDL receptor protein. Thirteen of the exons are homologous with functionally similar domains in other proteins. The central exon of globin genes encodes the haem-binding domain of the protein [43] and although the other exons do not encode such clear-cut domains as the above examples, they do divide the protein into separate compact structures [44]. Other examples such as lysozyme [45] and $\alpha$-foetoprotein [46] show that the domain structure of some proteins was derived from the amplification and subsequent divergence of exons within a gene.

Further support for the role of exons in the evolution of proteins comes from the study of gene families and structurally related genes. Thus, all known vertebrate globin and myoglobin genes have two introns at the same position within the coding sequence (Fig. 9.3), an arrangement that fits well with the concept that they evolved from a primitive globin gene by the mechanisms outlined above. The relationship even extends to the plant leg haemoglobins [47] which have three introns, two of which correspond to those of vertebrate globins. However, not all of the evidence supports the concept that introns arose during the evolution of genes. Insects are more closely related to vertebrates than either are to plants so the recent finding that insect globins contain no introns is difficult to explain [48]. Similarly, the structurally related $\alpha$1-antitrypsin and chicken ovalbumin genes have introns that are totally dissimilar in number, distribution and size [49]. This latter example has led to the suggestion that some introns may be introduced into genes and that, in the case of ovalbumin and $\alpha$1-antitrypsin, introductions occurred after the two genes diverged. Stone *et al.* [50] have built upon these ideas and, as a result of a comparison of the glyceraldehyde phosphatase gene in seven species (two bacteria, a yeast, a crustacean, a bird and two mammals including man), have suggested that introns are of two types. *Associative (type A)* introns are those created when two exons are brought together by genetic recombination. An organism possessing the splicing mechanism necessary for type A introns could also tolerate *divisive (type B)* introns. These are seen as DNA inserted into the gene by aberrant recombination, the insertion of a retrovirus or transposon or by retropositional insertion of DNA copies of cellular mRNA. Stone *et al.*[50] claim that both types of intron can be recognized in the chicken glyceraldehyde phosphatase gene, and the serine protease gene family [51] also provides examples of apparent intron insertion as well as intron loss and exon rearrangement. Very recently, a mechanism for the insertion of introns has been described in yeast mitochondria. Two groups [419, 420] have described an intron-encoded protein that has a transposase-like function and is required for spreading the intron of mitochondrial 21S rRNA gene between intron-plus and intron-minus variants.

All of these ideas imply that introns are functionless sequences and the fact that the two-intron and one-intron rat insulin genes are equally well expressed shows that the larger insulin insert can be lost with no ill effects. Furthermore, with the exception of the splicing regions, the base sequence of introns exhibit a much greater sequence divergence than that of exons, again indicating a low information content (compare the sizes of the introns and exons in the different $\beta$-globin genes in Fig. 9.3). On the other hand, experimental systems in which introns were deleted from recombinant genes are contradictory and show that in some but not all genes the splicing out of intron transcripts may be prerequisite for stable mRNA formation [52, 53, 54]. Furthermore, introns can be functional in cases of differential processing.

When this occurs, the product of one gene may be differentially spliced to yield more than one mRNA and the intron of one message may become the exon of another. This, and the internal maturase sequences of introns of mitochondrial genes, which control their own processing, are discussed elsewhere (Sections 10.6.2 and 9.6.4 respectively).

### 9.2.2  Gene families and gene clustering

#### (a)  *Multigene families*

As outlined in the preceding section, one way in which new genes may evolve is by the grouping of previously unrelated exons into a new transcriptional unit. Another mechanism, which appears to be common, is gene duplication. This may give rise to multicopy genes and/or allow the evolution of new genes. The mouse $\alpha$-foetoprotein and serum albumin genes probably arose by duplication of a common ancestral gene [55]. rRNA, tRNA, 5S RNA and histones are encoded by multiple copy genes. This has been rationalized in terms of the large quantities of the end product of these genes which the cell might need to synthesize over a short period of time. However, genetic manipulation has revealed that other genes, particularly those encoding abundant proteins, also often occur as members of structurally related gene families. Detection of such families usually follows $C_0t$ analysis (Section 3.2.5) or the use of a cloned cDNA copy of a mRNA as a probe for the gene from which the original mRNA was transcribed. A family is indicated when the cDNA hybridizes to more than one area of the genome. Subsequent analysis may reveal that some members of a family are not expressed (pseudogenes) or are only expressed in specific tissues. The mechanisms by which multigene families are thought to arise are described in Section 7.3.2.

Gene families have been detected for several structural proteins including collagen [56], actin [57], tubulin [57, 58], keratin [59], lens crystallins

[60] and insect eggshell chorion [61]. They also occur for other abundant proteins such as histones (see below) and ribosomal proteins which in higher eukaryotes have from seven to more than 20 gene copies per protein [62] although there is no evidence for more than one functional gene. Clearly, different isoenzymes may also, in some cases, represent the products of gene families that have evolved from a common ancestor. Multiple, structurally related, rat amylase genes encode enzymes that are disparately expressed in different tissues [63, 64]. However, isoenzymes, including those of amylase, may also be generated by differential processing of one gene product (see Section 10.6.2).

(b) *Gene clustering*

All the members of a gene family may occur closely coupled (arranged in tandem along the DNA) or they may be dispersed as individual genes or small clusters. Histone genes, which exhibit all three arrangements, will serve as an example [for reviews see 35, 65, 66].

Histones are small highly conserved proteins which organize DNA into nucleosomes (see Section 3.3.1), are required in large quantities by the cell and are often encoded in highly repetitive genes. They were first analysed in detail in the sea-urchin *Psammechinus miliaris* where they are so highly repeated and clumped that, in caesium chloride–actinomycin density gradients, the genes migrate as a satellite, separable from the main band of DNA [67]. The histone genes so obtained, when digested with the restriction enzyme *Eco*RI or *Hin*dIII give rise to 6.3-kb fragments of DNA which contain the genes for all five histones [67]. They were arranged in the order shown in Fig. 9.5.

The arrows indicate the polarity of the genes

and show that they are all transcribed in the same direction and therefore occur in sequence on the same strand of the DNA [68]. There is no evidence, however, that they are ever transcribed as a polycistronic mRNA and each gene has its own TATA box promoter in its 5′-flanking sequences (Section 8.3.1). Each of the genes is separated from the next by non-transcribed, AT-rich spacer and the quintets are arranged in tandem with 300–600 copies per haploid genome. Each cluster has one *Eco*RI and one *Hin*dIII cleavage site thus permitting the isolation and subsequent cloning of the clusters.

The availability of the sea-urchin histone genes allowed their use as probes for the isolation and cloning of histone genes from other species. In other cases, cDNA copies of purified histone mRNA were employed as a probe for the same purpose. It was soon found that the histone genes of two other distantly related sea-urchin species occur in very similar quintets [69] in which the main difference from *P. miliaris* is the size and sequence of the non-transcribed spacers. The spacers show some heterogeneity (*microheterogeneity*) even within a species.

The genes of the fruit fly *Drosophila melanogaster* are also in quintets [70] though here the order is different (Fig. 9.6). Furthermore, the *Drosophila* genes have a variable polarity within the cluster, some running from left to right and the others in the reverse direction. Since transcription is always in a 5′ to 3′ direction (Section 8.1), this means that the coding sequences of some genes are on one strand of the DNA while the rest are on the other. There are about 100 copies of the *Drosophila* histone genes.

At this point, it appeared that histone genes and their transcripts were characterized by a number of features that set them apart from other known protein-encoding genes. They were

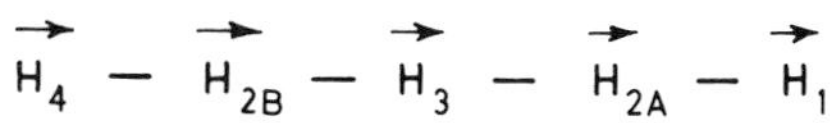

$$\overrightarrow{H_4} - \overrightarrow{H_{2B}} - \overrightarrow{H_3} - \overrightarrow{H_{2A}} - \overrightarrow{H_1}$$

**Fig. 9.5** The histone gene clusters of *Psammechinus*.

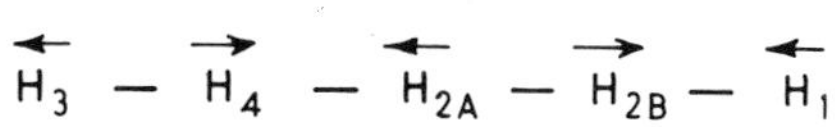

$$\overleftarrow{H_3} - \overrightarrow{H_4} - \overleftarrow{H_{2A}} - \overrightarrow{H_{2B}} - \overleftarrow{H_1}$$

**Fig. 9.6** The histone gene clusters of *Drosophila*.

highly reiterated, clustered into tandem repeated quintets, contained no introns and their transcripts were not polyadenylated. It is now known that none of these generalizations is universally true. A few histone genes do contain introns [35], some histone mRNAs are poly-adenylated (see Section 9.3.3) and many do not occur as highly reiterated, tandemly repeated quintets. Reiteration can vary from 1600-gene copies in *Axolotl* down to two copies in yeast [35]. Clustering also exhibits great variability between species and within a species. Thus, within the amphibian *Xenopus borealis* at least 70% of histone genes occur in a major cluster that has the same gene order as *Drosophila* but is in a large 16-kb segment of DNA for which tandem linkage has not been demonstrated. The closely related *Xenopus laevis*, on the other hand, has at least two major cluster types each of which shows considerable length and sequence heterogeneity [71, 72]. Mammals and birds show even less regularity and isolated clones indicate that histone genes are clustered in a largely random fashion except that H2A and H2B tend to be paired [36, 66]. Even in sea-urchins, the 80% of genes in quintets are the so-called 'early genes' that encode the histones made in very large quantities during the rapid early cleavages of the embryo. The 'late histone genes' which take over histone gene expression at the blastula stage of development are clustered but are not organized in quintets. Sea-urchins and other species also contain histone genes known as *orphons*, which are clearly related to those of tandem repeats but exist as solitary elements remote from the tandem clusters [73].

Why the pattern of arrangement of histone genes should be so variable; why they are clustered and what might be the value of their common but far from universal arrangement in quintets is unknown. It has been suggested that the arrangement of the early genes in sea-urchins is associated with the high rates of histone synthesis in the early embryonic cleavages but this does not explain why *Xenopus* quintets are

expressed throughout life. In many cells, histones are synthesized in a cell-cycle-dependent manner being closely coupled to S phase and DNA synthesis (see Sections 3.3.1 and 6.12). The coupling occurs, however, in both clustered and non-clustered genes. Presumably, clustering arose initially from gene duplication but it seems likely that some evolutionary pressure has been responsible for holding the genes together (these concepts are further discussed in [36]).

Other much-studied gene clusters are those for the $\alpha$- and $\beta$-globins (for a review see [74]).

In the adult mammal, haemoglobin consists of two alpha($\alpha$) and two beta($\beta$) polypeptides. In the foetus, the $\beta$ chain is replaced by gamma($\gamma$) chains which, since they give haemoglobin a higher affinity for oxygen, facilitate the passage of the gas across the placenta from maternal to foetal blood. In human embryos of less than 8 weeks two further polypeptides occur in haemo-globin. The $\beta$ chain is replaced by epsilon($\epsilon$) polypeptides and the $\alpha$ chains are replaced by zeta($\zeta$) polypeptides. From eight weeks the embryonic polypeptides are gradually replaced by foetal chains and hybrid molecules occur.

There is then a family of $\alpha$-like globins and a family of $\beta$-like globins and, at a genomic level, these are represented as clustered genes. The arrangement of the human globin genes is illustrated in Fig. 9.7. It can be seen that separate genes encode each of the embryonic and adult polypeptides and that some ($\alpha$, $\zeta$ and $\gamma$) are encoded in duplicated genes. All members of each family have two introns inserted at the same position in the coding sequence as $\alpha$ and $\beta$ respectively. In both families the genes are arranged in the order in which they are expressed. Both families also include pseudo-genes (identified with the symbol $\psi$) which arise by duplication but are defective and do not give rise to a functional polypeptide (pseudogenes are discussed in Sections 7.6 and 7.7). The globin genes of other primates are very similar to those of man and, while those of other vertebrates have their own embryonic variants, in mammals

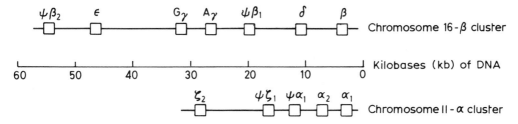

**Fig. 9.7** The human globin gene clusters.

at least, it appears that the active genes are arranged in a 5′ to 3′ direction in the order in which they are expressed.

As with histone genes, the advantage of having the globin genes in a strongly conserved familial cluster is unknown. It is tempting to speculate that the arrangement facilitates the control of their sequential expression but as yet there is little or no evidence to support such speculation. A group of gene families for which the mechanistic advantages of clustering are apparent are the immunoglobulin genes of which the gene arrangement is considered in Section 7.9.1.

## 9.3 TRANSCRIPTION AND PROCESSING OF EUKARYOTE PRE-MESSENGER RNA

Three characteristics of eukaryote mRNAs make them easier to study than those of prokaryotes. Firstly, they are more stable and have half-lives ranging from 1 to 24 h [75, 76]. Secondly, they have 3′-polyadenylate tails which allows their purification by affinity chromatography on columns of oligo(dT)-cellulose or poly(U)-Sepharose (see Section A.2.2). Thirdly, in some specialized cells, a high percentage of the total mRNA codes for one or a few abundant proteins. Thus 50% of the mRNA produced by the tubular gland cells of the magnum portion of the hen oviduct encodes the egg white protein ovalbumin [77] and as a result it was one of the first mRNA species isolated [78]. Similar tissue

or cell specific abundance led to the early isolation of globin mRNA from reticulocytes [79], silk fibroin mRNA from the silk glands of the larvae of the silk moth *Bombyx mori* [80], crystallin mRNA from chick lens [81] and immunoglobulin light chain mRNA from myeloma cells [82]. The studies with these and other abundant mRNA species led to the development of techniques whereby much less abundant species could be studied (see Appendix).

### 9.3.1 The nature of gene transcripts

It is now established that cytoplasmic mRNA of eukaryotic cells is the product of the maturation of longer precursors which form at least a part of the high-$M_r$ nuclear RNA known as heterogeneous nuclear RNA (hnRNA). The early kinetic evidence for the relationship between hnRNA and mRNA was ambiguous. Most of a pulse of radioactive precursor incorporated into hnRNA turns over within the nucleus. This, combined with the difficulties encountered in conducting effective chase experiments and in adequately fractionating hnRNA from mRNA and precursor to rRNA, led to conflicting data. The relationship between the two types of molecule was strengthened with the finding that both possess 3′-polyadenylated tails and modified 5′-termini known as caps. The greatest problem with the precursor–product hypothesis, however, was the severalfold difference in size between the two classes of molecule. hnRNA

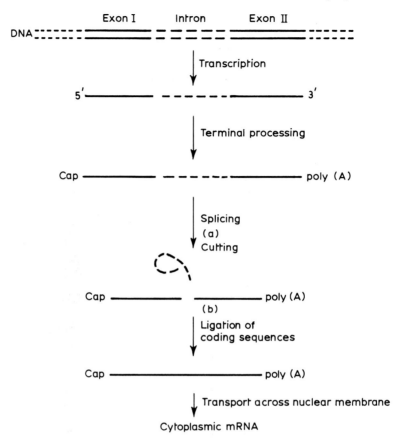

**Fig. 9.8**    A diagrammatic representation of the processing of the transcripts of a gene containing a single intron.

can have a sedimentation value of 100S indicating lengths of up to 50 000 nucleotides [83] while even a large mRNA, such as that of the avian egg yolk precursor protein, vitellogenin (6700 nucleotides) [32] is very much smaller. Aggregation does contribute to the apparent large size of hnRNA molecules but, even under severely denaturing conditions, the average hnRNA molecule is still several times the length of the average mRNA [84].

Various models put forward to account for this paradox included the possibility that an hnRNA might contain more than one mRNA sequence or that the hnRNA might be cleaved and digested to conserve an mRNA sequence within it. The former solution did not account for the fact that there is no evidence for polycistronic mRNA in eukaryotes or the fact that most hnRNA turns over within the nucleus. However, the alternative appeared to require the destruction of at least one protected end of the hnRNA and the synthesis, on the mRNA, of a new 5'-cap or 3'-polyadenylate. A third possibility, that hnRNA might be processed by the removal of internal portions of the molecule, would have received scant attention before the discovery that genes are themselves discontinuous (Section 9.2.1). In those cases where characterization has proceeded far enough, however, it is precisely this solution that solves the precursor–product paradox.

Figure 9.8 is a diagrammatic representation of

**Table 9.1** Small RNAs of eukaryotic cells.

| Notation of RNA components | Length (nucleotides) | Abundance (copies/cell) | Major subcellular location | Transcribing RNA polymerase | Nature of 5'-terminus | Reference |
|---|---|---|---|---|---|---|
| U1 or D | 165 | $1 \times 10^6$ | Nucleoplasm | II | | Section 9.3.4.c |
| U2 or C | 188–189 | $5 \times 10^5$ | Nucleoplasm | II | | Sections 9.3.3.c and 9.3.4.c |
| U3 or A | 210–215 | $3 \times 10^5$ | Nucleolus | II | 2',2'-Dimethyl-7-methylguanosine | Section 9.4.2.a |
| U4 or F | 142–146 | $1 \times 10^5$ | Nucleoplasm | II | | Section 9.3.3.b |
| U5 or G' | 107–121 | $2 \times 10^5$ | Nucleoplasm | II | | Ref. [85] |
| U6 or H | 107–108 | $3 \times 10^5$ | Nucleoplasm | II | *Modified 5'-triphosphate | Section 9.3.3.b |
| U7 | ~56–57 | $3 \times 10^4$ | Nucleoplasm | Presumed to be II | Uncertain [125, 423] | Section 9.3.3.c |
| U8 | 141 | $2.5 \times 10^4$ | Nucleolus | | 2',2'-Dimethyl-7-methylguanosine | Ref. [87] |
| U9 | ~130 | Not known | Not known | | | |
| U10 | ~65 | Not known | Not known | | | |
| 7S or L or 7SL | 195–300 | $5 \times 10^5$ | Cytoplasm | III | | Section 11.7.1 |
| 7.1 } 7SM | 260 | | Nucleolus | III | | Ref. [85] |
| 7.2 } | 290 | $1 \times 10^5$ | Nucleolus | III | | Ref. [85] |
| 7.3 or 7SK | 331 | $0.5$–$1 \times 10^5$ | Nucleoplasm | III | 5'-Triphosphate | Refs. [85], [86] and refs. therein |
| Y | 83–112 | $1 \times 10^5$ $10^5$–$10^6$ | Cytoplasm | III | | Refs. [85], [86] and refs. therein |
| 4.5S | 90–99 | $3 \times 10^5$ | Nucleoplasm | III | | Refs. [85], [86] and refs. therein |
| 5.8S | 158 | $5 \times 10^6$ | Cytoplasm | I | 5'-Monophosphate | Section 11.8.1 |
| 5S | 121 | $7 \times 10^6$ | Cytoplasm | III | 5'-Triphosphate | Section 11.8.1 |
| tRNA | 74–95 | $1 \times 10^8$ | Cytoplasm | III | 5'-Monophosphate | Section 11.1 |

*Nature of modification not known.

the processing of the transcripts of eukaryote genes. Each of its three main steps: 5'-capping, 3'-polyadenylation and the removal of intron transcripts (splicing) will be considered in turn in the following sections. Considerable evidence suggests that the members of a group of small nuclear RNA molecules, collectively known as *snRNA*, play a considerable role in the processing of gene transcripts. Table 9.1 lists some of the characters of these RNA molecules, compares them with more abundant small RNA species, such as tRNA and 5SRNA, and specifies the sections of the subsequent text in which their proposed function is discussed. For a review of

snRNA see Busch *et al.* [85], for a compilation of snRNA sequences see [86] and for recent data on U8 snRNA see [87].

### 9.3.2 Caps and 5'-leader sequences of eukaryotic mRNA

The mRNAs and pre-mRNAs of eukaryotes and most of their viruses are unique in having their 5'-termini blocked by the post-transcriptional addition of a methylated guanosine cap (for reviews see [88–90]). The guanosine is linked through a 5'-5' pyrophosphate bridge to the 5'-terminal nucleotide of the primary transcript

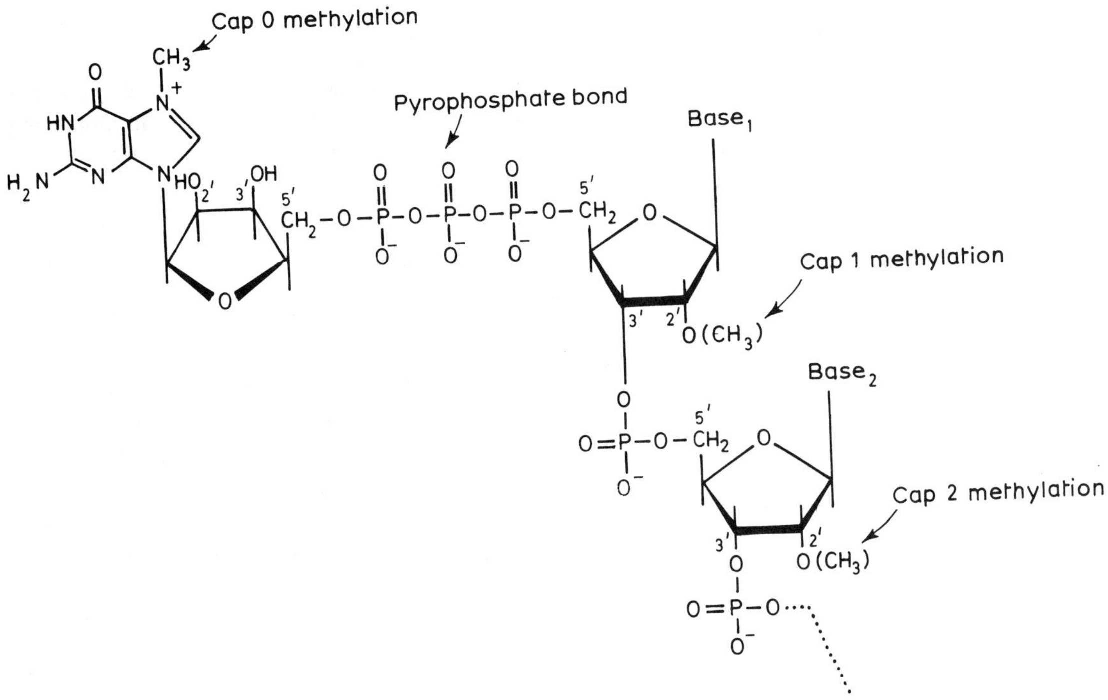

**Fig. 9.9** The 5′ cap of eukaryotic mRNA.

and is then methylated at carbon 7 to form cap 'zero'. Further modification may involve 2′-*O*-methylation of the ribose of the first or the first two nucleotides of the transcript thus forming 'cap one' and 'cap two' respectively (Fig. 9.9).

The relative abundance of the different cap structures changes with evolutionary complexity. Yeast mRNAs have cap zero; those of slime moulds are mainly cap zero but 20% have cap one; messages of the brine shrimp and sea-urchin are all terminated with cap one while mammals have high percentages of caps one and two [88]. The enzymic reactions of cap formation were first elucidated with purified reovirus [91], vaccinia virus [92] and HeLa cell nuclei [93] and are illustrated in Fig. 9.10.

Caps are added to the 5′-ends of incomplete transcripts [94] and, in those systems where sufficient data are available, the 5′-termini of both the mature mRNA and its precursors are

identical and map at the same point in the gene sequence. Even where microheterogeneity exists in the initiation site of a gene [95, 96] this appears to be reflected in similar heterogeneity in the cytoplasmic mRNA and there is little if any 5′ trimming. In any case the addition of caps to trimmed transcripts seems unlikely as it would require different enzymic reactions.

Between the cap and the translational initiation codon, AUG (Section 11.6.1), there is a length of non-translated RNA known as the *leader sequence* (Fig. 9.11). It varies in length from three nucleotides in the immunoglobulin kappa chain [97] to 256 nucleotides for liver α-amylase [98]. Leaders contain no universally conserved sequence elements but within those of some gene families there do seem to be some conserved nucleotides and even some similar elements of possible secondary structure [90].

Although it has been suggested that caps may

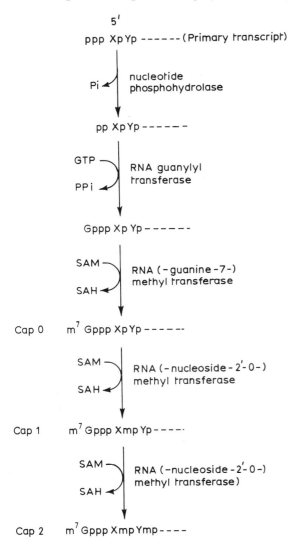

**Fig. 9.10** A mechanism for the biogenesis of 5′-terminal cap structures (Cap 0; Cap 1; Cap 2) in eukaryotic mRNA precursors.
SAM, S-adenosyl-1-methionine;
SAH, S-adenosyl-1-homocysteine.

function in the transport of mRNA from the nucleus there is no convincing evidence for such a role. The most important functions of the cap appear to be the protection of mRNA from nuclease attack and in facilitating ribosome attachment to mRNA (see Section 11.6.1).

Recent evidence has also indicated that the cap may play a role in splicing (see Section 9.3.4.c). The leader sequence [90] and a cap-binding protein [89] may also play a role in enhancing and stabilizing the interaction of the mRNA with the ribosome and translational initiation factors.

### 9.3.3 Polyadenylate tails, 3′-processing and 3′-non-coding sequences of eukaryotic mRNAs

Three features of the 3′-ends of the transcripts of protein-encoding genes warrant consideration. Firstly, it has recently become apparent that whether or not they are subsequently poly-adenylated, the 3′-terminus is often not generated by simple termination but by termination followed by processing (for a review see [99]). Secondly, the 3′-terminus of most but not all pre-mRNAs are modified by the addition of a polyadenylate tail and thirdly many messages have substantial 3′-non-coding trailer sequences (Fig. 9.11).

### (a) *Processing in preparation for polyadenylation*

As discussed in Section 8.3.4, the transcripts of eukaryotic genes often, though not always, terminate well beyond the apparent end of the gene. The primary transcript is then processed by nucleolytic action to within approximately 15 nucleotides of a hexanucleotide AAUAAA [100]. It seems likely that the nucleases concerned are endonucleases as the transcripts of adenoviruses are cleaved and polyadenylated before the transcript is complete, i.e. before the availability of a 3′-end for exonucleolytic attack [101]. Genes which give rise to multiple mRNAs with heterogeneity in their 3′-termini (see

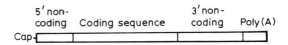

**Fig. 9.11** Diagrammatic structure of a polyadenylated mRNA.

Section 10.6.1) tend to have the hexanucleotide, or a close analogue, preceding each potential termination point [100]. The role of the AAUAAA sequence in the generation of the processed 3'-terminus has been demonstrated in the adenovirus early region 1A gene where its mutation to AAGAAA inhibits cleavage of transcripts but has no effect on polyadenylation [102]. Similarly, a form of α-thalassaemia (Saudi thal) has the hexanucleotide of the α-globin gene mutated to AATAAG and results in the production of mRNA with 3'-ends that extend well beyond their normal terminus [103]. The instability of these transcripts gives rise to thalassaemia (thalassaemias are discussed more generally in Section 9.3.4).

It seems unlikely that the AAUAAA is the sole recognition site for 3'-terminus formation. The sequence also occurs in internal regions of genes and is then not recognized as a termination signal [31, 104, 105]. Recently, Nevins and coworkers [106] have demonstrated that the generation of 3'-termini in adenovirus E2A mRNA required the AAUAAA hexanucleotide and a sequence occurring between 20 and 35 nucleotides downstream of the polyadenylation site. Similarly, Gil and Proudfoot [107] have shown that correct 3'-end formation in rabbit β-globin requires a sequence lying 16–50 bp downstream of the AAUAAA (+3 to −31 with respect to the polyadenylation site). The adenovirus and globin downstream sequences did not show marked homology but that of globin did include part of a CAYUG penta-nucleotide found just before or just after the AAUAAA in many vertebrate mRNAs. Berget [108] has drawn attention to the fact that the snRNA species known as U4 (Table 9.1) contains nucleotide sequences that would hybridize both to AAUAAA and to CAYUG. She suggests that U4 is involved in the selection of the termination site in the same way that U1 is thought to be involved in splicing the transcripts of introns (see Section 9.3.4). SnRNA U6 occurs base-paired to snRNA U4 in U4/U6 snRNP

[109, 110] so both RNA species could be involved in 3'-processing. Furthermore, *in vitro* polyadenylation systems are inhibited by anti-sera to snRNP particles containing U1 so this small RNA may also be involved in termination [112]. Another consensus sequence, YGTGTTYY, found downstream of 67% of genes examined, has a preferred location 24–30 bp downstream of the AATAAA signal and has also been proposed as a requirement for efficient formation of mRNA 3'-termini [111]. It is not known, however, whether it may play a role in termination (Section 8.3.4) or processing.

### (b) *Polyadenylation*

Nuclear poly(A) polymerase catalyses the addition of approximately 200 AMP residues onto the 3'-terminus of pre-mRNA 11 to 30 nucleotides after the hexanucleotide AAUAAA [113]. The extent to which polyadenylation is coupled to 3'-processing and whether the above sequence elements and U4 RNA might be involved in both processes is, at present, unclear. Fitzgerald and Shenk [113] showed that deletion of the AAUAAA sequence prevented polyadenylation but this could reflect a coupling of polyadenylation with processing rather than a requirement of the hexanucleotide for poly-adenylation. While a mutation of AAUAAA to AAGAAA inhibited most processing, the small percentage of transcripts that were processed were polyadenylated correctly [102]. A recently described *in vitro* polyadenylation system [114] accurately added approximately 200 adenosine nucleotides to the correct adenylation site of adenovirus L2 RNA but it would only use precursor RNA synthesized *in situ*. Exogenously added precursor was not polyadenylated thus suggesting that polyadenylation was coupled to transcription or processing or both. Conversely, the fact that polyadenylation but not processing was inhibited by the ATP analogue adenosine 5'-[α,β-methylene]-triphosphate [112] indicates that the two processes can be uncoupled. Poly-adenylation in this system was inhibited by antibodies raised against various snRNA-containing ribonucleoproteins but the inhibition was rather non-specific and was not shown to be limited to the snRNP fraction that included U4 RNA-containing particles. Polyadenylation may also require a sequence downstream of the coding sequence [115].

The balance of evidence supports a role for the poly(A) tail in extending the stability of mRNA (this is considered in Section 11.11.1). A further role that has been suggested for the polyadenylate tract is the modulation of mRNA transport. Some turnover of the poly(A) appears to take place during mRNA maturation and transport, and further polyadenylation, which is perhaps necessary for mRNA protection, occurs in the cytoplasm [116, 117].

(c) *Non-polyadenylated mRNA and its processing*

Up to 30% of mRNA is not polyadenylated (for a review see [118]) and appears to comprise some mRNA species which are not normally polyadenylated as well as mRNA species which are bimorphic in having poly(A)$^+$ and poly(A)$^-$ forms.

The most prominent mRNA species of which most, but not all, variants lack polyadenylation are the histones (for a compilation of histone mRNAs that are polyadenylated see [35]), and it has recently become clear that, like polyadenylated species, they are subject to 3'-processing. The 3'-ends of histone genes have a series of conserved elements. The first block is a GC-rich region of dyad symmetry which is usually hyphenated by a run of four Ts. This is

followed by the ACCA sequence the 3'-end of which corresponds to the 3'-end of the mRNA. A third, strongly conserved element, especially in sea-urchins, occurs about eight nucleotides downstream from the end of the message into the spacer of the gene quintets and has the sequence CAAGAAAGA. A consensus sequence for the 3'-end of histone genes is shown in Fig. 9.12.

The termination of transcription of histone genes appears to be at multiple sites and, in the case of sea-urchin histone H2A, occurs heterogeneously approximately 100 nucleotides downstream from the mRNA 3'-terminus [119]. The correct 3'-end of the message is generated by endonucleolytic cleavage of longer transcripts [120, 121] by a nuclear ribonucleoprotein enzymic complex which includes an RNA of approximately 60 nucleotides in length that has been called, snRNA U7 [122, 123]. The ribonucleoprotein is antigenically related to those carrying snRNAs, U1 to U6 (Table 9.1) and the RNA has a 5'-sequence which base-pairs with six of the nine nucleotides in the CAAGAAAGA element and 13 of the 16 nucleotides in the conserved palindrome [123]. It would appear from experiments with mutant genes that both of these elements, together with downstream elements, are essential for processing [124]. Furthermore, recognition of the palindrome is at the level of an RNA stem–loop in the transcript rather than the DNA cruciform [125].

A consensus sequence GTTTN$_{0-3}$AAA RNNAGA is a conserved element downstream of snRNA genes and is part of the sequence required to generate the 3'-end of U2 snRNA [126]. The sequence is clearly similar to the

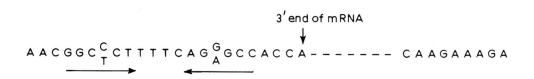

**Fig. 9.12** Histone gene 3' consensus sequence.

histone termination signal and may thus employ the same processing mechanism. This in turn implies that snRNAs could participate in their own processing. The recent advances in the study of 3'-processing have been the subject of a short review [127].

### (d) *The 3'-non-coding regions of eukaryotic mRNA*

Between the translational termination codon and the polyadenylation site, eukaryote mRNAs often have substantial segments of non-coding nucleotide sequence which varies in length from the 30 nucleotides of mouse $\alpha$-amylase [98] to the 637 nucleotides of chicken ovalbumin [128]. From the above discussion of polyadenylation sites and transcript-processing signals, it is clear that these segments contain some conserved elements. Otherwise a comparison of closely related mRNAs, such as the $\beta$-globin family [90], reveals little sequence conservation. Furthermore, apart from the above recognition sequences, they have no known function.

### 9.3.4 Removal of intron transcripts from pre-mRNA

#### (a) *Introns are transcribed into pre-mRNA*

The first demonstration that the precursors of mRNA contain transcripts of introns was provided by Tilghman *et al.* [129]. They followed their initial observation that the gene for $\beta$-globin contains at least one intron (see Section 9.2.1) with the demonstration that a $\beta$-globin pre-mRNA contains a transcript of the intron. Specifically, they showed that normal 10S globin mRNA, when hybridized to its gene, causes the formation of a double-stranded DNA R-loop indicative of an intron as illustrated in Fig. 9.2(b). Conversely, when a 15S pre-globin mRNA species is hybridized to the gene, no such loop is observed. A single continuous R-loop of

displaced single-stranded DNA (Fig. 9.2(a)) showed that the precursor could hybridize with both the exons and the introns of its gene. Tsai *et al.* [130] combined the technique of R-loop analysis with a kinetic approach in which finely minced oviduct was pulse-labelled to follow the processing of ovomucoid mRNA. They showed that all seven introns of this gene are transcribed into the largest pre-mRNA detected and that the seven intervening sequences are removed in a preferred but not obligatory order. Similar processing events were soon observed in the transcripts of other genes including a 7.8-kb ovalbumin pre-mRNA [130] a 10.6-kb precursor to immunoglobulin light chain mRNA [131] and a large precursor to amphibian vitellogenin mRNA [132].

#### (b) *Signals for the removal of intron transcripts*

The obvious place to look for clues on the mechanism by which intron transcripts are removed from pre-mRNA was the nucleotide sequence of the intron–exon boundary. It was soon noticed that these so-called *splice sites* were strongly conserved and a number of authors published consensus sequences from which any known sequence deviated only marginally. Most recently, Mount [437] catalogued the sequences of 139 *exon–intron boundaries* (the splice site at the 5'-end of the intron which is also known as the *donor site*) and 130 *intron–exon boundaries* (the splice site at the 3'-end of the intron which is also known as the *acceptor site*). He derived a consensus sequence which would give rise to the transcript as illustrated in Fig. 9.13.

The sequence immediately before the intron–exon splice site consensus sequence is always pyrimidine-rich and free of the dinucleotide AG. The most invariant aspect of the consensus sequence is the GU at the beginning of the intron transcript and the AG at its end (the so-called GT/AG rule when applied to the coding strand of the genomic DNA).

The role of this sequence in splicing has been

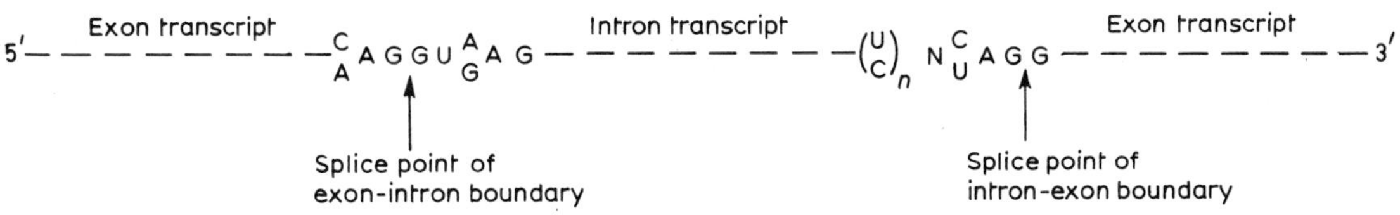

Exon transcript                Intron transcript                Exon transcript

$$5'\text{——————} \begin{smallmatrix}C\\A\end{smallmatrix}AGGU\begin{smallmatrix}A\\G\end{smallmatrix}AG\text{——————}\left(\begin{smallmatrix}U\\C\end{smallmatrix}\right)_n N\begin{smallmatrix}C\\U\end{smallmatrix}AGG\text{——————}3'$$

Splice point of                             Splice point of
exon-intron boundary                  intron-exon boundary

**Fig. 9.13** Intron splice site consensus sequence.

supported in numerous studies of the genes of eukaryotes and their viruses and is well exemplified in the $\beta$-thalassaemias, a group of hereditary anaemias in which the production of $\beta$-globin is either diminished ($\beta^+$-thalassaemia) or absent ($\beta^0$-thalassaemia). In one form of $\beta^0$-thalassaemia, lack of $\beta$-globin production was found to be due to a mutation in the GGT splice point at the exon–intron boundary of the large intron of the gene [133, 134]. As a result processing of $\beta$-globin transcripts was impaired. In one form of $\beta^+$-thalassaemia, defective globin synthesis was shown to be associated with a single G $\rightarrow$ A mutation which created a sequence CTATTAG within the large intron. The sequence closely resembled the intron–exon sequence CCGTTAG and competed with it as a splice point to the extent that 90% of transcripts were incorrectly spliced [136, 137]. The conclusions from these and similar studies are that mutations in the GT and AG splice site sequences severely affect splicing, and mutations near the GT may also have an effect. Interference with normal splicing may lead to the utilization of previously unused *cryptic, splice sites*. Thus, a mutation in, for instance, the 5'-splice site may result in abnormal splicing such that the exon at the 3'-splice site is ligated to a new 5'-site which may show strong homologies with the normal splice site but would not normally be used. The existence of sequences, often within coding sequences, that resemble the splice site consensus sequences indicates that other factors must be involved in the selection of the correct intron boundaries.

(c) *Mechanisms for the removal of pre-mRNA intron transcripts – splicing*

As well as apparently being the norm in the protein-encoding genes of the eukaryote nucleus, introns have now been identified in some rRNA and tRNA genes, in the genes of subcellular organelles such as mitochondria, in the thymidylate synthase gene of bacteriophage T4 [417] and in the genomes of archaebacteria [39]. Furthermore, there would appear to be at least three major mechanisms by which the transcripts of introns are removed. In the following section, the splicing of protein-encoding nuclear genes is discussed. The removal of intron transcripts, where they occur in rRNA, tRNA and mitochondrial gene transcripts, is discussed in Sections 9.4.3.b, 9.5.2.c and 9.6.4 respectively.

The first mechanism that was proposed for the removal of introns from the transcripts of protein-encoding genes arose from the independent observations of Lerner *et al.* [138] and Rogers and Wall [139]. They noticed that the 5'-terminus of the most abundant snRNA, U1 (Table 9.1), exhibited considerable sequence complementarity with the consensus sequence of both the exon–intron and intron–exon boundaries of gene transcripts. The most notable feature of this complementarity was the tetranucleotide ACCU which would hybridize with the invariant GU/AG splice points if these were brought together to form the sequence UGGA (Fig. 9.14). The authors suggested that U1 could hybridize with both ends of an intron thus drawing two exons into alignment such that they

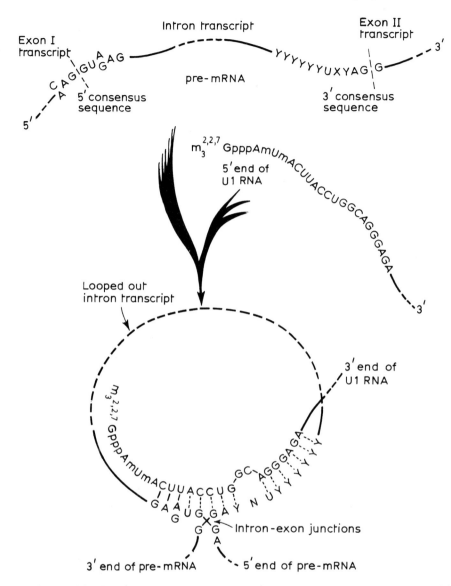

**Fig. 9.14**  An early model for the base pairing between pre-mRNA and U1 SnRNA in which base-pairing was envisaged between U1 and both the 5′ and 3′ consensus sequences [138, 139]. Y indicates a pyrimidine and X a variable base.

could be ligated were the intron transcript removed.

Neither snRNA nor pre-mRNA occur in the nucleus as free nucleic acids but are complexed with proteins as *small nuclear ribonucleoprotein particles (snRNP or snurps)* and *heterogeneous*

*nuclear ribonucleoproteins (hnRNP)* respectively [140]. Lerner *et al.* [138] provided further evidence for their model (Fig. 9.14) when they noticed that an observed association between snRNP and hnRNP did not occur if the 5′-end of the U1 RNA molecule was missing.

This 5′-end, i.e. the portion that shows complementarity with the intron consensus sequences, appears to be exposed on the surface of snRNP as it is very vulnerable to digestion by RNase. Furthermore, affinity columns that employ antibodies to the cap-like 2,2,7-trimethylguanine at the 5′-terminus of the RNA can be used to isolate snRNP [141].

The development of *in vitro* systems able correctly to splice viral and globin pre-mRNAs [142–145] provided more evidence for the role of U1 snRNP. The ribonucleoprotein copurifies with splicing activity [146] and the removal of the 5′-sequence of the U1 RNA inhibits the *in vitro* system [147] as does the inclusion of antibodies raised against snRNP [148]. The specific requirements and characteristics of these *in vitro* splicing systems are that they require U1 snRNP, ATP and $Mg^{2+}$. The kinetics are slow and there is a lag of 30 min during which time ATP has to be present. In some cases at least, the mRNA has to be capped, and splicing is inhibited by the cap analogue $^7$mGpppN [149]. Introns appear to have a minimum length of 25–30 nucleotides [150, 151].

The *in vitro* systems demonstrate that U1 RNA is involved in interaction with the 5′-splice site. So far, however, they have not supported a role for the RNA in interacting with the 3′-splice site as proposed in the model of Fig. 9.14 and its variants. Thus, after incubation of purified U1 snRNP with globin pre-mRNA, 80% of the globin sequences can be precipitated with anti-snRNP antibodies. Examination of the recovered complexes reveals that the U1 ribonucleoprotein particle protects from ribonuclease digestion a nucleotide fragment at the 5′ exon–intron boundary of the pre-mRNA. There is apparently no interaction at the 3′ intron–exon splice site.

More evidence that the model of Fig. 9.14 required revision came from the use of *in vitro* splicing systems to detect the intermediates of the process. The excised intervening sequence was shown to have an unusual structure which

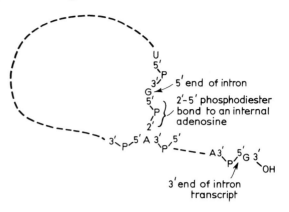

**Fig. 9.15** The lariat structure of excised intron transcripts.

has been described as a *lariat* (Fig. 9.15) and in which the 5′-terminal guanosine of the intron transcript is joined via a 2′–5′ phosphodiester bond to an internal adenosine residue of the intron [152].

Specifically, the 5′-terminal guanosine forms a branch with the adenine at −24 (for the adenovirus 2 major late transcription unit [153]) or the adenine at −37 (for the β-globin pre-mRNA [154]) with respect to the 3′ splice site. In both the adenovirus and globin splicing systems, there is a considerable degree of sequence complementarity between the 5′-end of the intron transcript and the nucleotides to the 5′-side of the branchpoint. This has led to a proposed scheme for the processing of the two precursors in which the formation of the 2′–5′ pyrophosphate branch is preceded by the formation of a lariat held together by base-pairing [153, 155]. Fig. 9.16 illustrates the model as applied to the β-globin precursor.

Analysis of further higher eukaryotic mRNA [421, 422] reveals that the sequence

$$\begin{array}{cccc} C & & A & C \\ & U & & \\ U & & G & U \\ & & \underline{A} & \end{array}$$

is weakly conserved around the internal adenosine branchpoint (adenosine of sequence underlined), and Keller and Noon [421] draw attention to the fact that U2 snRNA could base-pair with

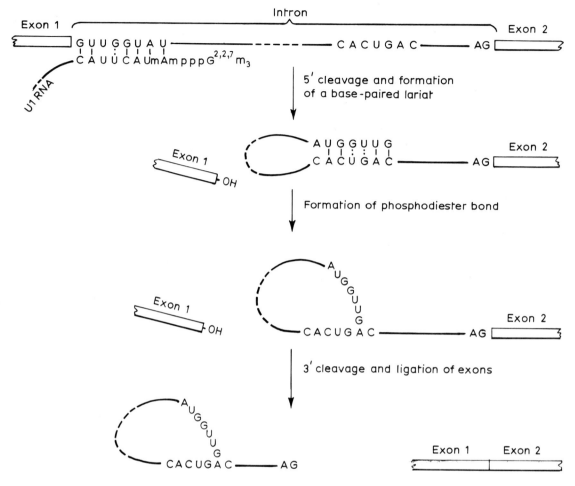

**Fig. 9.16**   A model for the splicing of exons in eukaryotic mRNA [153, 155].

the sequence. However, studies with mutated branchpoints indicate that while adenosine is important for the formation of the 2–5 phosphodiester bond, the surrounding nucleotides can be variable [423]. Ruskin *et al.* [423, 424] have suggested that branchpoint selection is based on a mechanism that measures distance from the 3' splice site and requires its presence 22–37 nucleotides upstream of the chosen adenosine.

Recent studies with *in vitro* splicing systems indicate that splicing occurs at a multicomponent splicing complex which has been called a *spliceosome* [425–427] and, in support of the above

observation, the complex includes U2 snRNA as well as U1 [428, 429]. Krainer and Maniatis [428] have identified six factors necessary for splicing *in vitro*. Splicing factor 1 (SF1), which includes U1 and U2 snRNA, together with two uncharacterized factors, SF2 and SF5B, plus ATP and $Mg^{2+}$, are apparently required for cleavage at the 5' splice site and lariat formation. Splicing factors 3 and 4A, which are also uncharacterized, together with potassium, are required for cleavage at the 3' splice site. The excised lariats are apparently processed to linear molecules by a 2'–5'-phosphodiesterase [430].

### (d) *Yeasts have a variant splicing mechanism*

The fact that the yeast, *Saccharomyces cerevisiae*, is unable to excise the intervening sequences of higher eukaryotes [156] has led to the discovery that yeast and perhaps other fungi have a variant of the above splicing mechanism. In comparison with higher eukaryotes, many yeast nuclear genes do not contain introns. Exceptions are the genes for actin, the mating-type proteins and some ribosomal proteins. Introns within these genes have a highly conserved sequence,

TACTAAC

which occurs four to 53 nucleotides before the AG at the 3′ splice site and is essential for splicing [157–159]. The conserved sequence shows considerable homology with the 5′-end of U1 snRNA and could act as the counterpart of U1 in yeast cells by base-pairing with the 5′ splice site [158]. This is supported by the finding that the sequence GTATGT, which would form four base pairs with TACTAAC is highly conserved at the 5′ splice site of yeast genes. Furthermore, lariat structures have now also been reported in yeast and the 5′–2′ phosphodiester bond appears to be between the 5′-G residue at the exon–intron boundary and the third adenosine of the TACTAAC base [160, 161]. In contrast to higher eukaryotes, the 3′ splice site of yeast introns is not required for efficient lariat formation [431].

The TACTAAC box has also been found in *Aspergillus* introns [162].

### 9.4 THE ARRANGEMENT OF rRNA GENES, THEIR TRANSCRIPTION AND PROCESSING

#### 9.4.1 The prokaryotic rRNA genes and their processing

In *E. coli* there are seven rRNA transcription units dispersed through the genome and arranged in operons that are termed *rrnA* to E, G and H. Each operon contains the genes for the three ribosomal RNAs in the order 16S, 23S, 5S together with one or more genes for tRNA. Each gene is separated from the next by a transcribed spacer. The number and location of the tRNA genes vary. In four of the seven operons the spacer between the 16S and 23S RNA contains a tRNA$^{Glu}$ gene while in the remaining three operons the spacer contains genes for tRNA$^{Ile}$ and tRNA$^{Ala}$ [163–166]. There may also be tRNA genes following the 5S RNA (trailer genes), and in the *E. coli* operon, *rrnC*, this region contains the only tRNA$^{Trp}$ gene of the genome together with a gene for tRNA$^{Asp}$ (Fig. 9.17). The operon *rrnD* contains two 5S RNA genes with a tRNA$^{Thr}$ gene sandwiched between them [172].

The *rrnB* operon has been completely sequenced [166]; as has the promoter region of five other operons [167–170] and the terminator regions of three other operons [171–173]. All units have two tandem promoters (P1 and P2) about 120 nucleotides apart each having an AT-rich region followed by the normal −35 and −10 region (see Section 8.2.1). Transcription is initiated eight nucleotides after P1 or seven nucleotides after P2 [168, 174] and a major fraction of cellular transcriptional activity is from these operons [175]. It is regulated by growth rate control [176, 177] and by the so-called stringent control in which limited availability of charged tRNA inhibits *rrn* operon transcription in *rel*A$^+$ (stringent strains) but not in relaxed strains (see Section 11.12). The terminators all have a stem–loop structure followed by a string of U residues characteristic of rho-independent termination signals (Section 8.2.3.a).

The transcription units are transcribed in their entirety into a single precursor with a sedimentation value of approximately 30S. This precursor does not normally accumulate but is subjected to processing while still a nascent transcript. The study of processing has depended on rRNA genes inserted in plasmids and on mutants defective in the RNA-processing enzymes involved (for reviews see [14, 15, 178, 179]).

**Fig. 9.17** A schematic representation of the *E. coli* ribosomal transcript unit, *rrnC*.

### (a) *Processing the 16S and 23S transcript*

The sequences flanking both the 16S and 23S gene transcripts form inverted repeats which are respectively 1700 and 2900 nucleotides apart. These sequences interact to form base-paired stems [180, 181] at the base of giant loops that contain the complete sequences of 16S and 23S RNA and which in the nascent transcript are already in the form of a ribonucleoprotein. The base-paired stems form specific catalytic sites for the enzyme ribonuclease III (Fig. 9.18) and precursors containing them accumulate in mutants lacking the enzyme [182]. Exactly how RNase III selects its cleavage site is unclear; there is little sequence homology between its primary catalytic sites in pre-ribosomal RNA or in its other much studied substrate, the early RNA of bacteriophage T7 [183, 179]. It cleaves

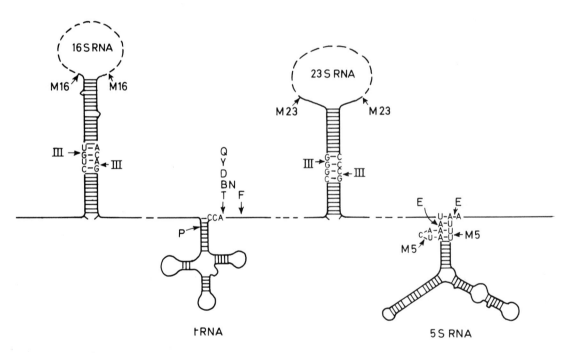

**Fig. 9.18** The processing of prokaryotic rRNA, tRNA and 5S RNA. The following symbols are employed for the ribonuclease activities involved: III, sites cleaved by RNase III in the double stranded regions flanking 16S and 23S RNA: P, site of cleavage by RNase P to generate the 5'-end of mature tRNA; M16, M23, maturation of 16S and 23S RNA by ill-characterized RNase activities M16 and M23; F, endonucleolytic cleavage to the 3' side of tRNA, possibly by RNase F; Q, Y, D, BN, T, generation by exonuclease action of the mature tRNA 3' end, possibly involving one or more of the nucleases Q, Y, D, BN and T; E, initial endonucleolytic cleavage in the generation of 5S RNA by RNase E; M5, maturation of 5S RNA by RNase M5.

double-stranded RNA at intervals of approximately 15 nucleotides with little apparent specificity [184] and, in *E. coli* pre-ribosomal RNA, produces staggered cuts in each stem (Fig. 9.18). The specificity is, however, such that even a small perturbation of the normal double stranded site strongly inhibits the enzyme [185]. The products of RNase III are precursor ribonucleoproteins containing RNAs slightly longer than the mature molecules and known as p16 and p23. They are processed to mature 16S and 23S RNA by ill-characterized enzymes which probably act endonucleolytically and are known as RNase M16 and RNase M23 respectively [186].

### (b) *Processing the 5S transcript*

Three enzymatic steps are required for the generation of 5S rRNA from the transcripts of *rrn* operons (Fig. 9.18). The first of these is the removal of the precursors to 23S rRNA from the growing nascent rRNA precursor chain. This is catalysed by RNase III as described above. The second enzyme, RNase E, has been purified as a $66\,000$-$M_r$ protein but perhaps requires additional factors [187]. It generates a precursor of 5S RNA known as p5S which, compared with the mature molecule (m5S), has three extra nucleotides at both the 5'- and 3'-termini [188]. In a temperature-sensitive, RNase E mutant, a 9S RNA accumulates which the added enzyme will convert to p5S [189]. Singh and Apirion [190] have analysed this molecule from the *rrnB* operon (does not have a trailer tRNA gene after the 5S gene) and have shown that it is 246 nucleotides long. It includes the 5S sequence, a rho-independent terminator and 81 highly conserved nucleotides lying between the 5S gene and the RNase III cleavage site of 23S RNA. The latter may be important for recognition of the molecule by both enzymes [190]. The catalytic site for RNase E occurs in an extension of the double-stranded stem, which is formed by the 5'- and 3'-ends of the 5S RNA sequence and known as a molecular stalk [191].

RNA I, a 100 nucleotide *E. coli* RNA which regulates plasmid DNA replication, is also a substrate for RNase E [192], and the cleavage site, in the middle of the nonanucleotide

$$\text{ACAG}^{\text{A}}_{\text{U}}\text{AUUG}$$

is identical to that of 9S RNA precursor.

The third enzyme, RNase M5, is responsible for the terminal maturation of 5S RNA and endonucleolytically cleaves the trinucleotide segments from both the 5'- and 3'-ends of the precursor, p5S [193]. Enzymic activity in *Bacillus subtilis* is dependent on two proteins $\alpha$ and $\beta$, of which $\alpha$ is the catalytic activity and $\beta$ is ribosomal protein BL16. The latter is probably required to lock the RNA into a conformation recognizable by the nuclease [194, 195].

### (c) *Processing tRNA genes in the rRNA transcript*

Spacer and trailer tRNA sequences in the transcripts of *rrn* operons are removed and processed by a series of RNases as described in Section 9.5.2.a.

## 9.4.2 The rRNA genes of eukaryotes

Eukaryotic ribosomes contain four species of RNA known from their sedimentation constants in higher animals as 18S, 28S, 5.8S and 5S. In fact their size is variable and in lower eukaryotes the major species may have sedimentation constants as low as 17S and 25S. Ribosomal RNA represents more than 80% of the cellular RNA and, in order that the cell can produce the quantities required, all species are encoded in multiple genes and are transcribed at a rapid rate. In most eukaryotes the genes for 5S RNA are located in a separate part of the genome and are transcribed by RNA polymerase III rather than polymerase I. They will be considered separately. (For reviews of the structure, expression and, where relevant, the amplification of rDNA from a range of species, see many of the chapers in ref. [196].)

(a) *The 18S, 5.8S, 28S rRNA transcription unit*
The ribosomal transcription unit contains the genes for 18S, 5.8S and 28S rRNA separated from each other by *internal transcribed spacers (ITS)* and flanked on its 5′-side by an *external transcribed spacer (ETS)*. The transcript units themselves are also clustered in *tandem repeats*, each unit being separated from the next by a *non-transcribed spacer (NTS)* (Fig. 9.19).

In the mature ribosome, 5.8S RNA is base-paired to the 5′-end of 28S RNA and in sequence and secondary structure the two are evolutionarily related to the 5′-end of prokaryotic 23S RNA. One can therefore think of the ITS between the 5.8S and 28S genes (ITS-2) as the equivalent of a piece of DNA inserted 150 nucleotides from the end of 23S RNA (for a review of 5.8S RNA see [197]).

The clusters of ribosomal genes occur at a few locations which are known as *nucleolar organizers* because of their ability to associate into nucleoli. It was the clustering of ribosomal genes and the high G + C content of the rDNA of some species that permitted its early isolation as satellites on CsCl density gradients [198]. This also explains why a single mutation in an anucleolate *Xenopus laevis* [198] or the so-called bobbed mutants in *Drosophila* [199, 200] can lead to the deletion of all or most rRNA gene copies.

Considerable variation occurs in the number, arrangement, size and chromosomal location of the rDNA cluster (for reviews see [201–203]). Man has five clusters located on chromosomes 13, 14, 15, 21 and 22 [104]. Maize contains all of its several thousand genes in one organizer on chromosome 6 and *Drosophila* has organizers on the X and Y chromosomes [200].

Most lower eukaryotes have several hundred tandem rDNA repeated genes; mammals have 100–300. Plants and amphibia have several thousand or even tens of thousands. There is, however, considerable species variation, and Long and Dawid [201] have catalogued numerous instances in which closely related species or even strains exhibit large variations in the repetitiveness of their rRNA genes. It would appear that there are often excess genes over the number required for viability. In addition, some organisms are able to amplify their ribosomal rRNA genes (see Section 7.5.1). Amplification is a common feature of oogenesis and, in *Xenopus*, occurs by the production of large circles of extrachromosomal DNA each containing hundreds or even thousands of rDNA repeating units [205]. The amplification allows the accumulation of ribosomes and prepares the oocyte for the massive increase in protein synthesis in early embryonic development. Amplification of rDNA is also a feature of *Tetrahymena* and other ciliate protozoans. The vegetative cell of *Tetrahymena* has two types of nucleus. The *micronucleus* is the germinal nucleus that undergoes meiosis in sexual conjugation. However, it is only a repository of genetic information and is not expressed. It contains a single ribosomal transcription unit [206]. The *macronucleus* is derived from the micronucleus, is *polyploid* and controls normal cell metabolism. In it, the rDNA is amplified to 400 copies per haploid genome equivalent to approximately 20 000 copies per polyploid nucleus [207, 208]. The mechanism of amplification is discussed in Section 7.5.3.

A further source of variability in rDNA is in the length and arrangement of both the coding and the non-coding regions of the transcription units. The genes for the major rRNA species are always arranged on the bacterial pattern, i.e.

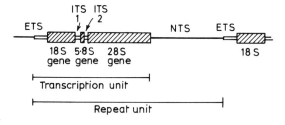

**Fig. 9.19**   The organization of eukaryotic rDNA (*Xenopus laevis*).

ETS–18S–ITS–28S, and the 5.8S gene is inserted into the ITS. However, there is considerable variation on this theme. Largely owing to differences in the transcribed spacers, the transcription unit can vary in length from 6 to 8 kbp in lower eukaryotes [209] and up to 13 kbp in mammals [210]. Furthermore, the non-transcribed spacers show even greater length diversity so that the total repeating unit of rDNA varies from 9 kb in yeast [211] to 34–44 kbp in mammals [212–214].

Other variations are found in the gene arrangement and are illustrated in Fig. 9.20. Thus, yeast of the genus *Saccharomyces* and the slime mould, *Dictyostelium*, are unique amongst those so far investigated in having their 5S RNA genes within the non-transcribed spacer of the ribosomal transcript unit [211, 215]. The transcribed spacer of *Drosophila* also contains an extra gene which encodes a 2S RNA species found hydrogen-bonded to the 26S RNA in the ribosomes of these species [216]. About half of the 250 26S rRNA genes of *Drosophila* also contain insertion sequences which vary in length between 0.5 and 6.0 kbp and fall into two classes. Those known as type I occur predominantly as 0.5, 1.0 and 5.0 kbp inserts and are confined to the nucleolar organizer of the X chromosome. Type II inserts occur in about 15% of the genes on both the X and Y chromosome and show no homology with type I inserts (for a review see [217]). Those *Drosophila* genes that contain these *ribosomal insertions* are not processed to rRNA to a significant extent and the fact that some type I inserts are flanked by duplicated sequences suggests that they arose as transposable elements [217]. Similar non-transcribed inserts, which should perhaps be called interruptions, have been found in a small percentage of the rDNA genes of a number of other insects (see for instance [218]). In other cases, however, such as several strains and species of *Tetrahymena*, the slime mould, *Physarum poly cephalum*, yeast mitochondria and *Chlamydomonas* chloroplasts, all of the rDNA tran

scription units contain introns which must therefore be transcribed [219]. The splicing of the intron transcripts of *Tetrahymena* has been subject to considerable study and is considered in Section 9.4.3.b.

A third unusual feature of *Drosophila* rDNA is gaps in the 26S RNA. The RNA has ribonuclease-sensitive sequences that are cleaved during processing so that the mature RNA consists of two halves ($\alpha$ and $\beta$ 26S RNA) held together by hydrogen-bonding [220].

The amplified rDNA of *Tetrahymena* also exhibits a further variation in rDNA organization. The extrachromosomal genes occur as palindromes each containing two transcriptional units usually in a linear molecule (Fig. 9.20) but in some cases as circles [221, 222].

Considerable sequence data now exist for the ribosomal RNA genes derived from species of widely varying phylogenetic complexity. They include genes from eukaryotic nuclei, mitochondria and chloroplasts as well as those of both eubacteria and archebacteria. Nelles *et al.* [223] have for instance listed references to the sequences of the small ribosomal subunit RNA species from 26 different sources, and they present one further sequence. Considerable homology is found in the small rRNA gene; so much so that, from the primary sequence of the DNAs, it is possible to derive primary and secondary structures for the RNAs which show that those of all species examined more or less conform to a common plan (this is further discussed in Section 11.8.1). From the more limited data available on the larger RNA species, it would appear that there are again strongly conserved nucleotide tracts and a similar overall secondary structure.

In contrast to the gene-coding regions, the transcribed spacers show little sequence homology. Although short regions of homology have been noted between the ITS of rat and *Xenopus* [224], in general the spacers show sequence and size heterogeneity even between closely related species and within the repeating

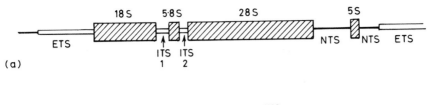

(a)

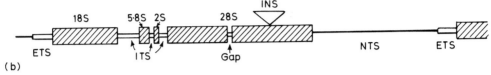

(b)

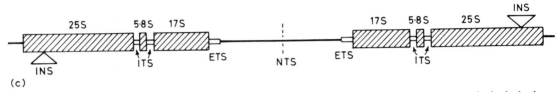

(c)

**Fig. 9.20** Some unusual eukaryotic ribosomal gene arrangements (a) *Saccharomyces* repeat units include the 5S genes; (b) *Drosophila* repeat units contain 2S genes, a gap in the 28S gene and some contain inserts; (c) extra chromososmal transcription units of *Tetrahymena* are palindromes and the 25S gene may contain an insert.

units of the same species. Both the ITS and ETS tend to have regions rich in one or more nucleotides, lengths of homopolymers, di-nucleotide polymers and repeated elements but again these are characteristic of the species and are not conserved between species (for a review see [209]).

The non-transcribed spacers show even greater heterogeneity and commonly vary in length in a single gene cluster (for a review see [225]). Much of this heterogeneity arises from repeated elements of which the most studied are those of *Xenopus* and *Drosophila*. In *Xenopus* the NTS contains four different types of repeated unit (see Fig. 9.21). They are known as repeated regions 0, 1, 2 and 3 and each is composed of a variable number of repeating elements (for a review see [226]). For instance, repeat region 0 may contain from three to nine copies of a 34-bp sequence while repeat region 1 may contain six to eight copies of a 97-bp repeat. The repeat regions 2 and 3 each contain a variable number of 60 and 81-bp repeats of which the 81-bp repeat is the same as the 60-bp element, but has an extra

21 added base pairs. Separating the repeat regions 1, 2 and 3 are from two to seven copies of an approximately 300-bp sequence which, because they contain sites for the restriction enzyme *Bam*H1, are known as *Bam* Islands [227]. It is the variable repetition of these, together with the repeat regions 2 and 3 (the so-called *super repeats*) that contribute most to the heterogeneity of the NTS.

The feature of the *Xenopus* NTS repeats that has attracted most attention is their content of promoter-like sequences. This is most obvious in the *Bam* Islands which contain a sequence that is 90% homologous to the region $-14$ to $+4$ at the transcription start site. They have led Moss [228] to propose that the islands act as 'sinks' for RNA polymerase I, increasing the affinity of the transcription unit for the enzyme. Other components of the NTS with a possible function in transcription are 42-bp sequences that form part of the 60/81-bp repeats of regions 2 and 3. These have imperfect homology with the $-100$ region of the RNA polymerase I promoter [229] and it has been suggested that they may function as

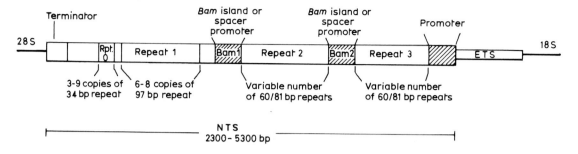

**Fig. 9.21** The non-transcribed spacer of *Xenopus laevis* rDNA.

enhancers providing entry points for RNA polymerase [226, 228]. Their similarity to enhancers extends to an ability to function in either orientation and in various positions [230]. (Enhancers are considered in detail in Section 10.4.2.)

Various *Drosophila* species also have, within the much simpler structure of their NTS, repeating elements that contain copies of the promoter [231]. Each unit also contains a site for the restriction endonuclease *Alu*I. Promoter-like sequences in the NTS of two such distant species suggests a general phenomenon. However, the repetitive elements in the NTS of mouse [232], rat [233] and human [234] rDNA do not appear to show homology with the promoters of their respective transcript units. Several groups have shown that the repetitive sequences in the latter three species also occur elsewhere in their respective genomes, and Mroczka *et al.* [233] have demonstrated that, in the rat, they include RNA polymerase III initiation sites and sequences that hybridize to small RNAs. They have put forward a model suggesting that the repetitive regions on either side of the transcription unit interact with each other via a small RNA or protein so causing the transcribed portion of the unit to be looped out.

(b) *Eukaryotic 5S RNA genes and their expression*

Eukaryotic 5S RNA genes are the subject of several chapters in [196] and are also reviewed in [235]. The multiple genes that encode 5S RNA are not normally linked to those of other ribosomal RNA species although they are in the yeast, *Saccharomyces* (see Section 9.4.2 and Fig. 9.20), and in the slime mould, *Dictyostelium discoidium* [211, 215]. In some lower eukaryotes, including *Neurospora* [236] and the yeast, *Schizosaccharomyces pombe* [237], they are dispersed elsewhere in the genome but in all higher eukaryotes so far investigated, including plants [238, 239], insects [240–242], amphibia [reviewed in 200] and mammals [243], they are clustered elsewhere in tandem arrays. Chromosomal location is quite variable. For instance, the approximately 160 repeats of the *Drosophila* 5S gene appear to be arranged in a single cluster on chromosome 2 [240–242] while those of *Xenopus* occur at the telomeres of most if not all chromosomes [244].

These two genera represent the earliest and most studied 5S genes. Those of *Drosophila* are simpler than *Xenopus* and the repeating unit consists of the 135-bp gene separated from its neighbour by an AT-rich spacer. This spacer varies in length depending on the number of copies of a heptameric base sequence [245], but in most units there are five copies giving a 238-bp spacer and a repeat length of 373 bp. The genes of *Xenopus* are more complex and occur as two major types encoding oocyte and somatic 5S RNAs. The oocyte genes are only expressed during oogenesis while the somatic type is expressed in oocytes and somatic cells [246]. The

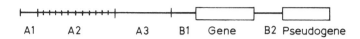

**Fig. 9.22**  Structure of the oocyte 5S rRNA gene repeat of *Xenopus laevis*. A1, first 40 nucleotides of AT spacer, constant in length and highly conserved; A2, a 15-nucleotide repeating sequence, variation in the number of repeats is the main reason for heterogeneity in the length of the gene repeat; A3, a 24/25-nucleotide repeating unit repeated eight times with some variation at nucleotides 4, 13, 14, 18 and 20 and in the presence or absence of nucleotide 25; B1, 49 nucleotides, GC-rich and much less repetitive than the remainder of the spacer but containing palindromes; Gene, 121 nucleotides containing an internal promoter for RNA polymerase III (see Section 8.3.2); B2, 73 nucleotides showing considerable homology with B1 plus the 3′ 24 nucleotides of A3; Pseudogene, 101 nucleotides, identical to the first 101 nucleotides of the gene in all but 10 nucleotides.

two RNA variants differ from each other in six nucleotides [247]. In *X. laevis* there are 24 000 copies of the oocyte gene [248] compared with several hundred of the somatic gene. It is assumed that the large number of the oocyte genes allows their transcription to keep pace with that of the amplified genes for 18S and 28S rRNA during oogenesis. The repeat units of *Xenopus laevis* oocyte 5S DNA are larger than those of *Drosophila* and each contains a pseudogene [249] as well as the 5S gene (for a discussion of pseudogenes see Sections 7.6, 7.7 and 9.2.2.b). They are also heterogeneous in length, again largely because of variation in the length of an AT-rich spacer [250–252]. Fig. 9.22 illustrates the organization of the repeating unit.

The 5S DNA of mammals is also heterogeneous in length, sequence and possibly function and has very long repeat lengths which range from 6 kbp in mouse, 2.2 kbp in hamster and 2.8 kbp in man [243, 250, 251]. Again the variation is due to sequences other than the gene which, like the RNA it encodes, is highly conserved across the phylogenetic spectrum [432].

The 5S genes are transcribed by RNA polymerase III and the enzyme, its promoters, need for trans-acting initiation factors and termination sites have been considered in Chapter 8. In contrast to other RNA species, 5S RNA does not undergo post-transcriptional modification [252].

### 9.4.3  The transcription and processing of eukaryotic ribosomal RNA

(a) *Transcription and nucleolytic processing*
The transcription of rRNA, together with most processing and assembly into ribosomal particles, occurs in the nucleolus (the transcribing enzyme RNA polymerase I, its structure initiation and termination have been considered in Chapter 8). Electron micrographs of oocyte rDNA engaged in rRNA transcription reveal the rDNA fibre forming the 'trunk' of a 'Christmas-tree-like' structure of which the 'bole of the trunk' is the non-transcribed spacer which separates the rRNA gene units [253]. The 'branches' on the transcription unit are represented by about 100 nascent transcripts which increase in length along the length of the fibre axis and have at their base a transcribing molecule of RNA polymerase (Fig. 9.23). The longest branches are complete transcripts of the transcription unit but they are only approximately one-twelfth its length because they are already partially wrapped into ribonucleoproteins.

RNA polymerase I produces a single long pre-rRNA molecule that contains the 18S, 5.8S and 28S rRNAs joined to each other by transcripts of the transcribed spacers. Where, as in yeast, the rDNA repeat includes the gene for 5S RNA, this is separately transcribed from the

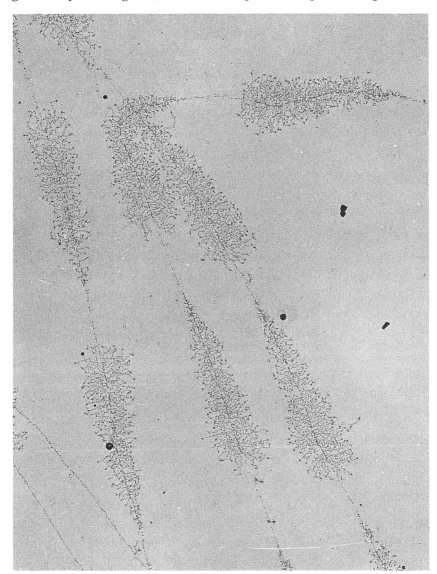

**Fig. 9.23** An electron micrograph of rRNA genes of the amphibian *Triturus*. The gradient of fibrils radiating from each gene are transcripts of pre-rRNA in successive stages of completion. They are produced by approximately 100 molecules of RNA polymerase I that are simultaneously transcribing the gene. The DNA fibre between each gene is the non-transcribed spacer. Reproduced, with permission, from [253] Copyright 1969 by the AAAS.

opposite DNA strand by RNA polymerase III. The initial transcript is subject to ribonucleolytic cleavage which produces the mature ribosomal species via a succession of intermediate pre-rRNA species which have been identified on sucrose density gradients, in polyacrylamide and agarose gels, by hybridization and by sequencing (for reviews see [201, 202]). Fig. 9.24 presents a generalized scheme for this maturation process. In step I the transcript of the ETS is removed.

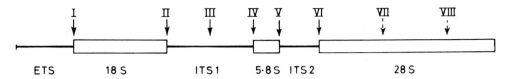

**Fig. 9.24**  The steps in the maturation of eukaryotic pre-rRNA. The numbered steps are explained in the text.

Steps II and III do not appear to have an obligatory order and release the mature 18S RNA while steps IV, V and VI give rise to mature 5.8S RNA which then remains hydrogen-bonded to the 5′-terminus of 28S RNA. Step VII is specific to *Drosophila* and other dipterous flies and gives rise to the break in their 26S rRNA that is discussed in Section 9.4.2.a. Step VIII is the splicing site in *Tetrahymena* and is discussed in the following section. Variation in the size of the initial transcript reflects the variation in the lengths of the transcription unit. Fig. 9.25 illustrates the maturation pathway of the 45S transcript of the mammalian rDNA transcription unit but similar pathways have been observed for the primary transcripts of many species [254–256].

Little is known of the activities that catalyse the maturation of rRNA in eukaryotes. Nucle-olar ribonuclease activities that process 45S pre-rRNA to 18S and 28S RNA have been described [257] but as yet both the enzymes and their product remain poorly characterized. One of these enzymes was reported to contain an RNA component which is of interest in view of the likely role of snRNA U3 in nucleolar RNA processing. U3, together with U8 and 7SM RNA (Table 9.1), is found only in the nucleolus and is found hydrogen-bonded to mammalian 32S pre-rRNA but not to mature 28S RNA [258, 259]. Two different groups have recently proposed models for U3 involvement in processing. One of these suggests that three short sequences in U3 base-pair with 5.8S RNA and its flanking sequences and is important in the secondary folding of pre-rRNA into the correct con-formation for processing [260]. The second model draws attention to the potential for

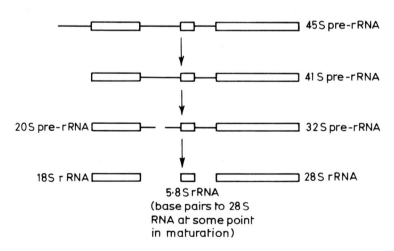

**Fig. 9.25**  Maturation of human pre-rRNA.

base-pairing between ITS-2 and U3 and suggests that U3 may be involved in the actual excision process [261]. As yet there is little evidence to support the models, although it is clear that the transcribed spacer can have considerable secondary structure.

### (b) The splicing of pre-rRNA

As discussed in Section 9.4.2.a, the extra-chromosomal rDNA of some strains of *Tetrahymena* contains introns, the transcripts of which are removed by splicing. Cech and his colleagues [262] demonstrated that this splicing occurred by a fundamentally new mechanism in which, at least *in vitro*, RNA catalyses its own splicing. No enzyme or any other protein is involved and all that is required is $NH_4^+$ and $Mg^{2+}$ ions and a guanosine cofactor.

Splicing occurs in a series of steps which are illustrated in Fig. 9.26 as follows:

(i)   Guanosine binds specifically at or close to the 5′ splice junction [263].

(ii)  The RNA behaves like an enzyme in that the binding of guanosine promotes the cleavage of the phosphodiester bond at the 5′ splice junction and the transfer of the 5′-phosphate to the 3′-OH of guanosine [264].

(iii) The 3′ splice junction is cleaved in a similar way except that the 5′-phosphate created in the break is joined through a normal phosphodiester bond to the upstream exon; so forming the spliced RNA.

(iv)  The released insertion sequence circularizes in a cleavage/ligation reaction in which the linear molecule is cleaved 15 nucleotides from its 5′-end and the A residue to the right of the cleavage site is joined to the G residue at the 3′-terminus [265]. Thus, a circular as well as a short linear molecule are produced both of which are rapidly degraded [266]. It is assumed that the formation of the circular molecule forces in a forward direction a reaction that would otherwise be reversible.

The evidence for this is not clear-cut, however, as the cyclization is itself reversible and the equilibrium *in vitro* favours the linear molecule [433].

Many of the events in the above scheme, for instance guanosine recognition, specific cleavage etc. would be likely to depend on the secondary and tertiary structure of the RNA. Two groups, those of Cech [267] and Davies [268], have shown that this is the case and the latter group in particular has shown that the insert of *Tetrahymena* can be drawn with a secondary structure very similar to that proposed for mitochondrial Class I introns [269, 270 and Section 9.6.4]. In particular, both introns share short conserved sequences, called E, P, Q, R, E′ and S that could form base-paired regions and they also have a so-called internal guide sequence (IGS) that could function in the alignment of splice points.

Class I inserts occur not only in the protein-encoding genes of the mitochondria of lower eukaryotes but also in their rDNA. Recent analysis of the *in vitro* splicing of the ribosomal precursor of *Neurospora* has shown that the insert in the 21S rRNA can again be removed autocatalytically [271]. Processing of pre-rRNA in yeast has also been shown to generate Tetrahymena-like intermediates [272]. However, at least some Class I mitochondrial introns, in both yeast and *Neurospora*, require trans-acting factors for efficient splicing. Some of these factors are encoded in nuclear genes and some (maturases) are encoded in open reading frames in the introns (they are discussed in Section 9.6.4). The *in vitro* splicing of *Tetrahymena* pre-rRNA is only one-sixtieth as efficient as splicing *in vivo*, so it is possible that in the intact nucleus protein factors are used. However, that the process can occur autocatalytically *in vivo* was demonstrated by Waring *et al.* [268] when they showed that it would occur in *E. coli* cells in which any trans-acting factor would presumably not be present. Maturases do not appear to be encoded in the introns of

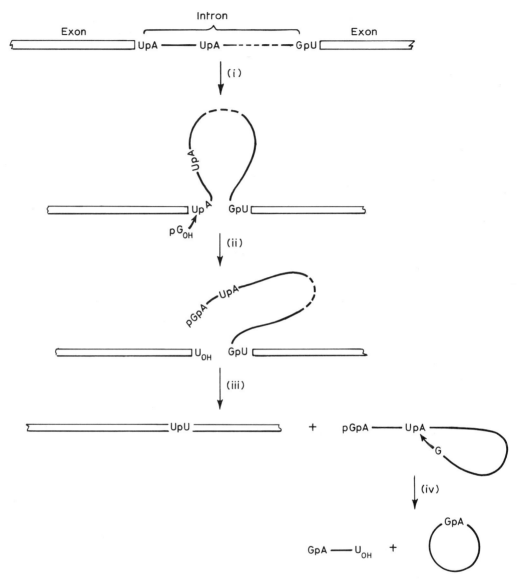

**Fig. 9.26**  Autocatalytic splicing of *Tetrahymena* rDNA. Based on Zaug *et al.* [265], reproduced with permission.

mitochondrial rDNA but a trans-acting nucleus-specified component required for excision of the intron has been identified in *Neurospora* [273].

(c) *Post-transcriptional modification of rRNA*
The analyses of the late-1950s established that rRNA contains methylated nucleotides [274] and it has since become apparent that it also contains pseudouridine [275] and in eukaryotes $N^4$-acetylcytosine [276] and the hypermodified nucleotide 3-(3-aminocarboxypropyl)-1-methyl-pseudouridine [277]. Some modified bases are strongly conserved and presumably have important functions. Thus, a ubiquitous sequence $G.m_2^6A.m_2^6A$ found at the 3'-ends of both prokaryote and eukaryote small subunit rRNA

seems to have a role in the initiation of protein synthesis (Section 11.8.3). Inhibition with the antibiotic kasugumycin of the methylases that catalyse this modification stops initiation in prokaryotes by inhibiting the binding of fMet-tRNA [278]. Conversely, some modifications are not shared between prokaryotes and eukaryotes. 5'-Methylcytosine is found only in prokaryotic rRNA [276] while the much larger methylated nucleotide content of eukaryotic rRNA is largely attributable to nucleotides that carry 2'-*O*-methyl groups on the ribose. Of 70 methyl groups on HeLa cell 28S rRNA, 65 are 2'-*O*-ribose methylations and the remainder are on bases. Similarly, there are 40 2'-*O*-methylriboses on 18S rRNA and six base methylations [279]. These methylations are strongly conserved between eukaryotic species [280] and their overall number appears to show a one-to-one relationship with the number of pseudouridine residues [281]. All of the 2'-*O*-methylations occur on the rRNA encoding region of 45S rRNA [279] but the conversion of uridine to pseudouridine occurs on conserved and non-conserved sequences [275]. Both of these modifications occur in the nucleus but base methylation occurs in the cytoplasm [279].

Analysis of the 18S rRNA of *Xenopus laevis* by Maden and coworkers [282, 283] established the location of all of the modified nucleotides, and more recently [284] they have located the methyl groups within the consensus secondary structure for the molecule. In general, they occur in non-helical regions, at hairpin loop ends or at helix boundaries and imperfections. The base methylations are all in the 3' domain of the molecule, and one cluster of 2'-*O*-ribose methylations occurs in an area of complicated secondary structure in the 5' third of the molecule.

(d) *Precursor ribonucleoproteins in the formation of ribosomes*

As already discussed in Section 9.4.1, the primary processing events in the generation of prokaryotic rRNA give rise to precursor forms of 16S, 23S and 5S rRNA that are known at p16, p23 and p5. Because many of the ribosomal proteins attach sequentially to the nascent rRNA during its transcription, each of these precursors occurs as a ribonucleoprotein particle [178]. Most of the *E. coli* genes for ribosomal proteins have been mapped and many of them are found to be organized into multicistronic operons [285, 286]. The regulation of prokaryotic ribosome synthesis is considered in Sections 10.2.2 and 11.10.2.

In lower eukaryotes, as in bacteria, there are only one or two copies of each ribosomal protein gene [287–289] but in mammals there are multiple copy genes. Those of mouse vary from 7 to 20 per protein species and they are scattered over more than one chromosome [62, 290]. The amphibian, *Xenopus*, appears to have an intermediate repetitiveness with two to five gene copies per protein [291]. Ribosomal protein genes contain introns [292 and refs. therein].

The assembly of new ribosomes in eukaryotic cells occurs predominantly in the nucleolus and the precursor rRNA species do not occur as free RNA but as ribonucleoprotein particles. Thus, the mammalian 45S pre-rRNA transcript and the 32S processing intermediate (Fig. 9.25) occur as 80S and 55S particles respectively (for a review see [293]). A 40S nucleolar precursor to the 40S subunit can also be isolated [294]. Similar maturation pathways have been followed in other organisms. The 37S initial transcript of yeast occurs in a 90S pre-ribosome which matures to 40S and 60S mature ribosomal subunits via 66S and 43S precursor particles [295].

A number of groups have demonstrated the sequential addition of ribosomal proteins during the formation of ribosomes. Todorov *et al.* [294] showed that the mammalian 80S pre-ribosome contained most of the small subunit proteins. Only two S3 and S21 are added during the formation of the 40S nucleolar precursor and four others S2, S19, S26 and S29 are added later, presumably in the cytoplasm. Similar sequential addition of the large subunit proteins has also been observed [296–298]. The nucleolar pre-ribosomal particles contain some proteins that are not present in the mature ribosome [296, 297].

## 9.5 THE ARRANGEMENT AND EXPRESSION OF tRNA GENES

### 9.5.1 tRNA genes

(a) *Prokaryotic tRNA*
The structure and organization of the tRNA genes of *E. coli* and *B. subtilis* are the subject of recent reviews [299, 300].

54 tRNA genes, encoding 35 subspecies and representing an estimated two-thirds of the total tRNA genes, have been mapped in *E. coli*. Many were located as a result of the mapping of suppressor mutations most of which result from changes in the anticodon-encoding region of tRNA genes. Others have been mapped by demonstrating the enhanced synthesis of specific tRNAs after the insertion of a prophage or an F′ factor causes the amplification of its expression. Still others have been characterized by the analysis of tRNA gene clusters and of rRNA operons that contain tRNA genes (see Section 9.4.1). Fournier and Ozeki [299] have produced a chromosome map of *E. coli* showing that the 54 genes are dispersed throughout the genome with no correlation between their position and that of the genes for aminoacyl-tRNA synthetases. Of the 54, however, 37 occur in polycistronic operons of which 14 are in the seven rRNA operons (Section 9.4.1 and Fig. 9.17). The remaining clustered genes occur in six operons that encode from three to seven tRNAs per cluster. Fournier and Ozeki [299] have catalogued the sequences of 15 *E. coli* tRNA genes.

The analysed genes of the coliform bacterium *Salmonella typhimurium* reveal its close relationship with *E. coli*. Thus, one four-tRNA gene operon is almost identical in sequence and chromosomal position to its counterpart in *E. coli* [301, 302]. The more distantly related *Bacillus subtilis*, however, differs considerably [300]. Its tRNA genes are heavily clustered into ten known operons which are concentrated in two regions of the chromosome. One operon contains 21 genes and another 16 genes. All *E. coli* tRNA genes encode the -CCA terminus, 12 of the 51 analysed *B. subtilis* genes do not. Repeated genes of a specific tRNA species occur in several *E. coli* operons but they have not been found in the operons of *B. subtilis*. Other differences between these examples of Gram-negative and Gram-positive eubacteria are listed by Vold [300].

(b) *Eukaryotic tRNA genes*
The structure and organization of eukaryotic tRNA genes has been reviewed by Clarkson [303].

While prokaryotes contain one or a few copies of the gene encoding any given tRNA species, eukaryotic tRNAs are encoded in multiple-copy genes. Presumably this is necessary in order that the cell can produce its required quantities of tRNA (20% of total cell RNA). Nevertheless, a tabulation of tRNA gene numbers by Long and Dawid [201] reveals that abundance is poorly correlated with genetic complexity. For instance, among the fungi, *Saccharomyces* has 360 genes for total tRNA while *Neurospora* has 2640.

The most striking feature of the arrangement of tRNA genes in eukaryotic cells is the species-to-species variation. Some, like many yeast genes, are dispersed throughout the genome [304, 305]. Others occur in clusters. The 600–800 *Drosophila* genes are arranged in approximately 60 such clusters spread over most of the chromosomes. The arrangement within a cluster is totally irregular, however, and any given cluster may contain several types of tRNA gene, some in multiple copies. Furthermore, they are irregularly spaced and transcribed from both strands of the DNA. One cluster that has been fully analysed contained eight tRNA$^{Asn}$ genes, four tRNA$^{Arg}$ genes, five tRNA$^{Lys}$ genes and one tRNA$^{Ile}$ gene [306]. The tRNA genes of the amphibian, *Xenopus*, occur in tandemly repeating clusters. One such cluster consisted of approximately 150 copies of a 3.18-kb repeating fragment that contained eight tRNA genes, including two copies of tRNA$^{Met}$, and six single-copy genes [307, 303]. In mammals, tRNA genes appear to be solitary or arranged in small clusters (1–2.5 kb of DNA) with each cluster separated by a larger region (8–20 kb) of non-tRNA-encoding DNA. A 13.4-kb rat DNA fragment containing two such clusters has been analysed by Rosen *et al.* [308]. One cluster contained tRNA$^{Leu}$, tRNA$^{Asp}$, tRNA$^{Gly}$ and tRNA$^{Glu}$ and the second cluster contained the same genes except that tRNA$^{Leu}$ was a pseudogene. Blot experiments showed that the clusters were part of a repeating unit that was copied about 10 times.

The enzyme that transcribes eukaryotic tRNA genes, RNA polymerase III, together with its promoters and termination signals, have been described in Chapter 8. With one known exception, each tRNA gene, whether solitary or part of a cluster, is organized as a single transcription unit. The exception is a pair of yeast genes, tRNA$^{Arg}$ and tRNA$^{Asp}$, which are separated by only 10 bp and are transcribed as a 170-nucleotide dimeric precursor which is processed to the mature tRNAs [309]. Transcription depends on the promoter of the 5′ gene [310].

### 9.5.2 The processing of tRNA

(a) *Excision from precursors and 5′ and 3′ trimming*

(i) *Prokaryotic.* As has been seen in previous sections, tRNA-containing precursor molecules of prokaryotes can be transcripts of single genes or of gene clusters. The transcripts may also contain mRNA or rRNA sequences (Section 9.4.1). The maturation of tRNA from these transcripts has been the subject of a number of reviews [14, 15, 311–313].

A key enzyme in the maturation process is RNase P. This is a complex ribonucleoprotein consisting of a polypeptide of approximately 18 000 $M_r$ and a RNA of 375–377 nucleotides known as M1 RNA [314, 315]. The M1 RNA contains a nucleotide sequence which is complementary to the GTψCPu loop of tRNA [316] and it has been suggested that RNA:RNA base pairing positions the enzyme complex on the precursor and is the reason that the enzyme appears to recognize the tRNA domain rather than specific sequences for cleavage. With some substrates, the RNA can effect catalytic cleavage in the absence of the protein of the enzyme [317, 434]. Together with *Tetrahymena* pre-rRNA (Section 9.4.3.b) it is thus the second known example of catalytic RNA. The mechanism of catalysis differs, however, from the ribosomal RNA self-splicing as it does not depend on nucleophilic attack by ribose 3′-OH groups [435].

Endonucleolytic catalysis by RNase P generates the mature 5′-terminus of the tRNA (Fig. 9.18). In monomeric precursors, this might be the primary cleavage. Thus, it specifically removes the 41-nucleotide sequence upstream of the tRNA gene of the *E. coli* tRNA$^{Tyr}$ transcript [318]. Multimeric precursors, however, may first be cleaved further to the 5′-side of the gene by other ribonucleases such as RNase III (Section 9.4.1).

The cleavage events that generate the 3′-end of the mature prokaryotic tRNA appear to be less precise and are not well understood. Endonucleolytic cleavage of precursors generates a 3′-end that still contains a few extra nucleotides. These are then removed exonucleolytically (Fig. 9.18). Ribonuclease F [319] is a possible contender for the endonucleolytic function although some of the properties of this enzyme are those associated with degradative activities [319, 320]. Of the possible exoribonuclease activities, Deutscher [313] has reviewed the evidence implicating *E. coli* ribonucleases Q, Y, D and BN and to these possibles must be added the recently characterized RNase T [321].

(ii) *Eukaryotic.* Although they are all mono-cistronic, the primary transcripts of tRNA genes in eukaryotic cells also have extra sequences at their 5′- and 3′-ends. In some instances they also contain insertion sequences (see Section 9.5.2.c). The lack of mutants in all but the lowest eukaryotes has slowed an understanding of the trimming reactions involved in processing and this is only partially offset by the use of cloned genes to develop *in vitro* systems in which precursors might accumulate. Such a system is the nucleus of *Xenopus* oocytes (Section A.10), and Melton *et al.* [322] made elegant use of this to study the expression of a plasmid-cloned yeast tRNA gene microinjected directly into the nucleus. They were able to detect mature tRNA, its precursors and the subcellular location of processing (Fig. 9.27). Recently, stable precursors for tRNA$^{Met}$ and tRNA$^{Leu}$ have been isolated from HeLa cells and may aid the investigation of processing [323]. They both contain four extra nucleotides at the 5′-terminus, nine extra nucleotides at the 3′-terminus and several nucleotides that are modified in the mature molecule are undermodified in the pre-cursors.

A RNase P-like activity, which may function in 5′ trimming, has been partially purified from yeast [324] and detected in chicken and human cells [325, 326]. At the 3′-end of tRNA-containing transcripts, the processing reactions of eukaryotic cells must precisely remove the extra sequence prior to the addition of the CCA terminus (Section 9.5.2.b). Activities have been detected in *Xenopus* oocyte nuclei which cleave the entire 3′-extension of a *Bombyx mori* pre-tRNA$^{Ala}$ as a single 22-nucleotide fragment [327, 328], and an activity which cleaves an 8–12-nucleotide trailer of human pre-tRNAi$^{Met}$ has been partially purified [436]. The latter enzyme

required a mature 5′-end to the tRNA precursor suggesting that processing followed an obligatory order of 5′ maturation followed by 3′ maturation.

(b) *Synthesis of the CCA 3′-terminus*
In prokaryotes, the 3′-terminal trinucleotide sequence CCA, which forms the amino acceptor stem of the mature molecule, is usually though not invariably encoded within the gene [329, 330]. Conversely, no known eukaryotic genes encode the CCA terminus, and this includes those of mitochondria and chloroplasts. The terminus is added post-transcriptionally by tRNA nucleotidyltransferase (EC 2.7.7.25) [331], a highly specific enzyme which, in the presence of ATP, CTP and tRNA lacking the 3′-terminal sequence, will catalyse the reaction:

$$tRNA + ATP + 2\,CTP \rightarrow tRNACCA + 3\,PP_i$$

The enzyme has been identified in a wide variety of organisms, subcellular particles, bacteria and viruses [331]. It is present in *E. coli* and required for its normal growth despite the fact that all known tRNA genes in this organism encode the CCA terminus. The explanation for this is that tRNA nucleotidyltransferase is also required for the repair of damaged tRNA molecules which arise by end turnover, mainly of the 3′-AMP residue, when tRNA molecules are not fully charged [332].

(c) *The mechanism of tRNA splicing*
The tRNA genes of eukaryotes [333–335] and archaebacteria [39] may contain introns and, in yeast nuclei, approximately 40 of the 400 tRNA genes are so interrupted [336]. The single intron may be from 14 to 60 nucleotides in length but in all known examples it is inserted one nucleotide to the 3′-side of the anticodon Fig. 9.27. Furthermore it includes a sequence which base-pairs with all or part of the anticodon and sometimes with a portion of the anticodon loop [336].

A yeast temperature-sensitive mutant, *rna1*, accumulates the precursors of nine tRNAs when

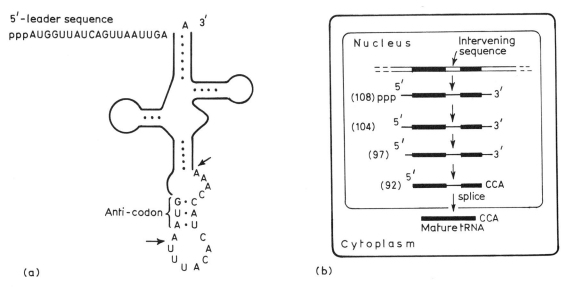

5'-leader sequence
pppAUGGUUAUCAGUUAAUUGA

(a)

(b)

**Fig. 9.27** Processing of yeast tRNA$^{Tyr}$ precursors into mature tRNA$^{Tyr}$. (a) schematic illustration of 108-base precursor with 10-nucleotide long 5'-leader sequence, 14-nucleotide intervening sequence (between arrows) and 5'-terminal triphosphate; (b) schematic diagram of processing steps. The numbers in brackets indicate the nucleotides in each of the intermediates. The 5'-leader is removed in three stages, the last of which is accompanied by the excision of the 3'-trailer end and addition of the 3'-CCA end group. Nucleotide modifications take place between 108 and 104, 97 and 92 and 92 and mature tRNA [322].

grown at the non-permissive temperature [337, 338]. These precursors have mature 5'- and 3'-termini but contain the intervening sequences. They can be correctly spliced *in vitro* in systems derived from yeast [339, 340], *Xenopus* germinal vesicle [341] and HeLa cells [342]. Their splicing can be separated into two stages [343]. In the first of these, the tRNA precursor is precisely cleaved at the intron boundaries, thus generating two half-tRNA molecules and a linear intervening sequence. In the second stage, the two half-molecules are ligated together to produce the mature tRNA.

The endonucleolytic cleavages are unusual in producing 5'-hydroxyl termini and 3'-phosphorylated termini, a feature that is common to degradative enzymes such as the RNases A and T1, but is not normally found in processing enzymes. Endonuclease activities that accurately cleave the yeast precursors have been

partially purified from oocyte germinal vesicles [344] and substantially purified from yeast [345]. The yeast enzyme has two unusual features. It appears to be a membrane protein and it produces a 2',3'-cyclic phosphorylated terminus rather than a 3'-phosphate (Fig. 9.28).

Ligation has also been studied in both yeast [346] and animal cell systems [342, 347] and again appears to differ between the two. In the yeast system, the 3'-half of the tRNA is phosphorylated and is then adenylated by an adenylated ligase. Condensation occurs between this activated 3'-half of the rRNA molecule and the 2', 3'-cyclic phosphate terminus of the 5'-half of the molecule. The phosphodiester linkage formed thus has the form

$$N \diagdown P \diagup P \diagup N$$

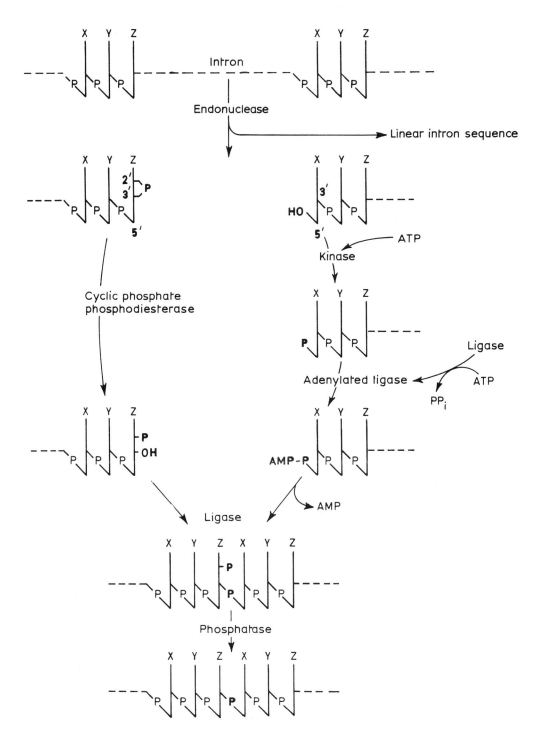

**Fig. 9.28** The ligation of tRNA exons in yeast.

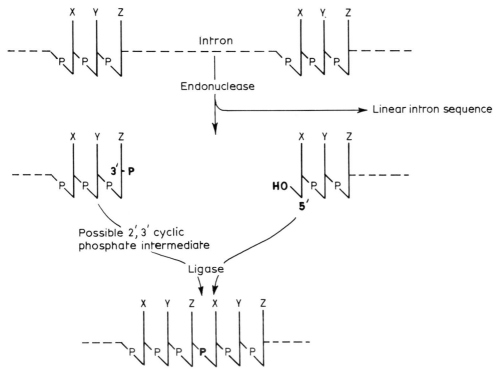

**Fig. 9.29** The ligation of tRNA exons in animal cells.

with the 3′,5′-bond derived from the 5′-terminal phosphate. This reaction mechanism is similar to that of bacteriophage T4 ligase [348]. The 2′-phosphate is subsequently removed by phosphatase activity (Fig. 9.28).

Animal cell ligation appears to involve a different strategy [342, 347]. The 3′-half of the tRNA molecule is not phosphorylated and the phosphate of the phosphodiester bond is derived from the 5′-half of the molecule (Fig. 9.29). A HeLa cell ligase has been purified and is dependent on $Mg^{2+}$, ATP or dATP and RNA substrates with 5′-hydroxyl and 2′,3′-cyclic phosphate termini [349].

### (d) tRNA base modification
Sprinzl and Gauss [350] list 43 different rare nucleosides that have been found in tRNA and

all of these arise by post-transcriptional base modification or base transposition. Much remains to be learned both of the enzyme activities catalysing the modification and of the function of the modified nucleosides. Both formation [351, 352] and function [353] have, however, been the subject of reviews and Table 9.2 lists some of the commonest modifications and the enzymes involved in their formation. One unusual form of modification, by which inosine and queuosine are inserted into the wobble position of the anticodon of a number of tRNA species, is a direct base replacement. This is discussed in Section 4.8. Some insight into substrate recognition by tRNA-modifying enzymes has come from the recent finding that the formation of pseudouridine in the anticodon of yeast tRNA$^{Tyr}$ depends on the presence of an

**Table 9.2** Base modification in tRNA.

| Modified base | tRNA nucleotide modified* | Enzyme | Ref. |
|---|---|---|---|
| 7-Methylguanosine | 46 | $m^7G$, tRNA methyltransferase | [355] |
| 2-Methylguanosine | 6, 26 | $m^2G$, tRNA methyltransferase | [356] |
| 1-Methyladenine | 1, 4, 22, 58 | $m^1A$, tRNA methyltransferase | [357] |
| 5-Methylcytidine | Various | $m^5C$, tRNA methyltransferase | [358] |
| Ribosylthymine | 54 | rT, tRNA methyltransferase | [359] |
| 5-Methylaminomethyl-2-thiouridine | 34 | $mam^5S^2U$ tRNA methyltransferase | [360] |
| Pseudouridine | Various | Pseudouridylate synthase | [354, 361] |
| 4-Thiouridine | 8 | $S^4U$ tRNA sulphurtransferase | [362] |
| Queuosine | 34 | tRNA guanine ribosyltransferase | Section |
| Inosine | 34 | tRNA hypoxanthine ribosyltransferase | 4.8 |

*The numbering of the modified nucleotides is based on the numbering system derived from yeast phenylalanine tRNA (Fig. 11.1). It indicates the usual location of the modified nucleotide and does not imply that the position is so modified in all tRNAs.

intron in the gene [354]. Presumably, pseudo-uridylate synthase recognizes an intron-induced secondary structure in the pre-tRNA.

Other modifications are less well understood and some, such as the formation of wyosine and 2-methylthio-$N^6$-isopentenyladenosine, require a sequence of enzymic steps [352]. The possible role of modified nucleotides in translation is considered in Sections 11.2.5, 11.9.2 and 11.13.2.

## 9.6 THE ARRANGEMENT AND EXPRESSION OF MITOCHONDRIAL AND CHLOROPLAST GENES

The genes of subcellular particles are organized and expressed in ways which differ both from those of prokaryotes and those of the eukaryote nucleus (for reviews see [363–368]). Even within themselves they are very variable. In general, animals have small circular mitochondrial genomes of 14–18 kb while plant mitochondria and chloroplasts have much larger circles (see Section 3.4).

Mitochondrial and chloroplast RNA poly-merases have been described in Section 8.1.2.b, their promoters considered in Section 8.3.5 and the transcription of mitochondrial DNA has been reviewed [365, 367]. Each strand of mammalian mitochondrial DNA is symmetric-ally transcribed. The light strand (L), which encodes the sense sequence of the rRNAs and most of the tRNAs and mRNAs, is transcribed 2–3 times faster than the heavy strand [369]. However, the genomic sequence reveals that virtually all of the H strand would need to be transcribed while transcription of the L strand may not be complete [370]. The only site on the mammalian mitochondrial genome without a coding function is the so-called *displacement loop (D loop)* (see Fig. 3.13) and all the available evidence indicates that this is the site of tran-scriptional initiation [365, 367]. *In vitro* tran-scription systems employing partially purified mitochondrial RNA polymerases that are specific for the D-loop initiation site have been developed [371, 372]. Furthermore, promoters of related but unusual sequences have been identified immediately adjacent to and over-

**Table 9.3**   A comparison of the structure and expression of the yeast (*Saccharomyces*) and human mitochondrial genomes.

| Genome structure and expression | Yeast | Human |
|---|---|---|
| Size | 73–78 kb | 16–57 kb |
| Shape | Circular | Circular |
| Encoded rRNA | 15S and 21S | 12S and 16S |
| Transcription | Asymmetric | Symmetric |
| Promoters | Five | One |
| Introns | In some genes | None |
| Processing | No polyadenylation | mRNAs polyadenylated |

lapping the transcriptional start sites of both the H and L strands of human mitochondrial DNA [373, 374]. The same sequences are not found, however, in the D loops of other vertebrates.

Most of the rRNA and protein-encoding genes of the mammalian mitochondrial genome are flanked by tRNA genes with no separating nucleotides. The processing of transcripts must therefore involve the recognition of the 5'- and 3'-ends of the tRNA and precise cleavage at these points. By analogy with prokaryotic RNase P, it has been suggested that the catalytic entity could be a ribonucleoprotein and it has been proposed that mitochondrial 7S RNA could be involved [359]. A 9S RNA has been similarly implicated in tRNA processing of yeast mitochondrial transcripts [375].

Transcription of yeast mitochondrial DNA is asymmetric and initiates from five different sites. Putative promoters have been identified at three of these, namely an 18-bp sequence to the 5'-side of both of the rRNA genes [376] and a promoter for subunit 3 of cytochrome oxidase which is also assumed to function for the tRNA$^{Val}$ gene [377]. These genes of yeast mitochondria are scattered and separated by AT-rich regions, the function of which, if any, is unknown. The above 9S RNA is encoded in one such region [375].

Table 9.3 compares the structure and expression of the mitochondrial genomes of yeast and man.

### 9.6.1  Protein-encoding genes of mitochondria and chloroplasts

Most mitochondrial proteins are synthesized from nuclear gene products [378]. Some, however, including all subunits of cytochrome oxidase, apocytochrome *b*, one or two subunits of the mitochondrial ATPase complex and the intron-encoded maturases (Section 9.6.4) are encoded in the mitochondrial genome. In addition, there are other open reading frames of unknown function that are known as *unassigned reading frames (URF)*. The variability within mitochondria from various sources extends to the individual genes. Thus, the so-called COB gene, encoding apocytochrome *b*, exists in two- and five-intron versions in yeast but in man contains no introns and is flanked by tRNA genes. The mRNA transcripts of mammalian mitochondria are polyadenylated (Table 9.3) and the economy of the genomic sequence is such that many of the gene transcripts require the poly(A) tail to complete the translation stop codon [367]. No obvious polyadenylation signals have been identified in mitochondrial transcripts, and mitochondrial mRNAs are not capped [379].

There are many more genes encoded in chloroplast genomes than in those of mitochondria. Apart from one to three copies of the rRNA genes and up to 40 tRNA genes, they encode part of the ATPase complex, cytochrome *f*, cytochrome $b_{599}$, elongation factors G and T, the large subunit of ribulose bisphosphate carboxylase, at least one ribosomal protein, aminoacyl-tRNA synthetases and undoubtedly others still to be identified [368, 380, 381].

### 9.6.2 Mitochondrial and chloroplast rDNA

Mitochondria and chloroplasts have the capacity to make some of their own proteins and to do so produce their own ribosomes, the RNAs for which are encoded in their genomes.

The sequences of the rRNA genes of both mitochondria and chloroplasts are known for a wide variety of species (for a list, see [233] and for a review see [382]). In general they support the concept that these organelles may have arisen by endosymbiotic capture of bacteria (for further discussion of these points see Section 11.9). The argument is complicated, however, by the very variable gene arrangement, particularly in mitochondria. The small 5 μm circular mitochondrial DNA of mammals have their

genetic information highly compressed. Their rRNAs are very small (12S and 16S) and the rRNA genes are interspersed with tRNA genes rather than spacers [383]. The large 25 μm circular DNA of yeast mitochondria, on the other hand, has the genes for the 15S and 21S rRNAs separated by a segment of DNA which is at least 25 kbp long and contains two cytochrome oxidase genes and most of the mitochondrial tRNA genes [384]. Furthermore, the gene for the 21S rRNA contains an insert [384] and the two ribosomal genes are separately transcribed from different promoters [385]. The mitochondrial genes of *Aspergillus* [386] and the linear mitochondrial genome of ciliate protozoans [387] appear to present an intermediate degree of complexity. Fig. 9.30 illustrates the gene arrangement of the mitochondrial rDNA of mouse (rat, human and bovine genes share the same arrangement) and *Aspergillus*.

The yeast mitochondrial rRNAs are both made as precursors with 3'-extensions. They are still made and processed in petite mutants that cannot make mitochondrial proteins so the

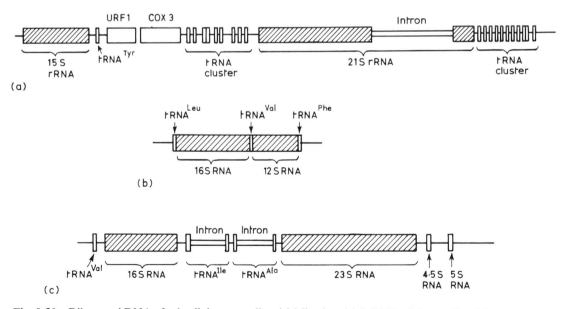

**Fig. 9.30** Ribosomal DNA of subcellular organelles. (a) Mitochondrial rDNA of *Aspergillus*; (b) Mitochondrial rDNA of mouse; (c) Chloroplast rDNA of maize.

nucleases involved may be nucleus-encoded and imported from the cytoplasm [388].

The genes for the four rRNAs of higher-plant chloroplasts are clustered in the order 16S, 23S, 4.5S and 5S [389, 390] and have intron-containing tRNA genes in their intervening sequences (Fig. 9.30c). In many angiosperms the chloroplast rDNA is part of a large (22–25 kb) inverted repeat and there are therefore two sets of genes [391 and refs therein]. That of the algae, *Chlamydomonas reinhardii*, however, contain no 4.5S gene, has an insertion in the 23S gene and contains two small genes encoding 3S and 7S RNA in the spacer between the 16S and 23S genes [392]. The 7S and 3S genes are homologous to the 5'-end of the chloroplast rRNA of other species as well as that of prokaryotes and they are hydrogen-bonded to 23S RNA in the ribosome.

### 9.6.3 Mitochondrial and chloroplast tRNA genes

The human, mouse and bovine mitochondrial genome each encode 22 tRNA genes. They are located as singles, doubles, triplets and one five-gene sequence such that they separate from each other nearly all of the other mitochondrial genes (for reviews see [364, 367] and several chapters in [393]). They occur on both strands of the DNA and their possible role in processing has already been discussed. The encoded tRNAs are smaller than their cytoplasmic counterparts (59–75 nucleotides) and are relatively under-methylated. The 59-nucleotide tRNA$^{Ser}$ has lost the D stem and loop portion of the clover leaf secondary structure (see Section 11.9.2).

The tRNA genes of fungal mitochondria are clustered [384], do not contain introns and do not encode the 3'–C–C–A terminus. The AT-rich gene-flanking sequences do not contain any obvious promoters or termination sites; however, one segment has been shown to encode a 9S, 450-bp RNA which plays a role in processing tRNA-containing transcripts [375]. Yeast

mitochondria contain 25–26 tRNA species all encoded in the organelle genome except one lysine isoacceptor that appears to be imported from the cytoplasm.

The chloroplast tRNA genes of *Euglena* and higher plants have been extensively mapped and some have been isolated and sequenced (for reviews see [394, 395, 396]). Four of the nine loci of *Euglena* have been examined in detail. One contains tRNA$^{Ile}$ and tRNA$^{Ala}$ in the 16S–23S rRNA spacer. The others are clusters of four, six and seven tRNA genes in which each gene is separated from the next by a short spacer of 2–175 bp [397 and refs therein]. A chloroplast transcription system has been developed that transcribes tRNA genes and processes the product to mature tRNA including the addition of a 3'CCA terminus [398]. Nothing is currently known of the promoters or regulatory sites for transcription, but it is assumed that the gene clusters are transcribed into multicistronic precursors. Introns occur in two tRNA genes in the ribosomal transcription unit of maize chloroplasts ([399] and Fig. 9.30).

### 9.6.4 The introns of mitochondrial genes and their splicing

Introns have not been demonstrated in the tightly packed genomes of animal cell mitochondria. Their protein-encoding genes are continuous coding sequences and tend not to be separated from each other by non-coding sequences but by tRNA genes. The much less tightly packed and larger genomes of fungal mitochondria do contain introns as well as intragenic non-coding sequences. These introns have, however, little in common with those of nuclear protein-encoding genes. They do not obey the GT/AG rule and indeed do not have strongly conserved splice site sequences. Some mitochondrial introns contain large open reading frames that are in phase with the preceding exon. Furthermore, because in mitochondria splicing and translation occur in the same subcellular compartment, these sequences could be translated into protein. Introns of this sort in *Saccharomyces* occur in three of the five introns of the cob gene that encodes apocytochrome *b*, in four introns in the gene for subunit 1 of cytochrome oxidase and in an intron of the gene of the large mitochondrial ribosomal RNA (for review see [400]).

Genetic analysis of splicing defective mutants of the cob gene introns first led Slonimski and colleagues to suggest that the open reading frames encoded trans-acting diffusible proteins responsible for the splicing and maturation of mitochondrial mRNA. They called these proteins, *maturases* [401]. Antibodies against conserved nucleotide sequences within the putative maturase-encoding sequences have recently been employed to confirm the existence of such proteins [402, 403] and their occurrence may not be limited to mitochondria as similar open frame coding sequences are found in introns of a ribosomal gene in the chloroplast of the algae *Chlamydomonas reinhardii* [404]. In splicing, the maturase is believed to recognize short nucleotide sequences within the introns from which they are translated so that, as well as catalysing the removal of the intron, they participate in the destruction of their own message and are autoregulating. The system also has features which allows for the concerted control of more than one gene product. The maturase produced from the fourth intron of the apocytochrome *b* gene transcript is believed to recognize identical signal sequences in its own intron and in an intron of the cytochrome oxidase gene transcript. Thus, it catalyses the processing of both gene products [405, 406]. A similar system has also been described in *Aspergillus* [407]. Several other trans-acting components that are required for mitochondrial splicing are specified in the nuclear genome [408, 409, 410].

It may prove artificial to separate the splicing of mitochondrial genes from the autocatalytic self-splicing of *Tetrahymena* ribosomal gene

introns (see Section 9.4.3.b). Indeed van der Horst and Tabak [411] have recently demonstrated self-splicing in several yeast mitochondrial precursor mRNAs. The introns of both *Tetrahymena* and fungal mitochondria share a highly conserved secondary structure and are known as class I introns [270, 406, 412]. Davies [413] has suggested that all class I introns may once have been self-splicing but in some mitochondrial genes, proteins appear to have evolved perhaps to assist in the alignment of precursor RNAs which then splice autocatalytically. The secondary structure of the intron RNA sequence certainly appears to be important, as the splicing-defective mutants in the yeast cob gene all have mutations which disrupt the proposed base-pairing of the intron sequence [412, 413].

Four of the thirteen introns in yeast mitochondria, namely introns 1, 2 and 5 of subunit one of cytochrome oxidase and intron 1 of the cob gene, are placed in a separate class (class II). They bear little or no resemblance to class I introns but have in common conserved sequence elements, a circular excision product and a highly conserved secondary structure [414]. Nothing is known of their processing but they may also encode maturase [415] and, incredibly, proteins related to the reverse transcriptases of retroviruses [416].

## 9.7 A POSTSCRIPT ON SPLICING

Eukaryotes would appear to have evolved a plethora of different splicing mechanisms, the best-known examples of which are summarized in Table 9.4. Our knowledge of all of the above systems is very incomplete and, with time, their differences may prove to be the evolutionary modification of one or two common themes. Similarly, the removal of the transcribed spacers from pre-rRNA may represent a modification in which the cleavage products are not religated.

The discovery of an intron-containing gene in a bacteriophage [417] raises the question of how

**Table 9.4** A summary of splicing mechanisms.

| Splicing system | Precursor-encoded splicing requirements | External requirements | Splicing excision product |
|---|---|---|---|
| Nuclear pre-mRNA (see Section 9.3.4) | Splice site consensus sequence | ATP U1 snRNP U2 snRNP | Lariat |
| Fungal pre-tRNA (see Section 9.5.2) | Nuclease-recognition sites | Specific nucleases and RNA ligase | Linear molecule |
| Tetrahymena pre-rRNA (see Section 9.4.3) | RNA catalyses its own excision | $Mg^{2+}NH_4$ guanosine | Circular and linear molecules |
| Mitochondrial precursors (see Section 9.6.4) | Maturases | Nucleus-encoded components | Circular molecules |

exons are joined in prokaryotic systems. Recent studies reveal that, here too, a pre-mRNA is processed to remove the intron transcript [418] but nothing is yet known of the mechanism.

REFERENCES

1 Hershey, A. D. (1953), *J. Gen. Physiol.*, **37**, 1.

2 Volkin, E. and Astrachan, L. (1956), *Virology*, **2**, 149.

3 Spiegelman, S., Hall, B. D. and Storck, R. (1961), *Proc. Natl. Acad. Sci. USA*, **47**, 1135.

4 Brenner, S., Jacob, F. and Meselson, M. (1961), *Nature (London)*, **190**, 576.

5 Gros, F., Hiatt, H., Gilbert, W., Kurland, C. G., Risebrough, R. W. and Watson, J. D. (1961), *Nature (London)*, **190**, 581.

6 Marmar, J., Greenspan, C. M., Policek, E., Kahan, F. M., Levine, J. and Mandel, M. (1963), *Cold Spring Harbor Symp. Quant. Biol.*, **28**, 191.

7 Tocchini-Valentini, G. P., Slodolsky, M., Aurisicchio, A., Sarnat, M., Fraziosi, F., Weiss, S. B. and Geiduschek, E. P. (1963), *Proc. Natl. Acad. Sci. USA*, **50**, 935.

8 Hayashi, M., Hayashi, M. N. and Spiegelman, S. (1963), *Science*, **140**, 1313.

9 Miller, O. L., Hamkalo, B. A. and Thomas, C. A. (1970), *Science*, **169**, 392.

10 Morse, D. E., Baker, R. F. and Yanofski, C. (1968), *Proc. Natl. Acad. Sci. USA*, **60**, 1428.

11 Leive, L. and Kollin, V. (1967), *J. Mol. Biol.*, **24**, 247.

12 Morikawa, N. and Imamoto, F. (1969), *Nature (London)*, **223**, 37.

13 Morse, D. E. and Yanofsky, C. (1969), *J. Mol. Biol.*, **41**, 317.

14 Gegenheimer, P. and Apirion, D. (1981), *Microbiol. Rev.*, **45**, 502.

15 Apirion, D. and Gegenheimer, P. (1981), *FEBS Lett.*, **125**, 1.

16 Allman, S., Guerrier-Takada, C., Frankfort, H. M. and Robertson, H. D. (1982), in *Nucleases* (ed. S. M. Linn and R. J. Roberts), Cold Spring Harbor Laboratory, New York, p. 243.

17 Dunn, J. J. and Studier, F. W. (1973), *Proc. Natl. Acad. Sci. USA*, **70**, 3296.

18 Dunn, J. J. and Studier, F. W. (1974), *Brookhaven Symp. Biol.*, **26**, 267.

19 Gitelman, D. R. and Apirion, D. (1980), *Biochem. Biophys. Res. Commun.*, **96**, 1063.

20 Tilghman, S. M., Tiemeier, D. C., Polsky, F., Edgell, M. H., Seidman, J. G., Leder, A., Enquist, L. W., Norman, B. and Leder, P. (1977), *Proc. Natl. Acad. Sci. USA*, **74**, 4406.

21 Tilghman, S. M., Tiemeier, D. C., Seidman, J. G., Peterlin, B. M., Sullivan, M., Maizel, J. V. and Leder, P. (1978), *Proc. Natl. Acad. Sci. USA*, **75**, 725.

22 Schaffner, W., Gross, K., Telford, J. and Birnstiel, M. (1976), *Cell*, **8**, 471.

23  Nagata, S., Mantei, N. and Weissmann, C. (1980), *Nature (London)*, **287**, 401.

24  McKnight, S. L. (1980), *Nucleic Acids Res.*, **8**, 5949.

25  Aleström, P., Akasjärvi, G., Perricaudet, M., Mathews, M. B., Klessing, D. F. and Petterson, U. (1980), *Cell*, **19**, 671.

26  Konkel, D. A., Tilghman, S. M. and Leder, P. (1978), *Cell*, **15**, 1125.

27  van Ooyen, A., van den Berg, J., Mantei, N. and Weissmann, C. (1979), *Science*, **206**, 337.

28  Lawn, R. M., Efstratiadis, A., O'Connell, C. and Maniatis, T. (1980), *Cell*, **21**, 647.

29  Nishioka, Y. and Leder, P. (1979), *Cell*, **18**, 875.

30  Stein, J. P., Catterall, J. F., Kristo, P., Means, A. R. and O'Malley, B. W. (1980), *Cell*, **21**, 681.

31  Woo, S. L., Beattie, W. G., Catterall, J. F., Dugaiczyk, A., Staden, R., Brownlee, G. G. and O'Malley, B. W. (1981), *Biochemistry*, **20**, 6437.

32  Wilks, A., Cato, A. C., Cozens, P. J., Mattaj, I. W. and Jost, J-P. (1981), *Gene*, **16**, 249.

33  Wozney, J., Hanahan, D., Tate, V., Boedtker, H. and Doty, P. (1981), *Nature (London)*, **294**, 129.

34  Lai, E. C., Stein, J. P., Catterall, J. F., Woo, S. L. C., Mace, M. L., Means, A. R. and O'Malley, B. W. (1979), *Cell*, **18**, 829.

35  Old, R. W. and Woodland, H. R. (1984), *Cell*, **38**, 624.

36  Gilbert, W. (1978), *Nature (London)*, **271**, 501.

37  Darnell, J. E. (1978), *Science*, **202**, 1257.

38  Lomedico, P., Rosenthal, N., Efstratiadis, A., Gilbert, W., Kolodner, R. and Tizard, R. (1979), *Cell*, **18**, 545.

39  Kaine, B. P., Gupta, R. and Woese, C. R. (1983), *Proc. Natl. Acad. Sci. USA*, **80**, 3309.

40  Sakano, H., Rogers, J. M., Hüppi, K., Brack, C., Traunecker, A., Maki, R., Wall, R. and Tonegawa, S. (1979), *Nature (London)*, **277**, 627.

41  Lonberg, N. and Gilbert, W. (1985), *Cell*, **40**, 81.

42  Südhof, T. C., Goldstein, J. D., Brown, M. S. and Russell, D. W. (1985), *Science*, **228**, 815.

43  Craik, C. S., Bachman, S. R. and Beychok, S. (1980), *Proc. Natl. Acad. Sci. USA*, **77**, 1384.

44  Gō, M. (1981), *Nature (London)*, **291**, 90.

45  Artymiak, P. J., Blake, C. C. F. and Sippel, A. E. (1981), *Nature (London)*, **290**, 287.

46  Eiferman, F. A., Young, P. R., Scott, R. W. and Tilghman, S. M. (1981), *Nature (London)*, **294**, 713.

47  Jensen, E. O., Paludan, K., Hyldig-Nielsen, J. J., Jorgensen, P. and Marcker, K. A. (1981), *Nature (London)*, **291**, 677.

48  Antoine, M. and Niessing, J. (1984), *Nature (London)*, **310**, 795.

49  Leicht, M., Long, G. L., Chandra, T., Kurachi, K., Kidd, V. J., Mace, M., Davie, E. W. and Woo, S. L. C. (1982), *Nature (London)*, **297**, 655.

50  Stone, E. M., Rothblum, K. N. and Schwartz, R. J. (1985), *Nature (London)*, **313**, 498.

51  Rogers, J. (1985), *Nature (London)*, **315**, 458.

52  Hamer, D. H. and Leder, P. (1979), *Cell*, **18**, 1299.

53  Gruss, P. and Khoury, G. (1980), *Nature (London)*, **286**, 634.

54  Gruss, P., Efstratiadis, A., Karathansis, S., König, M. and Khoury, G. (1981), *Proc. Natl. Acad. Sci. USA*, **78**, 6091.

55  Ingram, R. S., Scott, R. W. and Tilghman, S. M. (1981), *Proc. Natl. Acad. Sci. USA*, **78**, 4694.

56  Solomon, E. (1980), *Nature (London)*, **286**, 656.

57  Firtel, R. A. (1981), *Cell*, **24**, 6.

58  Gwo-Shu Lee, M., Lewis, S. A., Wilde, C. D. and Cowan, N. J. (1983), *Cell*, **33**, 477.

59  Fuchs, E. V., Coppock, S. M., Green, H. and Cleaveland, D. W. (1981), *Cell*, **27**, 75.

60  Piatigorsky, J. (1984), *Cell*, **38**, 620.

61  Sim, G. K., Kafatos, F. C., Jones, C. W., Koehler, M. D., Efstratiadis, A. and Maniatis, T. (1979), *Cell*, **18**, 1303.

62  Monk, R. J., Meyuhas, P. and Perry, R. P. (1981), *Cell*, **24**, 301.

63  MacDonald, R. J., Crerar, M. M., Swain, W. F., Pictet, R. L., Thomas, G. and Rutter, W. J. (1980), *Nature (London)*, **287**, 117.

64  Pittet, A-C. and Schibler, U. (1985), *J. Mol. Biol.*, **182**, 359.

65  Hentschel, C. C. and Birnstiel, M. L. (1981), *Cell*, **25**, 301.

66  Maxson, R., Cohn, R. and Kedes, L. H. (1983), *Annu. Rev. Genet.*, **17**, 239.

67  Schaffner, W., Gross, K., Telford, J. and Birnstiel, M. (1976), *Cell*, **8**, 471.

68  Gross, K., Schaffner, W., Telford, J. and Birnstiel, M. (1976), *Cell*, **8**, 479.

69  Wu, M., Holmes, D. S., Davidson, N., Cohn, R. H. and Kedes, L. H. (1976), *Cell*, **9**, 163.

70  Lifton, R. P., Goldberg, M. L., Karp, R. and Hogness, D. S. (1977), *Cold Spring Harbor Symp. Quant. Biol.*, **42**, 1047.

71  Old, R. W., Woodland, H. R., Ballantine, J. E. M., Aldridge, T. C., Newton, C. A., Bains, W.

A. and Turner, P. C. (1982), *Nucleic Acids Res.*, **10**, 7561.

72 Turner, P. C. and Woodland, H. R. (1983), *Nucleic Acids Res.*, **11**, 971.

73 Childs, G., Maxson, R., Cohn, R. H. and Kedes, L. (1981), *Cell*, **23**, 651.

74 Maniatis, T., Fritsch, E. F., Lauer, J. and Lawn, R. M. (1980), *Annu. Rev. Genet.*, **14**, 145.

75 Singer, R. H. and Penman, S. (1973), *J. Mol. Biol.*, **78**, 321.

76 Spradling, A., Hui, J. and Penman, S. (1975), *Cell*, **4**, 131.

77 Palmiter, R. D., Moore, P. B. and Mulvihill, E. R. (1976), *Cell*, **8**, 557.

78 Rosenfeld, G., Comstock, J. P., Means, A. R. and O'Malley, B. W. (1972), *Biochem. Biophys. Res. Commun.*, **47**, 387.

79 Williamson, R., Morrison, M., Lanyon, G., Eason, R. and Paul, J. (1971), *Biochemistry*, **10**, 3014.

80 Suzuki, Y. and Brown, D. D. (1972), *J. Mol. Biol.*, **63**, 409.

81 Williamson, R., Clayton, R. and Truman, D. E. S. (1972), *Biochem. Biophys. Res. Commun.*, **46**, 1936.

82 Swan, D., Aviv, H. and Leder, P. (1972), *Proc. Natl. Acad. Sci. USA*, **69**, 1967.

83 Greenberg, J. R. and Perry, R. P. (1971), *J. Cell Biol.*, **50**, 774.

84 Federoff, N., Welleuer, P. K. and Wall, R. (1977), *Cell*, **10**, 597.

85 Busch, H., Reddy, R., Rothblum, L. and Choi, Y. C. (1982), *Annu. Rev. Biochem.*, **51**, 617.

86 Reddy, R. (1985), *Nucleic Acids Res.*, **13**, r155.

87 Reddy, R., Henning, D. and Busch, H. (1985), *J. Biol. Chem.*, **260**, 10 930.

88 Shatkin, A. J. (1976), *Cell*, **9**, 645.

89 Shatkin, A. J., Darzynkiewicz, W., Furuichi, Y., Kroath, H., Morgan, M. A., Tahara, S. M. and Yamakawa, M. (1982), *Biochem. Soc. Symp.*, **47**, 129.

90 Baralle, F. E. (1983), *Int. Rev. Cytol.*, **81**, 71.

91 Faruichi, Y., Mulhukrishnan, S., Tomasz, J. and Shatkin, A. J. (1976), *J. Biol. Chem.*, **251**, 5043.

92 Moss, B., Gershowitz, A., Wei, C-M. and Boone, R. (1976), *Virology*, **72**, 341.

93 Venkatesan, S., Gershowitz, A. and Moss, B. (1980), *J. Biol. Chem.*, **255**, 2829.

94 Babich, A., Nevins, J. R. and Darnell, J. E. (1980), *Nature (London)*, **287**, 246.

95 Lai, E. C., Roop, D. R., Tsai, M. J., Woo, S. L.

and O'Malley, B. W. (1982), *Nucleic Acids Res.*, **10**, 5553.

96 Gertlinger, P., Krust, A., LeMeur, M., Perrin, F., Cochell, M., Gannon, F., Dupret, D. and Chambon, P. (1982), *J. Mol. Biol.*, **162**, 345.

97 Hamlyn, P. H., Gait, M. J. and Milstein, C. (1981), *Nucleic Acids Res.*, **9**, 4485.

98 Hagenbüchle, P., Tosi, M., Schibler, U., Bovey, R., Welleur, P. K. and Young, R. A. (1981), *Nature (London)*, **289**, 643.

99 Birnstiel, M. L., Busslinger, M. and Strub, K. (1985), *Cell*, **41**, 349.

100 Proudfoot, N. J. (1982), *Nature (London)*, **298**, 516.

101 Nevins, J. R. and Darnel, J. E. (1978), *Cell*, **15**, 1477.

102 Montell, C., Fisher, E. F., Caruthers, M. H. and Berk, A. J. (1983), *Nature (London)*, **305**, 600.

103 Higgs, D. R., Goodbourn, S. E. Y., Lamb, J., Clegg, J. B., Weatherall, D. J. and Proudfoot, N. J. (1983), *Nature (London)*, **306**, 398.

104 Fiers, W., Contrerus, R., Haegeman, G., Rogiers, R., Van de Voorde, A., Van Heuverswyn, H., Van Herreweyhe, J., Volckaert, G. and Ysebaert, M. (1978), *Nature (London)*, **273**, 113.

105 Perricaudet, M., Moullec, J-M., Tiolais, P. and Pettersson, U. (1980), *Nature (London)*, **288**, 174.

106 McDevitt, M. A., Imperiale, M. J., Ali, H. and Nevins, J. R. (1984), *Cell*, **37**, 993.

107 Gil, A. and Proudfoot, N. J. (1984), *Nature (London)*, **312**, 473.

108 Berget, S. M. (1984), *Nature (London)*, **309**, 179.

109 Hashimoto, C. and Steitz, J. A. (1984), *Nucleic Acids Res.*, **12**, 3283.

110 Rinke, J., Appel, B., Digweed, M. and Lührmann, R. (1985), *J. Mol. Biol.*, **185**, 721.

111 McLaughlan, J., Gaffney, D., Whitton, J. L. and Clements, J. B. (1985), *Nucleic Acids Res.*, **13**, 1347.

112 Moore, C. L. and Sharp, P. A. (1985), *Cell*, **41**, 845.

113 Fitzgerald, M. and Shenk, T. (1981), *Cell*, **24**, 251.

114 Moore, C. L. and Sharp, P. A. (1984), *Cell*, **36**, 581.

115 Waychick, R. P., Lyons, R. H., Post, L. and Rottman, F. M. (1984), *Proc. Natl. Acad. Sci. USA*, **81**, 3944.

116 Diez, J. and Brawerman, G. (1974), *Proc. Natl.*

*Acad. Sci. USA*, **71**, 4091.

117 Brawerman, G. and Diez, J. (1975), *Cell*, **5**, 271.

118 Katinakis, P. K., Slater, A. and Burdon, R. H. (1980), *FEBS Lett.*, **116**, 1.

119 Birchmeier, C., Schümperli, D., Sconzo, G. and Birnstiel, M. L. (1984), *Proc. Natl. Acad. Sci. USA*, **81**, 1057.

120 Krieg, P. A. and Melton, D. A. (1984), *Nature (London)*, **308**, 203.

121 Price, D. H. and Parker, C. S. (1984), *Cell*, **38**, 423.

122 Galli, G., Holfteller, H., Stannenberg, H. G. and Birnstiel, M. L. (1983), *Cell*, **34**, 823.

123 Strub, K., Galli, G., Busslinger, M. and Birnstiel, M. L. (1984), *EMBO J.*, **3**, 2801.

124 Georgiev, O. and Birnstiel, M. L. (1985), *EMBO J.*, **4**, 481.

125 Birchmeier, C., Folk, W. and Birnstiel, M. L. (1983), *Cell*, **35**, 433.

126 Yuo, C-Y., Ares, M. and Weiner, A. M. (1985), *Cell*, **42**, 193.

127 Turner, P. (1985), *Nature (London)*, **316**, 105.

128 McReynolds, L., O'Malley, B. W., Nisbet, A. D., Fothergill, J. E., Givol, D., Fields, S., Robertson, M. and Brownlee, G. G. (1978), *Nature (London)*, **273**, 723.

129 Tilghman, S. M., Curtis, P. J., Tiemeier, D. C., Leder, P. and Weissmann, C. (1978), *Proc. Natl. Acad. Sci. USA*, **75**, 1309.

130 Tsai, M., Ting, A., Nordstrom, J., Zimmer, W. and O'Malley, B. W. (1980), *Cell*, **22**, 219.

131 Herbert, M. G. and Wall, R. J. (1979), *J. Mol. Biol.*, **135**, 879.

132 Ryffel, G. U., Wyler, T., Muellenes, D. and Weber, R. (1980), *Cell*, **19**, 53.

133 Baird, M., Driscoll, C., Schreiner, H., Sciarratta, G. V., Sansone, G., Niazi, G., Ramirez, F. and Bank, A. (1981), *Proc. Natl. Acad. Sci. USA*, **78**, 4218.

134 Treisman, R., Proudfoot, N. J., Shander, M. and Maniatis, T. (1982), *Cell*, **29**, 903.

135 Busslinger, M., Maschonas, N. and Flavell, R. A. (1981), *Cell*, **27**, 289.

136 Weatherall, D. J. and Clegg, J. B. (1982), *Cell*, **29**, 7.

137 Mount, S. and Steitz, J. (1983), *Nature (London)*, **303**, 380.

138 Lerner, M. R., Boyle, J. A., Mount, S. M., Wolin, S. L. and Steitz, J. A. (1980), *Nature (London)*, **283**, 220.

139 Rogers, J. and Wall, R. (1980), *Proc. Natl. Acad. Sci. USA*, **77**, 1877.

140 Knowler, J. T. (1983), *Int. Rev. Cytol.*, **84**, 103.

141 Bringman, P., Rinke, J., Appel, B., Rueter, R. and Lührman, R. (1983), *EMBO J.*, **2**, 1129.

142 Goldenberg, C. J. and Raskus, H. J. (1981), *Proc. Natl. Acad. Sci. USA*, **78**, 5430.

143 Kole, R. and Weissman, S. M. (1982), *Nucleic Acids Res.*, **10**, 5429.

144 Goldenberg, C. J. and Hauser, S. D. (1983), *Nucleic Acids Res.*, **11**, 1337.

145 Padgett, R. A., Hardy, S. F. and Sharp, P. A. (1983), *Proc. Natl. Acad. Sci. USA*, **80**, 5230.

146 Di Maria, P. R., Kaltwasser, G. and Goldenberg, C. J. (1985), *J. Biol. Chem.*, **260**, 1096.

147 Krämer, A., Keller, W., Appel, B. and Luhrmann, R. (1984), *Cell*, **38**, 299.

148 Padgett, R. A., Mount, S. M., Steitz, J. A. and Sharp, P. A. (1983), *Cell*, **35**, 101.

149 Konarska, M. M., Padgett, R. A. and Sharp, P. A. (1984), *Cell*, **38**, 731.

150 Rautman, G., Matthes, H. W. D., Gait, M. J. and Breathnach, R. (1984), *EMBO J.*, **3**, 2021.

151 Wieringa, B., Hofer, E. and Weissman, C. (1984), *Cell*, **37**, 915.

152 Grabowski, J. T., Padgett, R. A. and Sharp, P. A. (1984), *Cell*, **37**, 415.

153 Konarska, M. M., Grabowski, P. J., Padgett, R. A. and Sharp, P. A. (1985), *Nature (London)*, **313**, 552.

154 Zeitlin, S. and Efstratiadis, A. (1984), *Cell*, **39**, 589.

155 Keller, W. (1984), *Cell*, **39**, 423.

156 Langford, C., Nellen, W., Niessing, J. and Gallwitz, D. (1983), *Proc. Natl. Acad. Sci. USA*, **80**, 1496.

157 Langford, C. J. and Gallwitz, D. (1983), *Cell*, **33**, 519.

158 Pikielny, C. W., Teem, J. L. and Rosbash, M. (1983), *Cell*, **34**, 395.

159 Langford, C. J., Klinz, F. J., Donath, C. and Gallwitz, D. (1984), *Cell*, **36**, 645.

160 Rodriguez, J. R., Pikielny, C. W. and Rosbash, M. (1984), *Cell*, **39**, 603.

161 Domedey, H., Apostol, B., Lin, R-J., Newman, A., Brody, E. and Abelson, J. (1984), *Cell*, **39**, 611.

162 Boel, E., Hansen, M. T., Hjort, I., Hoegh, I. and Fiil, N. P. (1984), *EMBO J.*, **3**, 1581.

163 Lund, E., Dahlberg, J. E., Lindahl, L., Jaskunas, S. R., Dennis, P. P. and Nomura, M. (1976), *Cell*, **7**, 165.

164 Lund, E. and Dahlberg, J. E. (1977), *Cell*, **11**, 247.
165 Young, R. A., Macklis, R. and Steitz, J. A. (1979), *J. Biol. Chem.*, **254**, 3264.
166 Brosius, J., Dull, T. J., Sleeter, D. D. and Noller, H. F. (1981), *J. Mol. Biol.*, **148**, 107.
167 deBoer, H. A., Gilbert, S. F. and Nomura, M. (1979), *Cell*, **17**, 201.
168 Young, R. A. and Steitz, J. A. (1979), *Cell*, **17**, 225.
169 Csordas-Toth, E., Boros, I. and Venetianer, P. (1979), *Nucleic Acids Res.*, **7**, 2189.
170 Shen, W-F., Squires, C. and Squires, C. L. (1982), *Nucleic Acids Res.*, **10**, 3303.
171 Sekiya, T., Mori, M., Takahashi, N. and Nishimura, S. (1980), *Nucleic Acids Res.*, **8**, 3809.
172 Duester, G. L. and Holmes, W. M. (1980), *Nucleic Acids Res.*, **8**, 3793.
173 Liebke, H. and Hatfall, G. (1985), *Nucleic Acids Res.*, **13**, 5515.
174 Gilbert, S. F., De Boer, H. A. and Nomura, M. (1979), *Cell*, **17**, 211.
175 Ryals, J., Little, R. and Bremer, H. (1982), *J. Bacteriol.*, **151**, 1261.
176 Gausing, K. (1982), *Trends Biochem. Sci.*, **7**, 65.
177 Nomura, M., Gourse, R. and Baughman, G. (1984), *Annu. Rev. Biochem.*, **53**, 75.
178 Dunn, J. J. (1982), in *The Enzymes* (ed. P. D. Boyer), Academic Press, New York, Vol. 15, p. 485.
179 Robertson, H. D. (1982), *Cell*, **30**, 669.
180 Young, R. A. and Steitz, J. A. (1978), *Proc. Natl. Acad. Sci. USA*, **75**, 3593.
181 Bram, R. J., Young, R. A. and Steitz, J. A. (1980), *Cell*, **19**, 393.
182 King, T. C., Sirdeshmukh, R. and Schlessinger, D. (1984), *Proc. Natl. Acad. Sci. USA*, **81**, 185.
183 Dunn, J. J. and Studier, F. W. (1983), *J. Mol. Biol.*, **166**, 477.
184 Schweitz, H. and Ebel, J-P. (1971), *Biochimie*, **53**, 585.
185 Stark, M., Gourse, R. L., Jemiolo, D. K. and Dahlberg, A. E. (1985), *J. Mol. Biol.*, **182**, 205.
186 Dahlberg, A. E., Dahlberg, J. E., Lund, E. Tokimatsu, H., Robson, A. B., Calvert, P. C., Reynolds, F. and Zahalak, M. (1978), *Proc. Natl. Acad. Sci. USA*, **75**, 3598.
187 Roy, M. K. and Apirion, D. (1983), *Biochim. Biophys. Acta*, **747**, 200.
188 Roy, M. K., Singh, B., Roy, B. K. and Apirion, D. (1983), *Eur. J. Biochem.*, **131**, 119.
189 Ghora, B. K. and Apirion, D. (1978), *Cell*, **15**, 1055.
190 Singh, B. and Apirion, D. (1982), *Biochim. Biophys. Acta*, **698**, 252.
191 Pieler, T. and Erdmann, V. A. (1982), *Proc. Natl. Acad. Sci. USA.*, **79**, 4599.
192 Tomcsányi, T. and Apirion, D. (1985), *J. Mol. Biol.*, **185**, 713.
193 Sogin, M. L., Pace, B. and Pace, N. R. (1977), *J. Biol. Chem.*, **252**, 1350.
194 Stahl, D. A., Pace, B., Marsh, T. and Pace, N. R. (1984), *J. Biol. Chem.*, **259**, 11 448.
195 Pace, B., Stahl, D. A. and Pace, N. R. (1984), *J. Biol. Chem.*, **259**, 11 454.
196 Busch, H. and Rothblum, L. (eds) (1982), *The Cell Nucleus*, Vols 10, 11 and 12, Academic Press, New York and London.
197 Walker, T. H. and Pace, N. R. (1983), *Cell*, **33**, 320.
198 Wallace, H. and Birnstiel, M. L. (1966), *Biochim. Biophys. Acta*, **114**, 296.
199 Marrakechi, M. (1974), *Mol. Gen. Genet.*, **135**, 213.
200 Ritossa, F. (1976), in *The Genetics and Biology of Drosophila* (ed. M. Ashburner and F. Novilski), Academic Press, New York, Vol. 1B, p. 801.
201 Long, E. D. and Dawid, T. B. (1980), *Annu. Rev. Biochem.*, **49**, 727.
202 Hadjiolov, A. A., in *Subcellular Biochemistry* (1980), (ed. D. B. Roodyn), Plenum Press, New York, Vol. 7, p. 1.
203 Mandal, R. K. (1984), *Prog. Nucleic Acid Res. Mol. Biol.*, **31**, 115.
204 Henderson, A. S., Warburton, D. and Atwood, K. C. (1972), *Proc. Natl. Acad. Sci. USA*, **69**, 3394.
205 Miller, O. L. and Beatty, B. R. (1969), *Genetics*, **61** (*Suppl.*), 133.
206 Yao, M-C. and Gall, J. G. (1977), *Cell*, **12**, 121.
207 Yao, M-C., Blackburn, E. and Gall, J. G. (1978), *Cold Spring Harbor Symp. Quant. Biol.*, **43**, 1293.
208 Yao, M-C., Blackburn, E. and Gall, J. (1981), *J. Cell Biol.*, **90**, 515.
209 Klootwijk, I., de Jonge, P. and Planta, R. J. (1979), *Nucleic Acids Res.*, **6**, 27.
210 Wellauer, P. K. and Dawid, I. B. (1973), *Proc. Natl. Acad. Sci. USA*, **70**, 2827.
211 Philippsen, P., Thomas, M., Kramer, R. A. and Davis, R. W. (1978), *J. Mol. Biol.*, **123**, 387.
212 Meunier-Rotival, M., Cortadas, I., Macaya, G.

and Bernardi, G. (1979), *Nucleic Acids Res.*, **6**, 2109.

213 Kominami, R., Urano, Y., Mishima, Y. and Maramatsu, M. (1981), *Nucleic Acids Res.*, **9**, 3219.

214 Wellauer, P. K. and Dawid, I. B. (1979), *J. Mol. Biol.*, **128**, 289.

215 Maizels, N. (1976), *Cell*, **9**, 431.

216 Jordan, B. R. and Glover, D. M. (1977), *FEBS Lett.*, **78**, 271.

217 Glover, D. M. (1981), *Cell*, **26**, 297.

218 Lecanidou, R., Eickbush, T. H. and Kafatos, F. C. (1984), *Nucleic Acids Res.*, **12**, 4703.

219 Beckingham, K. (1982), in *The Cell Nucleus* (ed. H. Busch and L. Rothblum), Academic Press, New York, Vol. 10A, p. 205.

220 Wellauer, P. K. and Dawid, I. B. (1977), *Cell*, **10**, 193.

221 Karrer, K. M. and Gall, J. G. (1976), *J. Mol. Biol.*, **104**, 421.

222 Engberg, J., Andersson, P. and Leick, V. (1976), *J. Mol. Biol.*, **104**, 455.

223 Nelles, L., Fang, B-L., Volckaert, G., Vandenberghe, A. and de Wachter, R. (1984), *Nucleic Acids Res.*, **12**, 8749.

224 Subrahmanyam, C. S., Cassidy, B., Busch, H. and Rothblum, L. I. (1982), *Nucleic Acids Res.*, **10**, 3667.

225 Treco, D., Brownell, D. and Arnheim, N. (1982), in *The Cell Nucleus* (ed. H. Busch and L. Rothblum), Academic Press, New York and London, Vol. 12, p. 106.

226 Reeder, R. H. (1984), *Cell*, **38**, 349.

227 Bosley, P., Moss, T., Mächler, M., Portmann, R. and Birnstiel, M. (1979), *Cell*, **17**, 19.

228 Moss, T. (1983), *Nature (London)*, **302**, 223.

229 Moss, T. (1982), *Cell*, **30**, 835.

230 Labhart, P. and Reeder, R. H. (1984), *Cell*, **37**, 285.

231 Simeone, A., La Volpe, A. and Boncinelli, E. (1985), *Nucleic Acids Res.*, **13**, 1089.

232 Kuehn, M. and Arnheim, N. (1983), *Nucleic Acids Res.*, **11**, 211.

233 Mroczka, D. L., Cassidy, B., Busch, H. and Rothblum, L. I. (1984), *J. Mol. Biol.*, **174**, 141.

234 Miesfeld, R. and Arnheim, N. (1982), *Nucleic Acids Res.*, **10**, 3933.

235 Miller, J. R. (1983), in *Eukaryotic Genes: Their Structure, Activity and Regulation* (ed. N. Maclean, S. P. Gregory and R. A. Flavell), Butterworths, London, p. 225.

236 Selker, E. U., Yanofsky, C., Driftmier, K., Metzenberg, R. L., Alzner-DeWeerd, B. and Raj Bhandary, U. L. (1981), *Cell*, **24**, 819.

237 Mao, J., Appel, B., Schaack, J., Sharp, S., Yamada, H. and Söll, D. (1982), *Nucleic Acids Res.*, **10**, 487.

238 Rafalski, J. A., Wiewiorowski, M. and Söll, D. (1982), *Nucleic Acids Res.*, **10**, 7635.

239 Goldbrough, P. B., Ellis, T. and Lomonossoff, G. P. (1982), *Nucleic Acids Res.*, **10**, 4501.

240 Hershey, N. D., Conrad, S. E., Sodja, A., Yen, P. H., Cohen, M. and Davidson, N. (1977), *Cell*, **11**, 585.

241 Procunier, J. D. and Dunn, R. J. (1978), *Cell*, **15**, 1087.

242 Artavanis-Tsakonas, S., Schedl, P., Tschudi, C., Pirrotta, V., Steward, R. and Gehring, W. J. (1977), *Cell*, **12**, 1057.

243 Emerson, B. M. and Roeder, R. G. (1984), *J. Biol. Chem.*, **259**, 7916.

244 Pardue, M. L., Brown, D. D. and Birnstiel, M. L. (1973), *Chromosoma*, **42**, 191.

245 Tschudi, C. and Pirrotta, V. (1980), *Nucleic Acids Res.*, **8**, 441.

246 Ford, P. J. and Southern, E. M. (1973), *Nature (London) New Biol.*, **241**, 7.

247 Ford, P. J. and Brown, R. D. (1976), *Cell*, **8**, 485.

248 Brown, D. D., Wensink, P. C. and Jordan, E. (1971), *Proc. Natl. Acad. Sci. USA*, **68**, 3175.

249 Jacq, C., Miller, J. R. and Brownlee, G. G. (1977), *Cell*, **12**, 109.

250 Hart, R. P. and Folk, W. R. (1982), *J. Biol. Chem.*, **257**, 11 706.

251 Krol, A., Gallinaro, H., Lazar, E., Jacob, M. and Branlant, C. (1981), *Nucleic Acids Res.*, **9**, 769.

252 Birkenmeier, E. H., Brown, D. D. and Jordon, E. (1978), *Cell*, **15**, 1077.

253 Miller, O. L. and Beatty, B. R. (1969), *Science*, **164**, 955.

254 Maden, B. E. H., Salim, M. and Summers, D. F. (1972), *Nature (London) New Biol.*, **237**, 5.

255 Wellauer, P. K. and Dawid, I. B. (1973), *Proc. Natl. Acad. Sci. USA*, **70**, 2827.

256 Hadjiolov, A. A. and Nikolaev, N. (1976), *Prog. Biophys. Mol. Biol.*, **31**, 95.

257 Denoya, C., Costa Giomi, P., Scodeller, E. A., Vasquez, C. and La Torre, J. L. (1981), *Eur. J. Biochem.*, **115**, 375.

258 Prestayko, A. W., Tonato, M. and Busch, H. (1970), *J. Mol. Biol.*, **47**, 505.

259 Wise, J. A. and Weiner, A. M. (1981), *J. Biol.*

*Chem.*, **256**, 956.

260 Crouch, R. J., Kanayu, S. and Earl, P. L. (1983), *Mol. Biol. Rep.*, **9**, 75.

261 Bachellerie, J-P., Michot, B. and Raynal, F. (1983), *Mol. Biol. Rep.*, **9**, 79.

262 Cech, T. R., Zaug, A. J. and Grabowski, P. J. (1981), *Cell*, **27**, 487.

263 Bass, B. L. and Cech, T. R. (1984), *Nature (London)*, **308**, 820.

264 Zaug, A. J. and Cech, T. R. (1982), *Nucleic Acids Res.*, **10**, 2823.

265 Zaug, A. J., Grabowski, P. J. and Cech, T. R. (1983), *Nature (London)*, **301**, 578.

266 Brehm, S. L. and Cech, T. R. (1983), *Biochemistry*, **22**, 2390.

267 Price, J. V., Kieft, G. L., Kent, J. R., Sievers, E. L. and Cech, T. R. (1985), *Nucleic Acids Res.*, **13**, 1871.

268 Waring, R. B., Ray, J. A., Edwards, S. W., Scazzocchio, C. and Davies, R. W. (1985), *Cell*, **40**, 371.

269 Waring, R. B., Scazzocchio, C., Brown, T. A. and Davies, R. W. (1983), *J. Mol. Biol.*, **167**, 595.

270 Davies, R. W., Waring, R. B., Ray, J. A., Brown, T. A. and Scazzocchio, C. (1982), *Nature (London)*, **300**, 719.

271 Garriga, G. and Lambowitz, A. M. (1984), *Cell*, **39**, 631.

272 Tabuk, H. F., Van der Horst, G., Osinya, K. A. and Arnberg, A. C. (1984), *Cell*, **39**, 623.

273 Bertrand, H., Bridge, P., Collins, R. A., Garriga, G. and Lambowitz, A. M. (1982), *Cell*, **29**, 517.

274 Smith, J. D. and Dunn, D. B. (1959), *Biochim. Biophys. Acta*, **31**, 573.

275 Jeanteur, P., Amaldi, F. and Attardi, G. (1968), *J. Mol. Biol.*, **33**, 757.

276 Thomas, G., Gordon, J. and Rogg, H. (1978), *J. Biol. Chem.*, **253**, 1101.

277 Brand, R. C., Klootwijk, J., Planta, R. J. and Maden, B. E. H. (1978), *Biochem. J.*, **169**, 71.

278 Helser, T. L., Davies, J. E. and Dahlberg, J. A. (1971), *Nature (London) New Biol.*, **233**, 12.

279 Maden, B. E. H. and Salim, M. (1974), *J. Mol. Biol.*, **88**, 133.

280 Khan, S. N., Salim, M. and Maden, B. E. H. (1978), *Biochem. J.*, **169**, 531.

281 Hughs, D. G. and Maden, B. E. H. (1978), *Biochem. J.*, **171**, 781.

282 Maden, B. E. H. (1980), *Nature (London)*, **288**, 293.

283 Salim, M. and Maden, B. E. H. (1980), *Nucleic Acids Res.*, **8**, 2871.

284 Almadja, J., Brimacombe, R. and Maden, B. E. H. (1984), *Nucleic Acids Res.*, **12**, 2649.

285 Isono, K. (1980), in *Ribosomes, Structure, Function and Genetics* (ed. G. Chambliss, G. R. Craven, J. Davies, L. Kahan and M. Nomura), University Park Press, Baltimore, p. 641.

286 Nomura, M. and Post, L. E. (1980), in *Ribosomes, Structure, Function and Genetics* (ed. G. Chambliss, G. R. Craven, J. Davies, L. Kahan and M. Nomura), University Park Press, Baltimore, p. 671.

287 Nomura, M., Morgan, E. A. and Jaskunas, S. P. (1977), *Annu. Rev. Genet.*, **11**, 297.

288 Woolford, J. C., Hereford, L. M. and Rosbash, M. (1979), *Cell*, **18**, 1247.

289 Fried, H. M., Pearson, N. J., Kim, C. H. and Warner, J. R. (1981), *J. Biol. Chem.*, **256**, 10 176.

290 D'Eustachio, P., Meyuhas, O., Ruddle, F. and Perry, R. O. (1981), *Cell*, **24**, 307.

291 Bozzoni, I., Beccari, E., Xun Luo, Z., Amaldi, F., Pierandrei-Amaldi, P. and Campioni, N. (1981), *Nucleic Acids Res.*, **9**, 1069.

292 Pikielny, C. W., Teem, J. L. and Rosbash, M. (1983), *Cell*, **34**, 395.

293 Maden, B. E. H. (1971), *Prog. Biophys. Mol. Biol.*, **22**, 127.

294 Todorov, I. T., Noll, F. and Hadjiolov, A. A. (1983), *Eur. J. Biochem.*, **131**, 271.

295 Planta, R. J. and Mager, W. H. (1982), in *The Cell Nucleus* (ed. H. Busch and L. Rothblum), Academic Press, New York and London, Vol. XII, p. 213.

296 Auger-Buendia, M. A. and Longuet, M. (1978), *Eur. J. Biochem.*, **85**, 105.

297 Fujisawa, T., Imai, K., Tanaka, Y. and Ogata, K. (1979), *J. Biochem. (Tokyo)*, **85**, 277.

298 Lastick, S. M. (1980), *Eur. J. Biochem.*, **113**, 175.

299 Fournier, M. J. and Ozeki, H. (1985), *Microbiol. Rev.*, **49**, 379.

300 Vold, B. S. (1985), *Microbiol. Rev.*, **49**, 71.

301 Bossi, L. (1983), *Mol. Gen. Genet.*, **192**, 163.

302 Hsu, L. M., Klee, H. J., Zagorski, J. and Fournier, M. J. (1984), *J. Bacteriol.*, **148**, 934.

303 Clarkson, S. G. (1983), in *Eukaryotic Genes: their Structure, Activity and Regulation* (ed. N. McLean, S. P. Gregory and R. A. Flavell), Butterworth Press, London, p. 239.

304 Olson, M. V., Montgomery, D. L., Hopper, A. K., Page, G. S., Horodyski, F. and Hall, B. D. (1977), *Nature (London)*, **267**, 639.

305 Valenzuela, P., Venegas, A., Weinberg, F., Bishop, R. and Rutter, R. J. (1978), *Proc. Natl. Acad. Sci. USA*, **75**, 190.

306 Yen, P. H. and Davidson, N. (1980), *Cell*, **22**, 137.

307 Müller, F. and Clarkson, S. G. (1980), *Cell*, **19**, 345.

308 Rosen, A., Sarid, S. and Daniel, V. (1984), *Nucleic Acids Res.*, **12**, 4893.

309 Schmidt, O., Mao, J., Ogden, R., Beckman, J., Sakano, H., Abelson, J. and Söll, D. (1980), *Nature (London)*, **287**, 750.

310 Kjellin-Straby, K., Engelke, D. R. and Abelson, J. (1983), *Int. tRNA Workshop*, Hakone, Japan.

311 Altman, S. (1981), *Cell*, **23**, 3.

312 Kole, R. and Altman, S. (1982), in *The Enzymes* (ed. P. D. Boyer), Academic Press, New York, Vol. 15, p. 470.

313 Deutscher, M. P. (1984), *CRC Crit. Rev. Biochem.*, **17**, 45.

314 Stark, B. C., Kole, R., Bowman, E. J. and Altman, S. (1978), *Proc. Natl. Acad. Sci. USA*, **75**, 3717.

315 Guthrie, C. and Atchison, R. (1980), in, *tRNA, Biological Aspects* (ed. D. Söll, J. Abelson and P. Schimmel), Cold Spring Harbor Press, New York, p. 83.

316 Reed, R. E., Baer, M. F., Guerrier-Takada, C., Donis-Keller, H. and Altman, S. (1982), *Cell*, **30**, 627.

317 Guerrier-Takada, C., Gardiner, K., March, T., Pace, N. and Altman, S. (1983), *Cell*, **35**, 849.

318 Robertson, H. D., Altman, S. and Smith, J. D. (1972), *J. Biol. Chem.*, **247**, 5243.

319 Watson, N. and Apirion, D. (1981), *Biochem. Biophys. Res. Commun.*, **103**, 543.

320 Gurevitz, M., Watson, N. and Apirion, D. (1982), *Eur. J. Biochem.*, **124**, 553.

321 Deutscher, M. P., Marlor, C. W. and Zaniewski, R. (1984), *Proc. Natl. Acad. Sci. USA*, **81**, 4290.

322 Melton, D. A., DeRobertis, E. M. and Cortese, R. (1980), *Nature (London)*, **284**, 143.

323 Harada, F., Matsubara, M. and Kato, N. (1984), *Nucleic Acids Res.*, **12**, 9263.

324 Kline, L., Nishikawa, S. and Söll, D. (1981), *J. Biol. Chem.*, **256**, 5058.

325 Bowman, E. J. and Altman, S. (1980), *Biochim. Biophys. Acta*, **613**, 439.

326 Koski, R. A., Bothwell, A. and Altman, S. (1976), *Cell*, **9**, 101.

327 Garber, R. L. and Gage, L. P. (1979), *Cell*, **18**, 817.

328 Hagenbüchle, O., Larsen, D., Hall, G. I. and Sprague, K. U. (1979), *Cell*, **18**, 1217.

329 Green, C. J. and Vold, B. S. (1983), *Nucleic Acids Res.*, **11**, 5763.

330 Wawrousek, E. F. and Hansen, J. N. (1983), *J. Biol. Chem.*, **258**, 291.

331 Deutscher, M. P. (1982), in *The Enzymes* (ed. P. D. Boyer), Academic Press, New York, Vol. 15, p. 183.

332 Deutscher, M. P., Lin, J. C. and Evans, J. A. (1977), *J. Mol. Biol.*, **117**, 1081.

333 Ogden, R. C., Knapp, G., Peebles, C. L., Johnson, J. and Abelson, J. (1981), *Trends Biochem. Sci.*, **6**, 154.

334 Müller, F. and Clarkson, S. G. (1980), *Cell*, **19**, 345.

335 Robinson, R. R. and Davidson, N. (1981), *Cell*, **23**, 251.

336 Kaine, B. P., Gupta, R. and Woese, C. R. (1983), *Proc. Natl. Acad. Sci. USA*, **80**, 3309.

337 Ogden, R. C., Lee, M-C. and Knapp, G. (1984), *Nucleic Acids Res.*, **12**, 9367.

338 Hopper, A. K., Banks, F. and Evangelidis, V. (1978), *Cell*, **14**, 211.

339 O'Farrell, P. Z., Cordell, B., Valezuela, P., Rutter, W. J. and Goodman, H. M. (1978), *Nature (London)*, **274**, 438.

340 Knapp, G., Ogden, R. C., Peebles, C. L. and Abelson, J. (1979), *Cell*, **18**, 37.

341 Ogden, R. C., Beckmann, J. S., Abelson, J., Kang, H. S., Söll, D. and Schmidt, O. (1979), *Cell*, **17**, 399.

342 Filipowicz, W. and Shatkin, A. J. (1983), *Cell*, **32**, 547.

343 Peebles, C. L., Ogden, R. C., Knapp, G. and Abelson, J. (1979), *Cell*, **18**, 27.

344 Otsuka, A., dePaolis, A. and Tocchini-Valentini, G. P. (1981), *Mol. Cell Biol.*, **1**, 269.

345 Peebles, C. L., Gegenheimer, P. and Abelson, J. (1983), *Cell*, **32**, 525.

346 Greer, C., Peebles, C. L., Gegenheimer, P. and Abelson, J. (1983), *Cell*, **32**, 537.

347 Filipowicz, W., Konarska, M., Gross, H. J. and Shatkin, A. J. (1983), *Nucleic Acids Res.*, **11**, 1405.

348 Uhlenbeck, O. C. and Gumport, R. I. (1981), in *The Enzymes* (ed. P. D. Boyer), Academic Press, New York, Vol. 15, p. 31.

349 Perkins, K. K., Furneaux, H. and Hurwitz, J. (1985), *Proc. Natl. Acad. Sci. USA*, **82**, 684.

350 Sprinzl, M. and Gauss, D. H. (1982), *Nucleic Acids Res.*, **10**, r1.

351 Söll, D. and Kline, L. K. (1982), in *The Enzymes* (ed. P. D. Boyer), Academic Press, New York, Vol. 15, p. 557.

352 Kline, L. K. and Söll, D. (1982), in *The Enzymes* (ed. P. D. Boyer), Academic Press, New York, Vol. 15, p. 567.

353 Kersten, H. (1984), *Prog. Nucleic Acids Res. Mol. Biol.*, **31**, 59.

354 Johnson, P. F. and Abelson, J. (1983), *Nature (London)*, **302**, 681.

355 Aschhoff, H. J., Elton, H., Arnold, H. H., Mahal, G., Kersten, W. and Kersten, H. (1976), *Nucleic Acids Res.*, **3**, 3109.

356 Izzo, P. and Gantt, R. (1977), *Biochemistry*, **16**, 3576.

357 Glick, J. M. and Leboy, P. S. (1977), *J. Biol. Chem.*, **252**, 4790.

358 Keith, J. M., Winters, E. M. and Moss, B. (1980), *J. Biol. Chem.*, **255**, 4636.

359 Delk, A. S., Nagle, D. P. and Rabinowitz, J. C. (1980), *J. Biol. Chem.*, **255**, 4387.

360 Taya, Y. and Nishimura, S. (1973), *Biophys. Biochem. Res. Commun.*, **51**, 1062.

361 Arena, F., Ciliberto, G., Ciampi, S. and Cortese, R. (1978), *Nucleic Acids Res.*, **5**, 4523.

362 Abrell, J. W., Kaufman, E. E. and Lipsell, M. N. (1971), *J. Biol. Chem.*, **246**, 294.

363 Attardi, G. (1981), *Trends Biochem. Sci.*, **6**, 86.

364 Borst, P., Tabak, H. F. and Grivell, L. A. (1983), in *Eukaryotic Genes: Their Structure, Activity and Regulation* (ed. N. McLean, S. P. Gregory and R. A. Flavell), Butterworths, Cambridge, p. 71.

365 Tabak, H. F., Grivell, L. A. and Borst, P. (1983), *CRC Crit. Rev. Biochem.*, **14**, 297.

366 Grivell, L. A. (1983), *Sci. Am.*, **248(3)**, 60.

367 Clayton, D. A. (1984), *Annu. Rev. Biochem.*, **53**, 573.

368 Bedbrook, J. R. and Kolodner, R. (1979), *Annu. Rev. Plant Physiol.*, **30**, 593.

369 Canatore, P. and Attardi, G. (1980), *Nucleic Acids Res.*, **8**, 2605.

370 Ojala, D., Merkel, C., Gelford, R. and Attardi, G. (1980), *Cell*, **22**, 393.

371 Yaginuma, K., Kobayashi, M., Taira, M. and Koike, K. (1982), *Nucleic Acids Res.*, **10**, 7531.

372 Walberg, M. W. and Clayton, D. A. (1983), *J. Biol. Chem.*, **258**, 1268.

373 Chang, D. D. and Clayton, D. A. (1984), *Cell*, **36**, 635.

374 Bogenhagen, D. F., Applegate, E. F. and Yoza, B. K. (1984), *Cell*, **36**, 1105.

375 Miller, D. L. and Martin, N. C. (1983), *Cell*, **34**, 911.

376 Locker, J. and Rabinowitz, M. (1981), *Plasmid*, **6**, 302.

377 Thalenfield, B. E., Hill, J. and Tzagoloff, A. (1983), *J. Biol. Chem.*, **258**, 610.

378 Douglas, M. and Takeda, M. (1985), *Trends Biochem. Sci.*, **10**, 192.

379 Grohmann, K., Amalric, F., Crews, S. and Attardi, G. (1978), *Nucleic Acids Res.*, **5**, 637.

380 Imbault, P., Colas, B., Sarantoglou, V., Boulanger, Y. and Weil, J-H. (1981), *Biochemistry*, **20**, 5855.

381 Bartsch, M., Kimura, M. and Subrumanian, A-R. (1982), *Proc. Natl. Acad. Sci. USA*, **79**, 6871.

382 Grant, D. M. and Lambowitz, A. M. (1982), in *The Cell Nucleus*, (ed. H. Busch and L. Rothblum), Academic Press, New York, Vol. 10, p. 387.

383 Van Etten, R. A., Walberg, M. W. and Clayton, D. A. (1980), *Cell*, **22**, 157.

384 Borst, P. and Grivell, L. A. (1978), *Cell*, **15**, 705.

385 Levens, D., Ticho, B., Ackerman, E. and Rabinowitz, M. (1981), *J. Biol. Chem.*, **256**, 5226.

386 Köchel, H. and Kuntzel, H. (1981), *Nucleic Acids Res.*, **9**, 5689.

387 Seithamer, J. J. and Cummings, D. J. (1981), *Nucleic Acids Res.*, **9**, 6391.

388 Tabak, H. F., van der Laan, J., Osinga, K. A., Schouten, J. P., van Boom, J. N. and Veeneman, G. H. (1981), *Nucleic Acids Res.*, **9**, 4475.

389 Bedbrook, J. R. and Kolodner, R. (1979), *Annu. Rev. Plant Physiol.*, **30**, 593.

390 Rochaix, J. D. and Darlix, J. L. (1982), *J. Mol. Biol.*, **159**, 383.

391 Palmer, J. D. and Thompson, W. F. (1982), *Cell*, **29**, 537.

392 Kossel, H., Edwards, K., Fritzche, E., Koch, W. and Schwarz, Z. (1983), in *Proteins and Nucleic Acids in Plant Systematics* (ed. U. Jensen and D. F. Fairbrothers), Springer Verlag, Berlin/Heidelberg, p. 36.

393 Slonimski, P. P., Borst, P. and Attardi, G. (eds) (1982), *Mitochondrial Genes*, Cold Spring Harbor Laboratory, Cold Spring Harbor, New York.

394 Hallick, R. B., Greenberg, B. H., Gruissem, W., Hollingsworth, M. J. and Karabin, M. J. (1983), in *Structure and Function of Plant Genomes* (ed. D. Ciferri and L. Dure), Plenum Press, New York, p. 155.

395 Bohnert, H-J., Crouse, E. J. and Schmitt, M. (1982), in *Encyclopedia of Plant Physiology* (ed. B. Pathier and D. Boulter), Springer-Verlag, New York, Vol. 14B, p. 475.

396 Whitfield, P. R. and Bottomley, W. (1983), *Annu. Rev. Plant Physiol.*, **34**, 279.

397 Karabin, G. D. and Hallick, R. B. (1983), *J. Biol. Chem.*, **258**, 5512.

398 Gruissem, W., Greenberg, B. M., Zurawski, G., Prescott, D. M. and Hallick, R. B. (1983), *Cell*, **35**, 815.

399 Koch, W., Edwards, K. and Kössel, H. (1981), *Cell*, **25**, 203.

400 Mahler, H. R. (1983), *Int. Rev. Cytol.*, **82**, 1.

401 Lazowska, J., Jacq, C. and Slonimski, P. P. (1980), *Cell*, **22**, 333.

402 Jacq, C., Banroques, J., Becam, A. M., Slonimski, P. P., Guise, N. and Danchin, A. (1984), *EMBO J.*, **3**, 1567.

403 Guise, N., Dreyfas, M., Siffert, O., Danchin, A., Spyridakis, A., Gargouri, A., Claisse, M. and Slonimski, P. P. (1984), *EMBO J.*, **3**, 1769.

404 Rochaix, J. D., Rahire, M. and Michel, F. (1985), *Nucleic Acids Res.*, **13**, 975.

405 De La Salle, H., Jacq, C. and Slonimski, P. P. (1982), *Cell*, **28**, 721.

406 Netter, P., Jacq, C., Carignani, G. and Slonimski, P. P. (1982), *Cell*, **28**, 733.

407 Waring, R. B., Davies, R. W., Scazzocchio, C. and Brown, T. A. (1982), *Proc. Natl. Acad. Sci. USA*, **79**, 6332.

408 Pillar, T., Lang, B. F., Steinberger, I., Vogt, B. and Kaudewitz, F. (1983), *J. Biol. Chem.*, **258**, 7954.

409 McGraw, P. and Tzagoloff, A. (1983), *J. Biol. Chem.*, **258**, 9459.

410 Simon, M. and Faye, G. (1984), *Proc. Natl. Acad. Sci. USA*, **81**, 8.

411 van der Horst, G. and Tabak, H. F. (1985), *Cell*, **40**, 759.

412 Michel, F. and Dujon, B. (1983), *EMBO J.*, **2**, 33.

413 Davies, R. W. (1984), *Biosci. Rep.*, **4**, 707.

414 Schmelzer, C., Schmidt, C., May, K. and Schwegen, R. J. (1983), *EMBO J.*, **3**, 2047.

415 Carignani, G., Groudinsky, O., Frezza, D., Schiavon, E., Bergantio, E. and Slonimski, P. (1983), *Cell*, **35**, 733.

416 Michel, F. and Lang, B. F. (1985), *Nature (London)*, **316**, 641.

417 Chu, F. K., Maley, G. F., Maley, F. and Belfort, M. (1984), *Proc. Natl. Acad. Sci. USA*, **81**, 3049.

418 Belfort, M., Pederson-Lane, J., West, D., Ehrenman, K., Maley, G., Chu, F. and Maley, F. (1985), *Cell*, **41**, 375.

419 Jacquier, A. and Dujon, B. (1985), *Cell*, **41**, 383.

420 Macreadie, I. G., Scott, R. M., Zinn, A. R. and Butow, R. A. (1985), *Cell*, **41**, 395.

421 Keller, E. B. and Noon, W. A. (1985), *Nucleic Acids Res.*, **13**, 4971.

422 Reed, R. and Maniatis, T. (1985), *Cell*, **41**, 95.

423 Ruskin, B., Greene, J. M. and Green, M. R. (1985), *Cell*, **41**, 833.

424 Ruskin, B. and Green, M. R. (1985), *Nature (London)*, **317**, 732.

425 Grabowski, P. J., Seiler, S. R. and Sharp, P. A. (1985), *Cell*, **42**, 345.

426 Frendewey, D. and Keller, W. (1985), *Cell*, **42**, 355.

427 Brody, E. and Abelson, J. (1985), *Science*, **228**, 963.

428 Krainer, A. R. and Maniatis, T. (1985), *Cell*, **42**, 725.

429 Black, D. L., Chabot, B. and Steitz, J. A. (1985), *Cell*, **42**, 737.

430 Ruskin, B. and Green, M. R. (1985), *Science*, **229**, 135.

431 Rymond, B. C. and Rosbash, M. (1985), *Nature (London)*, **317**, 735.

432 Erdmann, V. A., Wolters, J., Huysmans, E. and DeWachter, R. (1985), *Nucleic Acids Res.*, **13**, r105.

433 Sullivan, F. X. and Cech, T. R. (1985), *Cell*, **42**, 639.

434 Guerrier-Takada, C. and Altman, S. (1984), *Science*, **223**, 285.

435 Marsh, T. L. and Pace, N. R. (1985), *Science*, **229**, 79.

436 Castaño, J. G., Tobian, J. A. and Zasloff, M. (1985), *J. Biol. Chem.*, **260**, 9002.

437 Mount, S. M. (1982), *Nucleic Acids Res.*, **10**, 459.

438 Wells, D. and Kedes, L. (1985), *Proc. Natl. Acad. Sci. USA*, **82**, 2834.

# 10

# Control of transcription and mRNA processing

An organism must be able to modulate the expression of its genes. Some proteins, such as the enzymes of key metabolic pathways and the ribosomal proteins, will be required in large quantities. Thus, in an *E. coli* cell it has been calculated that there are 20 000 ribosomes so it follows that there will be at least 20 000 copies of each of their constituent proteins. Conversely, some *E. coli* proteins, such as $\beta$-galactosidase, are normally present in very low concentrations (approximately 5 molecules per cell) but, under circumstances to be described in the following sections, can be synthesized in much greater quantities. Clearly, there must be mechanisms for ensuring and controlling the selective utilization of the genome and those mechanisms that operate at a transcriptional and RNA processing level are the subject of this chapter. Those that operate at the level of protein synthesis are considered in Chapter 11.

## 10.1 THE REGULATION OF PROKARYOTIC RNA CHAIN INITIATION

The best-understood transcriptional control systems are those of prokaryotes and their viruses. Bacteria must be flexible enough to employ many substrates for their metabolic requirements but must combine flexibility with economy by not making enzymes unless they are needed. Their answer to this challenge is to link the genes for functionally related proteins, such as the enzymes of a metabolic pathway, into clusters which are subjected to co-ordinated control and are known as *operons*. Typically but not exclusively, the operon is under the control of a single promoter from which RNA polymerase transcribes all of the constituent genes as a *polycistronic mRNA*. The promoter is in turn controlled by a series of regulatory elements which respond to the bacterial environment and modulate the affinity of RNA polymerase for the promoter and hence the rate of RNA chain initiation.

### 10.1.1 Induction of the *lac* operon – a negative control system

The *E. coli lac* operon is the classic example of the negative regulation of a bacterial operon by a *repressor protein*. It provides the mechanism by which the bacterial cell does not normally make the enzymes of lactose metabolism (except in very small quantities) but retains the capacity to do so should it find itself in an environment in which glucose is unavailable but lactose is present as an alternative. Under such conditions, the concentration of $\beta$-galactosidase, the enzyme which catalyses the splitting of lactose to galactose and glucose, rapidly increases from approximately five to several thousand molecules per cell. The enzyme is described as *inducible*, as are the functionally related lactose-metabolizing enzymes, β-galactoside permease and β-galactoside transacetylase. Jacob and Monod [1] derived a series of mutant bacteria defective in the production of these enzymes. Some mutations affected the synthesis of one enzyme and were assumed to be in the gene for that enzyme. Others, known as *lacI* mutations, disturbed the regulation of all three genes such that, instead of being inducible in the presence of lactose or other β galactosides, they were synthesized *constitutively*; that is they were made at induced levels regardless of the presence or absence of an inducer. As a result of their studies [1], Jacob and Monod proposed the operon theory, a hypothesis which has since been proved by direct analysis.

The features of the *lac* operon, in its presently understood form, are illustrated in Fig. 10.1 and are the subject of a number of reviews [2, 3]. They may be summarized as follows:

(i)  The *structural genes* for the three enzymes are contiguous on the DNA and are transcribed from a single promoter into a polycistronic mRNA.

(ii) The I region of the DNA is a *regulator gene* which is transcribed from a separate promoter. The resulting I mRNA is translated into a 38 000-$M_r$ polypeptide which aggregates into a tetrameric *repressor* protein and, in the absence of inducers, binds to the *operator*. Binding of the

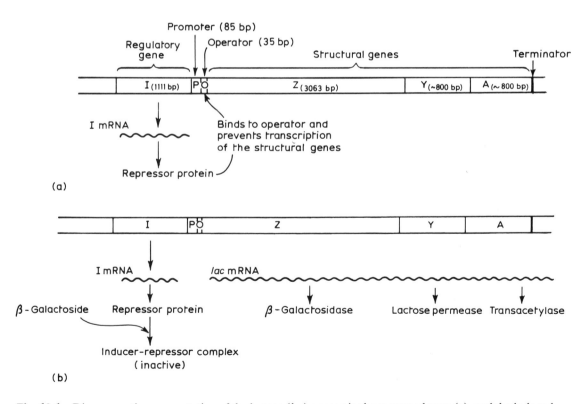

**Fig. 10.1** Diagrammatic representation of the lactose (*lac*) operon in the repressed state (a), and the induced state (b).

repressor protein to the operator blocks the transcription of the structural genes. The above *lacI* mutations of Jacob and Monod [1] produced mutant repressors that were unable to bind to the operator or block transcription. Similar constitutive mutants may arise from mutation of the operator.

(iii) In the presence of lactose, a complex is formed between the inducer (a metabolite of lactose) and the repressor protein rendering the latter unable to bind to the operator. Transcription of the structural genes can thus proceed (Fig. 10.1b). The natural inducer is not lactose but allolactose formed from lactose by transglycosylation. In studying this system induction is maximized by using a non-metabolizable or *gratuitous inducer*, isopropyl thiogalactoside (IPTG).

(iv) The 3′ portion of the promoter (Fig. 10.2) is more or less typical of prokaryotic promoters (bacterial promoters are discussed in Section 8.2.1 and the divergence of the *lac* promoter from that of the prokaryotic consensus sequence is reviewed by Reznikoff [3]). The 5′ portion of the promoter functions in catabolite repression, a further control mechanism which inhibits the induction of the *lac* operon when lactose and glucose are both available as substrates (catabolite repression is discussed in Section 10.1.3).

The first bacterial repressor protein to be isolated was that of the *lac* operon [4]. The repressor protein of wild-type *E. coli* represents approximately 0.001% of the total protein of the cell and its purification would have presented formidable problems had not Gilbert and Müller-Hill [4] increased its synthesis by genetic means. They first produced a strain of *E. coli* with a mutated promoter which overproduced the repressor tenfold. This mutated DNA they then incorporated into transducing phages which could occur at many copies per cell. The result was an overall thousandfold increase in the synthesis of the protein. Purification made use of the affinity of the repressor for the gratuitous inducer IPTG [4].

The *lac* repressor is an acidic protein with an $M_r$ of 150 000 and consists of four identical subunits. Each subunit binds one molecule of

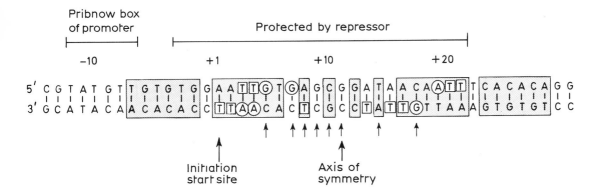

**Fig. 10.2** The *lac* repressor binding region. Shaded boxes indicate regions of symmetry. Open squares enclose thymidine residues which, when replaced by 5-bromouracil and exposed to ultraviolet light, become cross-linked to the repressor. Open circles enclose residues which the repressor protein protects from modification by dimethyl sulphoxide. Short arrows identify the position of point mutations which lead to constitutive expression of the structural genes.

inducer but only the tetramer binds to the *lac* operator [5]. When the inducer is bound to the repressor it causes an allosteric change in its configuration that greatly reduces its affinity for the operator (reviewed in [6]). Experiments with partially digested repressor, mutant repressors and analogy with other known repressors (for reviews see [3 and 6]) suggest that the binding of the protein to the operator involves the *N*-terminal 60 amino acids of the repressor and has led to models in which the DNA-binding *N*-termini of the four polypeptides protrude as arms from the core of the protein. More recent data, however, are not consistent with such a clear-cut segregation of activity [7] and the structure–function relationships in the binding of the *lac* repressor to DNA are not as well understood as is the case with some other repressor and activator proteins (see Section 10.1.6).

The binding of the repressor to the operator is very strong ($K_d \sim 10^{-13}$ M) and is about $4 \times 10^6$ times stronger than its affinity for non-specific DNA. This enabled the purification of DNA fragments carrying the operator region as a receptor–DNA complex which was retained on nitrocellulose filters while non-complexed DNA passed through. Isolation from these DNA fragments of sequences that were protected from DNase digestion by their association with the repressor yielded a 27-bp operator sequence of which 16 bases occurred as a hyphenated palindrome [8]. It has since been shown that the twofold symmetry extends beyond the protected region and involves 28 bp of a 35-bp sequence. The operator sequence is illustrated in Fig. 10.2 with the symmetrical sequences in boxes. It also indicates those nucleotides that have been shown to be important in repressor interaction. These include base pairs which if subject to mutation give rise to the constitutive phenotype, nucleotides which the receptor protects from *in vitro* methylation by dimethyl sulphate, and thymidines which, when replaced by 5-bromouracil and exposed to ultraviolet light, can be cross-linked to the repressor.

The 35-bp operator sequence runs from −7 to +28 with respect to the transcription start site of the β-galactosidase gene. It thus overlaps the gene and butts on to the Pribnow box of the promoter. Evidence was presented in Chapter 8 (Section 8.2.1) that RNA polymerase, when it binds to the promoter, covers up to 50 nucleotides on the 5′-side of the gene and up to 20 nucleotides into the gene. Thus, the repressor binding site at least partially overlaps the sequence recognized by RNA polymerase and would deny the enzyme access to its promoter.

The speed with which the repressor protein is able to locate its operator, which represents less than one millionth of the bacterial genome, is too fast to be accounted for by random association and dissociation with the DNA. Rather it suggests that there must be some mechanism for the facilitated transfer of the protein to its regulatory site. Evidence has been presented that the protein first binds non-specifically to DNA and is then transferred to the operator either by sliding along the DNA or by direct intersegment transfer [10].

---

*Cis-acting and trans-acting regulation*

These terms are in wide use to describe the two types of regulatory element seen in the *lac* operon and many other control systems. A *cis-acting regulator* is one which like the *lac* operator, controls genes with which it is contiguous. It could not control genes remote from its location. Thus, if an *E. coli* mutant was created that contained two *lac* operons of which one contained an operator with a constitutive mutation, the two operons would function independently. One would be constitutively expressed but its mutant operator would not influence the other operon which would still be responsive to the repressor.

The product of the regulator gene, the repressor protein, is an example of a *trans-acting regulator* because it is a diffusible substance, that is capable of interacting with any *lac* operator in the genome.

### 10.1.2 Repression of the *trp* operon

*Escherichia coli* has the capacity to synthesize all of the amino acids required for protein synthesis. In many cases, however, the necessary enzymes are only made in the absence of an adequate exogenous source of the amino acid. Perhaps the most studied such system is the *trp* operon [11]. This consists of five structural genes which encode the five polypeptides of the three enzymes required to synthesize tryptophan from chorismate (chorismic acid is the common precursor for the synthesis of the three aromatic amino acids). The operon is again under the control of a single promoter, is transcribed into a polycistronic mRNA and is controlled by a repressor protein which binds to the operator (Fig. 10.3). In this instance, however, the regulator gene (*trpR*) is remote from the rest of the operon. It encodes a repressor protein which is inactive and, in the absence of tryptophan, does not bind to the operator. When tryptophan is present, however, it binds to and activates the repressor which now binds to the operator and blocks transcription. Tryptophan is a *corepressor* in that, while lactose decreases the affinity of the *lac* repressor for the *lac* operator, tryptophan greatly increases the affinity of the *trp* repressor for its operator [320]. Both systems are, however, examples of *negative control* because the active repressor switches off transcription. The *trp* operon is also controlled by attenuation (see Section 10.2.1).

The twofold axis of symmetry of the *lac* operator has proved to be a common theme and the *E. coli trp* operator has a 20-bp sequence of diad symmetry that includes only one mismatch. This operator falls entirely within the promoter, and the region of symmetry includes the Pribnow box.

### 10.1.3 Catabolite repression – a positive control system

When grown in the presence of both lactose and glucose, *E. coli* selectively metabolizes glucose. This is achieved by a control of the *lac* operon that is distinct from repression by the repressor protein. It is called *catabolite repression* and is controlled by the *catabolite activator protein (CAP)* (also known as the *cyclic AMP receptor protein, CRP*). CAP is a positive activator of the C terminus which interacts with the 5′ region (−72 to −52) of the *lac* promoter thereby facilitating the formation of an RNA polymerase initiation complex. It is only active, however, in the presence of cyclic AMP (cAMP) and catabolite repression depends on the ability of glucose to reduce the cellular levels of this nucleotide (for reviews see [3, 12, 13]).

The control system is obviously of survival value to the bacteria as it is energetically desirable to metabolize glucose rather than lactose especially if by so doing it becomes unnecessary to express the *lac* operon. Furthermore, catabolite repression is not restricted to the *lac* operon as the binding of the same CAP–cAMP complex is also necessary for the expression of other bacterial operons including those for a number of alternative carbohydrate energy and carbon sources.

The DNA sequence to which the CAP–cAMP complex binds has been determined in a number of operons and, like those of operators, has been found to exhibit partial diad symmetry. Further-

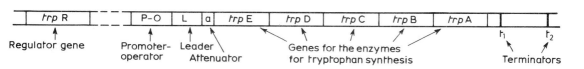

**Fig. 10.3** Diagrammatic representation of the tryptophan (*trp*) operon.

more, after an analysis of 18 such regions, Ebright *et al.* [14] derived a CAP-binding DNA consensus sequence. These workers also demonstrated the importance of the interactions between specific base pairs within the sequence and specific amino acid residues in the CAP protein. Thus, a mutant protein in which Glu-181 was replaced by Lys, Leu or Val could be compensated for by an altered DNA sequence in which G:C base pairs at two positions in the recognition sequence were replaced by A:T base pairs [14, 15]. The similarity between CAP and repressor proteins is discussed in Section 10.1.6.

Notwithstanding the above shared features of catabolite repression in different operons, there are sufficient differences between the target operons to make it difficult to envisage a common mode of action. Much of this relates to the location of the CAP-binding site. In the *lac* operon, at −72 to −52 with respect to the transcription start site, it is adjacent to the RNA-polymerase binding site. In the galactose (*gal*) operon, at −50 to −23, it appears to overlap the RNA polymerase-binding site while in the arabinose (*ara*) operon the major CAP-binding site is much further away at −107 to −78. A second minor site on this operon is still further away at −146 to −121. *In vitro* transcription assays of the *lac* operon [16] indicate that CAP acts to stimulate the formation of a closed RNA polymerase–promoter complex (the first stage in the formation of a transcription complex, see Section 8.2.1). A number of models have been proposed for the way in which this is achieved but there is no simple explanation that fits all of the available data (reviewed in [3]).

### 10.1.4 Other variations in the control of initiation at bacterial operons

(a) *The arabinose operon of E. coli encodes a dual-control protein and has dispersed genes*

The arabinose (*ara*) operon (for a review see [17]) allows *E. coli* to use arabinose as an energy source. It is induced by arabinose and subject to catabolite repression when glucose is also present. Three of its structural genes encode the enzymes required to convert arabinose to xylulose 5-phosphate, an intermediate in the pentose phosphate pathway. They are contiguous and are transcribed into a polycistronic mRNA. However, the control unit also includes two other structural genes that encode proteins involved in arabinose uptake into the bacterial cell and which are remote from the operon but are under the same regulation. Such dispersed control units are known as *regulons* (Fig. 10.4).

The regulatory gene of the *ara* operon (*araC*) is, like *lacI*, adjacent to its operator. It encodes a protein which is bifunctional and exists in two conformations variously known as $C^{rep}$ and $C^{ind}$ or $P_1$ and $P_2$. In the absence of arabinose, $C^{rep}$ binds to the operator and functions as a typical repressor protein in preventing transcription. Conversely, when arabinose is present, the conformational equilibrium of the regulatory protein is shifted to $C^{ind}$ which, together with the complex of CAP and cAMP, binds to a regulatory area of the DNA which appears to extend from inside the 3′-end of the operator to the RNA polymerase-binding site. The entire sequence of this regulatory element is known and its association with regulatory proteins has been studied by DNase-protection studies and oligonucleotide-directed mutation. Nevertheless, the precise sites of interaction, particularly those of the CAP–cAMP complex, remain controversial and the subject of differing models [18, 19, 20]. The binding of $C^{ind}$ and the CAP–cAMP complex enables RNA polymerase to initiate transcription. Thus, the protein product of the *araC* gene can function as both a positive and negative regulator.

(b) *The E. coli galactose operon has two differentially responsive promoters*

The three structural genes of the galactose (*gal*) operon encode enzymes responsible for metabolizing galactose to glucose 1-phosphate. They

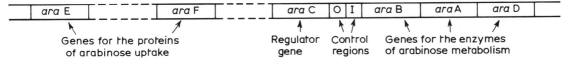

Fig. 10.4 Diagrammatic representation of the arabinose (*ara*) regulon.

are transcribed from either of two partially overlapping promoters, $P_{G1}$ and $P_{G2}$ [21]. The promoters are differentially responsive to the CAP–cAMP complex in that its binding to a single site activates $P_{G1}$ but inhibits $P_{G2}$ [22]. In the absence of CAP–cAMP, $P_{G1}$ is inactive but transcription can still occur from $P_{G2}$.

The *gal* promoters are also subject to negative regulation by the *gal* repressor protein which is the product of a non-contiguous gene (*galR*). Expression is repressed by the binding of repressor to both of two operators located on either side of the overlapping promoters [23].

### (c) *Autogenic control of bacterial genes*

The synthesis of *E. coli* alanyl-tRNA synthetase is under autogenic control. The synthetase represses the transcription of its own gene by binding to an operator-type palindromic sequence which flanks the transcription start site of the gene [24]. High concentrations of the cognate amino acid enhance the repression by causing an even tighter association between the synthetase and the DNA (see also Section 6.9.3b).

Several *E. coli* ribosomal proteins also autogenously control their own synthesis but this appears to be mainly a translational control mechanism and is considered in Section 11.10.2.

### (d) *Divergent transcription*

The biotin (*bio*) operon of *E. coli* contains five structural genes (A, B, F, C and D) that encode enzymes involved in biotin synthesis and which have an operator–promoter region located between genes *A* and *B*. Gene *A* is transcribed by leftwards transcription from one promoter while the remaining four genes are transcribed from a second promoter by rightwards tran-

scription of the other DNA strand. The operator between the two promoters appears to have a single binding site for the repressor–biotin complex that negatively regulates the operon. It exhibits partial diad symmetry and overlaps both promoters [25, 26]. Such a dual promoter system provides a mechanism for differential rates of expression of the different genes and is also found with several other genes including the arginine (*arg*) operon [27].

### 10.1.5 The repressors of the bacteriophage lambda (phage λ)

Other repressor-operator interactions that have been studied in considerable detail are those of the *E. coli* bacteriophage, lambda (λ) (for reviews see [28–31]). This, and related viruses, can exhibit two different life-styles. In the lysogenic state, the viral DNA, known as a prophage, is integrated into the host genome and is replicated with it. Conversely, in the lytic cycle, the genes of the bacteriophage are transcribed by the host RNA polymerase leading ultimately to the production of many progeny bacteriophage which are released with the lysis of the bacterial cell (see Section 3.6.8).

### (a) *Maintenance of the lysogenic state*

Maintenance of the lysogenic state of infection depends on a single protein, the lambda repressor. The way in which this protein suppresses the alternative lytic cycle of the phage is illustrated in Fig. 10.5 and can be summarized as follows.

(i) The repressor polypeptide can be conceived as a dumb-bell-shaped molecule in which *N*-terminal and *C*-terminal domains are

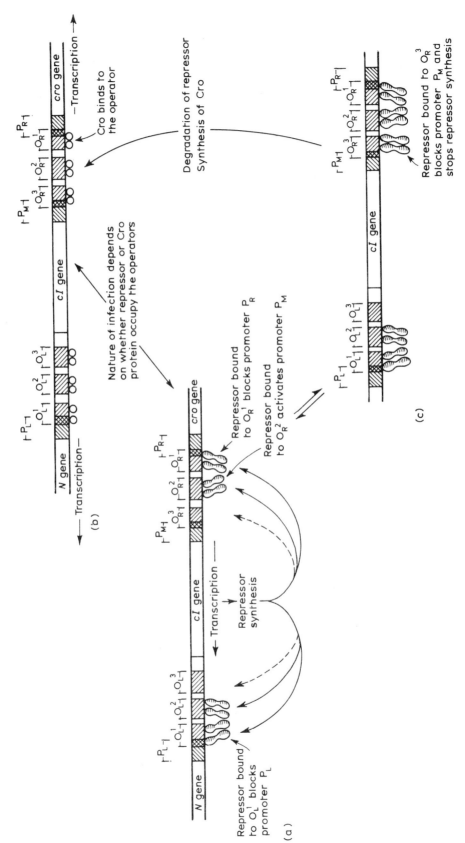

**Fig. 10.5** The mode of action of the lambda (λ) repressor. (a) lysogeny; (b) entry into the lytic cycle; (c) autogenic control of repressor synthesis.

joined by an intervening stalk. The active protein is a dimer of two of the 27 000-$M_r$ polypeptides joined through their carboxy domains.

(ii) The viral DNA has two operators $O_L$ and $O_R$ each of which has three 17-bp repressor-binding sites designated $O_L^1$, $O_L^2$, $O_L^3$ and $O_R^1$, $O_R^2$ and $O_R^3$ respectively. The binding sites conform to a consensus sequence and show partial diad symmetry. Each binding site will bind the *N*-terminal domains of a repressor dimer (as illustrated on the front cover).

(iii) The binding sites $O_L^1$ and $O_R^1$ exhibit the greatest affinity for the repressor and they overlap with the promoters $P_L$ and $P_R$. The preferential binding of the repressor to these sites thus inhibits transcription from the promoters and in so doing stops the expression of the proteins required in the lytic cycle of viral infection.

(iv) The presence of repressor bound at $O_L^1$ and $O_R^1$ greatly stimulates the binding of further repressor dimers at $O_L^2$ and $O_R^2$ and the adjacent repressors interact through their carboxyl domains (Fig. 10.5a). The presence of bound repressor at $O_R^2$ stimulates the interaction of RNA polymerase with the promoter $P_M$ and the transcription of the *cI* gene which encodes the repressor. The repressor thus exerts a negative influence on the promoters $P_R$ and $P_L$ but is a positive regulator of the promoter $P_M$ and of its own synthesis.

(v) The binding sites of $O_L^3$ and $O_R^3$ have a relatively low affinity for repressor and normally remain unoccupied. If, however, large quantities of repressor accumulate, the site $O_R^3$ will become occupied so blocking the promoter $P_M$ and shutting off repressor synthesis. The repressor thus exerts an *autogenic* control over its own synthesis (Fig. 10.5c). Repressor bound at $O_R^3$ does not interact with repressor bound at $O_R^1$ and $O_R^2$.

(b) *Switching from lysogenic to lytic infection*
The lysogenic prophage retains the capacity to switch to lytic infection and this is best achieved in the laboratory by damaging the bacterial DNA with ultraviolet light or chemical mutagens. This stimulates in the host a set of responses which are known collectively as SOS functions and which include the activation of a protease activity in the *recA* protein that normally functions in genetic recombination (see Section 7.4). The activation results from the binding of single-stranded DNA generated by the damage to the host cell genome. Activated *rec* A protein cleaves and inactivates the lambda repressor and thus derepresses the operator $P_R$ ($P_R$ is derepressed before $P_L$ as the latter has the higher affinity for the protein). Transcription from $P_R$ then leads to the production of the Cro protein, a dimer which like lambda repressor recognizes the three operator components $O_R^1$, $O_R^2$ and $O_R^3$ (Fig. 10.5b). Its binding is, however, non-co-operative and its affinity is the reverse of that of the lambda repressor. Thus, at low concentrations, Cro binds preferentially to $O_R^3$ so blocking the $P_M$ promoter and inhibiting the replacement of the degraded repressor. At higher concentrations it also occupies $O_R^2$ and $O_R^1$ turning down the transcription of the early genes; a step which by then is a necessary part of the lytic cascade (see Section 10.2.2).

(c) *The establishment of the lysogenic state versus entry into the lytic cycle*
Establishment of lysogeny requires the synthesis of two proteins, the repressor and a DNA topoisomerase called integrase, that catalyses the integration of the prophage into the host DNA (see Section 7.4.3). It was seen in Section 10.1.5a, however, that the repressor activates and controls its own synthesis from the binding sites of the $O_R$ operator. One might therefore question how sufficient repressor could be made to start the process. In fact the *cI* gene encoding the repressor is under the influence of a second promoter, $P_E$, which is functional during the

establishment of lysogeny while the $P_M$ promoter takes over for its maintenance.

When lambda first infects a bacterium, no repressor is present so the operators $O_R$ and $O_L$ are unblocked and transcription will initiate from $P_L$ and $P_R$. The first products are the proteins Cro and N and, as will be seen in Section 10.2.2, this permits the further transcription of the genes for the proteins cII and cIII. The combined effect of these two proteins is to stimulate the synthesis of repressor from its alternative promoter, $P_E$. They also stimulate the production of integrase.

It can be seen that whether bacteriophage infection results in lytic or lysogenic infection is under delicate balance (Fig. 10.6). The early products of infection include the proteins Cro, cII and cIII one of which blocks the action of the repressor while the other two stimulate its synthesis. Whether the infection becomes lytic or lysogenic will depend on whether Cro or repressor occupies the viral operators.

The control of the lytic cycle is considered in Section 10.2.2.

### 10.1.6 The interaction of repressor and activator proteins with DNA

Recent studies of the lambda repressor and Cro proteins have centred on their recognition of the operator-binding sites (for reviews see [31–33]). Both proteins can be made in large quantities from genes inserted in recombinant plasmids and both have been subjected to sequencing and crystallographic studies. The structure of the Cro protein is known in almost complete atomic detail [34] and the structure of the operator-binding N-terminal domain of the lambda repressor has also been determined by crystal diffraction [35] (see front cover). At first sight the two proteins are not very similar. The Cro protein consists of only 66 amino acids while the repressor protein contains 236 residues. However, both proteins have at their DNA-binding N-termini a pair of $\alpha$-helices separated by a $\beta$-turn. They are known as helix 2 and helix 3, are superimposable in space and contain conserved amino acids at key positions such as at the $\beta$-turn between the helices. Furthermore, hybrid

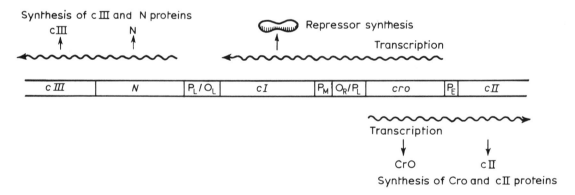

**Fig. 10.6** Initial events following the infection of *E. coli* with bacteriophage lambda: Cro competes with the repressor for occupancy of $O_R$ and $O_L$.

Whether the infection is lytic or lysogenic depends on the outcome; N is an antiterminator which permits the synthesis of cII and cIII proteins (see Section 10.2.2); cII and cIII stimulate repressor synthesis from the promoter $P_E$.

Note that the transcript includes the anti-sense strand of *cro*.

proteins in which the phage repressor helix 3 is replaced by the helix 3 of Cro exhibit identical interactions with DNA. Very similar models have been proposed for the interaction of both proteins with their operators [31–33] in which the $\alpha_3$ helix is seen as occupying the major groove of B-form DNA with the proteins' amino acids as side chains positioned to make base-specific interaction with the DNA. The $\alpha_2$ helix is thought to lie across the major groove and is presumed to interact with the DNA primarily through the phosphate backbone. The models are supported by genetic and chemical modification, protection studies, model building [319, reviewed in 15, 16, 17] and most recently by nuclear spin-labelling [36] and the resolution at 7 Å of crystals of the repressor operator complex [37]. The latter studies were of the close relative of lambda phage, coliphage 434, which was shown to have the same $\alpha_2$–$\alpha_3$ or helix–turn–helix, motif. Specific changes in the amino acid sequence of the $\alpha_3$ helix changed the specificity of the binding [38] as did ethylation of the contact phosphate of the DNA [39].

Of perhaps even greater significance is the accumulating evidence that other gene-regulating, sequence-specific, DNA-binding proteins share the helix–turn–helix motif. The structure of *E. coli* CAP (Section 10.1.3) has also been determined at a three-dimensional level [40]. In contrast to the phage proteins it interacts with DNA through its *C*-termini while its larger *N*-terminus interacts with cyclic AMP. Nevertheless, the $\alpha_D$, $\alpha_E$ and $\alpha_F$ helices of the *C*-terminus can be superimposed on the $\alpha_1$, $\alpha_2$ and $\alpha_3$ helices of Cro. The superimposition between $\alpha_E$–$\alpha_F$ and $\alpha_1$–$\alpha_2$ is particularly striking [41] and they appear to interact with DNA in the same way [14, 15].

Very recently the crystal structure of the *E. coli trp* repressor has been solved at a resolution of 2.6 Å [320]. It also has a helix–turn–helix DNA-binding motif which is stabilized by the interaction with tryptophan. Furthermore, a number of other DNA-binding proteins share

sequence homologies with the phage proteins and CAP, and several groups have predicted that they also employ the $\alpha_2$–$\alpha_3$ structure in their DNA interactions ([32, 33] and Section 10.3.1).

## 10.2 THE REGULATION OF THE TERMINATION OF TRANSCRIPTION IN PROKARYOTES

### 10.2.1 Attenuation

The expression of the *E. coli* tryptophan operon varies approximately 600-fold but this is not entirely due to the repressor–corepressor–operator interaction described in Section 10.1.2. Repression can reduce transcription 70-fold [42] while a controlled termination of transcription, known as *attenuation*, can reduce it by 8–10-fold [43] giving a combined effect of 600-fold. Attenuation is the subject of many reviews [44–47] and was discovered when mutants of the *trp* operon were isolated that gave rise to increased levels of all of the enzymes of the tryptophan-synthetic pathway and mapped not at the operator or repressor but at a site between the operator and the first structural gene, E (Fig. 10.3). This region of the DNA was found to consist of a sequence of 162 nucleotides lying downstream of the promoter and is thus transcribed by RNA polymerase. The transcript, which forms the 5′-end of the polycistronic mRNA, is believed to be translated but does not form a functional protein and is known as the *leader sequence*.

The leader has been sequenced, as has the entire tryptophan operon [48], and it was found that the transcript can undergo extensive base-pairing. The region numbered 1 in Fig. 10.7a can base-pair with region 2. This leaves region 3 free to base-pair with region 4 to form a typical rho-independent terminator site (see Section 8.2.3) of a G : C-rich hairpin followed by a string of uridine residues (Fig. 10.7c). However, the leader sequence has an alternative base-pairing strategy in which region 2 is base-paired with 3, a structure which would preclude the formation of

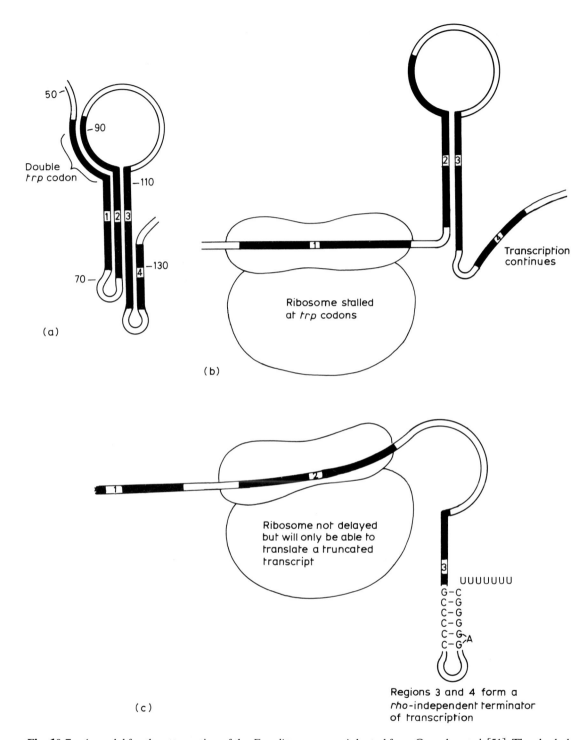

**Fig. 10.7** A model for the attenuation of the *E. coli trp* operon. Adapted from Oxender *et al.* [51]. The shaded areas 1–4 indicate regions of the leader able to undergo extensive base-pairing.
(a) base-pairing possibilities of the *trp* leader transcript; (b) base-pairing in low tryptophan concentrations; (c) base-pairing in high tryptophan concentrations.

the terminator site. The accepted model of attenuation [49–51] predicts that the leader transcript can adopt either of these two conformations and in one case will result in the termination of transcription at the end of the leader sequence while in the other will permit the transcription of the entire operon. The model also postulates that the conformation that is adopted will depend on the abundance of tryptophan. This latter concept arose from the finding that region 1 of the leader transcript contains two successive UGG codons for tryptophanyl-tRNA, a codon which in the genome as a whole is relatively rare. The model envisages that, if tryptophan is abundant, charged tryptophanyl-tRNA will be available to interact with UGG codons without delaying the translating ribosome. The leader sequence will be rapidly translated and the ribosomes will follow closely behind RNA polymerase such that the mRNA will be translated almost as fast as it is transcribed. Region 1 will have been translated and region 2 will be in the process of translation before a hybrid between regions 2 and 3 could form. Under these circumstances, there would be nothing to stop regions 3 and 4 base-pairing and terminating transcription (Fig. 10.7c). Thus, if tryptophan is abundant, those polymerase molecules that manage to initiate transcription (i.e. are not blocked by a repressor occupying the operator – Section 10.1.2) will only transcribe the 162 nucleotides of the leader. When, however, tryptophan is in short supply there will be a different outcome. The rarity of charged tryptophanyl-tRNA will slow the translating ribosome and cause it to pause at the dual *trp* codons. This will result in the ribosome being further behind RNA polymerase and will create time for the interaction of regions 2 and 3. This in turn precludes the hybridization of regions 3 and 4 into a terminator and allows RNA polymerase to continue past the leader and into the structural genes (Fig. 10.7b).

Clearly, the model depends on the close relationship between transcription and translation in prokaryotes. It also depends on the spatial relationship between the ribosome pause site and the regions 2 and 3. By making assumptions about the range over which a ribosome was likely to cause steric hindrance, Yanofsky [44] demonstrated that only tryptophan and, to a much lesser extent, arginine starvation can cause attenuation of the *trp* operon in *E. coli*. Only a ribosome stalled at these codons could allow the formation of the region 2 : region 3 base-paired stem.

Considerable evidence has accumulated that supports the above model of attenuation. Studies of the sensitivity of the leader to ribonuclease are consistent with the predicted base-pairing [51] as are studies with base change [44] and deletion mutants ([52] and references therein) which modify the ability of the different regions to base-pair with each other. The adjacent tryptophan codons in the *trp* leader are conserved in the *trp* operon of other bacterial species [44] but the best evidence for their importance in attenuation comes from the analysis of the operons for the biosynthesis of other amino acids. In every case investigated, the amino acids that control the expression of an operon have been found to have clustered codons in the leader of that operon. Often the clustering is more dramatic than that of the *trp* operon. Thus, the histidine (*his*) operon, which encodes the enzymes' for the biosynthesis of histidine and which is controlled by histidine levels, has six consecutive histidine codons in its leader sequence [53]. Similarly, the leader of the leucine (*leu*) operon has four consecutive leucine codons [50] and the phenylalanine (*phe*) operon leader has a sequence of nine codons of which seven specify phenylalanine [54]. Even more convincing are those operons regulated by more than one amino acid. The threonine (*thr*) operon is regulated by threonine and isoleucine and a sequence of thirteen codons in its leader includes eight for threonine and four for isoleucine [55]. Similarly, the *ilu* operon is regulated by leucine, valine and isoleucine and a stretch of seventeen codons in its leader includes four for leucine, five for isoleucine and six for valine [56]. The leaders of some of these operons have been shown to be capable of forming mutually exclusive secondary structures similar to that described for the *trp* leader [57–59].

Attenuation-like control also contributes to the controlled expression of the phe-tRNA synthetase operon [60] and the autogenic control of ribosomal protein synthesis [61] but the mechanism is less well understood.

### 10.2.2 Antiterminators of transcription

During lytic infection by bacteriophage lambda and related phage, the viral genes are expressed sequentially in what is known as the lytic cascade (Fig. 10.8). The primary control of this process is at the termination of transcription which is inhibited by antiterminator proteins (for reviews see [62–65]).

As described in Section 10.1.5.c, when the lambda phage first infects a cell, two genes, *cro* and *N*, are expressed by host RNA polymerase transcribing in opposite directions from the promoters $P_R$ and $P_L$ (rightward and leftward promoters). These two genes are known as the *immediate early genes* and they encode the antirepressor protein Cro, the function of which is described in Section 10.1.5.c, and an anti-terminator, the N protein. This latter protein permits the lytic cycle to continue with the expression of the next block of genes: *the delayed early genes*. It achieves this by allowing RNA polymerase to transcribe past *rho*-dependent terminators ($t_L$ and $t_{R1}$) located at the 3′-ends of the *N* gene and *cro* gene respectively (rho-dependent terminators are considered in Section 8.2.3). The delayed early genes encode the cII and cIII proteins discussed in Section 10.1.5.c, plus two proteins which are involved in the replication of the viral genome, a series of proteins which function in recombination and another antitermination factor, Q. The protein Q unlocks the final block of bacteriophage

genes, the so-called *late genes*, which encode the viral head and tail proteins as well as two proteins that bring about host cell lysis. They are transcribed from a separate promoter $P'_R$ from which RNA polymerase constitutively synthesizes RNA but normally terminates at an adjacent rho-dependent terminator ($t_R'$) with the release of a 194-nucleotide RNA known as 6S RNA. The Q protein suppresses termination at $t_R'$ and allows RNA polymerase to transcribe through it and into the late genes. The lytic cycle is illustrated in Fig. 10.8.

*N*-mediated antitermination of intermediate early genes is the most extensively studied antitermination system. The *N* gene encodes a $12\,000$-$M_r$ protein which is a highly specific antiterminator for lambda genes. Related bacteriophage have equivalent proteins but they are often not interchangeable. This specificity suggests that the protein recognizes unique sites on lambda DNA and such sites have been identified from mutations that abolish their function. They are known as N protein utilization sites (*nutL* and *nutR*) and were identified as 17-bp sequences lying between $P_L$ and $t_L$ on the leftward strand and between $P_R$ and $t_{R1}$ on the rightward strand of the lambda DNA (Fig. 10.8). They are identical for 16 of their 17 bp and contain a 5-bp region of inverted symmetry [66]:

$$
\begin{array}{l}
\text{A}|\text{G C C C T}|\text{G A A Pu A}|\text{A G G G C}|\text{A} \\
\text{ }\mid\text{ }\qquad\qquad\mid\qquad\qquad\mid\qquad\qquad\mid \\
\text{T}|\text{C G G G A}|\text{C T T Py T}|\text{T C C C G}|\text{T}
\end{array}
$$

The functional significance of this sequence was demonstrated when a cloned fragment including *nutR* caused termination-resistant transcription when placed between a non-lambda promoter and a termination signal [67]. The combination of N and *nut* will overcome any rho-dependent terminator and will also act at independent termination sites. Because it will function with both sorts of terminators it seems likely that the N protein acts on an element common to them

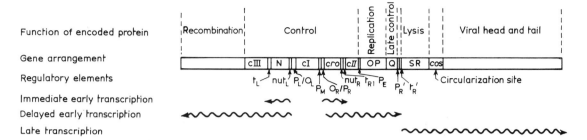

**Fig. 10.8** Diagrammatic representation of the events in the lytic cycle of bacteriophage lambda. Note that, although it is illustrated as a linear molecule, the lambda DNA is circularized on infection by the annealing of its single-stranded *cos* ends and ligation by host DNA ligase (see Section 3.6.6a). Some of the illustrated regulatory elements are discussed in Section 10.1.5.

both, namely RNA polymerase. This is supported by RNA polymerase mutations that prevent protein N activity and by compensatory N gene mutations which restore antiterminator function [68]. However, conclusive evidence that protein N acts by binding to RNA polymerase has not been presented.

Other *E. coli* mutations reveal that at least five host proteins are involved in antitermination. They are known as N utilization substances (nus) and are designated *nusA* to *E. NusD* maps to the gene that encodes rho protein and *nusC* maps at the *rpoB* gene that encodes the $\beta$ subunit of RNA polymerase. It has been suggested that protein N interacts with this portion of the polymerase. The importance of the product of the remaining three host genes, *nusA*, *nusB* and *nusE*, has been confirmed by the development in an *in vitro* transcription antitermination system dependent on all three proteins [69, 70]. However, little is known of their mode of action. The *nusA* gene encodes a 69 000-$M_r$ protein that binds to N protein [71]. It also binds with equimolar stoichiometry to RNA polymerase core enzymes [72] not, as previously suggested, by displacing sigma factor but after sigma is released [73]. It has a much lower affinity for RNA polymerase holoenzyme, but it has recently been demonstrated that it binds specifically to rho. It may thus provide a mechanism for

coupling rho to the elongating polymerase [74] and probably plays an essential role in bacterial termination. Putative DNA-binding sites have been identified for the protein. They are known as box A and have a consensus sequence:

$$5'\genfrac{}{}{0pt}{}{C}{T}\ GCTCTT(T)A3'$$

However, the evidence for a role for specific DNA sequences in the mode of action of the protein is contradictory (for a review see [65]) and it has recently been suggested that it binds to the $\lambda$ mRNA rather than the DNA. Such binding has been demonstrated *in vitro*, to a site immediately upstream of the box A signal [321]+ Even less is known of the function of the remaining host-encoded proteins. The *nusB* gene product is a 14 500-$M_r$ protein and *nusE* encodes the ribosomal protein S10 which can function in *in vitro* antitermination systems either as the purified protein or as 30S ribosomal subunits [75].

The other lambda antiterminator, Q, is a 23 000-$M_r$ protein which appears to depend on fewer host proteins. In an *in vitro* system it promotes readthrough of the $t_R'$ terminator with RNA polymerase and NusA protein as the only additional protein factors [76]. Grayhack *et al.*

[322] have shown that, in the presence of NusA, Q protein binds to RNA polymerase and causes it to read through a pause site early in the λ late gene operon.

A further recently described example of anti-termination is in the *E. coli* rRNA genes [77]. It was seen in the previous section that premature termination by attenuation results from the uncoupling of transcription and translation such that the nascent transcript was able to fold into a secondary structure that induced termination. Clearly, non-translated operons, such as those encoding rRNA, must either contain no potential early termination sites or must have some other antitermination mechanism. That the latter is the case is indicated by the fact that termination signals introduced into rDNA are ignored. Holben and Morgan [77] have presented evidence that the promoter–leader sequence of the rDNA is able to modify RNA polymerase so that it reads through transcription termination signals.

## 10.3 MODIFICATION OF PROKARYOTIC RNA POLYMERASE

### 10.3.1 Diversity in sigma factor

The specificity of transcription in bacteria is dependent on the association with RNA polymerase core enzyme of sigma factor; a subunit which allows the enzyme to recognize gene promoters but which leaves the ternary complex when transcripts are a few nucleotides in length (Section 8.2.1.b). The mechanics of the system suggests a potential control mechanism in which different classes of promoter could be recognized by a series of promoter-specific sigma factors. In general, such systems do not appear to be common although the number of known or suspected examples is growing.

The most studied system is the RNA polymerase of the spore-forming bacterium, *Bacillus subtilis* (for reviews see [78–80]). This enzyme associates with a series of sigma factors of which

the main equivalent of the *E. coli* factor has been called sigma[55] ($\sigma^{55}$) from an empirically determined molecular mass of 55 000 daltons. Recently, however, the protein has been sequenced and its 371 amino acids have an actual $M_r$ of 43 000 [81]. It has therefore been renamed $\sigma^{43}$ but the new notation is not yet universally used. It shares sequence homology with *E. coli* sigma [82] and recognizes a promoter sequence similar to the −10 and −35 regions [83].

In addition to $\sigma^{43}$, vegetative cells of *B. subtilis* contain at least three minor sigma factors known, again from their $M_r$, as $\sigma^{28}$, $\sigma^{32}$ and $\sigma^{37}$ (the status of several other reported factors is uncertain). Some of these factors become more abundant as the bacteria move out of logarithmic growth into stationary phase and a further factor, $\sigma^{29}$, appears when sporulation commences, in response to nutrient depletion. Each of these minor factors changes the promoter recognition specificity of RNA polymerase (Table 10.1) and thus dictates the expression of different sets of genes. Losick and Pero [78] proposed a cascade model for the sequential expression of genes during sporulation in which it was envisaged that regulation was based on the displacement from RNA polymerase of one sigma factor by another. The evidence for this is not yet conclusive but can be summarized as follows.

Sigma[37] appears to play a role in the stationary phase of bacterial growth. The abundance of $\sigma^{37}$-containing enzyme increases during late log phase and *in vitro* and *in vivo* it has been shown to direct the transcription of genes which are expressed only in the stationary phase [84–86]. Another gene which is expressed in both the growth and early-stationary phase was found to have promoters for both the major $\sigma^{43}$ and $\sigma^{37}$, suggesting that expression of this gene was controlled by $\sigma^{43}$ RNA polymerase during rapid growth and by $\sigma^{37}$ RNA polymerase as the bacteria prepares for sporulation [87]. Similarly, the gene *spoVG*, activated at the start of sporulation has been found to have two pro-

**Table 10.1** The sigma factors of *B. subtilis*.

| Sigma factor | Occurrence | Conserved promoter sequence | |
|---|---|---|---|
| | | −35 region | −10 region |
| $\sigma^{43}$ ($\sigma^{55}$) | Growing cells | TTGACA | TATAAT |
| $\sigma^{28}$ | Growing cells | CTAAA | CCGATAT |
| $\sigma^{32}$ | Growing cells | AAATC | TA–TG–TT–TA |
| $\sigma^{37}$ | Growing cells | AG––TT––A | GG–ATT–TTT |
| $\sigma^{29}$ | Sporulating cells | TT–AAA | CATATT |
| $\sigma^{gp28}$ | Infected cells | T–AGGAGA––A | TTT–TTT |
| $\sigma^{gp33–34}$ | Infected cells | CGTTAGA | GATATT |

moters, but in this case they are the recognition sequences for $\sigma^{37}$ and $\sigma^{32}$ [88]. *SpoVG* is under complex control and it is not known how this relates to the dual promoters. RNA polymerase containing $\sigma^{28}$ is another minor component of vegetative cells which disappears within an hour of the onset of sporulation and appears to be specific for the transcription of a small number of genes [89]. Gilman and Chamberlin [89] speculate that $\sigma^{28}$-transcribed genes may encode proteins that are involved in the transition from growth to sporulation. Conversely, $\sigma^{29}$ is a sporulation-specific component of RNA polymerase and it has been proposed that it is the specific transcription factor for genes transcribed early in sporulation [90].

The *B. subtilis* bacteriophage, SP01, capitalizes on the flexible RNA polymerase conformation of the bacteria by regulating it for its own reproduction [78, 80]. Three viral genes (genes 28, 33 and 34) encode sigma-like polypeptides that alter promoter recognition by the host cell polymerase (Table 10.1). As was seen with lambda bacteriophage (Section 10.2.2), the phage genes can be divided into three transcriptional blocks: those expressed early in infection, those turned on in the middle of the lytic cycle and those transcribed late in the cycle. The promoters for the phage early genes are recognized by the host polymerase with the $\sigma^{43}$ subunit. The product of one of these early genes, gene 28, is a sigma-like, 26 000-$M_r$ protein ($\sigma^{gp28}$)

[91] which directs the host RNA polymerase to recognize and initiate transcription of the middle genes [92]. Two of the middle genes, genes 33 and 34, encode the sigma-like proteins, $\sigma^{gp33}$ and $\sigma^{gp34}$, which direct the host polymerase to transcribe the late genes [78].

There is a growing number of additional examples in which multiple forms of RNA polymerase holoenzyme control the expression of specific sets of genes. Until recently it was thought that *E. coli* sigma factor[70] regulated the specificity of transcription initiation throughout the whole genome. It is now known, however, that the product of the *htpR* gene, that controls the heat-shock regulon (genes expressed in response to heat shock), is a sigma-like protein [93]. Similarly, the morphologically complex filamentous bacteria, *Streptomyces coelicolor*, has been shown to have at least two distinct RNA polymerase holoenzymes containing different sigma factors. These recognize different promoters and may be involved in the complex differentiation cycle of the bacteria [94]. It has also been proposed that the *nifA* and *ntrA* gene products, which control nitrogen assimilation in *Klebsiella pneumoniae* and *Rhizobium* respectively, could be sigma factors [95, 96].

Stragier *et al.* [82] have compared the sequences of the major $\sigma^{70}$ factor of *E. coli*, the major $\sigma^{43}$ factor of *B. subtilis*, the *E. coli* heat-shock sigma-like protein, Htpr, and a sporulation-induced *B. subtilis* protein spoIIG which may be identical to $\sigma^{29}$. They found a highly conserved internal region in all four proteins which they proposed bound the sigma factors to RNA polymerase. Furthermore, the *C*-terminal region of all four proteins exhibited the $\alpha$-helix–$\beta$-turn–$\alpha$-helix configuration previously described as the DNA-binding region in a number of repressor and activator proteins (see Section 10.1.6). They proposed that this region recognized the promoters of the relevant target genes.

### 10.3.2 Bacteriophage T4 modulation of host RNA polymerase

As with other bacteriophage, the development of the *E. coli* phage, T4, is characterized by the successive expression of several gene classes. All transcription is by the host RNA polymerase but the enzyme undergoes a series of phage-induced modifications which, it is believed, play a crucial role in the control of viral transcription (for a review see [97]). At least three classes of viral gene, early, middle and late, are known and each has a distinct promoter type. The early genes are at first transcribed by unmodified *E. coli* RNA polymerase from promoters indistinguishable from those of the host. However, immediately on infection, the polymerase undergoes its first modification carried out by the product of the T4 *alt* gene. This protein is a component of intact phage particles and is injected with the DNA. It brings about the addition to one of the host enzyme $\alpha$ subunits of an ADP-ribose residue. A few minutes later, a second ADP-ribosylation on the same residue of both $\alpha$ subunits (arginine 265) is catalysed by the product of one of the early genes (*mod*). Other T4-induced modifications include the association with the polymerase of several phage-encoded polypeptides, known from their relative masses as 10K, 12K, 15K and 22K proteins. Two of them, 12K and 22K, have been identified with the T4 control genes 33 and 55.

The precise role of all these modifications is uncertain. The 12K and 22K proteins are required for late-gene expression and the latter (also known as gp$^{55}$) has been shown to replace host sigma factor in their transcription. It may therefore function as an alternative sigma factor recognizing the late gene promoter TATAAATA [98]. In *in vitro* systems, the 10K protein inhibits host sigma factor [99] and the 15L protein elevates the melting temperature of T4 promoters [100] but their *in vivo* function is uncertain. T4-induced ADP-ribosylation appears to be involved in the shut-off of the early phage genes at the later stages of development [101].

## 10.4 THE CONTROL OF GENE EXPRESSION IN EUKARYOTES

Much less is known about the control of gene expression in eukaryotic cells than in prokaryotes. As well as genes that can be turned on and off, such as those regulated by hormones, a eukaryotic cell will have many permanently shut-down genes. Thus a liver cell will not express the genes encoding proteins found only in a brain cell or proteins made only during differentiation. Eukaryotic DNA exists in the nucleus as chromatin and it is assumed that the proteins of chromatin are intimately involved in the availability of genes for expression. The nucleus itself is a feature not found in prokaryotes and nuclear RNA processing and transport of RNA through the nucleoplasm and nuclear membrane are additional levels at which gene expression could be controlled.

There is no evidence that eukaryotes ever have operons of contiguous genes that are transcribed as polycistronic mRNA and controlled by closely associated regulatory genes. Eukaryotic gene expression is under the influence of specific, cis-acting, elements of DNA but these may be well away from the genes that they regulate (e.g. enhancers and enhancer-like elements). Furthermore, transcription may be influenced by diffusible substances such as proteins or steroid hormone–protein complexes collectively known as trans-acting substances. How these elements influence the transcription of genes, often from long distances, is little understood but increasing evidence suggests that gene activation is accompanied by changes in DNA conformation and modification.

All of these control mechanisms are subject to intense and continuing investigation and each is considered in the following sections. It should be emphasized, however, that, so far, it has not proved possible to weld the observed phenomena into proven mechanisms for gene control. For instance, trans-acting protein–hormone complexes control the expression of steroid hormone-responsive genes. They are known to interact with DNA elements near the genes that they activate and the activated genes have in many cases been shown to have changed chromatin conformation, changes in chromatin-associated proteins, relative hypomethylation and increased DNase sensitivity of the DNA in chromatin. Nevertheless, the mechanisms by which these changes are brought about and the ways that they influence the transcription of genes are largely unknown.

### 10.4.1 Promoters

In the preceding sections on control of prokaryotic gene expression, several examples were given of genes or operons transcribed from more than one promoter. These included the *E. coli gal* operon with two differentially responsive promoters (Section 10.2.4.c), the *B. subtilis spoVG* developmentally regulated genes with promoters sensitive to different RNA polymerase holoenzymes (Section 10.3) and the lambda repressor gene with different promoters for the establishment and maintenance of lysogeny (Section 10.1.5.c). There is growing evidence that eukaryotic genes can also have differentially responsive promoters.

The most-studied example is the mouse *Amy-1*[a] amylase gene [102–105]. This gene has two promoters of which one is exclusively active in the parotid gland while the other is active in both the parotid and the liver (Fig. 10.9). A difference in the relative strengths of the two promoters accounts for the fact that there are only 100 copies of amylase mRNA per liver cell compared with 10 000 copies per parotid gland cell [104].

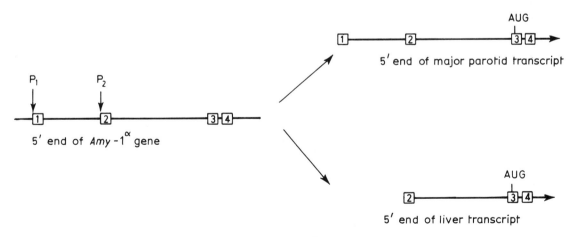

**Fig. 10.9** Use of the alternative promoters of the mouse *Amy-1*[a] amylase gene. The boxes 1–4 indicate the exons of the 5′-end of the gene and the intervening lines represent introns. $P_1$ is the strong parotid promoter from which transcription is initiated at 30 times the rate of that from $P_2$, the promoter active in liver cells. AUG is the translation start site of the mRNA.

The mouse parotid gland differentiates largely after birth and the parotid-specific promoter is activated between 14 and 21 days of age [105].

Similar findings to the mouse amylase gene have been reported in other systems although their significance is not understood. Thus, the transcript of the single alcohol dehydrogenase of *Drosophila melanogaster* has different 5'-ends in different tissues and at different stages of development. The differences arise from the use of alternative promoters [106]. Other examples include the invertase gene of yeast [107, 108] and the discoidin-1 gene of *Dictyostelium* [109].

The mechanism by which different promoters are selected is not understood but there is evidence for the involvement of cis-acting enhancers and trans-acting protein factors (the nature of enhancers and trans-acting substances is described in subsequent sections). Promoters may be enhancer-dependent or enhancer-independent [110] and Paul and coworkers have presented evidence that the human $\epsilon$-globin gene has both sorts [111]. Promoter-specific protein factors have been detected in SV40-infected HeLa cells. Of two isolated factors, one was required for transcription from a wide range of promoters while the second, called Sp1, was a promoter-specific factor required for the transcription from SV40 early promoters, one late promoter and several host cell promoters [112 and ref. therein].

### 10.4.2  Cis-acting control elements

#### (a)  *Enhancers*
The region of DNA required for the efficient initiation by RNA polymerase II includes at least three elements (see Fig. 8.6). The TATA box lying approximately 30 bp upstream of the transcriptional start site is involved in fixing the position of transcription initiation. A second, so-called upstream element, including CAAT boxes and $-100$ elements, is less strongly conserved, located a variable distance upstream of the start site ($-40$ to $-110$) and appears to be important in promoting efficient initiation without playing any role in its accuracy. These two regions have been considered in Section 8.3.1. A third region, the so-called *enhancer*, together with various *enhancer-like elements* are regulatory in their function and are considered here. (There is some evidence for a fourth kind of region which like the promoters of RNA polymerase III occur as *downstream elements*, that is they are downstream of the transcriptional start site.) The evidence for these has been reviewed [113].

The first enhancer discovered was a 72-bp tandem repeat located more than 100 bp upstream of the early SV40 genes but essential for their expression [114, 115]. Since then, enhancers have been discovered in many DNA tumour viruses [116 and refs therein] and a variety of other viruses [117, 118]. They have also been found in cellular genes notably those for immunoglobulins [119–121], insulin and chymotrypsin [122] (for review see [117, 123, 124] and for other enhancer-like elements see the following Section).

Enhancers are cis-acting DNA elements which strongly stimulate the transcription of adjacent genes. Thus, deletion of the SV40 enhancer reduces early gene expression by a factor of 100 and abolishes viral viability [114, 115]. Typically, enhancers can produce their effects regardless of their orientation and although they preferentially activate proximal promoter elements, they can, with decreased efficiency, act over considerable distances (several kilobases) and can even function in a downstream position [125]. Many, like the SV40 enhancer, can affect homologous or heterologous promoters of a wide variety of species and cell types [126]. Others, like those of polyoma virus, show host cell preference [127] while the immunoglobulin enhancers exhibit strict cell-type specificity [119–121], and the lymphotropic papovavirus enhancer is specifically active in human haemotopoietic cells [128].

No extensive homology exists among the viral

**Table 10.2**  Some enhancer-like elements in eukaryotic cells.

| Response | Stimulus | DNA sequence implicated | References |
|---|---|---|---|
| *Genes influenced by external stimuli* | | | |
| MMTV | Glucocorticoids | | 135–147 |
| Metallothionein | Glucocorticoids | ⎫ TGTTCT consensus sequence | 148 |
| Induced enzymes | Glucocorticoids | ⎬ | 149, 151 |
| Uteroglobin | Glucocorticoids | ⎭ | 150 |
| Growth hormone | Glucocorticoids | Within 500 nucleotides of 5'-side of gene | 152 |
| Ovalbumin | Progesterone | Within 300 nucleotides of 5'-side of gene | 153–154 |
| Vitellogenin II | Oestrogen | GCGTGACCGGAGCTGAAAGAACAC | 155 |
| Metallothionein | Cadmium | $\begin{smallmatrix}T\ G\\C\ T\end{smallmatrix}$CG-CCCGGC$\begin{smallmatrix}T\ C\\\ \ C\end{smallmatrix}$ | 148 |
| Heat-shock proteins | Heat-shock | CT-GAA--TTC-AG and other elements | 156, 157 |
| α- and β-interferon | virus | Region from −109 to −64 containing repetitious oligonucleotides | 158–160 |
| Ribulose 1,5-bisphosphate carboxylase | Light | 33 bp surrounding the TATA box and including the consensus CATTATATAG$\begin{smallmatrix}C\\A\end{smallmatrix}$ | 161 |
| Yeast gal genes | Galactose | 75-bp region between the divergently transcribed *gal1* and *gal-10* genes | 162, 163 |
| *Genes under negative regulation* | | | |
| Yeast a-specific genes | — | CATGTAA-T-C------G--A-TTACATG palindrome | reviewed in 164<br>165, 173 |
| *Genes under co-ordinate control* | | | |
| α- and β-globin | — | Regulating elements both 5' and 3' to translation initiation site | 166, 167 |
| Yeast ribosomal proteins | — | Conserved 12-bp homology in six genes | 168 |
| cAMP-regulated genes | — | 29-nucleotide conserved sequence | 169 |
| *Tissue/development-specific gene expression* | | | |
| Immunoglobulin | — | At least three sequence elements within and outwith the gene | 119–121, 170, 171 |
| Insulin | — | 5'-Flanking region of gene | 122 |
| Chymotrypsin | — | 5'-Flanking region of gene | 122 |
| Albumin | — | Within 400 nucleotides of 5'-side of gene | 172 |

enhancers although various consensus sequences have been pointed out. Of these, the core element

$$GTGG\frac{AAA}{TTT}G$$

is present in a number of viral enhancers [129] and the related sequence GTGGTTTT(T)GAA is present as a closely spaced repeat in the cell-specific immunoglobulin gene enhancer [119]. However, Weber *et al.* [130] and others (Table 10.2) have presented sequences with only limited homology with this consensus.

The mode of action of enhancers is unknown. One popular model proposes that they provide a bidirectional entry site for RNA polymerase. They alter the conformation of DNA in chromatin as judged by DNase sensitivity (Section 10.4.4.a) and electron microscopy [131] and several groups have demonstrated that the SV40 enhancer increases the number of RNA polymerase II molecules on the DNA [132, 133].

Enhancers, or elements that function in a similar way, are not confined to the genes transcribed by RNA polymerase II. The '*Bam* Islands' and associated enhancer-like elements in the non-transcribed spacer of *Xenopus* rDNA have been discussed in Section 9.4.2.a. They are very promoter-like in sequence and are presumed to act as sinks for RNA polymerase I so increasing the affinity between the enzyme and its transcription unit [134].

### (b) *Enhancer-like elements*

Many less-well-characterized DNA sequences have been identified which, in the absence of more specific evidence, will be classed here as enhancer-like elements. Typically they consist of short DNA sequences which have been associated with the tissue-specific expression of genes or with gene expression induced by specific stimuli such as hormones, heat shock, heavy metals or viral infection. Generally they have been identified in one of two ways. They may be isolated as a section of DNA which forms a binding site (*acceptor site*) for a trans-acting

substance such as a hormone–receptor protein complex. Alternatively, they may be identified as a fragment of DNA which when linked to a gene can bring that gene under the control of a hormone, heat shock or some other stimulus. The control of gene expression by glucocorticoid hormones will serve as an example (for review see [135]).

Glucocorticoids modulate the expression of a wide variety of genes which include enzymes, structural proteins, house-keeping proteins, membrane proteins, secretory proteins and, the very popular model, the mouse mammary tumour virus (MMTV). Transcription of the MMTV genome is stimulated by glucocorticoids, not only in mouse mammary tumours, but in a wide variety of tissue culture cells. Experiments in which cells were transfected with cloned MMTV confirmed that the signal for hormone responsiveness is within the viral DNA [136] and further studies with viral fragments linked to marker genes [137, 138] or plasmids [139] showed that the signal is in the viral long terminal repeat (LTR). However, as with typical viral enhancers, there would appear to be multiple enhancer-like elements in MMTV, and glucocorticoid–receptor complexes bind at a number of sites both within the LTR and elsewhere in the viral genome [140]. The LTR is 1328 bp long and also includes signals for viral replication. Enhancer-like elements within it have been identified by DNase I protection and foot printing studies (Section A.8.2) which define those regions of the DNA protected from enzyme digestion by the complex. They are backed up by further fragmentation studies in which portions of the LTR were linked to marker genes. A consensus of the many publications resulting from these studies appears to show that the major regulatory elements reside within a fragment of the LTR between −52 and −236. The precise boundaries are, however, contentious [141–144]. Within this sequence there are two strong and two weak binding sites for the glucocorticoid–receptor complex [145, 146]. All four of these contain a hexanucleotide sequence 5′TGTTCT3′ the guanosine of which is protected by the hormone–receptor complex from methylation by dimethyl sulphate [147]. Furthermore, the two strong binding sites share a more extended homology having the consensus sequence

$$\begin{smallmatrix}T\\C\end{smallmatrix}GGT\begin{smallmatrix}A\\T\end{smallmatrix}CA\begin{smallmatrix}AA\\CT\end{smallmatrix}TGT\begin{smallmatrix}T\\C\end{smallmatrix}CT$$

which they also share with the gluco-corticoid-responsive element of the human metallothionein gene [148]. Other gluco-corticoid-responsive genes support the indication that this sequence is important. Thus, the genes for tyrosine aminotransferase and tryptophan oxygenase have sequences related to the hexanucleotide at an equivalent position with respect to the transcriptional start site [149]. The uteroglobin gene has three copies of the consensus which binds the glucocorticoid receptor 2.6 kb upstream of the gene [150] and the gene for lysozyme has a strong hormone-binding site which includes the hexanucleotide in the reverse orientation [151]. The most enhancer-like property of all is the ability of the −52 to −236 region of the MMTV LTR to confer glucocorticoid responsiveness on the normally unresponsive thymidine kinase or α-globin genes when inserted at their 5′- or 3′-ends, in either orientation and when placed up to several hundred nucleotides from the gene [144].

There are many more examples of elements of DNA, usually located near the 5′-end of the gene, which have been implicated in the specific responses of that gene. Some, like those responding to glucocorticoids, respond to external stimuli. Others, like the immunoglobulin, insulin and chymotrypsin enhancers mentioned in the previous section, appear to be associated with tissue-specific expression. Still others, which repress gene activity and have been called *silencers*, appear to work against enhancers [173]. Table 10.2 presents a selection of sequences which have been identified or proposed as controlling elements.

### 10.4.3 Trans-acting factors

Much remains to be discovered about the mode of action of enhancers and the other cis-acting elements described above. One thing is, however, clear; many, probably all of them, function as targets for diffusible, DNA-binding molecules. Some of these, such as steroid hormone–receptor protein complexes, have been studied for many years and have been purified. Many others surely await discovery.

They are collectively known as trans-acting factors but this term also includes the protein factors which interact with the promoters for eukaryotic RNA polymerases. These have been considered elsewhere (Sections 8.3.2, 8.3.3 and 9.4.1) and include the proteins which interact with the internal promoters of genes transcribed by RNA polymerase III (Section 8.3.2) and those that bind to the TATA box of genes transcribed by RNA polymerase II [174]. They appear to be important for the formation of a stable initiation complex with the template and the accurate initiation of transcription. This does not preclude them also having a role in the control of gene expression; indeed there is evidence that the *Xenopus* 5S gene transcription factor, TF IIIA (see Section 8.3.2), exhibits altered cellular levels during differentiation [175]. Much of the preference for the expression of somatic 5S genes rather than oocyte 5S genes in somatic cells can be attributed to the concentration of this factor, as it binds four times more strongly to the internal promoter of somatic genes than to that of oocyte genes [323]. Also the two developmentally regulated promoters of the *Drosophila* alcohol dehydrogenase gene may be differentially controlled by multiple sequence-specific DNA-binding proteins [176], and in Ehrlich ascites tumour cells, growth-dependent regulation of rRNA synthesis is mediated by a transcription factor TIF-IA [324].

Of the known trans-acting regulatory proteins, those which perhaps most nearly approach the repressor–operator interaction of prokaryotic operons are found in yeast. Transcription of the *GAL* genes of yeast is induced more than 5000-fold by galactose, the induction being dependent on an enhancer-like element known as an *upstream activating sequence* (UAS) located about 250 bp from the start of each of the divergently transcribed $GAL_1$ and $GAL_{10}$ genes [162, 163]. Induction is also dependent on a positive regulatory protein, $GAL_4$, which has recently been shown to function as a trans-acting factor that binds to the upstream activating sequence [177]. Similarly, the upstream activating sequence which negatively regulates the expression of the yeast aspecific mating-type genes is recognized by a sequence-specific DNA-binding protein, $\alpha2$ [165].

Perhaps the greatest problem associated with the control of gene expression is how an embryonic cell comes to follow a specific pathway of differentiation. Most progress in analysing the regulation of this process has been in *Drosophila* in which nine genes involved in embryonic segmentation have each been shown to contain, within the protein-coding sequence, an element of approximately 180 bp known as the homeobox (reviewed in [332]). Using the *Drosophila* homeoboxes as probes, cross-hybridizing sequences have been found in a wide range of higher animals including man and here again the mammalian homeoboxes are expressed during embryogenesis [333]. How the homeobox-containing proteins function is unknown but they share homology with DNA-binding proteins and it has been suggested that they are trans-acting factors that control specific pathways of morphogenesis by regulating large batteries of genes. As such, they fulfil the requirements of the 'selector gene' hypothesis of Garcia-Bellido [334], in which it was proposed that small numbers of control genes might specify the development of body patterns.

The trans-acting factors which bind to enhancers would be expected to be different from those that bind to the other promoter elements

and in SV40 this has been shown to be the case [178]. However, the SV40 trans-acting factor, which binds to both the 5' and 3'-domains of the SV40 enhancer, will also bind to other enhancers. Emerson *et al.* [1979] have suggested that an enhancer may consist of multiple binding sites for different factors which would presumably form an enhancing complex. This is supported by the data of Wu [180] showing that enhancer-like elements associated with the activation of heat-shock genes in *Drosophila* are activated by the sequential binding of at least two protein factors.

What the protein-enhancer combination does to stimulate transcription is far from clear. Factors which bind specifically to the 5'-region of the chicken adult $\beta$-globin gene do so to a DNase I hypersensitive region in the chromatin, the sensitivity of which is correlated with the expression of the gene [179]. It has been suggested that the binding of specific proteins to DNA recognition sequences may occur when the DNA is free of nucleosomes during its replication [181, 182]. The binding of the factors at the expense of histones might in turn predispose the DNA in the chromatin to DNase sensitivity as well as placing it in the correct conformation to enhance transcription.

In concluding the brief résumé of our current knowledge of trans-acting factors it is worth mentioning that there is currently considerable interest in the concept that a number of viral oncogenes encode proteins which stimulate transcription and may function as trans-acting gene regulators (reviewed in [183]). There is also considerable current interest in the possibility that specific RNA species may play a role in the control of gene expression and one possibility is that they function as trans-acting factors. Regulation of gene expression by RNA is considered in Section 10.5.

### 10.4.4 The nature of active chromatin

Active eukaryotic genes tend to share a number of features which appear to be associated with, if not sufficient for, their transcription. These include an increased accessibility to DNase digestion, a tendency to be relatively under-methylated and other features which indicate that the DNA or chromatin is differently arranged when compared with inactive genes.

(a) *The hypomethylation of active genes*
The methylation of DNA has been considered in Section 4.6.1 and the point was made that a number of observations support the theory that changes in methylation are associated with the control of gene expression (for short reviews see [184, 185]). As mentioned in Section 4.5.3, pairs of restriction endonucleases are available which recognize the same nucleotide sequence but are differently affected by its methylation (iso-schizomeric enzyme pairs). *Msp*I, which cuts the sequence CCGG whether or not the second C is methylated, and *Hpa*II which cuts the same sequence provided it is unmethylated are commonly used to study methylation. Thus, if the methylation of this sequence within a gene varies under differing growth conditions, this will be manifested as a change in sensitivity to the *Hpa*II digestion. No change will occur in the sensitivity to digestion by *Msp*I.

Studies with these enzyme pairs have shown that the degree of methylation at specific sites does vary with gene expression. For example, some but not all of the CCGG sequences in globin genes are specifically undermethylated in those tissues, or stages of development, in which they are expressed [186–188]. Similarly, the genes for the egg white proteins are relatively undermethylated in oviduct when compared to tissues in which they are not expressed [189, 190]. Other examples include the loss of methyl groups from immunoglobulin heavy and light chain genes during lymphocyte differentiation [191–193], the relative hypomethylation of metallothionein genes when induced by cadmium or glucocorticoid [194] and the correlation between expression and hypo-methylation of $\alpha$-foetoprotein genes [195]. Many

other examples could be given and are considered in detail in [326].

The importance of methylation in the sup-pression of gene expression has been illustrated *in vivo* in experiments with mouse retroviruses. Early mouse embryos are non-permissive for retrovirus expression and Jähner *et al.* [196] showed that when Maloney murine leukaemia virus is introduced into pre-implantation mouse embryos, the viral DNA becomes methylated and is not expressed. When, however, the virus infects post-implantation embryos, it is not methylated and gives rise to viral infection. Even then, infection by cloned provirus can be in-hibited if it is methylated with rat liver DNA methyltransferase before administration [197].

In many of the studied systems the relative undermethylation of expressed genes occurs in the 5'-flanking sequences. The importance of this in the $\gamma$-globin gene was particularly well illustrated by Busslinger *et al.* [198]. They synthesized cloned globin genes in which certain segments were methylated and others were left unmodified. Methylation of the structural gene had no inhibitory effect on globin transcription while methylation of the 5'-sequences from $-760$ to $+100$ prevented transcription. The picture is not, however, clear-cut as the mouse dihydrofolate reductase gene [199] and the chicken $\alpha2$ collagen gene [200] have un-methylated 5'-ends regardless of the extent to which they are expressed. Furthermore, the ovalbumin gene domain [190] and the glucose 6-phosphate dehydrogenase gene [201] have clusters of methylation sites at their 3'-ends that are specifically undermethylated when the genes are expressed.

An example that casts doubt on the whole concept of relative demethylation as a pre-condition for transcription is the oestrogen-controlled vitellogenin gene. In chicken, specific demethylation of a CCGG sequence at the hormone-binding site of this gene is associated with, and apparently necessary for, gene expres-sion [202, 203]. In the *Xenopus* vitellogenin gene, however, all *Hpa*II-sensitive sites are methyl-ated whether or not the gene is expressed [204]. This finding may be explained by the fact that only 4% of potentially methylatable CG sequences are recognized by *Hpa*II. More difficult to rationalize is the almost total lack of methylation in the DNA of *Drosophila* and other insects. Indeed it is very difficult to assign a generally applicable role to DNA methylation when it occurs to such a variable extent across the phylogenetic tree (see Section 4.6.1).

Where, as in the above examples, specific demethylation is involved in gene expression, it appears that it is a prerequisite for expression but not enough to ensure it. Thus, the oestrogen-induced demethylation at the 5' end of the chicken vitellogenin gene occurs in two of the hormone's target tissues: the liver, where the gene is expressed and the oviduct where it is not expressed [202].

(b) *The DNase sensitivity of active chromatin*
Weintraub and Groudine [205] first demon-strated that active chromatin is preferentially digested by pancreatic DNase I. They found that the globin gene in chromatin from erythrocytes but not from other tissues was selectively digested by the enzyme. Similarly, the oval-bumin gene was shown to be preferentially digested in oviduct chromatin but again was relatively insensitive to the enzyme in tissues, such as erythrocytes, in which it was not expressed [206].

DNase sensitivity of active genes has since proved to be a general phenomenon (for a review see [207]). Typically, the area of sensitivity includes the region of transcription and may extend to many kilobases of sur-rounding DNA. Thus, the glycerophosphate dehydrogenase gene is located in a 12-kbp domain of increased DNase I sensitivity [208]. Furthermore, such domains may include gene clusters. The region of sensitivity of the β-globin gene domain includes the whole cluster of β-like genes (see Section 9.2.2) plus at least 8 kbp on the

3'-side and 6 kbp on its 5'-side [209, 210]. Likewise, the ovalbumin gene of chicken, to-gether with the related X and Y genes, is linked in a 100-kbp domain of DNA this is preferentially sensitive to DNase I digestion (Fig. 10.10).

High DNase sensitivity often, but not always [211], correlates with the relative under-methylation of the genes and like undermethyl-ation appears to be a prerequisite for tran-scription but is insufficient to ensure it. Thus, the whole human β-globin gene family domain is more DNase-sensitive than bulk DNA [210] but the genes within it are sequentially expressed through the early life of the individual. The ovalbumin gene of fowl oviduct and the vitell-ogenin gene of *Xenopus* liver both exhibit enhanced sensitivity to DNase I when oestrogen activates the gene. In each case, however, they retain their sensitivity when gene expression is shut down after the withdrawal of the hormone [212, 213].

Superimposed on this general DNase I sensi-tivity of active or potentially active genes, there are discrete, smaller regions which are even more sensitive to the enzyme and are known as *hypersensitive sites*. They are usually demon-strated by first very lightly digesting nuclei with DNase I. The DNA is then purified, digested with restriction endonucleases and the digestion products analysed with a gene-specific probe by Southern blotting (see Section A.3). Res-triction fragments smaller than normally expected indicate a preferential cleavage by DNase I and the size of the fragments can be used to locate the position of the cleavage. Hypersensitive sites are characteristically less than 200 bp in length and, usually but not always, occur in the 5' sequences of genes. They have been found associated with various promoter elements [214–217, 325], enhancers [214, 217], hormone–receptor complex binding sites [218], other enhancer-like elements [219], the LTR of a retrovirus [220] and even with negative control elements [217]. They have also been located apparently bracketing the 3'-end of genes [221]

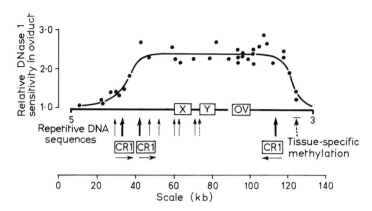

**Fig. 10.10** The ovalbumin gene domain. The solid circles indicate the relative DNase I sensitivity of various regions of the domain. The solid arrows denote the locations of repetitive DNA sequences within the domain and the bolder arrows point out the specific locations of CR1 family repetitive sequences. The orientations of the CR1 sequences are indicated by horizontal arrows. The dashed arrow indicates a region where distinctive methylation patterns are noted for different chicken tissues and where oviduct DNA is specifically undermethylated (reproduced, with permission, from Stumph *et al.* [190] copyright 1983 American Chemical Society).

and gene cluster domains [222]. The human β-globin gene locus is one of the most thoroughly investigated systems [210, 222]. As stated above, the entire locus including the β, γ, δ and ε genes are located in a domain which, in erythrocytes, is more sensitive to DNase I than bulk DNA. Within this region there are hypersensitive sites located at the 5′-ends of all the genes. Even these so-called minor hypersensitive sites are, however, relatively resistant to DNase I digestion compared with the major hyper-

sensitive sites which occur at what appears to be the boundary of the human β-globin domain (Fig. 10.11).

Like the generalized DNase I sensitivity of chromatin that is active or can be activated, some hypersensitive sites appear to reflect a potential for transcription rather than active transcription. Conversely, the appearance of other sites is strongly correlated with gene expression. The oestrogen-activated chicken vitellogenin gene seems to have three classes of hypersensitive site

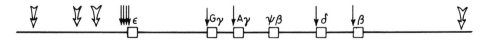

**Fig. 10.11** The DNase I hypersensitive sites of the human β-globin domain in adult erythrocyte chromatin. The boxes β, A$_\gamma$, G$_\gamma$, σ, ε and ψβ represent the genes of the globin gene family (see Section 9.2.2). Narrow arrows indicate minor hypersensitive sites at the 5′-ends of the genes. Broad arrows indicate major hypersensitive sites which appear to bracket the gene domain. Adapted from Tuan *et al.* [222].

all of which are confined to oestrogen target tissues [223]. Some of these exist prior to oestrogen treatment while three others appear in the 5′-flanking sequence in response to the hormone. Of these three, two are stable and retained after hormone withdrawal while the third, located 0.7 kbp upstream of the gene, is observed only during oestrogen-induced expression. The LTR of mouse mammary tumour virus contains a hypersensitive site which coincides with the glucocorticoid–receptor complex binding site and also disappears upon hormone withdrawal [218]. The chicken lysozyme gene is controlled by steroid hormones in chicken oviduct but is constitutively expressed in macrophages. Different sets of hypersensitive sites characterize these different functional states [224], as they do in malignant and non-malignant expression of the *c-myc* oncogene [217, 220].

(c) *The role of nucleosome phasing, supercoiling*
     *and Z-DNA in the activity of genes*
A number of studies have indicated that active genes exhibit an increased nucleosomal spacing [225] and that their DNase I hypersensitive sites may be nucleosome-free [226]. Furthermore, DNA regions, which on chromatin are hypersensitive to DNase I, often have sites sensitive to S1 nuclease [227, 228]. Jointly these data indicate that, if not actually present as single-stranded regions, then hypersensitive sites must be in a conformation on which nucleosomes cannot form and in which single-strand specific nucleases can gain access (for a review of the anatomy of hypersensitive sites see [229]). Nucleosomes may be located on DNA by specific factors interacting with sequences 80 bp apart [203, 335–337]. Such positioning may regulate the exposure of specific sequences to trans-acting factors [338]. Felsenfeld and colleagues [231] have recently shown that two proteins interact specifically with two discrete protected regions within the hypersensitive site of the 5′-flanking region of the chicken β-globin gene. The proteins are found in the nuclei of the erythrocytes of adult chicken but not in other tissues. The most 5′-binding site contains a 16-nucleotide sequence of $G + C$ base pairs which, when placed in supercoiled plasmids, is very sensitive to S1 nuclease. It also contains the 6-bp sequence found in SV40 and other promoters. The second binding site contains a partial inverted repeat in which 18 out of 25 nucleotides are complementary. One might therefore speculate that multiple specific trans-acting factors recognize sequences in the 5′-flanking sequences of genes so creating or further modifying hypersensitive sites. In so doing they allow the DNA to adopt a more open conformation that is accessible to RNA polymerase and to the formation of an initiation complex. The 5S RNA-specific transcription factor (see Section 8.3.2) is known to induce specific gyration in the promoter of 5S DNA so inducing a torsional strained conformation that is active in transcription [232]. Conversely, nucleosomes suppress the untwisting of DNA [233]. Villeponteau *et al.* [234] have shown that torsional stress promotes DNase I sensitivity and they propose that the altered structure around active genes is maintained by DNA supercoiling. Topoisomerase II induces the torsion and the topoisomerase inhibitor, novobiocin, reverses it and the DNase sensitivity. Recent evidence that topoisomerases might be responsible for establishing the DNase I-sensitive domains of chromatin has been reviewed by North [235] and the mode of action of the enzymes has been discussed in Section 6.4.6.

The most extreme form of conformational change which might be associated with transcriptional control is the left-handed helical Z-DNA (the structure of Z-DNA has been considered in Section 2.6.3). The SV40 enhancer has been shown to contain nucleotides which *in vitro* can form Z-DNA [236], and Z-DNA-binding proteins have been isolated which bind to the viral control region [237]. Regions with the potential to form Z-DNA have also been found flanking the rat somatostatin gene [238], near the element involved in heavy metal regulation of the metallothionein gene [148] and in a negative regulatory region of a $\beta$-globin gene [239]. Negative regulation was also observed when sequences able to form Z-DNA *in vitro* were inserted between the split promoters of a tRNA gene [240].

### (d) *The chromatin proteins of active genes*

Specific trans-acting factors involved in the control of eukaryotic gene expression were discussed in Section 10.4.3. Some of these, namely those associated with sites hypersensitive to DNase I or with regions of Z-DNA, were considered in the previous section. Also apparently involved in the differentiation between active and inactive chromatin are the major proteins of nucleosomes. These have already been discussed in Section 3.3, and their consideration here is limited to their possible role in the control of transcription (for reviews see [207, 241]).

Several studies have indicated that histone H1 is non-randomly deposited in the nucleus [242] and depleted in active chromatin [243]. Weintraub [244] has presented evidence that it crosslinks adjacent nucleosomes in inactive regions of the genome but not in active regions, where its loss may contribute to the more open structure and increased accessibility to nucleases and the trans-acting factors necessary for gene expression.

The development of sea-urchin embryos is accompanied by the successive incorporation into the chromatin of primary sequence histone variants [245, 246]. The sperm-specific histones (sp histones) are sequentially replaced with cleavage-stage or CS histones in the 16-cell embryo. Then, in the morula and blastula stages of development, the $\alpha$-histone variants predominate to be followed later by the $\beta$, $\gamma$ and $\delta$ variants. Simultaneously with these changes, the nucleosomal spacing of the chromatin is changing [247] and presumably the stage-specific replacements are associated with the derepres-sion of the transcriptionally inert sperm genome. The replacement of histones with protamines during spermatogenesis in salmonid fishes and mammals [248] and the associated condensation of chromatin into a transcriptionally inert configuration would appear to be akin to a reverse of the developmental changes in sea-urchins. The transcriptionally active macro-nucleus of *Tetrahymena* also contains histone variants not present in the transcriptionally inactive micronucleus [249].

There is considerable evidence for the enrichment of acetylated histones in transcriptionally active chromatin and for their association with increased RNA synthesis (see for instance [250, 251]). Their function in transcription is unknown but the best correlation is between transcriptional activity, the extent of contribute hyperacetylation of histone H4 and the turnover of acetyl groups in active chromatin [250]. Chalkley and colleagues [251] argue that the coupled process of acetylation and deacetylation generates a sequential uncoiling and repacking of the regions of the chromatin to be expressed.

Ubiquitin is a highly conserved 76-amino acid protein required for ATP-dependent non-lyso-somal intracellular protein degradation [327] but it is also found conjugated to histones, principally histone H2A. Ubiquitinated histones are present at higher concentration in actively transcribed chromatin regions and may therefore play a role in the regulation of gene expression [252].

Histone may also be phosphorylated, methylated and ADP-ribosylated and attempts have been made to link all of these modifications to transcriptional activation [207, 241]. Poly ADP-ribosylated histones have been shown to inhibit RNA synthesis [253] and histone phosphorylation [254].

The non-histone proteins HMG 14 and HMG 17 are also strongly associated with active chromatin [255, 256]. They are responsible, at least in part, for the DNase I sensitivity of active chromatin [257, 258] and can even be immobilized on an affinity column and used to isolate actively transcribed chromatin [259]. They are differentially phosphorylated during the cell cycle [260].

### (e)  *Active genes may be associated with the nuclear matrix*

The nuclear matrix, also known as the nuclear skeleton or lamina, is the remnant that remains when nuclei are sequentially extracted with high salt concentrations and detergent and are then treated with DNase and RNase. It consists of a network of thin proteinaceous fibres together with residual nucleolar structures, nuclear pores and connecting lamina (see Section 3.1). There is an accumulating mass of often conflicting data as to whether the matrix may be intimately associated with both transcriptional and post-transcriptional processes.

DNA appears to be organized into a series of supercoiled loops anchored to the matrix and when the loops are cleaved with endonucleases the DNA sequences which remain attached to the matrix can be analysed. In this way the active ovalbumin [261–263], conalbumin [262], globin [264, 265] and vitellogenin genes [266] have all been shown to be preferentially associated with the matrix. The data are not, however, universally accepted. Kirov *et al.* [265] have suggested that the apparent association of active genes with the matrix may be an artifact created by the association of active genes with proteins that are resistant to the high-salt methods of preparing the matrix. When DNase I was used in the absence of high salt no such association was observed. Mirkovitch *et al.* [328], also using a low-salt extraction method, have shown that histone gene clusters are attached to the matrix by an AT-rich region in the spacer between the clusters. They found, however, that the attachment was transcription-independent. Conversely, Jackson [267] used gentle isotonic methods to make preparations of nucleoskeletons and found that RNA polymerase, nascent transcripts and genes being transcribed

were closely associated with it.

Similar argument to those discussed above centre on whether, in the mitotic chromosome, potentially active genes are associated with the so-called chromosomal scaffold (see for instance [268]).

### 10.4.5  Multiple gene copies, amplification and gene rearrangement

#### (a)  *Multiple gene copies*

The mechanisms by which multi-gene families are thought to arise are described in Section 7.3.2 and the best-known examples are discussed in Section 9.2.2. It is worth repeating here, however, that in many cases, such as the genes for rRNA, 5S RNA, histones and many structural proteins, the existence of multicopy genes appears to provide a mechanism by which the cell can produce large quantities of the end products concerned.

#### (b)  *Gene amplification*

The mechanisms by which amplification is thought to occur are described in Section 7.5.3, and the amplified rDNA of amphibia and ciliate protozoa is discussed in Sections 7.5.1 and 9.4.2.a. Consideration here is limited to the relevance of gene amplification to the control of gene expression (for reviews see [269, 270]). In fact, relatively few examples are known where the increased accumulation of gene products results from gene amplification, the more common mechanisms being enhanced transcription or translation of an unaltered number of genes. Like genes which are carried as multiple copies in the germ-line DNA, amplified genes are found where expression must be faster than could be achieved by a single gene copy. In this case, however, the multiplicity arises in specific cell types during differentiation. The amplification of rDNA during oogenesis in amphibia serves to allow the rapid accumulation of rRNA in the oocyte. Similarly, the amplification of rDNA in the macronucleus of *Tetrahymena* vegetative cells compensates for the single copy which serves as a repository of genetic information in the micronucleus of germinal cells (Section 9.4.2.a). The genes which encode the peptides of the communal cocoon of the sciarid fly, *Rhynchosciara americana*, are amplified when they are required in large quantities [271] as are the chorion genes of *Drosophila* when their products are required for egg shell proteins [272–274]. The latter occurs only in the cells of the ovarian follicle and takes place about 18 h prior to the onset of chorion synthesis. De Cicco and Spradling [275] have demonstrated that the initiation of chorion gene amplification on chromosome 3 of *Drosophila* is controlled by elements within a 3.8-kbp length of DNA associated with the gene cluster.

Although not strictly a control mechanism, it is worth noting that a number of examples exist in which gene amplification has resulted in the acquisition of resistance to drugs and poisons [103, 276] (see Section 7.5.2). The best-known example of this is the emergence of amplified dihydrofolate reductase (*dhfr*) genes in tissue culture cells [270], tumours [277] and *Leishmania tropica* parasites [278] treated with the DHFR inhibitor, methotrexate. It should also be mentioned that amplification is not confined to eukaryotic cells but occurs commonly in prokaryotes in which it can be revealed by selective culture [279]. It is freely reversible in prokaryotes and probably plays an important role in allowing adaptation to environmental change.

(c) *Gene rearrangement*

Possible mechanisms of gene rearrangement as well as many of the best-known examples have been discussed in Sections 7.8 and 7.9. It suffices here merely to restate that genomic rearrangements are the means by which gene expression is controlled in a wide diversity of situations which range from the generation of antibody diversity in vertebrates, variation in the mating type of yeast, the generation of different cell surface antigens by trypanosomes and the elimination of portions of the genome in a wide variety of organisms.

## 10.5 REGULATION OF GENE EXPRESSION BY RNA

### 10.5.1 Antisense RNA

An antisense RNA, that is an RNA with a sequence complementary to that of a mRNA, clearly has the potential to modulate gene expression. The extent to which modulation happens *in vivo* is uncertain but several examples are now known in both prokaryotic and eukaryotic cells (for reviews see [280, 281]).

One of the best-characterized examples is the expression of a transposon Tn10 (transposable elements are considered in Section 7.7) which is inserted at random in the *E. coli* genome [282]. Tn10 is 9300 bp in length with inverted repeats of an insertion sequence (IS10) at its ends. The right-hand IS10 sequence encodes a transposase

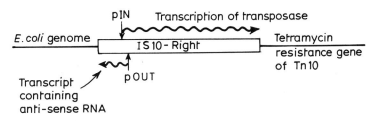

**Fig. 10.12** The transcription of the insertion sequence IS10-Right. pIN is the promoter for the transposase gene. pOUT is a strong promoter directing the synthesis of an RNA which includes 36 bp of antisense RNA covering the start codons for transposase.

protein which catalyses the insertion of the element into the chromosome. Transcription from one of the two transposase promoters produces the mRNA for the protein. Transcription from the second promoter, on the other hand, begins from just inside the coding region, proceeds in an upstream direction and produces an RNA complementary to the 5'-end of the transposase message (Fig. 10.12). It appears that the formation of a duplex between the mRNA and the antisense RNA negatively controls the translation of the transposase protein. A similar mechanism appears to regulate the expression of the *E. coli ompC* and *ompF* genes which encode two outer membrane proteins of the bacteria [283]. Initiation of replication of the plasmid Col E1 is also regulated by antisense RNA ([329] and Section 6.9.3.a).

A possible candidate for control by antisense RNA in eukaryotic cells is the dihydrofolate reductase (*dhfr*) gene of mouse. Farnham *et al.* [330] have characterized a transcription unit in the 5'-flanking region of the gene orientated in the opposite direction to the gene and giving rise to a 180–240 nucleotide opposite strand RNA. The RNA is abundant in the nuclei, is not polyadenylated but it is not known whether it functions in the control of *dhfr* gene expression.

Using the above indications of natural regulation of gene expression by antisense RNA, a number of groups have demonstrated artificial regulation of specific genes by antisense RNAs generated from inducible promoters on plasmids. These have included several *E. coli* [284, 285] and eukaryotic genes [286].

The mic RNA, which inhibits the expression of the *E. coli* outer membrane protein, OmpF, is complementary to the 5' leader region of the mRNA including the ribosome-binding sequence and the initiation codon. On this basis, the inhibition by antisense RNA would appear to operate at the level of translation [284]. A similar interpretation could be placed on the finding that artificial antisense transcripts of *β*-galactosidase mRNA were effective so long as they were complementary to the coding region of the message. However, in eukaryotic cells, duplexes between thymidine kinase mRNA and an antisense RNA were only detected in the nucleus, suggesting that duplex-containing transcripts failed to enter the cytoplasm [287]. It has been suggested that the medical use of antisense RNA might provide a mechanism for the artificial prevention of viral infection or the expression of harmful genes [288].

### 10.5.2 Identifiers

Sutcliffe *et al.* [289] have identified an 82-bp nucleotide sequence present in multiple copies in the rat genome and transcribed into 62% of RNA polymerase II and III transcripts made in brain nuclei. The sequence is present, however, in fewer than 4% of the transcripts from the nuclei of other tissues. The RNA sequences have been called *identifiers* or *ID sequences* and it has been suggested that they might be control elements governing brain-specific gene expression [289, 290]. The concept has aroused considerable interest. Minárovits *et al.* [291] for example have suggested that identifier sequences acquired by oncoviruses might determine the potential target cell for malignant transformation. Recently, however, the validity of the identifier theory has been challenged as, when assayed directly rather than as run-off transcription, brain identifier RNA was found not to be restricted to the brain but to occur in similar abundance elsewhere [292]. Whether identifiers really exist and whether they occur in other tissue types are questions that will only be resolved by further investigation.

### 10.6 THE CONTROL OF PRE-mRNA PROCESSING

#### 10.6.1 3'-Processing and poly-adenylation

As already discussed in Section 9.3.3, it has recently become apparent that the 3'-termini of

RNA polymerase II transcripts are not generated by simple termination but by termination followed by processing. It has also become apparent that this processing can, at least in some cases, follow alternative routes leading to the generation of mRNAs with differing 3'-ends. Furthermore, the nature of the 3'-end may govern the nature of further processing events, such as the pattern of splicing, and may lead to the production of different polypeptides.

Examples of genes in which alternative processing has been observed include those encoding the immunoglobulin heavy chain [293–295], calcitonin [296], tropomyosin [297], glycinamide ribotide transformylase [298] and the adenovirus late genes [299]. In each of these examples, the alternative processing results in the production of different polypeptides. Thus the product of the calcitonin gene in the thyroid gland is calcitonin but the alternative processing in neural tissue generates the so-called calcitonin gene-related peptide [296]. Similarly, mRNAs with different 3'-ends encode the different iso-

forms of tropomyosin [297] and the membrane-bound and secreted forms of the immunoglobulin $\mu$ chain [293–295] (Fig. 10.13).

In addition to the above, a number of examples are known where the gene encodes alternative polyadenylation sites which are used to differing exents but do not apparently affect the nature of the protein product [300–303]. That is, the 3' heterogeneity is in the 3'-non-coding region of the mRNA. The choice of polyadenylation site in these transcripts may, however, influence the overall extent of gene expression and this appears to be the case with the dihydrofolate reductase gene [300].

The factors which govern the choice of polyadenylation site are, at present, totally unknown.

### 10.6.2 Splicing of pre-mRNA

The complex series of immunoglobulin gene DNA rearrangements, which give rise to the variable gene regions and which replace one

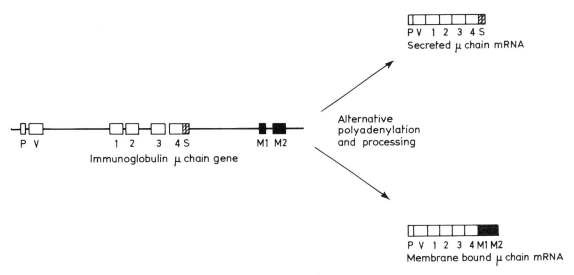

**Fig. 10.13** A schematic representation of the alternative processing of transcripts of the immunoglobulin $\mu$ chain gene. Boxes represent exons and the intervening lines, introns. P refers to the signal peptide exon and V to the rearranged variable gene (see Section 7.9.1). Open boxes 1 to 4 are shared exons. The hatched boxes, S, encode the 3' terminus of secreted $\mu$ chain mRNA and the shaded boxes M1 and M2 are unique to the membrane-bound $\mu$ chain in mRNA.

constant region gene with another, has been considered in Section 7.9.1. However, DNA rearrangement cannot explain all of the variation in immunoglobulin gene expression. It was seen in the previous section that differential 3′ processing determines whether lymphocytes produced the membrane-bound or secreted form of the $\mu$ heavy chain [293–295]. The change, during lymphocyte differentiation, from IgM expression to the co-expression of IgM and IgD also derives from differential processing [304, 305]. Here, the production of an IgD molecule appears to require RNA polymerase to transcribe a very long (25 kb) pre-mRNA-containing transcripts of the variable locus, the $\mu$ heavy gene and the $\delta$ heavy gene (Section 7.9.1d). Splicing then removes the transcript of the $\mu$ gene treating it as though it were an intron. Conversely, production of an IgM molecule results from a shorter transcript not containing the $\delta$ sequence and in which the $\mu$ transcript is seen as an exon [304, 305]. The simultaneous expression of $\mu$ and $\epsilon$ heavy chain also appears to involve similar differential processing [306].

The above examples are not clearly different from those in the previous section on alternative 3′ processing because all available evidence indicates that it is the location of the 3′-end of the pre-mRNA that determines the nature of the splicing. This is not the case in animal viruses such as SV40, polyoma and adenovirus, all of which provide examples of differential splicing that is independent of the polyadenylation site [307]. For example, the L1 RNA of adenovirus is produced at both early and late stages of infection. At early times, it is processed to produce a 4-kb early mRNA while, late in infection, the same transcript is subjected to several splicing strategies to produce three mRNAs of 4.3, 3.8 and 2.3 kb [308–310]. More recently, this type of termination site-independent differential processing has also been observed in cellular genes. In some examples the length heterogeneity appears to result from the use of alternative splice sites within a single intron. These include the three different fibronectin mRNAs that appear to arise by differential splicing of the products of a single gene [311] and the altern-

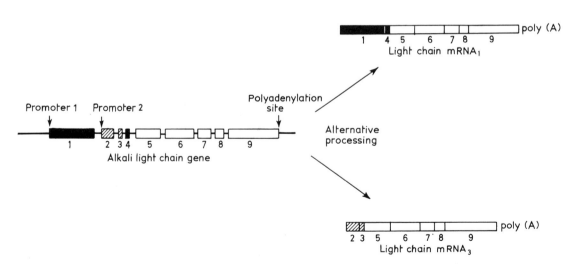

**Fig. 10.14** A schematic representation of the alternative processing of myosin alkali light chain gene transcripts. Boxes 1–9 represent exons and the intervening lines are introns. Shaded exons are unique to light chain mRNA$_1$, which is transcribed from promoter 1. Hatched exons are unique to light chain mRNA$_3$, which is transcribed from promoter 2.

ative processing products of murine $\alpha$ A-crystalin pre-mRNA [312]. In other examples, such as chicken myosin light chain [313] and mouse myelin basic protein [314], the alternative splicing brings together different exon combinations and results in very different polypeptides (Fig. 10.14).

Recently, trans-splicing, the joining of exons from two different mRNAs has been demonstrated *in vitro* [331]. It remains to be seen whether it occurs *in vivo* and whether it provides a mechanism for the distribution of a single exon to more than one mRNA.

The factors which control the splicing strategy are totally unknown.

### 10.6.3 Control of the rate of processing and the nuclear transport of mRNA

Very little is known as to the possible role that the rate of processing or transport might play in the control of gene expression. In many instances such control is indicated by the presence in the nucleus of transcripts which fail to appear as mRNA in the cytoplasm. Thus the RNA sequences which encode the embryonic proteins of sea-urchins are still found in the nuclear polyadenylated RNA of adults but they are not processed and do not enter the cytoplasm [315]. Similarly, histone mRNAs appear to be made throughout the cell cycle of HeLa cells but occur in the cytoplasm only during the period of DNA synthesis [316]. Analbuminaemic rats, the livers of which lack cytoplasmic albumin mRNA, do have albumin mRNA precursors in their liver nuclei [317].

In the above examples pre-mRNA accumulates in the nuclei, thus indicating a defect or control of processing. In rat liver deprived of glucocorticoids, however, there appears to be a defect in the nucleocytoplasmic transport of mature message. Thus, in adrenalectomized animals, mature $\alpha$2a-globulin mRNA accumulates in the nuclei [318].

## REFERENCES

1 Jacob, F. and Monod, J. (1961), *J. Mol. Biol.*, **3**, 318.
2 Miller, J. H. and Reznikoff, W. S. (eds) (1978), in *The Operon*, Cold Spring Harbor Laboratory, New York – eight articles are relevant to the *lac* operon.
3 Reznikoff, W. S. (1984), in *The Microbe* (part II) (ed. D. P. Kelly and N. G. Carr), Cambridge University Press, Cambridge, p. 195.
4 Gilbert, W. and Müller-Hill, B. (1966), *Proc. Natl. Acad. Sci. USA*, **56**, 1891.
5 Riggs, A. D. and Bourgeois, S. (1968), *J. Mol. Biol.*, **34**, 361.
6 Beyreuther, K. (1978), in *The Operon* (ed. J. H. Miller and W. S. Reznikoff), Cold Spring Harbor Laboratory, New York, p. 123.
7 Manly, S. P., Bennett, G. N. and Matthews, K. S. (1984), *J. Mol. Biol.*, **179**, 335.
8 Gilbert, W. and Maxam, A. (1973), *Proc. Natl. Acad. Sci. USA*, **70**, 3581.
9 Barkley, M. D. (1981), *Biochemistry*, **20**, 3833.
10 Winter, R. B., Berg, O. G. and von Hippel, P. H. (1981), *Biochemistry*, **20**, 6961.
11 Platt, T. (1978), in *The Operon* (ed. J. H. Miller and W. S. Reznikoff), Cold Spring Harbor Laboratory, New York, p. 263.
12 Reznikoff, W. S. and Abelson, J. H. (1978), in *The Operon* (ed. J. H. Miller and W. S. Reznikoff), Cold Spring Harbor Laboratory, New York, p. 221.
13 Ullman, A. and Danchin, A. (1983), *Adv. Cyclic Nucleotide Res.*, **15**, 1.
14 Ebright, R., Cossart, P., Gicquel-Sanzey, B. and Beckwith, J. (1984), *Nature (London)*, **311**, 232.
15 Ebright, R., Cossart, P., Gicquel-Sanzey, B. and Beckwith, J. (1984), *Proc. Natl. Acad. Sci. USA*, **81**, 7275.
16 McClure, W. R., Hawley, D. K. and Malan, T. P. (1982), in *Promoters: Structure and Function* (ed. R. L. Rodriguez and M. J. Chamberlin), Praeger, New York, p. 111.
17 Lee, N. (1978) in *The Operon* (ed. J. H. Miller and W. S. Reznikoff), Cold Spring Harbor Laboratory, New York, p. 389.
18 Ogden, S., Haggerty, D., Stoner, C. M., Kolodrubetz, D. and Schleif, R. (1980), *Proc. Natl. Acad. Sci. USA*, **77**, 3346.
19 Lee, N. L., Gielow, W. O. and Wallace, R. G. (1981), *Proc. Natl. Acad. Sci. USA*, **78**, 752.

20 Miyada, C. G., Stoltzfus, L. and Wilcox, G. (1984), *Proc. Natl. Acad. Sci. USA*, **81**, 4120.

21 Adhya, S. and Miller, W. (1979), *Nature (London)*, **279**, 492.

22 Busby, S., Irari, M. and de Crombrugghe, B. (1982), *J. Mol. Biol.*, **154**, 197.

23 Majumdar, A. and Adhya, S. (1984), *Proc. Natl. Acad. Sci. USA*, **81**, 6100.

24 Putney, S. D. and Schimmel, P. (1981), *Nature (London)*, **291**, 632.

25 Ketner, C. and Campbell, A. (1975), *J. Mol. Biol.*, **96**, 13.

26 Otsuka, A. and Abelson, J. (1978), *Nature (London)*, **276**, 689.

27 Kelker, N. E., Mass, W. K., Yang, H-L. and Zubay, G. (1976), *Mol. Gen. Genet.*, **144**, 17.

28 Ptashne, M., Jeffrey, A., Johnson, A. D., Maurer, R., Meyer, B. J., Pabo, C. O., Roberts, T. M. and Sauer, R. T. (1980), *Cell*, **19**, 1.

29 Adhya, S. L., Garges, S. and Ward, D. F. (1981), *Prog. Nucleic Acids Res. Mol. Biol.*, **26**, 103.

30 Johnson, A. D., Poteete, A. R., Lauer, G., Sauer, R. T., Ackers, G. K. and Ptashne, M. (1981), *Nature (London)*, **294**, 217.

31 Ptashne, M. (1984), *Trends Biochem. Sci.*, **9**, 142.

32 Takeda, Y., Ohlendorf, D. H., Anderson, W. F. and Matthews, B. W. (1983), *Science*, **221**, 1020.

33 Pabo, C. O. and Sauer, R. T. (1984), *Annu. Rev. Biochem.*, **53**, 293.

34 Anderson, W., Ohlendorf, D., Takeda, Y. and Matthews, B. W. (1981), *Nature (London)*, **290**, 754.

35 Pabo, C. and Lewis, M. (1982), *Nature (London)*, **298**, 443.

36 Metzler, W. J., Arndt, K., Tecza, E., Wasilewski, J. and Lu, P. (1985), *Biochemistry*, **24**, 1418.

37 Anderson, J. E., Ptashne, M. and Harrison, S. C. (1985), *Nature (London)*, **316**, 596.

38 Warton, R. P. and Ptashne, M. (1985), *Nature (London)*, **316**, 601.

39 Bushman, F. O., Anderson, J. E., Harrison, S. C. and Ptashne, M. (1985), *Nature (London)*, **316**, 651.

40 McKay, D. and Steitz, T. (1981), *Nature (London)*, **290**, 744.

41 Steitz, T., Ohlendorf, D. H., McKay, D. B., Anderson, W. F. and Matthews, B. W. (1982), *Proc. Natl. Acad. Sci. USA*, **79**, 3097.

42 Jackson, E. and Yanofsky, C. (1972), *J. Mol. Biol.*, **69**, 307.

43 Bertrand, K. and Yanofsky, C. (1976), *J. Mol. Biol.*, **103**, 339.

44 Yanofsky, C. (1981), *Nature (London)*, **289**, 751.

45 Platt, T. (1981), *Cell*, **24**, 10.

46 Watson, M. D. (1981), *Trends Biochem. Sci.*, **6**, 180.

47 Kolter, R. and Yanofsky, C. (1982), *Annu. Rev. Genet.*, **16**, 113.

48 Yanofsky, C., Platt, T., Crawford, I. P., Nichols, B. P., Christie, G. E., Horowitz, H., Van Cleemput, M. and Wu, A. M. (1981), *Nucleic Acids Res.*, **9**, 6647.

49 Zurawski, G., Elseviers, D., Stauffer, G. V. and Yanofsky, C. (1978), *Proc. Natl. Acad. Sci. USA*, **75**, 5988.

50 Keller, E. B. and Calvo, J. M. (1979), *Proc. Natl. Acad. Sci. USA*, **76**, 6186.

51 Oxender, D., Zarawski, G. and Yanofsky, C. (1979), *Proc. Natl. Acad. Sci. USA.*, **76**, 5524.

52 Kuroda, M. and Yanofsky, C. (1983), *Proc. Natl. Acad. Sci. USA*, **80**, 2206.

53 Barnes, W. M. (1978), *Proc. Natl. Acad. Sci. USA*, **75**, 4281.

54 Zurawski, G., Brown, K., Killingly, D. and Yanofsky, C. (1978), *Proc. Natl. Acad. Sci. USA*, **75**, 4271.

55 Gardner, J. F. (1979), *Proc. Natl. Acad. Sci. USA*, **76**, 1706.

56 Nargang, F. E., Subrahmanyam, C. S. and Umbarger, H. E. (1980), *Proc. Natl. Acad. Sci. USA*, **77**, 1823.

57 Keller, E. B. and Calvo, J. M. (1979), *Proc. Natl. Acad. Sci. USA*, **76**, 6186.

58 Johnston, H. M., Barnes, W. M., Chumley, F. G., Bossi, L. and Roth, J. R. (1980), *Proc. Natl. Acad. Sci. USA*, **77**, 508.

59 Hauser, C. A. and Halfield, G. W. (1983), *Nucleic Acids Res.*, **11**, 127.

60 Springer, M., Mayaux, J. F., Fayat, G., Plumbridge, J. A., Gaffe, M., Blanquet, S. and Grunberg-Manago, M. (1985), *J. Mol. Biol.*, **181**, 467.

61 Lindahl, L., Archer, R. and Zengel, J. M. (1983), *Cell*, **33**, 241.

62 Greenblatt, J. (1981), *Cell*, **24**, 8.

63 Ward, D. F. and Gottesman, M. E. (1982), *Science*, **216**, 946.

64 Friedman, D. and Gottesman, M. (1983), in *Lambda II* (ed. F. W. Stahl and R. A. Weisberg), Cold Spring Harbor Laboratory, New York, p. 21.

65 Friedman, D. I., Olsin, E. R., Georgopoulos, C., Tilly, K., Herskowitz, I. and Banuell, F. (1984), *Microbiol. Rev.*, **48**, 299.

66 Rosenberg, M., Court, D., Shimatake, H., Brady, C. and Wulff, D. L. (1978), *Nature (London)*, **272**, 414.

67 deCrombrugghe, B., Mudryj, M., Dihaura, R. and Gottesman, M. (1979), *Cell*, **18**, 1145.

68 Georgopoulos, C. P. (1971), *Proc. Natl. Acad. Sci. USA*, **68**, 2977.

69 Das, A. and Wolska, K. (1984), *Cell*, **38**, 165.

70 Ghosh, B. and Das, A. (1984), *Proc. Natl. Acad. Sci. USA*, **81**, 6305.

71 Greenblatt, J. and Li, J. (1981), *J. Mol. Biol.*, **147**, 11.

72 Greenblatt, J. and Li, J. (1981), *Cell*, **24**, 421.

73 Shimamoto, N., Kamiguchi, T. and Utiyama, H. (1983), in *Microbiology – 1983* (ed. D. Schlessinger), American Society for Microbiology, Washington, DC, p. 7.

74 Schmidt, M. C. and Chamberlin, M. J. (1984), *J. Biol. Chem.*, **259**, 15 000.

75 Das, A., Ghosh, B., Barik, S. and Wolska, K. (1985), *Proc. Natl. Acad. Sci. USA*, **82**, 4070.

76 Grayhack, E. J. and Roberts, J. W. (1982), *Cell*, **30**, 637.

77 Holben, W. E. and Morgan, E. A. (1984), *Proc. Natl. Acad. Sci. USA*, **81**, 6789.

78 Losick, R. and Pero, J. (1981), *Cell*, **25**, 582.

79 Doi, R. H. (1982), *Arch. Biochem. Biophys.*, **214**, 772.

80 Losick, R. and Youngman, P. J. (1984), in *Microbiol. Development* (ed. R. Losick and L. Shapiro), Cold Spring Harbor Laboratory, New York, p. 63.

81 Gitt, M. A., Wang, L. F. and Doi, R. H. (1985), *J. Biol. Chem.*, **260** (in press).

82 Stragier, P., Parsot, C. and Bouvier, J. (1985), *FEBS Lett.*, **187**, 11.

83 Moran, C. P., Lang, N., LeGrice, S., Lee, G., Stephens, M., Shonenshein, A. L., Pero, J. and Losick, R. (1982), *Mol. Gen. Genet.*, **186**, 339.

84 Moran, C. P., Lang, N. and Losick, R. (1981), *Nucleic Acids Res.*, **9**, 5979.

85 Moran, C. P., Lang, N., Banner, C. D. B., Haldenwang, W. G. and Losick, R. (1984), *Cell*, **25**, 783.

86 Wong, S. L., Price, C. W., Goldfarb, D. A. and Doi, R. H. (1984), *Proc. Natl. Acad. Sci. USA*, **81**, 1184.

87 Wang, P-Z. and Doi, R. H. (1984), *J. Biol. Chem.*, **259**, 8619.

88 Johnson, W. C., Moran, C. P. and Losick, R. (1983), *Nature (London)*, **302**, 800.

89 Gilman, M. Z. and Chamberlin, M. J. (1983), *Cell*, **35**, 285.

90 Haldenwang, W. G., Lang, N. and Losick, R. (1981), Cell, **23**, 615.

91 Costanzo, M. and Pero, J. (1983), *Proc. Natl. Acad. Sci. USA*, **80**, 1236.

92 Costanzo, M. and Pero, J. (1984), *J. Biol. Chem.*, **259**, 6681.

93 Grossman, A. D., Erickson, J. W. and Gross, C. A. (1984), *Cell*, **38**, 383.

94 Westpheling, J., Ranes, M. and Losick, R. (1985), *Nature (London)*, **313**, 22.

95 DeBruijn, F. J. and Ausubel, F. M. (1983), *Mol. Gen. Genet.*, **192**, 342.

96 Raibaud, O. and Schwartz, M. (1984), *Annu. Rev. Genet.*, **18**, 173.

97 Rabussay, D. (1983), in *Bacteriophage T4* (ed. C. K. Matthews, E. M. Kulter, G. Mosig and P. B. Berget), American Society of Microbiology, Washington, DC, p. 167.

98 Kassavetis, G. A. and Geiduschek, E. P. (1984), *Proc. Natl. Acad. Sci. USA*, **81**, 5101.

99 Stevens, A. and Rhoton, J. C. (1975), *Biochemistry*, **144**, 5074.

100 Malik, S. and Goldfarb, A. (1984), *J. Biol. Chem.*, **259**, 13 292.

101 Goldfarb, A. and Palm, P. (1981), *Nucleic Acids Res.*, **9**, 4863.

102 Hagenbüchle, O., Tosi, M., Schibler, U., Bovey, R., Wellauer, P. and Young, R. (1981), *Nature (London)*, **289**, 643.

103 Young, R. A., Hagenbüchle, O., Wellauer, P. K. K. and Pittet, A. (1983), *Cell*, **33**, 501.

104 Schibler, U., Hagenbüchle, O., Wellauer, K. and Pittet, A. (1983), *Cell*, **33**, 501.

105 Shaw, P., Sordat, B. and Schibler, U. (1985), *Cell*, **40**, 907.

106 Benyajati, C., Spoerel, N., Haymerle, H. and Ashburner, M. (1983), *Cell*, **33**, 125.

107 Carlson, M. and Bolstein, D. (1982), *Cell*, **28**, 145.

108 Perlman, D., Halvorson, H. D. and Cannon, L. E. (1982), *Proc. Natl. Acad. Sci. USA*, **79**, 781.

109 Jellinghaus, U., Schatzle, U., Schmid, W. and Roewekamp, W. (1982), *J. Mol. Biol.*, **159**, 623.

110 Humphries, R. K., Ley, T., Turner, P., Moulton, A. D. and Nienhuis, A. W. (1982), *Cell*, **30**, 173.

111 Allan, M., Zhu, J-d, Montague, P. and Paul, J.

(1984), *Cell*, **38**, 399.

112 Dynan, W. S. and Tjian, R. (1983), *Cell*, **35**, 79.

113 Reudelhuber, T. L. (1984), *Nature (London)*, **312**, 700.

114 Benoist, C. and Chambon, P. (1981), *Nature (London)*, **290**, 305.

115 Gruss, P., Dhar, R. and Khoury, G. (1981), *Proc. Natl. Acad. Sci. USA*, **78**, 943.

116 Laimins, L. A., Tsichlis, P. and Khoury, G. (1984), *Nucleic Acids Res.*, **12**, 6427.

117 Picard, D. (1985), in *Viral and Cellular Transcript Enhancers, Oxford Surveys on Eukaryotic genes* (ed. N. Maclean), Oxford University Press, Oxford, Vol. 2.

118 Boshart, M., Weber, F., Jahn, G., Dorsch-Häsler, K., Fleckenstein, B. and Schaffner, W. (1985), *Cell*, **41**, 521.

119 Gillies, S. D., Morrison, S. L., Oi, V. T. and Tonegawa, S. (1983), *Cell*, **33**, 717.

120 Banerji, J., Olson, L. and Schaffner, W. (1983), *Cell*, **33**, 729.

121 Queen, C. and Baltimore, D. (1983), *Cell*, **33**, 741.

122 Walker, M. D., Edland, T., Boulet, A. M. and Rutter, W. J. (1983), *Nature (London)*, **306**, 557.

123 Khoury, G. and Gruss, P. (1983), *Cell*, **33**, 313.

124 Gluzman, Y. and Shenk, T. (eds), (1983), in *Enhancers and Eukaryotic Gene Expression*, Cold Spring Harbor Laboratory, New York.

125 Augereau, P. and Wasylyk, B. (1984), *Nucleic Acids Res.*, **12**, 8801.

126 Sassone-Corsil, P., Dougherty, J. P., Wasylyk, B. and Chambon, P. (1984), *Proc. Natl. Acad. Sci. USA*, **81**, 308.

127 de Villiers, J., Olson, L., Tyndall, C. and Schaffner, W. (1982), *Nucleic Acids Res.*, **10**, 7965.

128 Mosthaf, L., Pawlita, M. and Gruss, P. (1985), *Nature (London)*, **315**, 597.

129 Weiher, H., Konig, M. and Gruss, P. (1983), *Science*, **219**, 626.

130 Weber, F., de Villiers, J. and Schaffner, W. (1984), *Cell*, **36**, 983.

131 Jongstra, J., Reudelhuber, T. L., Oudet, P., Benoist, C., Chae, C-B., Jeltsch, J-M., Mathis, D. J. and Chambon, P. (1984), *Nature (London)*, **307**, 708.

132 Treisman, R. and Maniatis, T. (1985), *Nature (London)*, **315**, 72.

133 Weber, F. and Schaffner, W. (1985), *Nature (London)*, **315**, 75.

134 Moss, T. (1983), *Nature (London)*, **302**, 223.

135 Roasseau, G. G. (1984), *Biochem. J.*, **224**, 1.

136 Buetti, E. and Diggelman, H. (1981), *Cell*, **23**, 335.

137 Lee, F., Mulligan, R., Berg, P. and Ringold, G. (1981), *Nature (London)*, **294**, 228.

138 Huang, A. L., Ostrowski, M. C., Berard, D. and Hager, G. L. (1981), *Cell*, **27**, 245.

139 Fasel, N., Pearson, K., Buetti, E. and Diggelman, H. (1982), *EMBO J.*, **1**, 1.

140 Payvar, F., DeFranco, D., Firestone, G. L., Edgar, B., Wrange, Ö., Okret, S., Gustafsson, J-A. and Yamamoto, K. R. (1983), *Cell*, **35**, 381.

141 Hynes, N. E., van Ooyen, A., Kennedy, N., Herrlich, P., Ponta, H. and Groner, B. (1983), *Proc. Natl. Acad. Sci. USA*, **80**, 3637.

142 Buetti, E. and Diggelman, H. (1983), *EMBO J.*, **2**, 1423.

143 Majors, J. and Varmus, H. E. (1983), *Proc. Natl. Acad. Sci. USA*, **80**, 5866.

144 Ponta, J., Kennedy, N., Skroch, P., Hynes, N. E. and Groner, B. (1985), *Proc. Natl. Acad. Sci. USA*, **82**, 1020.

145 Scheidereit, C., Geisse, S., Westphal, H. M. and Beato, M. (1983), *Nature (London)*, **304**, 749.

146 Pjahl, M., McGinnis, D., Hendrich, M., Groner, B. and Hynes, N. E. (1983), *Science*, **222**, 1341.

147 Scheidereit, C. and Beato, M. (1984), *Proc. Natl. Acad. Sci. USA*, **81**, 3029.

148 Karin, M., Haslinger, A., Holtgreve, H., Richards, R. I., Krauter, P., Westphal, H. M. and Beato, M. (1984), *Nature (London)*, **308**, 513.

149 Shinomiya, T., Scherer, G., Schmid, W., Zentgraf, H. and Schütz, G. (1984), *Proc. Natl. Acad. Sci. USA*, **81**, 1346.

150 Cato, A. C. B., Geisse, S., Wenz, M., Westphal, H. M. and Beato, M. (1984), *EMBO J.*, **3**, 2771.

151 Renkawitz, R., Schütz, G., von der Ahe, D. and Beato, M. (1984), *Cell*, **37**, 503.

152 Robins, D. M., Paek, I., Seebury, P. H. and Axel, R. (1982), *Cell*, **29**, 623.

153 Mulvihill, E., LePennec, J-P. and Chambon, P. (1982), *Cell*, **24**, 621.

154 Dean, D. C., Knoll, B. J., Riser, M. E. and O'Malley, B. W. (1983), *Nature (London)*, **305**, 551.

155 Jost, J-P., Seldran, M. and Geiser, M. (1984), *Proc. Natl. Acad. Sci. USA*, **81**, 429.

156 Pelham, H. R. B. and Bienz, M. (1982), *EMBO J.*, **1**, 1473.

157 Corces, V. and Pellicer, A. (1984), *J. Biol. Chem.*, **259**, 14 812.

158 Fujita, T., Ohno, S., Yasumitsu, H. and Taniguchi, T. (1985), *Cell*, **41**, 489.

159 Ryals, J., Dierks, P., Ragg, H. and Weissman, C. (1985), *Cell*, **41**, 497.

160 Goodbourn, S., Zinn, K. and Maniatis, T. (1985), *Cell*, **41**, 509.

161 Morelli, G., Nagy, F., Fraley, R. T., Rogers, S. G. and Chua, N. H. (1985), *Nature (London)*, **315**, 200.

162 Yocum, R., Hanley, S., West, R. and Ptashne, M. (1984), *Mol. Cell. Biol.*, **4**, 1985.

163 West, R., Yocum, R. and Ptashne, M. (1984), *Mol. Cell. Biol.*, **4**, 2467.

164 Brent, R. (1985), *Cell*, **42**, 3.

165 Johnson, A. D. and Herskowitz, I. (1985), *Cell*, **42**, 237.

166 Charnay, P., Treisman, R., Mellon, P., Chao, M., Axel, R. and Maniatis, T. (1984), *Cell*, **38**, 251.

167 Wright, S., Rosenthal, A., Flavell, R. and Grosveld, F. (1984), *Cell*, **38**, 265.

168 Teem *et al.* (1984), *Nucleic Acids Res.*, **12**, 8295.

169 Nagamine, Y. and Reich, E. (1985), *Proc. Natl. Acad. Sci. USA*, **82**, 4606.

170 Mason, J. D., Williams, G. T. and Neuberger, M. S. (1985), *Cell*, **41**, 479.

171 Grosschedl, R. and Baltimore, D. (1985), *Cell*, **41**, 885.

172 Ott, M-D., Sperling, L., Herbomel, P., Yaniv, M. and Weiss, M. C. (1984), *EMBO J.*, **3**, 2505.

173 Brand, A. H., Breeden, L., Abraham, J., Sternglanz, R. and Nasmyth, K. (1985), *Cell*, **41**, 41.

174 Davison, B. L., Egly, J-M., Mulvihill, E. R. and Chambon, P. (1983), *Nature (London)*, **301**, 680.

175 Honda, B. M. and Roeder, R. G, (1980), *Cell*, **22**, 119.

176 Heberlein, U., England, B. and Tjian, R. (1985), *Cell*, **41**, 965.

177 Giniger, E., Varnum, S. M. and Ptashne, M. (1985), *Cell*, **40**, 767.

178 Corsi, P. S., Wildeman, A. and Chambon, P. (1985), *Nature (London)*, **313**, 458.

179 Emerson, B. M., Lewis, C. D. and Felsenfeld, G. (1985), *Cell*, **41**, 21.

180 Wu, C. (1984), *Nature (London)*, **309**, 229.

181 Bogenhagen, D. F., Wormington, W. M. and Brown, D. D. (1982), *Cell*, **28**, 413.

182 Gottesfeld, J. and Bloomer, L. S. (1982), *Cell*, **28**, 781.

183 Kingston, R. E., Baldwin, A. S. and Sharp, P. A. (1985), *Cell*, **41**, 3.

184 Jaenisch, R. and Jähner, D. (1984), *Biochim. Biophys. Acta*, **782**, 1.

185 Bird, A. P. (1984), *Nature (London)*, **307**, 503.

186 Weintraub, H., Larsen, A. and Groudine, M. (1981), *Cell*, **24**, 333.

187 Shen, C. J. and Maniatis, T. (1980), *Proc. Natl. Acad. Sci. USA*, **77**, 6634.

188 van der Phloeg, L. and Flavell, R. A. (1980), *Cell*, **19**, 947.

189 Mandel, J. L. and Chambon, P. (1979), *Nucleic Acids Res.*, **7**, 2081.

190 Stumph, W. E., Baez, M., Beattie, W. G., Tsai, M-J. and O'Malley, B. W. (1983), *Biochemistry*, **22**, 306.

191 Rogers, J. and Wall, R. (1981), *Proc. Natl. Acad. Sci. USA*, **78**, 7497.

192 Yagi, M. and Koshland, M. E. (1981), *Proc. Natl. Acad. Sci. USA*, **78**, 4907.

193 Storb, U. and Arp, B. (1983), *Proc. Natl. Acad. Sci. USA*, **80**, 6642.

194 Compere, S. J. and Palmiter, R. D. (1981), *Cell*, **25**, 233.

195 Eiferman, E. A., Young, P. R., Scott, R. W. and Tilghman, S. M. (1981), *Nature (London)*, **294**, 713.

196 Jähner, D., Stahlman, H., Stewart, C. L., Harbers, K., Lohler, J., Simon, I. and Jaenisch, R. (1982), *Nature (London)*, **298**, 623.

197 Simon, D., Stuhlmann, H., Jähner, D., Wagner, H., Werner, E. and Jaenisch, D. (1983), *Nature (London)*, **304**, 275.

198 Busslinger, M., Hurst, J. and Flavell, R. A. (1983), *Cell*, **34**, 197.

199 Stein, R., Sciaky-Gallili, N., Razin, A. and Cedar, H. (1983), *Proc. Natl. Acad. Sci. USA*, **80**, 2423.

200 McKeon, C., Ohkubo, H., Pastan, I. and de Crombrugghe, B. (1984), *Nucleic Acids Res.*, **12**, 3491.

201 Battistuzzi, G., D'Urso, M., Toniolo, D., Persico, G. M. and Luzzatto, L. (1985), *Proc. Natl. Acad. Sci. USA*, **82**, 1465.

202 Wilks, A. F., Cozens, P. J., Mattaj, I. W. and Jost, J-P. (1982), *Proc. Natl. Acad. Sci. USA*, **79**, 4252.

203 Jost, J-P., Seldran, M. and Geiser, M. (1984), *Proc. Natl. Acad. Sci. USA*, **81**, 429.

204 Gerber-Huber, S., May, F. E., Westly, B. R., Felber, B. K., Hosbach, H. A., Andres, A. C.

and Ryffel, G. U. (1983), *Cell*, **33**, 43.

205 Weintraub, H. and Groudine, M. (1976), *Science*, **193**, 848.

206 Garel, A. and Axel, R. (1976), *Proc. Natl. Acad. Sci. USA*, **73**, 3966.

207 Weisbrod, S. (1982), *Nature (London)*, **297**, 289.

208 Alevy, M. C., Tsai, M-J. and O'Malley, B. W. (1984), *Biochemistry*, **23**, 2309.

209 Stalder, J., Larsen, A., Engel, J. D., Dolan, M., Groudine, M. and Weintraub, H. (1980), *Cell*, **20**, 451.

210 Groudine, M., Kohwi-Shigematsu, T., Gelinas, R., Slamatoyannopoulos, G. and Papayannopoulou, T. (1983), *Proc. Natl. Acad. Sci. USA*, **80**, 7551.

211 McKeon, C., Pastan, I. and de Crombrugghe, B. (1984), *Nucleic Acids Res.*, **12**, 3491.

212 Palmiter, R., Mulvihill, E., McKnight, S. and Senear, A. (1977), *Cold Spring Harbor Symp. Quant. Biol.*, **42**, 639.

213 Williams, J. L. and Tata, J. R. (1983), *Nucleic Acids Res.*, **11**, 1151.

214 Jongstra, J., Reudelhuber, T. L., Oudet, P., Benoist, C., Chae, C-B., Jeltsch, J-M., Mathis, D. J. and Chambon, P. (1984), *Nature (London)*, **307**, 708.

215 Becker, P., Renkawitz, R. and Schütz, G. (1984), *EMBO J.*, **3**, 2015.

216 Shimada, T. and Nienhuis, A. W. (1985), *J. Biol. Chem.*, **260**, 2468.

217 Siebenlist, U., Hennighausen, L., Battey, J. and Leder, P. (1984), *Cell*, **37**, 381.

218 Zaret, K. S. and Yamamoto, K. R. (1984), *Cell*, **38**, 29.

219 Tuan, D. and London, I. M. (1984), *Proc. Natl. Acad. Sci. USA*, **81**, 2718.

220 Schubach, W. and Groudine, M. (1984), *Nature (London)*, **307**, 702.

221 Garguilo, G., Razui, F., Ruberti, I., Mohr, I. and Worcel, A. (1985), *J. Mol. Biol.*, **181**, 333.

222 Tuan, D., Solomon, W., Li, Q. and London, I. M. (1985), *Proc. Natl. Acad. Sci. USA*, **82**, 6384.

223 Burch, J. and Weintraub, H. (1983), *Cell*, **33**, 65.

224 Fritton, H. P., Igo-Kemenes, T., Nowock, J., Strech-Jurk, U., Theisen, M. and Sippel, A. E. (1984), *Nature (London)*, **311**, 163.

225 Smith, R. D., Seale, R. L. and Yu, J. (1983), *Proc. Natl. Acad. Sci. USA*, **80**, 5505.

226 McGhee, J. D., Wood, W. I., Dolan, M., Engel, J. D. and Felsenfeld, G. (1981), *Cell*, **27**, 45.

227 Larsen, A. and Weintraub, H. (1982), *Cell*, **29**, 609.

228 Weintraub, H. (1983), *Cell*, **32**, 1191.

229 Elgin, S. C. R. (1984), *Nature (London)*, **309**, 213.

230 Kornberg, R. (1981), *Nature (London)*, **292**, 579.

231 Emerson, B. M., Lewis, C. D. and Felsenfeld, G. (1985), *Cell*, **41**, 21.

232 Kmiec, E. B. and Worcel, A. (1985), *Cell*, **41**, 945.

233 Morse, R. J. and Cantor, C. R. (1985), *Proc. Natl. Acad. Sci. USA*, **82**, 4653.

234 Villeponteau, B., Lundell, M. and Martinson, H. (1984), *Cell*, **39**, 469.

235 North, G. (1985), *Nature (London)*, **316**, 394.

236 Nordheim, A. and Rich, A. (1983), *Nature (London)*, **303**, 674.

237 Azorin, F. and Rich, A. (1985), *Cell*, **41**, 365.

238 Hayes, T. E. and Dixon, J. E. (1985), *J. Biol. Chem.*, **260**, 8145.

239 Gilmour, R. S., Spandidos, D. A., Vass, J. K., Gow, J. W. and Paul, J. (1984), *EMBO J.*, **3**, 1263.

240 Santoro, C., Costanzo, F. and Ciliberto, G. (1984), *EMBO J.*, **3**, 1553.

241 Cartwright, I. C., Abmayr, S. M., Fleischmann, G., Lowenhaupt, K., Elgin, S., Keene, M. A. and Howard, G. C. (1982), *Crit. Rev. Biochem.*, **13**, 1.

242 Leffak, M. and Trempe, J. P. (1985), *Nucleic Acids Res.*, **13**, 4853.

243 Pederson, T. (1978), *Int. Rev. Cytol.*, **55**, 1.

244 Weintraub, H. (1984), *Cell*, **38**, 17.

245 Newrock, K. M., Alfageme, C. R., Nardi, R. V. and Cohen, L. H. (1977), *Cold Spring Harbor Symp. Quant. Biol.*, **42**, 421.

246 Poccia, D., Salik, J. and Krystal, G. (1981), *Dev. Biol.*, **82**, 287.

247 Savic, A., Richman, P., Williamson, P. and Poccia, D. (1981), *Proc. Natl. Acad. Sci. USA*, **78**, 3706.

248 Marushige, K. and Dixon, G. H. (1971), *J. Biol. Chem.*, **246**, 5799.

249 Allis, C. D., Glover, C., Bowen, J. K. and Gorovsky, M. A. (1980), *Cell*, **20**, 609.

250 Oliva, R. and Mezquita, C. (1982), *Nucleic Acids Res.*, **10**, 8049.

251 Chalkley, R. and Shires, A. (1985), *J. Biol. Chem.*, **260**, 7698.

252 Levinger, L. and Varshavsky, A. (1982), *Cell*, **28**, 375.

253 Yu, F-L. (1985), *FEBS Lett.*, **190**, 109.

254 Ushiroyama, T., Tanigawa, Y., Tsuchiya, M.,

Matsuara, R., Ueki, M., Sagimoto, O. and Shimoyuma, M. (1985), *Eur. J. Biochem.*, **151**, 173.

255 Levy, W. B. and Dixon, G. (1978), *Nucleic Acids Res.*, **5**, 4155.

256 Albanese, I. and Weintraub, H. (1980), *Nucleic Acids Res.*, **8**, 2790.

257 Weisbrod, S. and Weintraub, H. (1979), *Proc. Natl. Acad. Sci. USA*, **76**, 631.

258 Weisbrod, S., Groudine, M. and Weintraub, H. (1980), *Cell*, **19**, 289.

259 Weisbrod, S. and Weintraub, H. (1981), *Cell*, **23**, 391.

260 Bhorjee, J. S. (1981), *Proc. Natl. Acad. Sci. USA*, **78**, 6944.

261 Robinson, S. I., Nelkin, B. D. and Vogelstein, B. (1982), *Cell*, **28**, 99.

262 Robinson, S. I., Small, D., Idzerda, R., McKnight, G. S. and Vogelstein, B. (1983), *Nucleic Acids Res.*, **11**, 5113.

263 Ciejek, E. M., Tsai, M-J. and O'Malley, B. W. (1983), *Nature (London)*, **306**, 607.

264 Hentzen, P. C., Rho, J. H. and Bekhor, I. (1984), *Proc. Natl. Acad. Sci. USA*, **81**, 304.

265 Kirov, N., Djondjurov, L. and Roumen, T. (1984), *J. Mol. Biol.*, **180**, 601.

266 Jost, J-P. and Seldran, M. (1984), *EMBO J.*, **3**, 2005.

267 Jackson, D. A. and Cook, P. R. (1985), *EMBO J.*, **4**, 919.

268 Kuo, M. T. (1982), *J. Cell. Biol.*, **93**, 278.

269 Stark, G. R. and Wahl, G. M. (1984), *Annu. Rev. Biochem.*, **53**, 447.

270 Schimke, R. T. (1984), *Cell*, **37**, 705.

271 Glover, D. M., Zaha, A., Stocker, A. J., Santelli, R. V., Pueyo, M. T., de Toledo, S. M. and Lara, F. J. S. (1982), *Proc. Natl. Acad. Sci. USA*, **79**, 2947.

272 Spradling, A. C. and Mahowald, A. P. (1980), *Proc. Natl. Acad. Sci. USA*, **77**, 1096.

273 Spradling, A. C. (1981), *Cell*, **27**, 193.

274 Ish-Horowicz, D. (1982), *Nature (London)*, **296**, 806.

275 de Cicco, D. V. and Spradling, A. C. (1984), *Cell*, **38**, 45.

276 Fox, M. (1984), *Nature (London)*, **307**, 212.

277 Horns, R. C., Dower, W. J. and Schimke, R. T. (1984), *J. Clin. Oncol.*, **2**, 2.

278 Coderre, J. A., Beverley, S. M., Schimke, R. T. and Santi, D. V. (1983), *Proc. Natl. Acad. Sci. USA*, **80**, 2132.

279 Anderson, R. P. and Roth, J. R. (1977), *Annu. Rev. Microbiol.*, **31**, 473.

280 Travers, A. (1984), *Nature (London)*, **311**, 410.

281 Laporte, D. C. (1984), *Trends Biochem. Sci.*, **9**, 463.

282 Simmons, R. W. and Kleckner, N. (1983), *Cell*, **34**, 683.

283 Mizuno, T., Chou, M-Y. and Inouye, M. (1984), *Proc. Natl. Acad. Sci. USA*, **81**, 1966.

284 Coleman, J., Green, P. J. and Inouye, M. (1984), *Cell*, **37**, 429.

285 Ellison, M. J., Kelleher, R. J. and Rich, A. (1985), *J. Biol. Chem.*, **260**, 9085.

286 Izant, J. G. and Weintraub, H. (1984), *Cell*, **36**, 1007.

287 Kim, S. K. and Wold, B. J. (1985), *Cell*, **42**, 129.

288 Coleman, J., Hirashima, A., Inokuchi, Y., Green, P. J. and Inouye, M. (1985), *Nature (London)*, **315**, 601.

289 Sutcliffe, J. G., Milner, R. J., Bloom, F. E. and Lerner, R. A. (1982), *Proc. Natl. Acad. Sci. USA*, **79**, 4942.

290 Sutcliffe, J. G., Milner, R. J., Gottesfeld, J. M. and Lerner, R. A. (1984), *Nature (London)*, **308**, 237.

291 Minárovits, J., Kovács, Z. and Földes, I. (1984), *FEBS Lett.*, **174**, 208.

292 Owens, G. P., Chaudhari, N. and Hahn, W. E. (1985), *Science*, **229**, 1263.

293 Alt, F. W., Bothwell, A. L., Knapp, M., Siden, E., Mather, E., Koshland, M. and Baltimore, D. (1980), *Cell*, **20**, 293.

294 Early, P., Rogers, J., Davis, M., Calame, K., Bond, M., Wall, R. and Hood, L. (1980), *Cell*, **20**, 313.

295 Nelson, K. J., Haimovich, J. and Perry, R. P. (1983), *Mol. Cell. Biol.*, **3**, 1317.

296 Rosenfeld, M. G., Mermod, J. J., Amara, S. G., Swanson, L. W., Sawchenko, P. E, River, J., Vale, W. W. and Evans, R. M. (1983), *Nature (London)*, **304**, 129.

297 Boardman, M., Basi, G. S. and Storti, R. V. (1985), *Nucleic Acids Res.*, **13**, 1763.

298 Henikoff, S., Sloan, J. S. and Kelly, J. D. (1983), *Cell*, **34**, 405.

299 Nevins, J. R. and Chen-Kiang, S. (1981), *Adv. Virus Res.*, **26**, 1.

300 Kaufman, R. J. and Sharp, P. A. (1983), *Mol. Cell. Biol.*, **3**, 1598.

301 Aho, S., Tate, U. and Boedtker, H. (1983), *Nucleic Acids Res.*, **11**, 5443.

302 Mason, P. J., Jones, M. B., Elkington, J. A. and Williams, J. G. (1985), *EMBO J.*, **4**, 205.

303 Tosi, M., Young, R. A., Hagenbüchle, O. and Schibler, U. (1981), *Nucleic Acids Res.*, **11**, 2313.

304 Maki, R., Roeder, W., Traunecker, A., Sidman, C., Wabi, M., Raschke, W. and Tonegawa, S. (1981), *Cell*, **24**, 353.

305 Mather, E. L., Nelson, K. J., Haimovich, J. and Perry, R. P. (1984), *Cell*, **36**, 329.

306 Yaoita, Y., Kamagai, Y., Okamura, K. and Honjo, T. (1982), *Nature (London)*, **297**, 697.

307 Ziff, E. B. (1980), *Nature (London)*, **287**, 491.

308 Chow, L. T., Broker, T. R. and Lewis, J. B. (1979), *J. Mol. Biol.*, **134**, 265.

309 Akasjarvi, G. and Persson, H. (1981), *Nature (London)*, **292**, 420.

310 Nevins, J. R. and Wilson, M. C. (1981), *Nature (London)*, **290**, 113.

311 Schwarzbauer, J. E., Tamkun, J. W., Lemischka, I. R. and Hynes, R. D. (1983), *Cell*, **35**, 421.

312 King, C. R. and Piatigorsky, J. (1983), *Cell*, **32**, 707.

313 Nabeshima, Y-i., Fujii-Kuriyama, Y., Muramatsu, M. and Ogata, K. (1984), *Nature (London)*, **308**, 333.

314 Takahashi, N., Roach, A., Teplow, D. B., Prusiner, S. B. and Hood, L. (1985), *Cell*, **42**, 139.

315 Wold, B. J., Klein, W. H., Hough-Evans, B. R., Britten, R. J. and Davidson, E. H. (1978), *Cell*, **14**, 941.

316 Melli, M., Spinelli, G. and Arnold, E. (1977), *Cell*, **12**, 167.

317 Esumi, H., Takahashi, Y., Sekiya, T., Sato, S., Nagase, S. and Sugimura, T. (1982), *Proc. Natl. Acad. Sci. USA*, **79**, 734.

318 Fulton, R., Birnie, G. D. and Knowler, J. T. (1985), *Nucleic Acids Res.*, **13**, 6467.

319 Nelson, H. C. M. and Sauer, R. T. (1985), *Cell*, **42**, 549.

320 Schevitz, R. W., Otwinowski, Z., Joachimiak, A., Lawson, C. L. and Sigler, P. B. (1985), *Nature (London)*, **317**, 782.

321 Tsugawa, A., Kurihara, T., Zuber, M., Court, D. L. and Nakamura, Y. (1985), *EMBO J.*, **4**, 2337.

322 Grayhack, W. J., Yang, X., Lau, L. F. and Roberts, J. W. (1985), *Cell*, **42**, 259.

323 Brown, D. D. and Schissel, M. S. (1985), *Cell*, **42**, 759.

324 Buttgereit, D., Pflugfelder, G. and Grummt, I. (1985), *Nucleic Acids Res.*, **13**, 8165.

325 Spinelli, G. and Ciliberto, G. (1985), *Nucleic Acids Res.*, **13**, 8065.

326 Adams, R. L. P. and Burdon, R. H. (1985), in *Molecular Biology of DNA Methylation*, Springer-Verlag, New York.

327 Finley, D. and Varshavsky, A. (1985), *Trends Biochem. Sci.*, **10**, 343.

328 Mirkovitch, J., Mirault, M-E, and Laemmli, U. K. (1984), *Cell*, **39**, 223.

329 Cesareni, G. and Banner, D. W. (1985), *Trends Biochem. Sci.*, **10**, 303.

330 Farnham, P. J., Abrams, J. M. and Schimke, R. T. (1985), *Proc. Natl. Acad. Sci. USA*, **82**, 3978.

331 Konarska, M. M., Padgett, R. A. and Sharp, P. A. (1985), *Cell*, **42**, 165.

332 Gehring, W. J. (1985), *Cell*, **40**, 3.

333 Manley, J. L. and Levine, M. S. (1985), *Cell*, **43**, 1.

334 Garcia-Bellido, A. (1977), *Ann. Zool.*, **17**, 613.

335 Wittig, S. and Wittig, B. (1982), *Nature (London)*, **297**, 31.

336 Rhodes, D. (1985) *EMBO J.*, **4**, 3473.

337 Strauss, F. and Varshavsky, A. (1984), *Cell*, **37**, 889.

338 Weintraub, H. (1980), *Nucleic Acids Res.*, **8**, 4745.

# 11

# The translation of mRNA: protein synthesis

The structural genes of cellular DNA contain the genetic information to specify proteins. The sequence of amino acids in proteins is determined by, and is colinear with [1, 2], the sequence of bases in the DNA. In the process of *transcription* (Chapter 8) the genetic information is transferred from the DNA to messenger RNA. It is from the mRNA that the four-letter language of the genetic information of the nucleic acids is *translated* into the twenty-letter language of the proteins. This process of protein biosynthesis occurs on the ribosome and involves the participation of ribosomal RNA and transfer RNA. Particular emphasis on the part played by the nucleic acids will be given in the following account of translation. More detailed treatment of this subject may be found elsewhere [3–9].

## 11.1 THE STRUCTURE OF tRNA

Transfer RNA plays the key role in decoding the genetic information of the sequence of bases in the mRNA into a sequence of amino acids in proteins. Although the elucidation of the general features of the genetic code preceded the discovery of tRNA [10, 11], it is convenient to deal with the structure of tRNA (see [12, 13] for reviews) before considering the details of the genetic code (Secction 11.2).

Transfer RNAs are small RNAs the essential common features of which are that each can covalently attach one specific amino acid to its 3′-end, and that each possesses a sequence of three bases, the *anticodon*, complementary to and able to interact by hydrogen-bonded base-pairing with a mRNA codon for this, so-called, cognate amino acid. The genetic code is thus determined by which aminoacyl-tRNAs can recognize which codons.

The primary structures of over 350 tRNAs from many different species have now been established. The tRNAs range in length from 73 to 94 nucleotides and are characterized by a relatively large proportion of modified or non-standard nucleosides (see Section 2.1.2 and 9.5.2). All these primary structures allow themselves to be arranged into a common secondary structure, specific and generalized examples of which are shown in Fig. 11.1(a) and (b), respectively. The common features of these structures are as follows.

(i) A stem, the acceptor or aminoacyl stem, containing the 5′ and 3′ extremities of the molecule. It consists of a helix of seven pairs of bases generally making Watson–Crick base pairs (i.e. A–U or G–C) but occasionally (e.g. Fig. 11.1(a)) with a G–U base pair (see also Fig. 11.4, below), together with an unpaired sequence of four bases at

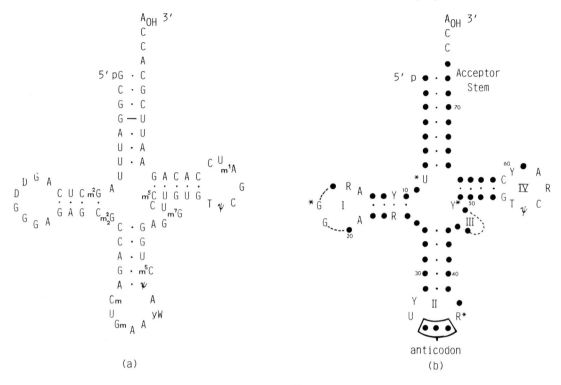

**Fig. 11.1** Secondary structure of tRNA. (a) Yeast tRNA[Phe]; (b) generalized structure in which only the invariant or semi-invariant bases are named, the others being represented by filled circles. An asterisk indicates that a base may be modified. Hydrogen bonds in the helical stems are indicated by dots, except for the G-U pair in (a) which is indicated by a line for emphasis. The dotted lines indicate a variable number of nucleotides. R and Y stand for purine and pyrimidine, respectively; other symbols are defined at the beginning of the book. An explanation of the numbering of the variable loops can be found in [12].

the end of the 3'-strand of the stem. The last three bases are always CCA, and it is to the 3'-terminal adenosine residue that the amino acid is attached (see Section 11.3).

(ii) An arm, the dihydrouridine (D) arm, comprising a stem of three or four base pairs together with the D loop (loop I) of eight to eleven nucleotides, some of which are invariant. Although loop I generally contains one or more dihydrouridine residues (hence one of its common names), several examples are known in which this base is absent.

(iii) An arm, the anticodon arm, comprising a helical stem of five base pairs together with

the anticodon loop (loop II) of seven nucleotides. It is worth pointing out that the base 5' to the centrally placed anticodon is always U, and that two other bases in the loop are semi-invariant (i.e. restricted to being purines or pyrimidines, respectively). The 5'-base of the anticodon, the base in the 'wobble' position (see Section 11.2.5) is frequently a modified or non-standard base.

(iv) An extra arm (III) of extreme variability, ranging from four to 21 nucleotides. This may be either a helical stem together with a loop of three or four nucleotides (13–21 nucleotides in all), or merely a loop of three to five nucleotides.

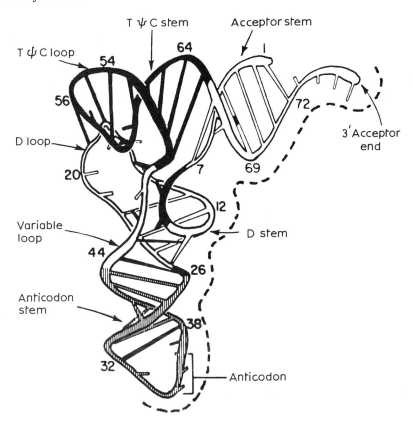

**Fig. 11.2** Tertiary structure of yeast tRNA$^{Phe}$. The sugar–phosphate backbone is shown as a coiled tube, the numbers referring to nucleotide residues starting from the 5'-end. Hydrogen bonding interactions between bases are shown as cross-rungs, tertiary interactions being shaded solid black. Bases not involved in hydrogen bonding are shown as shortened rods attached to the backbone (from [17] with permission).

(v) An arm, the TψC arm, comprising a helical stem of five base pairs, and a loop (loop IV) of seven nucleotides.

The X-ray crystal structures of four tRNAs have so far been determined [12–16], and that for yeast tRNA$^{Phe}$, the one for which greatest resolution has been obtained, is shown in Fig. 11.2. This has been variously described as L-shaped or T-shaped, and although both designations legitimately stress particular aspects of the structure, either alone is inadequate. The L-shape is perhaps most immediately apparent in the two angularly orientated domains of the molecule: one comprising the TψC stem and acceptor stem with the 3'CCA at its furthest extremity; and the other comprising the D-stem, the variable loop (III) and the anticodon arm, with the anticodon loop (II) at its furthest extremity. The manner in which the individual helical stems augment one another to form these two domains of extensive base stacking is perhaps the most fundamentally important feature of the structure. The corner made by the junction of the two domains comprises the D-loop (I) and the TψC-loop (IV). The dimensions of the domains are similar: approximately 60 Å (6 nm) from the TψC-loop at the corner to either anticodon or 3′ acceptor end; whereas the distance between these extremities is approximately 75 Å (7.5 nm).

The inner surface of the two domains (broken line in Fig. 11.2) rather belies the description, L-shaped, that applies to the outer surface. It comprises a more or less planar region that in two dimensions can be regarded as the top of a T, the base of which (at about nucleotide 56) is equivalent to the angle of the L. This inner surface contains many of the variable bases of the D-stem and variable loop (III), which, furthermore, are not involved in any of the tertiary interactions discussed below. They are thus available for interaction with other molecules, and may be of importance in the interaction of a tRNA with its aminoacyl-tRNA synthetase (Section 11.3).

The tertiary structure of yeast tRNA$^{Phe}$ is maintained by a large number of hydrophobic stacking interactions in the augmented helices, together with additional specific hydrogen-bonding interactions between nucleotide residues that are often widely separated in the secondary structure cloverleaf (Fig. 11.3). Many of these interactions are between invariant or semi-variant bases, suggesting a rationale for their restricted variance and a generality for the yeast tRNA$^{Phe}$ tertiary structure. These hydrogen bonding interactions are in no way confined to Watson–Crick base pairs, but include a variety of non-standard interactions, some involving three bases (Fig. 11.4). Indeed, some of the interactions involve the sugar–phosphate backbone. Although these hydrogen bonds may at first sight seem esoteric, their occurrence in tRNAs merely reflects the fact that the bases involved are not confined to the relative spatial orientations which they are forced to adopt in a standard RNA A-helix. The potential for the 2′-OH of the ribose to participate in hydrogen-bonding may be the basis of the ability of RNA molecules to form such non-helical structures, not possible in the case of DNA.

There is much evidence to support the

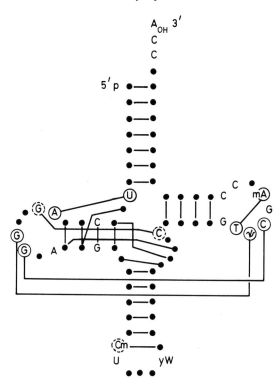

**Fig. 11.3** Relationship of tertiary hydrogen bonding interactions to cloverleaf secondary structure of yeast tRNA$^{Phe}$. The generalized convention of Fig. 11.1(b) has been applied to the structure in Fig. 11.1(a) to highlight the involvement of invariant and semi-invariant bases in the tertiary interactions (extended lines). (Adapted from [17] with permission.)

proposition that the crystal structure of yeast tRNA$^{Phe}$ also corresponds to that which it adopts in solution [13]. Furthermore the general base-stacked two-domain structure stabilized by tertiary interactions also obtains in yeast tRNA$^{Asp}$ [16], and in yeast and *E. coli* tRNA$_f^{Met}$ (see Section 11.2.4), suggesting that it is a general feature of tRNAs. Nevertheless it is possible that conformational changes in the structure of tRNA may occur during protein biosynthesis on the ribosome.

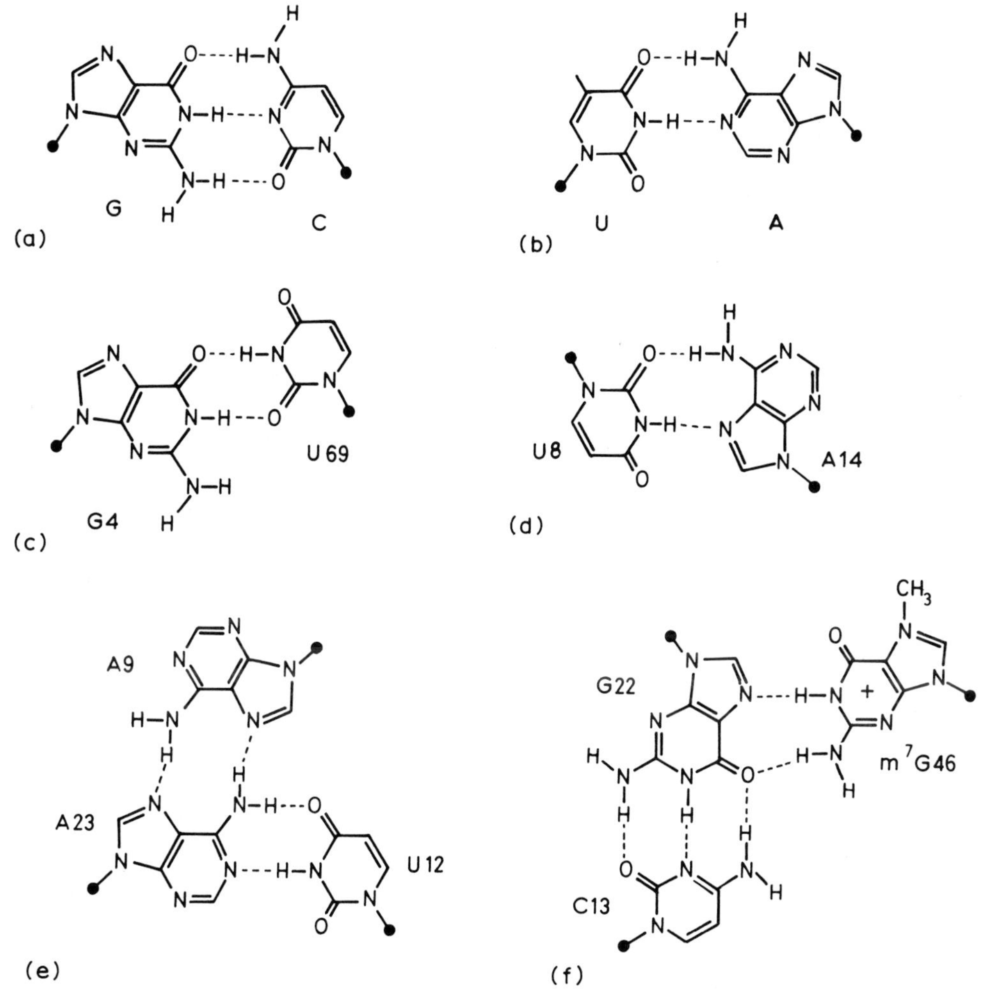

**Fig. 11.4** Secondary and tertiary structure hydrogen bonds in yeast tRNA$^{Phe}$. (a), (b) and (c) show the structure of the types of base pairs in the helical secondary structure; (d), (e) and (f) show examples of some of the unusual hydrogen-bonded interactions in the tertiary structure (cf Figs. 11.2 and 11.3). These latter involve bases in the D-stem. Numbering is as in Fig. 11.1(b). (Adapted from [12], with permission.)

## 11.2 THE GENETIC CODE

### 11.2.1 The codon as a nucleotide triplet

In proteins, 20 different kinds of amino acids are commonly found, whereas only four main kinds of base occur in the nucleic acids. The genetic code describes how a sequence derived from 20 or more units is determined by a sequence derived from four units of a different type (for reviews see [18–23]).

Since there are only four kinds of base but 20 kinds of amino acid, the correspondence cannot be a simple 1:1 relationship between bases and amino acids. Nor are there sufficient combinations of two bases ($4^2$, i.e. 16) to account for 20 amino acids. It had been suspected, therefore, that each amino acid would be determined by a

```
        C A T｜C A T｜C A T｜C A T｜C A T｜C A T｜..........
 + 1    C A T｜C A N｜T C A｜T C A｜T C A｜T C A｜T...........
+1 –1   C A T｜C A N｜T C T｜C A T｜C A T｜C A T｜..........
 + 2    C A T｜N C A｜N T C｜A T C｜A T C｜A T C｜A T.......
 + 3    C A N｜T N C｜A N T｜C A T｜C A T｜C A T｜C A T...
```

**Fig. 11.5** Hypothetical sequence of bases in a DNA strand showing genetic message in triplets. Addition of one base (second line) or two bases (fourth line) makes the code unreadable but it can be restored if one base is added and another removed nearby (third line). The message is still readable if three bases are inserted (last line).

sequence of at least three bases, which would give 64 combinations ($4^3$) – more than adequate for the coding of the 20 amino acids. Crick and his colleagues [18, 22, 23] produced fairly clear-cut evidence that the *triplet* theory was correct and that what they termed the *codon* is a sequence of three nucleotides. Their experiments were carried out on the A and B cistrons of the rII locus of bacteriophage T4 in which one particular region of the DNA determines whether or not the phage can attack strain K of *E. coli*; and they used proflavine (Section 7.2.3) to bring about either the insertion of an additional base into the DNA sequence or the deletion of a single base from this.

If we assume that the sequence of bases in a portion of DNA is as shown in the top line of Fig. 11.5, and that the message is read in groups of three from left to right starting at the first C, then the insertion by the mutagen of another base N in the second triplet from the left will upset the reading of all triplets to the right of the point of insertion (Fig. 11.5 second line). The mutant so produced will be seriously defective and will not infect strain K. However, if a further mutation can now be produced which brings about the removal of the third A from the left, the fourth, fifth and subsequent triplets will read correctly and only the second and third triplets will be faulty (Fig. 11.5, third line). Only two amino acids, corresponding to these two triplets, will be 'wrong', and, if the presence of these two amino acids does not affect the structure of the protein significantly, the bacteriophage will behave

normally and will infect strain K. In practice it was in fact found that bacteriophages with an insertion and a deletion close together behaved normally, whereas the chances of normal behaviour diminished as the distance between the insertion and the deletion increased.

It was, moreover, possible to combine mutants in other ways. When two (+) mutations were combined, the recombinants were defective (Fig. 11.5, fourth line) but recombinants with three (+) or three (–) mutations behaved normally and infected strain K (Fig. 11.5, bottom line).

These results could best be interpreted by assuming that coding takes place by consecutive triplets in the nucleic acid. The insertion of one or two bases at any point will so alter the sequence of triplets as to make the code unreadable, whereas if three bases are added (or if one base is added and another is deleted) the sequence of triplets is restored after the first two changes, and the original message on the DNA can be interpreted as before.

### 11.2.2 Assignment of amino acid codons

The problem of assigning triplets of bases to each of the 20 amino acids was attacked in several ways.

(a) *The use of biosynthetic messengers*
The first approach utilized protein-synthesizing systems prepared from cell-free extracts of *E. coli* [24]. Such extracts contained ribosomes,

tRNAs, aminoacyl-tRNA synthetases and other enzymes, and, in the crude state, also DNA and messenger RNA. When ATP was added together with GTP, $Mg^{2+}$, $K^+$ and amino acids, the amino acids were readily incorporated into an acid-insoluble protein product and the incorporation process could be followed by using amino acids labelled with $^{14}C$. When the DNA in such extracts was destroyed by DNase, protein synthesis ceased after the messenger RNA had been depleted but could be restored by adding RNA fractions from various sources. These latter were found to include synthetic polynucleotides produced by the action of polynucleotide phosphorylase (see Section 4.4). This led Nirenberg and his coworkers to examine which amino acids were incorporated into polypeptide when different synthetic polynucleotides were added.

In 1961, Nirenberg and Matthaei [25] observed that, when the synthetic polymer poly(U) was added to the cell-free system with mixtures of 20 amino acids, only one amino acid in each mixture being radioactive, the only amino acid to be incorporated into acid-insoluble protein-like material was phenylalanine, and the product was polyphenylalanine. When taken with the evidence [22] that the codon is a triplet, this established that phenylalanine may be coded for by UUU, the first codon to be deciphered. Similarly, poly(A) and poly(C) were found to direct the synthesis of polylysine and polyproline, respectively. Nirenberg and his colleagues, and Ochoa and his colleagues, extended this approach from homopolynucleotides to heteropolynucleotides. Such heteropolynucleotides could be synthesized with various defined base compositions. Thus, although they had random sequences, the statistical incidence of different triplet codons could be calculated, and correlated with the relative incorporation of different [$^{14}C$]amino acids [26, 27].

Such use of random heteropolynucleotides could never do more than indicate the base composition of the codons specifying different amino acids. Khorana and his colleagues were subsequently able to attack this problem more directly with copolymers of defined sequence, prepared by an elegant combination of organic chemical and enzymic syntheses. The enzyme DNA polymerase I (see Section 6.4.2) was used to extend a short double-helical oligodeoxyribonucleotide, e.g. poly d(AC)·poly d(TG), previously prepared by chemical means. This synthetic DNA was then transcribed using RNA polymerase (see Section 8.1) to yield a specific polyribonucleotide poly(U–G) containing U and G in a repeating sequence. This could then be used in a cell-free protein-synthesizing system in order to determine codon assignments (Table 11.1) [28, 29].

For example, a $(A-C)_n$ sequence will be read as ACA–CAC–ACA–CAC–A . . . and will yield a polypeptide containing two amino acids, alternating those coded by ACA and CAC. In fact, the amino acids incorporated were threonine and histidine when poly(AC) was used as messenger. Which of these codons corresponds to which amino acid? In this case the results of the translation of the polytrinucleotide, poly(CAA), provided the answer. From this sequence three homopolymers should be coded corresponding to the triplets CAA, AAC, and ACA. This is becuase the starting-point is not clearly defined and the message may be read in any of the three forms:

· · C–A–A–C–A–A–C–A–A–C–A · ·

· · C–A–A–C–A–A–C–A–A–C–A · ·

· · C–A–A–C–A–A–C–A–A–C–A · ·

In practice, the amino acids incorporated were glutamine, threonine and asparagine. As poly(CAA) only specifies one triplet, ACA, that is also found in poly(A–C), above, this triplet must code for threonine, the only amino acid incorporated in both cases. Thus the other triplet, CAC, specified by poly(A–C), must correspond to histidine.

**Table 11.1**   Amino acid incorporations stimulated by mRNA containing repeating nucleotide sequences [29].

| Messenger | Amino acids incorporated | Messenger | Amino acids incorporated |
|---|---|---|---|
| *Repeating dinucleotides* | | *Repeating trinucleotides* | |
| Poly(UC) | Ser-Leu | Poly(GUA) | Val, Ser |
| Poly(AG) | Arg-Glu | Poly(UAC) | Tyr, Leu, Ile, Ser |
| Poly(UG) | Val-Cys | Poly(AUC) | Ile, Ser, His |
| Poly(AC) | Thr-His | Poly(GAU) | Met, Asp |
| | | | |
| *Repeating trinucleotides* | | *Repeating tetranucleotides* | |
| Poly(UUC) | Phe, Ser, Leu | Poly(UAUC) | Tyr, Leu, Ile Ser |
| Poly(AAG) | Lys, Glu, Arg | Poly(GAUA) | none |
| Poly(UUG) | Cys, Leu, Val | Poly(UUAC) | Leu, Thr, Tyr |
| Poly(CAA) | Gln, Thr, Asn | Poly(GUAA) | none |

In other cases the fact that more than one triplet codon may specify a single amino acid prevented unambiguous assignments being made from these results alone. However, by taking into account the previous data obtained with random heteropolynucleotides, and the results of mutation experiments (see (c) below), the amino acids specified by about half the codons were established [28, 29].

(b)  *The ribosome-binding technique*
At about the same time as Khorana and his colleagues were performing the studies just described, Nirenberg and Leder were employing a different, and less ambiguous, approach to the problem. Instead of translating synthetic messengers, they made use of the specific codon–anticodon interaction between oligonucleotide triplets and aminoacyl-tRNA that they found could occur on ribosomes *in vitro* [30]. When a mixture containing ribosomes, aminoacyl-tRNA and a triplet is allowed to react under suitable ionic conditions, and is then poured on to a nitrocellulose filter, the free tRNA (and triplet) passes through the filter, whereas the ribosomes – and any bound tRNA – are retained on it. By using a series of 20 different amino acid mixtures, each containing one $^{14}$C-labelled

amino acid, it is possible to identify the amino acid corresponding to a triplet by means of the radioactivity adsorbed by the filter. For example, the trinucleotide GUU retains valyl-tRNA whereas UGU and UUG do not [31]. All 64 possible triplets were synthesized and tested, and more than 50 of them gave unambiguous results [32].

(c)  *The use of mutations*
The cell-free techniques, described in (a) and (b) above, established the amino acid assignments shown in Table 11.2. Confirmatory evidence was obtained from genetic mutations, and this was of especial value as it derived from intact cells.

The principle involved in the use of base substitution mutants may be illustrated by the results obtained with the aid of artificially induced mutants of tobacco mosaic virus (TMV) [33–35]. When RNA is treated with nitrous acid (Section 7.2) two changes are brought about: (i) cytosine is deaminated to uracil (C → U), and (ii) adenine is deaminated to hypoxanthine, which is equivalent to guanine in coding (A → G) [33, 34]. When TMV RNA treated with $HNO_2$ was used to infect tobacco plants, mutants were produced in which single amino acids in the viral protein were replaced by different amino

**Table 11.2**    The standard genetic code.

| | Second letter | | | | |
|---|---|---|---|---|---|
| | **U** | **C** | **A** | **G** | **Third letter (3')** |
| **U** | UUU } Phe<br>UUC }<br>UUA } Leu<br>UUG } | UCU ⎫<br>UCC ⎬ Ser<br>UCA ⎪<br>UCG ⎭ | UAU } Tyr<br>UAC }<br>UAA } Stop<br>UAG } | UGU } Cys<br>UGC }<br>UGA  Stop<br>UGG  Trp | U<br>C<br>A<br>G |
| **C** | CUU ⎫<br>CUC ⎬ Leu<br>CUA ⎪<br>CUG ⎭ | CCU ⎫<br>CCC ⎬ Pro<br>CCA ⎪<br>CCG ⎭ | CAU } His<br>CAC }<br>CAA } Gln<br>CAG } | CGU ⎫<br>CGC ⎬ Arg<br>CGA ⎪<br>CGG ⎭ | U<br>C<br>A<br>G |
| **A** | AUU ⎫<br>AUC ⎬ Ile<br>AUA ⎭<br>AUG  Met* | ACU ⎫<br>ACC ⎬ Thr<br>ACA ⎪<br>ACG ⎭ | AAU } Asn<br>AAC }<br>AAA } Lys<br>AAG } | AGU } Ser<br>AGC }<br>AGA } Arg<br>AGG } | U<br>C<br>A<br>G |
| **G** | GUU ⎫<br>GUC ⎬ Val<br>GUA ⎪<br>GUG ⎭ | GCU ⎫<br>GCC ⎬ Ala<br>GCA ⎪<br>GCG ⎭ | GAU } Asp<br>GAC }<br>GAA } Glu<br>GAG } | GGU ⎫<br>GGC ⎬ Gly<br>GGA ⎪<br>GGG ⎭ | U<br>C<br>A<br>G |

*First letter (5')* is indicated to the left of the rows.

Stop = termination codon
* Also usual initiation codon.

acids at certain positions in the polypeptide chain in such a way that the replacements could be correlated with the changes C → U or A → G (Fig. 11.6). Similar evidence was obtainable from the mutations affecting the A protein of tryptophan synthetase [36], and from the different varieties of human haemoglobin [20].

Protein sequence determination was also performed on 'frame shift' mutants of the type

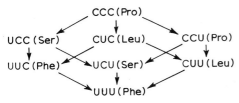

**Fig. 11.6**  Steps by which the triplet CCC which codes for proline may be changed by deamination to the triplet UUU which codes for phenylalanine. The amino acids corresponding to each triplet are shown on the right of the codon.

that had been used to establish the triplet nature of the genetic code (Fig. 11.5). The results were indeed consistent with the mutations having been caused by the insertion or deletion of a base producing a shift in the reading frame of the genetic message, according to the predicted genetic code [37].

### (d)  *Evidence from nucleic acid sequence determination*

Final confirmation of the genetic code *in vivo* came from determination of the sequences of mRNAs. The first such analysis was undertaken on the small RNA bacteriophages (see Section 11.10.1), the genome of which is also the mRNA. In 1969 Sanger and his colleagues published the nucleotide sequence of a segment of RNA from bacteriophage R17 which correlated exactly with the amino acids known to be in positions 81–99 of the coat protein [38].

Subsequently Fiers and his colleagues determined the complete sequence of the related bacteriophage MS2, and confirmed all the amino acid assignments of the genetic code [39].

The determination in 1981 [40] of the nucleotide sequence of the rII region of bacteriophage T4 can in a sense be seen as having closed the circle started in the late '50s.

### 11.2.3 Assignment of termination codons

The amino acid assignments derived by the methods described in Sections 11.2.2 (a), (b) and (c) accounted for 61 of the possible 64 triplets. The remaining three, UAA, UAG, and UGA, designated 'Stop' in Table 11.2, do not code for any amino acid but are signals for polypeptide chain termination.

The existence of specific termination signals was indicated by genetic studies with a class of mutants termed '*nonsense*', to distinguish them from the '*missense*' mutants in which one amino acid was replaced by another [41]. The identities of the termination codons responsible for two classes of these mutants ('amber' UAG, and 'ochre' = UAA) were deduced from mutagenesis experiments analogous to those described in Section 11.2.2.c above, but employing hydroxylamine and 2-aminopurine as mutagens [42]. A third class of 'nonsense' mutant (sometimes called 'opal' or 'umber') was subsequently discovered and shown to correspond to the termination codon UGA [43]. The designation of codons UAG and UAA as terminators explains why neither poly(GAUA) nor poly(GUAA) could be translated by Khorana and his colleagues (Table 11.1). The termination codon assignments have been confirmed by the nucleotide sequences of mRNAs.

Although tRNA is *not* involved in decoding the termination triplets (see Section 11.5.3 below), certain mutant tRNAs have been found to recognize them. These occur in the 'suppressor' strains of *E. coli*, so called for their ability to suppress particular classes of 'non-

sense' mutants. Nucleotide sequence analysis of such a mutant of a tRNA$^{Tyr}$ from an amber suppressor strain showed that its anticodon is changed from 3'AUG5' to 3'AUC5'. Thus it is able to insert tyrosine into a polypeptide chain in response to the termination codon UAG, rather than to the tyrosine codon UAC [44]. The reason that such suppressor mutations are viable, and do not cause premature termination of the bulk of normal proteins, is because they occur in the minor, otherwise redundant, representatives of certain pairs of iso-accepting tRNAs (see Section 11.2.5).

### 11.2.4 Assignment of initiation codons

It was originally thought that no specific codon served to signal the start of a polypeptide chain. This was because all 64 triplets were otherwise accounted for and no common triplet was required for synthetic polynucleotides to be translated *in vitro* (Table 11.1). However, in 1964 Marcker and Sanger discovered that, in *E. coli*, methionyl-tRNA could be formylated in the $\alpha$-amino position of the methionine residue [45]. The resulting species, *N*-formylmethionyl-tRNA, was thus only capable of forming a peptide bond with its $\alpha$-carboxyl group. When this was considered with earlier data showing that methionine occupied the *N*-terminal position of about 45% of *E. coli* proteins [46], it raised the possibility that bacterial proteins might all be initiated at a codon specifying formylmethionine, the formyl group and, sometimes, the methionyl residue being subsequently removed.

Direct evidence for this came from the use of cell-free systems to translate the RNA of bacteriophages R17 and f2. (This translation of natural mRNA requires a much lower $Mg^{2+}$ concentration than the promiscuous translation of artificial polynucleotides.) The coat proteins of the bacteriophages synthesized *in vivo* have *N*-terminal sequences commencing Ala-Ser . . . .

However, coat proteins were synthesized *in vitro* with *N*-terminal sequences commencing fMet-Ala-Ser [47, 48]. Presumably the cell-free system lacked the deformylase [49, 50] and aminopeptidase [50] activities that were subsequently demonstrated to be present in the intact cell.

The tRNA that inserts the initiating fMet into polypeptide chains, tRNA$_f^{Met}$, has a different nucleotide sequence from the tRNA that inserts Met internally: tRNA$_m^{Met}$ [51, 52]. Both species accept methionine, but only the Met-tRNA$_f$ can then be formylated by a *transformylase* enzyme that has $N^{10}$-formyltetrahydrofolic acid as a cofactor [53]. Although formylation of Met-tRNA$_f^{Met}$ is normally an absolute requirement for initiation in *E. coli* and many other bacteria, the specific recognition of the initiator tRNA in the initiation process (Sections 11.5.1 and 11.6.1) is most certainly influenced by the structure of the tRNA itself. Thus in eukaryotes the methionyl residue of the initiator tRNA is not formylated, and the initiator tRNA of at least one bacterium can compensate for lack of formylation by undergoing a base substitution in loop IV [54]. Comparison of the primary and secondary structures of tRNA$_m^{Met}$ and tRNA$_f^{Met}$ gives only limited clues to the basis of the specific recognition of the latter. However, the X-ray crystal structures of the initiator tRNAs from *E. coli* [55] and yeast [56] do reveal a significant difference from the elongator tRNAs in the folding of the anticodon loop, which results in the latter having an external-facing rather than internal-facing disposition.

In the ribosome-binding assay fMet-tRNA will recognize the triplets, AUG, GUG and UUG [53, 57]. Both AUG and, 3–4% as frequently, GUG have been found as natural initiation codons in those bacterial mRNAs the nucleotide sequences of which have so far been determined (usually from the DNA sequence of the gene). There are, in addition, some examples of UUG [58] and one of AUU [59] functioning as initiation codons in natural bacterial mRNAs,

and some examples of AUA functioning as a reinitiation codon in 'amber' mutants of phage T4 in which premature termination occurs [58]. All these minor initiation codons, like AUG, are recognized by fMet-tRNA.

### 11.2.5  The degeneracy of the genetic code

Since many of the 20 amino acids are encoded by more than one triplet (Table 11.2), the code is said to be degenerate.

Triplets coding for the same amino acid are not distributed at random, but are grouped together so that they generally share the same 5' and middle base. This has the consequence that mutations producing a change in the base at the 3' position of the codon often have no effect on the amino acid specified. Furthermore the different amino acids are segregated to a considerable extent on the basis of chemical similarity (hydrophobicity, hydrophilicity, acidity and basicity). Thus a mutation in the 5' base of any of the six leucine codons would give a codon specifying another hydrophobic amino acid. Such a change might not impair the function of a particular globular protein if the altered amino acid merely performed a structural role in the hydrophobic core of this protein. It has therefore been argued that the specific arrangement of codons in the genetic code serves to reduce the potentially harmful effect of possible mutations.

It might have been expected that each degenerate codon would require its own tRNA with a corresponding anticodon. Some such discrete *iso-accepting* tRNAs, recognizing the same amino acid, have been found; however, their number is less than the 61 required for all codon–anticodon interactions to involve three standard Watson–Crick base pairs. This situation is viable because there is a degree of latitude in the complementary base-pairing between the base in the 3'-position of the codon and that in the 5'-position of the anticodon. Such latitude

**Table 11.3**  Predicted and observed 'wobble' base-pairing.

Original wobble predictions

| 5'-Anticodon base | 3'-Codon bases read | Observed patterns of base-pairing in the anticodon wobble position |
|---|---|---|
| A | U | A never found as 5'-anticodon base. |
| C | G | Occurs as predicted without wobble. |
| G | $\left\{ \begin{array}{l} U \\ C \end{array} \right.$ | Wobble occurs as predicted. Also when G is replaced by $G_m$ or Q. |
| I | $\left\{ \begin{array}{l} U \\ C \\ A \end{array} \right.$ | Wobble occurs as predicted, but not where it would cause misreading of genetic code. |
| U | $\left\{ \begin{array}{l} A \\ G \end{array} \right.$ | U never found as 5'-anticodon base.* $mcm^5s^2U$, $mcm^5U$ and $mnm^5s^2U$ recognize A only. $mo^5U$, $cmo^5U$ recognize, A, G or U. |

*Except in mitochondrial tRNAs and in two isolated examples where it recognizes all four bases.

was called '*wobble*' by Crick [60] in the hypothesis he formulated to allow the two pairs of 3'-codon bases that are grouped together in degenerate codons, U and C, and A and G, to be recognized by a common 5'-anticodon base. From stereochemical considerations Crick suggested that the 5'-anticodon base, G (but not A), might be able to pair with *both* U and C in the 3'-position of the codon; and that the 5'-anticodon base, U (but not C), might be able to pair with *both* A or G in the 3'-position of the codon. Moreover it had already been found that several tRNA anticodons contained the nucleoside inosine (which pairs only with C in a double helix), and he suggested that this might be able to pair with A, U or C in the 3'-position of the mRNA codon. These proposals are shown in Table 11.3, and, with one exception where the prediction is exceeded (see below), have been verified by ribosome-binding experiments in the cases where the particular anticodon 5'-bases are found to occur. The experimentally observed pattern of anticodon–codon interactions for some of the modified bases subsequently found in the 5'-anticodon position of tRNAs is also indicated in Table 11.3. The genetic code itself

places constraints on the occurrence of the wobble base inosine in certain tRNAs. Thus, to prevent misreading of the related codon with a 3'-A, inosine must be excluded from the 5'-position of the anticodon of tRNAs reading Phe, Tyr, Cys, His, Asn and Asp, as well as from that of Ser coded by AGU or AGC. This is what is found.

Of the five 5'-anticodon bases considered by Crick, two (A and U) do not, in fact, occur in the vast majority of tRNAs, the structures of which have been determined. The non-usage of A to read 3'-Us in codons has been rationalized on the basis that it is more economical to use G, the reading of both U and C by which in no case violates the genetic code. The reason U almost never occurs in the wobble position is almost certainly because the wobble that could occur with it would exceed the limitation predicted by the hypothesis, and thus in two-codon amino acid families cause violation of the genetic code. In such two-codon families certain modified forms of U, which, in fact, can only base-pair with A, are employed (see Table 11.3). For four-codon amino acid families other modified forms of U ($mo^5U$ and $cmo^5U$) are often encountered, and these allow interaction with U in addition to that predicted to occur with A and G. The structural basis for the restriction or amplification of wobble recognition with modified U residues has been discussed [448, 454]. In the two known examples of tRNAs with an unmodified U in the wobble position all four bases are recognized [13].

Attempts have been made to relate the structure of tRNA (Section 11.1) to the phenomenon of wobble. Thus it is possible to accommodate a small conformational change into the tertiary structure of the anticodon of yeast tRNA$^{Phe}$, such that the 5'-anticodon wobble base Gm could move into an alternative position that would allow pairing with U rather than C. The reason that such a change is confined to the 5'-anticodon base seems to lie in the fact that the base immediately 3' to the anticodon, which is always a purine (Fig. 11.1) and frequently heavily modified, has strong base-stacking interactions with the two adjacent bases

of the anticodon (Fig. 11.7). This serves to anchor these bases in a helical conformation that only allows standard Watson–Crick base pairs to form. It is noteworthy that in *E. coli* tRNA$_f^{Met}$, in which there can be ambiguity in interaction of the anticodon with the initiation codon in the first position (see Section 11.2.4), the base 3' to the anticodon is not modified. However, in eukaryotic cytoplasmic tRNA$_f^{Met}$, which only recognizes AUG (see Section 11.6.1), this base is modified.

The essence of the wobble hypothesis is the possibility of alternative base-pairing interactions involving the 5'-base of the anticodon. Because of our lack of knowledge of the three-dimensional structure of the codon–anticodon complex it is not possible to say whether juxtaposition of bases other than allowed by wobble in this position would cause stereochemical or electrostatic repulsion, or merely result in a weaker interaction involving only two base pairs. Lagerkvist [61] has considered the latter possi-

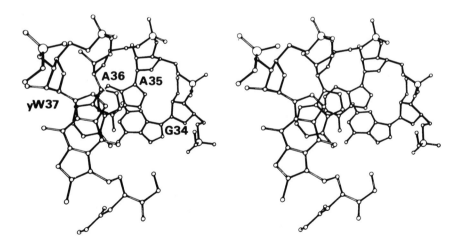

**Fig. 11.7** Stereoscopic view of the anticodon bases [34–36] of yeast tRNA$^{Phe}$ and the 3'-hypermodified base yW37 as viewed from the interior of the molecule. It can be seen that the stacking of residues yW37, A36 and A35 is greater than that of A35 to the 'wobble' base G34. (From [17] with permission.)

bility in his 'two-out-of-three' hypothesis, which proposes that a single tRNA that can recognize the first two codons of a four-codon family has the potential to decode the whole of such a family. The four-codon families for which such 'two-out-of-three' recognition is proposed are UCN, CUN, CCN, CGN, ACN, GUN, GCN and GGN (see Table 11.2). All of these involve at least one strong G–C base pair in the first two positions.

There is some experimental evidence to support the possibility of 'two-out-of-three' codon reading. This comes mainly from the successful translation of phage MS2 RNA in a cell-free system from *E. coli* in which the individual tRNAs that can normally translate particular codons have been omitted and replaced by tRNAs that would be predicted to translate them according to the 'two-out-of-three' hypothesis, but not according to the 'wobble' hypothesis [62]. It should be stressed that the efficiency of this process is such that it would be unable to compete with the normal tRNA that can make a 'three-out-of-three' interaction, and is thus most unlikely to be involved in contemporary codon–anticodon interactions. Nevertheless the hypothesis has interesting evolutionary implications and deserves at least a hearing in relation to the pattern of codon–anticodon interaction found in mitochondria (see Section 11.9.2).

Finally, in contrast to the *degenerate* tRNAs, which recognize more than one codon, it should be mentioned that there are, so-called, redundant, iso-accepting tRNAs, sharing exactly the same anticodon but differing at other points in their nucleotide sequences. (The normal iso-accepting tRNAs for an amino acid have, of course, different anticodons to allow decoding of all the degenerate codons.) These redundant iso-accepting tRNAs generally differ in the bases at only a very few positions [63], suggesting that they arose by gene duplication and subsequent mutation. Such gene duplication (which in other cases may produce genes coding

for identical tRNAs) is assumed to help protect the organism against catastrophic mutation in its tRNAs.

### 11.2.6 Variations in the code and codon usage

It was long thought that the genetic code of Table 11.2 was universal. This assumption derived from comparison between *E. coli* and higher mammals, where initially it was shown that the tRNAs recognize the same codon triplets *in vitro* [64], a result subsequently corroborated by comparison of protein and nucleic acid sequences. It is now clear that variations exist between the genetic code for the majority of cellular proteins of eukaryotes, expressed on the cytoplasmic ribosomes, and that for the small number of mitochondrial proteins specified by the mitochondrial DNA. (These mitochondrial genetic codes are discussed in more detail in Section 11.7.2.) It has recently emerged that deviations from the 'standard' genetic code even occur in certain lower eukaryotes and prokaryotes. Thus, in ciliated protozoa it appears that UAA and UAG code for glutamine rather than acting as chain-termination codons, whereas in the prokaryote, *Mycoplasma capricolum*, these codons are chain terminators but UGA codes for tryptophan [65]. It would seem that either the lines leading to these species diverged before the standard code became fixed, or at the time of divergence the termination system was primitive enough to allow further change.

The standard genetic code does apply to the vast majority of prokaryotes and eukaryotes studied, and, with the determination of the sequences of a large number of structural genes, it has been possible to discover how the usage of codons specifying the same amino acid or punctuation signal varies. The usage of such 'synonymous' codons is definitely non-random, and differs between prokaryotes and eukaryotes, viruses and their hosts. Furthermore the codon usage for different proteins of the same organism may be different. As regards the usage of amino acid codons, it has been suggested that this may be governed by considerations of the energies of codon–anticodon interaction and the relative abundance of different iso-acceptor tRNAs. The possible regulatory implications of this are discussed below (Section 11.10.3). Nevertheless host–virus differences are not easily accounted for in these terms and there may be other, e.g. structural, constraints on codon usage. Differences in the usage of termination codons may be related to the phenomenon of natural suppression (see Sections 11.5 and 11.13), whereas the greater promiscuity in selection of initiation codons in prokaryotes compared with eukaryotes (where only AUG is used) may reflect the differences in the initiation process between these kingdoms (see Sections 11.5.1 and 11.6.1).

## 11.3 AMINOACYLATION OF tRNA

Before a tRNA molecule can act as an adaptor by interacting with its corresponding anticodon

in the decoding process, it must first be 'charged' with its cognate amino acid. The enzymes responsible for this process are called *aminoacyl-tRNA synthetases* (*amino acid-tRNA ligases* EC 6.1.1.) and catalyse the reaction illustrated in Fig. 11.8.

The reaction occurs in two stages [11, 66], in the first of which the amino acid is *activated* by ATP to form an aminoacyl adenylate:

$$\text{ATP} + \text{amino acid}_1 + E_1 \rightleftharpoons$$
$$(\text{amino acid}_1 - \text{AMP})E_1 + PP_i$$

The aminoacyl–adenylate complex then reacts with the terminal adenosine moiety of the appropriate tRNA to form an aminoacyl-tRNA:

$$(\text{amino acid}_1 - \text{AMP})E_1 + \text{tRNA}_1 \rightleftharpoons$$
$$\text{tRNA}_1 - \text{amino acid}_1 + \text{AMP} + E_1$$

Although the amino acid is located at the 3'-OH of the ribose of the terminal adenosine moiety of tRNA during peptide bond formation, the initial point of attachment can be the 2'-OH, the 3'-OH, or either, depending on the amino

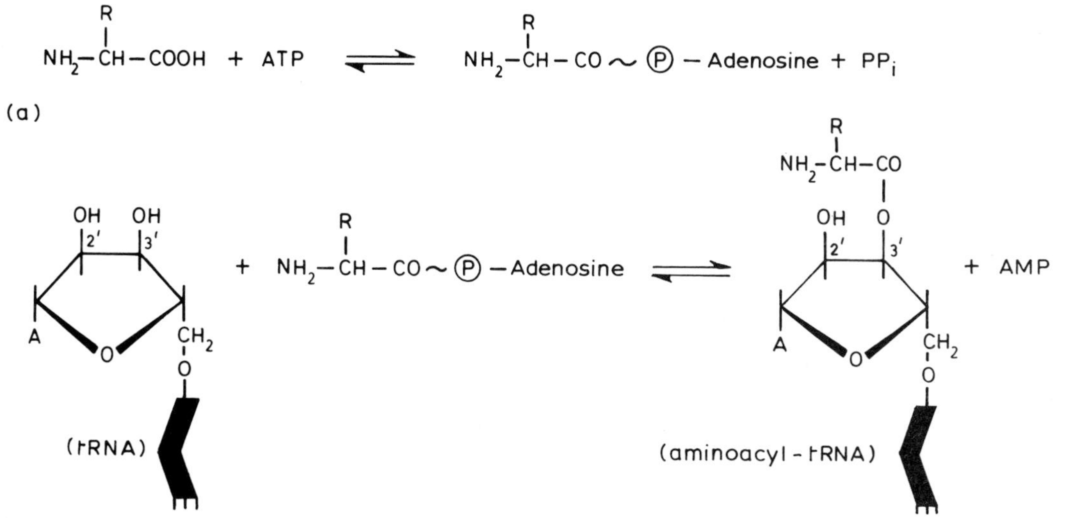

(a)

(b)

**Fig. 11.8** The activation of amino acids and their attachment to tRNA. The reaction occurs in two stages, (a) and (b), both of which are catalysed by the same enzyme, and to which the intermediate is bound (see text). The symbol ~ represents a bond with a relatively high standard free energy of hydrolysis.

acid [67]. After attachment, rapid migration between the two positions is possible [68]. The aminoacyl ester linkage has a relatively high standard free energy of hydrolysis, derived from the ATP hydrolysed during its activation. This is important as it provides the necessary energy for the subsequent peptide bond formation to occur [69].

Despite the fact that multiple species of tRNA (iso-accepting tRNAs) exist for a single amino acid (see Section 11.2.5), there appears to be only one aminoacyl-tRNA synthetase for each amino acid. Even the different initiating and elongating methionyl-tRNAs are recognized by the same enzyme. The different aminoacyl-tRNA synthetases have very different structures, some being large single-chain enzymes, some multi-chain enzymes of similar subunit size, and some multi-chain enzymes of different subunit size [70]. Furthermore they have very little primary structure homology. It appears that a portion of protein, comprising approximately 330 amino acids, is sufficient for the aminoacylation function of the synthetases, and that the remainder of the structure of the larger enzymes is concerned with other functions (e.g. DNA binding and gene regulation in prokaryotes, or synthesis of $Ap_4A$ in eukaryotes [71]). Although the three-dimensional structure of two of these enzymes has been determined by X-ray crystallography, and the amino acid–adenylate binding-sites located, the nature of the specific interaction of a tRNA with its cognate synthetase is still unclear because of disordered regions in the molecules [72]. There is evidence that the recognition involves molecular interactions at a number of points on the top of the 'T' (inside angle of the 'L') in the three-dimensional structure of the tRNA (the dotted line in Fig. 11.2), but crystallization and analysis of the complex between a tRNA and its synthetase will be required before the structural basis of the specific recognition of tRNA by synthetase is elucidated.

It has already been asserted that the specificity of decoding resides in the complementary base-pairing between the anticodon of the tRNA and the codon of the mRNA. Indeed, once the amino acid has been enzymically attached to the tRNA, it may be converted chemically to a different amino acid without altering the codon–anticodon interaction. This was shown in the classic experiment in which Cys-tRNA$^{Cys}$ was reduced with Raney nickel to give Ala-tRNA$^{Cys}$, which then incorporated alanine in response to codons for cysteine in a cell-free system [73]. The maintenance of the accurate decoding conferred by the specific codon–anticodon interaction is thus absolutely dependent on the aminoacyl-tRNA synthetases charging the tRNAs with their correct cognate amino acids. This requires the enzymes to be specifically able to recognize the correct amino acid as well as the correct nucleic acid. There are several pairs of amino acids differing in structure by no more than a single methyl group, and this poses a real problem in discrimination for the synthetases. Thus, it was observed that valine bound appreciably to Ile-tRNA synthetase [74]. There is good evidence to support the existence of a 'proof-reading' or 'editing' mechanism in those synthetases for which there are inappropriate isosteric or smaller amino acids with which the tRNA may be mischarged. This involves a hydrolytic site on the enzyme, close to but distinct from the acylation site, to discharge such inappropriate aminoacyl-tRNAs [75]. The accuracy of recognition at the hydrolytic site can, of course, be no greater than that at the initial acylation site; but by requiring the amino acid to be recognized twice, an error frequency of (e.g.) 1 in $10^2$ would be reduced to 1 in $10^4$.

## 11.4 GENERAL ASPECTS OF POLYPEPTIDE FORMATION

The interaction of the aminoacylated tRNA with the mRNA takes place on the ribosomes where successive codons are read in an ordered manner and the amino acids linked together to form a polypeptide chain. The direction of growth of the polypeptide chain is from the *N*-terminus to the *C*-terminus [76], and the direction of reading of the mRNA is $5' \rightarrow 3'$ [77, 37]. As this latter is also the direction in which the mRNA is synthesized, it is possible, in prokaryotes, for ribosomes to start translating a mRNA before its transcription is complete [78, 79] (see Section 9.1). The rates of protein synthesis in prokaryotes and eukaryotes appear to be quite similar. It has been reported that this is 15 amino acids $s^{-1}$ for β-galactosidase in *E. coli* [80], and seven amino acids $s^{-1}$ for globin chains in rabbit reticulocytes [81]. This is, however, much slower than the rates estimated for DNA or RNA synthesis (800 and 50 nucleotides $s^{-1}$, respectively, in prokaryotes).

The formation of peptide bonds is catalysed by an enzymic activity, *peptidyltransferase*, which is an integral part of the larger ribosomal subunit [82], and there are two physically distinct sites on the ribosome at which the reacting molecules of tRNA are bound [83]. One of these sites, the A-site, is occupied by successive molecules of aminoacyl-tRNA, whereas the other site, the P-site, is occupied prior to peptide bond formation by the tRNA carrying the growing polypeptide chain, peptidyl-tRNA. In addition it has been separately proposed that a third site, either for recognition [84], or for exit [85] exists. Although such a site(s) cannot be excluded, the evidence is not yet such as to have led to general acceptance of either of these ideas.

Fig. 11.9 is a schematic representation of the ribosome just before the aminoacyl ester bond of the peptidyl-tRNA is broken and the polypeptide chain transferred to the α-amino group of the aminoacyl-tRNA at the A-site. The events involved in the reading of successive codons are described in detail below (Sections 11.5 and 11.6) and involve relative movement of the 5'-portion of the mRNA and the ribosome away from one another.

Although Fig. 11.9 is a convenient type of representation for presenting these events on the

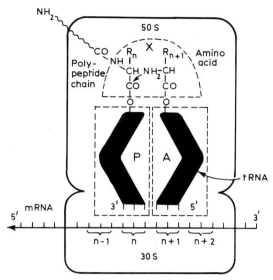

**Fig. 11.9** Diagrammatic representation of a prokaryotic ribosome. Two tRNA molecules are bound to the ribosome in response to the mRNA codons designated $n$ and $n + 1$. The tRNA bearing the growing polypeptide chain is occupying the peptidyl site (rectangular area, marked P), and the tRNA bearing an amino acid is occupying the aminoacyl site (rectangular area, marked A). The peptidyltransferase centre where the peptide bond formation is catalysed is represented by the semicircular area, marked X. The actual sizes of the various components are discussed in the text. Note also that the size of the amino acid (about 1 Å) is exaggerated in relation to that of tRNA (75 Å).

ribosome, comparison with Fig. 11.10 shows how it differs from physical reality. Not only are the overall shapes and sizes misleading, there is evidence to suggest that the tRNAs are, at least partially, sandwiched between the two subunits rather than decorating their surface. Nuclease protection experiments indicate that 35–45 nucleotides of mRNA (i.e. 12–15 codons) are in contact with the ribosome, this being equivalent to 135 Å (13.5 nm) of helical RNA. This compares with 200–300 Å (20–30 nm) for the diameter of the ribosome, and 75 Å (7.5 nm) for the 'length' of the tRNA. Analogous experi-

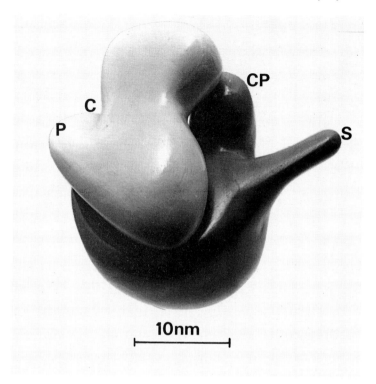

**Fig. 11.10**   Model of the 70S ribosome of *E. coli* based on the electron-microscopic studies of Lake and co-workers. The 30S subunit is light, the 50S subunit is dark. The cleft (C) and platform (P) of the 30S subunit, and the stalk (S) and central protuberance (CP) of the 50S subunit, referred to in the text, are indicated.

ments with proteases have indicated that 25–30 amino acid residues of nascent polypeptide chain are in contact with the ribosome. This would have a length of approximately 50 Å (5 nm) in an $\alpha$-helical conformation, or approximately 100 Å (10 nm) in the, perhaps more probable, extended conformation [86].

The overall length of the mRNA and the rate of initiation are usually such that the initiation codon can attach another ribosome before the first one completes its polypeptide chain. In fact several ribosomes are normally found on a given molecule of mRNA, translating different parts of it simultaneously, and such groups of ribosomes are termed *polyribosomes* or *polysomes* (see Fig. 11.11). These may be visualized by electron microscopy [87], and polysomes containing different numbers of ribosomes may be resolved by sucrose density gradient centrifugation [88]. The size of the polysomes increases with the size of the mRNA: polysomes synthesizing haemoglobin $\beta$-chains (molecular weight 16 000 approx.) contain four to five ribosomes [89], whereas those synthesizing myosin heavy chains (molecular weight, 200 000 approx.) contain about 50–60 ribosomes [90].

The ribosome is responsible only for synthesizing a single polypeptide chain comprising the 20 genetically coded amino acids. After peptide bond formation, amino acids may form additional inter- or intra-molecular covalent bonds, may be subject to specific proteolysis, or be chemically modified in any of a large number of different possible ways [91, 92], a description

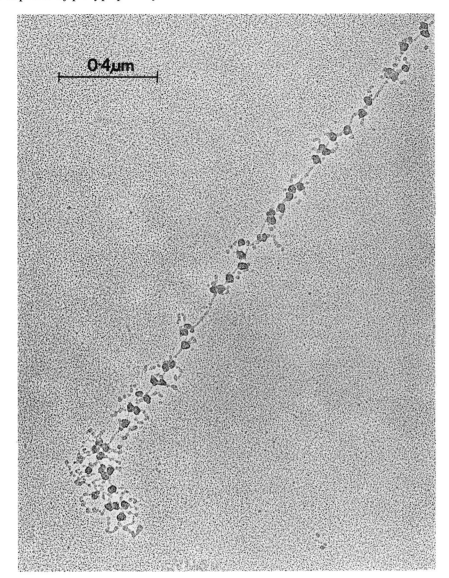

**Fig. 11.11** Electron micrograph showing the translation of silk fibroin mRNA on polysomes. The extended fibrous fibroin molecules can be seen emerging from the ribosomes (dark irregular particles). The length of the fibroin molecules increases from the top right to the bottom left of the frame, indicating that this is the $5' \rightarrow 3'$ direction along the mRNA (courtesy of Dr Steven L. McKnight and Dr Oscar L. Miller, Jr).

of which is beyond the scope of this book. It is worth pointing out, however, that modification of the *N*-terminal amino acid is especially frequent, particularly in eukaryotes. This is a point of practical relevance in relation to amino acid sequence determination. Finally it should be mentioned that in the case of certain small bacterial peptides, synthesis occurs in a manner not dependent on mRNA and ribosomes. The reader interested in this topic is directed elsewhere [93].

## 11.5 THE EVENTS ON THE BACTERIAL RIBOSOME

The following discussion uses the uniform nomenclature for bacterial protein-synthesis factors. The relationship of this to the various older nomenclatures that may be encountered in the original references is given in [94].

### 11.5.1 Chain initiation [95–98]

In polypeptide chain initiation fMet-tRNA is bound to the initiation codon of the mRNA on the 30S ribosomal subunit [99] and the resulting 30S initiation complex then reacts with the 50S ribosomal subunit to give a 70S initiation complex. This process requires GTP and the initiation factors, IF-1, IF-2 and IF-3 [100–102] and in polycistronic mRNAs can occur independently at several different initiation sites [103].

In bacterial-cell extracts the initiation factors are found associated with the 30S subunit, from which they have been extracted and purified [104]. As already mentioned (Section 11.2.4), the requirement for AUG and factors in initiation was only seen *in vitro* when salt-washed ribosomes and a suitably low concentration of

$Mg^{2+}$ (usually about 5 mM) were used [105, 106]. The exact sequence of events in initiation and the precise roles of all the factors is still uncertain, partly because of the concerted manner in which they act. For this reason, the initiation scheme presented in Fig. 11.12 avoids, for the most part, giving more detail than there is general agreement upon.

The initiating 30S subunit most probably has bound to it IF-3 and IF-1 (Fig. 11.12a) which are involved in generating free ribosomal subunits after polypeptide chain termination (see Section 11.5.3). Their role in initiation is distinct from this latter as they are required for the formation of a 30S initiation complex (Fig. 11.12b) even when 50S subunits are not present [107]. IF-3 (molecular weight 21 000) is primarily involved in binding mRNA to the ribosome [108]. Although it is needed for translation of natural mRNAs, there is no absolute requirement for IF-3 in either the AUG-dependent ribosome binding of fMet-tRNA or the translation of artificial polynucleotides such as $AUGA_n$ [109]. This suggests that, either directly or indirectly, IF-3 facilitates the recognition of the untranslated 'leader' sequences that precede the initiating AUGs of natural mRNAs. Such bacterial 'leader' sequences contain short polypurine regions complementary to a polypyrimidine sequence at the 3'-end of the 16S rRNA (Table 11.4); and it was suggested by Shine and Dalgarno [110] that base-pairing between these

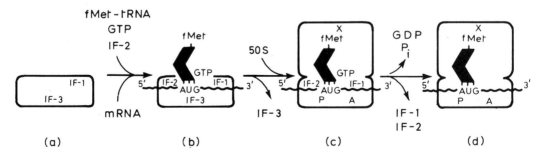

**Fig. 11.12** A schematic diagram of prokaryotic polypeptide chain initiation. The ribosome and tRNA are represented as in Fig. 11.9.

regions is the means by which bacterial ribosomes select the correct AUG codon for initiation. There is now overwhelming evidence (reviewed in [58]) that this hypothesis is correct, although other factors, such as secondary structure, may prevent every potential 'Shine and Dalgarno' sequence being used for initiation [111]. Three of four known exceptions in which initiation codons lack such a polypurine region in their 'leader' sequence nicely prove the rule, as these occur in mRNAs coding for extremely weakly expressed proteins: the $C_I$ repressor protein of phage λ, and the DNA primase and *trp* repressor of *E. coli*.

The primary role of IF-2 (molecular weight 97 000) is to bind fMet-tRNA to the ribosome in a reaction that requires GTP and is stimulated by the other initiation factors, especially IF-1 [112]. As it is possible *in vitro* to bind either mRNA or fMet-tRNA independently to 30S ribosomal subunits it is not certain which reaction occurs first *in vivo*, although the evidence, on balance, seems to favour binding of mRNA before fMet-tRNA [98]. Nor is it clear whether the unstable ternary complex between IF-2, fMet-tRNA and GTP, found *in vitro*, actually occurs free of the ribosome *in vivo*, or whether such a complex is found only on the 30S subunit.

Undoubtedly interaction between IF-2 and fMet-tRNA occurs at some point, and the factor must recognize some specific structural feature of the initiator tRNA as it will not interact with aminoacyl-tRNAs, including Met-tRNA$_m$ [113].

Once a 30S initiation complex, containing mRNA and fMet-tRNA, has been formed (Fig. 11.12b) the 50S ribosomal subunit can associate with it (Fig. 11.12c) causing the release of IF-3 [108]. The non-hydrolysable analogue of GTP, 5'-guanylylmethylene diphosphonate, has been used to show that hydrolysis of GTP is not required for this step, but GTP hydrolysis is required for the fMet-tRNA to become available for reaction with puromycin (an analogue of aminoacyl-tRNA, binding in the A site: Fig. 11.13) [114], at which stage IF-2 and (probably) IF-1 are released [115]. As GTP hydrolysis does not cause relative movement of mRNA and the ribosome, fMet-tRNA must be bound directly at the P-site [116, 117]. The role of the GTP hydrolysis, which is discussed in more detail below (Section 11.8), cannot, therefore, be the provision of energy for movement of the fMet-tRNA from A-site to P-site.

It will be evident from the foregoing that although IF-1 (molecular weight, 8000) is absolutely required for initiation, its role is not

**Table 11.4** Complementarity between pre-initiation regions of *E. coli* mRNAs and 16S rRNA.

| 16S rRNA | | 3' HOA U U C C U C C A C U A G . . . . . . . . 5' |
|---|---|---|
| MS2 coat | 5' . . . . . | U C A A C C G G A G U U U G A A G C A U G . . . 3' |
| MS2 replicase | 5' . . . . . | C A A A C A U G A G G A U U A C C C A U G . . . 3' |
| MS2 A protein | 5' . . . . . . | U C C U A G G A G G U U U G A C C U G U G . . 3' |
| λ Cro | 5' . . | A U G U A C U A A G G A G G U U G U A U G . . . . . . 3' |
| *gal* E | 5' . . . | A G C C U A A U G G A G C G A A U U A U G . . . . . 3' |
| β-lactamase | 5' . | U A U U G A A A A A G G A A G A G U A U G . . . . . . . 3' |
| Lipoprotein | 5' . . . . . . . | A U C U A G A G G G U A U U A A U A A U G . 3' |
| Ribosomal protein S12 | 5' . . . . | A A A A C C A G G A G C U A U U U A A U G . . . . 3' |
| RNA polymerase β | 5' . . . . | A G C G A G C U G A G G A A C C C U A U G . . . . 3' |
| *trp* E | 5' . . . . | C A A A A U U A G A G A A U A A C A A U G . . . . 3' |

Regions of complementarity are underlined, and the initiation codons are italicized.

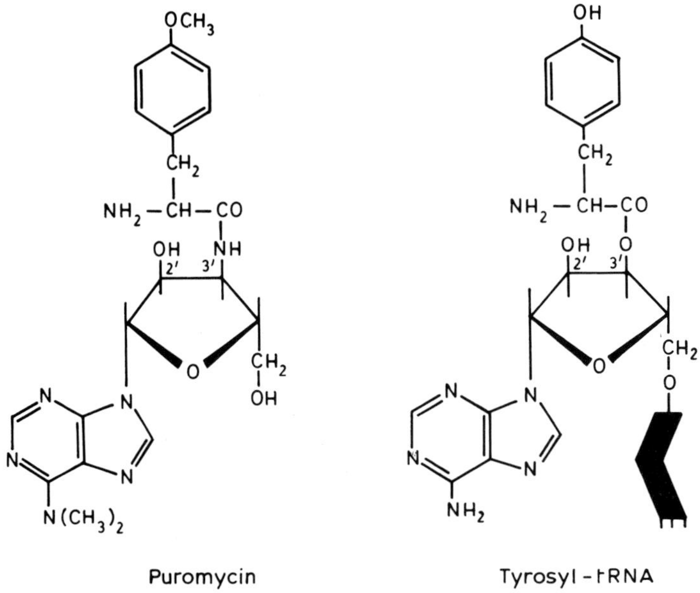

**Fig. 11.13**  The structure of puromycin compared with that of the terminal adenosine residue of Tyr-tRNA. Compare also Figs 11.8 and 11.9.

clearly defined. The fact that it cycles on and off the ribosome during initiation established that it is indeed an initiation factor, rather than a loosely bound ribosomal protein such as S1 (see Section 11.8). Although IF-1 seems especially to facilitate the action of IF-2, it is clear that it does not have a function analogous to that of EF-Ts in elongation (see Section 11.5.2). It may be involved in the expulsion of IF-2 from the ribosome after hydrolysis of GTP has occurred.

Lower-molecular-weight sub-species of IF-2 and IF-3 exist, lacking portions of the *N*-termini of the larger species. The smaller form of IF-2 (IF-2β), rather than being a proteolytic cleavage fragment of the larger form (IF-2α), may be the product of a second initiation on the mRNA [118]. Nevertheless, despite the fact that these sub-species are active, there is as yet insufficient evidence to support suggestions that they have properties that differ from their parents in a way that would allow them to perform regulatory functions.

### 11.5.2  Chain elongation [83, 119–121]

In polypeptide chain elongation an aminoacyl-tRNA binds to the A-site of the ribosome and reacts with the peptidyl-tRNA (Fig. 11.9) or fMet-tRNA (Fig. 11.14) in the P-site, accepting the growing polypeptide chain. The tRNA is then moved across to the P-site (translocated), with concomitant movement of the mRNA and expulsion of the deacylated tRNA, in order to make the A-site available for the next aminoacyl-tRNA [122–124]. Elongation requires three soluble factors, EF–Tu, EF–Ts, EF–G [125], and the hydrolysis of two molecules of GTP.

Elongation factor EF–Tu (molecular weight 43 000 [126]) is responsible for the ribosomal binding of the aminoacyl-tRNA corresponding to the mRNA codon in the A-site (arbitrarily designated as Ala in Fig. 11.14), prior to which it forms a soluble ternary complex with the tRNA and GTP [127, 128]. All elongator aminoacyl-tRNAs will form this complex, but fMet-tRNA will not [128]. The non-hydrolysable analogue of

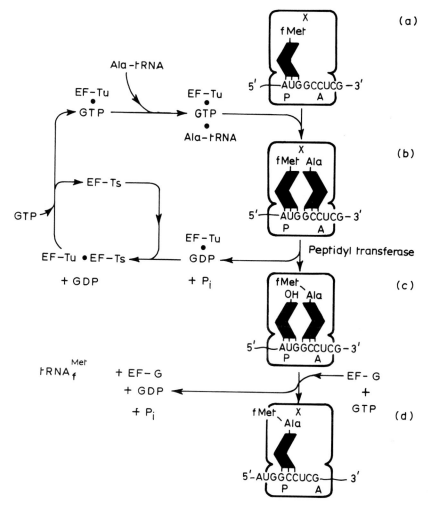

**Fig. 11.14**  A schematic diagram of prokaryotic polypeptide chain elongation. For convenience the 70S initiation complex of Fig. 11.12(d) has been taken as the starting-point, (a), although the scheme is equally true for 70S ribosomes bearing any peptidyl-tRNA in the P-site (e.g. (d)). Likewise the designations of the mRNA triplet in the A-site as coding for Ala, and the third triplet for Ser are purely arbitrary.

GTP, 5'-guanylylmethylene diphosphonate, will allow the aminoacyl-tRNA to bind to the 70S ribosome, but the GTP must be hydrolysed before peptide bond formation can occur [122]. The GTP hydrolysis is not required for the peptidyltransferase reaction itself and its possible role is discussed below. The EF–Tu and GDP are released from the ribosome as a complex. In this form the EF–Tu cannot react with GTP or aminoacyl-tRNA, and it is the function of EF–Ts (molecular weight, 30 000 [129]) to displace GDP from the EF–Tu·GDP complex [130, 131]. This results in the formation of an EF–Tu·EF–Ts complex from which the EF–Tu·GTP complex can be regenerated (Fig. 11.14).

The structure of EF–Tu has been studied by X-ray crystallography, but at present provides a detailed picture of only one of three domains [132]. It has overall dimensions similar to those of tRNA, and nuclease protection ('footprinting') and fluorescence studies indicate a primary interaction with the aminoacyl and TψC stems of tRNA (Fig. 11.1), leaving the anticodon loop exposed [133, 134]. An intriguing feature of EF–Tu that has recently emerged is that it has two binding sites for tRNA [447]. There are two forms of EF–Tu in *E. coli*, the products of separate genes (*tufA* and *tufB*), differing only in their *C*-terminal amino acids [126]. These two forms appear to be functionally equivalent. EF–Tu is extremely abundant, constituting some 5% of total bacterial cell protein, and occurring in approximately sixfold excess over ribosomes and other elongation factors. The significance of this is not known.

The aminoacyl-tRNA bound in the A-site (Fig. 11.14b) can now be linked to the carboxyl group of the fMet or nascent peptide, through the catalytic activity of the *peptidyltransferase* centre of the 50S ribosomal subunit [135, 82]. As already mentioned, the thermodynamic free energy for peptide bond formation comes from the hydrolysis of the 'energy-rich' acyl–ester bond of the fMet-tRNA, which in its turn is derived from the ATP hydrolysed during aminoacylation. The fact that GTP and supernatant factors are not required for the transpeptidation was perhaps most convincingly confirmed when it was discovered that, in the presence of ethanol (about 50%), a 3′-hexanucleotide fragment of fMet-tRNA could react with puromycin on the isolated 50S ribosomal subunit [136].

Extension of the 'fragment reaction' to even smaller oligonucleotide fragments has shown that CCA-fMet is the smallest species that can occupy the P-site of the peptidyltransferase [137], and other experiments with peptidyl-tRNA showed that puromycin can be replaced by CA–Gly at the A-site [138].

The translocation of the peptidyl-tRNA from the A-site to the P-site requires the elongation factor EF–G (molecular weight 77 000 [139, 140]) and GTP. This reaction has been shown to allow movement of the peptidyl end of the tRNA so that it becomes reactive towards puromycin [122, 141], movement of the mRNA relative to the ribosome [116, 117], and ejection of the deacylated-tRNA from the P-site [142]. The reaction requires hydrolysis of the GTP, the non-hydrolysable analogue being inactive [122, 123], although it will allow EF–G to bind to ribosomes [139]. The molecular mechanism underlying this process is perhaps the most intriguing and the least understood aspect of protein biosynthesis. It is clearly possible that a large part of the structural complexity of the ribosome, including even the division of the ribosome into subunits, may be a consequence of the need for this specific and concerted movement of macromolecules.

After translocation (Fig. 11.14d), one cycle of elongation has been completed (cf. Fig. 11.14a). The vacant A-site now contains a new mRNA codon, to which a corresponding aminoacyl-tRNA can bind, starting another round of elongation.

The function of GTP hydrolysis in the elongation reactions has long been a subject of debate. The earliest ideas involved the provision of energy for movement of molecules: the translocation of peptidyl-tRNA to the P-site in the case of EF–G, and the 'accommodation' of the aminoacyl-tRNA into the A-site (the most extreme form of which is from a separate entry site [84]) in the case of EF–Tu. However, the emergence of evidence for a single GTPase domain on the ribosome (see Section 11.8.3), to which EF–Tu and EF–G could not bind simultaneously [143–145], led to the suggestion of a unitary role for the GTP hydrolysis in expelling the factors (including IF-2) from the ribosome after they had fulfilled their functions [3, 4]. The presence of the factor on the ribosome was regarded as preventing the next reaction from occurring. The role of EF–G envisaged in this model would be to prime the ribosome for subsequent translocation which would *not*, in itself, require GTP. It is interesting in this regard that ribosomes can be made to perform a factor- and energy-independent translocation *in vitro*, albeit at rather a low rate [146, 147]. A totally different suggestion has, however, been made regarding the function of GTP hydrolysis. This is that it provides the energy for 'proof-reading' to ensure accuracy of the codon–anticodon inter- action [148, 149]. This is discussed further in Section 11.13.

### 11.5.3 Chain termination [150–152]

In polypeptide chain termination (Fig. 11.15), the ester linkage of the peptidyl-tRNA is hydro- lysed in response to one of three termination codons (see Section 11.2.3) in a reaction in- volving two of the three release factors, RF-1, RF-2 and RF-3 [153–156]. The deacylated tRNA

and the mRNA are expelled from the ribosome in the presence of release factor, RRF, and EF–G [157], liberating free ribosomal subunits. The subunits will associate to form 70S ribo- somes unless prevented from doing so by IF-3 and IF-1 (see Section 11.5.1).

In contrast with the other 61 codons, the three specific terminators are not read by tRNAs. This was shown using RNA from an amber mutant of bacteriophage R17 in which the first six codons of the coat protein are followed by a stop codon. Only the six appropriate aminoacyl-tRNAs and supernatant proteins were required for release of the hexapeptide [158]. This same system was subsequently used with purified elongation factors and release factors to show that release of the peptide required the peptidyl-tRNA to be at the P-site of the ribosome [159]. To study the factor requirements for termination at

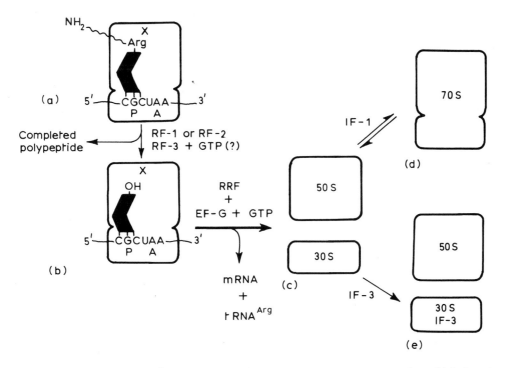

**Fig. 11.15** A schematic diagram of prokaryotic polypeptide chain termination. The amino acid designation in the P-site is purely arbitrary. Other possible termination codons in the A-site are UAG and UGA (see text).

all three codons, an assay was developed in which the termination codons could direct the release of fMet-tRNA, previously bound to ribosomes in the presence of the triplet AUG [160]. This led to the resolution [155] of two release factors, RF-1 (molecular weight 36 000 [161]) and RF-2 (molecular weight 38 000 [161]), of different codon specificities:

RF-1 for UAA or UAG

RF-2 for UAA or UGA

The third release factor, RF-3 (molecular weight 46 000), is not codon-specific and has no release activity in the absence of the other factors [156]. It stimulates the release of polypeptide promoted by the other factors and seems to stimulate both binding and release of these latter from the ribosome [162]. Its action is stimulated by GTP, but a requirement for GTP hydrolysis during termination in prokaryotes is not firmly established as GDP can replace GTP in the reaction *in vitro*.

There are quite strong grounds for thinking that the actual hydrolysis of the peptidyl ester linkage is catalysed by the peptidyltransferase centre of the ribosome, the reaction specificity of which has been modified by the binding of the release factors. This was suggested by the finding that the peptidyltransferase would catalyse the formation of an ester link to fMet-tRNA or its hexanucleotide fragment (see Section 11.5.2), if certain alcohols were presented to the ribosome instead of aminoacyl-tRNA [163]. If the hydroxyl groups of an alcohol could replace the $\alpha$-amino group of an aminoacyl-tRNA as a reactive nucleophile, it was possible that the hydroxyl group of water might do likewise. This suggestion was supported by the fact that a number of antibiotics (e.g. sparsomycin and chloramphenicol) and ionic conditions known to inhibit the peptidyltransferase reaction were also found to inhibit the termination reaction *in vitro* [155, 164].

After the release of the peptide, the mRNA

and deacylated tRNA are still attached to the ribosome (Fig. 11.15b) and must be removed before subunits can be regenerated for another round of protein synthesis. This process has been studied rather indirectly by assaying the release of ribosomes from polysomes after termination by puromycin [157], or using the amber coat protein mutant of bacteriophage R17, mentioned above [165]. A clear requirement was observed for GTP, EF–G and ribosome release factor, RRF (molecular weight 18 000 [157]). It has also been shown that RRF is required for synthesis of $\beta$-galactosidase in a coupled DNA-dependent transcription and translation system *in vitro* [166]. Although it might be expected that the primary role of EF–G in this process would be the expulsion of deacylated tRNA, there appear to be no data bearing on this question.

The released ribosomes can be in the form of 70S ribosomes or 30S and 50S subunits. It was originally thought that 70S ribosomes were released first and subsequently converted into subunits by a dissociation factor [167]. Although such an activity was identified [168], and later shown to be IF-3 [169], it seems most likely that it operates as an *anti-association factor*, preventing the 50S subunit from associating with the 30S·IF-3 complex [170]. Inactive 70S ribosomes do accumulate in cells, especially when inhibition of initiation results in a relative excess of 30S subunits over IF-3 [167]. To regenerate subunits when conditions improve there must be an equilibrium between 70S ribosomes and ribosomal subunits. Initiation factor IF-1 is thought to accelerate this reaction in both directions, without altering the position of its equilibrium [171]. Thus IF-1 will only stimulate dissociation if IF-3 is available to prevent the subunits reassociating [172].

Although all three termination codons are found to occur in actual mRNAs, the distribution of these is not random. The codon UAA is found most frequently, whereas the codon UAG is highly disfavoured in prokaryotes [173]. Furthermore there is a tendency for stop codons in prokaryotes to be followed by a U. Both these features appear to provide protection against read-through by natural suppressor tRNAs (see Section 11.13.2), rather than to contribute to the recognition of the codons by termination factors [174, 175].

## 11.6 THE EVENTS ON THE EUKARYOTIC RIBOSOME

Eukaryotic ribosomes catalyse essentially the same process as prokaryotic ribosomes. Although the details of eukaryotic protein synthesis are less well understood, it is clear that the differences from prokaryotic protein synthesis are relatively minor for elongation and termination, but much greater for initiation [9]. The following discussion relates to the nucleo-cytoplasmic protein-synthesizing system, the protein-synthesizing systems of mitochondria and chloroplasts being dealt with in Section 11.9.

### 11.6.1 Chain initiation

The first way in which it was realized that eukaryotic initiation differed from that in prokaryotes was in the initiating amino acid. Thus, although fMet-tRNA could be detected in yeast and rat liver mitochondria (in which it is now known to be the initiating amino acid) it was not found in the corresponding cytoplasms [176]. Nevertheless two distinct species of methionine tRNA were isolated from the cytoplasm: $tRNA_m^{Met}$ and $tRNA_f^{Met}$, the latter so designated because *in vitro* the Met-tRNA$_f$ could be formylated using *E. coli* transformylase [177]. Smith and colleagues [178, 179] used the Krebs II ascites tumour cell-free system of Mathews and Korner [180] to translate artificial polynucleotides, and obtained evidence that the methionine from the Met-tRNA$_f$ was preferentially incorporated at the *N*-terminus of the peptide products. Shortly afterwards it was demonstrated that methionine is transiently present at the *N*-termini of rabbit $\alpha$- and $\beta$-globins [181,

182] and trout protamine [183] synthesized *in vivo*.

However, the fundamental difference in eukaryotic initiation is the mode of selection of the initiating AUG, and it is clear that it is this aspect of initiation, although still incompletely understood, that accounts for the plethora of eukaryotic initiation factors [184–186]. In contrast with the situation in prokaryotes, internal initiations do not normally occur on eukarotic mRNAs [187], and the key CCUCC sequence involved in the 'Shine and Dalgarno' interaction in prokaryotes (Table 11.4) is conspicuously absent from the otherwise conserved 3'-end of the small (18S) rRNA of eukaryotes. According to Kozak, the principles of whose 'scanning model' [187, 58, 188] for eukaryotic initiation are widely accepted, the 40S ribosomal subunit binds to the 5'-end of the mRNA and moves along this until it encounters an appropriate (usually the first) AUG initiation codon, when attachment of the 60S ribosomal subunit can occur. There is good reason to believe that AUG is the only allowed initiation codon in eukaryotes [189], and the fact that the initiator tRNA binds to the 40S subunit before this attaches to the mRNA [190, 191] suggests that it is the CAU anticodon of the tRNA that scans for the AUG codon. In approximately 95% of cases the first AUG is the initiation codon, and most of the exceptions can be accommodated by postulating a context effect of the preceding bases, CCRCC*AUG*G representing the consensus for a strong initiation codon [188], AUG codons in weak contexts being bypassed. There are still some viral mRNAs which do not conform to this pattern, suggesting that our understanding here is still incomplete [449]. Nevertheless what is more striking than these exceptions is the extent to which viral mRNAs that are structurally polycistronic resort to strategies such as the production of polyproteins or the generation of nested mRNA subspecies, or both, that allow them to be functionally monocistronic (Fig. 11.16).

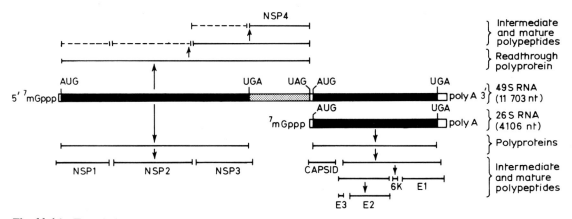

**Fig. 11.16**   Translation strategy of Sinbis virus [193]. Translation of the 49S genomic RNA occurs only from the first AUG, a second internal initiation site being recognized only in the sub-genomic 26S RNA, where it has now become the first AUG. The initial transcripts are polyproteins, which are proteolytically processed to give the mature non-structural and structural proteins identified. Readthrough (see Section 11.13.2) of the first stop codon produces a larger polyprotein from the 49S RNA, the processing of which generates NSP4 (and perhaps more of the other non-structural proteins: indicated by broken lines).

Turning now to the detailed mechanism of initiation (Fig. 11.17), the first stage is the formation of a (stable) ternary complex between Met-tRNA$_f^{Met}$, eIF-2 and GTP [191, 192], which then binds to the 40S ribosomal subunit, facilitated by the presence of eIF-3 and eIF-4C on the latter. The binding of this complex to the mRNA requires the secondary structure at its 5′-end to be melted, a process involving recognition of the 5′-cap structure by eIF-4F (a complex of the 24-kDa cap-binding protein (eIF-4E) [194], eIF-4A, and a third polypeptide of approximately 220 kDa [195]) together with eIF-4B. At some stage during or after this process ATP hydrolysis occurs [196, 197]. This is associated with eIF-4A activity and is thought to promote unwinding of the mRNA secondary structure [198]. The actual binding of the 40S pre-initiation complex to the melted mRNA requires eIF-1, and the joining of the 60S subunit after the 40S complex reaches the initiation codon requires eIF-5. The eIF-2 released at this stage is complexed to GDP which must be displaced by a factor eIF-2B (also called eRF and GEF) before it can continue to function (Fig. 11.30) [199]. This reaction, which is analogous to that between EF–Tu and EF–Ts (Fig. 11.14), is subject to regulation and is discussed in more detail in Section 11.11.2.

It should be mentioned that certain capped mRNAs (e.g. that of alfalfa mosaic virus 4) and uncapped mRNAs (e.g. that of polio virus) do not require eIF–4F (see also Section 11.11.3), and this is thought to be due to their having little secondary structure at their 5′-ends [200]. It may well be that the 5′-cap has two functions on cytoplasmic mRNA. The first may be to protect the mRNA from nucleolytic attack [201] (a function that may be sub-served in polio virus mRNA by the polypeptide attached to the 5′-end), and the second, as an attachment point for unwinding secondary structure, may not be required where such structure is absent.

Finally it should be pointed out that eIF–3 is an extremely complex factor, consisting of at least eight separate polypeptide chains with an

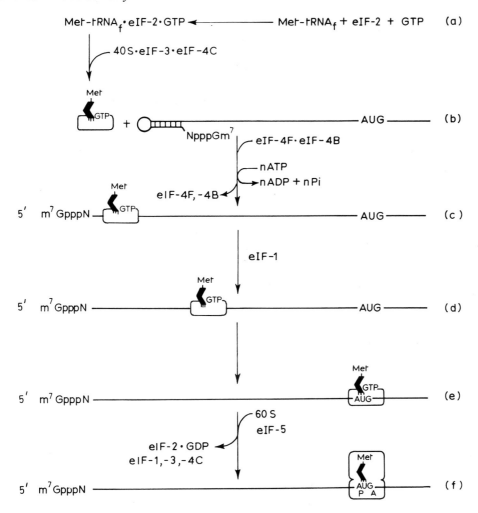

**Fig. 11.17** A schematic diagram of eukaryotic polypeptide chain initiation.

aggregate molecular weight of approximately 700 000 [202]. The reason for this complexity is unknown.

**11.6.2 Chain elongation**

At first it appeared that there was only a single aminoacyl-tRNA binding factor, EF–1, in eukaryotes, corresponding to EF–Tu + EF–Ts in prokaryotes [203]. However, the extreme tendency of EF–1 preparations to aggregate

[204] initially obscured the fact that the factor that resembles EF–Tu [205] and forms a ternary complex with aminoacyl-tRNA and GTP (now called EF–1$\alpha$), was associated with another activity [206, 207] (now called EF–1$\beta\gamma$). This latter factor has properties analogous to the prokaryotic EF–Ts [208, 209]. One perplexing feature of EF–1$\alpha$ is that, unlike EF–Tu, it can form a ternary complex with GTP and the initiator tRNA [210]. The mechanism that prevents unformylated Met-tRNA$_f$ in

eukaryotes being used for internal insertion of methionine residues [178] is unclear. Like prokaryotic EF–Tu, eukaryotic EF–1α is an extremely abundant cellular protein [211], and, at least in yeast, is encoded by two separate genes [212].

The elongation factor involved in the translocation reaction in eukaryotes, EF–2, has been purified [213, 214], and seems closely analogous to EF–G in prokaryotes. One point of interest is its specific inactivation by diphtheria toxin, which transfers ADP-ribose from $NAD^+$ to an unusual modified histidine residue in EF–2 [215, 216]. An enzyme with an analogous activity to that of diphtheria toxin has been detected in eukaryotic cells, suggesting a possible role for the modification of this residue of EF–2 in normal cellular regulation [445].

Eukaryotic, like prokaryotic, ribosomes possess an intrinsic peptidyltransferase activity, which has also been studied using the 'fragment' reaction [217]. Although the eukaryotic peptidyltransferase is inhibited by certain antibiotics (e.g. sparsomycin) that inhibit prokaryotic peptidyltransferase, it is resistant to the action of others (e.g. chloramphenicol), and hence must differ somewhat in its structure.

### 11.6.3 Chain termination

A single factor, RF, was found to catalyse the release of the completed polypeptide chain from eukaryotic ribosomes [218]. This appears to recognize all three termination codons, UAA, UAG and UGA, although it is necessary to use tetranucleotides to assay for this in the fMet-tRNA release reaction [219]. In contrast to the prokaryotic factors, the eukaryotic release factor shows a clear requirement for GTP hydrolysis for its action *in vitro*. In eukaryotes there seems to be a less biased usage of the three termination codons, and the preferred context differs from that in prokaryotes [173].

Although one imagines that there is a eukaryotic factor analogous to RRF, none has so far been described. Eukaryotic ribosomes have been shown to be liberated from polysomes as subunits, and a pool of inactive 80S monomers is present in eukaryotic cells [220, 221]. Anti-association activity (analogous to prokaryotic IF-3) was originally ascribed to eIF-3 [202] but has more recently been described as the property of a separate species designated eIF-6 [222, 223].

### 11.7 THE CONTROL OF THE CELLULAR LOCATION OF THE PRODUCTS OF TRANSLATION

Proteins synthesized on the cytoplasmic ribosomes of eukaryotic cells can be either retained in the cytoplasm, transferred to subcellular organelles, inserted into the plasma membrane, or secreted. Although, of course, prokaryotes do not have subcellular organelles, the other possibilities mentioned are also available to them, sequestration in the periplasmic space being formally similar to secretion in eukaryotes.

### 11.7.1 Secreted proteins

Most studied has been the synthesis of proteins destined for secretion (or periplasmic sequestration) and it has long been known that ribosomes synthesizing such proteins (unlike the majority of those synthesizing proteins to be retained intracellularly) are located on the membranes of the rough endoplasmic reticulum (in eukaryotes) or on the inner cell membrane (in prokaryotes). The proteins are extruded through the membrane as they are synthesized and, in the case of eukaryotes, pass from the cisternae of the endoplasmic reticulum, via the Golgi apparatus, to secretory vacuoles (see [224, 225] for reviews). The chemical basis for this segregation lies in a largely hydrophobic peptide, the 'signal peptide', at or near the *N*-terminus of the nascent protein. In most cases the signal peptide is an approximately 15–30 amino acid extension of the *N*-terminus of the mature protein (Table 11.5) that is later removed

**Table 11.5**  The structure of some signal peptides.

| Pre-protein | −25 | −20 | −15 | −10 | −5 | +1 |
|---|---|---|---|---|---|---|
| **Eukaryotic secreted** | | | | | | |
| Pre-proalbumin | | | M [K] W V T F L L | L L F I S G S | A F S \| | [R] |
| Pre-proparathyroid hormone | M M S A [K] (D) M V | [K] V M I V M L A | I C F L A [R] S | G (D) \| | | [K] |
| Pre-β-casein | | | M [K] V L I L A | C L V A L A \| | | [R] |
| Pre-ovomucoid | | M A M A G V F V L F | S F V L C G F L | P (D) A A F G ⌐ | A |
| Ovalbumin (uncleaved N-terminus) | M G S I G A A S M (E) | F C F (D) V F [K] (E) | L [K] V [H] I A ---- N | | | |
| **Eukaryotic membrane** | | | | | | |
| Pre-VSV glycoprotein | | | M [K] C L L Y L A | F L F I G V N C \| | | [K] |
| Pre-HLA 12.4 | | M A P [R] T L L L L | S G A L A L T Q | T W A \| | | [R] |
| **Prokaryotic periplasmic** | | | | | | |
| Pre-β-lactamase | M S I Q [H] F [R] V A | L I P F F A A F C | L P V F A \| | | | [H] |
| Pre-enterotoxin LT | | M [K] N I T F I F F | I L L A S P L | Y A ⌐ | | N |
| **Prokaryotic membrane** | | | | | | |
| Pre-lipoprotein | | M [K] A T [K] L V L G | A V I L G S T L | L A G \| | | C |
| Pre-maltose-binding protein | M [K] I [K] T G A [R] I | L A L S A L T T M | M F S A S A L A \| | | [K] |
| Pre-phage fD major coat protein | M [K] [K] S L V L [K] A | S V A V A T L V P | M L S F A ↑ | | A |
| | | | Sequence of signal peptide* | | | Cleavage point |

* Not for ovalbumin.
Basic residues are boxed, acidic residues are circled.

by proteolytic cleavage [226–228]. Although the signal peptide is hydrophobic, its first interaction, at least in eukaryotes, is not with the membrane, as might have been expected, but with an 11S signal recognition particle (SRP), which contains six polypeptide chains and a 7S RNA species, 260 nucleotides in length, to which the Alu family of repeated DNA sequences is related (see Fig. 7.24 and Section 7.7.3) [229]. This interaction may temporarily arrest protein synthesis, the proposed rationale for which is that the time taken for the signal peptide to 'find' the appropriate membrane site is greater than that needed for the growth of the nascent polypeptide to a size at which it would start folding into a tertiary structure in which the signal peptide would be inaccessible. Such elongational arrest does not, however, appear to occur in all cases [446]. The 'appropriate membrane site' is characterized by a specific membrane protein, 'docking protein' or 'SRP receptor' [230], the interaction of the arrested complex with which releases the block on translation. This allows the ribosome to attach to the membrane (presumably through specific proteins, perhaps those termed ribophorin I and II [231]) and the nascent peptide to be translocated into the intercisternal space. The signal peptide is, in most cases, cleaved by a peptidase on the luminal side of the endoplasmic reticulum [232] (Fig. 11.18). The protein then passes along towards the Golgi apparatus and thence to secretory vacuoles. The signal peptide has been shown in Fig. 11.18 as being led through the membrane in a hairpin configuration. This is partly by analogy with what is thought to occur in bacteria [233] (see below), and partly to accommodate the exceptional secreted protein, ovalbumin, where the signal sequence is not at the N-terminus (which is not cleaved – Table 11.5) but is thought to comprise the hydrophobic region in positions 26–55 [234].

The situation in prokaryotes is analogous, but not identical, to that just described [233]. Thus bacterial signal peptides can be recognized and cleaved by eukaryotic membranes. Nevertheless there is good evidence that in prokaryotes, but not in eukaryotes, the interaction of the negatively charged inner surface of the bacterial cell membrane with a positively charged residue near the N-terminus of the signal peptide is absolutely required for attachment to occur [235]. The existence of several uncharacterized genetic loci that affect export in *E. coli* (eg. *sec* Y, Fig. 11.29), and evidence for the involvement of a soluble activity in this process, suggest further complexity. Whether or not this involves a system similar to that involving the eukaryotic

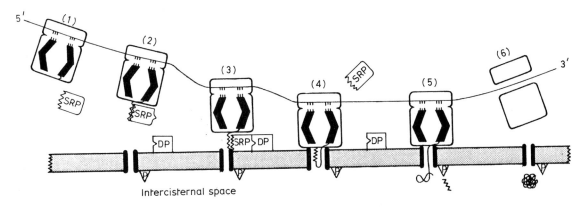

**Fig. 11.18** Model for the segregation of secretory proteins into the lumen of the endoplasmic reticulum: SRP, signal recognition particle; DP, docking protein; P, signal peptidase. For details see text.

signal recognition particle is at present an open question.

Secreted proteins are not the only proteins that are extruded into the intercisternal space of the endoplasmic reticulum in this way: the same is true for proteins destined for the lysozomes. In this case, manose 6-phosphate, added in the intercisternal space, is the signal that directs the proteins to the lysozomes and prevents their secretion [236].

### 11.7.2 Membrane proteins

Certain membrane proteins are also synthesized on membrane-bound ribosomes and are directed into the membrane cotranslationally via signal peptides which may or may not be subsequently cleaved [237, 451] (cf. Fig. 11.18). Unequivocal examples of these are proteins the *N*-termini of which are exposed to the external surface of the plasma membrane (which is believed to derive from the internal surface of the membrane of the endoplasmic reticulum). What stops such membrane proteins being secreted is a hydrophobic sequence, at or near the *C*-terminus, that anchors the protein into the membrane. This is especially well illustrated in the case of the membrane-bound and secreted forms of immunoglobulin heavy chain, which differ only in approximately 20 amino acid residues at their *C*-termini [238] (see also Section 10.6.1).

The situation regarding the synthesis of membrane proteins in which the *C*-terminus is orientated extracellularly, or of proteins that loop back and forth through the membrane, is less clear. Examples of both cotranslational and post-translational insertion have been reported for such proteins, and, in at least one case, an internal signal peptide is involved.

### 11.7.3 Organelle proteins

The question of the mode of transport to the mitochondrion of proteins synthesized in the cytoplasm is complicated by the different mitochondrial destinations of such proteins. These are the outer-membrane, inter-membrane space, inner membrane and matrix. Such mitochondrial proteins are synthesized on free cytoplasmic ribosomes, and those other than proteins destined for the outer membrane generally have *N*-terminal extensions that are cleaved on subsequent passage into the organelle, a process requiring a membrane potential (except in the case of cytochrome *c*) reminiscent of, but of opposite polarity to, that required for periplasmic sequestration in bacteria. These *N*-terminal extensions are more variable in size than those for secretory or membrane proteins, and, from currently available data, appear to be much more hydrophilic. It is thought that there are receptors to recognize different types of mitochondrial protein extension and direct the proteins to the right location (see [239, 452] for reviews).

The targeting of proteins to the lysozome is mentioned in Section 11.7.1 above, and targeting to other organelles is reviewed elsewhere [240, 241, 453].

### 11.8 THE RIBOSOME

Our most detailed knowledge of ribosomes comes from studies in *E. coli* [242–244], to which discussion will largely be confined, except as regards the structure of the individual ribosomal components, where something about the eukaryotic ribosome is known [245, 246]. The mitochondrial and chloroplast rRNAs will be included in discussion here but the ribosomes of these organelles are described in more detail in Section 11.9. In addition to the more recent references cited, two older volumes [247, 248] still contain much of value.

### 11.8.1 Structure of the components

Ribosomes comprise two subunits of dissimilar size that are held together by $Mg^{2+}$, and that, on

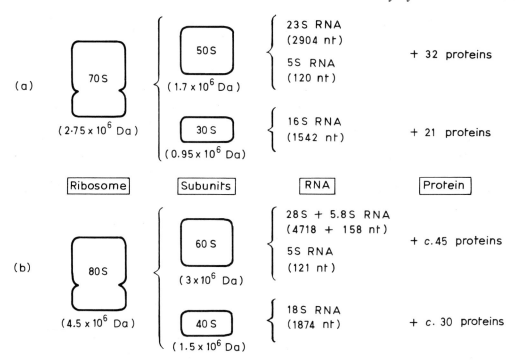

**Fig. 11.19**   A schematic diagram of the components of the ribosomes of (a) *E. coli* (b) rat. The molecular weights of the particles are mean values from physical determinations. The chemical molecular weights of 70S, 50S and 30S ribosomal particles from *E. coli*, based solely on RNA and protein contents, are $2.3 \times 10^6$, $1.45 \times 10^6$ and $0.85 \times 10^6$. The discrepancy between the two sets of values can be accounted for by the presence of metal ions and spermidine.

dissociation, sediment at 30S and 50S in the case of the 70S bacterial ribosome, and 40S and 60S in the case of the 80S eukaryotic ribosome of mammals. (The sizes of the ribosomes of other eukaryotes may differ from this.) They contain approximately 60% RNA and 40% protein. The smaller ribosomal subunit contains a single species of RNA (16S or 18S RNA for 30S and 40S subunits, respectively) together with ribosomal proteins; the larger ribosomal subunit contains a major species of RNA (23S or 28S RNA for 50S and 60S subunits, respectively) an additional small 5S RNA, together with a number of proteins exceeding that in the small subunit (Fig. 11.19). In eukaryotes the portion of the primary transcript that gives rise to the major species of rRNA in the 60S subunit undergoes

further nucleolytic cleavage after it has started adopting its secondary structure (see Section 9.4). This results in a 5.8S RNA species hydrogen-bonded to the 28S RNA in the case of mammals (Fig. 11.19). Similar 5'-post-transcriptional processing gives rise to a 2S RNA species in *Drosophila* and a 4.5S RNA species in plant chloroplasts (Sections 9.4.2, 9.4.3 and 9.6.2). As discussed further below, from the functional point of view these RNAs are better considered to be integral parts of the larger rRNA species.

The ribosomal RNAs contain a small number of specific modified (largely methylated) nucleotides. In the case of *E. coli* 10 base methylations have been identified in both 16S and 23S RNA; and the 23S RNA also contains three pseudouridine residues and a few ribothymidine residues and ribose methylations [249–250]. In mammals there are 46 and 71 methylated groups on the 18S and 28S RNA, respectively, most of which involve the 2′-O of the ribose moiety [251]. They are predominantly clustered in the 5′-half of 18S RNA and the 3′-half of 28S RNA [252]. Eukaryotic rRNAs also contain pseudo-uridine residues: approximately 37 and 60 for 18S and 28S rRNA respectively [251]. Neither eukaryotic nor prokaryotic 5S RNA contains modified nucleotides.

The ribosomes of eukaryotes, prokaryotes, mitochondria and chloroplasts catalyse similar reactions. It seems reasonable to assume that their rRNAs serve similar functions, whatever these might be. In comparing rRNAs that diverge markedly in size (from 597 to 1957 nucleotides for small subunit rRNAs, and 1152 to 4718 nucleotides for large subunit rRNAs [250]) attention has therefore centred on common structural features that may be crucial to such functions. In each class of rRNA there are a small number of relatively short, highly conserved, regions of primary structure. The confidence in the reality of such small similarities between molecules of such disparate size is strengthened by their occurrence at similar regions in a common type of secondary structure. An example of how small rRNAs can be assigned such similar secondary structures is shown in Fig. 11.20. It can be seen that the structure of the 12S RNA of the human mito-chondrion contains a core of features that are present in the progressively more elaborated 16S and 18S RNAs of *E. coli* and *X. laevis*, respectively. A similar situation obtains for the larger rRNAs, and this general model allows the 5.8S RNA to be accommodated, hydrogen-bonded to the 28S RNA, producing a structural feature similar to one present in 23S RNA [254].

These secondary structure models for the rRNAs have generally been derived from a combination of comparative sequence studies together with direct experimental analysis [249,

250]. As yet it has been difficult to apply computer prediction of secondary structure, based on thermodynamic considerations of helix stability, to these large molecules. In brief, secondary structure features have been predicted where these can be accommodated into many rRNAs of different phylogenetic origins. Especially useful are situations where there is poor conservation of primary structure but conservation of predicted secondary structure because of compensatory base changes (e.g. an A to G change in one strand of the RNA being compensated for by a U to C change in the complementary base of the other strand). Experimental probing of the secondary structure has been by reagents, the action of which requires RNA single-strandedness (certain nucleases, chemical reagents and oligonucleo-tide probes), and by methods that can indicate double-strandedness, such as RNA–RNA cross-linking and isolation of base-paired fragments. These experimental methods have generally tended to be used more confirmatively than predictively; however, they are especially important where either extreme conservation or divergence of primary structure makes the comparative approach inapplicable. In contrast, experimental methods have been of greater importance in proposals for the secondary structure of the 5S RNAs, about which, however, complete agreement has still not yet been reached [255, 256].

Although the proposed rRNA secondary structures are based, for the most part, on considerations of the RNAs outside the context of the ribosome, there is every reason to believe that they will be broadly valid for the RNAs within the ribosome. Clearly, there may be tertiary hydrogen-bonded interactions that the secondary structures take no account of, and it is also possible that interaction with ribosomal proteins (Section 11.8.2) might influence the structure. Our knowledge of the tertiary structure of the rRNAs is largely limited to the information obtained from RNA–RNA cross-

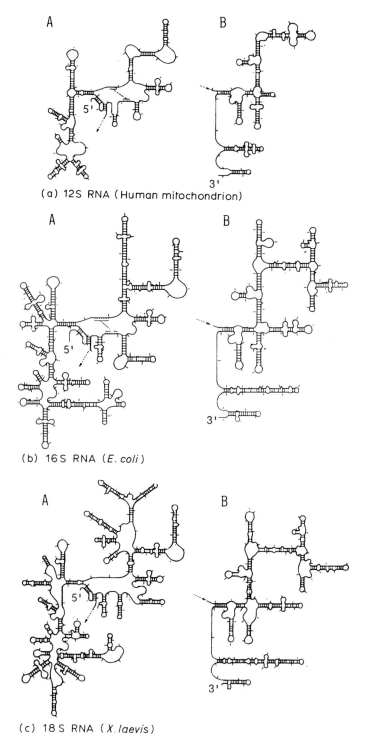

(a) 12S RNA (Human mitochondrion)

(b) 16S RNA (E. coli)

(c) 18S RNA (X. laevis)

**Fig. 11.20** Comparison of proposed secondary structures of the smaller ribosomal RNA from (a) a human mitochondrion, (b) a prokaryote and (c) a eukaryote. Each structure has been separated into two parts for clarity of presentation (after [253], with permission).

links, together with the location of a few specific features of the RNA on the ribosome as visualized by electron microscopy [249, 250].

In *E. coli* the 30S ribosomal subunit has 21 distinct proteins, designated S1–S21, one copy of each being present per 30S subunit. It was originally thought that there were 34 distinct proteins, L1–L34, on the 50S ribosomal subunit. However, L8 is in fact a complex of L7/L12 and L10, and L26 is identical to S20, there being on average 0.2 copies of L26 and 0.8 copies of S20 per 70S ribosome. Thus it is best regarded that there are 32 distinct proteins on the 50S subunit. All but two of these proteins are completely dissimilar and are present as single copies. The exceptions are L7 and L12, L7 being the *N*-acetylated form of L12. The two proteins together are present in a total of four copies per 50S subunit. Apart from proteins S1, S6, L7 and L12, the ribosomal proteins are all chemically basic, with molecular weights in the range 9 000 to 35 000. The total amino acid sequences of all of the ribosomal proteins of *E. coli* are known [242].

Only a few sequences are so far available for eukaryotic ribosomal proteins, but the data available, together with those obtained by immunochemical methods, have so far revealed only little structural homology to prokaryotic ribosomal proteins. Nevertheless evidence for common antigenic determinants has been obtained for certain eukaryotic ribosomal proteins and the *E. coli* proteins S6, L2, L11 and L7/L12 [257–259]. In the case of L7/L12 there has been a long controversy over the immunochemical cross-reactivity, which would not be expected from a comparison of the amino acid sequences of the corresponding proteins in *E. coli* and a number of eukaryotes. There is little doubt from their size, acidic nature and multicopy occurrence that the proteins are homologous. As far as primary structure is concerned an indirect link between the prokaryotic acidic proteins and the eubacterial L7/L12 is provided by the relatedness of both to the corresponding archaebacterial (see Section 11.9.1) ribosomal protein [260]. Two forms of this acidic ribosomal protein, with similar but distinct primary structures, occur in eukaryotes [260, 261]. In addition these proteins are phosphorylated to various extents, a modification found in another eukaryotic ribosomal protein (S6 – see Section 11.11.2) but not in prokaryotic ribosomal proteins [262].

### 11.8.2 Organization of components in the ribosome

The question of the mode of organization of RNA and ribosomal proteins in the ribosome has been approached from several standpoints. These are the relative proximity of the ribosomal proteins, the interaction of RNA and protein, and the position of the components in the ribosome as visualized by electron microscopy.

The most useful techniques for determining the relative positions of ribosomal proteins have been crosslinking with cleavable bifunctional reagents [263] and neutron scattering [264]. In the latter, extremely powerful, technique, ribosomes are reconstituted from their components, but with two of their proteins replaced by ones of a neutron density that contrasts with that of the rest of the particle. This is achieved by growth of *E. coli* in media with appropriately different proportions of $D_2O$. The results of such measurements for the proteins of the 30S ribosomal subunit are shown in Fig. 11.21, where they are superimposed on to the shape of the subunit derived from electron microscopy.

As regards the interaction of RNA and protein in the ribosome, certain proteins have been shown to be involved in the primary interaction with RNA during assembly *in vitro*, and probably also *in vivo*. For 16S RNA and 5S RNA these are fairly well defined as S4, S7, S8, S15, S17 and S20, and L5, L18 and L25 respectively [265]. In the case of 23S RNA an original list of 10 proteins (L1, L2, L3, L4, L6, L13, L16, L20, L23 and L24) has in more recent years been

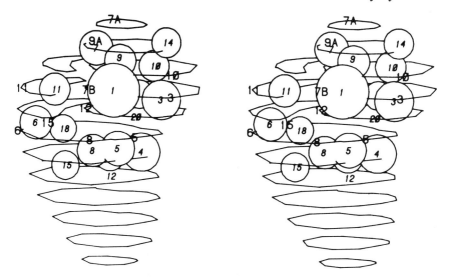

**Fig. 11.21** Stereoscopic diagram of the relative positions of the proteins (numbered balls) of the 30S ribosomal subunit of *E. coli* derived from neutron scattering studies. Superimposed on this are the contours of the subunit as deduced from the electron microscopic studies of Lake and coworkers, in the orientation that gives greatest agreement with the location of proteins (non-italic surface numbers) derived from immune electron microscopy. (From [264], with permission.)

extended to about half the proteins of the 50S ribosomal subunit [242, 244]. This may reflect a greater complexity of structure, but probably also illustrates the fact that most proteins in the assembled ribosome are in contact with RNA. The areas of RNA to which the proteins bind have been studied by nuclease-protection experiments, but only in a few cases has this enabled the binding sites to be defined relatively precisely [244]. It is interesting, however, that the RNA-binding site for protein L23 is structurally and functionally conserved in yeast, even though the corresponding yeast protein has little structural similarity to *E. coli* L23 [443]. Individual nucleotides in close proximity to particular ribosomal proteins have, however, been identified by RNA–protein crosslinking reagents [243, 249, 250].

The positions of the RNA and protein components in the structure of the ribosome deduced from electron microscopy (Fig. 11.10) have been

determined by examination of particles cross-linked by appropriate antibodies [266, 267]. This 'immune electron microscopy' has been applied mainly to ribosomal proteins, although the 5'- and 3'-ends of the rRNAs, and certain methylated bases, have also been 'visualized' in this way. The position of regions of rRNA shown to crosslink to particular proteins can, of course, be inferred indirectly. A summary of some of the results obtained for the 30S subunit is presented in Fig. 11.22.

### 11.8.3 Functional domains in the ribosome

The primary interest in elucidating the structure of the ribosome is the hope that this will help lead to an understanding of the way that this complex organelle catalyses protein biosynthesis. One can regard the individual components of the ribosome as having either structural or functional roles, and for a long time the general view

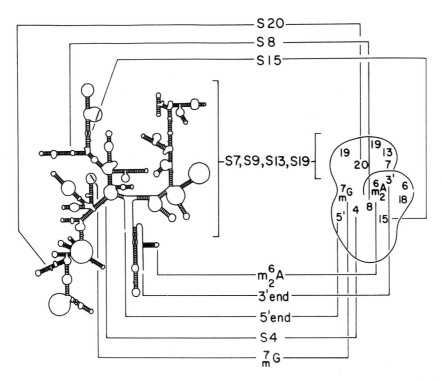

**Fig. 11.22** Location of protein and RNA components on the 30S ribosomal subunit of *E. coli* as deduced from immune electron microscopy and studies of RNA–protein interaction. (From [250], with permission.)

was that the rRNAs had a mainly structural role and that the catalytic functions of the ribosomes (e.g. peptidyltransferase) were sub-served by proteins. Although it was clear from reconstitution experiments that certain proteins may only be required for the assembly of the particles [268, 269], the idea of the functional pre-eminence of proteins was a natural extrapolation from what was known about enzymes. Much work was therefore performed to identify proteins at the functional sites of ribosomes, the main techniques being affinity labelling and crosslinking of protein substrates to proteins in the ribosome [270, 271].

A number of factors have contributed to a more recent change of emphasis from ribosomal proteins to ribosomal RNA. One was the failure to identify individual proteins containing the

catalytic activities associated with the ribosome: many of the likely candidates have been successively eliminated by the discovery of viable bacterial mutants in which these proteins are lacking [272]. Furthermore the success of the Shine and Dalgarno hypothesis (Section 11.5.1) awakened an interest in the possibility of a more extensive functional role for rRNA; and this interest was fuelled by the finding of accessible RNA at the active sites of ribosomes using appropriate affinity and crosslinking reagents for RNA [250]. This should, in fact, be regarded as a reawakening, for in 1968 Crick [273] had suggested that evolutionary considerations favoured a primitive ribosome consisting primarily of RNA. There has been a succession of hypotheses for rRNA functions involving base-pairing to other RNA molecules in or on the ribosome, but most of these have not survived the results of phylogenetic sequence comparison and experimental manipulation [250]. Rather than trying to assign functions to individual components of the ribosome, it seems at present more appropriate to discuss together the RNA and proteins present in different functional domains of the ribosome, bearing in mind that we are now aware of the existence of catalytic (Section 9.4.3) as well as informational RNA.

The *GTPase* domain on the 50S subunit [274, 275] will be considered first. There is clear evidence that the multicopy proteins L7/L12 are essential for the GTPase activity of the ribosome and the binding of elongation factors, even though these proteins are not labelled by affinity analogues of GDP. The protein that is labelled by such analogues, L11, does not, however, appear to be functionally indispensable, as mutants of *B. subtilis* resistant to thiostrepton (which prevents ribosomal binding of the EF–G·GTP complex) lack protein L11. The 23S RNA of the natural producer of thiostrepton, *Streptomyces azureus*, unlike that of *E. coli*, is methylated at nucleotide $A_{1067}$. This nucleotide is located in a sequence of RNA (1055–1081) to which EF–G can be crosslinked and also in the sequence (1052–1112) to which protein L11 binds (Fig. 11.23). The functional importance of this region of RNA is also supported by the fact that it contains highly conserved portions of primary structure which, incidentally, are not involved in secondary structure hydrogen-bonding.

The overall organization of the domain appears to involve a tetramer of L7/L12 complexed to L10, this latter protein being proximal to L11 and the RNA. The L7/L12 tetramer is highly elongated, and constitutes the 'stalk' seen in electron micrographs of the 50S subunit (cf Fig. 11.10). Whether or not the ribosome contains the GTPase activity is still unclear. This is because EF–Tu can exhibit a GTPase activity independent of ribosomes in the presence of

kirromycin or high concentrations of monovalent cations [276]. It is possible that the GTPase domain of the ribosome is necessary to trigger intrinsic GTPase activities of EF–Tu, EF–G [455] and IF–2, although there is no evidence for such activity in the latter case.

As regards the proteins and RNA implicated in the *peptidyltransferase* domain of the ribosome there is not such a clear picture of a physical complex, although immune electron-microscopic evidence places this domain near the central protuberance of the 50S subunit (Fig. 11.10). Several proteins are affinity-labelled by analogues of tRNA or by antibiotics such as chloramphenicol and puromycin that bind to this site, and some of these proteins have been shown to be necessary for peptidyltransferase activity [244, 271, 274]. Most strongly implicated, perhaps, has been protein L16, although there is good evidence that L2 and L27 are also required. The site of interaction, if any, of L16 and L2 with rRNA is not known, but protein L27 has been crosslinked to a nucleotide in the region 2332–2337 of 23S RNA. This is not too far away from the region of RNA thought to be at the peptidyltransferase centre (see below), but probably still far enough away to require tertiary folding to bring the regions together.

The region of RNA implicated in the peptidyltransferase domain is a non-hydrogen-bonded area that forms a 'junction' from which several hairpin stems emanate (Fig. 11.24). The primary evidence for the functional importance of this RNA is the fact that it contains bases that are the points of mutation that confer resistance to chloramphenicol or erythromycin (inhibitors of peptidyltransferase activity) to certain mitochondrial ribosomes [275]. Many nucleotides in this junction are conserved, and these include two that have been crosslinked to the amino acid analogue moiety of an affinity label derived from Phe-tRNA [277]. At the end of one of the duplexes that emanate from this junction is a phylogenically conserved loop that is the most likely site of a crosslink to a puromycin-

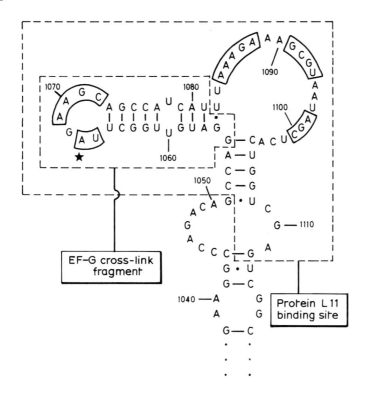

**Fig. 11.23** The proposed GTPase domain of the 50S ribosomal subunit of *E. coli* (after [275], with permission). ★, Nucleotide A $_{1067}$, the methylation of which endows resistance to thiostrepton in *S. azureus*. The boxed nucleotides are highly conserved in eubacterial and organelle ribosomes.

derivative [278]. The fact that the 3'-end of all tRNAs contains the sequence CCA led to the suggestion that this might base-pair to RNA at the peptidyltransferase centre. The most highly conserved exposed UGG sequence in rRNA comprises the nucleotides 807–809 in a completely separate region of the 23S RNA [277]. A crosslink between a nucleotide within the sequence 739–748 in this region and a nucleotide within the sequence 2609–2618 in the peptidyltransferase 'junction' [279] suggests that these regions may be close together in the tertiary structure, but direct evidence for the participation of this UGG sequence is at present lacking.

The basis of the catalytic activity is not yet clear. Although the idea of general acid–base catalysis involving a histidinyl residue is most favoured [244], the possibility of catalysis involving RNA has also been discussed [280].

Although not associated with a catalytic function, it is also useful to consider the *mRNA-decoding* domain on the 30S subunit where the codon–anticodon interaction occurs. This is close to the region where the mRNA interacts with the 3'-end of the 16S RNA (Section 11.5.1), and both are located in the cleft between the platform and main body of the 30S subunit. The mRNA-decoding domain has features that appear to be common to prokaryotes and eukaryotes, and involves the non-base-paired region from nucleotides 1392–1407 of the 16S RNA of *E. coli* which, unlike the extreme 3'-terminus involved in the 'Shine and Dalgarno'

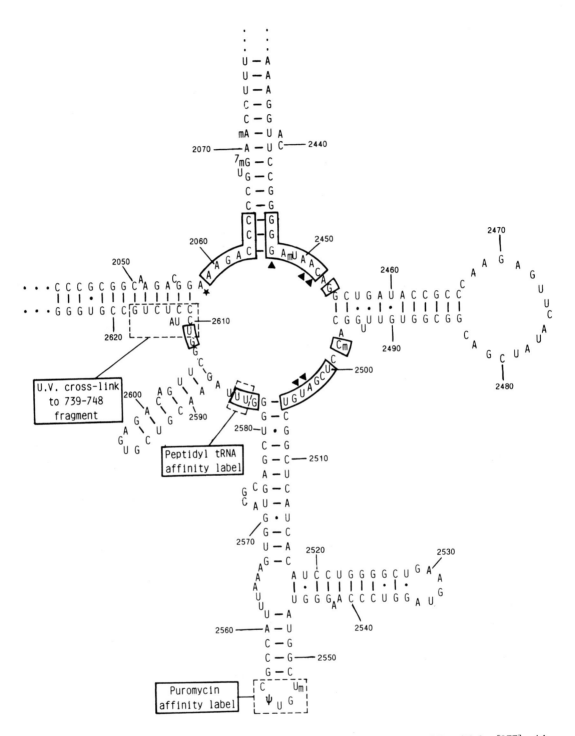

**Fig. 11.24** The proposed peptidyltransferase domain of the 50S ribosomal subunit of *E. coli* (after [277], with permission). ★, Site of base change conferring resistance to erythromycin in mitochondria; ▲, Site of base change conferring resistance to chloramphenicol in mitochondria. The boxed nucleotides are highly conserved throughout evolution.

interaction, is conserved in eukaryotic 18S RNA (Fig. 11.25). This region contains a nucleotide ($C_{1400}$) that can be crosslinked to the modified base, $cmo^5U$, at the wobble position of $tRNA^{Val}$ when this is bound to the P-site of *E. coli* or yeast ribosomes [281]. Furthermore, yeast mitochondrial ribosomes resistant to paromomycin show a base change at the adjacent nucleotide 1409 [282]. Also in this general area is the highly conserved sequence $m_2^6Am_2^6A$ (nucleotides 1518–1519), the functional importance of which was suggested by its loss of methylation in *E. coli* ribosomes resistant to the inhibitor of initiation, kasugamycin [283]. As well as inhibiting initiation, kasugamycin can cause misreading, as can streptomycin, resistance to which can involve mutation in protein S12, which is also implicated by other criteria in modulating the accuracy of translation (Section 11.13.1). Protein S12 has been crosslinked to $G_{1323}$ [284], which may mean that the stem and loop on which this is located are also associated with the decoding domain in the tertiary structure.

Two proteins that have long been implicated in the interaction of mRNA with the ribosome in *E. coli* are S1 and S21. It is thought that S21 makes the pyrimidine-rich 'Shine and Dalgarno' region accessible to the ribosome by affecting the secondary structure of the 16S RNA [285]. The involvement of protein S1, which can bind and be crosslinked to the 3′-end of 16S RNA, is suggested by the fact that, although absolutely required for translation of natural mRNAs, it can be dispensed with for translation of artificial polynucleotides. This protein has the ability to destabilize RNA helices, and one school of thought regards its function as to melt the purine-rich 'Shine and Dalgarno' region of the mRNA [286]. An alternative view lays stress on the extreme elongation of this atypical ribosomal protein and suggests that it may act as a flexible probe which searches the vicinity of the ribosome for mRNA [287].

In focusing on particular domains on the ribosome this discussion has neglected possible dynamic aspects of the ribosome, for example in relation to translocation of peptidyl-tRNA from the A- to the P-site. One intellectually appealing proposal is that this may involve 'switches' between alternative rRNA secondary structures [249].

## 11.9 OTHER PROTEIN-SYNTHESIZING SYSTEMS

### 11.9.1 Archaebacteria [288, 289]

In the discussion of protein-synthesizing systems so far a distinction has been made between the nucleocytoplasmic system of eukaryotes and the system found in those prokaryotes represented by the commonly studied bacteria (e.g. *E. coli*) and their viruses. There is good reason, however, to think that the prokaryotes comprise not one, but two, kingdoms. These are the *eubacteria*, into which kingdom the majority of bacteria fall, and the *archaebacteria*, a kingdom containing methanogens, extreme halophiles and certain thermoacidophiles. It is appropriate to discuss briefly the protein-synthesizing system of archaebacteria, not only because this shows certain differences from eubacteria and eukaryotes, but because analysis of the archaebacterial translation apparatus has provided the strongest evidence that the archaebacteria do in fact constitute a separate kingdom.

Archaebacteria have certain structural characteristics that distinguish them from eubacteria. Their cell walls (where these are found) lack peptidoglycan, and their lipids are unique in consisting of branch-chain fatty acids, and being linked to glycerol by an ether, rather than an ester, link. This, in itself, would hardly be sufficient to establish them as a separate kingdom. The evidence that does so comes from phylogenetic studies on ribosomal RNA. Woese and Fox [290] showed that the sequence divergence between the smaller (16S or 18S) rRNAs of eukaryotes and eubacteria clearly distinguished these two kingdoms. (This work

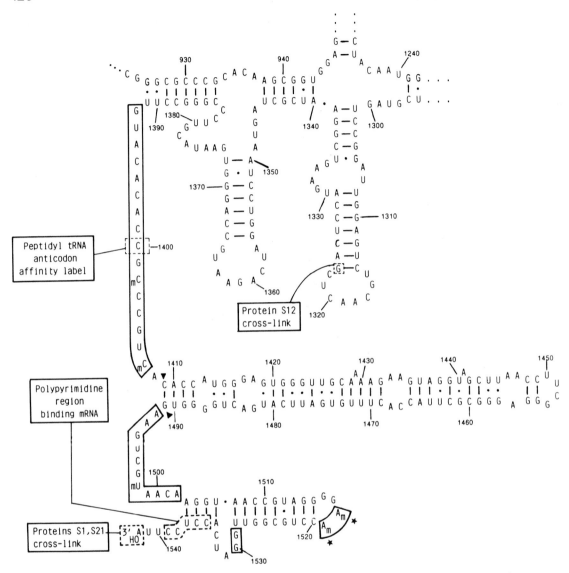

**Fig. 11.25** The proposed decoding domain of the 30S ribosomal subunit of *E.coli*. The RNA secondary structure is from [253], and shows part of the 3′ polypyrimidine region involved in four base-pairs not present in the structure suggested in Fig. 11.22. ★, Nucleotides m$_2^6$A$_{1518}$ and m$_2^6$A$_{1519}$, the demethylation of which endows resistance to kasugamycin in *E. coli*; ▲, Site of base change conferring resistance to paromomycin. The boxed nucleotides are those not involved in secondary structure that are highly conserved throughout evolution.

was actually done by analysis of oligonucleotides before complete sequence data were available.) When the rRNAs of the different archaebacteria were examined it was clear that they were related to one another, but were no more closely related to eubacteria than they were to eukaryotes.

The protein-synthesizing system of archaebacteria shows some similarities with that of eubacteria, some with that of eukaryotes, and some unique features. Although archaebacterial ribosomes are 70S, rather than 80S, they have an additional morphological feature (a 'bill') also present in eukaryotic ribosomes [291]. Most striking, however, is the sensitivity of archaebacterial ribosomes to antibiotics. Thus, they are insensitive to certain antibiotics (chloramphenicol and kanamycin) that had previously been regarded as specific for prokaryotic ribosomes, and sensitive to at least one inhibitor (anisomycin) that had been regarded as specific for eukaryotic ribosomes [292]. Further, the translocase enzyme from archaebacteria, like EF–2 from eukaryotes, has the modified histidine residue that can be ADP-ribosylated by diphtheria toxin [293]. The 3′-ends of the 16S RNAs of some archaebacteria do contain the polypyrimidine sequence that is complementary to the polypurine 'Shine and Dalgarno' sequence of eubacterial mRNAs. Nevertheless at least one archaebacterial mRNA (which, incidentally, employs the standard genetic code) lacks a corresponding polypurine sequence [294]. Another apparent difference from eubacteria in the initiation process of archaebacteria is the lack of formylation of the methionyl residue on the initiator tRNA [295]. To complete this brief survey, it should be mentioned that although no introns have yet been reported for archaebacterial mRNAs, there is strong evidence for introns in the precursors to archaebacterial tRNAs [296].

## 11.9.2 Mitochondria [297–300]

It has already been mentioned that the majority of mitochondrial proteins are encoded by mRNAs transcribed from nuclear genes, and that these mRNAs are translated on 80S ribosomes in the cytosol, the resulting proteins being transported into the mitochondrion (Section 11.7.3). The mitochondrion has its own genome (Section 3.4.1) and its own machinery for transcription (Section 9.6) and translation. Although the size and coding capacity of mitochondria vary quite widely between species (from *c.* 17 kbp in man, to *c.* 2500 kbp in some plants) all mitochondria appear to code for their own rRNAs and tRNAs, together with a variable sub-set of the proteins of the electron-transport and oxidative phosphorylation systems. The amount of coding DNA in the larger mitochondrial genomes is not much greater than that in the smaller ones, as much of the extra DNA in the former is simple-sequence DNA (see Chapter 3).

A widely held view is that the precursors of mitochondria were aerobic bacteria that were engulfed by, and evolved in a symbiotic relationship with, an anaerobic host cell. Despite there being differences between the translational machinery of eubacteria and mitochondria, there are a number of similarities that support this hypothesis. Although mitochondrial ribosomes other than those of plants differ from the ribosomes of prokaryotes and eukaryotes in lacking the otherwise essential 5S rRNA, their sensitivity to antibiotics (cf. Fig. 11.24) shows an extensive similarity to that of eubacteria. The number of mitochondrial ribosomal proteins (which are generally coded in the nucleus) seems to vary between species, but as yet there is insufficient structural information available to allow comparison with bacteria. In contrast, the yeast mitochondrial elongation factor responsible for binding aminoacyl-tRNA (also specified by the nucleus) is both functionally [301] and structurally [302] homologous to *E. coli* EF–Tu; and it has also been reported that the yeast translocase [300] (although not the bovine enzyme [303]) is functionally interchangeable with bacterial EF–G.

Mitochondria, like eubacteria, use fMet-tRNA to initiate protein synthesis [176]. However, the smaller rRNAs from mitochondria lack the bacterial polypyrimidine 3′-sequence; and in any case there is no opportunity for 'Shine and Dalgarno'-type base-pairing to mammalian mitochondrial mRNAs that start directly, or almost directly, with AUG [304]. It

**Table 11.6** Observed base-pairing pattern at 'wobble' position in mitochondria.

| 5'-Anticodon base | 3'-Codon bases read | Comments |
|---|---|---|
| C | { G | Normal |
| C | { A<br>{ G | Unique to mt tRNA$_m^{Met}$ |
| G | { U<br>{ C | Normal |
| U | ⎧ U<br>⎪ C<br>⎨ A<br>⎩ G | Unique[†] |
| U* | { A<br>{ G | Unique |

U* represents an unknown modification of U.
A, as in other tRNAs, is never found in the 'wobble' position.
[†]Except for two isolated examples (see text).

has been suggested that the untranslated 5'-regions of yeast mitochondrial mRNAs could base pair with a different sequence near the 3'-end of the smaller mitochondrial rRNA, but the variable location of the complementary sequence in the mRNA diminishes the likelihood of this possibility [58].

The most striking unique features of the mitochondrial translational system are the deviations from the standard genetic code, and the unique manner in which the codons are read by a restricted set of tRNAs. The actual number of mitochondrial tRNAs varies between species. Mammals seem to represent the minimum at 22; but even in the yeast mitochondrion, with its four-to five-times-larger genome, the 25 tRNAs are less than the 32 required to read the standard code (Table 11.2) according to the wobble hypothesis (Section 11.2.5). The wobble rules of prokaryotes and the eukaryotic cytoplasm are not obeyed by mitochondrial tRNAs, the groups of four synonymous codons being recognized by a single tRNA (Table 11.6). With UGA being read as tryptophan instead of acting as a

termination codon and UAA and UAG acting as stop codons (see below), 22 tRNAs will thus suffice in mammalian mitochondria – at least for elongation. What is the nature of the codon–anticodon interaction that allows this decoding pattern? All the tRNAs that decode a four-codon family have at the wobble position of their anticodons the unmodified U that is so much avoided elsewhere, but that results in the decoding of all members of a four-codon family in the two examples where it does occur (see Section 11.2.5). It is thought that this allows a relaxed pattern of base-pairing in which U-U and U-C are close enough together to form satisfactory hydrogen bonds [304] (Fig. 11.26). However, two-out-of-three decoding (Section 11.2.5) cannot yet be excluded. The recognition of synonymous codons with U or C in the 3'-position by a mitochondrial tRNA with a G in the anticodon wobble position (Table 11.6) poses no problems (Table 11.3). However, the ability to read only synonymous codons with A or G in the 3'-position is not possessed by any single bacterial or eukaryotic cytoplasmic tRNA. Mitochondria have, however, managed to devise such tRNAs by using a modified U of, as yet, unknown structure.

The mitochondrial tRNAs [305] do not conform to the general structural pattern described in Section 11.1 and summarized in Fig. 11.1, and it is possible that these deviations are important for their unique decoding properties. In particular the mitochondrial tRNAs violate some or all of the following: the conservation of the GT$\psi$CRA sequence in loop IV, the constant seven-nucleotide length of loop IV, the pattern of conserved bases in loop I. The most striking violation is by the mammalian mitochondrial tRNA$^{Ser}$ in which the whole of loop I and the corresponding stem is missing. It has, nevertheless, been proposed [306] that this tRNA can adopt a conformation similar to that of yeast tRNA$^{Phe}$ (Fig. 11.2).

Some of the variety of non-standard mitochondrial genetic codes are illustrated in Table

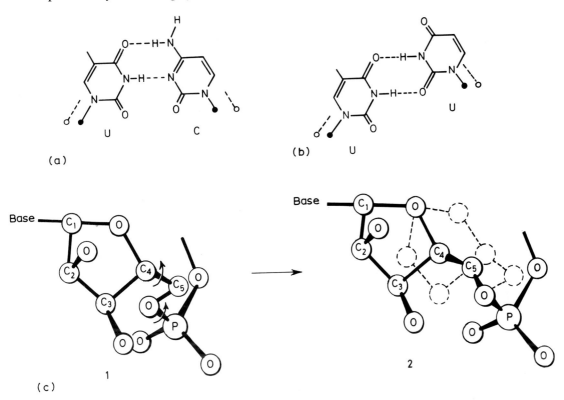

**Fig. 11.26** Possible structural basis for extended base-pairing by U in the 'wobble' position of mitochondrial tRNAs. (a) and (b) show how U could pair with C or U, respectively. The solid circles indicate the relative positions of the $C'_1$ carbons of the ribose rings necessary for the hydrogen-bonding, and the open circles the relative positions of these in the standard A-U or G-C base pairs of an RNA helix (cf Fig. 11.4, which also shows a G-U base pair). (c) shows how rotation of the $C'_4$–$C'_5$ and P–O bonds from their standard conformation (1) to an alternative conformation (2) would produce a movement of the ribose ring that might bring the $C'_1$ carbons close enough together to facilitate the base-pairing shown in (a) and (b). (After [304], with permission.)

11.7. These assignments are largely based on comparison of DNA sequences with protein sequences, together with deductions from tRNA sequences, assuming the mitochondrial pattern of wobble base-pairing described above. Because the corresponding protein sequences from the same species are seldom available (most have been determined for the bovine enzymes), it is not impossible that Table 11.7 contains errors. Furthermore, not all the possible codons in a given family may in fact occur. Thus complete sequencing of the mitochondrial genomes of man, beef and mouse show that, of the stop codons, UAG and AGG only occur in man, and AGA does not occur in mouse. Although AUN codons have been tentatively indicated as potential initiation codons for mammals because of the situation in mouse, AUC is not used in man or beef, and AUU is not used in beef. Despite these caveats, certain generalizations are possible. Firstly, there is no single mitochondrial genetic code, a result that suggests that different mitochondrial genomes have undergone separate evolution. Secondly, although some changes appear quite gratuitous (e.g. Leu to Thr for CUN in *S. cerevisciae*),

**Table 11.7**  Deviations of some mitochondrial genetic codes from the standard code.

| Codon | Standard code | Mammals | Maize | S. cerevisciae | S. pombe | D. melanogaster |
|---|---|---|---|---|---|---|
| CUU | | | | | | |
| CUC | } Leu | n.c. | n.c. | } Thr | n.c. | n.c. |
| CUA | | | | | | |
| CUG | | | | | | |
| AUU | | | | | | |
| AUC | } Ile | } Ile[†] | | } Ile | | } Ile |
| AUA | | | n.c. | | n.c. | |
| AUG | } Met* | } Met[†] | | } Met* | | } Met[§] |
| UGA | } Stop | | | | | |
| UGG | } Trp | } Trp | n.c. | } Trp | } Trp | } Trp |
| CGU | | | | | | |
| CGC | | | } Arg | | | |
| CGA | } Arg | n.c. | | n.c. | n.c. | n.c. |
| CGG | | | } Trp | | | |
| AGU | } Ser | } Ser | | | | |
| AGC | | | n.c. | n.c. | n.c. | } Ser |
| AGA | } Arg | } Stop | | | | |
| AGG | | | | | | |

*Also initiation codon(s) recognizing $tRNA_f^{Met}$.
[†]Also initiation codons, presumably recognising fMet-tRNA (no separate genes for $tRNA_f^{Met}$ and $tRNA_m^{Met}$).
[§]AUG and AUAA are initiation codons recognizing $tRNA_f^{Met}$.
n.c., no change from standard code.

others seem designed either to reduce the number of tRNAs required to decode the genome (e.g. changes from Arg for AGR) or to allow the mitochondrial wobble pattern for two codon families to operate (e.g. Trp for UGA).

The AUN family raises its own problems, however. Although one might have expected the use of AUA as a Met codon to involve a tRNA with a modified U recognizing both AUA and AUG, the anticodon found is 3'UAC5'. As the C, at least in yeast, is unmodified, one would expect decoding of only AUG. However, there are some grounds [307] for thinking that the unusual structure of the $tRNA_m^{Met}$ might allow a C–A base pair in the trans configuration. In the case of S. pombe, where AUA codes for Ile, a separate tRNA from that for AUY is required, and this must not decode AUG. This is thought

to be a tRNA with a C in the wobble position, and perhaps a modification of this C prevents it recognizing G. A further problem arises in the case of mammalian mitochondria where, although both Met-tRNA and fMet-tRNA are found, there are apparently no separate genes for $tRNA_f^{Met}$ and $tRNA_m^{Met}$. This is in contrast to the situation in mitochondria with larger genomes [297]. Unless one of these tRNAs has such an unusual structure that it has escaped detection in analysis of the DNA sequence, one must assume that the single gene gives rise to both tRNA species through differential modification. At least in mouse mitochondria, the initiator tRNA would appear to have to decode the whole AUN family.

The altered initiation and termination codons in these mitochondrial genetic codes raise interesting questions about the initiation and termination reactions. Unfortunately neither initiation nor termination factors have yet been identified in mitochondria to enable these questions to be answered.

### 11.9.3 Chloroplasts [298, 308]

The protein-synthetic apparatus of chloroplasts resembles that of bacteria in a much more straightforward way than does that of mitochondria. Indeed 5S RNA sequence comparison (cf. Section 11.9.1) shows clearly that chloroplasts are specifically related to cyanobacteria, with which they also share similarities in the structure of their chlorophylls and carotinoids [309]. The reason that this evolutionary relationship can be seen in the case of chloroplasts but not in the case of mitochondria (except in those of higher plants [444]) is that the larger genomes of the former (120–200 kbp DNA) contain a correspondingly greater amount of genetic information. Their rRNAs would not appear to have been under the same intense selection for size as have those of most mitochondria, and indeed 5S rRNA is found in all chloroplast ribosomes. The chloroplast genome is, in fact, able to code for much of its protein-synthetic apparatus besides coding for many of the proteins involved in photosynthesis and energy transduction. This includes the elongation factors and many of the ribosomal proteins, and these show typically 40–60% identity with clearly analogous proteins in *E. coli* [310, 311].

The chloroplast DNA codes for 35–40 tRNAs, enough to decode the standard genetic code according to the standard pattern of observed wobble shown in Table 11.3. Although it appears that chloroplasts use the standard genetic code their tRNAs do include one example of an unmodified U in the wobble position, which is presumably able to decode a family of four synonymous (Leu) codons [63]. The polypyrimidine sequence capable of 'Shine and Dalgarno' base-pairing is present at the 3'-end of

chloroplast 16S RNA, and it is likely that initiation codon selection involves a mechanism similar to that in eubacteria [58, 308].

### 11.10 THE REGULATION OF TRANSLATION IN PROKARYOTES

The regulation of protein synthesis in bacteria predominantly involves the modulation of transcription, as described in Chapter 10. Nevertheless there are some interesting examples of translational control involving the relatively stable *E. coli* bacteriophage RNA (Section 11.10.1), and certain features of these have also been found in both the less stable bacterial mRNAs (Section 11.10.2) and the mRNAs of eukaryotes. Another type of example (Section 11.10.3), also of more general validity, suggests that translational control may be more important in prokaryotes than is generally recognized.

### 11.10.1 Control of the translation of bacteriophage mRNA

The *E. coli* phages to be considered are those in the group which includes R17, MS2 and f2, and the related, but serologically distinct, bacteriophage Qβ. The single-stranded RNAs of these phages also act as mRNAs and have three clearly defined cistrons (Fig. 11.27). These code for a, so-called, A or maturation protein (which is present in one copy per virion, and one function of which is to ensure proper encapsulation of the viral RNA), a coat protein (180 copies of which are present per virion), and a subunit of the viral replicase (the other three subunits of which are EF–Tu, EF–Ts, and ribosomal protein S1) [312, 313]. Different amounts of these three proteins are synthesized during phage infection of *E. coli in vivo*, about 10–20 times as much coat protein as the other two proteins having been found at the end of infection [314]. There are, in addition, temporal differences in the synthesis of the proteins: synthesis of the replicase subunit and A-protein apparently predominating early

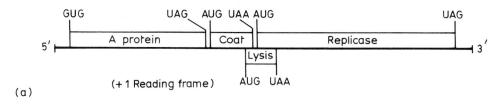

(a)

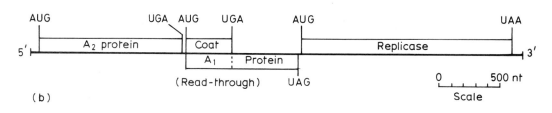

(b)

**Fig. 11.27** Maps of the genomes of phage MS2 (a) and phage Qβ(b). The central line from 5′ to 3′ represents the complete RNA, and the solid areas the regions that are translated. Only minor products are produced from the translation of the regions indicated below the central lines.

in the replicative cycle, whereas that of coat protein predominates at later times and continues after synthesis of the replicase subunit and A protein has declined.

Studies of the mechanisms underlying these phenomena utilized the fact that the phage RNAs can be translated in an *E. coli* cell-free system *in vitro* [315], and were assisted by knowledge of partial sequences of many of the RNAs and, ultimately, of the complete sequence of MS2 RNA [316]. The ratio of the amounts of the three proteins synthesized *in vitro* from phage f2 RNA is roughly comparable to that of those synthesized *in vivo* (coat protein:replicase subunit:A protein = 20:5:1 [315]). However, it should be noted that the temporal differences in translation *in vivo* are not mimicked during translation *in vitro*.

The determination of the gene order shown in Fig. 11.27, and the demonstration of independent initiation at the initiation codons of each cistron [317] disposed of early ideas that the cause of differential translation was analogous to

that in certain polar mutants of polycistronic mRNAs [318]. Nor were the sequences of the pre-initiation regions of the cistrons consistent with the relative extents of translation being governed by the extent of 'Shine and Dalgarno' base-pairing with 16S rRNA (see Table 11.4). Furthermore the claims that the differential translation was a result of differential efficiencies of cistron-specific sub-species of IF-3 [319, 320] have not been substantiated [321].

Instead it appears that secondary structure exerts a major influence over the quantitative control of initiation at the three sites. Lodish showed that destruction of the secondary structure of phage f2 RNA by mild formaldehyde treatment or by heating led to different relative amounts of the products translated from the RNA. The rate of synthesis of replicase subunit was increased to that of coat protein, and the rate of synthesis of A protein also increased substantially [322]. Furthermore, in a different type of experiment in which isolated oligonucleotides containing the initiation sites for the three cistrons were compared, ribosomal binding was greatest to the site for the A protein [323].

The proposed secondary structure for MS2 RNA shows the initiation site of the replicase subunit hydrogen-bonded to part of the coat protein cistron, the initiation site of the latter being well exposed (Fig. 11.28) [324]. This is consistent with the idea that part of the coat protein cistron must be translated before there can be initiation of the synthesis of replicase protein. This idea had been based on the fact that an amber mutation at the codon specifying amino acid position 6 of the phage f2 coat protein (see Section 11.5.3) severely repressed expression of the cistron for the replicase subunit, but that this effect was abolished if the phage RNA was treated with formaldehyde [322]. In contrast, amber mutations at the codons specifying the 50th, 54th and 70th amino acids did not have this effect (see [324]). It can be seen from inspection of Fig. 11.28 that translation of only the first six codons of the coat protein cistron of the related phage MS2 would not break the putative hydrogen bonds to the replicase subunit initiation site, but that translation to the 50th or subsequent mutant termination codons would.

It has been found that the A-protein cistron is translated *in vitro* more efficiently on the nascent RNA of replicative intermediates, and it has been suggested that translation from the mature RNA *in vivo* may be completely prevented by the secondary structure [325]. Although secondary structure is clearly a major factor in determining the extent of translation of the three cistrons, the fact that ribosomes from *B. stearthermophilus* will translate the A-protein cistron of the mature RNA of these *E. coli* bacteriophages [326] suggests that our understanding of this phenomenon may still be incomplete.

Although the temporal regulation of the translation of the three cistrons is not mimicked during simple translation of the phage RNA *in vitro*, experiments *in vitro* have provided some suggestions of how the temporal regulation may operate. Addition of coat protein inhibits trans-

lation of the replicase subunit cistron by binding specifically to the initiation site [327, 328], offering a persuasive explanation of the later decline in synthesis of the replicase subunit *in vivo*, as the coat protein builds up. If it is true that translation of the A protein can only occur on the replicative intermediate [325], the later decline in the synthesis of this protein *in vivo* can be explained quite well by the concomitant decline in RNA synthesis [314]. It is more difficult to see how the translation of replicase subunit can predominate over that of coat protein early in infection if the translation of coat protein is a pre-requisite for that of replicase subunit, as discussed above. A possible solution to this dilemma is afforded by the suggestion of an alternative non-hydrogen-bonded structure for the initiation site of the replicase subunit cistron [324]. It has been found that ribosomal protein S1 inhibits the synthesis of coat protein [329], and such an effect might explain the initial delay in the synthesis of the latter, if there were translation of replicase from the alternative RNA structure. This inhibition of coat protein translation would later be relieved as the newly synthesized replicase subunit associated with protein S1, which constitutes one of the host-specified subunits of the replicase [330].

An alternative explanation can be envisaged for the inhibition of coat protein synthesis by ribosomal protein S1 *in vitro*. This is that it reflects an inhibition which occurs *in vivo* with protein S1 integrated into the replicase. The purpose of this would be to prevent ribosomes starting along the coat protein in the $5' \rightarrow 3'$ direction as these would prevent the movement of the replicase in the $3' \rightarrow 5'$ direction. It has been suggested that the replicase would simultaneously attach to the $3'$-end of the RNA, which would start being replicated, but that the replicase would still remain attached to the coat protein initiation region until a considerable proportion of the transcription had occurred [331].

The coding potential of the RNA bacterio-

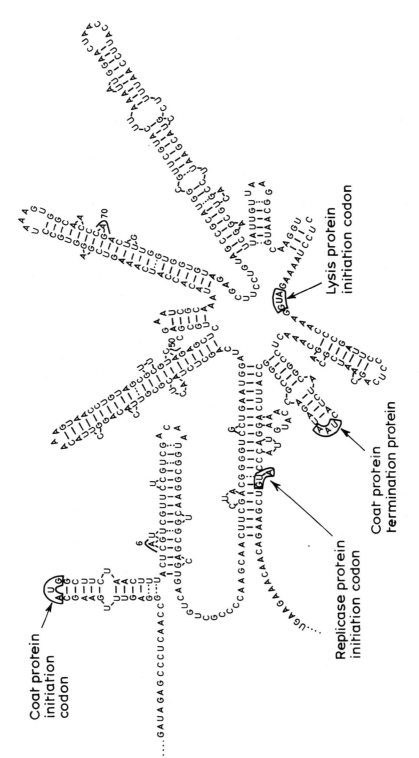

**Fig. 11.28** The 'flower-model' for the structure of the coat protein cistron of phage MS2. The positions of the coat protein initiation and termination codons are indicated, as well as the initiation codons for the replicase and for the minor lysis protein, translated in the + 1 reading frame. The numbers 6, 50 and 70 indicate codons which have been mutated in particular phage (see text) to produce premature termination. (After [324], with permission.)

phage is not restricted to the three clearly defined cistrons just discussed (Fig. 11.27). In the case of bacteriophage Qβ there is synthesis of minor amounts of a fourth protein, thought to be involved in release of progeny phage particles from the bacterium [332]. This protein (termed $A_1$ – the maturation protein is termed $A_2$) is composed of the coat protein and the product of translation of the subsequent intercistronic region and beyond, as a result of read-through of the single UGA termination codon [333]. This 'read-through', which recalls that in Sinbis virus (Fig. 11.16), is discussed in more detail in Section 11.13, but it is worth mentioning here that bacteriophages R17 and MS2, which do not show 'read-through', terminate their coat protein genes with a double stop, UAA UAG.

The fourth protein coded by bacteriophages f2 and MS2 is required for lysis of the host cell (a function fulfilled by the maturation protein, $A_2$, in Qβ). This lysis protein is initiated about 50 bases from the end of the coat protein cistron, but *out of phase* with the reading frame of the latter, and involves the translation of the intercistronic region and about 140 bases of the replicase cistron before finishing at an out-of-phase termination codon [334–336]. Although the amount of lysis protein translated from MS2 RNA *in vivo* is only very small, the initiation region of this cistron and the preceding 'Shine and Dalgarno' sequence appear to be exposed in the original secondary structure model (Fig. 11.28). However, this model has subsequently been revised to accommodate the results of experiments that indicate that secondary structure is the barrier to initiation of the lysis protein [337]. Other experiments imply that, in order for this barrier to be overcome, frame shifting must occur during the translation of the coat protein, followed by termination at an out-of-phase stop codon just preceding the lysis cistron. This disrupts the secondary structure and allows initiation of the synthesis of the lysis protein [338].

### 11.10.2 Autogenous control of ribosomal protein synthesis

The syntheses of the ribosomal proteins of *E. coli* are closely co-ordinated with one another and with that of rRNA. This regulation has been shown by Nomura and his colleagues to be translational and to involve protein interactions with mRNA, and, perhaps also, effects of mRNA secondary structure [339, 340]. The transcriptional control of ribosome synthesis, exerted by availability of amino acids, is discussed in Section 11.12.

There are approximately 25 transcriptional units (also referred to as operons) for the 52 ribosomal protein genes, some containing several ribosomal protein genes, whereas others contain only one. Genes for other proteins involved in macromolecular synthesis are also found in these units, some of which are shown in Fig. 11.29. The original evidence that the products of these transcription units were subject to feedback regulation came from experiments *in vivo* in which the copy number of several ribosomal genes was artificially increased by introducing into *E. coli* specialized transducing phages (Section 3.6.8) carrying several of these adjacent units. Although this had the effect of increasing the synthesis of the respective mRNAs, there was no increase in the synthesis of the corresponding ribosomal proteins. Subsequently recombinant DNA techniques allowed individual ribosomal protein genes to be introduced into multicopy plasmids and expressed under the control of an inducible promoter to allow overproduction of the protein. This demonstrated that only one protein in each transcription unit was able to repress the synthesis of the other members of the unit *in vivo*, as indicated in Fig. 11.29. Such specificity had previously been demonstrated *in vitro*, by adding individual purified proteins to a coupled system for transcription and translation from the DNA of the transducing phage. These results were broadly in agreement with those found *in*

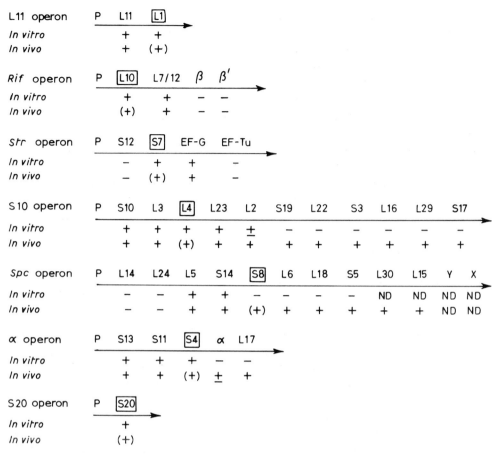

**Fig. 11.29** Autogenous regulation of ribosomal protein synthesis in *E. coli*. Each operon is indicated by an arrow, the direction of which is that of transcription. The promoters are indicated by P and the individual genes by the names of their protein products. Regulatory ribosomal proteins are boxed, and the effects of these on the synthesis of proteins in the same operon *in vivo* or *in vitro* are indicated (+, inhibition; −, no effect; (+), presumed to inhibit *in vivo*; ND, not determined). Y is the *Sec*Y gene product, X is an unknown protein α, β and β′ are subunits of RNA polymerase (from [340], with permission).

*vivo* (Fig. 11.29), although repression *in vitro* did not extend to all the distal genes affected *in vivo*. This is most likely an artifact of the system *in vitro*, but nevertheless these studies added substantially to those *in vivo*. This was because it was possible to dissociate transcription from translation and thus to demonstrate the effect of the ribosomal proteins on translation, and that it was not an effect on the stability of the mRNA. It was also possible to show that (at least in most cases) the repressor proteins also inhibited their own translation, something clearly not possible

in the experiments *in vivo*. It should be noted that some of the members of these transcription units (e.g. EF-Tu and the β and β′ subunits of RNA polymerase) were not subject to this feedback regulation.

A possible mechanism for the regulation was suggested by the fact that many of the proteins responsible for the feedback inhibition had been shown (originally also by Nomura [341]) to bind to rRNA early in the assembly of ribosomes (e.g. S4, S7 and S8 – see Section 11.8.2). Thus it was proposed that these proteins were also able to recognize a site on the polycistronic mRNA that resembled their rRNA binding site, and the binding to which would block translation. Binding to this site would be of lower affinity than that to rRNA so as not to interfere with assembly of ribosomes, but would occur when the ribosomal protein was produced in excess over the available rRNA. As the cotranscribed rRNAs would similarly regulate the uptake of all rRNA-binding proteins, the feedback of the excess of these would explain how the translation of the polycistronic mRNAs from different transcription units was also co-ordinately regulated.

It was possible using deletion mutants generated by both classical and recombinant DNA genetic techniques to demonstrate that the target of the translational feedback was generally a single site, located near the translational start of the first gene of the transcriptional unit (the *str* and *spc* operons are exceptional: Fig. 11.29). Furthermore, after the nucleotide sequences of the genes had been determined, it became possible to propose secondary structures in these regions of the mRNA that were similar to those in the rRNA-binding sites. Finally it was demonstrated that the appropriate rRNA could relieve the translational inhibition *in vitro*.

Although the model would therefore seem correct, it is still not certain how the binding of a ribosomal protein to a single site on a polycistronic mRNA can prevent translation starting at separate initiation sites on the same molecule. Present evidence favours translational coupling through effects on secondary structure of the type shown to occur for the coat protein and replicase genes of the RNA phages (Section 11.10.1). This suggests that those proteins that escape this feedback regulation (EF-Tu etc.) might do so by virtue of exposed translational initiation regions. This is not to pretend that no unsolved problems still remain: for example that of how four times as much protein L7/L12 (Section 11.8.1) is synthesized as the other ribosomal proteins.

Other RNA-binding proteins are also subject to autogenous regulation. Indeed this phenomenon was first discovered for the gene 32 protein of phage T4 and probably applies also to the *regA* protein of this phage [342]. Evidence has been presented for similar control for EF-Tu (from the *tufB* locus where it is cotranscribed with four tRNA molecules) [343], the β-subunit [344] and ρ-factor [345] of RNA polymerase and threonyl-tRNA synthetase (which is thought to be cotranscribed with IF-3 and two other tRNA synthetases) [346]. Nucleic acid interactions of ribosomal proteins have already been indicated for S1 (Section 11.10.1) and S10 (Section 10.2.2).

### 11.10.3 Codon usage and translational control

As already mentioned (Section 11.2.6) the usage of synonymous codons is non-random. This is true for codon usage within a gene, but differences are also found in the pattern of usage between genes in the same organism. In *E. coli* it was observed that certain codons which were rarely used in the genes for highly expressed proteins (e.g. ribosomal proteins) were found at a significantly greater frequency in the genes for weakly expressed proteins (e.g. the *lac* and other repressors) [347]. Although strength of expression can clearly be regulated transcriptionally, the necessity for translational regulation is nicely seen in the case of DNA primase. The gene for this protein is in the same transcription unit as those for ribosomal protein S21 and the σ factor of RNA polymerase, proteins that are approx. 800 and 50 times more strongly expressed respectively. Although the lack of a 'Shine and Dalgarno' sequence in the DNA primase gene (previously alluded to in Section 11.5.1) is undoubtedly important here, the difference in codon usage between this gene and its co-transcribed neighbours is most striking [348].

It has been suggested that the pattern of codon usage can regulate translation in two possible ways. The first suggestion, for which the indirect evidence is strongest, is that the speed of translation of synonymous codons varies with the abundance of the corresponding iso-accepting t-RNAs. A study of the relative abundance of the iso-accepting tRNAs in *E. coli* showed a strong correlation between tRNA abundance and codon choice in the genes of strongly expressed proteins [349].

The difference in usage of codons of the type NMU and NMC cannot be explained in this way, as they are both decoded by a tRNA with G or I in the 'wobble' position (Table 11.3). It was observed in such pairs that the usage correlated well with the predicted codon–anticodon inter-action energy, those with either maximum (e.g. CGC) or minimum (e.g. AUU) energies being less frequent in the genes for highly expressed proteins than those (e.g. CGU or AUC) with intermediate energies [347]. It was argued that intermediate codon–anticodon interaction energies would produce more efficient decoding by allowing the optimal balance between binding and release of tRNA. Although experiments *in vitro* have not provided support for this model [350], it is striking that a similar pattern of preference in NMY codons has been reported in yeast [351].

The studies in yeast also indicated a cor-relation between codon usage in highly expressed genes and the relative abundance of iso-accepting tRNAs, as had previously been observed for fibroin and the tRNAs of the silkworm, *B. mori* [352]. Although the situation for weakly expressed genes is less clear, the difference in patterns of relative abundance of iso-accepting tRNAs in yeast and *E. coli* does suggest that this method of translational regulation may extend to lower eukaryotes, at least.

## 11.11 THE REGULATION OF TRANSLATION IN EUKARYOTES

In eukaryotes the sites of synthesis and usage of mRNA are physically separated, and this, and

the relative stability of most eukaryotic mRNAs, provide scope for the translational control of protein synthesis. The following account deals with some well-defined examples of translational control, as well as touching upon topics which, although of interest or importance, are much less completely understood.

### 11.11.1 The role of mRNA structure

Both the 5'-end of the eukaryotic mRNA (with its 'cap') and the 3'-end (with its poly(A) tail) have structural features not found in prokaryotic mRNAs. The general significance of these feat-ures will be discussed here, the specific regu-latory strategies of viruses with uncapped mRNAs being dealt with in Section 11.11.3.

It is thought that the 3'-poly(A) segment of mRNAs may influence their degradation, the loss of this segment or its reduction to a certain minimum size leaving it vulnerable to exo-nuclease action. This proposition is supported by the following evidence. The poly(A) sequences of cytoplasmic mRNAs become progressively shorter with time [353, 354]; histone mRNAs, which lack poly(A), have shorter half-lives than most eukaryotic mRNAs [355]; and deadenyl-ated globin mRNA is translated less efficiently than normal globin mRNA when injected into *Xenopus* oocytes [356], but not when added to cell-free systems that only sustain translation for short periods [357]. Furthermore it has been possible to increase the stability of histone mRNAs by artificially polyadenylating them [358], and to decrease the stability of adenovirus mRNAs by deadenylating them in intact cells through administration of 3'-deoxyadenosine [359]. The exponential decay of poly(A)-con-taining mRNA implies that the susceptibility of such mRNA to degradation does not increase with decreasing size [355]. Rather, there is evidence that when the size of the poly(A) segment falls below a certain threshold value the mRNA suddenly becomes extremely susceptible to degradation [356].

Despite the convincing nature of this evidence it is important to stress that other factors may also regulate the stability of mRNAs, and that certain mRNAs with little or no poly(A) are nevertheless stable [360]. It appears that a region of the mRNA near the junction of the poly(A) segment with the rest of the mRNA is particularly accessible to nuclease [361], and it could be that this region can be protected either by proteins associated with the poly(A) segment [362] or by the secondary structure of the untranslated 3'-region of the mRNA itself.

Extension as well as decrease in the length of the poly(A) segment of mRNAs can occur in the cytoplasm following fertilization of clam oocytes [363], and to a considerable extent these changes correlate with changes in stability of the mRNAs. However, other developmental changes in the stability of mRNAs occur without concomitant changes in polyadenylation [364].

The complexity of the process of initiation of the translation of mRNA (Section 11.6.1) suggests that the structure at the 5'-end of the mRNA might easily affect the recognition of the 5'-cap by the eIF-4F·eIF-4B complex, and the ease with which the secondary structure is 'melted'. Certainly different capped viral mRNAs differ in their requirement for certain components of this complex [365, 366]. That this might influence the translational rate of cellular mRNAs is suggested by the case of the α- and β-globin mRNAs. It has been demonstrated that β-globin mRNA is translated more efficiently than α-globin mRNA both *in vivo* [367] and *in vitro* [368]. (It appears that more α-globin mRNA is present in reticulocyte cells to maintain production of similar amounts of both chains [369].) *In vitro* it has been found that this difference reflects different rates of initiation [368], and it has been reported that addition of eIF-4A and eIF-4B (although these factors may have been cross-contaminated with others) to a reticulocyte cell-free system synthesizing more β- than α-globin can relieve this imbalance [370].

It is a vexed question whether there are cellular systems for modulating the initiation of translation of specific mRNAs by interaction with the 5'-end of the mRNA. There have been repeated reports of species of RNA able specifically or non-specifically to inhibit translation [371, 372]. Although judgment should be reserved until those RNAs have been better characterized, the precedent of antisense mRNA in bacteria ([373] and Section 10.5.1) should not be forgotten.

### 11.11.2 The role of protein phosphorylation

Perhaps the best-characterized mechanism of translational regulation in eukaryotes is that involving phosphorylation of eIF-2. This has been most intensively studied in reticulocyte lysates lacking haem, the consequence of which is a rapid inhibition of protein synthesis. (Haem is rapidly converted in aqueous solution into an oxidized form, haemin, which is what is actually used in such studies.) Although one can rationalize this effect in terms of the co-ordination of the synthesis of the predominant reticulocyte protein, globin, with the availability of its prosthetic group, haem, it must be emphasized that, in fact, the synthesis of all reticulocyte proteins is similarly affected [374]. Early work established that it was the initiation of protein synthesis that was inhibited in cells deprived of haemin, and that the reaction involved was the binding of Met-tRNA$_f$ to the 40S ribosomal subunit [375], the reaction that requires eIF-2 (see Fig. 11.17). At the same time the actual inhibitor was partially purified from the lysates, and found to have properties consistent with it being a cytoplasmic protein, pre-existing in an inactive form in normal cells [376]. This inhibitor (known as HCR, haem-controlled repressor; or HRI, haem-regulated inhibitor) was subsequently found to be a cyclic AMP-independent protein kinase that specifically phosphorylates the α-subunit of eIF-2 [377]. Initially it was not possible to demonstrate an effect of the phosphorylation on the function of eIF-2 *in vitro* because systems lacking eIF-2B, and thus

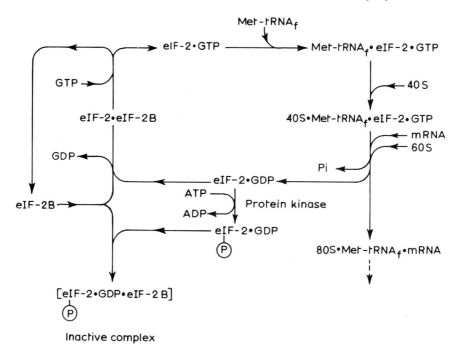

**Fig. 11.30**  A scheme for the role of factors eIF-2 and eIF-2B in the initiation of eukaryotic protein synthesis, and the effect of phosphorylation of the α-subunit of eIF-2. Other initiation factors (see Fig. 11.17) have been omitted for clarity. Note that the point at which the phosphorylation of eIF-2 has been indicated is purely arbitrary, and that some workers [401] believe eIF-2B is not released from the complex with eIF-2 until the stage at which Met-tRNAf is bound to the 40S subunit.

catalysing only a single round of initiation, were used. It is now clear that phosphorylation of eIF-2 prevents the liberation of GDP from the eIF-2·GDP complex released from the ribosome and causes the sequestering of eIF-2B in this complex [199, 378–380] (Fig. 11.30). This latter point is important as it provides an explanation of why phosphorylation of only approximately 20% of eIF-2 can cause complete inhibition of initiation. This is because there are thought to be only 20–25% as many molecules of eIF-2B as eIF-2.

Other stimuli can also provoke the phosphorylation of eIF-2 and the inhibition of protein synthesis in reticulocyte lysates. These include low concentrations of double-stranded RNA, oxidized glutathione and even the elevated pressure associated with ultracentrifugation. In

the case of double-stranded RNA, a different protein kinase inhibitor is activated [377]. The fact that haem is not the only possible stimulus for the phosphorylation of eIF-2 suggested that this phosphorylation might be a more general control mechanism. As is discussed in Section 11.11.3, nucleated cells treated with interferon synthesize a protein kinase which, on activation by double-stranded RNA, catalyses the phosphorylation of the α-subunit of eIF-2. However, the low endogenous concentration of eIF-2α kinase in such cells raised problems for the postulated rapid control of protein synthesis by this mechanism in other situations.

The quantification of the extent of phosphorylation of eIF-2 by two-dimensional gel electrophoresis combined with immunoblotting has recently established that eIF-2α does in fact become phosphorylated in other circumstances in nucleated cells. These include heat-shock [381] and deprivation of serum [382], conditions which are known to inhibit translation. Whether or not the double-stranded RNA-dependent protein kinase or a separate enzyme is responsible for this phosphorylation is unknown at present. The former possibility is not, however, incompatible with previous findings (above), because it is now known that the amount of this kinase in mouse fibroblasts increases with cell density [383].

The control of protein synthesis in cells deprived of serum is further complicated, however, by the finding that the activity of eIF-4B is impaired [384]. This is probably due to dephosphorylation of this factor [382], which also occurs during heat shock [381]. At the moment studies of the control of the phosphorylation of eIF-4B are still in an early stage.

It should also be mentioned that under certain conditions the extent of phosphorylation of another component of the 40S initiation complex varies in concert with changes in the protein-synthetic activity of the cell. This is a basic ribosomal protein, designated (eukaryotic) S6, which can accept up to five phosphoryl residues per molecule *in vivo*. The phosphorylation of this protein is greatest in rapidly growing cells [262, 385] and there is currently much interest in identifying the protein kinase(s) responsible for this phosphorylation because of a likely role in the transduction of extracellular signals for growth inside the cell [386]. Nevertheless, as far as the function of ribosomal protein S6 itself is concerned, it should be stressed that, as yet, no convincing direct effect of phosphorylation on the protein-synthetic ability of ribosomes has been demonstrated. This may well be because the phosphorylation modulates an activity of the ribosomes that has not been assayed *in vitro*. The recruitment of mRNA from storage particles (Section 11.11.4) or an effect on the association of ribosomes with the cytoskeleton come to mind as possibilities.

### 11.11.3 Regulation in cells infected with viruses

As with bacteria, the effects of viral infection provide some striking examples of translational control in eukaryotic cells. The latter, however, are able to respond in a common manner to interfere with the translation of viral mRNAs. This aspect will be considered first.

Animal cells infected with viruses produce a glycoprotein called interferon, which acts on adjacent uninfected cells, enabling them to resist infection by inhibiting the replicative cycle of the virus [387]. Interferons are species-specific but are effective against a wide spectrum of animal viruses. They also have inhibitory effects on cell growth which may be of importance in uninfected cells. It appears probable that the transcription as well as the translation of viral mRNAs is specifically inhibited in cells treated with interferon, but only the control of the latter process will be discusssed here. The synthesis of interferon by cells may be induced experimentally with double-stranded RNA, which is most probably involved in the initial production of interferon which the virus induces. When double-stranded RNA is added to lysates of interferon-treated cells there is an inhibition of protein synthesis which is much greater than that provoked by the interferon treatment or the double-stranded RNA alone. One consequence of this treatment is the activation of a protein kinase that phosphorylates eIF-2 [388]. A second consequence is the activation of an enzyme that catalyses the synthesis of the trinucleotide, pppA2′p5′A2′p5′A [389]. This nucleotide activates a relatively non-specific ribonuclease (Fig. 11.31) [390].

Because these inhibitors of protein synthesis induced by interferon are clearly not specific for viral mRNA, efforts have been made to determine whether they play a central role in the effects of interferon. The main approach has been to use cells which, although still possessing interferon receptors, are resistant to the effects of the molecule. The results of such studies

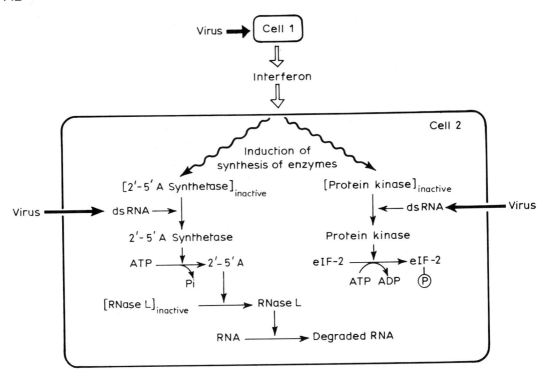

**Fig. 11.31** Pathways by which double-stranded RNA and interferon resulting from infection of cells by viruses may cause the inhibition of viral protein synthesis. $2'-5'$A is an abbreviation for $ppp(A2'p5'A2'p5'A)_n$. For other details, see text.

provide evidence for the prime importance of the protein kinase (e.g. [391]), on the one hand, and the nuclease (e.g. [392]), on the other. The conflict between the different studies seems likely to be due to differences in the cells and interferons used, and the most reasonable conclusion to be drawn would seem to be that both systems can be important in different circumstances. Indeed, perhaps this is why there are two systems. The ability of the double-stranded RNA-dependent protein kinase (whether or not this pre-exists in the cell or is induced by interferon) to inhibit viral protein synthesis *in vivo* has been clearly shown for adenoviral late mRNAs [393], and it is interesting to note that vaccinia virus produces an inhibitor of the kinase during infection [394].

There are now two main views regarding the problem of how these non-specific inhibitors could specifically affect viral translation. One concerns the action of the nuclease and suggests that the $2'-5'$-oligoadenylate required for its activation is rapidly degraded away from its site of synthesis, the double-stranded RNA. Thus its action would be largely restricted to this viral double-stranded RNA, thought to consist of replicative intermediates in the case of RNA viruses, or overlapping mRNA transcripts in the case of DNA viruses [395]. The other point of view is that the inhibition need not, in fact, be specific. The cell, unlike the virus, may be able to compensate for the effect of the nuclease by increasing its synthesis of RNA [396], and by resuming protein synthesis on reversal of the phosphorylation after the virus is destroyed.

We turn now from the antiviral defences of the cell to the specific inhibition of host protein synthesis which many, although not all, eukaryotic viruses produce. It is clear that different viruses operate by different mechanisms, and this will be illustrated in the case of two small RNA viruses.

Poliovirus is unusual in that its mRNA lacks the 5'-'cap' structure and does not require a functional eIF-4F complex for activity (see Section 11.6.1). Its mode of operation is to inactivate this complex and hence prevent the translation of capped host mRNAs [397]. It now appears that the mechanism of this inactivation is the proteolytic cleavage of the 220 000-Da component of eIF-4F [398]. Surprisingly, however, the virally coded protease involved in cleavage of the polyprotein (cf. Sinbis virus – Fig. 11.16) does not appear to be responsible for this [399]. The specific nature of the strategy adopted by poliovirus is illustrated by the fact that the related uncapped encephalomyocarditis (EMC) virus does not operate in the same manner, although rhinovirus 14 may [400]. A greater ability of EMC viral RNA to compete with cellular mRNA for the eIF-4F complex would appear important here [365, 366].

The other virus to be considered, Semliki Forest virus, is very similar in structure to Sinbis virus (Fig. 11.16). Here the inhibition of the translation of host mRNA also extends to the early viral mRNA (corresponding to the 49S species in Fig. 11.16) and allows preferential translation of the 26S mRNA which encodes the 'late' structural proteins. An inhibitor of translation of host and 49S RNA was isolated from infected cells and found to be the capsid protein [401]. The mechanism of this inhibition is unclear, but also appears to involve the cap-binding complex. This may appear surprising as the 26S mRNA is also capped (Fig. 11.16); however, this mRNA seems to have a low requirement for the factor complex [402] (see also Section 11.6.1).

### 11.11.4 Translationally inactive mRNA

The reversibility of the examples of translational inhibition described in Section 11.11.2 implies that during the inhibition the cells contain untranslated mRNA. It would appear that the untranslated mRNA is not associated with ribosomes, but is complexed to certain proteins in the form of messenger ribonuclear protein (mRNP) particles of sedimentation coefficient ranging from about 20 to 120S [403, 404]. Such mRNP particles (sometimes termed 'informasomes') are also found and in fact were first described in normal cells [405, 406].

It would be of great interest to know the mechanisms which regulate the uptake of mRNA into these particles and its re-entry into polysomes. Although there are hints that small RNA species may be involved, no system has yet been established to allow study of this transfer *in vitro*. Recently it has been suggested that the phosphorylation of a protein associated with mRNA may be involved in the inactivation [405].

A more extreme example of the mobilization of inactive mRNA is seen following fertilization in sea-urchin eggs [407]. Here there is rapid mobilization of a sequestered sub-set of maternal mRNAs which until fertilization are maintained in an inactive form. Despite studies of this system over many years, the mechanisms of the translational repression and its reversal are still not understood.

### 11.12 THE COUPLING OF TRANSCRIPTION AND TRANSLATION

When a bacterium, such as *E. coli*, is deprived of amino acids that are required for growth, not only is protein synthesis decreased, but so also is RNA synthesis [408]. This phenomenon is known as the 'stringent response' (see [409, 410] for reviews) and affects the synthesis of different types of RNA selectively. Specific inhibition of the synthesis of rRNA and tRNA, but not of the bulk of mRNA, was first to be recognized. Later it was found that the synthesis of certain mRNAs (e.g. those coding for ribosomal proteins) is also inhibited, although that of others (e.g. *lac* mRNA) is in fact stimulated. It is clearly advantageous for the cell to be able to avoid wasting valuable resources in synthesizing more ribosomes etc., which a deficiency of amino acids prevents being utilized. Likewise it is of benefit for the cell to retain the ability to synthesize essential proteins with the diminished supply of amino acids it can provide for itself through protein breakdown. Studies of the mechanism of the stringent response have been particularly assisted by two findings. One was that the lack of a single amino acid, or even of a functional aminoacyl-tRNA synthetase, would provoke the response. This latter was most clearly demonstrated with bacterial mutants having a temperature-sensitive aminoacyl-tRNA synthetase [411]. The other finding was that bacterial mutants could be obtained that did not show the stringent response. These mutant strains are known as 'relaxed' strains, and the initial strains examined all had mutations that mapped at what is now termed the *relA* locus [412]. The breakthrough in understanding the initial events in the stringent response came with the discovery by Cashel and Gallant that when stringent, but not relaxed, strains of *E. coli* were starved of amino acids two unusual nucleotides (or 'magic spots') appeared [413]. These were later found to be guanosine 5'-diphosphate 3'-diphosphate (ppGpp) and guanosine 5'-triphosphate 3'-diphosphate (pppGpp) [414]. A third nucleotide, guanosine 5'-diphosphate 3'-monophosphate, was subsequently found [415]. The presence of these nucleotides (particularly ppGpp) generally correlates very well with inhibition of RNA synthesis and, despite certain exceptions to this correlation [410], there is no doubt that these molecules play a key role in stringent control. The fact that the absence of an active aminoacyl-tRNA synthetase could provoke the stringent response had led to the

suspicion that deacylated tRNA might be the inducing molecule. Evidence supporting this came when it was found that deacylated tRNA stimulated the synthesis of ppGpp and pppGpp in an 'idling' reaction occurring on ribosomes that had been isolated from cells undergoing the stringent response [416]. The reaction by which pppGpp is synthesized utilizes both ATP and GTP (see Fig. 11.32) and requires that a codon cognate to the deacylated tRNA be present in the A-site of the ribosome. The reaction is catalysed by a protein known as 'stringent factor' (molecular weight 75 000), which has been shown to be the product of the *relA* gene [416]. It is perhaps worth mentioning that ribosomal protein L11, the absence of which from the ribosomes of *B. megaterium* does not prevent protein synthesis occurring (see Section 11.8.3), is nevertheless required for the stringent response [417].

Although ribosomes can utilize GDP to synthesize ppGpp *in vitro*, it is thought that the reaction *in vivo* mainly gives rise to pppGpp, which is converted into ppGpp by the gene product of the *gpp* locus [410]. This latter nucleotide is further degraded to ppG (GDP) in a reaction catalysed by the gene product of the *spoT* locus [418] (Fig. 11.32). The control of the degradation of ppGpp, although incompletely understood, is of some interest. One reason for this is that accumulation of ppGpp (but not pppGpp) also occurs when bacterial cells are deprived of an energy source, and, as this can occur in *relA⁻* strains, must involve a non-ribosomal mechanism. Although there may also be a non-ribosomal pathway for the synthesis of ppGpp, there is evidence for slow ribosomal synthesis of pppGpp (and hence ppGpp) in the absence of the stringent factor, and it is thought that one mechanism by which such ppGpp accumulates in these conditions is by inhibition of its degradation. It should be added that the stringent response also results in the inhibition of other metaoblic processes, such as fat synthesis, that are more directly related to energy charge.

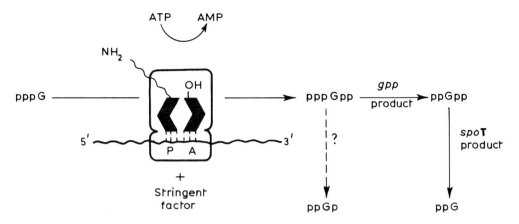

**Fig. 11.32** Pathways for the synthesis and degradation of guanine nucleotides involved in 'stringent control'.

However, the discussion here is confined to the mechanism of the control of RNA synthesis.

Many studies have investigated the effects of ppGpp on RNA synthesis *in vitro*. In some cases there is good correlation between the effects *in vitro* and the 'stringent' response *in vivo*. This is especially so for mRNAs, the synthesis of some (e.g. those for ribosomal proteins) being inhibited, whereas that of others (e.g. those for *lac* mRNA and *his* mRNA) is stimulated by the nucleotide [410]. In the case of rRNA and tRNA the situation has been confused by the fact that some workers have found specific inhibition of rRNA and tRNA synthesis, whereas others have found only non-specific effects. More recently analysis of the promoters for the rRNA and tRNA genes of *E. coli* has provided a basis for clarification [340, 419]. It has been found that rRNA genes have two tandem promoters, P1 and P2, P1 being the stronger and the member of the pair subject to stringent control [420]. It is thus especially convincing that low concentrations of ppGpp have now been shown specifically to inhibit transcription from the P1 promoter *in vitro* [421].

Travers [422] has identified a conserved consensus sequence, GCGCCNC, in the promoter region of genes coding for stable RNAs and also, interestingly, in the promoters of ribosomal protein genes. A four-base-pair substitution of this region in the promoter for tRNA$^{Tyr}$ abolished its susceptibility to the stringent response *in vivo* [423]. It is still not known how ppGpp exerts its effect on these promoters, and the question of whether or not it interacts with RNA polymerase is a vexed one. However, the isolation of a mutant RNA polymerase from a relaxed strain of *E. coli* [424] may help resolve this problem.

It appears that ppGpp has an additional role during amino acid starvation. This is to reduce translational errors (see Section 11.13), which are clearly more likely to occur if a particular aminoacyl-tRNA is in short supply [410].

Finally it should be emphasized that ppGpp does not occur in higher eukaryotes [425], in which RNA and protein synthesis appear to be less tightly coupled, at least in the short term [426].

## 11.13 TRANSLATIONAL ACCURACY AND MISREADING

### 11.13.1 Ribosomal optimization of translational accuracy

We have already seen how a mechanism exists to decrease the misreading of the genetic message that occurs as a result of mischarging of tRNA (Section 11.3). It can easily be imagined how misreading might also occur as the result of incorrect codon–anticodon recognition, and it was an early observation that certain antibiotics (streptomycin is the most studied example), when used at a concentration lower than that at which protein synthesis was completely inhibited, significantly increased the frequency of translational errors. The study of ribosomes from bacteria resistant to streptomycin showed that the error frequency of ribosomes is modulated by ribosomal proteins, and that ribosomes with different error frequencies are possible. Thus many streptomycin-resistant mutants have alterations in ribosomal protein S12 that confer greater accuracy to the ribosomes; whereas revertants from streptomycin-dependence (also involving S12) have mutations in ribosomal protein S4 that cause increased misreading [427]. Indeed a role for ribosomal protein S12 in modulating translational accuracy is indicated by the fact that ribosomes lacking S12 are more accurate than those from the wild-type [428].

If it is possible to select for ribosomes that have increased accuracy, the question arises as to why Nature has not already done this. The initial answer was that perfect accuracy is only possible if translation is infinitely slow, and hence the actual error frequency is the best compromise between accuracy and speed. Although the premise may be correct in the extreme case, the fact remains that ribosomal ambiguity mutants have been found in which there is no change in the speed of translation despite large changes in accuracy. An attractive alternative explanation [429] is that the actual error frequency in wild-type bacteria is, instead, a compromise between accuracy and the energetic cost of the type of proof-reading proposed by Hopfield [148] and Ninio [149]. Such proof-reading, as with that involved in charging (Section 11.3), would involve enhancement of the accuracy of a process by having several stages at which recognition occurs; but differs from that in the stress it

places on the need for an energy requirement to drive the process of discarding mismatched tRNAs. The proposal was that this energy requirement was provided by, and is indeed the reason for, the hydrolysis of GTP during protein synthesis (see Section 11.5.2). Kurland and Ehrenberg [429] regard this driving force as the irreversible free-energy loss associated with the displacement of the GTP:GDP ratio from equilibrium through the resynthesis of GTP, rather than the reversible standard free-energy change resulting from the hydrolysis of GTP. They regard the ability of ppGpp to increase translational accuracy (Section 11.12) to be due to the complex that forms between EF-Tu and ppGpp exerting a direct effect on the proof-reading of the ribosome.

Although the model described above is independent of structural considerations in respect to the actual recognition process, other workers have addressed themselves to this latter question. It has been argued that recognition involves more than the anticodon of the tRNA [430], an extreme example of the evidence in support of this idea being the finding of a suppressor tRNA with an unaltered anticodon but a base change in loop I [431]. As already mentioned, Lake [84] has proposed a separate recognition (R) site on the ribosome to which aminoacyl-tRNA binds before being transferred to the A site. Such a two-step process could, as he points out, have proof-reading as its rationale.

### 11.13.2 Suppression

The discussion of translational accuracy in Section 11.13.1, above, made the implicit assumption that errors would occur in a more-or-less random manner, and would lead to the synthesis of aberrant proteins that in many cases would be non-functional. However, we have already encountered examples of the misreading of specific codons of viral genomes in which new functional proteins are produced (see Sections

11.6.1 and 11.10.1). This misreading is modulated by suppressor tRNAs which insert amino acids at stop codons or which alter the reading frame of amino acid codons. In this section we shall confine consideration to the physical basis for readthrough suppression as our understanding of natural frame-shifting is still very limited [432, 433].

For suppression to occur at a specific codon in the mRNA there must be something in the surrounding sequence to distinguish this codon. As regards nonsense suppression, it has long been known to bacterial geneticists that amber mutations (i.e. artificially induced UAG stop codons) are particularly inclined to be 'leaky' (i.e. to be subject to readthrough suppression) but that the degree of 'leakiness' varied with the position of the mutation. We have already seen how UAG codons are less frequent than other stop codons in prokaryotes (Section 11.5.3), and it has been shown that the 3' context of stop codons is non-random, a U being most frequent after a prokaryotic stop codon and an A or G being most frequent after a eukaryotic stop codon [434]. Experiments have been performed which bear upon the question of whether the preferred prokaryotic context correlates inversely with the likelihood of misreading. The efficiency of suppression was measured for a variety of UAG and UGA codons of known location introduced by mutation into the *lacI* gene of *E. coli*. Clear context effects were seen, and, although exceptions were found, UAG codons followed by a purine – especially A – were well suppressed, whereas those followed by a pyrimidine were weakly suppressed [435, 436]. This is consistent with the preference for U in this position found in natural stop codons.

The codon context of natural prokaryotic stop codons subject to readthrough – the coliphage coat protein terminator and those of certain other phage and host genes – follows a clear pattern. This is a UGA codon followed by an A [437]. It appears that a normal *E. coli* tRNA$^{Trp}$ with a 3'ACC5' anticodon is responsible

for this suppression, and, as the wobble base C would not be allowed to recognize the 3'-A of the codon (Table 11.3), it has been suggested that the (universal) U, 5' to the anticodon, makes a base pair with the A, 3' to the codon, to allow this suppression to occur [437]. This would nicely explain the termination codon context effect in *E. coli*.

The situation in eukaryotes is less simple. The UGA codon subject to readthrough suppression in Sinbis virus (Fig. 11.16) is followed by a C, in murine leukaemia virus there is readthrough of a UAG followed by a G, whereas a UAG suppressed in tobacco mosaic virus is followed by a C [438]. It was shown in *E. coli* that two different suppressors having the same anticodon (as well as the universal U, 5' to the wobble base) differed markedly in efficiency [436], and this suggests that other features of the natural suppressor tRNAs in the different cells that these viruses infect need also to be taken into consideration. In the case of tobacco mosaic virus the UAG suppressor in tobacco leaves has been shown to be a normal tRNA$^{Tyr}$ with an anticodon 3'A–G 5' [439], even though G–G wobble should not occur (Table 11.3). This tRNA$^{Tyr}$ is also found in a modified form with the 5'-anticodon G replaced by a Q, but this species does not act as a suppressor [440]. A similar pair of tRNA$^{Tyr}$ species, with similar properties, has been found in *Drosophila* [441]. Although one might regard the primary physiological significance of readthrough as being confined to viruses (which need to obtain the maximum use of small genomes) the fact that the ratio of Q to G in tRNA$^{Tyr}$ changes during development in *Drosophila* (see [441]) has prompted the suggestion that suppression may also have a functional role in the cells themselves. Certainly a number of other natural suppressor tRNAs have been isolated from mammalian cells [438], perhaps the most curious of which is a bovine and chicken UGA suppressor, a minor tRNA$^{Ser}$ with an anticodon, 3'ACCm5', violating the 'wobble' hypothesis (Table 11.3) in that it does not appear to recognize any codon other than UGA [442]. This, the fact that the tRNA has other unusual structural features and the unique property it possesses of carrying phosphoserine suggest that its normal function may not be in translation. Nevertheless, this would not preclude it from having a subsidiary function in allowing readthrough of certain host mRNAs. It will be interesting to observe future developments in this area.

## REFERENCES

1 Sarabhai, A. S., Stretton, A. O. W., Brenner, S. and Bolle, A. (1964), *Nature (London)*, **201**, 13.
2 Yanofski, C., Carlton, B. C., Guest, J. R., Helinski, D. R. and Henning, V. (1964), *Proc. Natl. Acad. Sci. USA*, **51**, 266.
3 Lengyel, P. (1974), in *Ribosomes* (ed. M. Nomura, A. Tissières and P. Lengyel), Cold Spring Harbor Laboratory Monograph Series, p.13.
4 Weissbach, H. and Pestka, S. (1977), *Molecular Mechanisms of Protein Biosynthesis*, Academic Press, New York.
5 Weissbach, H. (1979), in *Ribosomes: Structure, Function and Genetics* (ed. G. Chambliss, G. R. Craven, J. Davies, K. Davis, L. Kahan and M. Nomura), University Park Press, Baltimore, p. 377.
6 Pérez-Bercoff, R. (1982), *Protein Biosynthesis in Eukaryotes*, Plenum, New York.
7 Hunt, T., Prentis, S. and Tooze, J. (1983), *DNA Makes RNA Makes Protein*, Elsevier Biomedical, Amsterdam.
8 Clark, B. F. C. and Petersen, H. U. (1984), *Gene Expression: The Translational Step and its Control*, Munskgaard, Copenhagen.
9 Moldave, K. (1985), *Annu. Rev. Biochem.*, **54**, 1109.
10 Hoagland, M. B., Zamecnik, P. C. and Stephenson, M. L. (1957), *Biochim. Biophys. Acta*, **24**, 215.
11 Hoagland, M. B., Stephenson, M. L., Scott, J. F., Hecht, L. I. and Zamecnik, P. C. (1958), *J. Biol. Chem.*, **231**, 241.
12 Goddard, J. P. (1977), *Prog. Biophys. Mol. Biol.*, **32**, 233.
13 Offengand, J. (1982), in *Protein Biosynthesis in Eukaryotes* (ed. R. Pérez-Bercoff), Plenum, New York, p. 1.
14 Schevitz, R. W., Podjarny, A. D., Krishnamachari, N., Hughes, J. J. and Sigler, P. B. (1979), *Nature (London)*, **278**, 188.
15 Woo, N. H., Roe, B. A. and Rich, A. (1980), *Nature (London)*, **286**, 346.
16 Westhof, E., Dumas, P. and Moras, D. (1985), *J. Mol. Biol.*, **184**, 119.
17 Quigley, G. J. and Rich, A. (1976), *Science*, **194**, 796.
18 Crick, F. H. C. (1967), *Proc. R. Soc. London, Ser. B*, **167**, 331.
19 Woese, C. R. (1967), *The Genetic Code. The Molecular Basis for Genetic Expression*. Harper and Row, New York.
20 Jukes, T. H. and Gatlin, L. (1971), *Prog. Nucleic Acid Res. Mol. Biol.*, **11**, 303.
21 Jukes, T. H. (1977), in *Comprehensive Biochemistry* (ed. M. Florkin and E. H. Stotz), Elsevier, Amsterdam, Vol. 24, p. 235.
22 Crick, F. H. C., Barnett, L., Brenner, S. and Watts-Tobin, R. J. (1961), *Nature (London)*, **192**, 1227.
23 Crick, F. H. C. (1966), *Cold Spring Harbor Symp. Quant. Biol.*, **31**, 3.
24 Matthaei, J. H. and Nirenberg, M. W. (1961), *Proc. Natl. Acad. Sci. USA*, **47**, 1580.
25 Nirenberg, M. W. and Matthaei, J. H. (1961), *Proc. Natl. Acad. Sci. USA*, **47**, 1588.
26 Nirenberg, M. W., Matthaei, J. H., Jones, O. W., Martin, R. G. and Barondes, S. H. (1963), *Fed. Proc.*, **22**, 55.
27 Speyer, J. S., Lengyel, P., Basilio, C., Wahba, A. J., Gardner, R. S. and Ochoa, S. (1963), *Cold Spring Harbor Symp. Quant. Biol.*, **28**, 559.
28 Khorana, H. G. (1965), *Fed. Proc.*, **24**, 1473.
29 Khorana, H. G., Büchi, H., Ghosh, H., Gupta, N., Jacob, T. M., Kössel, H., Morgan, R., Narang, S. A., Ohtsuka, E. and Wells, R. D. (1966), *Cold Spring Harbor Symp. Quant. Biol.*, **31**, 39.
30 Nirenberg, M. and Leder, P. (1964), *Science*, **145**, 1399.
31 Leder, P. and Nirenberg, M. (1964), *Proc. Natl. Acad. Sci. USA*, **52**, 420.
32 Nirenberg, M., Caskey, C. T., Marshall, R., Brimacombe, R., Kelly, D., Doctor, B., Hatfield, D., Levin, J., Rottman, F., Pestka, S., Wilcox, F. and Anderson, W. F. (1966), *Cold Spring Harbor Symp. Quant. Biol.*, **31**, 11.

33 Wittmann, H. G. (1962), *Z. Vererbungslehre*, **93**, 491.

34 Wittmann, H. G. and Wittmann-Liebold, B. (1963), *Cold Spring Harbor Symp. Quant. Biol.*, **28**, 589.

35 Fraenkel-Conrat, H. (1964), *Sci. Am.*, **211**(4), 46.

36 Yanofsky, C. (1967), *Sci. Am.*, **216**(5), 80.

37 Terzaghi, E., Okada, Y., Streisinger, G., Emrich, J., Inouye, M. and Tsugita, A. (1966), *Proc. Nat. Acad. Sci. USA*, **56**, 500.

38 Adams, J. M., Jeppesen, P. G. N., Sanger, F. and Barrell, B. G. (1969), *Nature (London)*, **223**, 1009.

39 Fiers, W., Contreras, R., Duerinck, F., Hageman, G., Iserentant, D., Merregaert, J., Min Jou, W., Molemans, F., Raeymaekers, A., Van den Berghe, A., Volckaert, G. and Ysebaert, M. (1976), *Nature (London)*, **260**, 500.

40 Pribnow, D., Sigurdson, C., Gold, L., Singer, B. S., Napoli, C., Brosius, J., Dull, T. J. and Noller, H. F. (1981), *J. Mol. Biol.*, **149**, 337.

41 Benzer, S. and Champe, S. P. (1962), *Proc. Natl. Acad. Sci. USA*, **48**, 1114.

42 Brenner, S., Stretton, A. O. W. and Kaplan, S. (1965), *Nature (London)*, **206**, 994.

43 Brenner, S., Barnett, L., Katz, E. R. and Crick, F. H. C. (1967), *Nature (London)*, **213**, 449.

44 Goodman, H. M., Abelson, J., Landy, A., Brenner, S. and Smith, J. D. (1968), *Nature (London)*, **217**, 1019.

45 Marcker, K. A. and Sanger, F. (1964), *J. Mol. Biol.*, **8**, 835.

46 Waller, J. P. (1963), *J. Mol. Biol.*, **10**, 319.

47 Adams, J. M. and Capecchi, M. R. (1966), *Proc. Natl. Acad. Sci. USA*, **55**, 147.

48 Webster, R. E., Engelhardt, D. L. and Zinder, N. S. (1966), *Proc. Natl. Acad. Sci. USA*, **55**, 155.

49 Adams, J. M. (1968), *J. Mol. Biol.*, **34**, 571.

50 Takeda, M. and Webster, R. E. (1968), *Proc. Natl. Acad. Sci. USA*, **60**, 1487.

51 Dube, S. K., Marcker, K. A., Clark, B. F. C. and Cory, S. (1968), *Nature (London)*, **218**, 232.

52 Cory, S., Marcker, K. A., Dube, S. K. and Clark, B. F. C. (1968), *Nature (London)*, **220**, 1039.

53 Marcker, K. A., Clark, B. F. C. and Anderson, J. S. (1966), *Cold Spring Harbor Symp. Quant. Biol.*, **31**, 279.

54 Delk, A. S. and Rabinowitz, J. C. (1974), *Nature (London)*, **252**, 106.

55 Woo, N. H., Roe, B. A. and Rich, A. (1980), *Nature (London)*, **286**, 346.

56 Shevitz, R. W., Podjarny, A. D., Krishnamachari, N., Hughes, J. J., Sigler, P. B. and Sussman, J. L. (1979), *Nature (London)*, **278**, 188.

57 Ghosh, H. P., Söll, D. and Khorana, H. G. (1967), *J. Mol. Biol.*, **25**, 275.

58 Kozak, M. (1983), *Microbiol. Rev.*, **47**, 1.

59 Sacerdot, C., Fayat, G., Dessen, P., Springer, M., Plumbridge, J. A., Grunberg-Manago, M. and Blanquet, S. (1982), *EMBO J.*, **1**, 311.

60 Crick, F. H. C. (1966), *J. Mol. Biol.*, **19**, 548.

61 Lagerkvist, U. (1978), *Proc. Natl. Acad. Sci. USA*, **75**, 1759.

62 Samuelsson, T., Axberg, T., Boren, T. and Lagerkvist, U. (1983), *J. Biol. Chem.*, **258**, 13 178.

63 Sprinzl, M., Moll, J., Meissner, F. and Hartmann, T. (1985), *Nucleic Acids Res.*, **13**, r1.

64 Marshall, R. E., Caskey, C. T. and Nirenberg, M. (1967), *Science*, **155**, 820.

65 Fox, T. D. (1985), *Nature (London)*, **314**, 132.

66 Lagerkvist, U., Rymo, L. and Waldenström, J. (1966), *J. Biol. Chem.*, **241**, 5391.

67 Julius, D. J., Fraser, T. H. and Rich, A. (1979), *Biochemistry*, **18**, 604.

68 Griffin, B. E., Jarman, M., Reese, C. B., Sulston, J. E. and Trentham, D. R. (1966), *Biochemistry*, **5**, 3638.

69 Zachau, H. G. and Feldman, H. (1965), *Prog. Nucleic Acid Res. Mol. Biol.*, **4**, 217.

70 Schimmel, P. R. and Söll, D. (1979), *Annu. Rev. Biochem.*, **48**, 601.

71 Jasin, M., Regan, L. and Schimmel, P. (1983), *Nature (London)*, **306**, 441.

72 Blow, D. M. and Brick, P. (1985), in *The Structure of Biological Macromolecules and Assemblies: Vol. 2. Nucleic Acids and Interactive Proteins* (ed. F. A. Jurnak and A. McPherson), Wiley, New York, p. 442.

73 Chapeville, F., Lipmann, F., von Ehrenstein, G., Weisblum, B., Ray, W. J., Jr and Benzer, S. (1962), *Proc. Natl. Acad. Sci. USA*, **48**, 1086.

74 Baldwin, A. N. and Berg, P. (1966), *J. Biol. Chem.*, **241**, 839.

75 Fersht, A. R. (1979), in *Transfer RNA: Structure, Properties and Recognition* (ed. P. R. Schimmel, D. Söll and J. N. Abelson), Cold Spring Harbor Laboratory Monograph Series, p. 247.

76 Dintzis, H. M. (1961), *Proc. Natl. Acad. Sci. USA*, **47**, 247.

77 Thach, R. E., Cecere, M. A., Sundrararajan, T.

A. and Doty, P. (1965), *Proc. Natl. Acad. Sci. USA*, **54**, 1167.

78 Byrne, R., Levin, J. G., Bladen, H. A. and Nirenberg, M. W. (1964), *Proc. Natl. Acad. Sci. USA*, **52**, 140.

79 Miller, O. L., Jr, Hamkalo, B. A. and Thomas, C. A. (1970), *Science*, **169**, 392.

80 Lacroute, F. and Stent, G. (1968), *J. Mol. Biol.*, **35**, 165.

81 Knopf, P. M. and Lamfrom, H. (1965), *Biochim. Biophys. Acta*, **95**, 398.

82 Maden, B. E. H., Traut, R. R. and Monro, R. E. (1968), *J. Mol. Biol.*, **35**, 333.

83 Harris, R. J. and Pestka, S. (1977) in *Molecular Mechanisms of Protein Biosynthesis* (ed. H. Weissbach and S. Pestka), Academic Press, New York, p. 413.

84 Lake, J. A. (1977), *Proc. Natl. Acad. Sci. USA*, **74**, 1903.

85 Rheinberger, H.-J., Sternbach, H. and Nierhaus, K. H. (1981), *Proc. Natl. Acad. Sci. USA*, **78**, 5310.

86 Rich, A. (1974) in *Ribosomes* (ed. N. Momura, A. Tissières and P. Lengyel), Cold Spring Harbor Monograph Series, p. 871.

87 Warner, J., Knopf, P. M. and Rich, A. (1963), *Proc. Natl. Acad. Sci. USA*, **49**, 122.

88 Wettstein, F. O., Staehelin, T. and Noll, H. (1963), *Nature (London)*, **197**, 430.

89 Lodish, H. F. and Jacobsen, M. (1972), *J. Biol. Chem.*, **247**, 3622.

90 Heywood, S. M., Dowben, R. M. and Rich, A. (1967), *Proc. Natl. Acad. Sci. USA*, **57**, 1002.

91 Wold, F. and Moldave, K. (1984), *Methods Enzymol.*, **106**.

92 Wold, F. and Moldave, K. (1985), *Methods Enzymol.*, **107**.

93 Kleinkauf, H. and von Döhren, H. (1983), *Trends Biochem. Sci.*, **8**, 281.

94 Caskey, T., Leder, P., Moldave, K. and Schlessinger, D. (1972), *Science*, **176**, 195.

95 Grunberg-Manago, M. and Gros, F. (1977), *Prog. Nucleic Acid Res. Mol. Biol.*, **20**, 209.

96 Revel, M. (1977), in *Molecular Mechanisms of Protein Biosynthesis* (ed. H. Weissbach and S. Pestka), Academic Press, New York, p. 245.

97 Grunberg-Manago, M. (1980) in *Ribosomes: Structure, Function and Genetics* (ed. G. Chambliss, G. R. Craven, J. Davies, K. Davis, L. Kahan and M. Nomura), University Park Press, Baltimore, p. 445.

98 Maitra, U., Stringer, E. A. and Chaudhuri, A. (1982), *Annu. Rev. Biochem.*, **51**, 869.

99 Guthrie, C. and Nomura, M. (1968), *Nature (London)*, **219**, 232.

100 Stanley, W. M., Salas, M., Wabha, A. J. and Ochoa, S. (1966), *Proc. Natl. Acad. Sci. USA*, **56**, 290.

101 Iwasaki, K., Sabol, S., Wabha, A. J. and Ochoa, S. (1968), *Arch. Biochem. Biophys.*, **125**, 542.

102 Revel, M., Lelong, J. C., Brawerman, G. and Gros, F. (1968), *Nature (London)*, **219**, 1016.

103 Steitz, J. A. (1969), *Nature (London)*, **224**, 957.

104 Hershey, J. W. B., Yanov, J., Johnston, K. and Fakunding, J. L. (1977), *Arch. Biochem. Biophys.*, **182**, 626.

105 Nakamoto, T. and Kolakofsky, D. (1966), *Proc. Natl. Acad. Sci. USA*, **55**, 606.

106 Lucas-Lenard, J. and Lipmann, F. (1967), *Proc. Natl. Acad. Sci. USA*, **57**, 1050.

107 Sabol, S., Sillero, M. A. G., Iwasaki, K. and Ochoa, S. (1970), *Nature (London)*, **228**, 1269.

108 Vermeer, C., Van Alphen, W., Van Knippenberg, P. H. and Bosch, L. (1973), *Eur. J. Biochem.*, **40**, 295.

109 Salas, M., Hille, M. B., Last, J. A., Wabha, A. J. and Ochoa, S. (1967), *Proc. Natl. Acad. Sci. USA*, **57**, 387.

110 Shine, J. and Dalgarno, L. (1974), *Proc. Natl. Acad. Sci. USA*, **71**, 1342.

111 Steege, D. A. (1977), *Proc. Natl. Acad. Sci. USA*, **74**, 4163.

112 Chae, Y.-B., Mazumder, R. and Ochoa, S. (1969), *Proc. Natl. Acad. Sci. USA*, **63**, 828.

113 Lockwood, A. H., Chakraborty, P. R. and Maitra, U. (1971), *Proc. Natl. Acad. Sci. USA*, **68**, 3122.

114 Anderson, J. S., Dahlberg, J. E., Bretscher, M. S., Revel, M. and Clark, B. F. C. (1967), *Nature (London)*, **216**, 1072.

115 Benne, R., Ebes, F. and Voorma, H. O. (1973), *Eur. J. Biochem.*, **38**, 265.

116 Thach, S. S. and Thach, R. E. (1971), *Proc. Natl. Acad. Sci. USA*, **68**, 1791.

117 Kuechler, E. (1971), *Nature (London) New Biol.*, **234**, 216.

118 Plumbridge, J. A., Deville, F., Sacerdot, C., Petersen, H. U., Cenatiempo, Y., Cozzone, A., Grunberg-Manago, M. and Hershey, J. W. B. (1985), *EMBO J.*, **4**, 223.

119 Miller, D. L. and Weissbach, H. (1977), in *Molecular Mechanisms for Protein Biosynthesis*

(ed. H. Weissbach and S. Pestka), Academic Press, New York, p. 324.

120 Bermek, E. (1978), *Prog. Nucleic Acid Res. Mol. Biol.*, **21**, 63.

121 Clark, B. F. C. (1983), in *DNA Makes RNA Makes Protein* (ed. T. Hunt, S. Prentis and J. Tooze), Elsevier Biomedical, Amsterdam, p. 213.

122 Haenni, A.-L. and Lucas-Lenard, J. (1968), *Proc. Natl. Acad. Sci. USA*, **61**, 1363.

123 Erbe, R. W., Nau, M. M. and Leder, P. (1969), *J. Mol. Biol.*, **39**, 441.

124 Gupta, S. L., Waterson, J., Sopori, M. L., Weissman, S. and Lengyel, P. (1971), *Biochemistry*, **10**, 4410.

125 Lucas-Lenard, J. and Lipmann, F. (1966), *Proc. Natl. Acad. Sci. USA*, **55**, 1562.

126 Jones, M. D., Petersen, T. E., Nielsen, K. M., Magnusson, S., Sottrup-Jensen, L., Gausing, K. and Clark, B. F. C. (1980), *Eur. J. Biochem.*, **108**, 507.

127 Ravel, J. M., Shorey, R. L. and Shive, W. (1967), *Biochem. Biophys. Res. Commun.*, **29**, 68.

128 Ono, Y., Skoultchi, A., Klein, A. and Lengyel, P. (1968), *Nature (London)*, **220**, 1304.

129 An, G., Bendiak, D. S., Mamelak, L. A. and Friesen, J. D. (1981), *Nucleic Acids Res.*, **9**, 4163.

130 Weissbach, H., Redfield, B. and Brot, N. (1971), *Arch. Biochem. Biophys.*, **144**, 224.

131 Beaud, G. and Lengyel, P. (1971), *Biochemistry*, **10**, 4899.

132 Rubin, J. R., Morikawa, K., Nyborg, J., La Cour, T. F. M., Clark, B. F. C. and Miller, D. L. (1981), *FEBS Lett.*, **129**, 177.

133 Wilkman, F. P., Siboska, G. E., Petersen, H. U. and Clark, B. F. C. (1982), *EMBO J.*, **1**(9), 1095.

134 Adkins, H. J., Miller, D. L. and Johnson, A. E. (1983), *Biochemistry*, **22**, 1208.

135 Traut, R. R. and Monro, R. E. (1964), *J. Mol. Biol.*, **10**, 63.

136 Monro, R. E. (1967), *J. Mol. Biol.*, **26**, 147.

137 Monro, R. E., Černa, J. and Marcker, K. A. (1968), *Proc. Natl. Acad. Sci. USA*, **61**, 1042.

138 Rychlík, I., Chládek, S. and Zěmlička, J. (1967), *Biochim. Biophys. Acta*, **138**, 640.

139 Parmeggiani, A. and Gottchalk, E. M. (1969), *Cold Spring Harbor Symp. Quant. Biol.*, **34**, 377.

140 Leder, P., Skogerson, L. E. and Nau, M. M. (1969), *Proc. Natl. Acad. Sci. USA*, **62**, 454.

141 Brot, N., Ertel, R. and Weissbach, H. (1968), *Biochem. Biophys. Res. Commun.*, **31**, 563.

142 Lucas-Lenard, J. and Haenni, A-L. (1969), *Proc. Natl. Acad. Sci. USA*, **63**, 93.

143 Richman, N. and Bodley, J. W. (1972), *Proc. Natl. Acad. Sci. USA*, **69**, 686.

144 Cabrer, B., Vázquez, D. and Modellel, J. (1972), *Proc. Natl. Acad. Sci. USA*, **69**, 733.

145 Miller, D. L. (1972), *Proc. Natl. Acad. Sci. USA*, **69**, 752.

146 Pestka, S. (1969), *J. Biol. Chem.*, **244**, 1533.

147 Gavrilova, L. T. and Spirin, A. S. (1971), *FEBS Lett.*, **17**, 324.

148 Hopfield, J. J. (1974), *Proc. Natl. Acad. Sci. USA*, **71**, 4135.

149 Ninio (1975), *Biochimie*, **57**, 587.

150 Tate, W. P. and Caskey, C. T. (1974), in *The Enzymes* (ed. P. D. Boyer), Academic Press, New York, Vol. 10, p. 87.

151 Caskey, C. T. (1977), in *Molecular Mechanisms of Protein Biosynthesis* (ed. H. Weissbach and S. Pestka), Academic Press, New York, p. 443.

152 Tate, W. P. (1984), in *Peptide and Protein Reviews* (ed. M. T. W. Hearn), Dekker, New York, Vol. 2, p. 173.

153 Ganoza, M. C. (1966), *Cold Spring Harbor Symp. Quant. Biol.*, **31**, 273.

154 Capecchi, M. R. (1967), *Proc. Natl. Acad. Sci. USA*, **58**, 1144.

155 Scolnick, E., Tompkins, R., Caskey, T. and Nirenberg, M. (1968), *Proc. Natl. Acad. Sci. USA*, **61**, 768.

156 Milman, G., Goldstein, J., Scolnick, E. and Caskey, T. (1969), *Proc. Natl. Acad. Sci. USA*, **63**, 183.

157 Hirashima, A. and Kaji, A. (1972), *Biochemistry*, **11**, 4037.

158 Bretscher, M. S. (1968), *J. Mol. Biol.*, **34**, 131.

159 Capecchi, M. R. and Klein, H. A. (1969), *Cold Spring Harbor Symp. Quant. Biol.*, **34**, 469.

160 Caskey, C. T., Tomkins, R., Scolnick, E., Caryk, T. and Nirenberg, M. (1968), *Science*, **162**, 135.

161 Craigen, W. J., Cook, R. G., Tate, W. P. and Caskey, C. T. (1985), *Proc. Natl. Acad. Sci. USA*, **82**, 3616.

162 Goldstein, J. L. and Caskey, C. T. (1970), *Proc. Natl. Acad. Sci. USA*, **67**, 537.

163 Fahnestock, S., Neumann, H., Shashoua, V. and Rich, A. (1970), *Biochemistry*, **9**, 2477.

164 Vogel, Z., Zamir, A. and Elson, D. (1969), *Biochemistry*, **8**, 5161.

165 Ogawa, K. and Kaji, A. (1975), *Eur. J. Biochem.*, **58**, 411.

166 Kung, H.-F., Treadwell, B. V., Spears, C., Tai, P.-C. and Weissbach, H. (1977), *Proc. Natl. Acad. Sci. USA*, **74**, 3217.

167 Davis, B. D. (1971), *Nature (London)*, **231**, 153.

168 Subramanian, A. R., Ron, E. Z. and Davis, B. D. (1968), *Proc. Natl. Acad. Sci. USA*, **61**, 761.

169 Subramanian, A. R. and Davis, B. D. (1970), *Nature (London)*, **228**, 1273.

170 Kaempfer, R. (1972), *J. Mol. Biol.*, **71**, 583.

171 Naaktgeboren, N., Roobol, K. and Voorma, H. O. (1977), *Eur. J. Biochem.*, **72**, 49.

172 Miall, S. H. and Tamaoki, T. (1972), *Biochemistry*, **11**, 4826.

173 Kohli, J. and Grosjean, H. (1981), *Mol. Gen. Genet.*, **182**, 430.

174 Miller, J. H. and Albertini, A. M. (1983), *J. Mol. Biol.*, **164**, 59.

175 Bossi, L. (1983), *J. Mol. Biol.*, **164**, 73.

176 Smith, A. E. and Marcker, K. A. (1968), *J. Mol. Biol.*, **38**, 241.

177 Caskey, C. T., Redfield, B. and Weissbach, H. (1967), *Arch. Biochem. Biophys.*, **120**, 119.

178 Smith, A. E. and Marcker, K. A. (1970), *Nature (London)*, **226**, 607.

179 Brown, J. C. and Smith, A. E. (1970), *Nature (London)*, **226**, 610.

180 Mathews, M. B. and Korner, A. (1970), *Eur. J. Biochem.*, **17**, 328.

181 Jackson, R. and Hunter, T. (1970), *Nature (London)*, **227**, 672.

182 Wilson, D. B. and Dintzis, H. M. (1970), *Proc. Natl. Acad. Sci. USA*, **66**, 1282.

183 Wigle, D. T. and Dixon, G. H. (1970), *Nature (London)*, **227**, 676.

184 Anderson, W. F., Bosch, L., Cohn, W. E., Lodish, H., Merrick, W. C., Weissbach, H., Wittmann, H. G. and Wool, I. G. (1977), *FEBS Lett.*, **76**, 1.

185 Thomas, A. A. M., Benne, R. and Voorma, H. O. (1981), *FEBS Lett.*, **128**, 177.

186 Shatkin, A. J. (1985), *Cell*, **40**, 223.

187 Kozak, M. (1978), *Cell*, **15**, 1109.

188 Kozak, M. (1984), *Nucleic Acids Res.*, **12**, 857.

189 Sherman, F., McKnight, G. and Stewart, J. W. (1980), *Biochim. Biophys. Acta*, **609**, 343.

190 Darnborough, C., Legon, S., Hunt, T. and Jackson, R. (1973), *J. Mol. Biol.*, **76**, 379.

191 Schreier, M. H. and Staehelin, T. (1973), *Nature (London) New Biol.*, **242**, 35.

192 Chen, Y. C., Woodley, C. L., Bose, K. K. and Gupta, N. K. (1972), *Biochem. Biophys. Res. Commun.*, **48**, 1.

193 Strauss, E. G., Rice, C. M. and Strauss, J. H. (1984), *Virology*, **133**, 92.

194 Sonenberg, N., Rupprecht, K. M., Hecht, S. M. and Shatkin, A. J. (1979), *Proc. Natl. Acad. Sci. USA*, **76**, 4345.

195 Edery, I., Hümbelin, M., Darveau, A., Lee, K. A. W., Milburn, S., Hershey, J. W. B., Trachsel, H. and Sonenberg, N. (1983), *J. Biol. Chem.*, **258**, 11 398.

196 Marcus, A. (1970), *J. Biol. Chem.*, **245**, 955.

197 Trachsel, H., Schreier, M. H., Erni, B. and Staehelin, T. (1977), *J. Mol. Biol.*, **116**, 727.

198 Grifo, J. A., Abramson, R. D., Satler, C. A. and Merrick, W. C. (1984), *J. Biol. Chem.*, **259**, 8648.

199 Safer, B. (1983), *Cell*, **33**, 7.

200 Gehrke, L., Auron, P. E., Quigley, G. J., Rich, A. and Sonenberg, N. (1983), *Biochemistry*, **22**, 5157.

201 Furuichi, Y., La Fiandra, A. and Shatkin, A. J. (1977), *Nature (London)*, **266**, 235.

202 Trachsel, H. and Staehelin, T. (1979), *Biochim. Biophys. Acta*, **565**, 305.

203 McKeehan, W. L. and Hardesty, B. (1969), *J. Biol. Chem.*, **244**, 4330.

204 Schneir, M. and Moldave, K. (1968), *Biochim. Biophys. Acta*, **166**, 58.

205 Van Hemert, F. J., Amons, R., Pluijms, W. J. M., Van Ormondt, H. and Möller, W. (1984), *EMBO J.*, **3**, 1109.

206 Iwasaki, K., Mizumoto, K., Tanaka, M. and Kaziro, Y. (1973), *J. Biochem. Tokyo*, **74**, 849.

207 Prather, N., Ravel, J. M., Hardesty, B. and Shive, W. (1974), *Biochem. Biophys. Res. Commun.*, **57**, 578.

208 Slobin, L. I. and Möller, W. (1978), *Eur. J. Biochem.*, **84**, 69.

209 Nagata, S., Motoyoshi, K. and Iwasaki, K. (1976), *Biochem. Biophys. Res. Commuun.*, **71**, 933.

210 Richter, D. and Lipmann, F. (1970), *Nature (London)*, **227**, 1212.

211 Slobin, L. I. (1980), *Eur. J. Biochem.*, **110**, 555.

212 Nagata, S., Nagashima, K., Tsunetsugu-Yokota, Y., Fujimura, K., Miyazaki, M. and Kaziro, Y. (1984), *EMBO J.*, **3**, 1825.

213 Galasinski, W. and Moldave, K. (1969), *J. Biol. Chem.*, **244**, 6527.

214 Raeburn, S., Collins, J. F., Moon, H. M. and Maxwell, E. S. (1975), *J. Biol. Chem.*, **250**, 720.

215 Collier, R. J. (1975), *Bacteriol. Rev.*, **39**, 54.

216 Van Ness, B. G., Howard, J. B. and Bodley, J. W. (1980), *J. Biol. Chem.*, **255**, 10 710.

217 Neth, R., Monro, R. E., Heller, G. Battaner, E. and Vázquez, D. (1970), *FEBS Lett.*, **6**, 198.

218 Goldstein, J. L., Beaudet, A. L. and Caskey, C. T. (1970), *Proc. Natl. Acad. Sci. USA*, **67**, 99.

219 Beaudet, A. L. and Caskey, C. T. (1971), *Proc. Natl. Acad. Sci. USA*, **68**, 619.

220 Hogan, B. L. and Korner, A. (1968), *Biochim. Biophys. Acta*, **169**, 139.

221 Kaempfer, R. (1969), *Nature (London)*, **222**, 950.

222 Russell, D. W. and Spremulli, L. L. (1979), *J. Biol. Chem.*, **254**, 8796.

223 Valenzuela, D. M., Chaudhuri, A. and Maitra, U. (1982), *J. Biol. Chem.*, **257**, 7712.

224 Palade, G. (1975), *Science*, **189**, 347.

225 Leader, D. P. (1983), in *DNA Makes RNA Makes Protein* (ed. T. Hunt, S. Prentis and J. Tooze), Elsevier Biomedical, Amsterdam, p. 257.

226 Milstein, C., Brownlee, G. G., Harrison, T. M. and Mathews, M. B. (1972), *Nature (London) New Biol.*, **239**, 117.

227 Blobel, G. and Dobberstein, B. (1975), *J. Cell Biol.*, **67**, 835.

228 Blobel, G., Walter, P., Chang, C. N., Goldman, B. N., Erickson, A. H. and Lingappa, V. R. (1979), *Symp. Soc. Exp. Biol.*, **33**, 9.

229 Walter, P. and Blobel, G. (1982), *Nature (London)*, **299**, 691.

230 Meyer, D., Krause, E. and Dobberstein, B. (1982), *Nature (London)*, **297**, 647.

231 Kreibich, G., Freienstein, C. M., Pereyra, P. N., Ulrich, B. C. and Sabatini, D. D. (1978), *J. Cell Biol.*, **77**, 488.

232 Walter, P., Gilmore, R. and Blobel, G. (1984), *Cell*, **38**, 5.

233 Randall, L. L. and Hardy, S. J. (1984), *Microbiol. Rev.*, **48**, 290.

234 Meek, R. L., Walsh, K. A. and Palmiter, R. D. (1982), *J. Biol. Chem.*, **257**, 12 245.

235 Inouye, S., Soberon, X., Franceschini, T., Nakamura, K., Itakura, K. and Inouye, M. (1982), *Proc. Natl. Acad. Sci. USA*, **79**, 3438.

236 Fischer, H. D., Gonzalez-Noriega, A., Sly, W. S. and Morré, D. J. (1980), *J. Biol. Chem.*, **255**, 9608.

237 Palade, G. E. (1983), *Methods Enzymol.*, **96**, xxix.

238 Rogers, J., Early, P., Carter, C., Calaine, K., Bond, M., Hood, L. and Wall, R. (1980), *Cell*, **20**, 303.

239 Schatz, G. and Butow, R. A. (1983), *Cell*, **32**, 316.

240 Strauss, A. W. and Boime, I. (1982), *CRC Crit. Rev. Biochem.*, **12**, 205.

241 De Robertis, E. (1983), *Cell*, **32**, 1021.

242 Wittmann, H. G. (1982), *Annu. Rev. Biochem.*, **51**, 155.

243 Wittmann, H. G. (1983), *Annu. Rev. Biochem.*, **52**, 35.

244 Nierhaus, K. H. (1982), *Curr. Top. Microbiol. Immunol.*, **97**, 81.

245 Wool, I. G. (1979), *Annu. Rev. Biochem.*, **48**, 719.

246 Bielka, H. (1982), *The Eukaryotic Ribosome*, Springer-Verlag, Berlin.

247 Nomura, M., Tissières, A. and Lengyel, P. (1974), *Ribosomes*, Cold Spring Harbor Monograph Series, New York.

248 Chambliss, G., Craven, G. R., Davies, J., Davis, K., Kahan, L. and Nomura, M. (1980), *Ribosomes: Structure, Function and Genetics*, University Park Press, Baltimore.

249 Brimacombe, R., Maly, P. and Zwieb, C. (1983), *Prog. Nucleic Acid Res.*, **28**, 1.

250 Noller, H. F. (1984), *Annu. Rev. Biochem.*, **53**, 119.

251 Maden, B. E. H., Khan, M. S. N., Hughes, D. G. and Goddard, J. P. (1977), *Biochem. Soc. Symp.*, **42**, 165.

252 Maden, B. E. H. (1980), *Nature (London)*, **288**, 293.

253 Brimacombe, R. (1984), *Trends Biochem. Sci.* **9**, 273.

254 Walker, T. A. and Pace, N. R. (1983), *Cell*, **33**, 320.

255 Singhal, R. P. and Shaw, J. K. (1983), *Prog. Nucleic Acid Res. Mol. Biol.*, **28**, 177.

256 Delius, N., Andersen, J. and Singhal, R. P. (1984), *Prog. Nucleic Acid Res. Mol. Biol.*, **31**, 161.

257 Chooi, W. Y., Sabatini, L. M. and Macklin, M. (1984), *Biochem. Genet.*, **22**, 749.

258 Juan-Vidales, F., Sánchez Madrid, F., Saenz-Robles, M. T. and Ballesta, J. P. G. (1983), *Eur. J. Biochem.*, **136**, 275.

259 Schmid, G., Strobel, O., Stöffler-Meilicke, M., Stöffler, G. and Böck, A. (1984), *FEBS Lett.*, **177**, 189.

260 Lin, A., Wittmann-Liebold, B., McNally, J. and Wool, I. G. (1982), *J. Biol. Chem.*, **257**, 9189.

261 Amons, R., Pluijms, W., Krick, J. and Möller,

W. (1982), *FEBS Lett.*, **146**, 143.

262 Leader, D. P. (1980), in *Molecular Aspects of Cellular Regulation* (ed. P. Cohen), Elsevier/North-Holland, Amsterdam, Vol. 1, p. 203.

263 Traut, R. R., Lambert, J. M., Boileau, G. and Kenny, J. W. (1980), in *Ribosomes: Structure, Function and Genetics* (ed. G. Chambliss, G. R. Craven, J. Davies, K. Davis, L. Kahan and M. Nomura), University Park Press, Baltimore, p. 89.

264 Ramakrishnan, V., Capel, M., Kjeldgaard, M., Engelman, D. M. and Moore, P. B. (1984), *J. Mol. Biol.*, **174**, 265.

265 Zimmerman, R. A. (1974), in *Ribosomes* (ed. M. Nomura, A. Tissières and P. Lengyel), Cold Spring Harbor Monograph Series, p. 225.

266 Lake, J. (1985), *Annu. Rev. Biochem.*, **54**, 507.

267 Stöffer, G. and Stöffler-Meilicke, M. (1984), *Annu. Rev. Biophys. Bioeng.*, **13**, 303.

268 Held, W. A. and Nomura, M. (1973), *Biochemistry*, **12**, 3273.

269 Nomura, M. and Held, W. A. (1974), in *Ribosomes* (ed. M. Nomura, A. Tissières and P. Lengyel), Cold Spring Harbor Monograph Series, p. 193.

270 Kuechler, E. and Ofengand, J. (1979), in *Transfer RNA: Structure, Properties and Recognition* (ed. P. R. Schimmel, D. Söll and J. N. Abelson), Cold Spring Harbor Monograph Series, p. 413.

271 Cooperman, B. (1980), in *Ribosomes: Structure, Function and Genetics* (ed. G. Chambliss, G. R. Craven, J. Davies, K. Davis, L. Kahan and M. Nomura), University Park Press, Baltimore, p. 531.

272 Dabbs, E. R., Hasenbank, R., Kastner, B., Rak, K-H., Wartusch, B. and Stöffler, G. (1983), *Mol. Gen. Genet.*, **192**, 301.

273 Crick, F. H. C. (1968), *J. Mol. Biol.*, **38**, 367.

274 Liljas, A. (1982), *Prog. Biophys. Mol. Biol.*, **40**, 161.

275 Garrett, R. (1983), *Trends Biochem. Sci.*, **8**, 189.

276 Fasano, O., De Vendittis, E. and Parmeggiani, A. (1982), *J. Biol. Chem.*, **257**, 3145.

277 Barta, A., Steiner, G., Brosius, J., Noller, H. F. and Kuechler, E. (1984), *Proc. Natl. Acad. Sci. USA*, **81**, 3607.

278 Branlant, C., Krol, A., Machatt, M. A., Pouyet, J., Ebel, J.-P., Edwards, K. and Kossel, H. (1981), *Nucleic Acids Res.*, **9**, 4303.

279 Stiege, W., Glotz, C. and Brimacombe, R. (1983), *Nucleic Acids Res.*, **11**, 1687.

280 Garrett, R. A. and Woolley, P. (1982), *Trends Biochem. Sci.*, **7**, 385.

281 Ehresmann, C., Ehresmann, B., Millon, R., Ebel, J.-P., Nurse, K. and Ofengand, J. (1984), *Biochemistry*, **23**, 429.

282 Li, M., Tzagoloff, A., Underbrink-Lyon, K. and Martin, N. C. (1982), *J. Biol. Chem.*, **257**, 5921.

283 Helser, T. L., Davies, J. E. and Dahlberg, J. E. (1971), *Nature (London) New Biol.*, **233**, 12.

284 Chiaruttini, C., Expert-Bezançon, A. and Hayes, D. (1982), *Nucleic Acids Res.*, **10**, 7657.

285 Van Duin, J., Ravensbergen, C. J. C. and Doornbos, J. (1984), *Nucleic Acids Res.*, **12**, 5079.

286 Thomas, J. O. and Szer, W. (1982), *Prog. Nucleic Acid Res. Mol. Biol.*, **27**, 157.

287 Subramanian, A. R. (1984), *Trends Biochem. Sci.*, **9**, 491.

288 Woese, C. R. (1981), *Sci. Am.*, **244**(6), 98.

289 Woese, C. R. and Wolfe, R. S. (1985), *The Bacteria, Vol. 8, Archaebacteria*, Academic Press, New York.

290 Woese, C. R. and Fox, G. E. (1977), *Proc. Natl. Acad. Sci. USA*, **74**, 5088.

291 Lake, J. A. (1983), *Cell*, **33**, 318.

292 Elhardt, D. and Böck, A. (1982), *Mol. Gen. Genet.*, **188**, 128.

293 Kessel, M. and Klink, F. (1980), *Nature (London)*, **287**, 250.

294 Dunn, R., McCoy, J., Simsek, M., Majumder, A., Chang, S. H., Raj Bhandary, U. L. and Khorana, H. G. (1981), *Proc. Natl. Acad. Sci. USA*, **78**, 6744.

295 White, B. N. and Bayley, S. T. (1972), *Biochim. Biophys. Acta*, **272**, 583.

296 Daniels, C. J., Gupta, R. and Doolittle, W. F. (1985), *J. Biol. Chem.*, **260**, 3132.

297 Slonimski, P., Borst, P. and Attardi, G. (1982), *Mitochondrial Genes*, Cold Spring Harbor Laboratory Monograph Series.

298 Grivell, L. A. (1983), *Sci. Am.*, **248**(3), 60.

299 Borst, P., Grivell, L. A. and Groot, G. S. P. (1984), *Trends Biochem. Sci.*, **9**, 48.

300 Attardi, G. (1985), *Int. Rev. Cytol.*, **93**, 93.

301 Richter, D. and Lipmann, F. (1970), *Biochemistry*, **9**, 5065.

302 Nagata, S., Tsunetsugu-Yokota, Y., Naito, A. and Kaziro, Y. (1983), *Proc. Natl. Acad. Sci. USA*, **80**, 6192.

303 Denslow, N. D. and O'Brien, T. W. (1979), *Biochem. Biophys. Res. Commun.*, **90**, 1257.

304 Montoya, J., Ojala, D. and Attardi, G. (1981), *Nature (London)*, **290**, 465.

305 Grosjean, H. J., de Henau, S. and Crothers, D. M. (1978), *Proc. Natl. Acad. Sci. USA*, **75**, 610.

306 De Bruijn, M. H. L. and Klug, A. (1983), *EMBO J.*, **2**, 1309.

307 Sibler, A. P., Dirheimer, G. and Martin, R. P. (1985), *Nucleic Acids Res.*, **13**, 1341.

308 Whitfeld, P. R. and Bottomley, W. (1983), *Annu. Rev. Plant Physiol.*, **34**, 279.

309 Küntzel, H., Heidrich, H. and Piechulla, B. (1981), *Nucleic Acids Res.*, **9**, 1451.

310 Subramanian, A. R., Steinmetz, A. and Bogorad, L. (1983), *Nucleic Acids Res.*, **11**, 5277.

311 Montandon, P.-E. and Stutz, E. (1983), *Nucleic Acids Res.*, **11**, 5877.

312 Hindley, J. (1973), *Prog. Biophys. Mol. Biol.*, **26**, 269.

313 Zinder, N. D. (1975), *RNA Phages*, Cold Spring Harbor Monograph Series.

314 Nathans, D., Oeschger, M. P., Polmar, S. K. and Eggen, K. (1969), *J. Mol. Biol.*, **39**, 279.

315 Lodish, H. F. (1968), *Nature (London)*, **220**, 345.

316 Fiers, W., Contreras, R., Duerinck, F., Hageman, G., Iserentant, D., Merregaert, J., Min Jou, W., Molemans, F., Raeymaekers, A., Van den Berghe, A., Volkaert, G. and Ysebaert, M. (1976), *Nature (London)*, **500**, 260.

317 Jeppesen, P. G. N., Steitz, J. A., Gesteland, R. F. and Spahr, P. F. (1970), *Nature (London)*, **226**, 230.

318 Ohtaka, Y. and Spiegelman, S. (1963), *Science*, **142**, 493.

319 Yoshida, M. and Rudland, P. S. (1972), *J. Mol. Biol.*, **68**, 465.

320 Groner, Y., Pollack, Y., Berissi, H. and Revel, M. (1972), *FEBS Lett.*, **21**, 223.

321 Lodish, H. F. (1976), *Annu. Rev. Biochem.*, **45**, 39.

322 Lodish, H. F. (1970), *J. Mol. Biol.*, **50**, 689.

323 Steitz, J. A. (1973), *Proc. Natl. Acad. Sci. USA*, **70**, 2605.

324 Min Jou, W., Haegeman, G., Ysebaert, M. and Fiers, W. (1972), *Nature (London)*, **237**, 82.

325 Robertson, H. D. and Lodish, H. F. (1970), *Proc. Natl. Acad. Sci. USA*, **67**, 710.

326 Lodish, H. F. (1969), *Nature (London)*, **224**, 867.

327 Sugiyama, T. and Nakada, D. (1968), *J. Mol. Biol.*, **31**, 431.

328 Bernardi, A. and Spahr, P. F. (1972), *Proc. Natl. Acad. Sci. USA*, **69**, 3033.

329 Inouye, H., Pollack, Y. and Petre, J. (1974), *Eur. J. Biochem.*, **45**, 109.

330 Groner, Y., Scheps, R., Kamen, E., Kolakofsky, D. and Revel, M. (1972), *Nature (London) New Biol.*, **239**, 19.

331 Kolakofsky, D. and Weissmann, C. (1971), *Nature (London) New Biol.*, **231**, 42.

332 Winter, R. B. and Gold, L. (1983), *Cell*, **33**, 877.

333 Weiner, A. M. and Weber, K. (1973), *J. Mol. Biol.*, **80**, 837.

334 Model, P., Webster, R. and Zinder, N. D. (1979), *Cell*, **18**, 235.

335 Atkins, J. F., Steitz, J. A., Anderson, C. W. and Model, P. (1979), *Cell*, **18**, 247.

336 Beremand, M. N. and Blumenthal, T. (1979), *Cell*, **18**, 257.

337 Kastelein, R. A., Berkhout, B. and van Duin, J. (1983), *Nature (London)*, **305**, 741.

338 Kastelein, R. A., Remaut, E., Fiers, W. and van Duin, J. (1982), *Nature (London)*, **295**, 35.

339 Dean, D. and Nomura, M. (1982), *Cell Nucleus*, **12**, 185.

340 Nomura, M., Gourse, R. and Baughman, G. (1984), *Annu. Rev. Biochem.*, **53**, 75.

341 Mizushima, S. and Nomura, M. (1970), *Nature (London)*, **226**, 1214.

342 Campbell, K. M., Stormo, G. D. and Gold, L. (1983), in *Gene Function in Prokaryotes* (ed. J. Beckwith, J. Davies and J. A. Gallant), Cold Spring Harbor Monograph Series, p. 185.

343 Van der Meide, Kastelein, R. A., Vijgenboom, E. and Bosch, L. (1983), *Eur. J. Biochem.*, **130**, 409.

344 Peacock, S., Brot, N. and Weissbach, H. (1983), *Biochem. Biophys. Res. Commun.*, **113**, 1018.

345 Kung, H-F., Bekesi, E., Guterman, S. K., Gray, J. E., Traub, L. and Calhoun, D. H. (1984), *Mol. Gen. Genet.*, **193**, 210.

346 Lestienne, P., Plumbridge, J. A., Grunberg-Manago, M. and Blanquet, S. (1984), *J. Biol. Chem.*, **259**, 5232.

347 Grosjean, H. and Fiers, W. (1982), *Gene*, **18**, 199.

348 Konigsberg, W. and Godson, G. N. (1983), *Proc. Natl. Acad. Sci. USA*, **80**, 687.

349 Ikemura, T. (1981), *J. Mol. Biol.*, **146**, 1.

350 Andersson, S. G. E., Buckingham, R. H. and Kurland, C. G. (1984), *EMBO J.*, **3**, 91.

351 Bennetzen, J. L. and Hall, B. D. (1982), *J. Biol. Chem.*, **257**, 3026.

352 Garel, J.-P. (1976), *Nature (London)*, **260**, 805.

353 Sheiness, D. and Darnell, J. E. (1973), *Nature*

(*London*) *New Biol.*, **241**, 265.

354 Brawerman, G. (1976), *Prog. Nucleic Acid Res. Mol. Biol.*, **17**, 117.

355 Perry, R. P. and Kelley, D. E. (1973), *J. Mol. Biol.*, **79**, 681.

356 Marbaix, G., Huez, G., Soreq, H., Gallwitz, D., Weinberg, E., Devos, R., Hubert, E. and Cleuter, Y. (1978), *FEBS Symp.*, **51**, 427.

357 Bard, E., Efron, D., Marcos, A. and Perry, R. P. (1974), *Cell*, **1**, 101.

358 Huez, G., Marbaix, G., Gallwitz, D., Weinberg, E., Devos, R., Hubert, E. and Cleuter, Y. (1978), *Nature* (*London*), **271**, 572.

359 Zeevi, M., Nevins, J. R. and Darnell, J. E. (1982), *Mol. Cell. Biol.*, **2**, 517.

360 Krowczynska, A., Yenofsky, R. and Brawerman, G. (1985), *J. Mol. Biol.*, **181**, 231.

361 Bergmann, I. E. and Brawerman, G. (1980), *J. Mol. Biol.*, **139**, 439.

362 Blobel, G. (1973), *Proc. Natl. Acad. Sci. USA*, **70**, 924.

363 Rosenthal, E. T., Tansey, T. R. and Ruderman, J. V. (1983), *J. Mol. Biol.*, **166**, 309.

364 Goto, S., Buckingham, M. and Gros, F. (1981), *Biochemistry*, **20**, 5449.

365 Blair, G. E., Dahl, H. H. M., Truelsen, E. and Lelong, J. C. (1977), *Nature* (*London*), **265**, 651.

366 Ray, B. K., Brendler, T. G., Adya, S., Daniels-McQueen, S., Miller, J. K., Hershey, J. W. B., Grifo, J. A., Merrick, W. C. and Thach, R. E. (1983), *Proc. Natl. Acad. Sci. USA*, **80**, 663.

367 Hunt, R. T., Hunter, A. R. and Munro, A. J. (1968), *Nature* (*London*), **220**, 481.

368 Lodish, H. F. (1971), *J. Biol. Chem.*, **246**, 7131.

369 Phillips, J. A., Snyder, P. G. and Kazazian, H. H. (1977), *Nature* (*London*), **269**, 442.

370 Kabat, D. and Chappell, M. R. (1977), *J. Biol. Chem.*, **252**, 2684.

371 McCarthy, T. L., Siegel, E., Mroczkowski, B. and Heywood, S. M. (1983), *Biochemistry*, **22**, 935.

372 Sarkar, S. (1984), *Prog. Nucleic Acid Res. Mol. Biol.*, **31**, 267.

373 Travers, A. (1984), *Nature* (*London*), **311**, 410.

374 Mathews, M. B., Hunt, T. and Brayley, A. (1973), *Nature* (*London*) *New Biol.*, **243**, 230.

375 Legon, S., Jackson, R. J. and Hunt, T. (1973), *Nature* (*London*) *New Biol.*, **241**, 150.

376 Gross, M. and Rabinovitz, M. (1972), *Biochim. Biophys. Acta*, **287**, 340.

377 Farrell, P. J., Balkow, K., Hunt, T. and Jackson,

R. J. (1977), *Cell*, **11**, 187.

378 Panniers, R. and Henshaw, E. C. (1983), *J. Biol. Chem.*, **258**, 7928.

379 Ochoa, S. (1983), *Arch. Biochem. Biophys.*, **223**, 325.

380 Salimans, M., Goumans, H., Amesz, H., Benne, R. and Voorma, H. O. (1984), *Eur. J. Biochem.*, **145**, 91.

381 Duncan, R. and Hershey, J. W. B. (1984), *J. Biol. Chem.*, **259**, 11 882.

382 Duncan, R. and Hershey, J. W. B. (1985), *J. Biol. Chem.*, **260**, 5493.

383 Petryshyn, R., Chen, J.-J. and London, I. M. (1984), *J. Biol. Chem.*, **259**, 14 736.

384 Salimans, M. M. M., van Heugten, H. A. A., van Steeg, H. and Voorma, H. O. (1985), *Biochim. Biophys. Acta*, **824**, 16.

385 Thomas, G., Martin-Pérez, J., Siegmann, M. and Otto, A. M. (1982), *Cell*, **30**, 235.

386 Parker, P. J., Katan, M., Waterfield, M. and Leader, D. P. (1985), *Eur. J. Biochem.*, **148**, 579.

387 Metz, D. H. (1975), *Cell*, **6**, 429.

388 Kimchi, A., Zilberstein, A., Schmidt, A., Shulman, L. and Revel, M. (1979), *J. Biol. Chem.*, **254**, 9846.

389 Kerr, I. M. and Brown, R. E. (1978), *Proc. Natl. Acad. Sci. USA*, **75**, 256.

390 Lengyel, P. (1982) in *Protein Biosynthesis in Eukaryotes* (ed. R. Pérez-Bercoff), Plenum, New York, p. 459.

391 Nilsen, T. W., Maroney, P. A. and Baglioni, C. (1982), *J. Biol. Chem.*, **257**, 14 593.

392 Salzberg, S., Wreschner, D. H., Oberman, F., Panet, A. and Bakhanashvili, M. (1983), *Mol. Cell. Biol.*, **3**, 1759.

393 Siekierka, J., Mariano, T. M., Reichel, P. A. and Mathews, M. B. (1985), *Proc. Natl. Acad. Sci. USA*, **82**, 1959.

394 Rice, A. P. and Kerr, I. M. (1984), *J. Virol.*, **50**, 229.

395 Baglioni, C., de Benedetti, A. and Williams, G. J. (1984), *J. Virol.*, **52**, 865.

396 Nilsen, T. W., Maroney, P. A. and Baglioni, C. (1983), *Mol. Cell. Biol.*, **3**, 64.

397 Ehrenfeld, E. (1982), *Cell*, **28**, 435.

398 Lee, K. A. W., Edery, I. and Sonenberg, N. (1985), *J. Virol.*, **54**, 515.

399 Lloyd, R. E., Etchison, D. and Ehrenfeld, E. (1985), *Proc. Natl. Acad. Sci. USA*, **82**, 2723.

400 Mosenkis, J., Daniels-McQueen, J., Janovec, S., Duncan, R., Hershey, J. W. B., Grifo, J. A.,

Merrick, W. C. and Thach, R. E. (1985), *J. Virol.*, **54**, 643.

401 Van Steeg, H., Kasperaitis, M., Voorma, H. O. and Benne, R. (1984), *Eur. J. Biochem.*, **138**, 473.

402 Van Steeg, H., van Grinsven, M., van Mansfeld, F., Voorma, H. O. and Benne, R. (1981), *FEBS Lett.*, **129**, 62.

403 Hershey, J. W. B. (1982), in *Protein Biosynthesis in Eukaryotes* (ed. R. Pérez-Bercoff), Plenum, New York, p. 157.

404 Schmid, H. P., Akhayat, O., de Sa, C. M., Pavion, F., Kochler, K. and Scherrer, K. (1984), *EMBO J.*, **3**, 29.

405 Perry, R. P. and Kelley, D. E. (1968), *J. Mol. Biol.*, **35**, 37.

406 Spirin, A. S. (1969), *Eur. J. Biochem.*, **10**, 20.

407 Woodland, H. (1982), *Biosci. Rep.*, **2**, 471.

408 Sands, M. K. and Roberts, R. B. (1952), *J. Bacteriol.*, **63**, 505.

409 Edlin, G. and Broda, P. (1968), *Bacteriol. Rev.*, **32**, 206.

410 Gallant, J. A. (1979), *Annu. Rev. Genet.*, **13**, 393.

411 Neidhart, F. C. (1966), *Bacteriol. Rev.*, **30**, 701.

412 Stent, G. S. and Brenner, S. (1961), *Proc. Natl. Acad. Sci. USA*, **47**, 2005.

413 Cashel, M. and Gallant, J. (1969), *Nature (London)*, **221**, 838.

414 Cashel, M. and Gallant, J. (1974), in *Ribosomes* (ed. M. Nomura, A. Tissières and P. Lengyel), Cold Spring Harbor Monograph Series, p. 733.

415 Pao, C. C., Dennis, P. P. and Gallant, J. A. (1980), *J. Biol. Chem.*, **255**, 1830.

416 Block, R. and Haseltine, W. A. (1974), in *Ribosomes* (ed. M. Nomura, A. Tissières and P. Lengyel), Cold Spring Harbor Monograph Series, p. 747.

417 Stark, M. J. R. and Cundliffe, E. (1979), *Eur. J. Biochem.*, **102**, 101.

418 Richter, D. (1980), in *Ribosomes: Structure, Function and Genetics* (ed. G. Chambliss, G. R. Craven, J. Davies, K. Davis, L. Kahan and M. Nomura), University Park Press, Baltimore, p. 743.

419 Lamond, A. I. (1985), *Trends Biochem. Sci.*, **10**, 271.

420 Sarmientos, P., Sylvester, J. E., Contente, S. and Cashel, M. (1983), *Cell*, **32**, 1337.

421 Glaser, G., Sarmientos, P. and Cashel, M. (1983), *Nature (London)*, **302**, 74.

422 Travers, A. A. (1984), *Nucleic Acids Res.*, **12**, 2605.

423 Lamond, A. I. and Travers, A. A. (1985), *Cell*, **40**, 319.

424 Nene, V. and Glass, R. E. (1983), *FEBS Lett.*, **153**, 307.

425 Silverman, R. H. and Atherly, A. G. (1979), *Microbiol. Rev.*, **43**, 27.

426 Warner, J. R., Tushinski, K. J. and Wejksnora, P. J. (1980), in *Ribosomes: Structure, Function and Genetics* (ed. G. Chambliss, G. R. Craven, J. Davies, K. Davis, L. Kahan and M. Nomura), University Park Press, Baltimore, p. 889.

427 Gorini, L. (1974), in *Ribosomes* (ed. M. Nomura, A. Tissières and P. Lengyel), Cold Spring Harbor Monograph Series, p. 791.

428 Ozaki, M., Mizushima, S. and Nomura, M. (1974), *Nature (London)*, **222**, 333.

429 Kurland, C. G. and Ehrenberg, M. (1984), *Prog. Nucl. Acid Res. Mol. Biol.*, **31**, 191.

430 Yarus, M. (1982), *Science*, **218**, 646.

431 Hirsch, D. (1971), *J. Mol. Biol.*, **58**, 439.

432 Atkins, J. F., Gesteland, R. F., Reid, B. R. and Anderson, C. W. (1979), *Cell*, **18**, 1119.

433 Mardon, G. and Varmus, H. E. (1983), *Cell*, **32**, 871.

434 Kohli, J. and Grosjean (1981), *Mol. Gen. Genet.*, **182**, 430.

435 Miller, J. H. and Albertini, A. M. (1983), *J. Mol. Biol.*, **164**, 59.

436 Bossi, L. (1983), *J. Mol. Biol.*, **164**, 73.

437 Engelberg-Kulka, H. (1981), *Nucleic Acids Res.*, **9**, 983.

438 Ryoji, M., Hsia, K. and Kaji, A. (1983), *Trends Biochem. Sci.*, **8**, 88.

439 Beier, H., Barciszewska, M., Krupp, G., Mitnacht, R. and Gross, H. J. (1984), *EMBO J.*, **3**, 351.

440 Beier, H., Barciszewska, M. and Sickinger, H.-D. (1984), *EMBO J.*, **3**, 1091.

441 Bienz, M. and Kubli, E. (1981), *Nature (London)*, **294**, 188.

442 Diamond, A., Dudock, B. and Hatfield, D. (1981), *Cell*, **25**, 497.

443 El-Baradi, T. T. A. L., Raué, H. A., de Regt, V. C. H. F., Verbee, E. C. and Planta, R. J. (1985), *EMBO J.*, **4**, 2101.

444 Yang, D., Oyaizu, Y., Oyaizu, H., Olsen, G. J. and Woese, C. R. (1985), *Proc. Natl. Acad. Sci. USA*, **82**, 4443.

445 Lee, H. and Iglewski, W. J. (1984), *Proc. Natl. Acad. Sci. USA*, **81**, 2703.

446 Siegel, V. and Walter, P. (1985), *J. Cell. Biol.*, **100**, 1913.

447 Van Noort, J. M., Kraal, B. and Bosch, L. (1985), *Proc. Natl. Acad. Sci. USA*, **82**, 3212.

448 Hillen, W., Egert, E., Lindner, H. J. and Gassen, H. G. (1978), *FEBS Lett.*, **94**, 361.

449 Hunt, T. (1985), *Nature (London)*, **316**, 580.

450 Dearsly, A. L., Johnson, R. M., Barrett, P. and Sommerville, J. (1985), *Eur. J. Biochem.*, **150**, 95.

451 Rapoport, T. A. and Wiedmann, M. (1985), *Curr. Top. Membr. Transp.*, **24**, 1.

452 Reid, G. A. (1985), *Curr. Top. Membr. Transp.*, **24**, 295.

453 Feldherr, C. M. (1985), *BioEssays*, **3**, 52.

454 Yokoyama, S., Watanabe, T., Marao, K., Ishikura, H., Yamaizumi, Z., Nishimura, S. and Miyazawa, T. (1985), *Proc. Natl. Acad. Sci. USA*, **82**, 4905.

455 De Vendittis, E., Masullo, M. and Bocchini, V. (1986), *J. Biol. Chem.*, **261**, 4445.

# Appendix:
# methods of studying
# nucleic acids

It was not possible in the main text of this book to describe the experimental basis of all the conclusions presented. The objective of this appendix is to try partially to remedy this deficiency by describing some of the most important general methods used in the study of nucleic acids. For some of the methods it was necessary to provide a full description earlier, and in these cases only a brief reference is made here for completeness. It should be stressed that the emphasis in this section is on the principles underlying the methods, and no attempt is made to provide laboratory protocols, which are now available in abundance, and to which reference will be made. It is inevitable that some of the methods dealt with here (especially those involving cloning), although pertinent to the work already described in this text, will in many cases be superseded in the lifetime of this edition. The reader wishing information about current methodologies is therefore advised to consult the most recent texts. However, at the time of writing, the cited references are recommended for more detailed information on principles [1–5] and practice [6–10].

## A.1 OCCURRENCE AND CHEMICAL ANALYSIS

### A.1.1 Chemical determination of nucleic acids in tissues

The usual approach to the determination of nucleic acids in tissues involves fractionation to remove lipids and proteins etc., followed by assay on the basis of either phosphorus, ribose or deoxyribose, or purine and pyrimidine. The tissue is treated with cold dilute trichloroacetic or perchloric acid to precipitate nucleic acid and protein, and lipid removed by extraction with organic solvent. In the procedure of Schneider [11] total nucleic acid (DNA + RNA) can be separated from protein by hydrolysis in hot acid, the liberated products being soluble (see Section 2.10.3). These latter are then assayed for ribose and deoxyribose. In the procedure of Schmidt and Thannhauser [12] the DNA and RNA are separated by alkaline hydrolysis of the latter followed by acidification (Fig. A.1).

Analysis of nucleic acid phosphorus in such fractions usually employs a colour reaction involving the formation of a phosphomolybdate

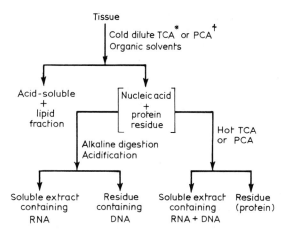

**Fig. A.1** Extraction and fractionation of nucleic acids from tissues. *TCA − trichloroacetic acid, †PCA – perchloric acid.

complex [13, 14]. Following depurination (Section 2.10.2), deoxyribose can be determined by a colour reaction with diphenylamine (Section 2.10.1) [15]. This reaction is specific for DNA. In contrast, the orcinol colour reaction employed for depurinated RNA also occurs to a

lesser extent with DNA [16]. The spectral properties of the purine and pyrimidine bases of such hydrolysed nucleic acids may also be used as a basis of nucleic acid assay. These bases have absorption maxima in the region of 260 nm (Fig. A.2). The spectral properties of double-stranded and single-stranded DNA have already been discussed (Section 2.7.1), as has the effect of pH thereon (Section 2.10.4). Protein also absorbs at 260 nm, although to a much lesser extent, dependent on the content of the aromatic amino acids phenylalanine, tryptophan and tyrosine. Because the latter two of these have an absorption maximum at 280 nm it is possible to obtain an approximate estimation of total nucleic acid and protein in a mixture by employing the following formulae based on [17]:

$$\text{Nucleic acid (mg/ml)} = 0.064\,A_{260} - 0.031\,A_{280}$$
$$\text{Protein (mg/ml)} = 1.45\,A_{280} - 0.74\,A_{260}$$

Typical values for the nucleic acid and protein contents of some rat tissues are given in Table

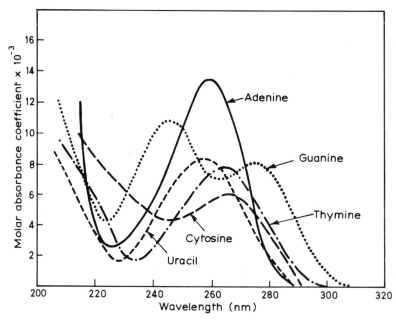

**Fig. A.2** Ultraviolet absorption spectra for purine and pyrimidine bases at pH7.

**Table A.1** Typical values for the concentrations of nucleic acids in different rat tissues.

| Tissue | RNA (mg/g protein) | DNA (mg/g protein) | RNA/DNA |
|---|---|---|---|
| Brain | 40 | 15 | 2.7 |
| Kidney | 43 | 18 | 2.4 |
| Liver | 54 | 12 | 4.5 |
| Skeletal muscle | 62 | 13 | 4.8 |
| Spleen | 58 | 58 | 1.0 |
| Thymus | 30 | 93 | 0.3 |

**Table A.2** Molar percentages of bases in RNAs from various sources.

| Type | Species | Adenine | Guanine | Cytosine | Uracil |
|---|---|---|---|---|---|
| Ribosomal[†] | | | | | |
| _E. coli_ | 16S rRNA | 25.2 | 31.6 | 22.9 | 20.3 |
| | 23S rRNA | 26.2 | 31.4 | 22.0 | 20.4 |
| | 5S rRNA | 19.2 | 34.2 | 30.0 | 16.7 |
| Rat | 18S rRNA | 22.5 | 29.2 | 26.5 | 21.8 |
| | 28S rRNA | 16.6 | 35.5 | 31.6 | 16.2 |
| | 5S rRNA | 18.3 | 32.5 | 27.5 | 21.7 |
| Messenger[*] | | | | | |
| _E. coli_ | Outer-membrane lipoprotein | 28.3 | 24.2 | 22.7 | 24.8 |
| Rabbit | _β_-globin | 23.6 | 27.2 | 23.0 | 26.2 |
| Rat | _α_-actin | 21.7 | 26.4 | 30.6 | 21.3 |
| Viral[*] | | | | | |
| Bacteriophage MS2 | | 23.4 | 26.0 | 26.1 | 24.5 |
| Poliomyelitis virus | | 29.6 | 23.0 | 23.3 | 24.1 |
| Rous sarcoma virus | | 23.8 | 28.8 | 25.3 | 22.1 |
| Tobacco mosaic virus | | 29.1 | 24.2 | 19.1 | 27.6 |

[†]Ignoring base modifications.
[*]Ignoring 3′ poly(A) tails.

A.1. These may be expressed per gram tissue wet weight, but it is often useful to express them in terms of tissue protein, which may be easily estimated [18, 19]. Values of haploid DNA content for various organisms are given in Table 3.1.

### A.1.2 Analaysis of base composition and nearest-neighbour frequency

The molar proportions of bases in nucleic acids are determined by hydrolysis, separation and spectral quantitation, as described for DNA in Section 2.4. Typical values for DNA are given in Table 2.3 and the variation in percentage GC content in different genomes can be found in Table 2.6. The principles of determining the base composition of RNA molecules are similar to those for DNA, and values for some types of RNA (actually based on sequence analysis) are given in Table A.2. Transfer RNA is not included because of the large proportion of modified bases (see Sections 2.1.2 and 9.5.2), which, in the case of rRNA (see Section 11.8.1), have been ignored in Table A.2. It may be remarked that although it was stated in Chapter 11 that rRNA is GC-rich, this need only be true in relation to the overall base composition of the genome, rather than in absolute terms. The examples of mRNAs in Table A.2 are rather arbitrary, and dramatic variations in the base compositions of different mRNAs are seen in the same organism.

The determination of nearest-neighbour frequencies (i.e. the relative frequency of occurrence of the sixteen different dinucleotides, NpM) in DNA, although to a considerable extent superseded by nucleotide sequence determination (see Section A.5), still has applications. It is described in Section 6.4.2.

### A.1.3  Estimation of the molecular weight of DNA

Methods of estimating the size of small DNA molecules (up to 10 kbp) using *electrophoresis* on acrylamide and agarose gels are described in Section A.2.1. Such methods can be accurately calibrated using DNA molecules of known sequence.

Although measurement of the *sedimentation rate* can be used to estimate the molecular weight of an RNA molecule (see Section A.2.2), this method is not readily applicable to DNA molecules because of their extremely large size and asymmetric shape. The theory behind the calculations depends on a knowledge of diffusion coefficients but these are equally difficult to obtain either by sedimentation analysis or optical mixing spectroscopy [130–132]. However, a series of empirical formulae relating sedimentation coefficient ($S^o_{20,w}$) and molecular weight (M) have been derived by comparing the sedimentation rates of molecules of known size. For example, the equation for linear, double-stranded DNA is

$$S^o_{20,w} = 2.8 + 0.00834\,M^{0.479} \quad [133, 134].$$

On sedimentation of DNA to *equilibrium* in

density gradients (usually of CsCl – see Section A.2.1) the width of the band of DNA is inversely proportional to the square root of the molecular weight. As with light scattering (see below) the result is dependent on the concentration of the DNA and a series of values must be obtained and extrapolated to zero concentration [135]. This is the most widely used, fundamental method of estimating the size of DNA molecules and is applicable to molecules of up to $10^8$ daltons (150 kbp).

*Light scattering* was widely used in the 1950s to measure molecular weights of DNA up to 5000 kbp, but only since the introduction of the laser and the production of machines to enable study of scattering at low angles has the method been extended to molecules of larger size [130–133]. However, at low angles the scattering by dust particles becomes a major problem and with large molecules different regions of the same molecule serve as scattering centres.

The above methods all suffer from the disadvantage that they are badly affected by any heterogeneity in the size of the DNA molecules. It is extremely difficult to prepare high molecular-weight DNA intact from chromosomes and these methods tend to reflect the size of the smaller molecules. The problem is even greater when attempts are made to relate molecular weight to *viscosity* as this involves subjecting the DNA solution to shear forces which tend to break long DNA molecules. However, Zimm developed a new, low-shear, viscometer consisting of a slowly rotating tube (the Cartesian diver) immersed in a DNA solution under pressure, and a relationship has been established between intrinsic viscosity ($\eta$) and molecular weight (M) of linear duplex DNA similar to that between sedimentation coefficient and molecular weight:

$$0.665 \log M = 2.863 + \log (\eta + 5) \;.[136].$$

A more important observation is that after the Zimm viscometer has been used to measure the size of very large DNA molecules the Cartesian diver rotates in the reverse direction for a while after the power has been switched off. This *viscoelastic* effect is caused by the stretched DNA molecules relaxing to the unstressed configuration. The theory derived by Zimm to explain this effect has been used to measure the size of the DNA in *Drosophila* chromosomes showing the method to be applicable to DNA molecules of molecular weight up to $79 \times 10^9$ (over $10^5$ kbp). The method has the major advantage that it measures the size of the largest DNA molecules in the solution [137].

Two 'visual' methods of measuring the size of DNA molecules involve electron microscopy and autoradiography. With *electron microscopy* the Kleinschmidt technique involves spreading the DNA on grids coated with polylysine or polystyrene-4-vinylpyridine [138]. A drop of DNA solution in 1M ammonium acetate containing 0.1 mg/ml denatured cytochrome *c* is applied to the surface of a solution of 0.2M ammonium acetate. A film of cytochrome *c* spreads across the surface, binding to the DNA which is disentangled and spread out. A grid is touched to the surface and the sample is dehydrated in ethanol. It may be stained with uranyl acetate or by evaporation of platinum on to the surface. In the latter process the grid is rotated so as to cause the metal to pile up against all surfaces of the DNA, which stands out from the grid [131].

Double-stranded DNA is readily visualized by this technique but single-stranded DNA (or RNA) molecules require the presence of formamide to prevent their collapse [139]. Even so they appear as rather wispy threads and may be stretched out to different extents. The incorporation of proteins which bind strongly to single-stranded DNA (e.g. phage T4 gene 32 protein – see Section 6.4.4) causes these regions to assume a profile thicker even than the native DNA regions and they now have a regular mass per unit length.

The presence, on the same grid, of marker DNA molecules of known length (e.g. SV40

DNA in the relaxed circular form) allows the size of an unknown DNA molecule to be found by measuring the relative contour lengths on electron micrographs. This method can readily be used with DNA molecules of up to about $10^3$ kbp, but larger molecules cannot be accommodated in a single frame.

In denaturation mapping (see Section 6.8.1) double-stranded DNA is partially denatured (usually by alkali treatment) and the unpaired bases reacted with formaldehyde to prevent their reannealing. The duplex DNA molecule now exhibits single-stranded bubbles in the (A + T) rich regions to give a characteristic map.

When DNA is labelled *in vivo* with ($^3$H)thymidine and the cells lysed on a slide to release the intact DNA molecules their position can be visualized by *autoradiography*. The radioactive DNA on the slide is covered with a layer of photographic film which is activated by the β-particles produced by the decay of the tritium to produce black grains. As with electron micrographs the size of the DNA can be obtained by measuring the contour length of the autoradiographic image. This method uses light microscopy and so the field size is much larger than with electron microscopy. Cairns applied it to the study of the *E. coli* chromosome, which is 3000 kbp long [140].

## A.2. ISOLATION AND SEPARATION OF NUCLEIC ACIDS

### A.2.1 Isolation and separation of DNA

The main problems to be overcome in isolating high molecular weight chromosomal DNA from bacterial [20] or eukaryotic cells [21] are the removal of protein (especially deoxyribonucleases) and RNA, and the avoidance of mechanical shearing. For eukaryotic cells a combination of pronase or protease K, a detergent and phenol extraction can be used directly to disrupt the cells and remove protein. For the cells of Gram-negative bacteria such as *E. coli* gentle lysis can be effected using lysozyme. Phenol extraction is used widely in the isolation of both DNA and RNA, which are retained in the aqueous phase while denatured protein collects at the interface between this and the phenol phase, which contains the lipids. RNA may be removed from the DNA by using pure pancreatic ribonuclease, or by isopycnic ultra-centrifugation in a gradient of CsCl, which is, in any case, a useful step for further purification.

In *isopycnic ultracentrifugation* a density gradient is established which encompasses the buoyant densities of the molecules to be sep-arated. Molecules will assume a position on the gradient corresponding to their own buoyant density, after which no further separation can occur. In this method an equilibrium is est-ablished, and the method is sometimes referred to as 'equilibrium ultracentrifugation' to distinguish it from the rate-zonal method (see Section A.2.2). As already mentioned (Section 2.7.3), the buoyant density of double-stranded DNA varies with the mole fraction of (G + C). In general, however, RNA has a much higher buoyant density (about 1.9 g/ml) than double-stranded DNA (about 1.7 g/ml), which, in turn, is higher than that of protein (about 1.3 g/ml). Single-stranded DNA is of slightly higher buoyant density than double-stranded DNA (e.g. for a double-stranded DNA of buoyant density 1.703 g/ml, the value for the single-stranded form is 1.717 g/ml).

Isopycnic ultracentrifugation is also used to separate small bacterial plasmids from chromo-somal DNA. The basis of separation in this case is not differing GC contents (these are, in general, similar) but the fact that the chromo-somal DNA is linear (normal methods do not extract the large circular bacterial chromosome intact) whereas the plasmids are circular. This difference is exploited by the addition of saturating concentrations of the intercalating fluorescent dye, ethidium bromide. The inter-calation requires that the DNA strands be forced apart, with concomitant decrease in buoyant

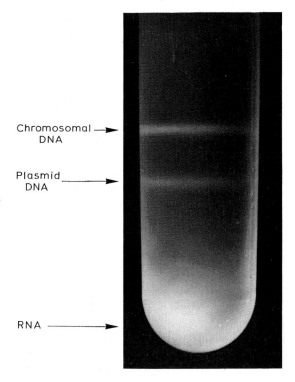

Chromosomal → DNA

Plasmid DNA

RNA →

**Fig. A.3** Separation of closed-circular DNA of plasmid pBR322 from *E. coli* chromosomal DNA by isopycnic ultracentrifugation in a CsCl density gradient in the presence of ethidium bromide. The band marked 'chromosomal DNA' may also contain nicked plasmid DNA molecules.

density and partial unwinding of the double helix. This latter process is hindered in the closed-circular plasmid molecules with the result that they bind less ethidium bromide and have a higher buoyant density than the linear chromo-somal (and open-circular and linearized plasmid) molecules (Fig. A.3). Before attempting to separate plasmid and chromo-somal DNA it is advisable to enrich the preparation in plasmid DNA. The most fre-quently employed method for achieving this in preparative work is the use of an alkaline pH during lysis and extraction [22]. This denatures the DNA, causing strand-separation in the case of linear molecules. On neutralization the latter will not renature, whereas the closed-circular plasmid DNA will and can be separated from the insoluble denatured DNA. For small-scale isolation of plasmid DNA, a boiling step is usually used in place of the treatment with alkali [23].

Ethidium/CsCl gradients are also used in the purification of small animal viruses (e.g. SV40) which contain cyclic genomes. In this case the initial purification involves precipitation of most of the high molecular weight cellular DNA adsorbed onto a precipitate of sodium dodecyl sulphate to leave behind the so-called Hirt extract [166].

Although most preparations of chromosomal DNA consist of sheared fragments, these do not in general separate on isopycnic ultracentrifugation because the fragmentation is random and the base compositions of individual fragments are generally similar. Exceptions do occur in the case of regions of repetitive DNA in eukaryotes (see Sections 3.2.5 and 7.7) which may be generated as quite homogeneous species in relatively large amounts if the fragments are of the appropriate size. Should the base composition of such a repeated DNA be markedly different from the average base composition of the chromosomal DNA, resolution from the bulk of the latter will occur. Such DNA was originally given the descriptive name 'satellite DNA' (see Section 3.2.5(b)), a designation which is sometimes applied to all repetitive DNA, regardless of base-composition or function. Such loose terminology is to be discouraged. It is also possible to isolate the tandemly repeated rDNA and histone genes of certain organisms by this method [24].

These exceptions aside, separation of DNA by isopycnic ultracentrifugation is largely confined to the different chromosomal species one might find in a given cell. Precise fragments of such DNAs can be generated by restriction endonucleases (see Section A.6) and the separation of such fragments is a common requirement in recombinant DNA manipulations. Most

frequently this is achieved by horizontal *agarose gel electrophoresis* [25]. This separates different linear DNA molecules according to size (the mobility is inversely proportional to the $\log_{10}$ of the molecular weight), as the charge per unit length of different molecules is identical and the larger molecules will be retarded by the gel matrix (Fig. A.4). The percentage of agarose can be increased to separate smaller-sized DNA fragments, but for very small species *polyacrylamide gel electrophoresis* (which is often used preparatively for fragments up to 1 kb) must be used [26]. This method is also employed in rapid nucleic acid sequencing methods (Section A.5), in which case 8M urea is included to ensure denaturation.

As agarose gel electrophoresis is frequently used to analyse preparations of plasmid DNA which may be slightly nicked, it is worth mentioning that the closed-circular supercoiled form of the plasmid migrates more rapidly than the less compact relaxed-circular form produced by single-stranded nicks. Any linearized plasmid molecules generally migrate at an intermediate position between these two, a result that is perhaps a little unexpected (Fig. A.4).

Nucleic acids may be precipitated with ethanol, 2.5 volumes in the presence of 0.3 M sodium acetate, being frequently employed.

### A.2.2 Isolation and separation of RNA

In principle the methods used to extract RNA from cells are generally similar to those for DNA. Degradation of RNA by ribonuclease is a greater threat than that of deoxyribonuclease degradation of DNA, especially in the case of mRNA, but there are a variety of methods to counter this. The major separation problem is posed by the large number of different RNA species in eukaryotes, and to study a particular species it is necessary to separate it from the others.

One preliminary that may be useful in particular cases is *subcellular fractionation* [27].

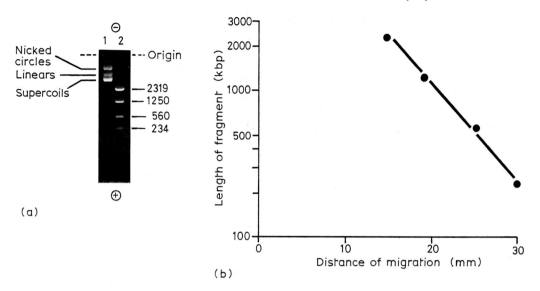

**Fig. A.4** Agarose gel electrophoresis of DNA. (a) Separation of: 1, different forms of DNA of plasmid pBR322; 2, fragments of DNA (lengths indicated in kbp) derived from plasmid pBR322 by double-digestion with restriction endonuclease *Bam* HI and *Bgl* I; (b) Plot of length of DNA fragment (log scale) against distance of migration (linear scale) of data from (a) 2, illustrating linear relationship.

Thus the nuclei may be removed by low-speed centrifugation (700 g for 5 min) after gentle disruption of the cells in the presence of sucrose (to preserve the osmolarity), the mitochondria may be removed by further centrifugation (10 000 g for 10 min), and the ribosomes separated from soluble RNA species such as tRNA by ultracentrifugation (e.g. 100 000 g for 90 min). It may of course be necessary further to purify the subcellular fraction in which one is interested, and methods for nuclei [28], nucleoli [29], mitochondria [30], and ribosomes [31] are available. It may be remarked that the most appropriate method for isolating an organelle, such as the nucleus, for subsequent RNA extraction (where purity is of greatest importance) may be quite different from the method which will yield the most biologically active material [32].

One of the most widely used methods of separating RNA molecules is *rate-zonal ultra-centrifugation* employing *sucrose density*

*gradients* [33]. In this method the separation of RNA is based primarily on size (in principle, molecular shape can also have an influence, but with RNAs differences in shape make no significant contribution to the separation). It should be emphasized that the density range of sucrose solutions does not extend to that of RNA species and, in contrast to isopycnic ultracentrifugation involving CsCl gradients (Section A.2.1), no equilibrium is reached. The main purpose of the sucrose gradient is to prevent diffusion of the zones of individual species that are separated from the narrow zone applied initially to the top of the gradient (Fig. A.5). Isopycnic ultracentrifugation in gradients of dense sucrose (often employing a discontinuous or 'step gradient') may be used to separate 'free' ribosomes from those bound to the membrane of the endoplasmic reticulum (see Section 11.7.1) [34].

Rate-zonal ultracentrifugation in sucrose density gradients is also used to separate

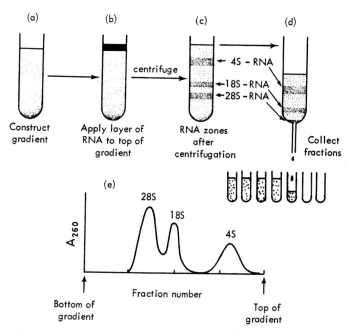

**Fig. A.5** Rate zonal centrifugation of RNA through a sucrose density gradient. A sucrose density gradient is constructed in a centrifuge tube (a) and the RNA solution applied as a layer on top (b). During ultracentrifugation the main components of the RNA separate into zones, primarily on the basis of molecular weight (c). These zones may be recovered by puncturing the bottom of the tube and collecting different fractions in separate tubes (d). The separated RNAs may be visualized and quantitated by measurement of the absorbance at 260 nm (e). Steps (d) and (e) may be conveniently combined by pumping the gradient through the flow-cell of a recording spectrophotometer.

different size classes of polyribosomes (see Section 11.4) and to separate the large and small subunits of ribosomes after their dissociation (Section 11.8.1).

Rate-zonal ultracentrifugation is clearly inappropriate for the separation of different species of *tRNA*, which are broadly similar in size. Methods involving separation on the basis of charge and hydrophobicity differences (e.g. chromatography on BD-cellulose and RPC5) may be employed for this purpose [35].

The major problem to be overcome in the isolation of *mRNA* is purification from other species of RNA, which are approximately 20 times more abundant. Affinity chromatography using oligo(dT)-cellulose [73] or poly(U)-Sepharose [74] – materials that bind the poly(A)

tails of mRNAs – is the basis of the separation of mRNA from other species of RNA. The minimum size of poly(A) tail required for binding to poly(U)-Sepharose is smaller than that for oligo(dT)-cellulose (about 6–10 residues compared with 15 residues), and although the latter material is in more widespread use, the former has applications to mRNAs having very short poly(A) tails. A mixture of different mRNAs can be separated to a certain extent on the basis of size, either by sucrose density gradient centrifugation or, on a smaller scale, by agarose gel electrophoresis – basically as for DNA but generally under denaturing conditions (see Section A.2.1). Both agarose gel electrophoresis and polyacrylamide gel electrophoresis have been applied to other types of RNA [36, 37]. The isolation of individual mRNA species is usually achieved by recombinant DNA techniques (see Section A.7).

## A.3 HYBRIDIZATION OF NUCLEIC ACIDS

The denaturation and renaturation of homoduplex DNA molecules has been dealt with in Section 2.7, where the effects of base composition, temperature and ionic strength were mentioned. The application of DNA renaturation to determining the copy number of different portions of eukaryotic genomes ($C_0t$ *value analysis*) has been described in Section 3.2.5. An important technique in molecular biology is the formation of heteroduplexes between different DNA molecules or between DNA and RNA. The application of electron microscopic visualization of such heteroduplexes formed in solution to the determination of the positions of introns (*R-loop mapping*) has already been described (Section 9.2.1), and the electron microscopic analysis of heteroduplexes can be most useful in a variety of other situations. However, for most routine work heteroduplexes are detected by the radioactivity of one of their components (methods for the radioactive labelling of nucleic

acids – 'probes'– are described in Section A.4). It is operationally convenient to have one component of the hybridizing pair immobilized on a solid 'paper' support in order that the hybrids may be easily visualized by autoradiography.

Before the introduction of heteroduplex formation involving immobilized nucleic acids analysis involved the time-consuming inconvenience of multiple solution-hybridizations, when the experimental protocol involved identifying which component of DNA fractionated by agarose gel electrophoresis hybridized to a particular radioactively-labelled probe. To overcome this problem Southern [38] devised a method of transferring the DNA from the fragile gel to nitrocellulose paper by means of capillary action (see Fig. A.6). The DNA must be denatured with alkali before transfer, and afterwards it is baked at 80°C (in a vacuum oven to prevent it igniting) in order to fix the DNA permanently to the nitrocellulose. This method of transfer is commonly called *Southern blotting*, after its originator. The nitrocellulose paper is usually hybridized with the radioactive DNA probe in a minimum volume of solution in a sealed polythene bag at 68°C and conditions of relatively high ionic strength that promote heteroduplex formation. (If 50% formamide is

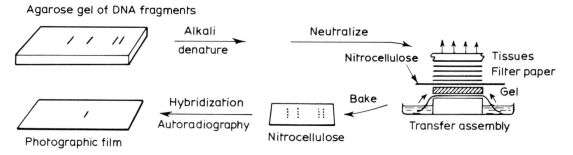

**Fig. A.6** Southern blotting. Fragments of DNA are fractionated by electrophoresis in agarose, the DNA denatured in alkali and, after neutralization, transferred (blotted) from the gel to nitrocellulose paper by capillary action at relatively high ionic strength in the transfer assembly shown. The transferred DNA is baked on to the nitrocellulose, after which it can be hybridized in solution with a radioactive probe and the hybridizing bands visualized by autoradiography. In the hypothetical example illustrated only one of the original four bands has hybridized to the probe.

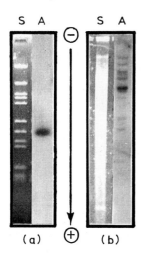

S    A         S    A

(a)          (b)

**Fig. A.7** Examples of results of Southern blotting experiment. (a) Cloned DNA from a mouse/bacteriophage lambda recombinant (cf. Fig. A.19), or (b) chromosomal DNA from mouse liver, were digested with (different) restriction endonucleases, subjected to electrophoresis on agarose gels and Southern blotting performed as in Fig. A.6. Hybridization was to a ($^{32}$P)-labelled mouse actin cDNA clone (cf. Fig. A.18). S: ethidium bromide stained gels photographed under illumination with ultraviolet light. A: autoradiographs of the nitrocellulose.

included the hybridization is performed at 42°C.) The membrane is washed under suitable conditions (see below), dried, and subjected to autoradiography. This method is most easily applied to detecting which restriction fragment of a larger piece of cloned DNA contains sequences homologous to a particular probe (see Fig. A.7(a)). However, the sensitivity of the method is such that it can detect individual fragments in the continuum produced by restriction enzyme digestion of total genomic DNA (see Fig. A.7(b)), and in fact the method was originally applied to uncloned genomic and sub-genomic DNAs. Such genomic Southern blotting has become a routine tool in screening for human genetic disorders by the detection of restriction fragment length polymorphisms (see

also Section A.8.3), for full details of which the reader is directed elsewhere [39].

The conditions of washing nitrocellulose membranes (nylon membranes may also be used) depend on the degree of homology between the two members of the heteroduplex. (The homology need not be 100% as hybridization may be between a variety of related but non-identical sequences.) When the two members of the heteroduplex are identical the washing is usually performed 12°C below the melting temperature ($T_m$), which in a solution 0.2 M in $Na^+$ is related to the mole percentage of $(G + C)$ by the equation:

$$T_m = 69.3 + 0.41\,(G + C) \quad [40]$$

The effect of ionic strength ($\mu$) on the melting temperature is given by the equation [41]:

$$T_{m2} - T_{m1} = 18.5\,\log_{10}\frac{\mu_2}{\mu_1}$$

Thus, to allow for the fact that the Tm of duplex DNA decreases by 1°C per 1–1.5% mismatch [42, 129] the 'stringency' of the hybridization can be decreased by lowering the temperature of the wash and/or by increasing the ionic strength of the washing buffer.

Southern's method for the transfer of DNA from gels to paper has been extended to RNA, where it has acquired the somewhat ridiculous name of *Northern blotting*. (There is a third point on the compass, 'Western blotting', the name sometimes given to the electrophoretic transfer of proteins from polyacrylamide gels to nitrocellulose.) The methodology of Northern blotting is different from that of Southern blotting because RNA does not bind to nitrocellulose paper under conditions in which DNA does, and is hydrolysed by alkali. To overcome this problem, transfer can be made to diazobenzyloxymethyl (DBM) paper, which binds RNA (and DNA) [43]. RNA can, in fact, be transferred to nitrocellulose paper from agarose gels provided that it is denatured (e.g. by formaldehyde) [44].

Hybridization of immobilized nucleic acids on paper supports is not restricted to material transferred from agarose gels. It can be used for detecting recombinant DNA in bacterial colonies or bacteriophage plaques after transfer from petri dishes (see Section A.7.2) or for multiple samples of total cellular RNA applied directly to the nitrocellulose (the so-called *Dot-blot* technique).

## A.4 METHODS OF LABELLING NUCLEIC ACIDS

There is a variety of circumstances in which the detection of DNA is only possible if the DNA is labelled in some way. Furthermore, certain techniques require DNA labelled specifically at one end. By far the most common method of achieving this is with radioactivity (usually $^{32}$P; less commonly $^{35}$S or $^{3}$H), and the discussion below will be exclusively in terms of, this. However, non-radioactive labelling methods are available. Currently the most prominent of these involves the use of biotin-labelled derivatives of dNTPs [45], which are detected by using an anti-biotin antibody. A variety of systems has been devised to visualize this, for example with a second antibody coupled to an enzyme such as horseradish peroxidase that catalyses a colour reaction.

### A.4.1 General labelling methods

When the objective of labelling a nucleic acid is merely that of allowing its detection, methods that cause the incorporation of radioactivity throughout the molecule are suitable, as well as the specific end-labelling methods described in Section A.4.2. Historically, RNA and DNA were labelled from ($^{3}$H), ($^{14}$C) or ($^{32}$P)-labelled precursors *in vivo*, but methods involving the incorporation of radioactive label *in vitro* are more convenient for recombinant DNA work and produce material of higher specific activity.

The most widespread method of labelling

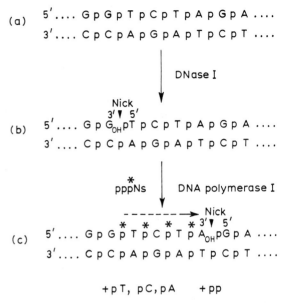

**Fig. A.8** Nick translation. Double-stranded DNA(a) is treated with a low concentration of pancreatic DNase producing occasional nicks with 3′-OH groups (b) on which DNA polymerase can commence polymerization. The use of $\alpha$ ($^{32}$P)dNTPs produces a radioactive phosphodiester bond p̂ and the 5′ → 3′ exonuclease action of the enzyme successively cleaves further non-radioactive phosphodiester bonds which are replaced by radioactive ones (c). In the course of this replacement the position of the nick undergoes vectorial translation, as indicated.

double-stranded DNA is by the method of '*nick translation*' (Fig. A.8) using $\alpha$($^{33}$P)dNTPs [46]. In this method nicks are introduced into the DNA with DNase and the 5′-phosphate ends produced can serve as substrates for the 5′ to 3′ exonucleolytic activity of *E. coli* DNA polymerase I, which at the same time repairs the gaps by addition of the $\alpha$($^{33}$P)dNTPs to the 3′-OH end of the 'nick' (see Section 6.4.2). Another replacement synthesis method involving T4 DNA polymerase has certain advantages over nick-translation, but is less widely used [47]. Single-stranded labelled DNA probes may be prepared by copying the DNA after cloning into the single-stranded phage vector M13 [48] (see Section A.5.1). More recently a method of producing highly radioactively-labelled RNA copies of a piece of DNA has been introduced, which involves transcription with bacteriophage SP6 RNA polymerase from the DNA cloned into a special vector that puts it under the control of a bacteriophage SP6 promoter [49]. Another method of producing very high specific activity DNA probes also involves copying but does not require preliminary subcloning. The Klenow fragment of DNA polymerase is employed and random oligonucleotides are used as primers with the denatured DNA as template [50].

It is possible to label mRNA directly *in vitro* by certain end-labelling methods (see Section A.4.2) but a copying method is generally preferred because it produces a more stable product of higher specific activity. This involves the use of $\alpha(^{32}P)$dNTPs and reverse-transcriptase to produce a $(^{32}P)$-labelled single-stranded cDNA copy (see Section A.7).

The labelling of tRNA *in vitro* is important in sequence determination in cases where insufficient material is available for spectral analysis and it is not possible to label with $^{32}P$ *in vivo*. As well as enzymic end-labelling methods (see below) chemical labelling using $(^{3}H)$borohydride has been employed [65, 66].

## A.4.2 End-labelling methods

These are particularly important in preparing DNA for sequence analysis by the method of Maxam and Gilbert (see Section A.5) and for certain other analytical purposes (see Sections A.8.1 and A.8.2). For these purposes the label must be covalently incorporated and the products must be homogeneous. The DNA to be labelled is generated by cleavage with a restriction endonuclease (see Section A.6) and the method of labelling depends on whether flush ends, or 'sticky' ends (in which there is a 5'-P or 3'-OH overhang) are produced. Ends with a 5'-P overhang are easily labelled by either of two

methods (Fig. A.9(a) and (b)), in one of which the 5'-P is removed by alkaline phosphatase and replaced by the $\gamma$-phosphate of $\gamma(^{32}P)$ATP in a reaction catalysed by bacteriophage-T4 polynucleotide kinase (see Section 4.7.1) [51]. In the other the Klenow fragment of *E. coli* DNA polymerase I (Section 6.4.2) is usually used to fill the gap in the strand with the recessed 3'-OH end, using an appropriate $\alpha(^{32}P)$dNTP, although bacteriophage T4 DNA polymerase may also be used (Fig. A.9(b)). (The 5' to 3' exonuclease of the complete *E. coli* enzyme would result in degradation of the template for this reaction.) Thus these two methods complement each other in labelling different strands.

The polynucleotide kinase method may be adapted for labelling the 5'-P of flush ends by effecting local denaturation of the latter, but cannot be applied efficiently where there is a 3'-OH overhang. The 3' to 5' exonuclease and 5' to 3' polymerase activities of the Klenow fragment of DNA polymerase I can be used to label flush ends or 3'-OH protruding ends by a replacement synthesis method (Fig. A.9(c)), although bacteriophage T4 DNA polymerase (which has a more powerful 3' to 5' exonuclease activity and also no 5' to 3 exonuclease) is more effective [52]. Currently, the most effective method of end-labelling 3'-OH overhangs for sequencing involves the use of terminal transferase [53] (see Section 6.4.2(e)), but with $\alpha(^{33}P)2',3'$dideoxyATP (see Section A.5.1(b)) to ensure addition of only a single nucleotide (Fig. A.9(d)).

A number of methods for end-labelling are available when the nucleic acid is only to be used as a probe. With DNA, extension of 3'-overhanging ends using terminal transferase and $\alpha(^{32}P)$dNTP [53] is an alternative to the use of DNA polymerase. For mRNA, extension of the 3'-end with *E. coli* poly(A) polymerase can be used for labelling with $\alpha(^{32}P)$ATP, or $\alpha(^{32}P)$ cordecypin (3'-deoxy ATP) if the addition of only a single nucleotide is required [54]. An alternative to covalent incorporation as a

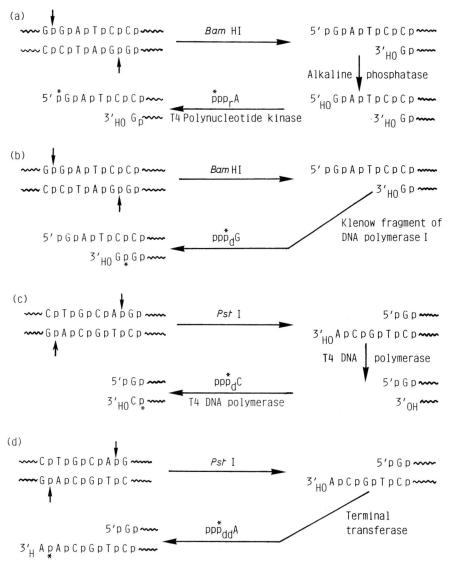

**Fig. A.9** Methods of end-labelling DNA fragments. (a) Use of polynucleotide kinase and $\gamma(^{32}\text{P})$ ATP for 5′-labelling of cohesive end with 5′-overhang produced by restriction endonuclease such as *Bam* HI; (b) use of Klenow fragment of DNA polymerase I and $\alpha(^{32}\text{P})$dNTPs for 3′-labelling of cohesive end with 5′-overhang produced by restriction endonuclease such as *Bam* HI. In this case any of the $\alpha(^{32}\text{P})$dNTPs could have been used for labelling if the other non-radioactive dNTPs had been included to allow complete fill-in; (c) use of T4 DNA polymerase and $\alpha(^{32}\text{P})$dNTPs for 3′-labelling of cohesive ends with 3′-overhang produced by restriction endonuclease such as *Pst* I. An excess of the other three non-radioactive dNTPs prevents the exonuclease activity proceeding further; (d) use of terminal transferase and $\alpha(^{32}\text{P})$ddATP for 3′-labelling of cohesive ends with 3′-overhang produced by restriction endonuclease such as *Pst* I. Methods (a), (c) and (d) may also be applied to blunt ends produced by restriction endonucleases such as *Sma* I, and method (d) can be applied to ends with a 5′-overhang. N.B. The other end of the fragment (not illustrated) will also be labelled if it is similar, and a second restriction endonuclease digestion or strand-separation will be required to obtain a fragment of DNA labelled at only one end.

method of end-labelling, useful in certain restricted circumstances, is hybridization of a labelled fragment to suitable single-stranded regions. In the case of eukaryotic mRNAs, ($^3$H)poly(U) can be hybridized to the poly(A) tails[55].

## A.5 DETERMINATION OF NUCLEIC ACID SEQUENCES

### A.5.1 Determination of DNA sequences

There are two main methods currently in use for sequencing large fragments of DNA. These methods (that of Maxam and Gilbert [56] and that of Sanger [57]) both involve the generation of a ladder of fragments of different sizes but with one common end. The basis of these methods is to generate specific sets of radio-actively labelled fragments, each set terminating at a particular base (in the ideal case), and the use of high-resolution polyacrylamide gels [58] allows fragments differing by only a single nucleotide to be resolved and the sequence to be deduced. In both cases cloned DNA is normally used, although the direct application of the Maxam and Gilbert method to genomic DNA has been described [59]. A third method of sequencing (the 'Wandering Spot' or 'Mobility Shift' method), although no longer used for large DNA molecules, still finds application to small oligonucleotides (especially those produced synthetically – Section A.8), which are difficult to sequence by the other methods. An outline of earlier methods, now superseded, can be found in [81].

### (a) *The method of Maxam and Gilbert* [56]
The essence of this method is the chemical fragmentation of either single- or double-stranded DNA by base-specific reactions. The reactions most commonly used are those absolutely specific for guanine residues and for cytosine residues respectively, and those that are specific only for purines or pyrimidines, res-

pectively. Cleavage at guanine residues is effected by methylation with dimethyl sulphate at the N7 position, leading to instability of the glycosidic linkage which is then hydrolysed by piperidine, followed by $\beta$-elimination of both phosphates from the sugar (Fig. A.10(a)). Purine nucleotide linkages are hydrolysed with acid (see Section 2.10.2), again followed by piperidine treatment. Pyrimidine residues are hydrolysed by hydrazine (Fig. A.10(b)), a reaction which, in the case of thymine, can be inhibited by 2 M NaCl, thus allowing specific cleavage at cytosine residues. (Under the conditions normally used, cleavage does not occur at 5-methylcytosine residues, causing a gap in the ladder.)

Conditions of chemical cleavage are generally adjusted to try to obtain a single scission per DNA molecule. Even so, each scission would produce two fragments. In order to visualize on polyacrylamide gels only fragments of increasing length emanating from one end of the DNA, a single end of the DNA must be radioactively labelled using one of the methods described in Section A.4.2. Such methods, in fact, generally label both ends of a piece of duplex DNA, and fragments with a single-labelled end are generated by restriction endonuclease digestion of this labelled DNA followed by separation on polyacrylamide gels. Less commonly, single end-labelled molecules are obtained by strand separation of the DNA. An example of a Maxam-Gilbert sequencing gel and its interpretation is given in Fig. A.11.

As sequence data accumulate, the use of computer programs to handle them and identify restriction endonuclease sites greatly facilitates sequencing by the Maxam-Gilbert method [60].

### (b) *The Sanger dideoxy method* [57]
The essence of this method is the primed synthesis of partial copies of the DNA to be sequenced, with random base-specific premature termination of the copying producing ladders of different length fragments terminating in each of

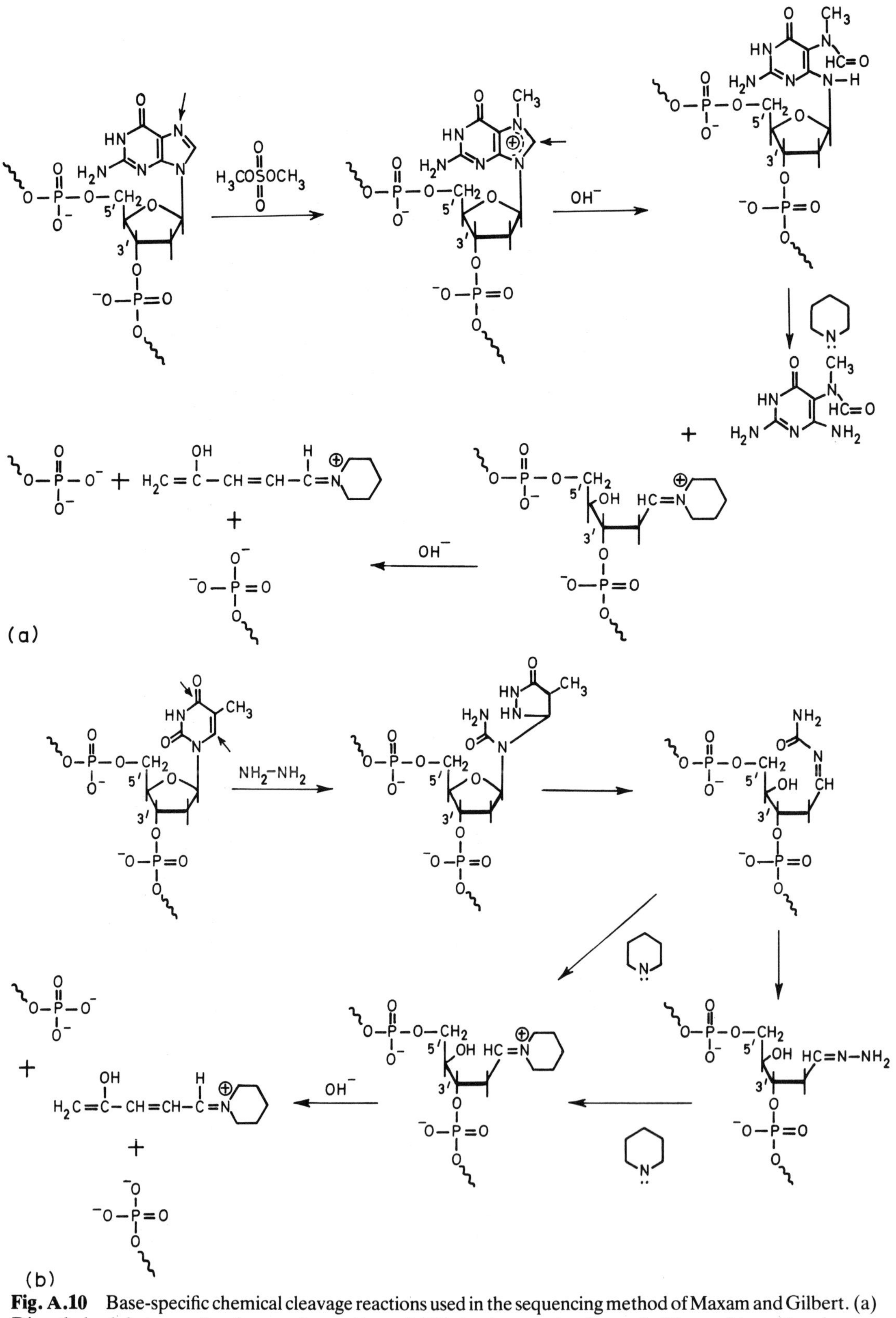

**Fig. A.10** Base-specific chemical cleavage reactions used in the sequencing method of Maxam and Gilbert. (a) Dimethyl sulphate reaction for guanine residues; (b) Hydrazine reaction for pyrimidine residues (thymine illustrated). For further details see text and reference [56], from which this figure is adapted, with permission.

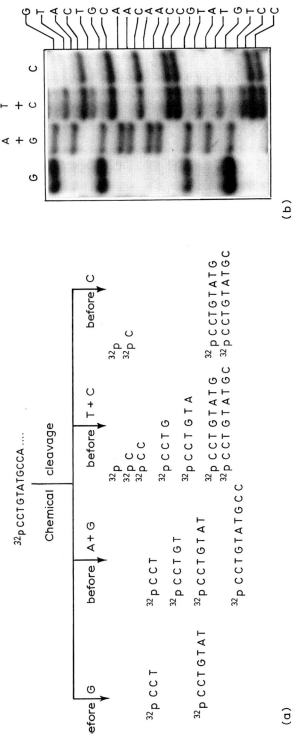

(b)

(a)

**Fig. A.11**  Example of DNA sequence determined by the method of Maxam and Gilbert. (a) The ³²P-end-labelled fragments derived from chemical cleavage of 5'-end-labelled DNA. These are set out in order of increasing size to allow comparison with (b) the autoradiograph of the part of the gel on which these (and longer fragments) were separated on the basis of size. It can be seen that bands appearing in both G and G + A tracks are assigned to G whereas those appearing only in the G + A track are assigned to A. C and T are similarly differentiated. (This is, in fact, a simplified example, and the extreme 5'-fragments are normally quite difficult to read.)

the four different bases. The copying is catalysed by the Klenow fragment of *E. coli* DNA polymerase I, which requires a primer and that the DNA to be copied be in the single-stranded form. (The Klenow fragment, lacking the 5′ to 3′ exonuclease, is used to prevent attack on the 5′-end of the primer.) The copies are made radioactive by inclusion of an $\alpha(^{32}P)$dNTP or a $\alpha(^{35}S)$dNTP in the reactions, and base-specific chain termination is effected by the addition of the appropriate 2′3′dideoxynucleotide triphosphate (ddNTP), which, lacking a 3′-OH, cannot be extended further. An example of a Sanger sequencing gel and its interpretation is given in Fig. A.12.

Initially, the widespread application of this method was prevented by the requirement for a single-stranded template and the need for a separate oligonucleotide primer for each piece of DNA to be sequenced. However, these limitations were overcome when the single-stranded bacteriophage, M13, was adapted for use as a sequencing vector [61, 62]. The DNA to be sequenced may now be subcloned into one of a variety of adjacent sites in the double-stranded replicative form of the bacteriophage vector, and can be sequenced in the easily prepared single-stranded DNA using oligonucleotide primers complementary to regions of the bacteriophage DNA flanking the cloning sites (see Fig. A.17b). This development has led to a dramatic increase in the use of the Sanger method. The random subcloning of fragments from the DNA to be sequenced has resulted in the need for computer programs to order the sequences obtained [63].

(c)  *The wandering-spot method* [64]
The Maxam-Gilbert and Sanger methods described above, although extremely powerful for large restriction fragments of DNA, are not well suited to smaller oligodeoxynucleotides. The sequence of these can be determined following end-labelling (Section A.4.2) by the chromatographic separation of a non-base-specific partial

enzymic digest of the oligodeoxynucleotide in which all the partial digestion products having the same labelled end are represented. Digestion is by venom or spleen phosphodiesterase and two-dimensional thin-layer homochromatography on DEAE-cellulose is employed. The shortening of the oligodeoxynucleotide by a single nucleotide causes a change in mobility that is characteristic of the nucleotide removed and which allows the sequence to be read (Fig. A.13).

### A.5.2  Determination of RNA sequences

Before the advent of rapid DNA sequencing methods considerable effort was expended on developing techniques for sequencing RNA. The availability of partially or totally base-specific endonucleases allowed employment of approaches formally rather similar to those used for sequencing proteins, but requiring different methods for separating and analysing the oligoribonucleotides. Although the sequencing of a DNA copy must now be the method of choice for sequencing any RNA, this approach may not be applicable to small RNAs or oligonucleotides, and in any case does not identify modified nucleotides. In these cases (especially for tRNAs) direct methods (including the 'wandering-spot' method, mentioned above) are still employed [67].

Rapid sequencing methods involving base-specific cleavage and separation of fragments on polyacrylamide gels have been introduced for 5′-end-labelled RNA [68, 69]. The cleavage is effected enzymically with ribonucleases (see Section 4.3), the following reactions being currently in use:

| | |
|---|---|
| Ribonuclease $T_1$ | : cleaves 3′ of G |
| Ribonuclease $U_2$ | : cleaves 3′ of A |
| Ribonuclease Phy M | : cleaves 3′ of A and U |
| *B. cereus* ribonuclease | : cleaves 3′ of U and C |

In other respects the method and the interpret-

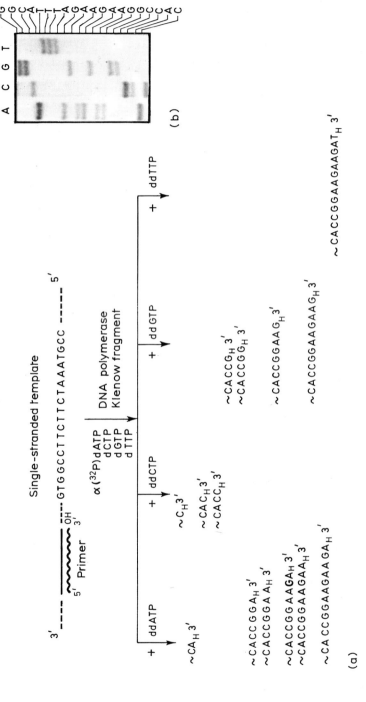

**Fig. A.12**  Example of DNA sequence determined by the dideoxy method of Sanger. (a) The various products of extension of the primer (~) by DNA polymerase are set out in order of increasing size, as in Fig. A.11. The terminating ddNTP is represented as $N_H$. Chains are labelled internally by the incorporation of the $\alpha(^{32}P)$- or $\alpha(^{35}S)dATP$; (b) autoradiograph of the part of the gel on which these (and longer) species were separated on the basis of size. Note that the sequence obtained is the complement of the original priming strand.

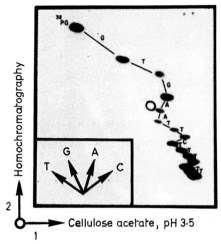

**Fig. A.13** Example of oligodeoxynucleotide sequence determined by the 'wandering spot' method. A partial pancreatic DNase digest of GGTGAATTCTTTCTT labelled with $^{32}$P at its 5'-end was subjected to two-dimensional separation. The inset shows the effect of removal of the four different 3'-end deoxynucleotides on the mobility of the resulting fragment. The identity of the 5'-nucleotide is determined by comparison with standards. The circle adjacent to the second A indicates the position of a marker dye. Adapted from [142], with permission.

ation of the results are formally similar to that of Maxam and Gilbert, described in A.5.1(a).

## A.6 RESTRICTION MAPPING OF DNA

When studying specific regions of genomic or cloned DNA it is necessary to have some sort of 'map' to differentiate one area from another. There are two different types of map. A *genetic map* can be constructed (in suitable organisms) in which the positions of functional genes contributing to particular phenotypes can be inter-related by studying the frequency of genetic recombination between normal and mutated alleles. A *physical map*, on the other hand, can be constructed in the absence of genetic data or knowledge of any function the DNA might have, and uses purely physical techniques. These include the formation of heteroduplexes

between standard DNA molecules and the DNA under investigation, and restriction mapping, the subject of this section.

A restriction map indicates the positions along a piece of DNA at which there are recognition sites for particular type II restriction endonucleases (see Section 4.5.3). The method in its simplest form involves digesting the DNA separately and in combination with different restriction endonucleases, separating the resulting fragments by agarose (or sometimes polyacrylamide) gel electrophoresis (see Section A.2.1), visualizing these under ultraviolet light after staining with ethidium bromide, and estimating their size in relation to standards [70]. If there are not too many recognition sites for the enzymes it is possible to deduce the position of these as illustrated in Fig. A.14.

It can be appreciated that for enzymes with many recognition sites it is both difficult and tedious to construct a restriction map by this procedure. In the case of large pieces of DNA (e.g. those cloned in bacteriophage lambda or cosmid vectors – Sections A.7.2(b) and (c)) most restriction endonucleases will have many recognition sites and it is difficult to circumvent this problem by choice of enzyme. In any case it is often useful to have a restriction map for enzymes that cleave a piece of DNA at many positions. An alternative approach was devised involving end-labelling of the DNA and *partial* digestion with *single* restriction endonucleases [71]. In essence the strategy is similar to that for Maxam and Gilbert DNA sequencing using a single base-specific reaction, except that analysis of the larger fragments usually employs the standard agarose gels. Although the pattern of stained fragments is complex, the autoradiograph of the dried gel reveals a ladder of fragments, the increasing sizes of which represent increasing distances from the labelled end. This is illustrated schematically in Fig. A.15. A modification of this strategy, for application to DNA cloned into bacteriophage or cosmid vectors, replaces end-labelling by

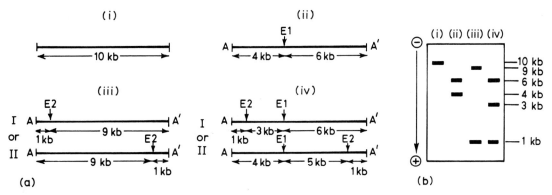

**Fig. A.14** Example of restriction mapping. It is desired to determine the relative positions of the recognition sites for two restriction endonucleases, E1 and E2, along a hypothetical 10 kb fragment of DNA. (a) If the 10 kb fragment (i) is cleaved by E1 to give 4 kb and 6 kb fragments the two ends of the DNA can be distinguished as A and A′ (ii). This then defines two possible positions of cleavage (I and II) of a second enzyme E2, which generates 1 kb and 9 kb fragments (iii). The two alternatives I and II may be distinguished by the different sized fragments they predict in the case of double-digestion with E1 and E2 (iv); (b) diagram of the results of agarose gel electrophoresis of (i)–(iv). The result in lane (iv) is only consistent with alternative I, and hence (iv)/I represents an (albeit limited) restriction map of the 10 kb fragment.

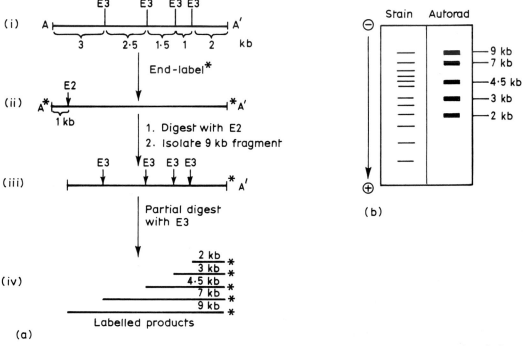

**Fig. A.15** Example of partial digestion restriction mapping. (a)(i) The objective is to determine the relative positions of the four recognition sites for the restriction endonuclease E3 on the hypothetical DNA fragment of Fig. A.14. This is end-labelled (ii), digested with enzyme E2, which is known to cleave close to one end, and the larger (9 kb) fragment isolated (iii). This is subjected to partial digestion with E3 giving a mixture of end-labelled fragments together with non-labelled fragments not illustrated (iv); (b) diagram of the results of electrophoresis of the partial digestion products. The distances of the E3 recognition sites from end A′ can be read off the autoradiography ladder using molecular size markers (cf. Fig. A.4(b)). This allows the restriction map of (a)(i) to be deduced.

hybridization of a specific ($^{32}$P)-labelled oligonucleotide to either the right or left cohesive end of the molecule [72]. This avoids the need for secondary cleavage and gel fractionation to produce a fragment of DNA labelled at one end.

## A.7 CLONING DNA

### A.7.1 The principles

The study of individual mRNAs and of particular regions of genomic DNA were hampered historically by two problems. The first was the difficulty of devising chemical methods to isolate one particular nucleic acid species from, perhaps, tens of thousands of chemically similar distinct species. The second was that, even if a chemical separation method had been found, the amount of nucleic acid isolated would in most cases have been too small to allow study. The cloning of DNA employs a biological, rather than a chemical, strategy to overcome both these problems. If one wishes to isolate a single bacterium from a million others one spreads them out on an agar plate so thinly that each is separated from its neighbour. Each bacterium gives rise to a visible colony which is, in fact, composed of identical bacteria derived from the division of the original one. Such an assembly of genetically identical individuals is called a *clone*. If the bacterium that gives rise to that clone harbours a particular mutation that distinguishes it from the wild type one could say that one had cloned the mutant DNA. Even though this mutant DNA originally existed as a single chromosome in minute amount, one could obtain large quantities of the DNA by using the colony to seed a bulk liquid culture. The basis of the purification and isolation of an mRNA or a chromosomal DNA segment using a biological strategy is to introduce them into individual bacteria so that they may be cloned in an analogous way to the hypothetical mutant chromosomal DNA described above.

There are several problems to be overcome before this strategy can be realized in practice. The foreign DNA must be presented as part of a molecule that can replicate along with the bacterial (or other host) chromosome. Such vehicles for introducing the DNA are called *vectors* (see Section A.7.2), and the chimeric DNA produced by the insertion of the DNA to be cloned is a *DNA recombinant*. The DNA must be inserted into the vector, or the mRNA be converted into a form in which it can be inserted (Section A.7.2), and the vector must be introduced into the bacterium, the process of *transformation* (Section 3.5.3). Methods must be available to distinguish the transformed bacteria from the untransformed ones, and bacteria containing recombinant DNAs from those containing non-recombinant DNAs from those containing non-recombinant vector (*selection*). Finally (see Section A.7.3) it is necessary to be able to identify which one (if any) of thousands of clones contains the particular DNA of interest (*screening*). The ways in which these problems may be overcome are illustrated below, almost exclusively in relation to *E. coli* as a biological host. For details of cloning in other bacteria, or in eukaryotic animal or plant cells, the reader is directed elsewhere [3].

### A.7.2 The construction of recombinants and their introduction into bacteria

As already stated, the cloning of a piece of foreign DNA in *E. coli* requires its introduction into a suitable vector. There are three main types of vector in use for primary cloning – plasmids, bacteriophage lambda and cosmids – and these will be discussed in turn, together with the way that DNA for cloning into them is prepared. It is worth mentioning at the outset that in all these vectors foreign DNA is inserted at specific sites by the use of the highly-specific restriction endonucleases. Although it is possible in theory to clone foreign DNA into vectors using cleavage by non-specific endonucleases such as DNase I, in actual practice it was the discovery of restriction endonucleases that made it practicable both to clone and subsequently analyse DNA. It was not only the specificity of restriction endonucleases that was important, but the fact that many of them give rise to DNA fragments with cohesive ends that can be far more easily joined with DNA ligase than can fragments with blunt ends.

### (a) *Cloning in plasmid vectors and bacteriophage M13*

As already described in Section 3.7, plasmids are self-replicating double-stranded circular DNA molecules found in certain bacteria. Especially useful are those that occur in multiple copies per cell. Natural plasmids have been extensively modified to serve as cloning vehicles, the most well known of which, pBR322 [75], is illustrated in Fig. A.16(a). The essential features are an origin of replication, unique restriction endonuclease sites into which foreign DNA may be inserted without disabling the plasmid, and selectable markers for the presence of the plasmid in a bacterial cell. These latter are genes endowing resistance to the antibiotics tetracycline and ampicillin (a derivative of penicillin). For clarity they are abbreviated as $Tc^R$ and $Ap^R$ in Fig. A.16, although the $Ap^R$ gene is referred to by its correct genetic designation (*bla*) in Chapter 7.

In the simplest case a fragment of foreign DNA generated from a larger molecule by digestion with a restriction endonuclease (e.g. *Bam* HI – see also Fig. A.9) is ligated with DNA ligase (see Section 6.4.3) to pBR322 DNA that has been digested with the same enzyme, generating the chimeric molecule shown in Fig. A.16(b). To enable the plasmid DNA to enter the *E. coli* cells (*transformation*) these are pretreated with $CaCl_2$ which, together with a short heat shock, appears to act by altering the structure of the cell wall [76]. Transformed cells are selected by plating the bacteria on agar containing the appropriate antibiotic (ampicillin in this case) which will prevent the growth of cells that do not contain any plasmid. Although re-ligation of the linearized plasmid to itself (but

not to foreign DNA) can be prevented by pretreating it with alkaline phosphatase, in many cases some of the transformed bacteria will contain plasmid with no insert. Colonies of such bacteria may be distinguished by the fact that they will also grow on tetracycline, unlike those containing foreign DNA inserted in the *Bam* HI site. (A 'replica' plate of the bacterial colonies is prepared for testing on tetracycline; the colonies of interest corresponding to those that will not grow on tetracycline being maintained on the ampicillin 'master' plate.) If, however, it had been necessary to clone into a restriction site such as *Eco* RI, which is not located in an antibiotic-resistance gene, the presence of inserts can only be confirmed by physical analysis of the plasmid DNA.

The problem of identifying transformed bacteria containing plasmids with inserted DNA is overcome in the more recently developed pUC vectors [77], which also have the advantage of containing a large number of useful cloning sites (Fig. A.17(a)). These vectors contain the ampicillin-resistance gene of pBR322 to allow selection of transformants, in addition to the *lac* operator and promoter regulatory regions and the N-terminal portion ($\alpha$-fragment) of the $\beta$-galactosidase gene. If the plasmid *lac* gene is induced with IPTG (isopropylthio-$\beta$-galactoside) in an *E. coli* strain with a $\beta$-galactosidase gene lacking the N-terminal portion (actually carried on the F'episome), complementation of the two peptides will occur giving enzymically-active $\beta$-galactosidase. This can be detected by the blue colour produced when 'X-gal' (5-bromo-4-chloro-3-indolyl-$\beta$-D-galactoside) is present as a substrate. In the pUC vectors the extreme N-terminal region of the $\alpha$-fragment is interrupted by a region of multiple cloning sites. This does not alter the reading frame or result in inactivation of the $\alpha$-fragment of $\beta$-galactosidase. However, in most cases the insertion of foreign DNA into these sites will cause inactivation (so-called 'insertional inactivation'). Thus if the transformed bacteria are plated on agar

(a)

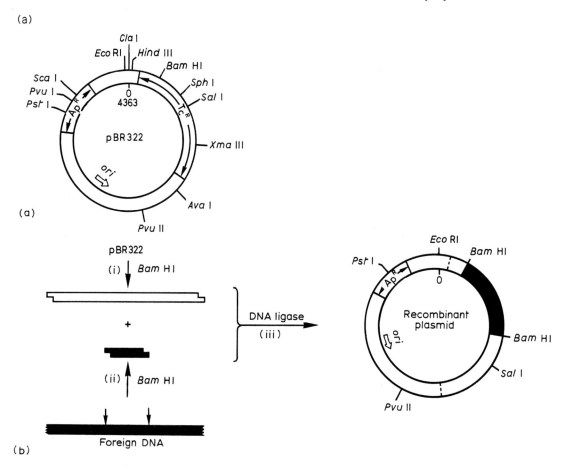

(b)

**Fig. A.18** Cloning cDNA. Poly(A)$^+$ mRNA is primed with oligo(dT) (i) and a cDNA copy synthesized (ii). After and tetracycline resistance (Tc$^R$), the origin of replication (*ori*), and the positions of some of the restriction endonucleases that have a single recognition site in this DNA: (b) to clone into the *Bam* HI site of pBR322 the DNA is digested with this enzyme (i), as is the foreign DNA (ii), and these are ligated together (iii), among the products of the ligation being the desired recombinant plasmid shown. Note that although this latter still retains the gene for ampicillin resistance, the gene for tetracycline resistance has been inactivated by the insertion of the foreign DNA. (N.B. For clarity, not all the original restriction sites are shown in the recombinant.)

containing IPTG and X-gal (as well as ampicillin) the white colonies will generally contain plasmid with inserts, whereas the blue colonies will generally not.

In Fig. A.17(b) is illustrated the cloning vector, M13mp18, a member of the family of M13mp vectors, for which the *lac* fragment of the pUC vectors was initially engineered [78].

Although this is technically a single-stranded phage vector, the replicative form illustrated is formally analogous to a plasmid (the ability to form 'plaques' replaces the ampicillin resistance). It can be used in a similar way, although, as already mentioned (Section A.5.1), the main purpose of cloning DNA fragments into M13 vectors is to facilitate the determination of their

(a)                                                                    (b)

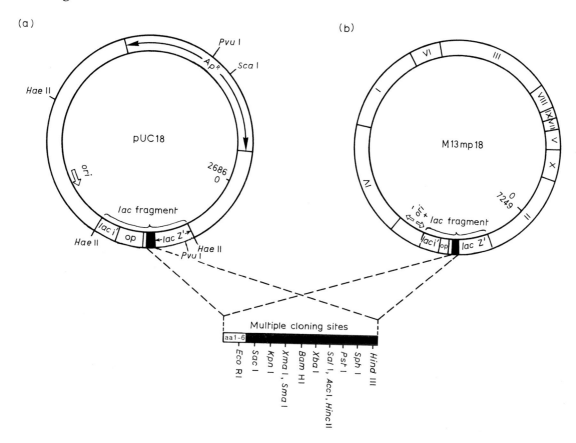

**Fig. A.17**   pUC18 and M13mp18. (a) The plasmid vector pUC18. This is derived from a fragment of pBR322 into which a fragment of the *lac* gene has been inserted at *Hae* II sites. The multiple cloning sequence interrupts the *lac* Z' (truncated β-galactosidase) gene. The *Pvu* I and *Sca* I sites of the original Ap^R gene are shown, but note that the *Pst* I site has been removed so that there is only a single *Pst* I site, as for the other sites in the multiple cloning region; (b) the double-stranded replicative form of single-stranded phage vector M13mp18. The *lac* fragment has been inserted into the phage M13, the genes of which are indicated I-X.

sequences by the Sanger dideoxy method. This is possible because the single-stranded form of the recombinant can easily be generated and, as cloning with different enzymes is into the same region, the same primers (corresponding to flanking regions of the *lac* Z' gene) can be used with different recombinants.

One of the major uses of plasmid vectors has been for *cloning cDNA* – DNA copies of mRNA. We shall therefore now briefly describe this procedure as it applies to polyA-containing

(poly(A)^+) mRNA (Fig. A.18). The poly(A)^+ mRNA is usually purified on an affinity column of oligo(dT)-cellulose (see Section A.2.2) and reverse-transcribed using retroviral reverse transcriptase (see Section 6.9.10) with oligo(dT) hybridized to the poly(A) tail of the mRNA as a primer. This results in a single-stranded cDNA, complementary to the mRNA. For a number of years the usual procedure for synthesizing the second strand relied on the self-priming that occurs when (after removing the mRNA by

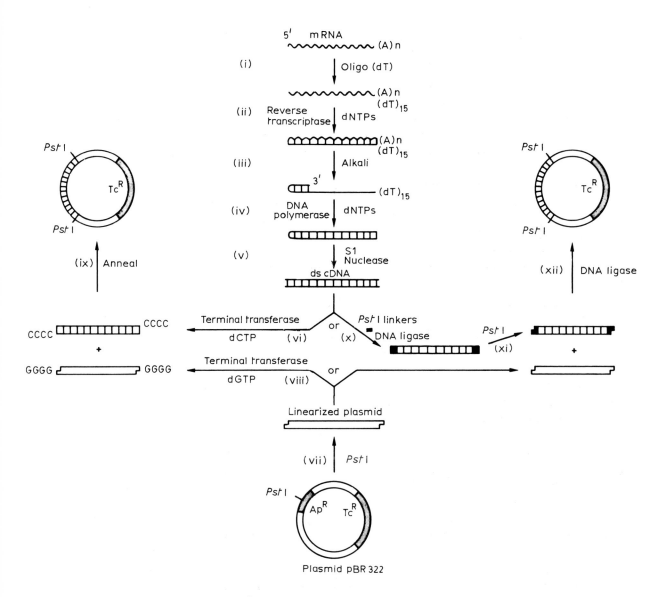

**Fig. A.18** Cloning cDNA. Poly(A)$^+$ mRNA is primted with oligo(dT) (i) and a cDNA copy synthesized (ii). After removing the mRNA (iii) this self-primes second-strand synthesis (iv) and the loop and any ragged ends are then removed with nuclease S1(v). One strategy is now to tail the double-stranded cDNA with dCTP using terminal transferase (vi), add this to linearized plasmid (vii) that has been tailed with dG (viii) and allow these to anneal (ix) to form the recombinant. Alternatively, linkers can be ligated onto the cDNA (x), these cleaved with the appropriate restriction endonuclease (xi) and then ligated (xii) to the plasmid linearized with the same enzyme. Other linkers may be used for step (x) but for tailing (vi and viii) a site such as *Pst* I, that on cleavage gives the necessary 3′-OH overhang for terminal transferase is required (cf. Fig. A.9(d)).

alkali or heat denaturation) the single-stranded cDNA folds back on itself. *E. coli* DNA polymerase I was generally used to generate a double-stranded cDNA by copying the first strand, and S1 nuclease (Section 4.2.1) had to be used to destroy the loop and digest any protruding ends on the double-stranded cDNA.

The ends of such double-stranded DNA must be modified so that they can anneal to the termini of a plasmid linearized by restriction endonuclease digestion. One way of achieving this is to 'blunt-end' ligate 'linkers' to them using DNA ligase (Section 6.4.3). Such linkers are double-stranded oligodeoxynucleotides containing a restriction endonuclease site which is cleaved after the ligation. This generates 'cohesive ends' that may be specifically ligated to the corresponding cohesive ends in the digested plasmid. (This is more efficient than blunt-end ligation directly into a blunt-ended site. Blunt-end ligation is much less efficient *per se* than ligation of fragments with cohesive ends. However, the high molar concentration of the small linkers can drive this reaction in a manner impossible with the larger molecules.) An alternative strategy – perhaps more widely used – has been to add homopolymeric tails of complementary oligonucleotides to the 3'OHs in both cDNA and linearized plasmid using the enzyme terminal transferase (see Section 6.4.2(e)). The cDNA and vector are 'tailed' with complementary oligonucleotides (generally C and G), and, after allowing these to anneal to one another, the hybrid can be used to transform *E. coli* without the need for prior ligation (which is subsequently effected by bacterial DNA repair mechanisms). *Pst*I is usually used as the cloning site in this strategy: not only does it generate the necessary 3'OH overhang for terminal transferase, but its recognition site and cleavage position (CTGCA'G) are such that 'tailing' with G will regenerate it.

The use of S1 nuclease in cDNA cloning has the disadvantage that it destroys the extreme 5' end of the cDNA. For this reason it is increasingly being superseded by methods that allow production of full-length cDNA copies of mRNAs. The problem is clearly how to prime the second strand. One solution has been to 'tail' the single-stranded cDNA with oligo(dC), allowing the second strand to be primed with oligo(dG) [79]. However, a more sophisticated solution, with an additional advantage of producing full-length cDNA with high efficiency, has been devised by Okayama and Berg [80]. The basic strategy is to use a plasmid vector tailed with oligo(dT) at one end as primer for the first cDNA strand, and, after tailing the free 5' end of the cDNA with oligo(dC), replacing the other end of the vector (which was also tailed with C) with an adaptor DNA previously tailed with oligo(dG) to allow completion of the second strand. The interested reader should consult the original reference [80] for the full details of how this is achieved in practice. However, one aspect of this rather sophisticated method has been adapted to allow almost complete second-strand copying of a simple complex of mRNA and first strand cDNA. This is the use of *E. coli* RNase H (Section 4.3.4) to nick the mRNA in the hybrid, and DNA polymerase (the Klenow fragment or the enzyme from bacteriophage T4) to perform replacement synthesis of DNA from the 3'OHs generated.

Finally, it should be mentioned that cDNA copies of poly(A)⁻ mRNAs can be obtained by polyadenylating with *E. coli* poly(A) polymerase (see Section A.4.2), or, more usually, by priming with a random mixture of oligonucleotides (cf. the use of oligonucleotides to prime labelling [50] – Section A.4.1), although in this case DNA corresponding to the 3' end of the mRNA may be lacking. Furthermore, in the method of *primer-extension* a single-stranded oligonucleotide fragment derived from the 5' end of a cDNA clone that is less than full length may be used as a primer in an attempt to obtain a cDNA clone corresponding to the area that is lacking (cf. the use of this method in mapping transcripts [82] – Section A.8.1.)

(b) *Cloning in bacteriophage lambda and cosmid vectors*

The plasmid and bacteriophage M13 vectors described above are most suitable for the cloning of relatively small pieces of DNA (usually up to about 3 kb), either as cDNA copies of mRNAs or as portions of larger pieces of cloned DNA (subcloning). However, other vectors are required when the objective is to prepare a eukaryotic genomic library – a collection of clones in which the whole DNA of the genome of a particular eukaryotic organism or cell is represented. The main reason for this is the large size of eukaryotic genomes. For example, to be 99% certain that a library of 15 kb inserts is representative of the $3 \times 10^6$ kbp mouse genome, $10^6$ clones are required. The reason for the unsuitability of plasmid vectors for such a library is the limited efficiency of the transformation of calcium-treated cells by DNA. This is true even with inserts of 1 or 2 kb, but the efficiency falls dramatically as the size of insert increases.

The use of derivatives of bacteriophage lambda as vectors for constructing genomic libraries rests on the fact that a large portion (approximately 16 kb) of the central region of the 49 kbp bacteriophage genome codes for lysogenic and other functions that are not required for the lytic infective cycle of the bacteriophage (see Sections 3.6.8 and 10.1.5). Thus if this region is replaced by foreign DNA the bacteriophage will still be viable. Furthermore, after constructing DNA recombinants of bacteriophage and foreign DNA it is possible to 'package' them into phage particles with high efficiency *in vitro*, using an extract from infected cells containing the necessary enzymes [83]. Cells may then be infected with the packaged recombinant, resulting in plaques. The size of DNA that can be cloned depends on the construction of the vector (see below), but is governed by the fact that there is a size restriction of 38.5–52 kb for the DNA to be packaged into phage heads.

In general, bacteriophage lambda vectors can be classed as 'insertion' or 'replacement' vectors. In *insertion vectors* (e.g. Charon 16A [84] – Fig. A.19(a)) there is a single unique restriction site in the central non-essential region into which the foreign DNA is ligated. The non-essential region is engineered so that the size of the vector DNA is towards the lower limit of packageability, and this type of vector can therefore accept foreign DNA up to a size that will produce a recombinant near the upper limit of packaging. In Charon 16A this is approximately 10 kb. Different vectors employ different genetic devices for identifying recombinants [85], but it will suffice to mention that Charon 16A employs as a cloning site a portion of the *E. coli lac* region containing the complete β-galactosidase gene (*lac* Z) and upstream operator and promoter regulatory elements, allowing colour selection of recombinant phage plaques using X-gal when plated on *E. coli* with a deletion in the *lac* Z gene (cf. above).

Lambda *replacement vectors* are generally employed when it is necessary to clone larger pieces of foreign DNA. An advanced vector of this type, EMBL 3 [86], is illustrated in Fig. A.19(b). This has been engineered so that restriction endonuclease digestion with *Bam* HI, *Eco* RI or *Sal* I cuts out a non-essential piece of DNA of 14 kb, which can be replaced by foreign DNA (up to about 23 kb) with compatible cohesive ends. To prevent re-ligation of the lambda 'centre' when using replacement vectors, the lambda 'arms' can be purified first. However, vectors such as EMBL 3 allow simpler methods to be employed (see legend to Fig. A.19(b)). It may be mentioned that the compatible cohesive ends generated in the fragments of genomic DNA need not be (and generally are not) produced using the same restriction endonuclease as used in cleaving the vector. Thus, cleavage of DNA with the enzyme *Sau* 3AI (which has the 4-base recognition sequence 'GATC) produces identical cohesive ends to cleavage with *Bam* HI (which has the 6-base recognition sequence G'GATCC). If high

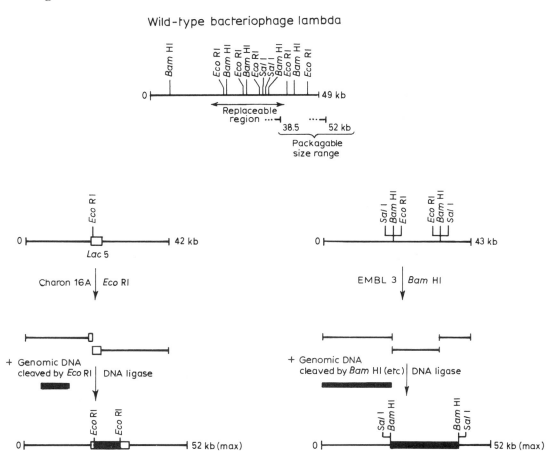

**Fig. A.19** Cloning vectors derived from bacteriophage lambda. (a) Charon 16A, an example of an insertion vector. *Lac* 5 is a fragment of the *E. coli lac* region containing all of the β-galactosidase (*lac* Z) gene and associated control elements. Only the restriction site for *Eco* RI is shown; (b) EMBL 3, an example of an insertion vector. The central fragment may be digested with *Eco* RI to prevent it being re-ligated to the '*Bam* arms'. If genomic DNA digested with *Sau* 3A rather than *Bam* HI is used (see text) the recombinant will generally lack the *Bam* HI sites indicated. However, the insert can be recovered using *Sal* I. Only the desired products of ligation are illustrated.

molecular weight chromosomal DNA is subjected to partial digestion with *Sau* 3AI one is more likely to obtain clonable fragments of any particular area of the genome than by complete (or partial) digestion with *Bam* HI, for which there are fewer recognition sites.

Because of the large size of the introns in many eukaryotic genes, a particular lambda genomic clone may not contain the whole of a given gene of interest. It is then necessary to perform the operation known as *chromosome walking*, in which a fragment of unique sequence DNA near the extremity of the clone is used as a probe to re-screen the genomic library for an overlapping

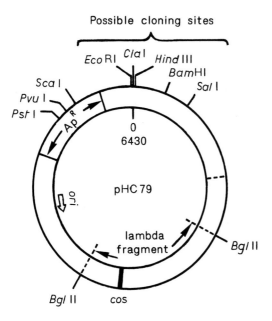

**Fig. A.20**  Cosmid vector pHC79. This simple cosmid vector contains the ampicillin resistance gene, origin of replication and useful cloning and manipulation sites of pBR322, together with a *Bgl* II fragment of circularized lambda DNA containing the *cos* site recognized by the lambda packaging system, and another fragment of DNA. The cleaved vector is ligated to large restriction fragments of genomic DNA under conditions that allow the formation of linear assemblies containing insert flanked by two molecules of linearized cosmid to provide the two *cos* sites needed for packaging *in vitro*.

clone containing the adjoining area of the genome.

An alternative to chromosome walking (or a way of speeding it up) is to construct the genomic library in a *cosmid vector* (e.g. pHC79 [87] – Fig. A.20) which can accommodate larger pieces of DNA than lambda replacement vectors. Cosmids are based on bacteriophage lambda but retain only the cohesive ends (*cos*) of the bacteriophage, two of which are sufficient to allow packaging *in vitro*. This may allow insertion of up to 40 kb of foreign DNA, but the packaged recombinant DNA is no longer infective after it penetrates the host cell. For this reason the cosmid vector has an origin of replication and antibiotic resistance gene that allows it to behave like a plasmid, and selection of transformed colonies can be made on antibiotic plates. Further discussion of cloning in cosmids is beyond the scope of this book.

### A.7.3  The identification of recombinant DNA clones

The objective of cloning DNA is, as already stated, to isolate a particular piece of DNA or mRNA copy in a pure state. However, the methods described above will, in general, produce a large number of clones which together represent many different mRNA copies or genomic segments. The solution to the problem of how to identify clones of the DNA of interest depends on its nature. We shall consider the most common case, that of DNA which encodes a protein, or part thereof. One might be able to identify this protein immunologically, by the biological properties it possesses, by its amino acid sequence or merely by its electrophoretic behaviour in polyacrylamide gels. In order to use these properties to identify the clone of interest it is usually necessary to express the genetic information in some way (random sequencing to search for cDNA clones corresponding to abundant proteins of known sequence [88] is an exception here). In general, direct expression may not occur and indirect expression is necessary.

By 'indirect expression' is meant the expression of the mRNA after selection by DNA-RNA hybridization. The cloned DNA is immobilized on a nitrocellulose filter, and an mRNA preparation (consisting of a mixture of species) from an appropriate tissue or cell is passed through [90]. Only the mRNA complementary to the particular immobilized DNA clone will form a hybrid and be retained on the filter. It can then be released (washed off at low ionic strength) and translated in a cell-free system [89]. This is known as *hybrid-selection* or *hybrid-release translation*. Alternatively, the denatured cloned DNA can be mixed with the mRNA preparation, the mixture added to the cell-free system and the loss of a particular product detected (*hybrid-arrest translation* [91]). In both cases the product may be identified immunologically or electrophoretically as appropriate (or, in particular cases, by its biological activity [92]). Screening a large number of clones by these methods is a formidable task, and it is usual to try to reduce this number by some means. For example, one can fractionate the mRNAs initially on sucrose density gradients and clone only those mRNAs in the relevant size range. If an antibody is available enrichment of the mRNA by immunological isolation of polysomes containing nascent peptides of the protein can be attempted [104]. Preliminary screening is often employed by comparing the hybridization of ($^{32}$P)-labelled single-stranded cDNA probes derived from mRNA populations enriched with or depleted of the species of interest. (This is possible if the mRNA is differentially expressed in certain tissues or circumstances.) This '*colony hybridization*' is due to Grunstein and Hogness [93] who devised a method of transferring duplicate copies of colonies from a 'master' agar plate to nitrocellulose filters (replica plating) on which the plasmid DNA could be immobilized after lysis of the cells (Fig. A.21). If the sequence of the protein, or part thereof, is known it is possible to screen the colonies with a ($^{32}$P)-labelled *oligonucleotide probe* corresponding to the nucleotide sequence predicted by a part of the amino acid sequence [94]. (Indeed, such an oligonucleotide may be used to prime cDNA synthesis as an initial cloning strategy [103].) Because of the degeneracy of the genetic code (see Section 11.2.5) such nucleotide sequences cannot be predicted with certainty. It is necessary to select sequences containing amino

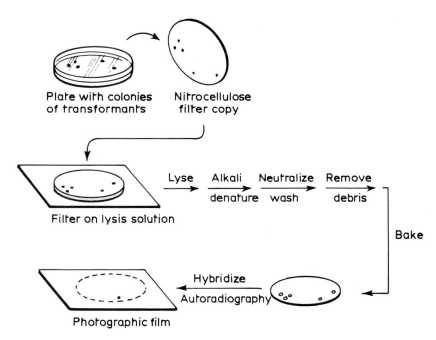

**Fig. A.21** Colony hybridization. Plasmid DNA in transformed bacteria is immobilized on nitrocellulose and hybridization of this to a suitable $^{32}$P-labelled probe (see text) allows identification of colonies carrying plasmids with inserts of interest. In the diagram the number of colonies has been greatly decreased for clarity.

acids of minimal codon degeneracy, to consider what bias in codon usage might exist in the organism from which the mRNA is derived (Section 11.10.3), and to employ sufficient different oligonucleotides of sufficient length to allow for mistakes. Nevertheless, this method is extremely powerful and has been increasingly employed as protein microsequencing technology has improved.

Identification of cloned DNA by direct expression of the protein product has the major advantage that thousands of clones can be screened directly. This is particularly useful for low-abundance mRNAs for the protein product of which no sequence data are available but against which antibodies have been raised. The problem with direct expression in *E. coli* is that a bacterial promoter and Shine and Dalgarno sequence (see Section 11.5.1) are required. In the case of bacterial genes this is no problem, and genetic selection can often be effected by using suitable auxotrophs [95], but eukaryotic DNAs lack such sequences. Although analogous genetic selection in eukaryotic cells can be effected in favourable instances where mutants are available [96], the more general solution to this problem has been to put the eukaryotic DNA under the influence of bacterial regulatory signals. In fact the reader may have realized that if an incomplete eukaryotic cDNA clone (i.e. lacking sequences corresponding to the 5′ untranslated portion of the mRNA) is cloned into the pUC or M13 vectors (Section A.7.2(a)), there is a one-in-six chance of this being inserted in the right orientation and with maintenance of the reading frame, and of this leading to the synthesis of a fusion protein comprising the N-terminal amino acids of β-galactosidase and a part of the protein corresponding to the cDNA clone. Such an expressed fusion protein would most likely have antigenic determinants that would be recognized by a polyclonal antibody to the eukaryotic protein.

If the objective is to screen a cDNA library by expression of a fusion protein in *E. coli* it is

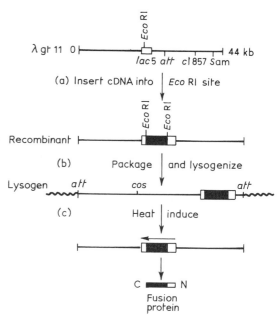

**Fig. A.22** Expression vector λgt11. Relevant differences from the vector Charon 16A (Fig. A.19) are indicated: the *att* site and other functions required for lysogenic integration into *E. coli* (absent in Charon 16A) are present, *cI*857 indicates a *ts* mutant in the *cI* gene responsible for maintaining lysogeny, *S*am indicates an amber mutation in the *S* gene coding for one of the cell lysis functions. (a) The vector is cut with *Eco* RI and ligated to a library of cDNA to which *Eco* RI linkers (cf. Fig. A.18) have been attached; (b) the recombinants are packaged *in vitro* and transfected into an *E. coli* strain carrying a mutation (*hfl*A) that ensures lysogeny rather than lysis; (c) subsequent heat induction causes release of the lambda recombinants, but these do not lyse the bacteria so that, in one case out of six, colonies contain large amounts of fusion proteins comprising the N terminal portion of the β-galactosidase gene joined to part of the protein encoded by the cDNA.

better to construct the library in a purpose-built expression vector rather than in the pUC or M13 vectors. Such expression vectors have features that maximize expression of the fusion protein to increase the chances of immunochemical detection. One such vector that is currently in wide use is the bacteriophage lambda insertion vector, λgt11 [97] (Fig. A.22). This has an *Eco* RI cloning site in the *lac* Z gene like the vector Charon 16A (Fig. A.19(a)) but has a mutation (*c*I 857) that allows thermal induction of the phage recombinants after lysogeny (Section 3.6.8) into a suitable *E. coli* host strain. Another mutation (*S*100) prevents lysis of the cells, resulting in the accumulation of large quantities of fusion protein inside the bacterial colonies. The protein is eventually made accessible to the antibody by treatment of the bacteria with chloroform.

In describing methods of screening recombinant DNA clones eukaryotic genomic clones have so far been ignored. This is because it is completely impracticable to apply indirect expression methods (mRNA selection) to massive genomic libraries, and direct expression in *E. coli* is generally impossible. In fact genomic clones for a particular protein are normally identified by plaque hybridization methods [98], analogous to colony hybridization (Fig. A.21) but using the ($^{32}$P)-labelled cloned cDNA insert as a probe. An alternative approach using genetic selection by recombination between the plasmid with the cDNA insert and the corresponding bacteriophage lambda [99] or cosmid [100] clone has been devised for more rapid screening of genomic libraries.

## A.8 ANALYSIS AND MANIPULATION OF CLONED DNA

Once a nucleic acid sequence has been cloned an investigator may wish to subject it to a wide variety of manipulations. In this section we restrict ourselves to describing the principles underlying the most common of these.

### A.8.1 Mapping of RNA transcripts

After cloning a piece of genomic DNA thought to encode a protein it is generally necessary to determine *in vivo* precisely which parts of the DNA are represented in the final processed RNA transcript. The positions of intron splice sites and the points at which transcription of mRNA starts and finishes are usually determined by the technique known as *nuclease S1 mapping* [101]. The principle of this method is to form a DNA/RNA hybrid between the mRNA encoded by the cloned DNA and the complementary single strand of the latter. This is possible by judicious choice of hybridization temperature, because of the greater stability of the DNA/RNA hybrids compared with the original DNA/DNA hybrid. The endonuclease S1 (Section 4.2.1(c)) is used to digest the non-hybridized, single-stranded regions of the DNA, leaving the regions of DNA complementary to the mRNA intact. The original method of characterizing the undigested DNA was by size after electrophoresis on alkaline agarose gels (to hydrolyse the RNA) and identification by hybridizing to an appropriate ($^{32}$P)-labelled probe. A more precise determination of the transcription boundaries can be obtained if an appropriate ($^{32}$P)-end-labelled restriction fragment of the cloned DNA is used in the hybridization. This can then be subjected to electrophoresis on a polyacrylamide sequencing gel alongside suitable standards, often produced by Maxam/Gilbert chemical cleavage (Section A.5.1(a)) of the restriction fragment in question (see Fig. A.23).

In determining the position of introns it can be helpful to employ exonuclease VII (see Section 4.5.2(g)) in addition to the nuclease S1 [102]. Exonuclease VII will digest single-stranded ends protruding from a mRNA/DNA hybrid but, unlike the endonuclease S1, cannot digest a single-stranded intron looped out from such a hybrid.

While discussing the topic of mapping RNA transcripts it is worth mentioning a situation that may arise with cDNA clones. As such clones may not reflect the full extent of the mRNA to its 5'-end one may wish to determine how far it lies from the position corresponding to the 5'-end of the cDNA. The technique of *primer extension* [82] utilizes a small restriction fragment near the

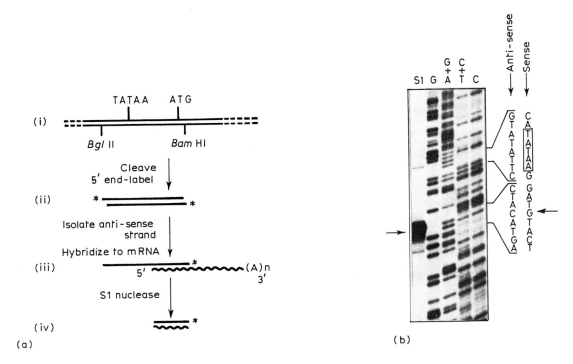

**Fig. A.23** Example of nuclease S1 mapping to determine transcriptional start point. (a) A cloned fragment of genomic DNA, hybridizing to a particular mRNA has been sequenced and the likely initiation codon and TATAA box identified (i). A restriction fragment that is likely to contain the transcriptional start point is isolated from this region and 5′-end-labelled (ii), the anti-sense strand isolated and hybridized to mRNA (iii) and the single stranded regions of the DNA (and RNA) not involved in the hybrid digested with S1 nuclease leaving the hybrid (iv); (b) the size of the protected DNA in (iv) is determined on an acrylamide sequencing gel. Maxam-Gilbert sequencing reactions are performed on the undigested anti-sense strand of (iii), and electrophoresed in parallel, allowing exact definition of the transcriptional start point in the sequence (the G indicated by the arrow). The position of the TATAA box is also indicated on the sequencing gel. (Data relating to the small subunit of HSV-2 ribonucleotide reductase mRNA, kindly provided by Dr Barklie Clements.)

5′ end of the cDNA clone which is denatured and then used to prime further copying of the mRNA with reverse transcriptase. If the primer is ($^{32}$P)-labelled at its 5′ end and non-radioactive dNTPs are used in the extension reaction, the size of the product can be determined by polyacrylamide gel electrophoresis, as for nuclease S1 mapping. Clearly, this technique is limited by the ability of the reverse transcriptase to overcome any secondary structure barriers that may exist at the 5′ end of the mRNA.

### A.8.2 Identification of regions of DNA that interact with proteins

The regulation of gene expression is achieved through the interaction of proteins with DNA, hence when a particular piece of genomic DNA has been cloned it is often of interest to identify the regions of DNA involved in such interactions. One general approach is to form a complex between the DNA and protein and then subject this to digestion with bovine DNase I

(Section 4.5.1). The region of DNA in contact with the protein is protected by the latter from digestion. Originally it was necessary to isolate this protected fragment and determine its structure; quite a difficult undertaking. This was much simplified by the adaptation of the Maxam-Gilbert sequencing strategy to the problem in the technique known as *DNA footprinting*.

In this approach [105] a fragment of double-stranded DNA is generated, radioactively end-labelled (Section A.4.2) at one end only, as for Maxam-Gilbert sequencing (Section A.5.1(a)). The protein is bound to the fragment and the complex subjected to digestion with DNase I under conditions that on average result in a single endonucleolytic cut per molecule (cf. the use of similar conditions for chemical cleavage). The DNA is then isolated and subjected to electrophoresis in the type of denaturing poly-acrylamide gel used for sequencing. If no protein is present during the DNase digestion a ladder of end-labelled oligonucleotides encompassing all possible sizes is produced on autoradiography. However, when the DNA–protein complex is subjected to DNase treatment no scission occurs in the protected region and no end-labelled oligonucleotides terminating in this region will be generated. This appears as a hole or gap on the sequencing gel (see Fig. A.24). Unprotected DNA subjected to chemical cleavage is sub-jected to electrophoresis on the same gel to allow the easy identification of the region of the DNA protected. When comparing the ladders pro-duced by chemical cleavage and cleavage with DNase I it must be remembered that chemical cleavage produces oligonucleotides with a 3'-phosphate group (see Fig. A.10) whereas DNase I produces oligonucleotides with a 3'-OH group (Section 4.5.1). This causes a difference in charge, and hence electrophoretic mobility, which must be taken into account.

Reagents other than DNase I have been used to cleave DNA, or to change its sensitivity to subsequent cleavage. Neocarzinostatin (a double-strand specific endodeoxyribonuclease)

has been used in place of DNase I because it generates products with 3'-phosphate groups that allow direct comparison with the chemical cleavage ladder [106]. Chemical methylation and 5-bromodeoxyuridine substitution with sub-sequent cleavage by ultraviolet light have been used to address the role of specific purines and thymine residues, respectively [107].

More recently the development of sequencing *in vivo* (see Section A.5.1) has allowed the footprinting technique to be extended to the study of interactions between uncloned DNA and protein *in vivo*. One approach is to irradiate nuclei with ultraviolet light, producing pyri-midine dimers (Section 7.2.5). The resulting saturation of the 5,6-double bond allows ring opening through reduction with $NaBH_4$ [108]. Alternatively, it is possible to apply the use of dimethyl sulphate for the methylation of purines to intact systems such as *E. coli* cells [109] and mammalian nuclei [110].

This section should not be concluded without mentioning that the most precise identification of the features of DNA–protein interactions has been achieved by X-ray crystallography of com-plexes formed between proteins and chemically synthesized oligonucleotides corresponding to their binding sites (e.g. [111]).

### A.8.3 Allocation of cloned genes to specific chromosomes

Three methods in which cloned genes can be assigned to specific chromosomes have been described in Section 3.2.3. These are the formation of somatic cell hybrids, hybridization to fixed chromosomes *in situ*, and hybridization to DNA from individual chromosomes sep-arated by a fluorescence-activated cell sorter. A fourth method, recently developed for the mouse, is useful for determining whether two genes are close together or distant, and will be described here as it also illustrates the use of restriction fragment length polymorphisms (RFLPs), mentioned in Section A.3.

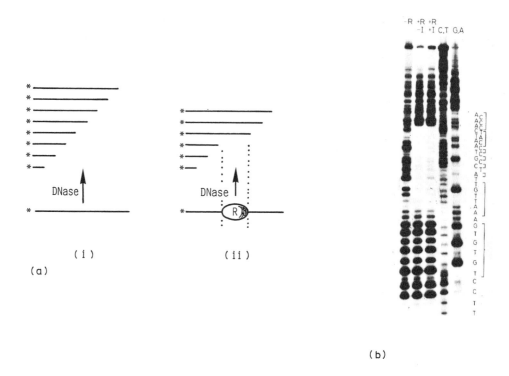

**Fig. A.24** Example of DNase I footprinting. (a) Partial digestion of a piece of end-labelled DNA to which a protein (R) is bound (ii) results in the absence of the end-labelled fragments cleaved in this region but found in the digest (i) of the unprotected DNA; (b) the results of separation of the products of such an experiment on a polyacrylamide sequencing gel. The example is with *lac* operator DNA and the *lac* repressor (R). I is IPTG (0.3 M), which does not prevent the binding of a mutant repressor used in this study. C,T and G,A represent the results of Maxam-Gilbert C + T and A + G reactions on the undigested end-labelled fragment. Adapted from [105], with permission.

This approach combines recombinant DNA with classical mouse genetics. The classical genetic approach to study the chromosomal 'linkage' of two gene loci is to determine whether markers co-segregate or segregate independently during genetic crosses between two different inbred strains that show polymorphism for the marker. In higher organisms externally observable phenotypes associated with polymorphisms are relatively infrequent; however, genetic analysis has been aided by the more numerous cases in which a polymorphism results in differences in electrophoretic mobility of an easily identifiable enzyme. (The assignment of certain markers to particular chromosomes is possible when they are the site of a visible chromosome translocation. Other markers can be assigned by linkage to these.) To assign cloned genes to particular chromosomes the above principle is taken one step further by directly observing the genotype rather than the phenotype. Advantage is taken of polymorphisms in the nucleotide sequence of the gene that affects sites for restriction endonucleases. Such polymorphisms will produce different-sized restriction fragments in the two cases and these can be visualized by hybridization of a radioactive probe to a genomic Southern blot of the DNA

(Section A.3). The best chance of finding suitable RFLPs in mouse is by comparing an inbred strain of the common laboratory mouse, *Mus musculus domesticus*, with an inbred strain of the distantly related *Mus spretus*. Crosses between these strains have been used to study the linkage of mouse myosin genes [112], as illustrated in Fig. A.25.

The use of RFLPs in the prenatal diagnosis of human hereditary diseases and the detection of carriers falls outside the terms of reference of this Appendix. Nevertheless, the utility of RFLPs generated by the mutation responsible for a particular genetic defect should be obvious. In addition, however, RFLPs outside the disease locus (which may be unknown) can be diagnostically useful if they show strong genetic linkage to this locus. The genetic analysis needed to use such RFLPs, although understandably more complex, is in principle similar to that just described [39].

### A.8.4 Expression in eukaryotic cells

Although bacterial systems are extremely well suited to cloning eukaryotic genes they cannot be used to study their transcription. In this section we shall discuss briefly some of the systems that can be used to detect expression of cloned eukaryotic genes and study factors important in their regulation. One can classify systems for expression in eukaryotes into three categories: those in which the exogenous DNA is neither replicated nor integrated into the host genome but is subject to *transient expression*, those in which the DNA is *stably integrated* into the host genome, and those in which the DNA is expressed on a *eukaryotic vector* based on an animal virus. This last category, although of considerable interest, is too wide to consider here, especially as expression in eukaryotic vectors is not yet a technique generally applied to cloned genes. (Yeast plasmid vectors are unfortunately unsuitable for expression of the genes of higher eukaryotes. This is because of fundamental differences in the signals for transcription and processing.)

One way of obtaining transient transcription of a cloned gene is *microinjection* into the nucleus of frog oocytes [113]. The large size of the nucleus in the oocyte facilitates this operation, but it is still a specialized technique requiring special equipment and considerable practice. More generally accessible is the technique of *calcium phosphate co-precipitation* of DNA on to tissue culture cells [114] which take up the precipitate, apparently by a process of phagocytosis. The DNA remains in the cells for several days and expression is usually studied about 48 h after transfection.

In transient expression experiments the problem is how to detect the relatively small amounts of gene products of the transcripts and distinguish them from the gene products of the host cell. In particular instances specific antisera may be available, but a more general approach is possible when, as is often the case, it is the regulatory rather than the coding regions of the gene that are being studied. In such cases constructs are made in which the 5'-non-coding and untranslated regions of the gene are linked to the coding region for the bacterial enzyme chloramphenicol acetyltransferase (CAT). *CAT assays*, as they are called, are extremely sensitive and are performed using [$^{14}$C]-chloramphenicol as a substrate, separating the acetylated chloramphenicol products from unreacted substrate by thin-layer chromatography [115].

The calcium phosphate co-precipitation technique also results in the stable integration of DNA into the chromosomal DNA of a small proportion of the cells. If such 'transformed' cells can be selected from the vast majority of untransformed cells, cell lines can be obtained in which to study the expression of such stably-integrated exogenous genes. This is achieved by co-precipitation with a piece of DNA containing a selectable marker, as it transpires that the form of integration is as long tandem concatamers,

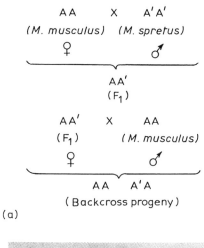

(a)

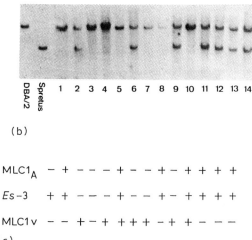

(b)

MLC1$_A$     − + − − − + − − + − + + + +

Es-3        + + − − − + − − + − + + + +

MLC1v       − − + − + + + + − + + − − −

(c)

**Fig. A.25** Allocation of genes to mouse chromosomes and study of linkage using restriction fragment length polymorphisms. (a) The mating strategy is illustrated. Because the F₁ male progeny are infertile the females must be back-crossed to one of the parental strains to allow the segregation of the two alleles A and A′ to be studied; (b) an example of the segregation of RFLPs for the MLCI$_A$ (myosin alkali light chain, cardiac atrial form) gene. The results of *Bam* HI digestion and electrophoresis of the DNA of the backcross offspring are shown. The different sized fragments in *Spretus* and DBA/2 (a *Musculus* strain) are formally equivalent to alleles A and A′; (c) comparison of the segregation of genes for MLCI$_A$, MLCI$_V$ (myosin alkali light chain, cardiac ventricle form), and *Es*-3 (an esterase enzyme exhibiting electrophoretic polymorphism) the gene for which is known to be located on chromosome 11. Homozygotes have been marked − , and heterozygotes + . These results illustrate the lack of linkage between the genes for MLCI$_A$ and MLCI$_V$, and that MLCI$_A$ lies on chromosome 11. (More offspring than illustrated were, in fact, analysed.) Adapted from [112], with permission.

and inclusion of a second DNA into the co-precipitate results in its inclusion in the integrated unit.

Many experiments have employed the herpes simplex viral thymidine kinase (tk) gene as a selectable marker. Cellular thymidine kinase is not required for the normal synthesis of dTTP from CDP, but allows salvage of thymidine produced by breakdown of DNA by converting it to dTMP (see Section 5.8). It is possible to select tk⁻ mutants by growing cells on bromo-deoxyuridine (BrUdr), which can only be incorporated into DNA in tk⁺ cells in which thymidine kinase converts BrUdr to BrdUMP. Ultraviolet light will fragment the DNA of those cells in which thymine has been replaced by bromouracil (cf. Section A.8.2), killing them, and hence allowing selection of spontaneous tk⁻ mutants. Tk⁻ cells that have been transformed to tk⁺ by the integration of exogenous DNA can be selected by growth on 'HAT' medium [116], containing hypoxanthine, aminopterin and thymidine. Aminopterin inhibits the activity of dihydrofolate reductase, a key enzyme in both purine and pyrimidine biosynthesis. Hypoxanthine can act as a preformed precursor of purines, but thymidine can only act as a precursor of pyrimidines if thymidine kinase is present (see Section 5.7).

Other selectable markers have been developed that do not require mutant cell lines such as tk⁻. Such dominant selectable markers include: dihydrofolate reductase (resistance to methotrexate) [117], the multifunctional CAD enzyme of uridine biosynthesis (resistance to *N*-phosphonacetyl-L-aspartate) [118], *E. coli* xanthine-guanine phosphoribosyltransferase (resistance to mycophenolic acid) [119], and bacterial neomycin phosphotransferase (resistance to neomycin) [120].

A limitation of systems in which cloned genes are stably integrated into tissue culture cells arises when the major interest is in the tissue-specificity of expression of a particular gene. A technique in which this problem is overcome involves the microinjection of cloned DNA into fertilized mouse eggs to produce so-called *trans-genic mice* [121]. The foreign DNA is integrated at such an early stage that it is generally present in all daughter cells, and certainly in the progeny of those mice that pass it on through the germ line to the next generation. It is thus possible, using suitable constructs, to study whether a particular tissue-specific gene is expressed in a tissue-specific manner in its new location, and to investigate the regions of DNA which confer tissue-specificity of expression [122].

### A.8.5 Mutagenesis *in vitro*

In studies of cloned genes of the type described in Sections A.8.2 and A.8.4, conclusions regarding the importance of regulatory regions of DNA can be tested and extended if it is possible to produce mutations in the regions of interest. Such mutations are difficult to produce in higher eukaryotes *in vivo*, but powerful techniques are available to produce them in cloned DNA *in vitro*. The technique of mutagenesis *in vitro*, especially when directed to specific nucleotides (see below) can also be applied to the coding regions of genes the products of which it is possible to express, purify and subject to structural or functional studies [123]. A variety of strategies has been used to produce mutations in cloned DNA. Discussion here will be restricted to two of the currently most widely used and powerful techniques, and the reader interested in a thorough treatment of this topic is directed elsewhere [5, 124].

*Deletion mutagenesis* is frequently performed using the enzyme *Bal* 31 nuclease (Section 4.2.1(f)). This enzyme has exonuclease activity against double-stranded DNA. Hence if a plasmid containing the cloned gene is linearized by restriction endonuclease digestion at a point in the region of interest, *Bal* 31 nuclease will digest it in a way that produces a deletion emanating from this point in both directions. The size of the deletion is determined by the time allowed for digestion, after which the blunt ends (treatment with the Klenow fragment of *E. coli* DNA polymerase and dNTPs is employed to fill in any overhangs) are ligated to linkers (see Section A.7.2(a)), cleaved, and ligated together with DNA ligase (see Fig. A.26) [125].

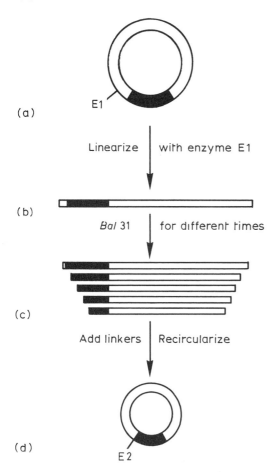

(a)

Linearize | with enzyme E1

(b)

*Bal* 31 | for different times

(c)

Add linkers | Recircularize

(d)

E 2

**Fig. A.26** Use of nuclease *Bal* 31 to generate deletions. A hypothetical plasmid (a) is illustrated in a region of which (solid shading) it is desired to obtain deletions. A neighbouring restriction endonuclease site is used to linearize the plasmid (b) and this is digested for different times generating a spectrum of deleted molecules (c). These are recircularized using linkers for a restriction endonuclease, E2, giving plasmids such as (d).

The most powerful technique at present available to produce mutations at specific sites is the so-called *site-directed mutagenesis using synthetic oligonucleotides* [126]. For this the DNA must be subcloned into a bacteriophage M13 (Fig. A.17(b)) or related vector that allows the generation of a single-stranded circular recombi-

nant DNA. An oligonucleotide (e.g. 10–12 mer) is synthesized, corresponding to the target of mutation and surrounding nucleotides, except that it contains the desired base-change. This is annealed to the single-stranded circular DNA, which is then converted to the replicative double-stranded form *in vitro* using the Klenow fragment of DNA polymerase I and DNA ligase. When *E. coli* are transfected with this DNA, phage containing both wild-type and mutant single-strands will result and give rise to 'plaques'. These can be distinguished by blotting on to nitrocellulose and hybridization to the ($^{32}$P)-labelled oligonucleotide originally used, under conditions that enable a perfect match (the desired mutant) to be distinguished from a single mismatch (the wild type). This is shown diagrammatically in Fig. A.27.

A modification of the procedure just described (and in fact preceding it historically [127]) may be used to generate short, specific deletions of a type hardly possible with *Bal* 31 nuclease. In this case the oligonucleotide synthesized is a hybrid of the sequences directly flanking the area to be deleted. This can hybridize to the single-stranded circular DNA if the area to be deleted loops out, and the desired mutant second strand can then be synthesized *in vitro*.

## A.9 CHEMICAL SYNTHESIS OF OLIGONUCLEOTIDES

The importance of chemically synthesized oligonucleotides in cloning cDNAs (Section A.7.2(a)), in labelling DNA (Section A.4.1), as primers for sequencing in M13 (Section A.5.1(b)), for producing mutagenesis *in vitro* (Section A.8.5) and in studying specific interactions between nucleic acids and proteins (Section A.8.2) has already been mentioned. Custom synthesis of oligonucleotides is now available commercially, as are machines for 'in house' synthesis. The chemistry involved derives from that developed by Khorana, the application of which culminated in the synthesis of a tRNA

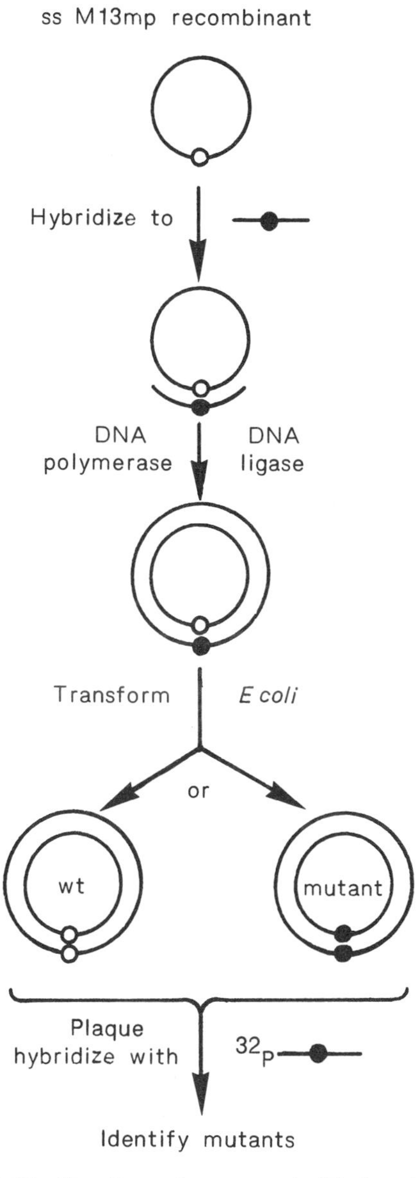

ss M13mp recombinant

**Fig. A.27** Site-directed mutagenesis. The base to be altered is shown as an open circle and the mismatch (mutation) in the chemically synthesized oligonucleotide as a solid circle. For other details, see text.

gene [128]. The methods Khorana used have now been superseded, and therefore the principles of the two methods at present in greatest use will be outlined here. Further details can be found elsewhere [141, 142].

In both methods successive diester bonds are formed between the 5'-OH group of one nucleotide derivative and 3'-OP of a second nucleotide derivative. The chemical environment of the P determines the chemistry of the condensation and is different in the two cases. Both methods share the common feature that other potentially reactive groups elsewhere in the molecules are 'protected' by reversible chemical modification, and such protected nucleotide derivatives are used as the starting building blocks. The groups protecting the bases are not removed until the end of the overall synthesis, and the same is true of the protecting group for the P–OH. However, as the oligonucleotide chain is chemically synthesized in a $3' \rightarrow 5'$ direction, the 5'-position (protected in both methods by a dimethoxytrityl group) must be deprotected after the addition of each monomer to allow further reaction. In order to simplify the synthetic procedure the first nucleotide is directly linked to a solid support (e.g. silica gel) packed in a column. This allows the excess reagents to be washed off the column after each step. It is usually necessary to block or 'cap' any unreacted 5'-OH groups after each step, using acetic anhydride. At the end of the synthesis the oligonucleotide must be chemically released from the column and deprotected.

In the *phosphotriester* method (Fig. A.28) the nucleotide with the reactive 3'-moiety is a derivative of the nucleoside 3'-monophosphate in which one P-OH is blocked by a 2-chlorophenyl group, and hence is in fact a phosphodiester. It is generated by a suitable base from a precursor in which the remaining hydroxyl on the P is blocked by a $\beta$-cyanoethyl group, and reaction in the presence of a coupling reagent and catalyst results in the formation of a phosphotriester. The 5'-O dimethoxytrityl protecting group is removed by suitable acid treatment and another activated monomer added to allow the next round of condensation. At the end of the whole reaction sequence the 2-chlorophenyl

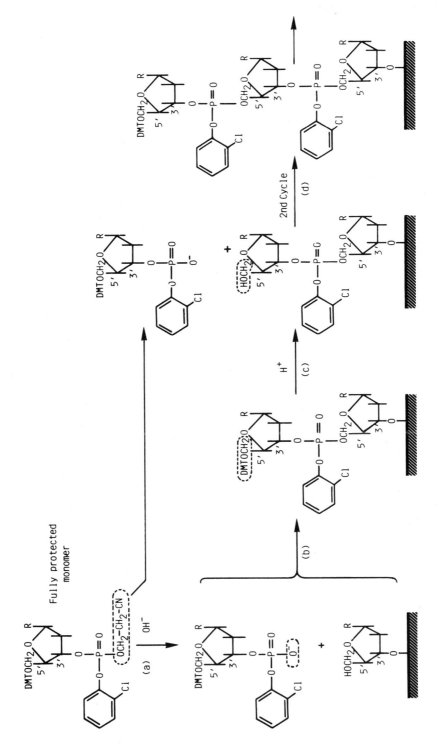

**Fig. A.28** The phosphotriester method of deoxyoligonucleotide synthesis. (a) The fully protected monomer is unblocked at the 3′-phosphate position by treatment with a base such as triethylamine allowing (b) condensation with a second molecule of monomer attached *via* its 3′-O to a solid support and having its 5′-OH unprotected. The dinucleotide produced must be (c) treated with acid to remove the dimethoxytrityl group (DMT) protecting the 5′-OH before (d) condensation with another molecule of 3′-activated monomer can occur. R = benzoyl adenine, benzoyl cytosine, isobutyryl guanine, or thymine.

groups are removed by tetramethylguanidine-aromatic aldoxime in aqueous dioxane, the acyl blocking groups of the base by concentrated ammonium hydroxide, and the dimethoxytrityl group by 80% acetic acid. Removal of the oligonucleotide chain from the solid support depends, of course, on the nature of the linkage. Ester linkage via succinyl groups is quite common, in which case cleavage occurs at the same time as deprotection of the bases.

The phosphotriester method is useful for synthesizing oligonucleotides containing up to about 50 bases, but is limited by the fact that each condensation takes about 1 h to go to completion, even in the presence of catalysts. The *phosphite triester* method (Fig. A.29) employs a more reactive form of the phosphorus which allows the condensation to be completed in about 2 min, and has been used to synthesize oligonucleotides containing more than 150 bases. The 5'-OH of the monomer is protected with the same blocking group as in the phosphotriester method but the reactive 3'-component is a dialkyl phosphoamidite (one of a variety of the possible dialkyl substitutions of the N is illustrated) and the product is a phosphite triester, which must be oxidized (e.g. with aqueous iodine) to a stable phosphotriester before the next synthetic step. A methyl group is usually employed to block the hydoxyl on the P. This so-called phosphoamidite method is now widely regarded as the method of choice for oligonucleotide synthesis, and most automatic synthesizers ('gene machines') use this chemistry.

## A.10 CELL-FREE SYSTEMS FOR TRANSCRIPTION AND TRANSLATION [9]

In Section A.8.4 a number of systems for studying the *transcription* of cloned genes in eukaryotic cells were mentioned. All of these are quite sophisticated and involve the use of intact cells. For many purposes the availability of cell-free systems which would allow transcription of exogenous DNA would offer considerable advantages; however, progress in this area has been slow. The state of development of such systems at the time of writing is described in Section 8.3.

The importance of systems in which to assay the *translation* of eukaryotic mRNAs has also been indicated in relation to cloning (Section A.7.3) and the study of protein synthesis in general (Chapter 11). Although the intact cell system of microinjection into frog oocytes (or eggs) has the advantage that translation may proceed for several days [143], cell-free systems are of greater utility to the non-specialized laboratory. Such systems are not as simple to establish in eukaryotes as in *E. coli* (see below) and activity is easily lost during cell fractionation. Necessary components of the system may be lost with nuclei and mitochondria or adhere to intercellular structures. Furthermore, the disruption of the cytoskeleton may be a factor contributing to loss of activity. It is therefore perhaps not coincidental that the two most widely used systems for cell-free translation are derived from rather specialized cells. By far the most popular system is that derived from rabbit reticulocytes using hypotonic lysis followed by low-speed centrifugation to remove the cell membranes. This *rabbit reticulocyte lysate* has been further modified to allow more efficient utilization of exogenous mRNA. The endogenous globin mRNA is degraded by the $Ca^{2+}$-dependent micrococcal nuclease that can subsequently be inactivated by chelation with EGTA [89]. Such nuclease-treated lysates are available from several commercial sources, obviating the need for the user to maintain and bleed his own anaemic rabbits. The other cell-free system still in general use is that derived from the post-microsomal supernatant of wheat-germ [144], which has an intrinsically low content of endogenous mRNA.

One limitation of the reticulocyte lysate and

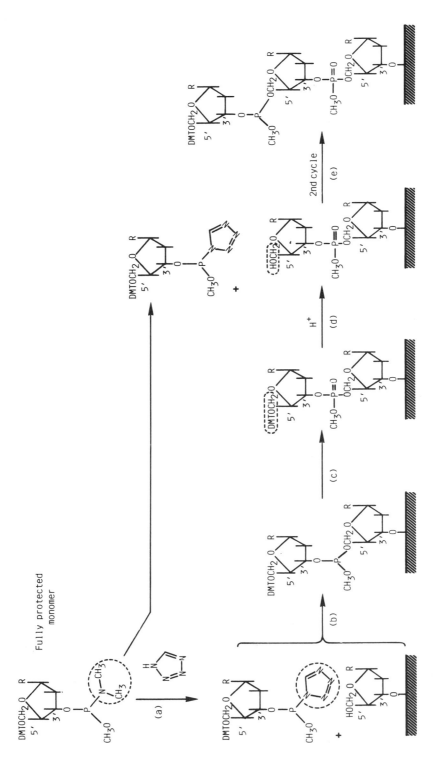

**Fig. A.29**   The phosphite triester method of deoxyoligonucleotide synthesis. (a) The protected phosphoramidite monomer is activated in the 3'-P position by a suitable weak acid (e.g. tetrazole) allowing condensation (b) with 5'-activated monomer attached to a solid support, as in Fig. A.28. The phosphite is then oxidized to a phosphate (c). Removal of the 5'-protecting DMT group of the dinucleotide with, for example, dichloroacetic acid (d) prepares this for the next cycle of condensation (e). R is as in Fig. A.28.

**Table A.3**  Some inhibitors of DNA synthesis.

| Category | Examples | Reference |
|---|---|---|
| **1. Reagents interacting with DNA*** | | |
| (a) Alkylating agents | Nitrogen and sulphur mustards | Section 7.2.2 |
| | Dimethyl sulphate | |
| | NNMG and NMS | |
| | Mitomycin C | |
| (b) Intercalating agents | Ethidium bromide | Section 7.2.3 |
| | Propidium diiodide | |
| | Acridine dyes | |
| | Actinomycins | |
| | Anthracenes | |
| | Benzpyrene | |
| | Adriamycin | [153] |
| (c) Intertwining agents | Netropsin | [154] |
| | Distamycin | [154] |
| **2. Analogues of bases, nucleosides etc.** [†] | 5-fluorouracil | Section 7.2.1 |
| | 6-azauracil | Section 7.2.1 |
| | 8-azaguanine | Section 7.2.1 |
| | 6-mercaptopurine | Section 7.2.1 |
| | diazaoxynorleucine | Section 5.2 |
| | azaserine | Section 5.2 |
| | 5-bromodeoxycytidine | [152] |
| | 5-fluorodeoxycytidine | [152] |
| | 2-aminopurine | Section 7.2.1 |
| | Cytosine $\beta$-1-D-arabinoside | Section 7.2.1 |
| | Adenine $\beta$-1-D-arabinoside | Section 7.2.1 |
| | Dideoxynucleosides | Section 6.4.2(e) and A.5.1(b) |
| | Aphidicolin | Section 6.4.2(e) |
| | Aminopterin | Sections 5.2 and 5.7 |
| | Amethopterin | Sections 5.2 and 5.7 |
| | 5-fluorodeoxyuridine | Section 5.2 |
| | Hydroxyurea | Section 5.5 |
| | Acyclovir | Section 3.6.7(e) |
| | 2'azido-2'-deoxynucleosides | [156] |
| **3. Inhibitors of topoisomerases** | Coumermycin | Section 6.4.6 |
| | Novobiocin | |
| | Nalidixic acid | |
| | Oxolinic acid | |
| **4. Inhibitors of cell division** | Colcemid | [157] |
| | Colchicine | |
| | Vinblastine | |
| | Vincristine | |

*These may be mutagens and/or agents which interfere with replication or transcription.
[†]These may interfere with nucleotide biosynthesis or, following incorporation into DNA, block further chain elongation or transcription.

wheatgerm systems (but not of the frog oocyte) is the lack of endoplasmic reticulum for correct processing of translation products that contain signal peptides (see Section 11.7.1). This may be overcome by the addition of a microsomal membrane fraction, e.g. from dog pancreas [145].

Prokaryotic cell-free systems for translation were developed very early on; the ribosomes, tRNAs and soluble factors for translation of exogenous mRNA being present in the supernatant after centrifugation of disrupted cells at 30 000 g for c 30 min. This system was originally developed by Nirenberg and Matthaei in their work on the genetic code (see Section 11.1.1) and was subsequently developed for use in studying the translation of the stable mRNAs of the RNA coliphages (Section 11.10.1) [146]. The low endogenous mRNA activity of such systems is a reflection of the very short half-lives of bacterial mRNAs, which make it very difficult to isolate bacterial mRNAs and study their translation directly. To overcome this problem a cell-free system for translation was devised in which the mRNA could be continually generated by simultaneous transcription of the gene. This coupled transcription–translation system [147] was originally applied to genes cloned by more traditional genetic techniques using specialized transducing phages.

## A.11 THE USE OF INHIBITORS IN THE STUDY OF GENE EXPRESSION

In the course of this book reference has been made to the use of antibiotics and other inhibitors of macromolecular synthesis. For the convenience of the reader the most common of these are summarized in Tables A.3–A.5. Many inhibitors of transcription and translation are specific for eukaryotic or prokaryotic systems, and this specificity is indicated in Tables A.4 and A.5. Most inhibitors of DNA synthesis (Table A.3) do not discriminate between eukaryotes and prokaryotes; however, the dideoxynucleo-

**Table A.4** Some inhibitors of RNA synthesis.

| Inhibitor | Remarks | Reference |
|---|---|---|
| Rifampicin and Streptovaricin | Bind to $\beta$ subunit of *E. coli* RNA polymerase, inhibiting initiation | Section 8.2.2 |
| Streptolydigin | Binds to $\beta$ subunit of *E. coli* RNA polymerase, inhibiting elongation | Section 8.2.2 |
| Actinomycin D | Inhibits both prokaryotic and eukaryotic transcription by complexing with deoxyguanosine residues. Hence transcription of rRNA is more sensitive than that of mRNA | Section 8.2.2 |
| $\alpha$-amanitin and other fungal amatoxins | Inhibits eukaryotic RNA polymerase II specifically (also RNA polymerase III at very high concentration) | Section 8.1.2 |
| Cordycepin (3'-deoxyadenosine) | Inhibits poly(A) polymerase responsible for 3'-polyadenylation of eukaryotic mRNAs | Section 9.3.3 |
| Dichlororibofuranosyl benzimidazole | Inhibits appearance of eukaryotic mRNA in the cytoplasm, although only partially inhibiting synthesis of hnRNA | [158] |

**Table A.5** Some inhibitors of protein synthesis.

| Inhibitor | Prokaryotic/ eukaryotic | Remarks | Reference |
|---|---|---|---|
| Streptomycin | P | Binds to 30S subunit to cause misreading and (at higher concentrations) inhibition of initiation | Section 11.13.1 |
| Paromomycin | P | Inhibits initiation. Resistant mitochondria have altered small rRNA | Section 11.8.3 |
| Kasugamycin | P | Inhibits initiation. $m_2^6 A m_2^6 A$ sequence of 16S rRNA involved | Section 11.8.3 |
| Pactamycin | P, E | Inhibits initiation, was used to determine gene order of proteins derived from picornaviral polyproteins | [159] |
| Neomycin | P, E | Causes miscoding and inhibits initiation and elongation. Used as a selectable marker in transfection of eukaryotic cells | Section A.8.4 |
| Edeine A | P, E | Inhibits initiation at either mRNA- or initiator tRNA-binding step | [150, 151] |
| Aurintricarboxylic acid | P, E | Inhibits initiation by preventing binding of mRNA to ribosome | [150, 151] |
| Showdomycin | P, E | Inhibits initiation at the stage of ternary complex formation | [150] |
| Fluoride ion | (P?) E | Inhibits initiation in intact cells | [160] |
| Puromycin | P, E | Inhibits elongation by binding to A-site and reacting with peptidyl-tRNA | Section 11.5 |
| Tetracycline | P | Inhibits elongation by blocking binding of aminoacyl-tRNA to the A-site on the 30S subunit | [161] |
| Chloramphenicol | P | Inhibits elongation at peptidyl transferase. Resistant mitochondria have altered large rRNA | Section 11.8.3 |
| Erythromycin | P | Inhibits elongation at transpeptidation step | Section 11.8.3 |
| Sparsomycin | P, E | Inhibits elongation at peptidyl transferase step | [150, 151] |
| Spectinomycin | P | Inhibits elongation at transpeptidation. Resistant ribosomes have altered protein S5 | [150, 151] |
| Thiostrepton | P | Inhibits elongation, preventing binding of EF-G·GTP complex to ribosome | Section 11.8.3 |

**Table A.5**  *(contd)*

| Inhibitor | Prokaryotic/ eukaryotic | Remarks | Reference |
|---|---|---|---|
| Fusidic acid | P, E | Inhibits elongation, preventing release of EF-G·GDP complex from ribosome. Resistant bacteria have altered EF-G | [150, 151] |
| Kirromycin | P | Inhibits elongation, preventing release of EF-Tu·GDP complex from ribosome | Section 11.8.3 |
| Kanamycin | P, E | Inhibits elongation and causes misreading | [150, 151] |
| Cycloheximide | E | Inhibits elongation and initiation, 'freezing' ribosomes on polysomes. Resistant yeast have altered 60S subunits | [150] |
| Emetine | E | Inhibits elongation at translocation step. Resistant hamster cells have altered protein S14 | [150, 162] |
| Diphtheria toxin | E | Inhibits elongation by inactivating EF-2 by ADP ribosylation | Section 11.6.2 |
| Ricin and abrin | E | Inhibit elongation, affecting 60S subunit | [150, 152] |
| Colicin E3 | P | Specific nuclease for site on 16S rRNA | [152, 163] |
| $\alpha$-sarcin | (P), E | Specific nuclease for site on 23S and 28S rRNA in ribosomes, although inactive against intact *E. coli*. Cleaves many sites in naked rRNAs | [164] |
| Guanylyl methylene diphosphonate and Guanylylimidodiphosphate | P, E | Non-hydrolysable analogues of GTP | Section 11.5.1 and 11.5.2 |
| Trimethoprim | P | Prevents formation of fMet-tRNA by inhibition of synthesis of $N^{10}$-formyl-$H_4$-folate | [165] |
| 5-fluorotryptophan | P, E | Prevents activation of tRNA$^{Trp}$ by competitive inhibition of synthetase | [165] |
| Norvaline | P, E | Prevents activation of tRNA$^{Val}$ | [165] |
| Ethionine | P, E | Causes synthesis of abnormal proteins with ethionine instead of methionine | [165] |
| O-methyl threonine | P, E | Causes synthesis of abnormal proteins with O-methyl threonine replacing threonine | [165] |

sides, on conversion to the triphosphates, do not inhibit eukaryotic DNA polymerase α. In contrast, a fairly specific inhibition of eukaryotic DNA polymerase α is obtained with aphidicolin. Further details may be found elsewhere [149–152].

## A.12 *E. COLI* GENES RELEVANT TO NUCLEIC ACID METABOLISM

Much of the study of nucleic acid metabolism has been conducted in *E. coli*, and many of the components involved were identified by mutations in their genes. Because of this, and in view of the widespread use of specialized *E. coli* strains among non-geneticists involved in recombinant DNA work, we felt it might be useful to provide a summary of the names given to some selected genetic loci of *E. coli* (Table A.6).

A few words of caution are necessary regarding Table A.6. Because genetic loci were often defined before the corresponding gene product was known, there exist alternatives to some of the names listed. In some cases names have merely been rationalized to conform to a pattern such as *dna*, *rpo*, *rps*; and older nomenclature based on, e.g. antibiotic resistance may still be found especially in description of genotypes. Thus bacteria with a totally inactive gene for ribosomal protein S12 may be designated *rps*L⁻, whereas bacteria with the mutation in the *rps*L gene that conveys dependence on the antibiotic streptomycin (Section 11.13.1) are generally designated $Sm^D$, the gene being originally called *str*A. Also worth mentioning are the *sup* genotypes: These define mutations in tRNA genes (e.g. *sup*F is in *tyr*T) which cause a particular termination codon to be read as an amino acid codon (suppression – see Section 11.13.2). For fuller details of these and other gene symbols the reader is advised to consult the *E. coli* linkage map published periodically in *Microbiological Reviews*. At the time of writing the 1983 edition [148] is the latest, but a new edition can be expected by 1987.

**Table A.6** Some *E. coli* genes relevant to nucleic acid metabolism.

| Gene symbol | Explanation | Reference |
|---|---|---|
| *ada* | Repair enzyme for removal of methyl groups | Section 7.3.1 |
| *ala*S | Alanyl-tRNA synthetase (similarly: *gln*S, *phe*S, etc.) | Section 11.3 |
| *ala*T (U, V etc.) | Alanyl-tRNAs (similarly: *gln*T, U, V etc.) | Section 9.5.1 and 11.1 |
| *att*λ | Integration site for bacteriophage lambda (similarly *att* φ80, etc.) | Section 7.4.3 |
| *cca* | tRNA nucleotidyl transferase | Section 9.5.2 |
| *dam* | DNA adenine methylase | Section 4.6.1 |
| *dcm* | DNA cytosine methylase | Section 4.6.1 |
| *dna*A (B, C etc.) | Enzymes of DNA biosynthesis | Tables 6.4 and 6.6 |
| *dut* | (*dna*S) dUTPase | Section 6.3.2 |
| *end*A | DNA endonuclease I | Table 4.3 |
| *fus*A | Protein synthesis factor, EF-G | Section 11.5.2 |
| *gpp* | Guanosine pentaphosphatase | Section 11.12 |
| *gyr*A,B | (*cou*, *nal*A) DNA gyrase (two subunits) | Section 6.4.6 |

**Table A.6** *1(contd)*

| Gene symbol | Explanation | Reference |
|---|---|---|
| *hfl* | High frequency of lysogenization by bacteriophage lambda | Section A.7.3 |
| *hsd*M, R, S | Host restriction/modification system (methylase, endonuclease and specificity components of *EcoB* and *EcoK* in B and K strains, respectively). | Section 4.5.3 |
| *inf*A, B, C | Protein synthesis factors, IF-1, $-2$ and $-3$ | Section 11.5.1 |
| *lac*A, I, Y, Z | Components of *lac* operon | Section 10.1.1 |
| *lep* (B) | Signal peptidase I | Section 11.7.1 |
| *lex*A | Repressor protein involved in response to DNA damage | Section 7.3.4 |
| *lig* | (*dna*L) DNA ligase | Table 6.6 |
| *lsp*A | Signal peptidase II (specific for lipoprotein) | Section 11.7.1 |
| *mutD* | (*dna*Q) DNA polymerase III ($\epsilon$) | Table 6.6 |
| *mut*H, L, R, S | Uncharacterized proteins involved in DNA repair | Section 7.3.3 |
| *nrd*A, B | (*dna*F) ribonucleotide reductase (two subunits) | Table 6.6 |
| *nus*A, B | Proteins involved in termination of transcription | Section 10.2.2 |
| *nus*C, D, E | = *rpo*B, *rho* and *rps*J, respectively | Section 10.2.2 |
| *ori*C | Origin of DNA replication | Section 6.9.4 |
| *phr* | Photolyase | Section 7.3.1 |
| *pnp* | Polynucleotide phosphorylase | Section 4.4 |
| *pol*A, B, C | DNA polymerases I, II and III (*dna*E) | Tables 6.3 and 6.6 |
| *pr1*A–D | Uncharacterized proteins affecting protein export | Section 11.7.1 |
| *prm*A, B | Ribosomal protein methylases | – |
| *pur*A–M | Enzymes of purine biosynthesis | Section 5.2 |
| *pyr*A–I | Enzymes of pyrimidine biosynthesis | Section 5.4 |
| *rec*A | Protein involved in recombination and repair | Section 7.4.1 |
| *rec*B, C | DNA exonuclease V (two subunits) | Table 4.4 and Section 7.4.1 |
| *rec*F | DNA exonuclease VIII | Section 4.5.2h |
| *rel*A | Stringent factor | Section 11.12 |
| *rep* | DNA helicase | Section 6.4.5 |
| *rho* | Transcription termination factor rho | Section 8.2.3 |
| *rim*B–L | Ribosomal protein modification enzymes (methylases, acetylases etc.) | – |
| *rna-rnh* | Ribonucleases I–III, D, E and H | Table 4.1 |
| *rnp*A, B | Ribonuclease P (protein and nucleic acid components) | Section 9.5.2 |

**Table A.6** (*contd*)

| Gene symbol | Explanation | Reference |
|---|---|---|
| *rpl*A–Z, *rpm*A–H | 50S subunit ribosomal proteins L1–L34 | Section 11.8.1 |
| *rpo*A–D | RNA polymerase subunits $\alpha$, $\beta$, $\beta'$ and $\sigma$, respectively | Section 8.1.1 |
| *rps*A–U | 30S subunit ribosomal proteins S1–S21 | Section 11.8.1 |
| *rrf*A–E, G, H | 5S rRNA genes of operons *rrn*A–H | Section 9.4.1 |
| *rrl*A–E, G, H | 23S rRNA genes of operons *rrn*A–H | Section 9.4.1 |
| *rrn*A–E, G, H | rRNA operons | Section 9.4.1 |
| *rrs*A–E, G, H | 16S rRNA genes of operons *rrn*A–H | Section 9.4.1 |
| *sbc*B | DNA exonuclease I | Table 4.4 |
| *sec*A, B, Y | Uncharacterized proteins affecting protein export (*sec*Y = *prl*A) | Section 11.7.1 |
| *spo*T | Guanosine tetraphosphatase | Section 11.12 |
| *ssb* | DNA-binding protein | Table 6.6 |
| *thy*A | Thymidylate synthase | Section 5.6 |
| *tdk* | Thymidine kinase | Section 5.8 |
| *top*A | Topoisomerase I | Table 6.6 |
| *tra*I | Helicase I (plasmid encoded) | Section 6.4.5 |
| *trm*A–D | tRNA methylases | Table 9.2 |
| *tsf* | Protein synthesis factor EF–Ts | Section 11.5.2 |
| *tuf*A, B | Protein synthesis factor EF–Tu (two genes) | Section 11.5.2 |
| *ung* | Uracil-DNA-glycosylase | Section 6.3.2 |
| *uvr*A–C | AP endonuclease (three components) | Table 4.3 and Section 7.3.2 |
| *uvr*D | Helicase II | Section 6.4.5 |
| *uvr*E | Protein involved in mismatch repair | Section 7.3.3 |
| *xth*A | DNA exonuclease III | Section 4.5.2c |
| *xse*A | DNA exonuclease VII | Table 4.4 |

## REFERENCES

1 Cantoni, G. L. and Davies, D. R. (1971), *Procedures in Nucleic Acid Research*, vol. 2, Harper and Row, New York.

2 Birnie, G. D. and Rickwood, D. (1978), *Centrifugal Separations in Molecular and Cell Biology*, Butterworths, London.

3 Old, R. W. and Primrose, S. B. (1985), *Principles of Gene Manipulation* (3rd edition), Blackwell, Oxford.

4 Walker, J. M. and Gaastra, W. (1983), *Techniques in Molecular Biology*, Croom Helm, London.

5 Glover, D. M. (1984), *Gene Cloning: the Mechanics of DNA Manipulation*, Chapman and Hall, London.

6 Habel, K. and Salzman, N. P. (1969), *Fundamental Techniques in Virology*, Academic Press, New York.

7 Maniatis, T., Fritsch, E. F. and Sambrook, J. (1982), *Molecular Cloning, a Laboratory Manual*, Cold Spring Harbor Laboratory, New York.

8 Osterman, L. A. (1984), *Methods of Protein and Nucleic Acid Research* vols 1 and 2, Springer, Berlin.

9 Hames, B. D. and Higgins, S. J. (1984), *Transcription and Translation: a Practical Approach*, IRL Press, Oxford.

10 Glover, D. M. (1985), *DNA Cloning: a Practical Approach*, vols 1 and 2, IRI Press, Oxford.

11 Schneider, W. C. (1945), *J. Biol. Chem.*, **161**, 293.

12 Schmidt, G. and Thannhauser, S. J. (1945), *J. Biol. Chem.*, **161**, 83.

13 Fiske, C. and Subbarow, Y. (1929), *J. Biol. Chem.*, **81**, 629.

14 Berenblum, I. and Chain, E. (1958), *Biochem. J.*, **82**, 286.

15 Burton, K. (1956), *Biochem. J.*, **62**, 315.

16 Ceriotti, G. (1955), *J. Biol. Chem.*, **214**, 59.

17 Warburg, O. and Christian, W. (1942), *Biochem. Z.*, **310**, 384.

18 Lowry, O. H., Rosebrough, N. J., Farr, A. L. and Randall, R. J. (1951), *J. Biol. Chem.*, **193**, 265.

19 Bradford, M. M. (1976), *Anal. Biochem.*, **72**, 248.

20 Marmur, J. (1961), *J. Mol. Biol.*, **3**, 208.

21 Blin, N. and Stafford, D. W. (1976), *Nucleic Acids Res.*, **3**, 2303.

22 Birnboim, H. C. and Doly, J. (1979), *Nucleic Acids Res.*, **7**, 1513.

23 Holmes, D. S. and Quigley, M. (1981), *Anal. Biochem.*, **114**, 193.

24 Wallace, H. and Birnsteil, M. L. (1966), *Biochim. Biophys. Acta*, **114**, 296.

25 Aaji, C. and Borst, P. (1972), *Biochim. Biophys. Acta*, **269**, 192.

26 Maniatis, T., Jeffrey, A. and Kleid, D. G. (1975), *Proc. Natl. Acad. Sci. USA*, **72**, 1184.

27 Birnie, G. D., Fox, S. M. and Harvey, D. R. (1972), in *Subcellular Components* (ed. G. D. Birnie), Butterworths, London, p. 235.

28 Roodyn, D. B. (1972) in *Subcellular Components* (ed. G. D. Birnie) Butterworths, London, p. 13.

29 Muramatsu, M., Hayashi, Y., Onishi, T., Sakai, M., Takai, K. and Kashiyama, T. (1974), *Exp. Cell Res.*, **88**, 345.

30 Estabrook, R. W. and Pullman, M. E. (1967), *Meths Enzymol.*, **10**.

31 Moldave, K. and Grossman, L. (1979), *Meths Enzymol.*, **59**.

32 Knowler, J. T., Moses, H. L. and Spelsberg, T. C. (1973), *J. Cell. Biol.*, **59**, 685.

33 McConkey, E. H. (1967), *Meths Enzymol.*, **12A**, 620.

34 Adelman, M. R., Blobel, G. and Sabatini, D. D. (1973), *J. Cell. Biol.*, **56**, 191.

35 Moldave, K. and Grossman, L. (1971), *Meths Enzymol.*, **20**.

36 DeWachter, R. and Fiers, W. (1971), *Meths Enzymol.*, **21**, 167.

37 Lehrach, H., Diamond, D., Wozney, J. M. and Boedtker, H. (1977), *Biochemistry*, **16**, 4743.

38 Southern, E. (1980) *Meths Enzymol.*, **69**, 152.

39 Botstein, D., White, R. L., Skolnick, M. and Davis, R. W. (1980), *Am. J. Hum. Genet.*, **32**, 314.

40 Marmur, J. and Doty, P. (1962), *J. Mol. Biol.*, **5**, 109.

41 Dove, W. F. and Davidson, N. (1962), *J. Mol. Biol.*, **5**, 467.

42 Bonner, T. I., Brenner, D. J., Neufeld, B. R. and Britten, R. J. (1973), *J. Mol. Biol.*, **81**, 123.

43 Alwine, J. C., Kemp, D. J. and Stark, G. R. (1977), *Proc. Natl. Acad. Sci. USA*, **74**, 5350.

44 Thomas, P. S. (1980), *Proc. Natl. Acad. Sci. USA*, **77**, 5201.

45 Langer, P. R., Waldrop, A. A. and Ward, D. C. (1981), *Proc. Natl. Acad. Sci. USA*, **78**, 6633.

46 Rigby, P. W. J., Dieckmann, M., Rhodes, C. and Berg, P. (1977), *J. Mol. Biol.*, **113**, 237.

47 O'Farrell, P. (1981), *Focus* **3**, 1.

48 Ricca, G. A., Taylor, J. M. and Kalinyak, J. E. (1982), *Proc. Natl. Acad. Sci. USA*, **79**, 724.

49 Green, M. R., Maniatis, T. and Melton, D. A. (1983), *Cell*, **32**, 681.

50 Feinberg, A. P. and Vogelstein, B. (1983), *Anal. Biochem.*, **132**, 6.

51 Chaconas, G. and van de Sande, J. H. (1980), *Meths. Enzymol.*, **65**, 75.

52 Challberg, M. D. and Englund, P. T. (1980), *Meths. Enzymol.*, **65**, 39.

53 Roychoudhury, R. and Wu, R. (1980), *Meths Enzymol.*, **65**, 43.

54 Gething, M. J., Bye, J., Skehel, J. and Waterfield, M. (1980), *Nature (London)*, **287**, 301.

55 Bishop, J. O., Rosbash, M. and Evans, D. (1974), *J. Mol. Biol.*, **85**, 75.

56 Maxam, A. M. and Gilbert, W. (1980), *Meths Enzymol.*, **65**, 499.

57 Sanger, F., Nicklen, S. and Coulson, A. R. (1977), *Proc. Natl. Acad. Sci. USA*, **74**, 5436.

58 Sanger, F. and Coulson, A. R. (1978), *FEBS Lett.*, **87**, 107.

59 Church, G. M. and Gilbert, W. (1984), *Proc. Natl. Acad. Sci. USA*, **81**, 1991.

60 Staden, R. (1977), *Nucleic Acids Res.*, **4**, 4037.

61 Gronenborn, B. and Messing, J. (1978), *Nature (London)*, **272**, 375.

62 Yanisch-Perron, C., Vieira, J. and Messing, J. (1985), *Gene*, **33**, 103.

63 Staden, R. (1979), *Nucleic Acids Res.*, **6**, 2607.

64 Fu, C-P. D. and Wu, R. (1980), *Meths Enzymol.*, **65**, 620.

65 Rich, A. and RajBhandary, U. L. (1976), *Annu. Rev. Biochem.*, **45**, 805.

66 Randerath, K., Gupta, R. C. and Randerath, E. (1980), *Meths Enzymol.*, **65**, 638.

67 Brownlee, G. G. (1972), *Determination of Sequences in RNA*, North Holland, Amsterdam.

68 Donis-Keller, H., Maxam, A. M. and Gilbert, W. (1977), *Nucleic Acids Res.*, **4**, 2527.

69 Donis-Keller, H. (1980), *Nucleic Acids Res.*, **8**, 3133.

70 Danna, K. J., Sack, G. H. and Nathans, D. (1973), *J. Mol. Biol.*, **78**, 363.

71 Smith, H. O. and Birnstiel, M. L. (1976), *Nucleic Acids Res.*, **3**, 2387.

72 Rackwitz, H-R., Zehetner, G., Frischauf, A-M. and Lehrach, H. (1984), *Gene*, **30**, 195.

73 Aviv, H. and Leder, P. (1972), *Proc. Natl. Acad. Sci. USA*, **69**, 1408.

74 Adesnik, M., Solditt, N., Thomas, W. and Darnell, J. E. (1982), *J. Mol. Biol.*, **71**, 21.

75 Bolivar, F. (1978), *Gene* **4**, 121.

76 Lederberg, E. M. and Cohen, S. N. (1974), *J. Bacteriol.*, **119**, 1072.

77 Vieira, J. and Messing, J. (1982), *Gene*, **19**, 259.

78 Messing, J. (1983), *Meths Enzymol.*, **101**, 20.

79 Land, H., Grey, M., Hanser, H., Lindenmaier, W. and Shutz, G. (1981), *Nucleic Acids Res.*, **9**, 2251.

80 Okayama, H. and Berg, P. (1982), *Mol. Cell. Biol.*, **2**, 161.

81 Sanger, F. (1981), *Biosci. Rep.*, **1**, 3.

82 Ghosh, P. K., Reddy, V. B., Swinscoe, J., Lebowitz, P. and Weissman, S. M. (1978), *J. Mol. Biol.*, **126**, 813.

83 Hohn, B. (1979), *Meths Enzymol.*, **68**, 299.

84 Blattner, F. R., Williams, B. G., Blechl, A. E., Denniston-Thompson, K., Faber, H. E., Furlong, L-A. *et al.* (1977), *Science*, **196**, 161.

85 Murray, N. E., Brammar, W. J. and Murray, K. (1977), *Molec. Gen. Genet.*, **150**, 53.

86 Frischauf, A-M., Lehrach, H., Poustka, A. and Murray, N. (1983), *J. Mol. Biol.*, **170**, 827.

87 Hohn, B. and Collins, J. (1980), *Gene*, **11**, 291.

88 Constanzo, F., Castagnoli, G., Dente, L., Arcari, P., Smith, M., Costanzo, P. *et al.* (1983), *EMBO J.*, **2**, 57.

89 Pelham, M. R. B. and Jackson, R. J. (1976), *Eur. J. Biochem.*, **67**, 247.

90 Parnes, J. R., Velan, B., Felsenfeld, A., Ramanathan, U., Ferrini, U., Appella, E. and Sideman, J. G. (1981), *Proc. Natl. Acad. Sci. USA*, **78**, 2253.

91 Paterson, B. M., Roberts, B. E. and Kuff, E. L. (1977), *Proc. Natl. Acad. Sci. USA*, **74**, 4370.

92 Nagata, S. Taira, H., Hall, A., Johnsrud, L., Streuli, M., Ecsödi, J. *et al.* (1980), *Nature (London)*, **284**, 316.

93 Grunstein, M. and Hogness, D. (1975), *Proc. Natl. Acad. Sci. USA*, **72**, 3961.

94 Montgomery, D. L., Hall, B. D., Gillam, S. and Smith, M. (1978), *Cell*, **14**, 673.

95 Cameron, J. R., Panasenko, S. M., Lehman, I. R. and Davis, R. W. (1975), *Proc. Natl. Acad. Sci. USA*, **72**, 3416.

96 Chang, A. C. Y., Nunberg, J. H., Kaufman, R. K., Ehrlich, H. A., Schimke, R. T. and Cohen, S. N. (1978), *Nature (London)*, **275**, 617.

97 Young, R. A. and Davis, R. W. (1983) *Proc. Natl. Acad. Sci. USA*, **80**, 1194.

98 Benton, W. D. and Davis, R. W. (1977), *Science*, **196**, 180.

99 Seed, B. (1983), *Nucleic Acids Res.*, **8**, 2427.

100 Poustka, A., Rackwitz, H-R., Frischauf, A. M., Hohn, B. and Lehrach, H. (1984), *Proc. Natl. Acad. Sci. USA*, **81**, 4129.

101 Berk, A. J. and Sharp, P. A. (1977), *Cell*, **12**, 721.

102 Berk, A. J. and Sharp, P. A. (1978), *Cell*, **14**, 695.

103 Noyes, B. E., Mevarech, M., Stein, R. and Agarwal, K. L. (1979), *Proc. Natl. Acad. Sci. USA*, **76**, 1770.

104 Shapiro, S. Z. and Young, J. R. (1981), *J. Biol. Chem.*, **256**, 1495.

105 Galas, D. J. and Schmitz, A. (1978), *Nucleic Acids Res.*, **5**, 3157.

106 Ross, W., Landy, A., Kikuchi, Y. and Nash, H. (1979), *Cell*, **18**, 297.

107 Siebenlist, U., Simpson, R. B. and Gilbert, W. (1980), *Cell*, **20**, 269.

108 Becker, M. M. and Wang, J. C. (1984), *Nature (London)*, **309**, 682.

109 Nick, H. and Gilbert, W. (1985), *Nature (London)*, **313**, 795.

110 Church, G. M., Ephrussi, A., Gilbert, W. and Tonegawa, S. (1985), *Nature (London)*, **313**, 798.

111 Frederick, C. A., Grable, J., Melia, M., Samudzi, C., Jen-Jacobson, L., Wang, B. C. *et al.* (1984), *Nature (London)*, **309**, 327.

112 Robert, B., Barton, P., Minty, A., Daubas, P., Weydert, A., Bonhomme, F. *et al.* (1985), *Nature (London)*, **314**, 181.

113 Mertz, J. E. and Gurdon, J. B. (1977), *Proc. Natl. Acad. Sci. USA*, **74**, 1502.

114 Graham, F. L. and van der Eb, A. J. (1973), *Virology*, **52**, 456.

115 Gorman, C. M., Moffat, L. F. and Howard, B. (1982), *Mol. Cell. Biol.*, **2**, 1044.

116 Wigler, M., Sweet, R., Sim, G. K., Wold, B., Pellicer, A., Lacy, E. *et al.* (1979), *Cell*, **16**, 777.

117 Wigler, M., Perculo, M., Kurz, D., Dana, S., Pellicer, A., Axel, R. and Silverstein, S. (1980), *Proc. Natl. Acad. Sci. USA*, **77**, 3567.

118 De Saint Vincent, B. R., Delbruck, S., Eckhart, W., Meinkoth, J., Vitto, L. and Wahl, G. (1981), *Cell*, **27**, 167.

119 Mulligan, R. C. and Berg, P. (1981), *Proc. Natl. Acad. Sci. USA*, **78**, 2072.

120 Colbère-Garapin, F., Horodniceanu, F., Kourilsky, P. and Garapin, A. C. (1981), *J. Mol. Biol.*, **150**, 1.

121 Gordon, J. W. and Ruddle, F. H. (1981), *Science*, **214**, 1244.

122 Palmiter, R. D. and Brinster, R. L. (1985), *Cell*, **41**, 343.

123 Winter, G., Fersht, A. R., Wilkinson, A. J., Zoller, M. and Smith, M. (1982), *Nature (London)*, **299**, 756.

124 Lathe, R. F., Lecocq, J. P. and Everett, R. (1983), *Genetic Engineering*, (ed. R. Williamson), Academic Press, London, Vol. 4.

125 Panayatos, N. and Truong, K. (1981), *Nucleic Acids Res.*, **9**, 5679.

126 Zoller, M. J. and Smith, M. (1983), *Meths Enzymol.*, **100**, 468.

127 Gillam, S., Astell, C. R. and Smith, M. (1980), *Gene*, **12**, 129.

128 Khorana, H. G. (1979), *Science*, **203**, 614.

129 Laird, C. D., McConaughy, B. L. and McCarthy, B. T. (1969), *Nature (London)*, **224**, 149.

130 Eason, R. and Campbell, A. M. (1978), in *Centrifugal Separations in Molecular and Cell Biology* (eds G. D. Birnie and D. Rickwood) Butterworths, p. 251.

131 Bloomfield, V. A., Crothers, D. M. and Tinoco, I. (1974), *Physical Chemistry of Nucleic Acids*, Harper and Row, New York.

132 Cantor, C. R. and Schimell, P. R. (1980), *Biophysical Chemistry* Part II, W. H. Freeman, San Francisco.

133 Freifelder, D. (1982), *Physical Biochemistry* (2nd edn) W. H. Freeman, San Francisco.

134 Freifelder, D. (1970), *J. Mol. Biol.*, **54**, 567.

135 Schmid, C. and Hearst, J. (1969), *J. Mol. Biol.*, **44**, 143.

136 Zimm, B. H. and Crothers, D. M. (1962), *Proc. Natl. Acad. Sci. USA*, **48**, 905.

137 Kavenoff, R., Klotz, L. C. and Zimm, B. H. (1973), *Cold Spring Harbor Symp. Quart. Biol.*, **38**, 1.

138 Kleinschmidt, A. (1968), *Meths Enzymol.*, **12B**, 361.

139 Inman, R. B. (1974), *Meths Enzymol.*, **12E**, 451.

140 Cairns, J. (1963), *Cold Spring Harbor Symp. Quart. Biol.*, **28**, 44.

141 Crockett, G. C. (1983), *Aldrichim. Acta*, **16**, 47.

142 Gait, M. J. (1984), *Oligonucleotide Synthesis, a Practical Approach*, IRL Press, Oxford.

143 Gurdon, J. B., Lane, C. D., Woodland, H. R. and Marbaix, G. (1971), *Nature (London)*, **233**, 177.

144 Marcu, K. and Dudock, B. (1974), *Nucleic Acids Res.*, **1**, 1385.

145 Shields, D. and Blobel, G. (1978), *J. Biol. Chem.*, **253**, 3753.

146 Robertson, H. D. and Lodish, H. F. (1969), *J. Mol. Biol.*, **45**, 9.

147 Zubay, G., Chambers, D. A. and Cheong, L. C. (1970), in *The Lactose Operon* (eds J. R. Beckwith and D. Zipser), Cold Spring Harbor Monograph Series, p. 375.

148 Bachmann, B. J. (1983), *Microbiol. Rev.*, **47**, 180.

149 Kersten, H. and Kersten, W. (1974), *Inhibitors of Nucleic Acid Synthesis, Biophysical and Biochemical Aspects*, Chapman and Hall, London.

150 Vázquez, D. (1979), *Antibiotic Inhibitors of Protein Biosynthesis*, Springer, Berlin.

151 Cundliffe, E. (1980), in *Ribosomes: Structure, Function and Genetics* (eds G. Chambliss, G. R. Craven, J. Davies, K. Davis, L. Kahan and M. Nomura), University Park Press, Baltimore, p. 555.

152 Cohen, P. and Van Heyningen, S. (1982), *Molecular Action of Toxins and Viruses*, Elsevier, Amsterdam.

153 Fritzsche, H., Triebel, H., Chaires, J. B., Dattagupta, N. and Crothers, D. M. (1982), *Biochemistry*, **21**, 3940.

154 Kopka, M. L., Yoon, C., Goodsell, D., Pjura, P. and Dickerson, R. E. (1985), *Proc. Natl. Acad. Sci. USA*, **82**, 1376.

155 Adams, R. L. P. and Burdon, R. H. (1985), *Molecular Biology of DNA Methylation*, Springer-Verlag, New York.

156 Sjoberg, B-M., Graslund, A. and Eckstein, F. (1983), *J. Biol. Chem.*, **258**, 8060.

157 Adams, R. L. P. (1980), *Cell Culture for Biochemists*, Elsevier, Amsterdam.

158 Seghal, P. B., Darnell, J. E. and Tamm, I. (1976), *Cell*, **9**, 473.

159 Summers, D. F. and Maizel, J. V. (1971), *Proc. Natl. Acad. Sci. USA*, **68**, 2852.

160 Marks, P. A., Burke, E. R., Conconi, F. M., Perl, E. and Rifkind, R. A. (1965), *Proc. Natl. Acad. Sci. USA*, **53**, 1437.

161 Bodley, J. W. and Zieve, F. T. (1969), *Biochem. Biophys. Res. Commun.*, **36**, 463.

162 Madjar, J-J., Nielsen-Smith, K., Frahm, M. and Roufa, D. J. (1982), *Proc. Natl. Acad. Sci. USA*, **79**, 1003.

163 Boon, T. (1972), *Proc. Natl. Acad. Sci. USA*, **69**, 549.

164 Wool, I. G. (1984), *TIBS* **9**, 14.

165 Nierhaus, K. H. and Wittmann, H. G. (1980), *Naturwiss.*, **67**, 234.

166 Hirt, B. (1967), *J. Mol. Biol.*, **26**, 365.

# Index